36th Riso International Symposium on Materials Science 2015

IOP Conference Series: Materials Science and Engineering
Volume 89

Riso, Denmark
7 - 11 September 2015

ISBN: 978-1-5108-0999-4
ISSN: 1757-8981

Printed from e-media with permission by:

Curran Associates, Inc.
57 Morehouse Lane
Red Hook, NY 12571

Printed by Curran Associates, Inc. (2015)

For permission requests, please contact the Institute of Physics
at the address below.

Institute of Physics
Dirac House, Temple Back
Bristol BS1 6BE UK

Phone: 44 1 17 929 7481
Fax: 44 1 17 920 0979

techtracking@iop.org

Additional copies of this publication are available from:

Curran Associates, Inc.
57 Morehouse Lane
Red Hook, NY 12571 USA
Phone: 845-758-0400
Fax: 845-758-2634
Email: curran@proceedings.com
Web: www.proceedings.com

Table of contents

Volume 89

36th Risø International Symposium on Materials Science
7–11 September 2015, Risø, Denmark

Accepted papers received: 29 June 2015
Published online: 7 August 2015

Preface

Papers

Invited Papers

Contributed Papers

PREFACE

Proceedings of the 36th Risø International Symposium on Materials Science

S. Fæster, N. Hansen, H. Hemmingsen, C. Hong, X. Huang, D. Juul Jensen, O. V. Mishin, J. Sun, T. Yu, Y.B. Zhang

The 36th Risø Symposium focuses on the effects of deformation-induced structural variations on annealing mechanisms. Although it is widely recognized that the processes occurring during annealing of deformed metals are determined by the local environment in which they occur, much of the current understanding, analysis and modelling is based on larger scale considerations.

Recent detailed investigations of deformation microstructures have led to a paradigm shift in the way these structures are characterized and analyzed. It is now clear that deformation microstructures are hierarchical, with dislocations and deformation-induced boundaries subdividing the original grains. This subdivision means that there are variations in the crystallographic orientations and in the distribution of stored energy on the scale of the subdivision, which typically is on the micrometer, sub-micrometer or nanometer scale. Structural variations in this subdivision may also be present from grain to grain in polycrystalline materials, thereby introducing variations on the grain scale. Finally, processing may also introduce structural variations on even larger scales. There are thus structural variations at many length scales, all of which play an essential role in subsequent annealing processes and in property optimization.

Recent advances in incorporating these structural variations into the understanding of annealing mechanisms and of how they affect the mechanical and physical properties of annealed metals and alloys are addressed in these Proceedings. The Proceedings contain 15 key-note and 46 contributed papers.

The 36th Risø International Symposium on Materials Science is organized by the Section for Materials Science and Advanced Characterization, Department of Wind Energy, Technical University of Denmark (DTU). We would like to thank all those at DTU who assisted in the preparations for the Symposium. We appreciate additionally the help from the international advisory committee consisting of: R.D. Doherty, USA; F.J. Humphreys, UK; E Pereloma; T. Sakai, Japan; M. Sugiyama, Japan; Q. Liu, China; R.A. Vandermeer, USA; Y.T. Zhu, USA. We gratefully acknowledge financial support from the following foundations: Civilingeniør Frederik Leth Christiansens Almennyttige Fond, Fabrikant Mads Clausens Fond, Knud Højgaards Fond, Kraks Fond and Otto Mønsteds Fond.

IOP Conf. Series: Materials Science and Engineering **89** (2015) 011002 doi:10.1088/1757-899X/89/1/011002

Peer review statement

All papers published in this volume of *IOP Conference Series: Materials Science and Engineering* have been peer reviewed through processes administered by the proceedings Editors. Reviews were conducted by expert referees to the professional and scientific standards expected of a proceedings journal published by IOP Publishing.

Microstructural evolutions and stability of gradient nano-grained copper under tensile tests and subsequent storage

W Chen[1], Z S You[1,2], N R Tao[1] and L Lu[1]

[1] Shenyang National Laboratory for Materials Science, Institute of Metal Research, Chinese Academy of Sciences, 72 Wenhua Road, Shenyang 110016, People's Republic of China
[2] Herbert Gleiter Institute of Nanoscience, Nanjing University of Science and Technology, 200 Xiaolingwei Street, Nanjing 210094, People's Republic of China

E-mail: llu@imr.ac.cn

Abstract. A gradient nano-grained (GNG) surface layer is produced on a bulk coarse-grained Cu by means of a surface mechanical grinding treatment. Homogeneous grain coarsening induced by mechanical deformation is observed in the GNG Cu layer under tensile tests at both 300 K and 123 K. The concurrent grain coarsening during tensile deformation is proven to be also thermally activated, because the extent of grain coarsening of the GNG Cu layer is less significant at 123 K than at 300 K, although a higher flow stress is achieved at 123 K. During the subsequent storage at 258 K after tensile tests, no obvious change can be found for the grain size in the GNG Cu layer deformed at 300 K. In contrast, widespread abnormal grain coarsening is frequently observed in the GNG Cu layer deformed at 123 K and stored for 100 days, which may be caused by the higher stored energy in the non-equilibrium grain boundary structures.

1. Introduction

Nano-grained (NG) metals with grain sizes smaller than 100 nm are generally unstable. Their grains could grow rapidly either under thermal [1] or mechanical loads [2,3]. The former is called thermally-induced grain coarsening relating to the thermal stability of the NG structures whereas the latter is referred to as mechanically-induced grain coarsening associated with the mechanical stability and imposed plastic deformation. Both the thermally- and mechanically-induced grain coarsening are of intimate relevance to the presence of a large volume fraction of grain boundaries (GBs), which possess considerable excess free energy and thus may provide a large driving force for the inherent structural instability.

In regards to the thermal stability of NG metals, there have been many literature reports, but the data seem to be inconsistent. For instance, nano-grains in pure Cu [1] grow even at room temperature (RT) (~0.2 T_m, where T_m is the absolute melting temperature), whereas NG Ni is relatively stable up to 473 K (~0.3 T_m) [4]. Indeed, a large number of NG metals exhibit remarkable thermal stability [5], manifested by a relatively high onset temperature (sometimes as high as 0.6 T_m) of grain coarsening. This unexpected thermal stability of NG metals could be attributed to the kinetic stabilization or thermodynamic stabilization. The kinetic stabilization is achieved by reducing the grain boundary (GB) mobility by solute drag, second-phase particle pinning (Zener pinning), or other mechanism such as porosity drag [5]. The thermodynamic stabilization is achieved by reducing the driving force for grain coarsening through reducing the GB energy via GB segregation [6].

Grain coarsening induced by mechanical deformation at RT has been reported in plenty of NG samples under diverse deformation modes such as indentation [3,7], rolling [8], compression [9] and tensile loading [10]. Specifically, Zhang *et al* [3] reported a significant grain coarsening in NG Cu (10-100 nm) after indentation at both RT and liquid-nitrogen temperature. Even more surprising is the observation that the extent of grain coarsening at liquid-nitrogen temperature is much greater than that at RT [11,12]. They proposed that such a grain coarsening is a primarily stress-induced and athermal process because the appreciably higher stress would be achieved at cryogenic temperature and conductive to the grain coarsening. Afterwards, Gurao and Suwas [8] reported contradictory results that grain coarsening is observed in NG Ni (~20 nm) during rolling at RT, rather than in a cryo-rolled sample, although a higher flow stress level is achieved in the NG Ni at the lower deformation temperature.

More recently, by means of a surface mechanical grinding treatment (SMGT) [13], a spatial gradient nano-grained (GNG) layer was produced on the surface of a bulk coarse-grained (CG) Cu rod. The surface NG layer exhibits a 10 times higher yield strength and a uniform tensile plasticity comparable to that of the CG substrate. The sustained tensile true strain could be as high as 100%, and had not induced any cracking or delamination in the GNG layer, which is starkly different from the behavior observed in free-standing NG metals [14]. The authors pointed out that the plastic deformation of the GNG structure is dominated by mechanically-induced GB migrations which concomitantly cause substantial grain coarsening [15].

For a better understanding of the mechanism of mechanically-induced grain coarsening of GNG Cu, in this study, the mechanical stability including the microstructural evolutions and grain coarsening of GNG Cu was systematically investigated under tensile tests at different temperatures (300 K and 123 K). The investigation was also undertaken with specific objective of studying the thermal stability of GNG Cu after tensile tests, i.e., the effect of the so-called "storage annealing" on the microstructure of GNG Cu during the subsequent storage at 258 K, i.e., −15 °C, for different numbers of days. Both the mechanisms related to the mechanical stability and thermal stability of the GNG Cu layer will be discussed.

2. Experimental

2.1. Sample fabrication

Commercial-purity copper rods (99.97 wt.%) consisting of well-annealed equiaxed coarse grains (with an average diameter of 25 μm) were machined into dog-bone tensile samples with a gauge length of 20 mm and a gauge diameter of 6 mm. Then both the gauge section and the arc transitions of the samples were subjected to SMGT. The basic principle and detailed procedure of the SMGT process have been described elsewhere [13]. In this study, the applied SMGT processing parameters are as follows: rotation speed υ_1 = 600 rpm, sliding velocity of the tool tip υ_2 = 3 mm s^{-1}, penetration depth of the tool tip into the sample a_p = 40 μm, and the Cu rod was cooled to a cryogenic temperature of ~173 K with liquid nitrogen during the SMGT process. In order to effectively refine the grains in the surface layer and increase the thickness of the nanostructured layer, the SMGT process was repeated six times with the same parameters.

2.2. Tensile tests

Uniaxial tensile tests of the SMGT-processed samples (hereafter referred to as the GNG/CG Cu) were performed on an Instron 5982 testing machine (Instron, UK) with a constant strain rate of 1×10^{-2} s^{-1} at 300 K and 123 K, respectively. The tensile tests at 123 K were carried out in an Instron environment chamber, which was cooled by liquid-nitrogen flow. An Instron 2620-601 clip-on extensometer was used to measure strain at RT, while an Epsilon model 3542 clip-on extensometer was used to measure strain at 123 K.

Several GNG/CG Cu samples were stretched to two predefined true strains of ε_T = 4% (shortly after the yield point) and 23% (well-developed in the uniform plastic deformation stage) and then unloaded for further microstructural examination at both temperatures. All the samples before and after tensile deformation were refrigerated at 258 K. To avoid the possibility of spontaneous grain coarsening in the GNG Cu layer during the storage, all the examinations including mechanical tests and microstructural characterising of the samples were finished within 5 days, unless otherwise specified.

2.3. Microstructural characterization

In order to reveal the microstructural evolutions, the longitudinal sections of the as-treated and tensile strained GNG/CG Cu samples were examined by a FEI NanoSEM Nova 430 scanning electron microscope (SEM) and by a JEOL-2010 transmission electron microscope (TEM) operated at a voltage of 200 kV. To protect the surface NG layer, a pure Cu layer was electro-deposited onto the surface. Then longitudinal samples were cut by using an electric spark cutting machine (ESM). Samples for SEM observation were polished first mechanically and then electrochemically in a solution of phosphoric acid (25%), alcohol (25%), and deionized water (50%) at RT. Samples for TEM observation were first mechanically polished to a final thickness of ~40 μm, then thinned by twin-jet polishing in an electrolyte of 33% phosphoric acid and 67% deionized water (by volume) at about −10 °C.

The grain size distributions of the top GNG Cu surface layer at different conditions were determined on bright-field TEM images. Over 800 grain sizes were measured from about 60 TEM images for each data set.

3. Results

3.1. Microstructure of the as-SMGT (GNG/CG) samples

The longitudinal cross-sectional SEM image in figure 1a shows a typical hierarchical microstructure of as-SMGT Cu samples, which apparently varies with the depth along the radial direction. This spatial microstructural gradient is formed by the SMGT technique that simultaneously imposes a high strain rate and a strong strain gradient at the sample surface. After the treatment, the surface is very flat and shining with a minor roughness ($R_a \approx 0.3$ μm), and crack-free.

Closer observations by TEM indicate that the top 20 μm thick surface layer is completely composed of nano-sized grains with an average grain size of 50 nm (the average transversal and longitudinal axis grain sizes are 42 nm and 60 nm, respectively) as shown in figure 1b. The selected-area electron diffraction (SAED) pattern (inset of figure 1b) indicates that the nano-grains are randomly oriented. The majority of the nano-sized grains are separated by indistinct curved or wavy GBs, which are apparently in a non-equilibrium state, as generally observed in nanostructured metals refined by severe plastic deformation [16]. These deformation distorted GBs are characterized by the presence of a high-density of extrinsic GB dislocations and hence of relatively high stress concentrations around GBs.

The average grain size gradually increases to about 350 nm as the depth increases from 20 to 100 μm (figures 1c and d). The grains are still nearly equiaxed. Beneath 100 μm, a typical deformation structure can be observed, which is characterized by dislocation tangles and dislocation cells with a size also gradually increasing from submicrometres to micrometres with the depth. The thickness of the overall deformed surface layer is about 700 μm. Therefore, the spatial gradient structure introduced by SMGT consists of a GNG layer, a deformed CG layer and an undeformed CG core. In this study, since only the evolution of NG structure is concerned, particular attention is paid to the top GNG layer (20 μm thick) with grain sizes smaller than 100 nm.

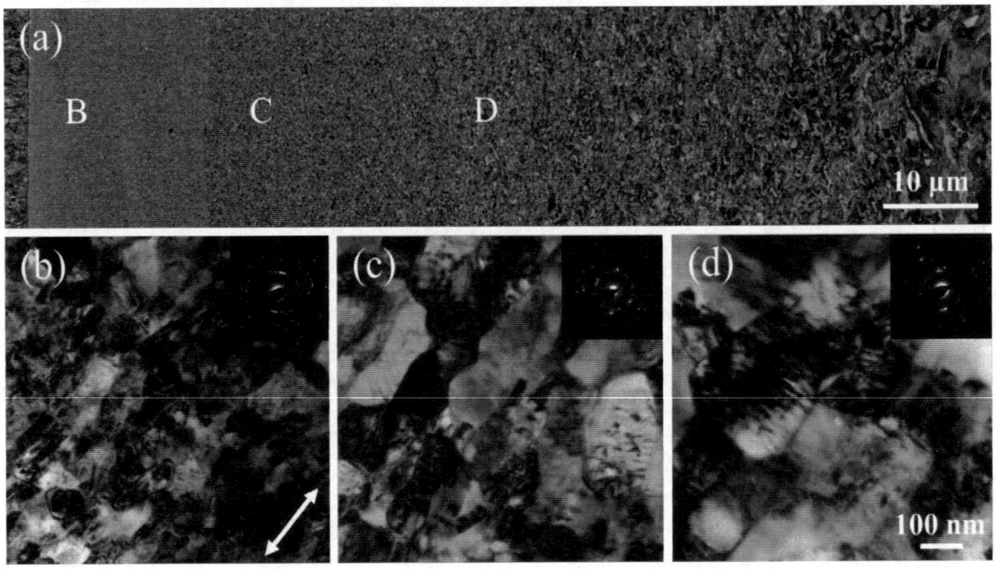

Figure 1 (a) Typical longitudinal section SEM image of a SMGT Cu sample. (b, c and d) Bright-filed TEM images at positions B, C and D indicated in (a), respectively. The double-head arrow indicates the tensile loading direction (the same as in the following Figures).

3.2. Microstructural evolutions of GNG Cu layer during tensile deformation

Tensile tests indicated that the GNG/CG Cu exhibits an obvious temperature dependence of the yield strength (0.2% offset). When the temperature decreases from 300 to 123 K, the yield strength of GNG/CG Cu elevates from 129 to 149 MPa, while a negligible enhancement of yield strength is observed in CG Cu. The relative stronger temperature dependence of the yield strength of the GNG/CG Cu sample suggests that the GNG Cu layer sustains a much higher tensile stress at 123 K than at 300 K. This is consistent with the literature report that the hardness of NG Cu with a grain size of 11 nm increased from 2.1 to 4 GPa, as the temperature decreased from 300 to 123 K [17].

Postmortem TEM observations (figures 2a-d) on samples tensioned to different strains under diverse temperatures provide an insight into the microstructural evolutions of the GNG Cu. Compared with the as-SMGT state (figure 1b), homogeneous grain coarsening is observed in the top GNG Cu layer after tensile deformation at 300 K (figures 2a and b). Moreover, the dislocation density in the microstructure is also greatly reduced, as indicated by the more uniform image contrast in the grains. As the strain increases from ε_T = 4% (figure 2a) to 23% (figure 2b) at 300 K, grains in the GNG Cu layer become larger in size and elongated slightly (the aspect ratio at ε_T = 23% is 1.9). Accompanying the grain coarsening, most of the GBs become straighter and sharper, which implies a more equilibrium state.

Homogeneous grain coarsening is also detected in the GNG Cu layer after tensile deformation at 123 K (figures 2c and d), although the extent of grain coarsening is less significant than that at 300 K. Furthermore, a high density of dislocations is retained in the microstructure due to the suppressed recovery at the lower deformation temperature. As the strain increases from ε_T = 4% (figure 2c) to 23% (figure 2d), grains in the top GNG Cu layer become larger but still remain equiaxed (the aspect ratio at ε_T = 23% is 1.5). Most of the GBs in the GNG Cu layer deformed at 123 K are indistinct and curved, indicating that they are still in a non-equilibrium state.

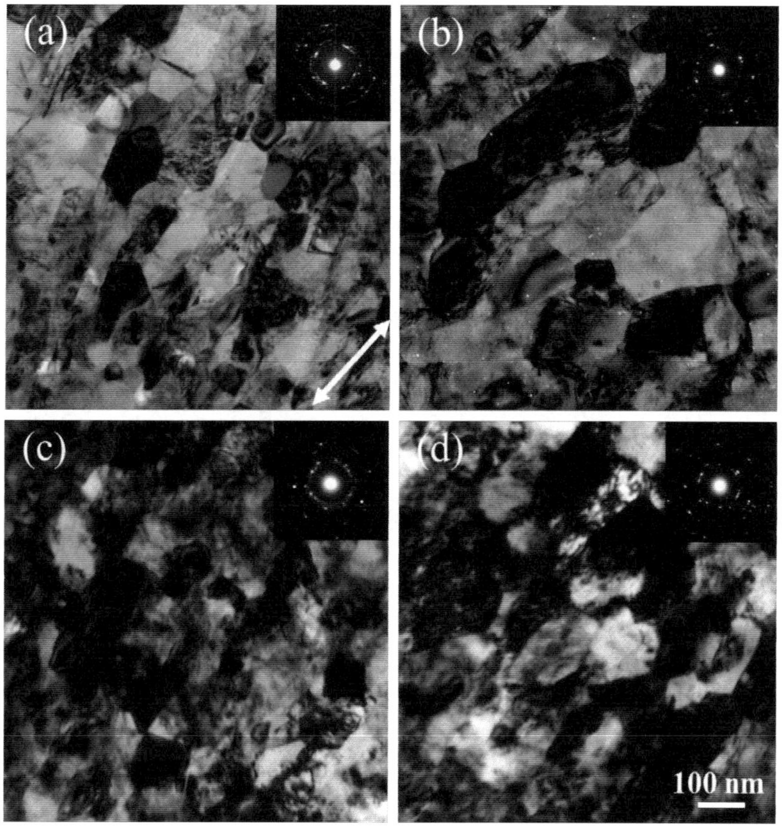

Figure 2 Bright-field TEM images showing the typical microstructures of the top GNG surface layer (2.5 μm below the sample surface) at different tensile temperatures and different strains. (a) $\varepsilon_T = 4\%$, 300 K. (b) $\varepsilon_T = 23\%$, 300 K. (c) $\varepsilon_T = 4\%$, 123 K. (d) $\varepsilon_T = 23\%$, 123 K.

Grain coarsening during tensile deformation is described more quantitatively in the form of transversal grain size distribution plots of the grain area fraction versus grain size at various temperatures and strains (figures 3a-d). The transversal grain size distribution of the as-SMGT sample is quite narrow ranging from 20 to 80 nm. With increasing tensile strains to $\varepsilon_T = 4\%$ and 23%, the transversal grain size distributions become broader at both 300 K (figures 3a and b) and 123 K (figures 3c and d). Moreover, the small grains in the GNG layer have gradually disappeared. It is noted that the transversal grain size distributions at 123 K (figures 3c and d) are much narrower and concentrating at the small grain size regime compared with that at 300 K (figures 3a and b), which indicates a less significant grain coarsening at 123 K. Examination of the longitudinal grain size distributions also gives the same but more obvious evidence that the extent of grain coarsening of the GNG Cu layer is less significant at 123 K (figures 4c and d) than that at 300 K (figures 4a and b). Furthermore, the unimodal grain size distributions at different temperatures and different strains also confirm that the grain coarsening accompanying tensile deformation is in a uniform mode.

It is interesting to find that although the GNG Cu layer sustains a much higher mechanical stress at 123 K than that at 300 K, the extent of grain coarsening induced by tensile deformation is less significant at 123 K. This suggests that the mechanical stability of the GNG Cu layer increases as the deformation temperature decreases.

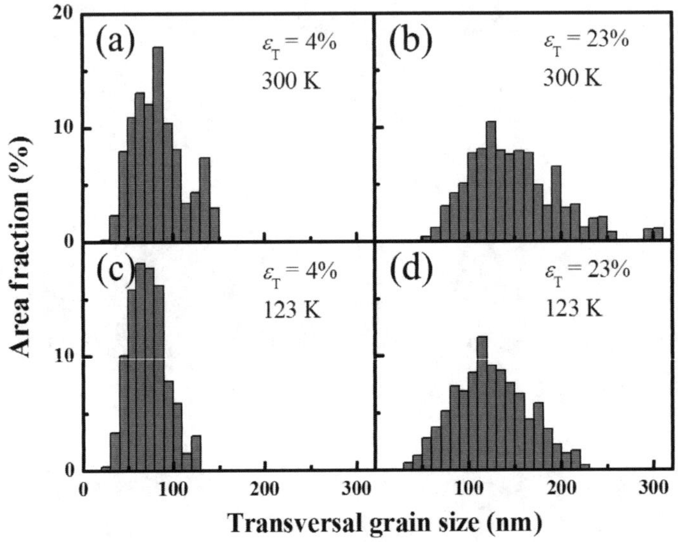

Figure 3 Area-weighted transversal grain size distributions in the top GNG surface layer at different temperatures and strains, as indicated.

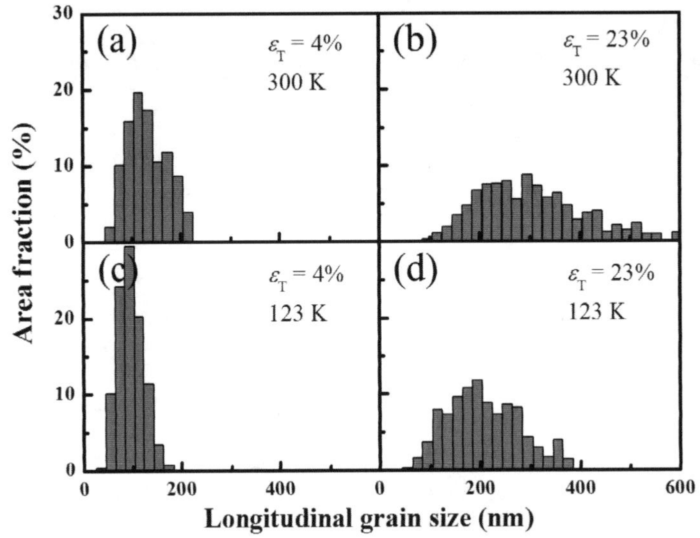

Figure 4 Area-weighted longitudinal grain size distributions in the top GNG surface layer at different temperatures and strains, as indicated.

3.3. Microstructural evolution of GNG Cu during storage at 258 K

After tensile tests, the GNG Cu samples deformed to ε_T = 23% at different temperatures were further stored at 258 K for different numbers of days. No obvious change in grain size can be found in the GNG Cu deformed at 300 K with increasing storage time under the resolution of the SEM observations (figures 5a-c). TEM images show more details for the microstructures of the GNG Cu layer with ε_T = 23% and stored for 10 days (figure 6a) and 30 days (figure 6b). Compared with the microstructure of the GNG Cu layer just after tension (figure 2b), no obvious change can be found in the microstructure during the storage (figures 6a and b), except for a slight decrease in the dislocation density.

Figure 5 SEM images of the top GNG Cu layers with ε_T = 23% deformed at 300 K and stored at 258 K for 10 days (a), 30 days (b) and 100 days (c), respectively.

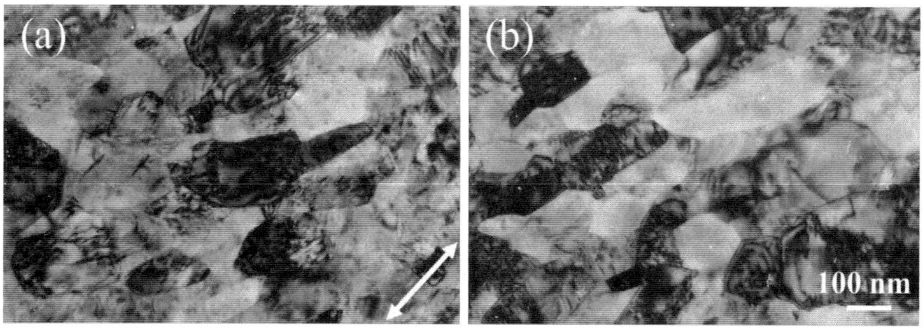

Figure 6 TEM images showing the typical microstructures of the top GNG Cu layers with ε_T = 23% deformed at 300 K and stored at 258 K for 10 days (a) and 30 days (b). No significant difference between the microstructures can be detected.

In contrast to the stable microstructure of the GNG Cu deformed at 300 K and stored at 258 K, the storage of the 123 K deformed GNG Cu at 258 K leads to the formation of very large grains in the matrix of fine grains, as clearly shown in the SEM observations (figure 7b for 30 days, and figure 7c for 100 days). These abnormally large grains are formed during the storage rather than in the tensile deformation processes, because only homogeneously distributed small grains in the GNG Cu are detected just after being stored for a short period of 3.5 hours (figure 7a). Moreover, it is worth noting that the abnormal grain coarsening is not an accident, but a common phenomenon. For example, widespread abnormally large grains with a grain size larger than a few micrometres are frequently observed in the GNG Cu layer stored for 100 days as shown in figure 7d.

To make sure whether the grain size is stable in the areas apart from the abnormally large grains as indicated by the square in figure 7b, closer TEM observations were applied to examine the fine grain areas in the GNG Cu layer deformed at 123 K and stored for 30 days. Obviously, even in the small grain size areas, abnormal grain coarsening had also occurred (figure 8a). It is noted that the dislocation density in the large grain (with a longitudinal grain size of 800 nm) is rather low, which is in sharp contrast to the high density of dislocations in the surrounding small grains. The GBs of the large grains are also much sharper than that of the small ones, which means a more equilibrium state. It is interesting to find that some clean and straight twin boundaries can be observed in some large grains (with a grain size usually larger than 1 μm) as shown in figure 8b, although the sample is stored at low temperatures.

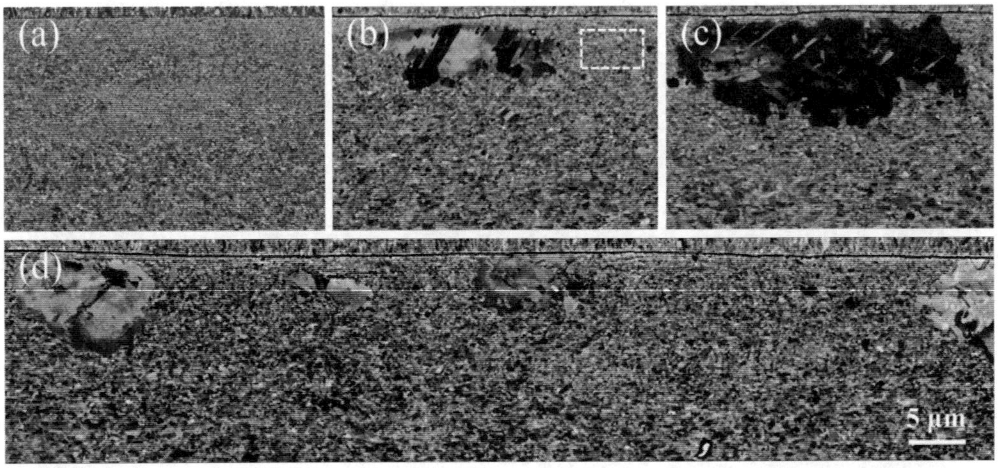

Figure 7 SEM images of the top GNG Cu layers which were deformed to ε_T = 23% at 123 K and stored at 258 K for 3.5 hours (a), 30 days (b) and 100 days (c), respectively. (d) An SEM picture with a large field of view of the sample stored for 100 days displays the widespread abnormal grain coarsening.

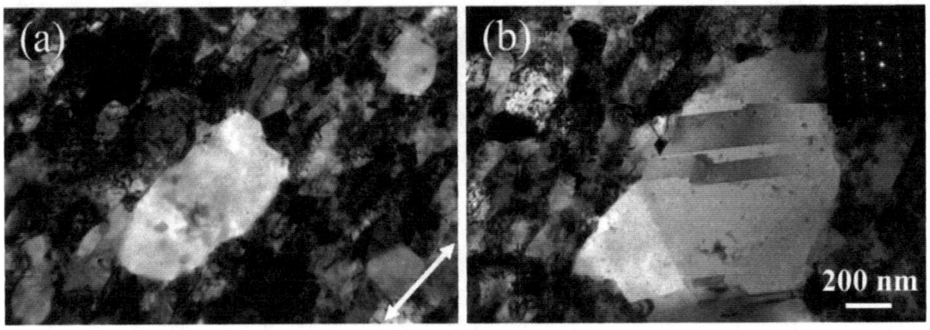

Figure 8 TEM images showing the relative small grain size areas of the GNG Cu layers deformed to ε_T = 23% at 123 K and subsequently stored at 258 K for 30 days. (a) A large dislocation-free grain is found to be embedded in the small grains. (b) An abnormally large grain is divided by clear (1 1 1) twin boundaries as indicated by the SAED patterns.

4. Discussion

Deformation-induced GB migration and associated grain coarsening have been experimentally verified to dominate the plastic deformation of various NG systems [10,15,18]. However, the understanding of detailed underlying mechanism is still far from mature. Computer molecular dynamics simulations have also been performed aiming at atomistically revealing the GB migration behavior [19-21], but the ultra-high strain rate and limited available time scale inevitably rule out some processes which may operate under realistic laboratory conditions. On the other hand, the grain coarsening mechanism can also be indirectly investigated by experimentally exploring the influence of such parameters as deformation temperature [3,8,22], strain rate [23,24], stress state [2,25] on the degree of grain growth, or by scrutinizing the character and evolution of the microstructure, especially the GB structure.

In this study, the grain coarsening behavior of the GNG Cu layer is investigated by tensile tests at different temperatures (300 K and 123 K) to reveal the temperature effect. The microstructural observations reveal a quite homogeneous normal grain coarsening accompanying tensile deformation at both temperatures, as shown in figures 2. Furthermore, the extent of grain coarsening increases with increasing tensile strain. However, at a given plastic strain, the grains grow less significantly when the sample is deformed at 123 K than at 300 K, indicating that reducing temperature appears to suppress the deformation-induced grain coarsening. Taking account of the fact that the surface GNG Cu layer sustains a higher flow stress at low temperatures, this result is clearly an indication that the grain coarsening during the deformation of NG structures is not exclusively controlled by the stress.

Zhang *et al* [3,11,12] reported that the grain coarsening in NG Cu under indentation at liquid-nitrogen temperature is more significant than at RT, suggesting that the athermal stress-induced GB migration and grain growth mediates the plastic deformation. The contrary in the temperature effect on the grain coarsening between the present result and the literature may be associated with the difference in the stress or strain state of the samples under mechanical deformation. Below the indentation, a larger strain gradient and complex stress state may change the dominant process controlling grain coarsening by facilitating the stress-induced GB migration. However, in the GNG Cu layer supported by a CG core that suppresses its strain localization, the nano-grains deform homogeneously under uniaxial tensile stresses. Although the detailed process of the mechanically-induced grain coarsening in the GNG Cu is still not well understood, it is reasonable to speculate that in order to accommodate the homogeneous grain growth, some GB activities other than the athermal stress-induced GB migration must also be triggered together. The more significant grain coarsening of the GNG Cu at higher deformation temperatures (RT versus 123 K) irrespective of being under lower applied stresses is a thermally activated process.

In addition to the extent of grain coarsening, the deformation temperature also exerts a strong influence on the deformation microstructure of the GNG Cu layer. It is noted that the GBs in the 300 K tensioned GNG Cu are in a much more equilibrium state, and appear straight and sharp, in stark contrast to the as-SMGT state where the GBs occupied by high-density extrinsic dislocations are indistinct. This suggests that mechanically-induced GB migration in GNG Cu deformed at RT has consumed the accumulated dislocations nearby, leading the GBs to a low energy state. By contrast, in the 123 K deformed GNG Cu, most GBs are still curved and blurred, in a non-equilibrium state. Inside the grains, there are a high-density of dislocations and strong stress contrast. The diverse GB states for GNG Cu deformed at 300 K and 123 K probably reflect different thermally activated grain coarsening behavior, and can also rationalize the thermal stability of deformed samples as discussed below.

The excess free energy due to GBs can be estimated by the Gibbs-Thomson relationship [26]:

$$\Delta G = \frac{4\Omega\gamma}{d} \tag{1}$$

where Ω is atomic volume, γ is the GB energy and d is the mean diameter of the grains. In the 123 K tensioned GNG Cu, the GBs are in a high-energy state due to the accumulation of high density defects. The energy of such non-equilibrium GBs can be two or more times the equilibrium GB energy [27]. Furthermore, the grain size in the 123 K tensioned GNG Cu is smaller than that in the GNG Cu deformed at 300 K. Thus, according to equation (1) and in view of both the smaller grain size and the higher GB energy, the GNG Cu deformed at 123 K should have higher excess free energy compared to the one deformed at 300 K. During the subsequent storage at 258 K, obvious abnormal grain growth is observed in the GNG Cu layer deformed at 123 K, but not in the GNG Cu layer deformed at 300 K. This suggests that the deformation at 300 K has completely eliminated the defects of and hence reduced the energy of nanoscale GBs in the course of grain coarsening.

It is noted that the grain growth in the GNG Cu deformed at 123 K and stored at 258 K for several tens of days is in an abnormal mode, which is usually observed for the low temperature grain coarsening of the NG metals [28]. The traditional hypothetical factors linked to the occurrence of abnormal grain coarsening such as inhomogeneous purities drag and pores drag seem impossible to apply to our samples. Several other factors may contribute to the abnormal grain growth in this study. Firstly, the stored energy in the GNG Cu layer introduced by plastic deformation may not be homogeneous and will cause nucleation and growth of defect-free grains preferentially at localized areas with the highest stored energy. Secondly, some larger dislocation-free grains induced by the mechanical deformation may directly act as nuclei for the abnormal grain growth.

The present results suggest that the thermal stability of NG metals is strongly dependent on its deformation history. All the factors related to the stored energy, final grain size as well as GB energy or morphology will have a direct impact on the microstructural stability during subsequent annealing treatment or storage.

5. Summary

The microstructural evolutions and mechanical/thermal stability of GNG Cu under tensile tests (at 300 K and 123 K) and subsequent storage at 258 K were systematically investigated by means of SEM and TEM. The GNG Cu layer sustains a higher stress at 123 K but the resulting grain coarsening is less significant compared with that caused by deformation at 300 K. The mechanically-induced grain coarsening can also be thermally activated in the GNG Cu. During subsequent storage at 258 K for tens of days, an obvious abnormal grain coarsening is observed in the GNG Cu layer deformed at 123 K, but not at 300 K under the same storage condition. The lower thermal stability of the sample deformed at 123 K may be caused by the larger stored energy at GBs. The insight into the stabilities of nanostructures gained in this work could be valuable for understanding the microstructural stability of NG metals under service process and thermomechanical treatment.

Acknowledgement

The authors acknowledge financial support from the National Basic Research Program of China (973 Program, 2012CB932202), the National Science Foundation of China (Grant Nos. 51420105001, 51171181, 51371171, 51471172 and 51401211). LL thanks the financial support of the "Hundreds of Talents Project" by the Chinese Academy of Sciences. The authors thank Mr. X. Si for SMGT sample preparation.

References

[1] Gertsman V Y and Birringer R 1994 *Scripta Metall. Mater.* **30** 577
[2] Rupert T J, Gianola D S, Gan Y and Hemker K J 2009 *Science* **326** 1686
[3] Zhang K, Weertman J R and Eastman J A 2004 *Appl. Phys. Lett.* **85** 5197
[4] Thuvander M, Abraham M, Cerezo A and Smith G D W 2001 *Mater. Sci. Technol.* **17** 961
[5] Suryanarayana C and Koch C C 2000 *Hyperfine Interact.* **130** 5
[6] Chookajorn T, Murdoch H A and Schuh C A 2012 *Science* **337** 951
[7] Jin M, Minor A M, Stach E A and Morris J W 2004 *Acta Mater.* **52** 5381
[8] Gurao N P and Suwas S 2009 *Appl. Phys. Lett.* **94** 191902
[9] Pan D, Kuwano S, Fujita T and Chen M W 2007 *Nano Lett.* **7** 2108
[10] Gianola D S, Van Petegem S, Legros M, Brandstetter S, Van Swygenhoven H and Hemker K J 2006 *Acta Mater.* **54** 2253
[11] Zhang K, Weertman J R and Eastman J A 2005 *Appl. Phys. Lett.* **87** 061921
[12] Gai P L, Zhang K and Weertman J 2007 *Scripta Mater.* **56** 25
[13] Li W L, Tao N R and Lu K 2008 *Scripta Mater.* **59** 546
[14] Meyers M A, Mishra A and Benson D J 2006 *Prog. Mater Sci.* **51** 427
[15] Fang T H, Li W L, Tao N R and Lu K 2011 *Science* **331** 1587
[16] Li Y S, Tao N R and Lu K 2008 *Acta Mater.* **56** 230
[17] Huang Z, Gu L Y and Weertman J R 1997 *Scripta Mater.* **37** 1071
[18] Fan G J, Fu L F, Choo H, Liaw P K and Browning N D 2006 *Acta Mater.* **54** 4781

[19] Schiøtz J 2004 *Mater. Sci. Eng.*, A **375** 975
[20] Farkas D, Mohanty S and Monk J 2008 *Mater. Sci. Eng.*, A **493** 33
[21] Monk J and Farkas D 2007 *Phys. Rev.* B **75** 045414
[22] Prasad M and Chokshi A H 2012 *Scripta Mater.* **67** 133
[23] Gianola D S, Warner D H, Molinari J F and Hemker K J 2006 *Scripta Mater.* **55** 649
[24] Bachmaier A, Hafok M and Pippan R 2010 *Mater. Trans.* **51** 8
[25] Brandstetter S, Zhang K, Escuadro A, Weertman J R and Van Swygenhoven H 2008 *Scripta Mater.* **58** 61
[26] Gibbs J W and Bumstead H A 1993 *Thermodynamics* Vol 1 (Longmans, Green and Company)
[27] Nazarov A A, Romanov A E and Valiev R Z 1993 *Acta Metall. Mater.* **41** 1033
[28] Ames M, Markmann J, Karos R, Michels A, Tschope A and Birringer R 2008 *Acta Mater.* **56** 4255

Uncovering the true nature of deformation microstructures using 3D analysis methods

M Ferry[1], MZ Quadir[1,2], N Afrin[1], W Xu[1], A Loeb[3], B Soe[3], C McMahon[3], C George[3] and L Bassman[3]

[1]School of Materials Science and Engineering, the University of New South Wales, NSW 2052, Australia
[2]Electron Microscope Unit, the University of New South Wales, NSW 2052, Australia
[3]Department of Engineering, Harvey Mudd College, Claremont, CA 91711, USA

Email: m.ferry@unsw.edu.au

Abstract. Three-dimensional electron backscatter diffraction (3D EBSD) has emerged as a powerful technique for generating 3D crystallographic information in reasonably large volumes of a microstructure. The technique uses a focused ion beam (FIB) as a high precision serial sectioning device for generating consecutive ion milled surfaces of a material, with each milled surface subsequently mapped by EBSD. The successive EBSD maps are combined using a suitable post-processing method to generate a crystallographic volume of the microstructure. The first part of this paper shows the usefulness of 3D EBSD for understanding the origin of various structural features associated with the plastic deformation of metals. The second part describes a new method for automatically identifying the various types of low and high angle boundaries found in deformed and annealed metals, particularly those associated with grains exhibiting subtle and gradual variations in orientation. We have adapted a 2D image segmentation technique, fast multiscale clustering, to 3D EBSD data using a novel variance function to accommodate quaternion data. This adaptation is capable of segmenting based on subtle and gradual variation as well as on sharp boundaries within the data. We demonstrate the excellent capabilities of this technique with application to 3D EBSD data sets generated from a range of cold rolled and annealed metals described in the paper.

1. Introduction

The ability to observe the structure of an opaque solid in three dimensions (3D) is important for understanding its true nature as it eliminates speculation about the spatial distribution of features associated with conventional 2D imaging techniques. This is an important step in exploring microstructure-property relationships relating to many metallurgical problems, since most practical materials have a complex, polycrystalline structure and most features affecting material properties are three dimensional. There are several 3D analysis techniques for determining the spatial distribution of microstructural features, each with varying resolution and specific capabilities [1]. For example, in situ 3D XRD has emerged as a powerful non-destructive 3D analysis tool for directly analysing microstructural changes subject to heating, loading etc. such as the time evolution of the size and shape of grains during recrystallization [2-5]. Schmidt *et al.* [2] used the technique to study recrystallization in deformed aluminium single crystals; the work showed that the orientation of new grains can be different to those present in the local deformation substructure as well as the growth fronts of the recrystallizing grains being highly irregular, which undermines the classic assumption of smooth and spherical growth of grains during recrystallization [2].

Over the past few years, considerable progress has been made in the development of another 3D technique, termed 3D electron backscatter diffraction (3D EBSD), for understanding various metallurgical phenomena [7-20]. Using a typical DualBeam™ or similar platform, sub-100 nm thick layers of a material can be milled sequentially by the focused ion beam (FIB) and each surface mapped by EBSD (figure 1a). The successive EBSD maps are then combined using a suitable post-processing method to generate crystallographic volumes of the microstructure (figure 1b). 3D EBSD is particularly suited to crystalline materials where the smallest resolvable crystallographic features are ~ 50 nm and encompasses many types of deformation and annealing phenomena.

The authors are using 3D EBSD as a tool for better understanding the geometric and crystallographic features of microstructures generated by deformation and annealing and phase transformation. In our particular 3D EBSD experiments, we achieve EBSD pattern indexing rates for each FIB-milled surface of greater than 85% using a 1nA FIB current for ion milling and 30kV SEM operating voltage. Since the reconstructed data represents a volume of the material, such information is useful for better understanding the mechanisms of formation of these features. For example, a significant structural feature associated with many types of plastically deformed metal is the characteristic subgrain/cellular structures that are generated.

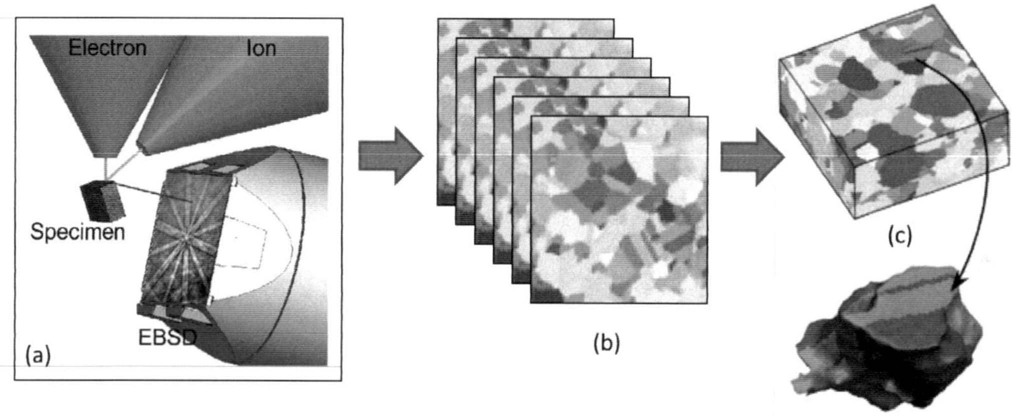

Figure 1. (a) Sequential focused ion beam milling of a sample and concurrent EBSD mapping to generate (b) consecutive EBSD maps. (c) EBSD maps are combined to generate a volume of microstructure up to 100 × 100 × 100 μm^3 that can be interrogated further for investigating specific microstructural features (figure adapted from ref. [6]).

FIB/EBSD serial sectioning readily generates hundreds of consecutive layers of material with each EBSD map often containing over 10^6 data points. Hence, a volumetric data set may easily contain 10^7-10^9 points. While preliminary reconstruction of such a data set may be done using commercial software tools, they do not generate the accuracy needed for low angle grain boundary (LAGB) identification, skew correction, fitting of surfaces or local surface analysis needed for this particular problem. For example, while the misorientation thresholding capabilities of commercial software can be used to determine high angle grain boundaries (HAGBs), subgrain/cell features generated by plastic deformation usually have gradual boundaries with pixel-to-pixel misorientations of a few degrees or less. A novel algorithm for repeatable identification of subgrain boundaries that addresses all of these foregoing issues is outlined in section 3. The algorithm automatically generates both 2D and 3D EBSD data sets that can be scrutinised to further our understanding of the nature of the deformed and annealed states.

2. Application of 3D EBSD to investigating various boundaries in deformed metals and alloys

There are various microstructural features generated by plastic deformation including, for example, cells/subgrains, microbands, deformation bands, shear bands and deformation zones at coarse, non-deformable particles [21]. Such features can have a pronouced influence on recrystallization and subsequent texture development and, hence, on the mechanical and physical properties of metals and alloys [21]. They are often analysed by standard 2D techniques and TEM. This section illustrates how 3D EBSD can be used as a tool for better understanding some key microstructural features associated with cold deformed metals.

2.1. Crystallographic nature of microbands in cold deformed metals and alloys

It is well-known that microbands (MBs) are generated in a wide range of deformed metals and alloys (Fe, Al, Cu, Ni etc.) in processes such as rolling and are recognizable by their characteristic lamellar appearance inclined at 25-35° to the rolling direction (RD) when viewing the normal direction (ND)-RD section (see *e.g.* figures 2c and d). In recent years there have been major efforts to determine the crystallographic nature of MBs with the aim of modelling their development during plastic straining and subsequent annealing. This has proven to be a challenge as shown by the recent controversy over the nature of their alignments, in particular for Al deformed in plane strain compression. The Risø group [22-24], based on detailed TEM observations, argue that subgrain boundaries or microbands lie mostly on the active {111} slip planes in fcc alloys. This is supported by the recent work using EBSD [25]. In contrast, the Manchester group [26-28], using EBSD microtexture analysis, asserts that the deformed microstructures are independent of crystallography and the MBs are aligned with the planes of maximum shear stress.

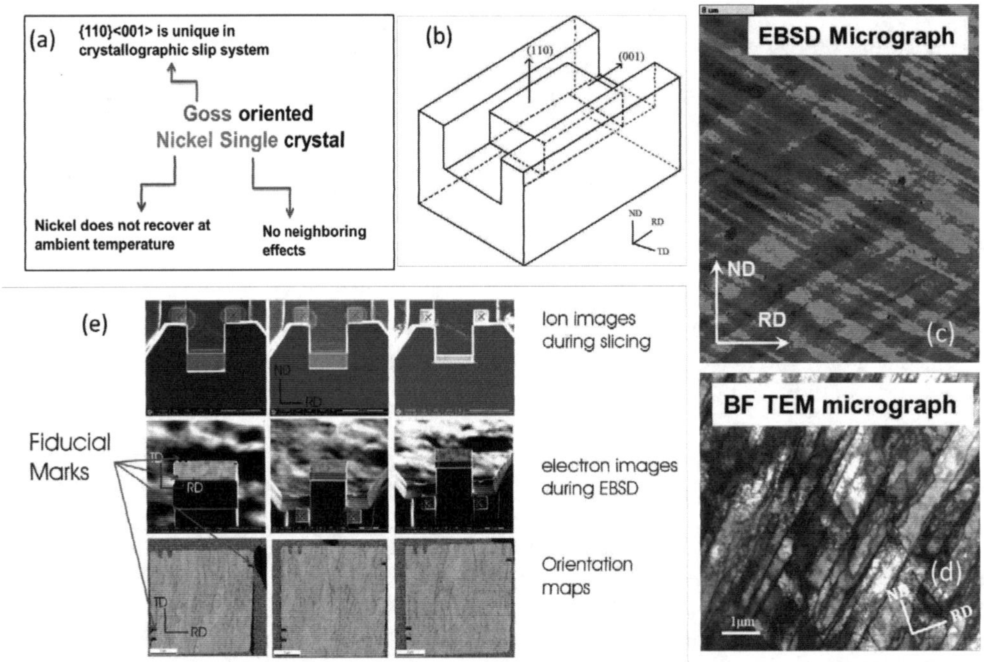

Figure 2. (a) Reasons for using Goss-oriented single crystals in investigations of plastic deformation of metals. (b) Plane strain compression (PSC) rig relative to the orientation of the Goss crystal. (c) 2D EBSD micrograph showing two non-coplanar sets of microbands. (d) Bright field TEM micrograph showing the dislocation structures associated with these microbands. (e) Standard method of FIB serial sectioning and EBSD mapping of our samples for 3D EBSD.

3D EBSD is being used for understanding the crystallographic nature of the surface of MBs in several types of material: cold rolled IF steel (BCC) [29-30]; plane strain compressed Ni (FCC) single crystals of the Goss orientation [31-33], and cold rolled commercial purity aluminium (FCC) [34]. These materials generate classic MBs. The actual analysis of individual MBs in these materials is challenging due to the subtle orientation gradients present in these features and the low angle nature of their boundaries. To address these challenges, an automated method has been developed for delineating the various types of boundaries found in deformed and annealed metals and alloys (see section 3). Serial sectioning and data reconstruction has enabled us to determine the general structure and surface topography of the MBs in these materials. The present section will focus on the formation of MBs in Ni single crystals (figure 2).

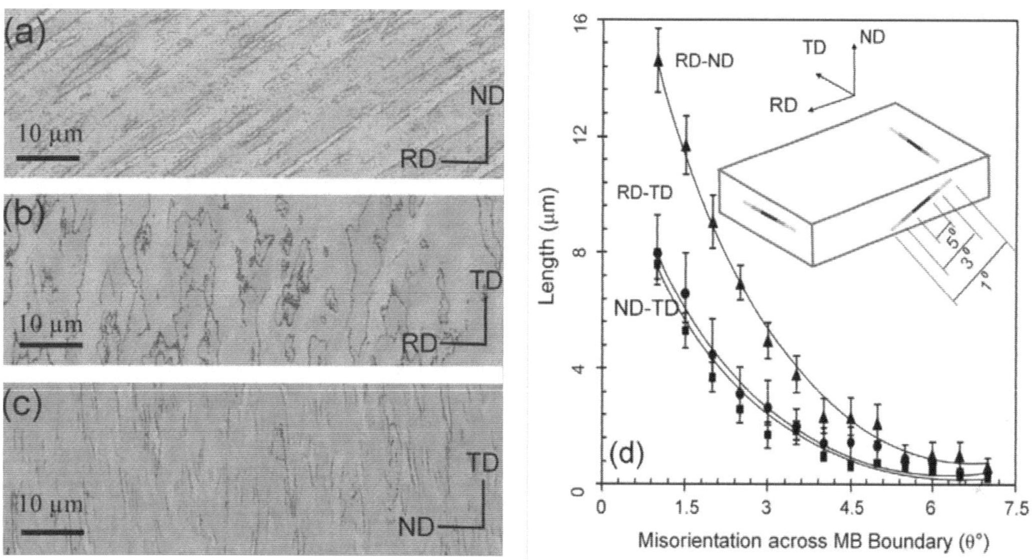

Figure 3. (a-c) EBSD micrographs of three orthogonal surfaces of the PSC Ni single crystal showing the 1.5° MB boundaries as black lines. (d) Plot of the MB boundary length as a function of misorientation showing that their boundary lengths with higher misorientations are short. In the TD section, the average MB length, for a given misorientation, is twice that of the ND section, after [32].

Figure 3 shows the three orthogonal sections of a 30% PSC Ni crystal, whereby MB traces are classically aligned in the transverse direction (TD) section at an acute angle from the rolling direction (RD), but appear wavy in the ND section. The latter observation may lead to the conclusion that microband boundaries are non-crystallographic. 3D-EBSD was used to reconstruct individual microbands in a deformed volume that was found to reveal significant new information about their structure [32]. Figure 4 shows the microband surfaces to be largely planar over large distances but frequently interrupted by local distortions and undulations due to interactions between intersecting, non-coplanar microbands (circled in figure 4). These findings are supported by detailed TEM analysis of general microband structure and interacting boundaries [31-33]. The combined EBSD/TEM investigation clearly reveals that MB boundaries are aligned close to an active {111} slip plane (i.e. they are crystallographic) but the bumps and distortions they contain are non-crystallographic in the sense that they deviate away from an active slip plane. The non-crystallographic features of microbands (as revealed by their wavy structure in the ND section, figure 3b) may be explained by the crystallographic oscillations of up to ±7.5° towards RD that occur during plastic deformation. Such oscillations result in varying fractions of slip on a given {111} plane, thereby resulting in varying degrees of interaction between the two sets of non-coplanar microbands. These local and intense microband interactions result in their deviation away from their active slip planes.

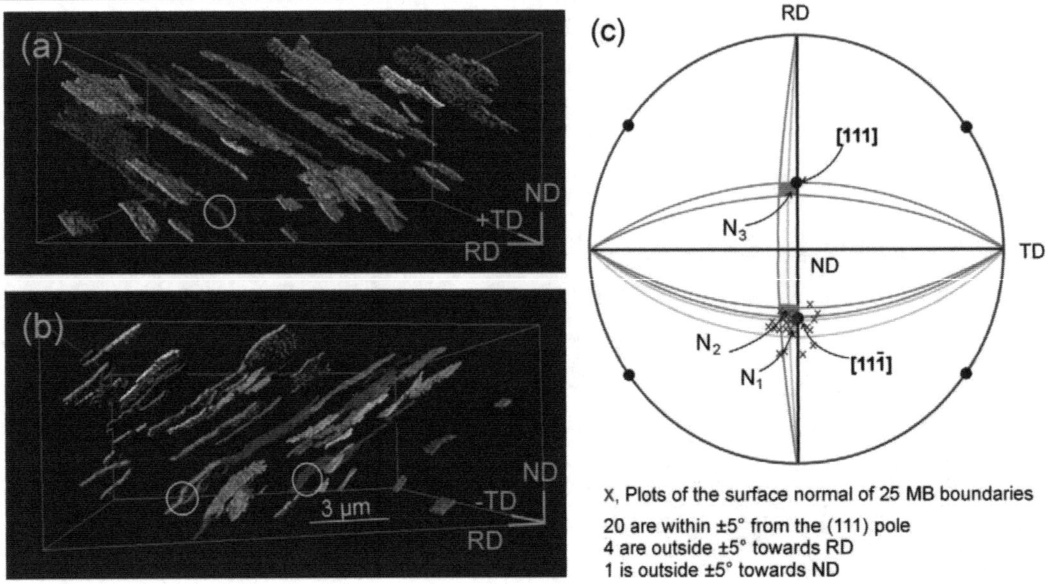

Figure 4. 3D EBSD reconstruction of 25 MB boundaries in the PSC Ni crystal, showing their inclinations and curvatures when viewed from +TD (a) and -TD (b). (c) <111> pole figure showing the plot of the average normal of these boundaries (x symbols) showing their close proximity to the $[11\bar{1}]$ pole of the Goss orientation ((110)[001]). The average normal of the flat segments of the MB boundaries are also plotted from their traces as N_1, N_2 and N_3 (in the shaded ranges), to show their close correspondence with the poles of the potential {111} slip planes, after [32].

2.2. Uncovering highly misoriented fragments at hotband grain boundaries in cold rolled IF free steel

3D EBSD was used to investigate the deformation heterogeneties that form in the vicinity of prior hotband grain boundaries in a 75% cold rolled IF steel [30]. This study focused on the actual structure of the boundary region between two near γ-fibre grains since it is generally reported for IF steels that nucleation of recrystallzation takes place in the highly fragmented substructures of these γ-fibre grains [35-36]. Within the γ-fibre grains (<111>//ND), shear bands [36-37] and deformation bands [35, 38] form profusely and these are arguably the principal sites for recrystallization. Nevertheless, considerable nucleation also occurs at the prior HAGBs of the hot band [39].

Figure 5 shows a representative EBSD micrograph from one of the 45 FIB milled sections showing three deformed grains N, O and P. Grain N contains a parallel set of LAGBs inclined at an acute angle to RD. While the total orientation spread within this grain is large (~28°) (inset), the average orientation is close to one of the prominent γ-fibre orientations ((1 1 1)[1 $\bar{2}$ 1]) of the BCC rolling texture. Grain P contains two sets of intersecting microbands at ±20° with RD and its average orientation (inset) is close to another prominent γ-fibre orientation (1 1 1)[0 $\bar{1}$ 1]. Within grain O, there is a very large orientation spread and the grain is highly fragmented, as revealed by the presence of numerous thick black lines in the EBSD micrograph.

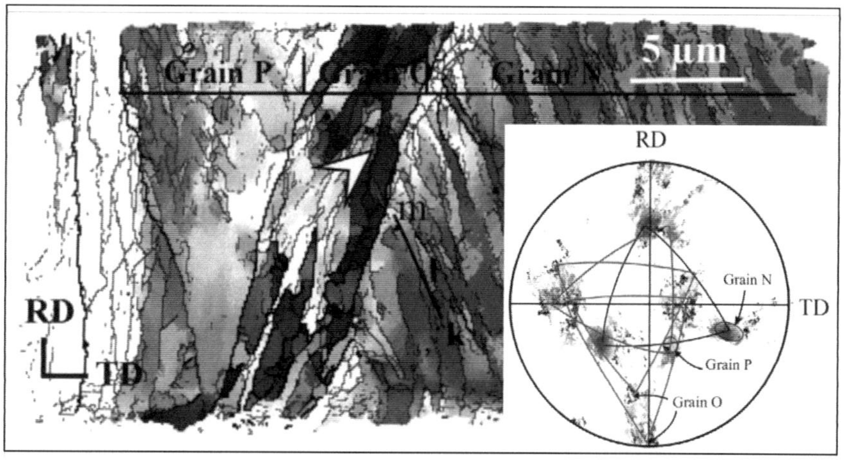

Figure 5. (a) EBSD micrograph (and inset showing the corresponding <200> pole figure) showing the orientation spread of deformed grains N, O and P. Here, HAGBs (>15°) and LAGBs (2-15°) are plotted as thick and thin black lines, respectively, after [30].

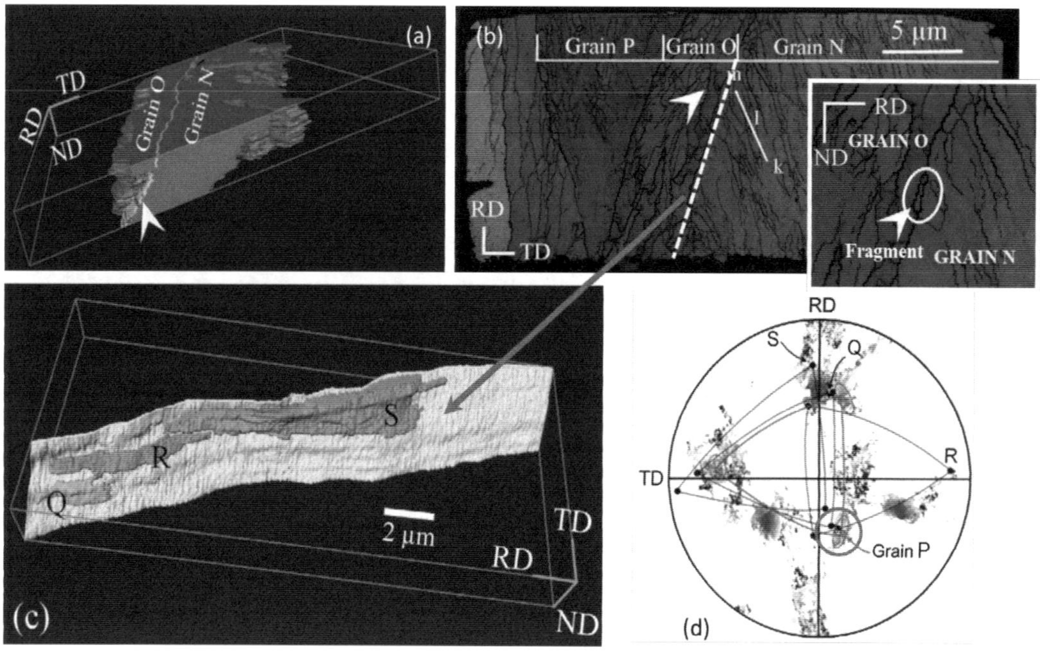

Figure 6. (a) 3D reconstruction of the N/O grain boundary affected region showing the serrated appereance of the microband-containing surface corresponding to grain N and irregular nature of the HAGB corresponding to grain O. (b) 2D EBSD section of the N/O grain boundary and (inset) elongated fragment on the boundary (circled). (c) 3D reconstruction of the actual N/O boundary showing substantial crystal fragments. (d) <200> pole figure showing the orientations of the boundary fragments in (c) superimposed on the local spread in orientations of grains N, O and P (figures adapted from ref. [30]).

Figure 6a shows the 3D reconstruction containing the N/O boundary (yellow line) with figure 6b showing one of the EBSD sections of the reconstructed volume. A close inspection of this boundary (dashed white line) reveals the presence of a thin, elongated fragment (circled in the inset in figure 6b). The actual shape of this feature is almost impossible to reveal in a single 2D EBSD micrograph, and such features can be easily overlooked.

Figure 6c shows the N/O boundary extracted from the volumetric EBSD data set. The boundary contains several thin, elongated fragments (Q, R, S fragments are shown herein but there are others). While these fragments may have small and large orientation variations along their length and slicing directions (see figure 7), their central orientations are plotted in the <200> pole figure in figure 6d, and compared with the orientations of grains N, O and P. The identified fragments sit on the left side of the N/O boundary adjacent to grain O, but their orientations are neither close to any of the orientations within these adjoining deformed grains (figure 6d) nor are they close to the α- and γ-fibre orientations [30]. However, these fragments have an average orientation close to grain P (figure 6d). Hence, it is likely that these boundary fragments have originated from an earlier disintegration event of grain P leaving remnant orientations of this grain as a Q-R-S film on the N/O boundary. The evidence also suggests that grains O and P were actually a single grain that eventually fragmented during deformation. Hence, these fragments do not have orientations within the local orientation spread near the N/O boundary. The present findings are consistent with the 2D EBSD work of Ushioda *et al.* [40], who reported the occurrence of numerous thin fragments on γ/γ boundaries in their 75% cold rolled IF steel, and these fragments were also found to have different orientations to their adjoining grains.

Figure 7 shows a series of EBSD slices highlighting a thin grain boundary fragment spanning many EBSD slices. This particular fragment has a large orientation variation within the slice length (see relevant pole figure with slice 22 in figure 7a). Identical orientation plots from the subsequent slices reveal (in figures 7b-e) a gradual rotation in orientation in the slicing direction (compare the pole figures associated with slice 22 to 30).

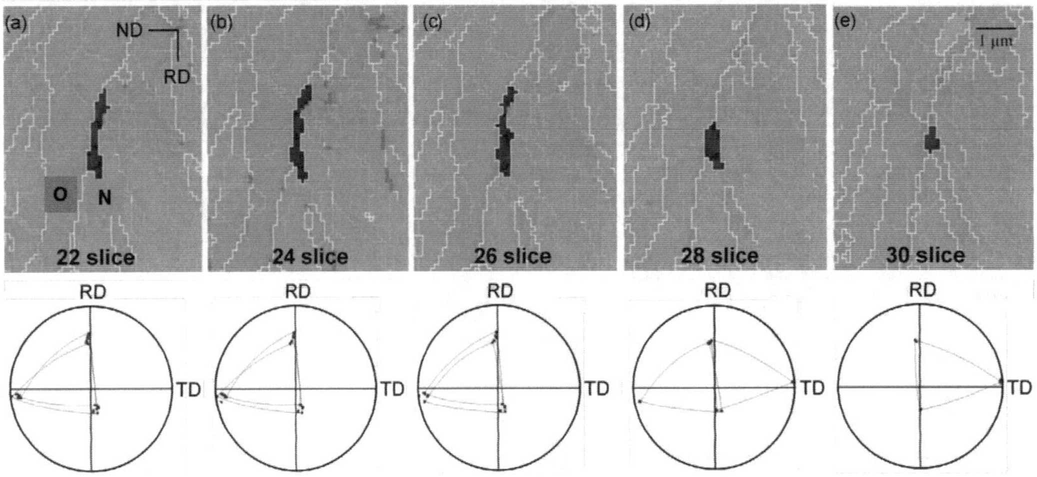

Figure 7. Series of EBSD micrographs (with corresponding <002> pole figures) of a thin, elongated fragment at the N/O grain boundary showing a large orientation spread within the length of the fragment and along the slicing direction in slices 22, 24, 26, 28 and 30. In the EBSD maps, the yellow lines represent misorientation boundaries over 5°, after [30].

The steep orientation gradients near HAGBs, deformation-induced interfaces and grain boundary fragments generated in the cold rolled IF steel are all potential sites for nucleation of recrystallization [21]. The steep orientation gradient near the N/O boundary in figure 5 is likely to be an ideal site for recrystallization. The large width of this region within the reconstructed volume (see figure 6a) also indicates that multiple nucleation events are likely to occur near a grain boundary face, as evidenced by the common occurrence of strings of grains nucleating along prior grain boundaries after deformation and subsequent annealing [30]. The high angle interface generated within grain O near the N/O boundary (figure 5) is also a potential site for recrystallization. Juul Jensen *et al.* [41], Hansen *et al.* [42] and Godfrey *et al.* [43] reported the early growth of subgrains at irregular boundary interfaces and concluded that local variations in texture (orientation) and microstructure (stored energy) play a major role in recrystallization.

Finally, but most interestingly, thin fragments left on the N/O boundary are bounded entirely by HAGBs and, hence, are also expected to be potent recrystallization nuclei. Despite these fragments not having a size advantage in the thickness direction over the average subgrain size of the deformation substructure (0.5 to 1.0 μm) in cold rolled steels [40,44-45]), their orientations are indeed found in low intensities in the recrystallization textures of IF steels. For example, the {110}<110> component (close to Q, R and S in figure 6d) is found in small intensities in the annealing textures of cold/warm rolled IF steels, despite these orientations not being identified in the rolling textures [36-38]. Therefore, it is possible that such orientations originate from these types of boundary fragments generated during rolling. Further, the discovery of such grain boundary fragments in the deformation microstructure provides a plausible explanation for the origin of recrystallized grains with orientations other than those found within the adjoining deformed grains in the vicinity of grain boundaries; this phenomenon has been commonly observed in texture data for many years and observed experimentally [21] but remained largely unexplained.

3. An automated method for reconstructing various types of boundaries and interfaces in 3D EBSD data sets

As discussed in section 2, plastic deformation generates complex deformation substructures that have a pronounced influence on subsequent recovery and recrystallization processes [21]. Unlike HAGBs, the variations between subgrain/cell structures are more subtle in magnitude (low total difference in orientation between subgrains) and spatially gradual (low point-to-point differences). This makes the automatic identification of their boundaries challenging and, hence, creates a significant obstacle to generating reliable data of the spatial distribution of these features within the overall deformation substructure. The ability to reliably segment a volumetric EBSD data set into substructure features is important for understanding the mechanisms of substructure formation by plastic deformation through to nucleation of recrystallization, and the overall relationship between microstructure and bulk material properties.

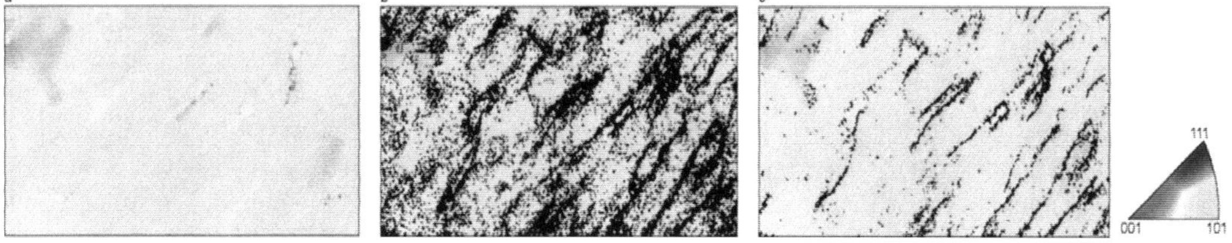

Figure 8. Misorientation thresholding of gradual boundaries in an EBSD data set containing microbands in a single grain of a commercial purity aluminium alloy loaded in uniaxial tension to a strain of 33%. (a) IPF colour map. The black points indicate locations of pixel-to-pixel misorientation greater than (b) 1° and (c) 2° (figure adapted from ref. [46]).

The identification of features in a typical EBSD data set may be achieved by grouping points that are not separated by a boundary with misorientation higher than a user-defined threshold. This is a common method in commercial software but fails when applied to subgrain/cell data since there is no threshold angle that identifies the boundaries of interest without being dominated by noise in the sample. As an example, figure 8 shows thresholding on EBSD data from a deformed aluminium sample containing microbands. Noise also causes problems, as mis-indexed pixels are usually classified as boundaries. In addition, volumetric reconstructions require closed boundaries, something the pixel-to-pixel method does not guarantee.

An approach to solve this thresholding problem is to use an edge-preserving lowpass filter to smooth the data. Humphreys *et al.* [47] have implemented a Kuwahara filter, which reduces the noise in a data set by statistically analysing neighbourhoods surrounding every point. Each pixel is iteratively replaced with the average of the neighbourhood with the least variance. This technique is capable of reducing orientation noise by approximately a factor of ten, but it may also lead to an oversegmentation of microstructures. Noise and random differences in variation can compound over successive iterations, creating microstructural artifacts, such as boundaries that should not exist. Furthermore, the results are unpredictable over spatially gradual boundaries such as those present in a typical deformation microstructure.

The authors have recently developed a robust boundary identification method, mathematically based on the fast multiscale clustering (FMC) image processing algorithm [46,48], which overcomes the major limitations of thresholding. The method can reliably reconstruct both 2D [46] and 3D [48] EBSD data sets to reveal the myriad boundaries found in deformed and annealed metals. Such boundaries can range from LAGBs through to HAGBs, the latter often associated with prior grain boundaries and those associated with nucleation of recrystallization. In our method, FMC was adapted to work with a quaternion representation of orientation data in 3D. The identification of features in a volumetric EBSD data set may be achieved by grouping points that are not separated by a boundary with a misorientation higher than a user-defined threshold. However, as noted above, this common method in commercial software is not applicable to subgrain data. Figure 9a-c shows an example of an artificially-constructed volume of orientation data and thresholding with two angles that illustrates this problem. We present a robust solution to segmenting 3D EBSD data, which identifies the actual boundary geometry, as shown in figure 9d.

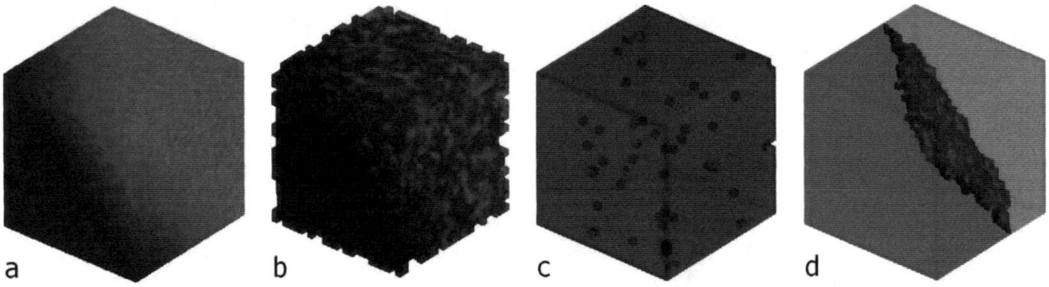

Figure 9. (a) A constructed data set with 25 voxels per side and 9° corner-to-corner misorientation. The transition occurs linearly along the middle third of the cube diagonal. Noise with average misorientation of 0.5° is added to make the data more realistic. A 0.5° threshold angle finds many superfluous boundaries (b), whereas a 0.9° angle finds very few boundaries (c). In both cases noise dominates. (d) Segmentation using FMC produces two clusters with a diagonal boundary.

3.1. Fast multiscale clustering (FMC)

This segmentation technique was originally developed for 2D image processing [49]. The method creates a hierarchy of scales from groupings of data, starting with individual data points at the finest scale and iteratively coarsening to combine groups. At each scale in the hierarchy every group is then scored based on the similarity of points it contains and distinction from neighbouring groups. The highest-scoring groups form the final segmentation of the data. Determining similarity of data requires a metric for measuring distance. For simple colour images, this can be a scalar difference between colours, but for application to EBSD data, the metric is based on misorientation. Hence, we have adapted FMC to work with a quaternion representation for orientation data. A detailed description of FMC and its modifications for both 2D and 3D EBSD data sets is provided elsewhere [46, 50].

At each scale in the hierarchy, groups inherit properties from the finer scale below, down to the individual data points [50]. This allows FMC to segment an entire data set with awareness of local and global trends. Another advantage is the freedom to decide neighbours of points, since the method does not depend on spatial distribution of the data. Thus FMC is not only applicable for 2D situations but easily applied in 3D. Furthermore, since there are fewer groups at coarser scales, FMC has a runtime linear with the number of data points, a necessity for methods applied to large data sets with more than two dimensions.

Misorientation is not useful on its own when comparing coarse groups containing many data points. Thus, Mahalanobis distance, where the distance between the average aggregated values is divided by the group's variance, is used. This gives groups with low variation in orientation naturally less similarity to their neighbours than groups with high variance. We developed a novel variance aggregation method since the elements of quaternions are not linearly correlated [46, 48]. While this Mahalanobis distance lacks physical significance, it can be scaled at the aggregation step to give sense to closeness and farness, depending on the nature of the data set. The scaling parameter, C_{Maha}, determines the sensitivity of the segmentation [49]. FMC requires several other user-defined values, but fixed values for the rest have been effective for all data sets examined. Hence, FMC has the additional advantage that it is controlled with a single parameter for application to different data, analogous to the cut-off angle for a thresholding method.

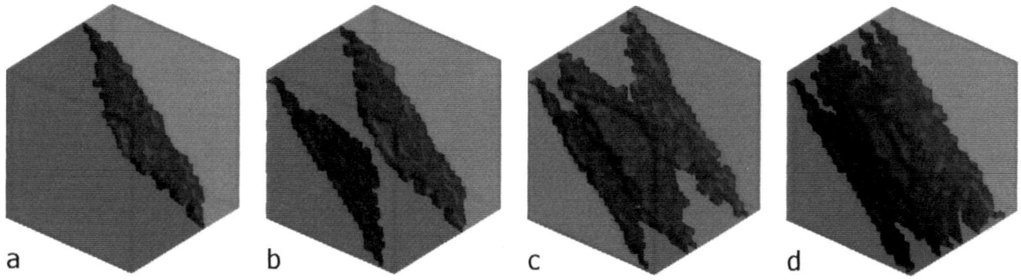

Figure 10. Segmentation of the data in figure 9 with C_{Maha} of: (a) 2.5, (b) 5, (c) 7.5, and (d) 10.

3.2. Determination of the optimal sensitivity parameter

The sensitivity of the segmentation, controlled by C_{Maha}, should be different based on the nature of the data. Boundaries between clusters in polycrystalline data should be on the distinct grain boundaries without consideration for variation within grains, but subgrain features are only captured with a more sensitive segmentation. There is no ground truth available for such a segmentation, so the user is responsible for deciding which segmentation to use. Thus, an automated method for determining the optimal value of C_{Maha} has been developed, inspired by similar clustering problems in machine learning [51]. Oversegmentation is characterized by low marginal variance change from splitting existing clusters. Therefore, an "elbow" in a plot of a global variance measure as a function of C_{Maha} indicates a shift from creating clusters that should be distinct towards erroneously breaking these clusters further. Once this elbow is identified, the segmentation created using the associated value of C_{Maha} is selected as optimal. The sensitivity of the segmentation controlled with C_{Maha} is demonstrated in figure 10.

3.3. Surface extraction

When analysing grain or subgrain volumes, it is often useful to isolate separate faces of an irregular feature. These faces may be defined as regions having similar normal vectors and, as the FMC method developed operates on orientations, only a small further modification is needed for application to normal vectors. With this modification, a second use of FMC identifies coherent faces from the set of points on the exterior of a cluster [48]. The voxel locations from a 3D segment are used as the input for a volume reconstruction using tentrahedra. The triangulated boundary surface is used to estimate local normal vectors, and these are used in the FMC process. As with the main use of FMC, the single C_{Maha} parameter controls the sensitivity of the segmentation of the surface, and the same elbow criterion is used to find the optimal segmentation that separates distinct faces without oversegmenting. It is pertinent to note that it is a straightforward task to identify all of the large clusters in a volume and all of the largest faces on those clusters, making study of many subgrain surfaces possible.

It is important to note that some spatial pre-processing must be carried prior to segmentation out since 3D EBSD data sets are collected in 2D sections generated by FIB. Measurement skew and instrument drift are corrected with a bilinear interpolation on each slice [52]. The data are then globally realigned, skewing the points off their regular grid. Points then may be interpolated to a rectangular grid for computational efficiency, but FMC can be performed with any spatial distribution of data points.

3.4. Automated reconstruction of complex boundaries in 3D EBSD data sets

We demonstrate the excellent capabilities of our method with application to 3D EBSD data sets generated from cold rolled aluminium containing well-defined microbands, channel-die plane strain compressed Goss-oriented nickel crystal containing microbands with very subtle changes in orientation, and cold rolled and partly recrystallized IF steel microstructure containing three magnitudes of boundary misorientations.

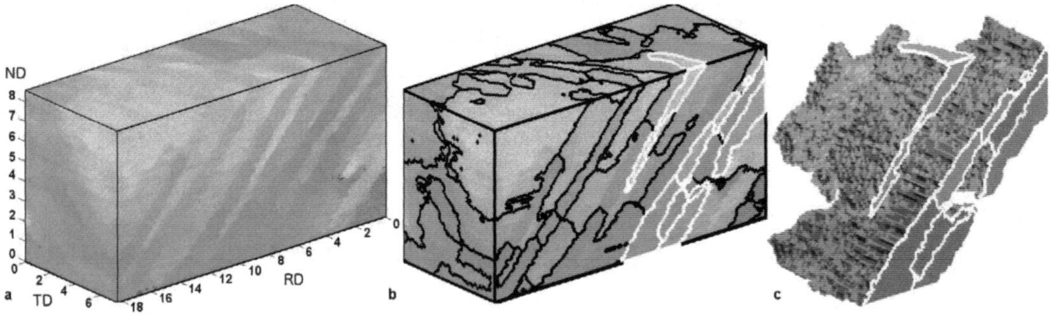

Figure 11. 3D reconstruction (1.1 million data points) of the cold rolled aluminium data set: (a) EBSD data using the IPF colour map of figure 8; (b) black and white lines show the boundaries of segments found with the optimal C_{Maha} value of 4.4, and (c) eight selected clusters, indicated with white boundaries in (b), showing the underlying subgrain structure, including a boundary plane that extends through the entire data set (Scale is in micrometres), after [48].

3.4.1. Microbands in cold rolled aluminium. We first apply FMC to segment microbands in a single grain of cold-rolled (22% reduction) commercial purity aluminium [34]. Figure 11 shows the EBSD data with inverse pole figure colouring, in which the microbands are visible, as well as its segmentation. The microband structure is captured by the FMC segmentation. The highlighted microbands in figure 11b are isolated in 3D in figure 11c to show the nature of the interior of the segmentation. This data set was also segmented with the Kuwahara filter method [47]. It was found that, while this type of filter partially aligns with the visible microband structure, no threshold value was found to be appropriate for the entire data set, which was evidenced by the presence of both significant oversegmentation and undersegmentation. This difficulty is overcome by FMC since its segments are determined with aggregate orientation and variance differences, not strictly local information.

The FMC surface extracting method was applied to extract microbands from the EBSD data set. Figure 12 shows the top microband of figure 11c, with the entire top and bottom surfaces identified. These surfaces can be used for further geometric and crystallographic analysis. FMC is a powerful tool for this application since it can form a full surface from similar components that are separated by a sharp step in the estimated normal vectors.

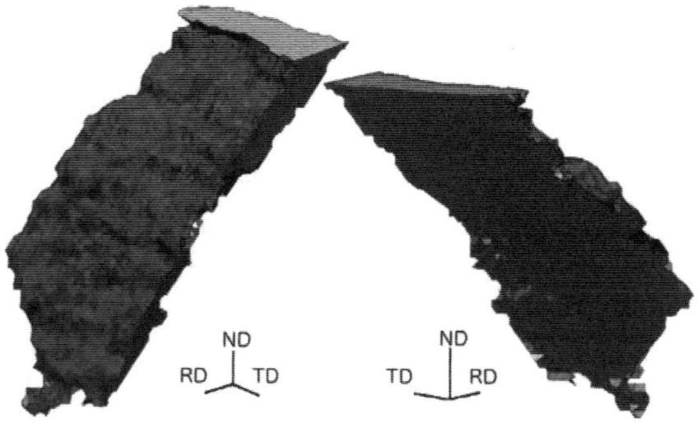

Figure 12. Two views of a cluster in the segmentation of cold rolled aluminium data. Colours indicate separate extracted surfaces. The top and bottom surfaces are isolated, as well as flat surfaces on the top and side from the boundary faces of the data set, after [48].

3.4.2. Microbands in deformed nickel single crystal. Microbands have been successfully isolated in a 3D EBSD data set from a channel die plane strain compressed (35% strain) Goss-oriented nickel single crystal [33]. The data have a standard deviation from the average orientation of only 0.7°. Figure 13 shows the segmented microbands. Identification of these subgrain features would not be attainable with conventional thresholding at any angle. The segmentation in figure 13b is performed with the optimal C_{Maha}. To show how even more subtle features may be identified, figure 13c shows a segmentation with a higher C_{Maha} value. This comes with the consequence of oversegmenting the data overall, but specific features of interest may still be used for further analysis.

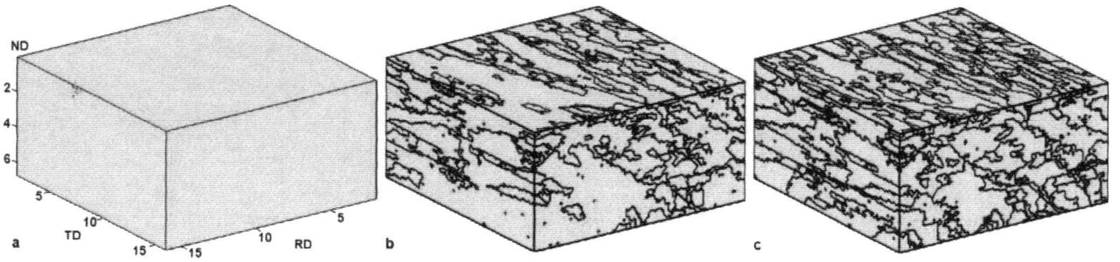

Figure 13. 3D reconstruction (> 1.3 million data points) of the nickel data set: (a) EBSD data using the IPF colour map of figure 8; (b) boundaries of segments found with the optimal C_{Maha} value of 6, and (c) boundaries of segments found with a C_{Maha} value of 7, showing the identification of more subtle features, after [48].

3.4.3. Recrystallizing grains in cold rolled and partly recrystallized IF steel. To further illustrate the capability of our method in identifying the broad range of misoriented boundaries that are often present in the same data set, the 3D EBSD data set of the partly recrystallized IF steel described earlier was analysed. Figure 14a shows the raw EBSD data of the reconstructed data set. Figures 14b and c reveal the subtle subgrain features, distinct grain boundaries, and a HAGB separating α- and γ-fibre regions; such boundaries are almost impossible to delineate using conventional thresholding techniques. Further, the presence of the HAGB separating the regions does not affect the detection of boundaries with a smaller misorientation, even those adjacent to it. This is because variance comparisons are done between each neighbouring group, rather than with a global standard over the entire data set. Finally, figure 14d shows the volume of local stored energy using the subgrain method of Choi and Jin [53] for 2D EBSD data sets, which has been extended to 3D by modifying their method to use all neighbouring EBSD points in 3D. In conjunction with FMC, this will be a powerful method for scrutinising how variations in stored energy coupled with grain boundary character can influence the nucleation of recrystallization through to the early and latter stages of recrystallization in deformed metals.

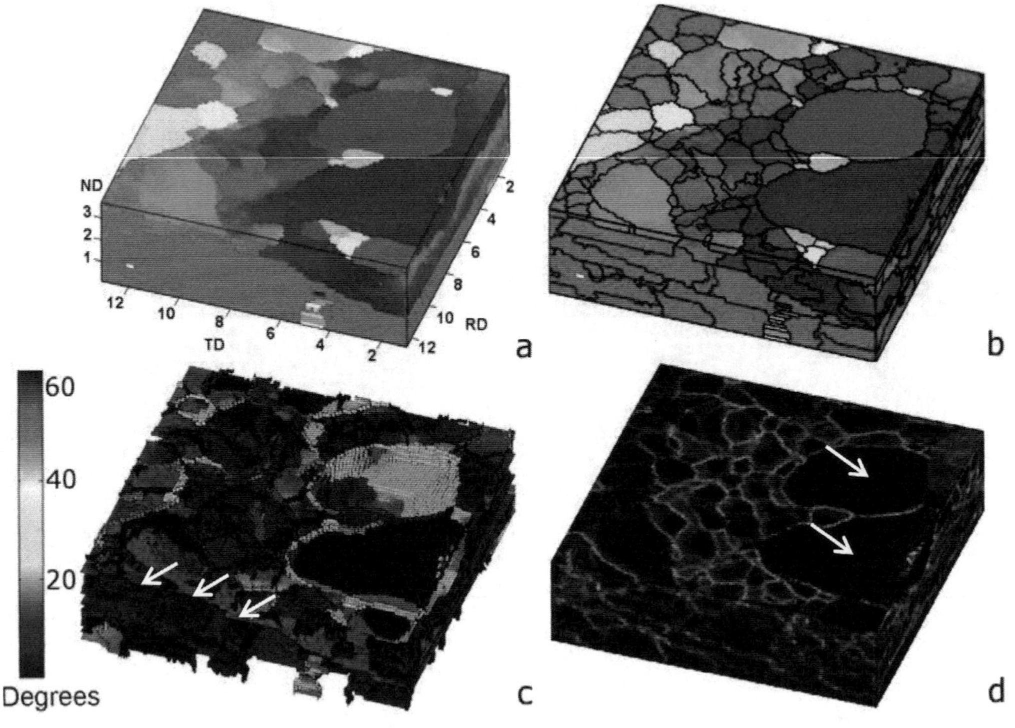

Figure 14. 3D reconstruction (350,000 data points) of the partly recrystallized IF steel data set [54]: (a) EBSD data using the IPF colour map of figure 8; (b) segmentation with the optimal C_{Maha} value of 3.4; (c) boundaries from the segmentation coloured by local misorientation, showing the HAGB between α- and γ-fibre regions (arrowed), and (d) map of local stored energy computed with the so-called subgrain method described in [53] showing the high energy substructure surrounding the low energy recrystallizing grains (arrowed), whereby black indicates regions of low energy and white indicates regions of high energy in relative terms (Scale is in micrometres).

The foregoing examples illustrate the effectiveness of FMC as a tool for segmenting 3D EBSD data sets containing subtle and gradual variations in orientation. We have incorporated FMC modified to operate on orientation data into the free open source texture analysis software package MTEX [8], thereby allowing users to take advantage of its existing methods for importing and manipulating data. The ability of the method to segment data based on a wide range of boundary magnitudes with a linear runtime make FMC a robust method for both 2D and 3D segmentations.

4. Concluding summary

By combining FIB with FEGSEM in a DualBeam™ or similar platform, it is possible to sequentially mill sections of a material by FIB and map the crystallographic features of each newly created surface by EBSD. This allows crystallographic volumes of microstructure to be generated and analyzed by this serial sectioning technique. 3D EBSD was used to investigate various phenomena associated with the plastic deformation of metals and alloys where new information it consistently revealed not clearly evident using standard 2D analysis techniques. For example, the structure of microbands in FCC Ni was highlighted and the new observation of grain boundary fragments that are potential sites for nucleation of recrystallization with orientations not present in the surrounding local substructure. We also illustrate an effective tool, based on fast multiscale clustering (FMC), for segmenting 3D EBSD data sets containing subtle and gradual variations in orientation. The ability of the method to segment data based on a wide range of boundary magnitudes with linear runtime make FMC a robust method for both 2D and 3D segmentations.

Acknowledgements
The authors gratefully acknowledge the Australian Research Council (ARC) and US National Science Foundation (NSF) for supporting much of our 3D EBSD work through the ARC Discovery grants program (grants no. DP0556927, DP0986600), ARC Centre of Excellence for Design in Light Metals (CE05611574), and NSF Materials World Network program (DMR-0807240). The 3D EBSD reconstruction method described in Section 3 was supported by the Laspa Fellowship in Applied Mechanics, the US NSF (DMR-0807240, DGE-1144153), ARC Centre of Excellence (CE05611574), and the Harvey Mudd College Department of Engineering. We thank Prof. Julian Driver and Dr Adeline Albou (Ecole des Mines des St Etienne, France) for preparing, cutting and deforming the Ni crystal, Emma Liu and Donald Bolton for contributions to collection of the Al and Ni 3D EBSD data sets, and Ayyappa Vemulkar and David Golay for helpful discussions. The authors also would like to acknowledge the support of the UNSW node of the Australian Microscopy and Microanalysis Research Facility (AMMRF).

References
[1] Möbus G and Inkson BJ 2007 *Materials Today* **10** 18
[2] Schmidt S, Nielsen SF, Gundlach C, Margulies L, Huang X and Juul Jensen D 2004 *Science* **305** 229
[3] Larsen AW, Poulsen H F, Margulies L, Gundlach C, Xing Q, Huang X and Juul Jensen D 2005 *Scripta Mater.* **53** 553
[4] Schmidt S, Olsen U L, Poulsen H F, Sørensen H O, Lauridsen E M, Margulies L, Maurice C and Juul Jensen D 2008 *Scripta Mater.* **59** 491
[5] West S S, Schmidt S, Sørensen H O, Winther G, Poulsen H F, Margulies L, Gundlach C and Juul Jensen D 2009 *Scripta Mater.* **61** 875
[6] www.edax.com/Products/EBSD
[7] Xu W, Ferry M, Mateescu N, Cairney JM and Humphreys FJ 2007 *Mater. Charact.* **58** 961
[8] Xu W, Ferry M, Cairney J M and Humphreys F J 2007 *Acta Mater.* **55** 5157
[9] Konrad J, Zaefferer S and Raabe D 2006 *Acta Mater.* **54**, 1369
[10] Mateescu N, Ferry M, Xu W and Cairney J M 2007 *Mater. Chem. Phys.* **106** 142
[11] Zaafarani Z, Raabe D, Singh RN, Roters F and Zaefferer S 2006 *Acta Mater.* **54** 1863
[12] Calcagnotto M, Ponge D, Demir E and Raabe D 2010 *Mater. Sci. Eng* A **527** 2738
[13] Robin L, Laws K J, Kurniawan G, Xu W, Privat K and Ferry M 2010 *Metall. Mater. Trans* A **41** 1691
[14] Rowenhorst D J, Gupta A, Feng C R and Spanos G 2006 *Scripta Mater.* **55** 11
[15] Lewis AC and Geltmacher A B 2006 *Scripta Mater.* **55** 81
[16] Lin FX, Godfrey A, Juul Jensen D and Winther G 2010 *Mater. Charact.* **61** 1203
[17] Bachmann F, Hielscher R and Schaeben H 2011 *Ultramicroscopy* **111** 1720
[18] Groeber M, Ghosh S, Uchic M D and Dimiduk DM 2008 *Acta Mater.* **56** 1257
[19] Xu W, Ferry M and Humphreys F J 2009 *Scripta Mater.* **60** 862
[20] Xu W, Quadir M Z, Ferry M and Humphreys F J 2009 *Metall. Mater. Trans* A **40** 1547

[21] Humphreys F J and Hatherly M 2004 *Recrystallization and Related Annealing Phenomena* (UK: Elsevier)
[22] Godfrey A, Juul Jensen D and Hansen N 1998 *Acta Mater.* **46** 823
[23] Winther G, Huang X and Hansen N 2000 *Acta Mater.* **48** 2187
[24] Winther G, Huang X, Godfrey A and Hansen N 2004 *Acta Mater.* **52** 4437
[25] Lin F X, Godfrey A and Winther G 2009 *Scripta Mater.* **61** 237
[26] Hurley P J and Humphreys F J 2003 *Acta Mater.* **51** 1087
[27] Hurley P J, Bate P S and Humphreys F J: *Acta Mater.* 2003 **51** 4737
[28] Humphreys F J and Bate P S 2007 *Acta Mater.* **55** 5630
[29] Quadir M Z, Mateescu N, Bassman L, Xu W and Ferry M 2007 *Scripta Mater.* **57** 977
[30] Afrin N, Quadir M Z and Ferry M 2015 *Metall. Mater. Trans* A (DOI: 10.1007/s11661-015-2878-4)
[31] Afrin N, Quadir M Z and Ferry M 2014 *Metall. Mater. Trans B* **45** 345
[32] Afrin N, Quadir M Z, Xu W and Ferry M 2012 *Acta Mater.* **60** 6288
[33] Afrin N, Quadir M Z, Bassman L, Albou A, Driver J H and Ferry M 2011 *Scripta Mater.* **64** 221
[34] George C, Soe B, King K, Quadir M Z, Ferry M and Bassman L 2013 *Mater. Charact.* **79** 15
[35] Duggan B J and Tse Y Y 2004 *Acta Mater.* **52** 387
[36] Barnett M R 1998 *ISIJ Int.* **38** 78
[37] Barnett M R and Jonas J J 1997 *ISIJ Int.* **37** 697
[38] Quadir M Z and Duggan B J 2006 *Acta Mater.* **54** 4337
[39] Quadir M Z 2003 *PhD thesis The University of Hong Kong*
[40] Ushioda K, Nakanishi S, Morikawa T, Higashida K, Suwa Y and Murakami K 2013 *Mater. Sci. Forum* **753** 58
[41] Juul Jensen D, Lin F X, Zhang Y B and Zhang Y H 2013 *Mater. Sci. Forum* **753** 37
[42] Hansen N and Juul Jensen D 2011 *Mater. Sci. Tech.* **27** 1229
[43] Godfrey A, Juul Jensen D, Hansen N 2001 *Acta Mater.* **49** 2429
[44] Dillamore I L, Smith C J E and Watson T W 1967 *Metal Sci.* **1** 49
[45] Goodenow R H 1966 *Trans ASM* **59** 804
[46] McMahon C, Soe B, Loeb A, Vemulkar A, Ferry M and Bassman L 2013 *Ultramicroscopy* **133** 16
[47] Humphreys F, Bate P and Hurley P 2001 *J. Microsc.* **201** 50
[48] Loeb A, Ferry M and Bassman L 2015 *Ultramicroscopy* (to be submitted)
[49] Kushnir D, Galun M, Brandt A, 2006 *Pattern Recogn.* **39** 1876
[50] Sharon E, Meirav G, Sharon D, Basri R and Brandt A 2006 *Nature* **442** 810
[51] Sugar C A, Lenert L A, Olshen R A 1999 *Technical Report. Stanford University*
[52] Soe B, McMahon C, Golay D, Quadir M Z, Ferry M and Bassman L 2012 *1st Int. Congress on 3D Materials Science* (Seven Springs, Pennsylvania, USA, 8-12 July 2012) eds M De Graef, H F Poulsen, A Lewis, J Simmons and G Spanos, pp 189-194
[53] Choi S and Jin Y 2004 *Mater. Sci. Eng A* **371** 149
[54] Loeb A, Soe B, McMahon C, Ferry M and Bassman L 2014 *2nd Int. Congress on 3D Materials Science* (Annecy, France, 29 June-2 July 2014) eds D Bernard, J-Y Buffiere, T Pollock, H F Poulsen, A Rollett and M Uchic, pp 9-14

IOP Conf. Series: Materials Science and Engineering **89** (2015) 012003 doi:10.1088/1757-899X/89/1/012003

Characterization and influence of deformation microstructure heterogeneity on recrystallization

A Godfrey[1], O V Mishin[2] and T Yu[2]

[1]Key Laboratory of Advanced Materials (MOE), School of Material Science and Engineering, Tsinghua University, Beijing 100084, P.R. China
[2]Danish-Chinese Center for Nanometals, Section for Materials Science and Advanced Characterization, Department of Wind Energy, Technical University of Denmark, 4000 Roskilde, Denmark

E-mail: awgodfrey@mail.tsinghua.edu.cn

Abstract. The microstructure resulting from plastic deformation of metals typically contains heterogeneity on several length scales. This is also true for samples deformed to large strains, where an important form of heterogeneity is in the variation in microstructural refinement by high angle boundaries. A methodology for quantifying this type of heterogeneity based on the identification of areas classified as low misorientation regions (LMRs) is described, and some parameters for quantification of both the extent and length scale of LMRs are presented. It is then shown how this approach can be used to investigate the early stages of recrystallization in samples deformed to large strains, by direct comparison of electron backscatter diffraction (EBSD) maps of the same area before and after annealing. Methods for estimation of the stored energy of deformation from EBSD data are also surveyed and the problems of each for quantification of the local variation in stored energy are discussed, where it is concluded that a method based on the summation of the contributions from individual boundary segments is considered to be the best suited at present for characterization of the local variation in stored energy on the scale of the dislocation boundary features.

1. Introduction

For many metals plastic deformation results in a process of microstructural subdivision on several length scales by the formation of dislocation boundaries [1-3]. This in turn leads to a refinement in the microstructural scale, which in some cases can extend down to the level of just a few nanometres [4,5]. A large number of studies have been carried out to investigate the evolution of deformation microstructure with increasing strain, and have resulted in the establishment of a well-defined framework to describe this process in terms of the types of dislocation boundaries formed, their spacing and misorientation angle, and their relationship to the underlying slip pattern [6-8].

An important characteristic of such deformation microstructures is the presence of heterogeneity, or local variation, in the microstructure. This heterogeneity can take several forms, over different length scales. For samples deformed from low to medium strains heterogeneity is present on the grain scale, as a result of the orientation dependence of the deformation microstructure [9,10]. Such an effect can in fact also be seen in samples deformed by rolling to large strains if the initial grain size is sufficiently large [11,12]. Inside individual grains heterogeneity can also exist in the form of micro-shear bands and localized glide bands [13-15], where such features are most commonly observed at medium to high strains. Sample scale heterogeneity can also be important, either in the form of macroscopic shear banding [16-18] or due to the nature of the applied deformation (for example, the effect of friction and roll gap geometry during rolling [19], or dead zones during compression).

The importance of the deformation microstructure for understanding recrystallization during annealing of deformed metals is well accepted. It is also worth pointing out, however, the high sensitivity of recrystallization to local variations in the deformation microstructure. This is true both for nucleation, for example, deformation zones around hard inclusions [20,21] or localized glide bands [22] as locations for potential nuclei, but also for growth of recrystallizing grains, where it is now recognized that protrusions and retrusions can form along a migrating recrystallization interface as a response to local variations in the deformed microstructure, and that the resulting interface profile can directly affect the average migration velocity of the interface [23]. A clear example of this high sensitivity to local variations in the deformation microstructure can be seen in the recrystallization behaviour of aluminium single crystals of {112}⟨111⟩ ("Copper") and {123}⟨634⟩ ("S") orientations deformed to a strain of $\varepsilon = 1.5$ by channel-die compression. For both crystals the average misorientation angles and spacings of the dislocation boundaries in the deformed microstructure, and hence the stored energy of deformation, are very similar [24], but for the Copper orientation crystal recrystallization starts at a temperature of 200°C after just 2 minutes, whereas for the S orientation crystal recrystallization is only observed at a temperature of 300°C after 30 minutes of annealing [25].

These observations highlight the importance of characterization of the deformed microstructure, and therefore estimation of the stored energy, on a scale that can capture such local variations, whether for analysis of recrystallization behaviour, or for use as an input to models of recrystallization. At present the two choices available for collecting such microstructure-based data are transmission electron microscope (TEM) observations and electron backscatter diffraction (EBSD) in the scanning electron microscope. TEM studies have the advantage of high spatial resolution and good angular resolution in measurement of the deformation microstructure, but only allow relatively small areas to be investigated even when making use of semi-automated or automated methods for measurement of crystal orientations [26-27]. In contrast, EBSD offers the possibility to collect microstructural data over several length scales (from tens of nanometres to millimetres), allowing heterogeneity within a deformed microstructure to be more fully characterized. The standard EBSD technique does however suffer from a limited angular resolution, such that boundary misorientations of a less than a certain angle (typically 1.5 – 2.0° for deformed samples) cannot be reliably identified [28-30]. Some consequences of this limitation in angular resolution for estimation of stored energy of deformation are considered later in this work. First we give an example of the use of the EBSD technique to quantify both the extent and length scale of deformation microstructure heterogeneity in samples deformed to large plastic strains, and show how this can be used to analyze recrystallization of such samples.

2. Characterization of heterogeneity in refinement at large plastic strains

2.1. Extent of refinement by high angle boundaries

There has been much interest in recent years in deformation to large plastic strains (so-called severe plastic deformation) by techniques such as equal channel angle pressing/extrusion (ECAP/ECAE), high pressure torsion, accumulated roll-bonding and heavy cold-rolling [31-33]. It is often assumed that the microstructure is uniform after deformation to large plastic strains, and in some cases the motivation for the use of such techniques is given as the preparation of samples with an ultra-fine grain size. However, even in samples deformed to large plastic strains, heterogeneity in the microstructure is present in the form of significant local variations in the refinement by high angle boundaries (HABs).

An example is given in figure 1, showing an EBSD map of a sample of aluminium AA1050 (99.5% purity) with an initial grain size of 45 μm after deformation by 8 passes of ECAE route Bc (90° sequential rotations between each pass) using a conventional split die with a sharp 90° corner. Further details of the processing conditions can be found in [34]. Note that all EBSD data in this paper were collected using the Oxford Instruments HKL Technology Channel 5 system. Some evidence of heterogeneity in the microstructure can be seen in figure 1(a), in which all boundary misorientations >2° are marked by black lines. This heterogeneity is more clearly seen though in figure 1(b), where HABs (misorientation angle >15°) are marked by thick black lines and low misorientation angles (between 2° and 15°) are marked by thin black lines. The map shows that in many places there are regions predominantly containing only low angle boundaries (LABs). Based on this characteristic, these have been termed low misorientation regions (LMRs).

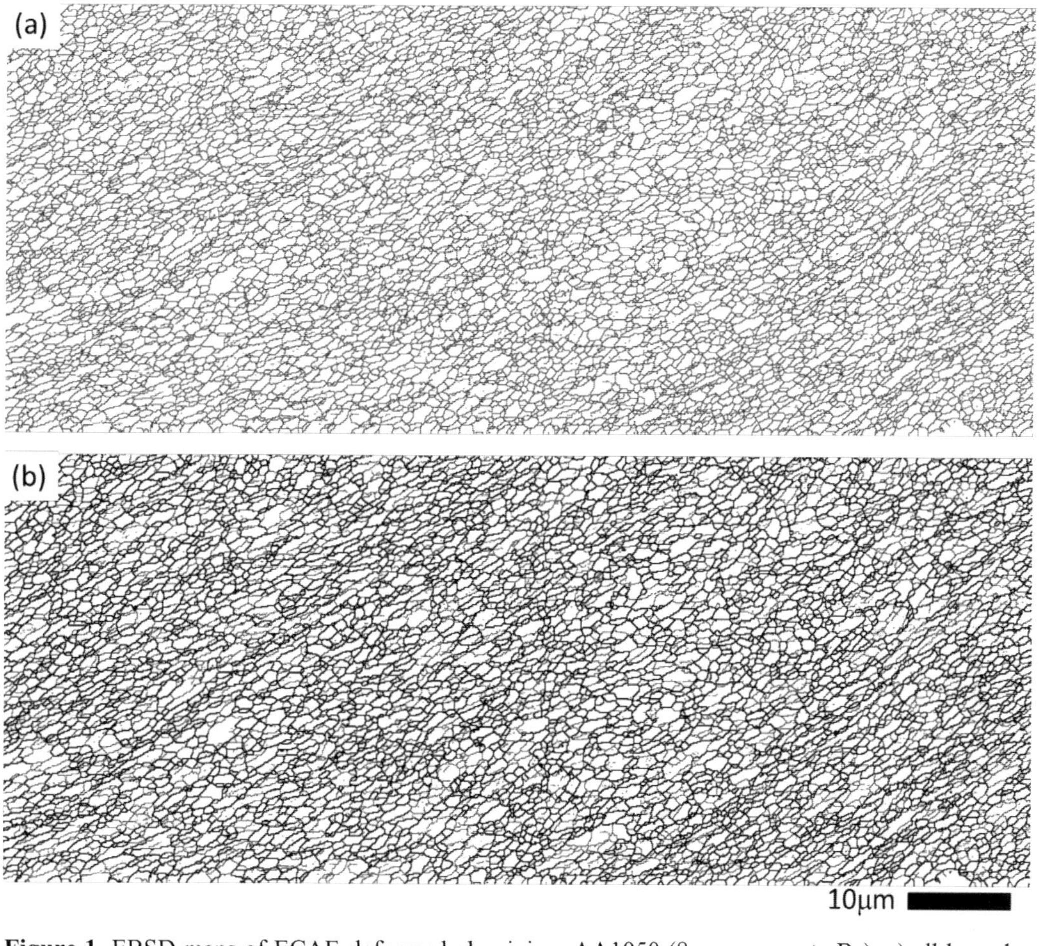

Figure 1. EBSD maps of ECAE-deformed aluminium AA1050 (8 passes; route Bc): a) all boundary misorientations > 2°; b) boundary misorientations >15° and 2° – 15° marked by thick and thin black lines, respectively.

A methodology to quantify the extent (volume or area fraction) of these LMRs using functions available in standard EBSD post-processing software has been proposed [35], as described in the following text. Firstly a "grain reconstruction" is performed using a misorientation angle definition of $\theta^G = 5°$. The result of such a procedure is shown in figure 2(a). Here each detected "grain" is coloured at random, and boundary misorientations of >5° and 2 – 5° are shown by thick and thin black lines, respectively. It can be seen that in some cases "grains" are detected, inside which no boundaries of lower misorientation angle are seen. In other cases, the detected "grains" contain boundaries of lower misorientation angle, and in many cases consist of clusters of dislocation cells. The use of quotation marks here is, therefore, a reminder that the detected areas are not necessarily grains in a conventional sense, but are areas of the deformed microstructure bounded by misorientation angles of > 5°.

A subset of these "grains" with an area greater than a specified threshold value (A^*) can then be defined, as shown in figure 2(b) for $A^* = 2.5\ \mu m^2$. The choice of this value for A^* is based on a multiple (10-times in this case) of d_{EBSD}^2, where d_{EBSD} is the average boundary spacing for the entire mapped area, measured via linear intercept measurement using the lowest appropriate misorientation threshold possible (typically 2° for such samples). In this way the LMRs identify places where the microstructure is predominantly subdivided by only <5° boundaries over a continuous area equivalent to at least 10 average-size dislocation cells/grains.

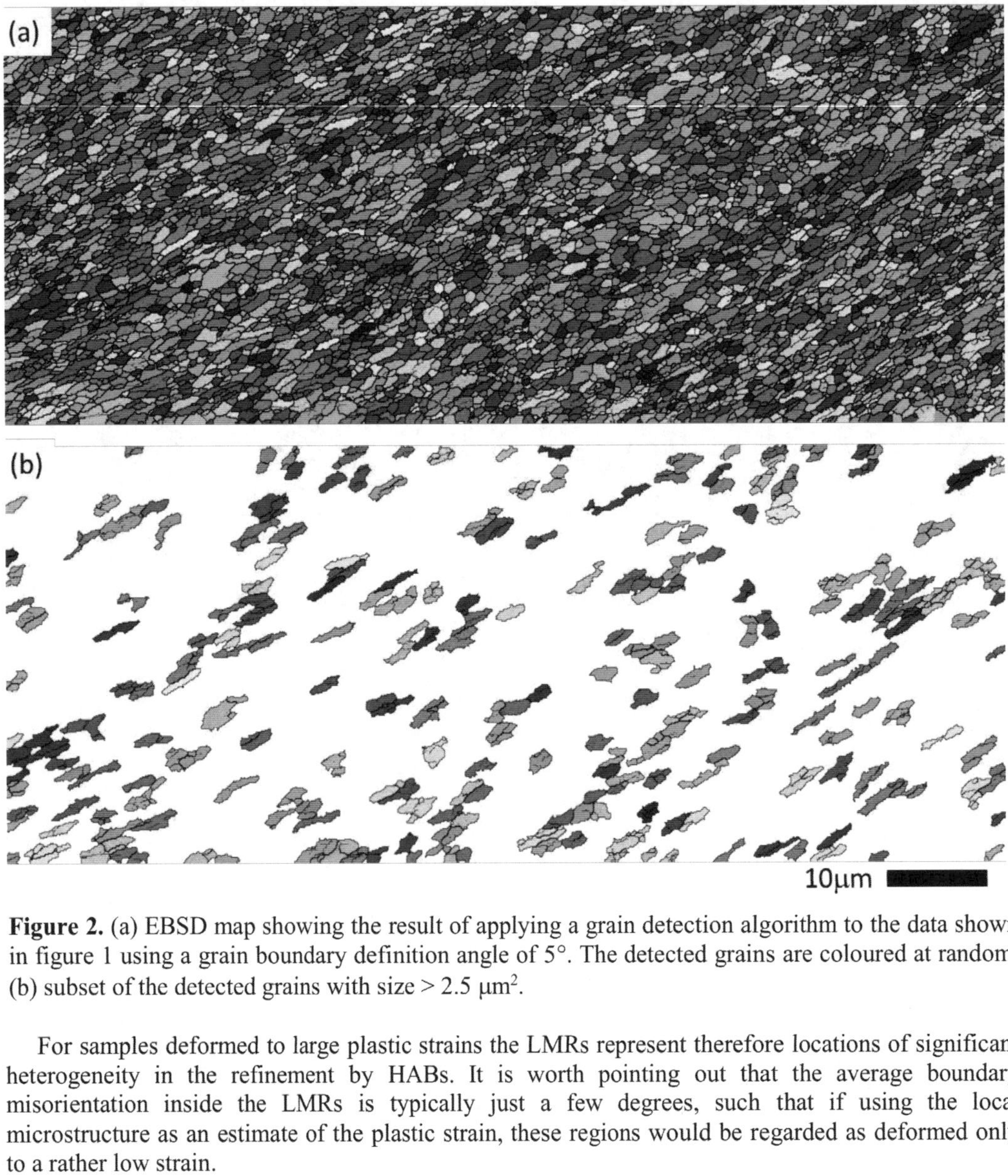

Figure 2. (a) EBSD map showing the result of applying a grain detection algorithm to the data shown in figure 1 using a grain boundary definition angle of 5°. The detected grains are coloured at random; (b) subset of the detected grains with size > 2.5 μm².

For samples deformed to large plastic strains the LMRs represent therefore locations of significant heterogeneity in the refinement by HABs. It is worth pointing out that the average boundary misorientation inside the LMRs is typically just a few degrees, such that if using the local microstructure as an estimate of the plastic strain, these regions would be regarded as deformed only to a rather low strain.

A quantitative measure of the extent of such heterogeneity is given by the area fraction of the mapped area covered by the LMRs, f_{LMR}. For the area shown in figure 2(b) f_{LMR} = 19%, i.e. approximately 1/5 of this highly deformed sample has a microstructure corresponding to a significantly lower level of plastic deformation. It can be noted also that the LMRs are in some cases isolated but can also be found in clusters. In the latter case they may be separated by HABs, but nevertheless the microstructure in these areas is predominantly subdivided only by LABs. The remaining area of the map is, by default, designated as high misorientation regions (HMRs) – although in this case the HMRs in fact form a single percolating region. The value of f_{LMR} will of course depend on the choices of both A* and of the misorientation angle used for the initial "grain" detection (θ^G). A discussion of the motivations for the choices of the values for these parameters is given elsewhere [35]. The values used here have, however, been found to be good choices for comparing heterogeneity in refinement by HABs between samples deformed under a number of different processing conditions [34-41].

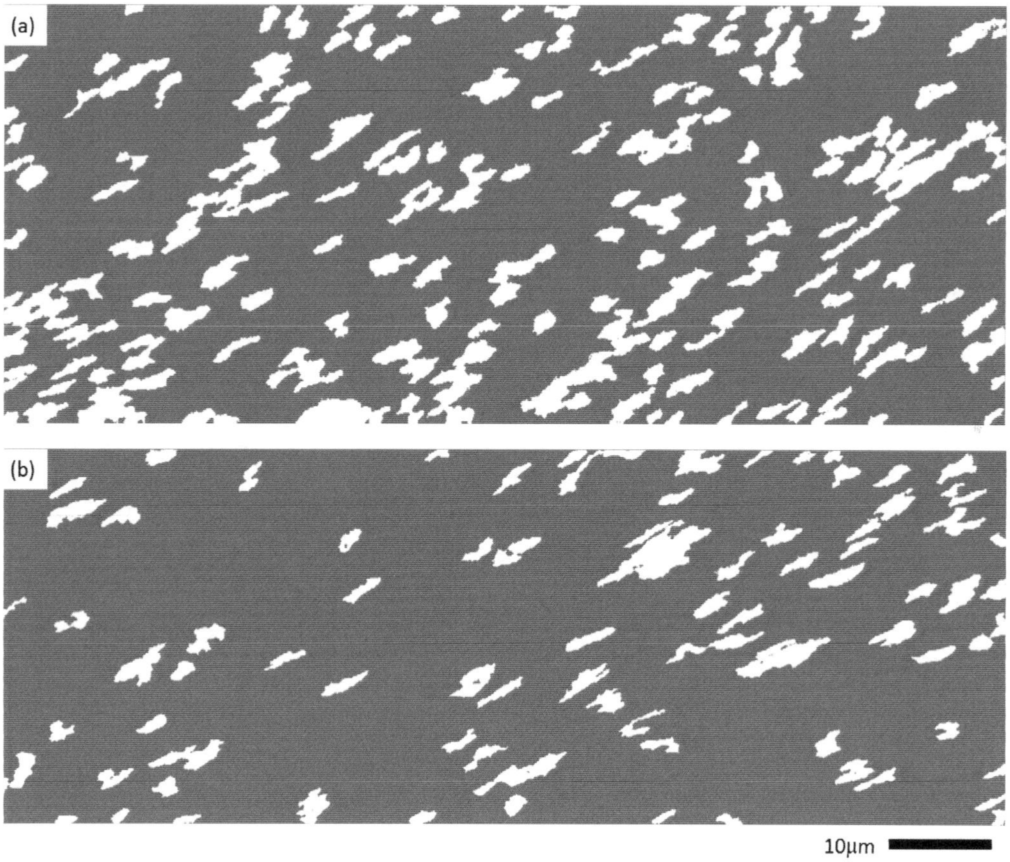

Figure 3. Phase-maps for two ECAE-deformed aluminium AA1050 samples showing LMRs in white, and the HMRs in grey: (a) sample described in [34]; (b) sample described in [40]. In each case $\theta^G = 5°$ and $A^* = 10d_{EBSD}^2$.

Such a comparison is illustrated in figure 3, where LMR/HMR "phase maps" (showing the regions identified as LMRs in white and the remaining HMRs in grey) are given both for the ECAE-processed AA1050 sample described above, and also for another sample of the same material (AA1050) with a similar initial grain size (50 μm), deformed also by 8 passes via route Bc, but using a different 90°

ECAE die with a sliding floor [40]. It is evident that the extent of heterogeneity in refinement by HABs is lower for the latter sample, where f_{LMR} is calculated as 10%.

2.2. *Estimation of length scale of heterogeneity*

In addition to quantifying the extent of heterogeneity in refinement by HABs it is also of interest to determine the length scale of this heterogeneity. One obvious measure that is readily available is the average area of the LMRs, $\langle A \rangle_{LMR}$. This parameter is influenced directly by the choice of A^*, in that the area of any individual LMR by definition cannot be smaller than this value. Another perhaps obvious choice to quantify the length scale of heterogeneity would be average spacing between the LMRs. This, however, is difficult to measure in the standard way (e.g. measuring the distance between LMRs along randomly placed test lines) as the often highly convoluted nature of the LMR-HMR interface results in unreasonably low values for the LMR spacing [35].

An important observation to consider in designing a parameter to measure the length scale of the LMRs is the fact that, as seen in figure 2(b), the LMRs are frequently found in clusters. A schematic illustration of some extreme cases for different spatial distributions of LMRs is shown in figure 4. For each case the area fraction (f_{LMR}) is identical. In figure 4(a) the LMRs are non-clustered and randomly distributed, whereas in figure 4(b) they are also non-clustered, but are found in dense pockets. For the case of figure 4(c) the LMRs are strongly clustered, but the clusters can be regarded as being randomly distributed. Ideally any length scale measure should be able to distinguish between these different types of spatial distribution, as well as providing an estimate of the relevant length scale for each.

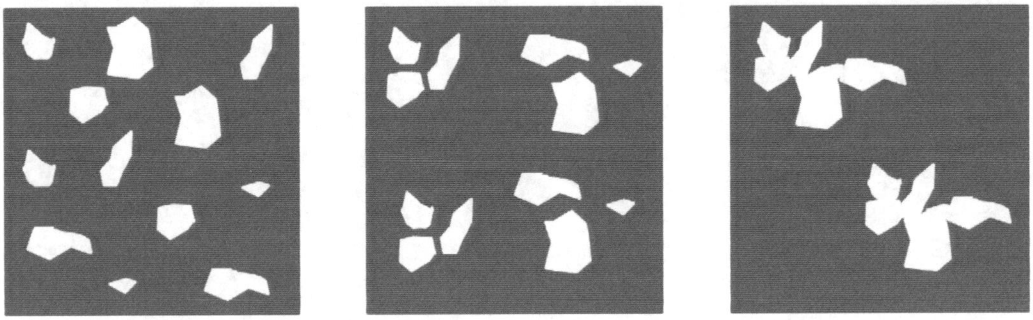

(a) (b) (c)

Figure 4. Schematic illustration showing some extreme cases for different spatial heterogeneity of LMRs: (a) non-clustered, random; (b) non-clustered, dense pockets; (c) clustered, with clusters randomly dispersed.

A simple parameter to estimate the extent of LMR clustering is the ratio of the average area of LMR clusters $\langle A_{LMR(cl)} \rangle$ to the average area of the individual LMRs $\langle A_{LMR} \rangle$. The difference between these two definitions is illustrated in figure 5, showing an enlarged view of a region of the EBSD map used in figures 1-3. In figure 5(a) each individual cluster is identified by a single colour, whereas in figure 5(b) each single colour identifies a LMR cluster. Note that not all LMRs are found in clusters; such LMRs are nevertheless included for the calculation of $\langle A_{LMR(cl)} \rangle$, assuming in effect a cluster size of one. Values for the ratio of $R_A = \langle A_{LMR(cl)} \rangle / \langle A_{LMR} \rangle$ for the two data sets described previously are 2.1 and 1.3 for the ECAE samples deformed using a conventional split-die (hereafter Sample 1) and a sliding-floor die (Sample 2), respectively. More sophisticated measures can certainly be extracted, for example by considering the distribution of areas, but this simple measure already captures the smaller extent of LMR clustering in Sample 2.

As a measure of the length scale in the spatial distribution of the LMRs (and LMR clusters) a method has been proposed based on a statistical analysis of representative test areas of increasing size. The basic idea is to determine f_{LMR} as a function of map area, i.e., $f_{LMR}(A_{map})$ and use this data to determine the minimum statistically representative area. For very small values of A_{map} (e.g. "maps" of area smaller than the average LMR area) a very large scatter in f_{LMR} will be found. As the value of A_{map} approaches the length scale of the heterogeneity, the scatter in f_{LMR} will decrease and approach a constant value. The idea is illustrated in figure 6, again using the data for Sample 1. As mentioned above, for the chosen θ^G and A^* the value of f_{LMR} for the entire map is 19%. If instead the map is divided into four sub-maps of 1/4 the map area, values of f_{LMR} = 18, 20, 22, and 23% are obtained. Similarly for eight sub-maps of 1/8 the map area, a still larger spread in f_{LMR} is obtained (in this case from 15% to 31%).

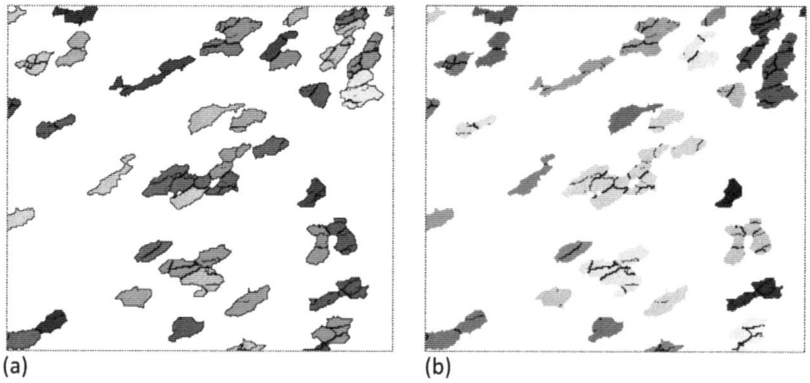

Figure 5. EBSD images, showing a subset of the map data for figures 1-3, illustrating the definition of individual LMRs and clusters of LMRs. In (a) each individual LMR is colored at random; in (b) LMR clusters are colored at random. Average areas are calculated based on the area of all coloured features.

In this case the original map in figure 3(a) has been divided into smaller non-overlapping areas (figure 6). In practice the calculation is best carried out by moving test-boxes of increasing area across a single large map, with some overlap between each test box, and then calculating the value of f_{LMR} for each position. Graphs showing the data following such an analysis of the EBSD maps of the ECAE-processed Sample 1 and Sample 2 are shown in figure 7, where the boxes were taken with a 50% overlap in both the x- and y-directions. As expected, for very small test box sizes the scatter in f_{LMR} is very large. With increasing test box size the spread in f_{LMR} decreases, and the values converge towards the average for the entire map area. An estimate for the length scale corresponding to the spatial distribution of the LMRs can then be obtained by determining the test box size where CV, the coefficient of variation (the relative standard deviation), falls below a pre-determined value. Here we use a value of $CV = 0.2$, giving representative test box lengths for the LMRs in Samples 1 and 2 of $L_{TestBox}$ = 15 μm and 25 μm, respectively. One requirement for such an approach is that the initial EBSD is large enough to fully capture the heterogeneity in the spatial distribution of the LMRs. With modern EBSD systems this is generally not a problem, but is something that should be remembered in planning a data collection strategy.

A summary of the comparison between the microstructural parameters for the two EBSD maps representing Samples 1 and 2 is given in Table 1. It is interesting to note that for both these samples the parameter most commonly used as a measure of the extent of microstructural refinement, namely the fraction of HABs (f_{HAB}), is almost identical. As such f_{HAB} does not provide any differentiation between the two microstructures, which is perhaps not surprising as this parameter does not contain any length scale information. The LMR-based parameters show that in addition to the smaller fraction of LMRs in Sample 2, the LMRs in this sample are of similar size, but significantly less clustered. The larger value of L_{TB} for Sample 2 indicates however a larger mesoscale heterogeneity for this sample, which is supported by the visual impression from comparing figures 3(a) and 3(b).

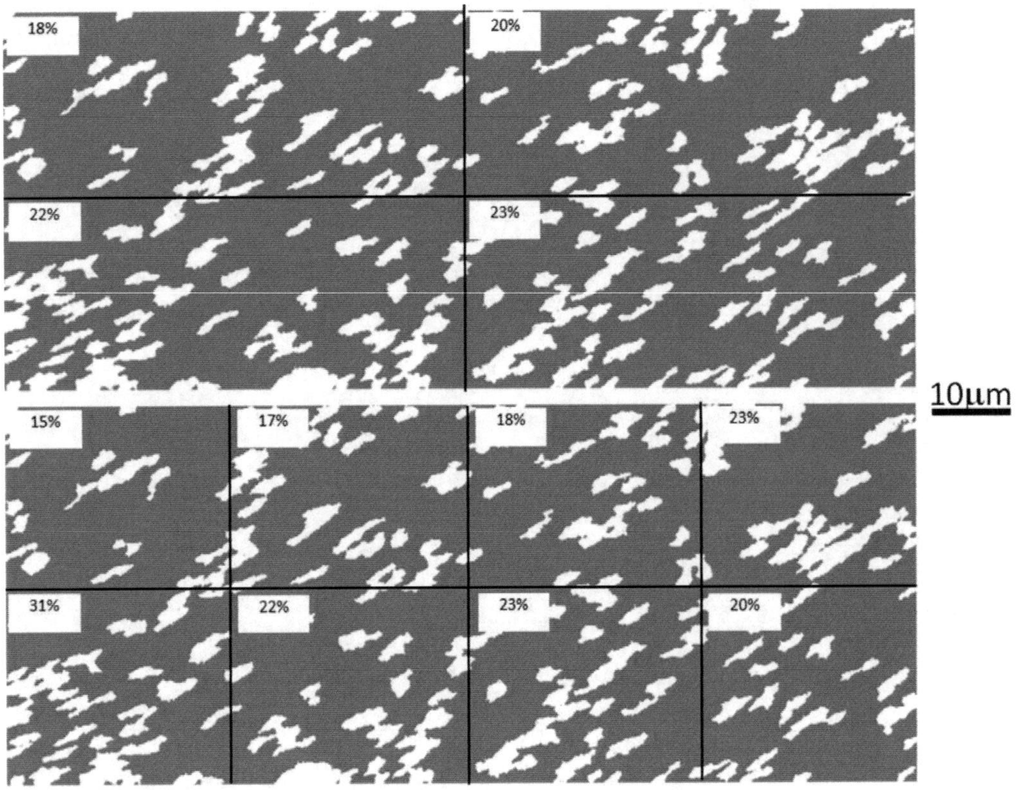

Figure 6. EBSD images illustrating length scale characterization by considering the variation in f_{LMR} with decreasing map area. The value of f_{LMR} for the entire map is 19%. Numbers in white boxes give the value of f_{LMR} for the sub-regions marked by black lines.

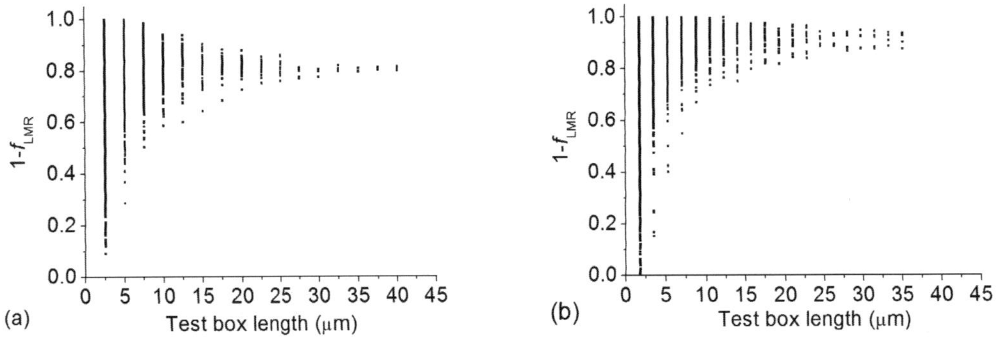

Figure 7. Variation of f_{LMR} (plotted as 1-f_{LMR}) as function of increasing test-box size, calculated using test boxes with a 50% overlap: (a) Sample 1 - corresponding to data in figure 3(a); (b) Sample 2 - corresponding to data in figure 3(b).

Table 1. LMR characteristics for the EBSD maps representing two different ECAE-processed (8 passes via route Bc) samples of aluminium AA1050

Sample	f_{HAB}	f_{LMR}	$<A>_{LMR}$ (μm²)	R_A (μm²)	$L_{TestBox}$ (μm)
1	68%	19%	4.0	2.1	15
2	69%	10%	3.6	1.3	25

3. Influence of heterogeneity on recrystallization

The presence of heterogeneity in the form of regions of significantly different microstructural characteristics can be expected to have direct influence on the recrystallization behaviour of metals after deformation to large plastic strains. The annealing of such metals is of particular interest, as there exists some debate concerning how uniformly, or otherwise, microstructural coarsening takes place in such samples. Clearly, however, the presence of significant heterogeneity in the microstructure can be expected to have some influence on the annealing behaviour.

As an illustration of the relationship between nucleation of recrystallization and heterogeneity in refinement by HABs, an investigation of the annealing behaviour of a sample of copper, deformed by ECAE is presented here. The copper (99.98%) sample was deformed by 15 passes using route A (no rotation of the billet between ECAE passes) using a 120° die angle (further details can be found in [37]). The longitudinal section of this sample was polished mechanically followed by electropolishing. One region in this sample was mapped using EBSD first in the as-deformed state, and the same region was mapped again after annealing for 30 minutes at 160°C. This annealing treatment was chosen to yield material at the start of recrystallization. In order to allow a direct comparison between locations of the recrystallization nuclei and the deformation microstructure, the EBSD map after annealing was collected without mechanical re-polishing of the sample surface (the surface was only quickly electropolished to remove the oxide layer formed during annealing in air). The two maps were aligned using clearly recognizable features.

An EBSD map showing the examined region of the ECAE copper sample is given in figure 8(a), in which an inverse pole-figure (IPF) colouring is used. The microstructure is predominantly lamellar, as typical for ECAE samples processed using route A, and consists of bands of fine subgrains separated by bands of less refined material. This heterogeneity is highlighted in figure 8(b) where LMRs, identified using parameters of $\theta^G = 5°$ and $A^* = 10d_T^2$ (= 2.1 μm²), are shown. For this map the LMR area fraction (f_{LMR}) is 32%. In this case the remainder of the map is formed of several large HMRs (in contrast to the cases shown above where the LMRs are distributed in a single fully connected HMR).

Figure 8(c) shows an IPF map of the same area after annealing at 160°C for 30 minutes. A number of small recrystallization nuclei are visible, both isolated and some found in clusters, confirming that the annealing treatment leads to an early stage of recrystallization. Figure 8(d) shows the locations of all the recrystallization nuclei in the mapped area. These were identified by first applying a grain detection procedure to the data set, and then using the average internal misorientation of each detected grain, as well as the size to identify a subset of the map containing the recrystallization nuclei. This automated procedure was able to pick-up most of the nuclei, though some minor manual adjustment of the subset was needed for cases where a nuclei retained some segment of low misorientation angle boundary to the adjacent deformed matrix.

Figure 8(e) shows the recrystallization nuclei superimposed onto the LMR/HMR phase-map for the as-deformed sample. Nuclei are seen predominantly in the HMRs, with 80% of the total nuclei area overlapping these regions and many nuclei fully inside HMRs. Very few nuclei are found fully inside LMRs. However, in several places nuclei overlap both LMR and HMR areas. In each of these cases the nuclei share a HAB with the adjacent LMR region. Although the LMRs represent regions of lower stored energy compared to HMRs, the fairly similar spread of orientations within the LMRs may provide good conditions for growth of recrystallization nuclei.

It should be remembered of course that there are always complications in interpretation of recrystallization mechanisms using only surface observations [42]. In the present case, however, the recrystallization nuclei are small and of comparable size to the LMRs and HMRs. As such even if the nuclei are formed below the sample surface, the location of the nuclei can still be related to the LMRs and HMRs. Nevertheless, it will be advantageous to use in-situ heating (to improve alignment between "before" and "after" data sets) and shorter annealing times (to confirm nucleation locations).

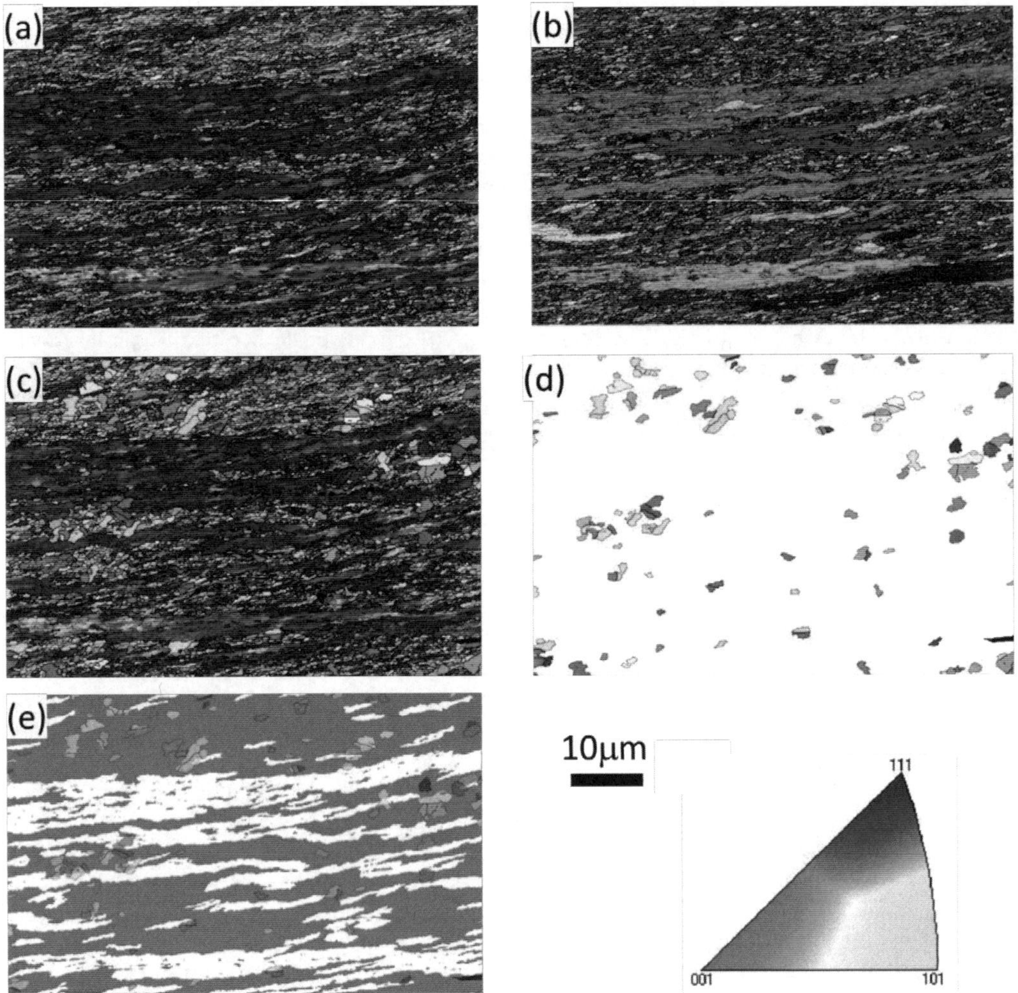

Figure 8. EBSD images of the same area of an ECAE-deformed copper sample before and after annealing at 160°C for 30 minutes: (a) before annealing (IPF colouring); (b) LMRs coloured at random and HMRs in grey; (c) after annealing (IPF colouring); (d) recrystallization nuclei; (e) recrystallization nuclei superposed on the LMR/HMR map of the as-deformed sample. For (a)-(d) thick and thin black lines represent misorientations of >15° and 2-15°, respectively.

4. Challenges in estimation of local stored energy

The stored energy of deformation provides the driving force for recrystallization, and is an important parameter for understanding both nucleation and growth of recrystallizing grains. In this regard the presence of heterogeneity in deformed microstructures requires the use of microstructure-based measurements of stored energy, in order to capture important local variations. It is clear, for example, that the LMRs and HMRs presented above will have different local stored energy, but variations between the stored energy of individual LMRs may also be important. In the case of metals with medium-to-high stacking fault energies, such as nickel and aluminium, the majority of dislocations are arranged in the form of low-energy dislocation boundaries [1-8]. The stored energy, E_s, can therefore be expressed by the equation

$$E_s = \gamma S_V, \tag{1}$$

where S_V is the boundary area per unit volume, and γ is the energy per unit area of the boundary. Here the small contribution of loose dislocations between these boundaries is ignored. Detailed analysis [24,43] of stored energy may be carried out in the TEM based on equation (1) and stereology. However, such measurements are difficult and can only be made over relatively small areas. In contrast, the EBSD technique can provide quantitative microstructural data over large areas and has been used by a number of authors, e.g. [44-48], for estimation of the stored energy of deformation. It is important to remember, however, that the EBSD technique has limited angular and spatial resolution [49]. Care must therefore be taken when EBSD data are used for stored energy calculations, whether for examining local variations or for calculation of average values. In the following we describe some of the methods that have been proposed/used for estimating the stored energy of deformation from EBSD data sets, and then apply each of these methods to a data set collected from a sample of deformed nickel, with a view to highlighting some of the problems in making such local stored energy calculations. Note that in the following text for convenience we refer to any volume enclosed wholly or partly by dislocation boundaries as a subgrain, For accuracy, however, it should be remembered, however, that for dislocation cell-forming metals the term "subgrain" should be interpreted to mean "subgrain or dislocation cell".

4.1. Calculation methods based on EBSD data

4.1.1 The average method (Method A). The most straightforward method [45,46,50] for estimating the stored energy (E_s^A) is to calculate γ using the average misorientation angle θ_{av} based on the Read-Shockley equation and to take $S_V = \alpha/D_{av}$, where D_{av} is the average subgrain size and $\alpha = 3$ for equiaxed structures. A subgrain is taken as the smallest area fully surrounded by misorientation angles greater than the low cut-off misorientation angle (θ_{cut}) as shown in figure 9(a), and the equivalent circular diameter, d_{ECD}, [49] is then taken as the subgrain size. The average misorientation angle can be chosen in one of two ways: 1) $\theta_{H,av}$, the mean of misorientation angles above θ_{cut} (therefore considering the stored energy in all the boundaries); and 2) $\theta_{L,av}$, the mean of misorientation angles between θ_{cut} and 15° (considering only the deformation-induced LABs).

4.1.2 The modified average method (Method B). The modified average method is similar to the average method, using the average subgrain size and the average misorientation angle (which can again be chosen using either of the above two definitions). However, for this method only the misorientation angles along subgrain boundaries are considered in the calculation. In other words, the misorientation angles inside subgrains (dashed lines in figure 9(b)) are excluded in the calculation of the stored energy (E_s^B).

4.1.3 The segment method (Method C). In this method the stored energy (E_s^C) is calculated from each boundary segment in the EBSD map [44]. For each individual boundary segment (figure 9(c)), which is represented by the misorientation of two adjacent pixels, the boundary length per unit area is $L_{A,\mathrm{EBSD}} = \Delta/A$, where Δ and A are the step size and the map area, respectively. In equiaxed structures, the boundary area per unit volume (S_V) and boundary length per unit area (L_A) have the relationship of $S_V = (4/\pi)L_A$. Additionally the stepped nature of EBSD maps requires a correction of $L_A = (\pi/4)L_{A,\mathrm{EBSD}}$ [51]. Substituting these relationships into equation (1), the stored energy contribution from each boundary segment, $E_{s,i} = (\Delta/A)\gamma_i$, can be calculated, where the boundary energy γ_i is estimated from the Read-Shockley function. The total stored energy over any examined area can then be determined as a sum of the contributions from all boundary segments as follows

$$E_S = \frac{\Delta}{A}\sum_i^N \gamma(\theta_i), \tag{2}$$

where N is the total number of misorientation segments above θ_{cut}.

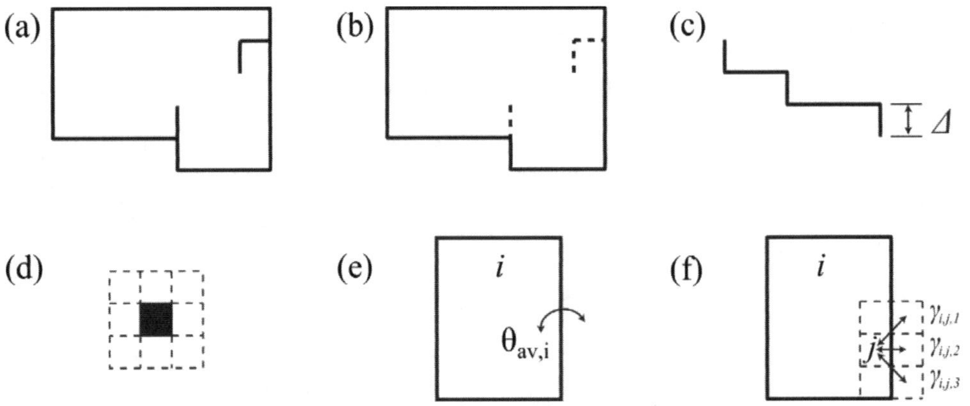

Figure 9. Illustration of methods for calculation of stored energy from EBSD data. (a) The average method includes all misorientation segments; (b) the modified average method excludes interior misorientation segments shown in dashed lines; (c) the segment method calculates the stored energy from each boundary segment; (d) the KAM method calculates the stored energy from each pixel; (e) the subgrain method performs boundary misorientation averaging for each subgrain; and (f) the modified subgrain method performs boundary energy averaging for each subgrain.

4.1.4 The kernel average misorientation (KAM) method (Method D). As an alternative to equation (1), the stored energy can also be expressed as the product of dislocation line energy (E_D) and the dislocation density (ρ) [52]

$$E_s = \rho E_D \approx \frac{\alpha\theta G b}{2d}, \tag{3}$$

where G is the shear modulus, *b* is the Burgers vector, *d* is the boundary spacing, θ is the boundary misorientation, and α is a constant and 3 for mixed boundary type.

The KAM here is defined for a given pixel (figure 9(d)) as the average misorientation of that pixel with all its immediate neighbours, which is calculated with the proviso that misorientation angles above a high cut-off (5° or 15° as used [47,53]) are not counted in the averaging process, and misorientation angles below θ_{cut} are considered as zero degrees, but are still counted during averaging. This approach differs from other KAM definitions [47,53], where no low cut-off misorientation angle is used. If the boundary misorientation angle θ and the boundary spacing *d* in equation (3) are replaced by the average KAM value, θ_{KAM}, and the step size Δ, respectively, the stored energy (E_s^D) can be expressed in the following form

$$E_s = \frac{3Gb}{2\Delta}\theta_{KAM}, \tag{4}$$

4.1.5 The subgrain method (Method E). After subgrain reconstruction with a certain value of θ_{cut}, the subgrain size D_i and the mean misorientation $\theta_{av,i}$ with its neighbours can be determined for each subgrain (figure 9(e)). Analogous to the average method, the stored energy associated with each subgrain can be obtained as

$$E_{s,i} = \alpha \frac{\gamma(\theta_{av,i})}{D_i} \tag{5}$$

The average stored energy (E_s^E) of the entire measured area can then be estimated by using the average (weighted by area fraction) for all subgrains.

4.1.6 The modified subgrain method (Method F). This method is similar to the subgrain method; however, in the modified subgrain method the average boundary energy is calculated [48] instead of the average misorientation angle. For each subgrain, the boundary pixels are first selected, and then the misorientation angles of each of these pixels with its n neighbours (n ≤ 8 for a square-grid data set) on the other side of the boundary are converted into boundary energies based on the Read-Shockley equation. Figure 9(f) shows the situation for pixel *j* of subgrain *i*. The boundary energy of each subgrain can then be calculated using the following equation

$$\gamma_{av,i} = \frac{1}{N}\sum_{j=1}^{N}\left(\frac{1}{n}\sum_{k=1}^{n}\gamma_{i,j,k}\right), \tag{6}$$

where N is the number of boundary pixels of subgrain *i*, and $\gamma_{av,i}$ is the average boundary energy of this subgrain. The area weighted average stored energy (E_s^F) can then be calculated by considering the whole map as for the subgrain method.

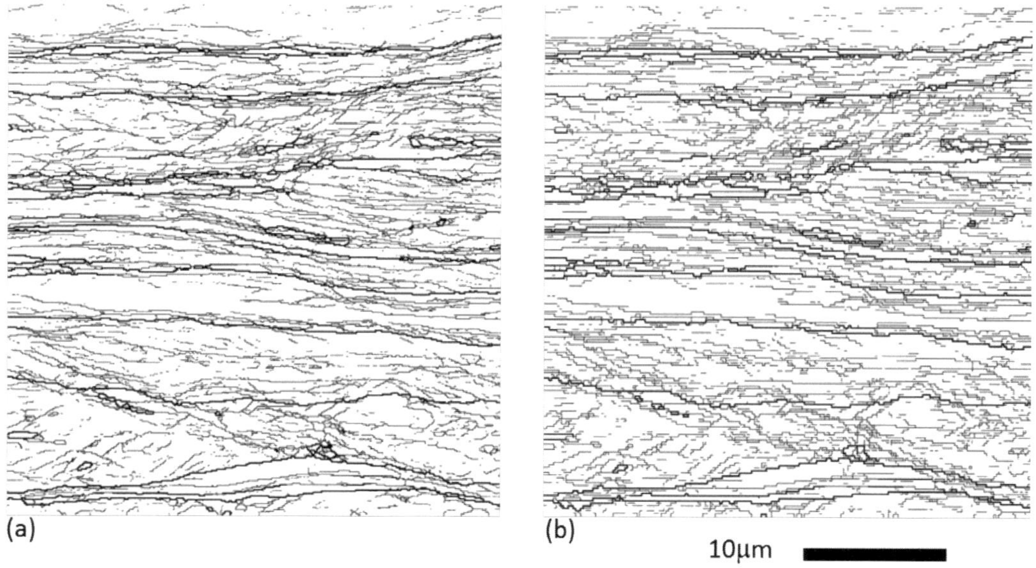

Figure 10. EBSD images of cold-rolled nickel for use in stored energy calculations: (a) experimental data at step size 0.05 μm; (b) resampled with a step size of 0.1 μm. Thick and thin black lines indicate misorientation angles of >15° and 2 – 15°, respectively.

4.2. Application to deformed nickel

High purity nickel (99.99%), cold-rolled to a thickness reduction of 90%, was examined using EBSD in the longitudinal section. A 400×400 pixel map, collected with a step size of 0.05 μm, was chosen for demonstration of the stored energy estimation. The average lamellar boundary spacing measured (θ_{cut} = 1.5°) in this map is 0.204 μm, which is very close to that measured from TEM images (0.21 μm) [54], and thus similar estimates of the stored energy (4.9 MJ m^{-3}) [43] should be expected. The effect of step size was investigated by artificially creating a 200×200 pixel map from the experimentally collected data set (effective map step size of 0.1 μm). For the stored energy calculations values of G = 79 GPa, and b = 0.249 nm were used.

The experimental EBSD data were first cleaned using the Channel 5 post-processing software. To further reduce the orientation noise, a maximum sub-grid minimum variance (MSMV) version of a modified Kuwahara filter [55] was applied to the data set. After data filtering, only a few interior misorientations were left. However, a large number of artificial single-pixel subgrains were still present along many LABs (see figure 10). The stored energy was then estimated for both maps using the methods described above, though it should be noted that single-pixel subgrains were excluded from the calculations involving subgrain reconstruction. The results are summarized in Table 2.

It is shown in Table 2 that both the step size and θ_{cut} affect the estimated stored energy significantly. A higher value of θ_{cut} results in a lower stored energy estimate for all methods examined, whereas the dependence of the stored energy estimate on the step size varies from method to method. For the average method and modified average method, the stored energy estimates (E_s^A and E_s^B) decrease with increasing step size, as larger mean ECDs are obtained with a larger step size. In contrast, for the segment method, the result (E_s^C) is almost independent of the step size used. Unexpected results were obtained for the subgrain method and the modified subgrain method, where the estimates (E_s^E and E_s^F) increase with increasing step size (discussed further in section 4.3.3).

The estimated stored energy is also affected by the choice of the average misorientation angle in the method A (the average method) and method B (the modified average method), with higher estimates obtained using $\theta_{H,av}$, and lower estimates using $\theta_{L,av}$. These two methods are very similar, but the latter ignores misorientation angles within subgrains, most of which are of very low misorientation angle, and thereby gives higher estimates than those calculated from the average method. Methods E and F (the subgrain method and the modified subgrain method) are grain-area weighted, and thus the results are strongly dependent on large subgrains, which are of low stored energy. Therefore much smaller estimates are obtained using these two methods.

Table 2. Stored energy (MJ m^{-3}) of cold rolled nickel estimated from experimentally measured EBSD data.

Step size (μm)	θ_{cut} (°)	d_{ECD} (μm)	E_s^A $\theta_{H,av}$	E_s^A $\theta_{L,av}$	E_s^B $\theta_{H,av}$	E_s^B $\theta_{L,av}$	E_s^C	E_s^D 5°	E_s^D 15°	E_s^E	E_s^F
0.05	1.5	0.33	7.1	4.9	7.8	5.7	3.5	3.6	7.5	1.9	1.4
0.05	2.0	0.41	6.5	4.7	6.8	5.6	3.0	2.8	6.7	1.3	1.0
0.10	1.5	0.38	6.1	4.2	6.4	4.5	3.4	3.9	7.5	2.9	2.0
0.10	2.0	0.46	5.4	3.8	5.7	4.3	3.0	3.2	6.8	1.9	1.4

4.3. Assumptions in the stored energy calculations and sensitivity to calculation parameters

In assessing the accuracy of the EBSD-based estimates of stored energy described above it should be remembered that the calculations all involve a number of important assumptions. A key assumption is that the majority of the energy left after deformation is stored in boundaries, whose energy can be described by the Read-Shockley equation. This is reasonable for metals with medium to high stacking fault energy, as both subgrain interior dislocations and extrinsic (redundant) dislocations within boundaries only contribute a small fraction to the stored energy [43,56]. However it should be still noted that this assumption results in underestimation of the stored energy as shown by comparison of DSC and TEM investigations [43]. A second common assumption is that subgrains in deformed metals are of equiaxed morphology. This assumption facilitates estimation, however, this will introduce some error in the results. A third key assumption is that θ_{cut} separates the misorientation angles of real boundaries from artefacts arising from orientation noise. This assumption, however, is only weakly satisfied due to the limited angular resolution of the EBSD technique, especially in highly deformed metals.

Some assumptions specific to particular stored energy calculations are also worth noting. For methods A and B the mean subgrain size and the mean misorientation angle are used. Therefore it is implicitly assumed that the subgrain size and the mean boundary misorientation angle have a narrow distribution. In highly deformed metals, however, both the subgrain size and the boundary misorientation angle have a wide distribution due to structural heterogeneity, and thus the mean values cannot well represent the microstructure. For these methods there are two choices for the mean misorientation angle: (1) $\theta_{H,av}$, the mean of all detected misorientation angles, considering therefore stored energy in all boundaries; and (2) $\theta_{L,av}$, which only considers LABs and assumes that all the HABs (>15°) are initial grain boundaries. It follows that $\theta_{H,av}$ may be suitable at high strains and $\theta_{L,av}$ may be preferred at low strains, where there are few deformation induced high angle boundaries.

The KAM method is based on equation (3), which is derived from the idea that LABs are built up by dislocations. An implicit assumption in this method is that each pixel in the map can be regarded as a subgrain, whose misorientation angle with its neighbouring subgrains is represented by the KAM. This assumption may be reasonable when the EBSD step size is close to the real subgrain size, however in such a case the subgrain size will be estimated poorly. For fine step sizes the assumption is weak and lacks physical meaning. The use of a high cut-off misorientation angle, to eliminate the effect of initial grain boundaries, introduces another weakly supported assumption of this method, namely that the deformation induced dislocation boundaries all have misorientation angles below the chosen high cut-off misorientation angle.

4.3.1 Influence of the misorientation threshold. A low cut-off misorientation angle θ_{cut} is always required in estimation of stored energy from EBSD data due to the influence of orientation noise. It is clear therefore that this parameter plays a key role in the calculations, especially for regions containing many LABs (such as the LMRs in highly deformed samples) where orientation noise will have a larger effect. A suitable value of θ_{cut} should be chosen as a compromise of including as many real boundaries as possible, while introducing as few artefacts as possible.

The conventionally-used KAM method does not include a low cut-off misorientation angle [47,53], and as a consequence orientation noise can strongly influence estimated stored energy values. In the calculations described above a θ_{cut} is introduced into the calculation such that misorientation angles within the pixel kernel below this value are taken as zero degrees in calculating the KAM. This gives an improvement in the stored energy estimates, although the values still depend strongly on the choice of the high cut-off misorientation angle. For an EBSD map with step size Δ, map area *A*, total boundary length *L*, and mean boundary misorientation θ, approximately 2L/Δ pixels will lie along subgrain boundaries. Furthermore, assuming a square-grid map, the mean KAM of these boundary pixels is approximately 3θ/8, since there are on average approximately three neighbours for all boundary pixels of a given grain, and the KAM of interior pixels is by definition zero degrees. Therefore, the global mean KAM is given by

$$\theta_{KAM} = \frac{(2L/\Delta)}{(A/\Delta^2)}\frac{3\theta}{8} = \frac{3}{4}L_A\theta\Delta \tag{7}$$

Combining this equation with the stereological relationship $(1/d) = (2/\pi)L_A$ (for an equiaxed structure) and inserting into equation (4) gives the following formulation of stored energy:

$$E_s = \frac{3Gb\theta}{2d}\frac{3\pi}{8} \tag{8}$$

This has the same form as equation (3), but with an additional coefficient of 3π/8. This explains why the KAM method gives reasonable estimates in some cases despite its various limitations. It should be recalled also, however, that equation (3) assumes that the energy of subgrain boundaries increases linearly with misorientation angle, which according to the Read-Shockley function for boundary energy is only true for boundaries of very low misorientation angles (where the effects of orientation noise will be the greatest).

4.3.2 Subgrain reconstruction. In contrast to grain reconstruction in fully recrystallized samples, subgrain reconstruction in deformed samples is challenging, due both to the smaller structural scale and more importantly to the lower misorientation angles of many of the boundaries. Subgrain reconstruction for an EBSD map of a deformed sample using a reasonable value of θ_{cut} invariably results in the detection of some huge subgrains, as even a single missing segment of a real boundary (due to the effect of orientation noise) leads to bleeding between neighbouring subgrains. At the same time orientation noise can also result in a large number of artificial single-pixel subgrains. Orientation filtering of the data does help to reduce the influence of the orientation noise [57,58], but only to a limited extent.

The effect of orientation noise on accurate subgrain detection is exacerbated by the fact that an apparent gradual change of orientation is often found as the electron beam is moved across boundaries of very low misorientation angle [59]. Preliminary studies show that such artificial gradients are found for boundaries with misorientation angle less than approximately 4°. Further work to establish the origin and extent of such effects is currently underway. It is assumed at present though that such artificial gradients arise as a result of the limited ability of the standard automated indexing procedures to distinguish overlapping EBSD patterns from volumes with very similar orientation. For these reasons it is clear that standard reconstruction of subgrains from EBSD data using low values of θ_{cut} based directly on orientation data is likely to be unreliable. Improvements may be possible by utilizing the observation of small cumulative misorientations either in pattern detection or for boundary detection, or alternatively by combining the orientation data with the pattern quality data [60], which has been shown in some cases to reveal clearly boundaries of low misorientation angle.

4.3.3 The step size. In order to correctly estimate the average subgrain size, a step size of less than 1/5 of the average subgrain size is required [49]. A larger step size gives larger mean values of d_{ECD} (Table 2), resulting in smaller estimates of the stored energy for the methods A and B (the average method and the modified average method). However, contrary to expectations, a smaller estimate of the stored energy is obtained for the smaller step size for the methods E and F (the subgrain method and the modified subgrain method). Closer examination of the data reveal that smaller step sizes give an increase in the size and number of very large subgrains for this data set as a result of the artificial changes in orientation measured across very low angle boundaries described above. The finer step size results therefore in a smoother artificial change in orientation, making it even harder to detect accurately boundaries of very low misorientation angle. This observation is a reminder that for the case of EBSD data, the presence of orientation noise means that a smaller step size is not always better, and that an appropriate step size (ideally close to 1/5 of the average size) should be chosen to estimate accurately the stored energy of deformation.

The data in Table 2 suggest that the segment method is not sensitive to the step size in the examined range. However, this method is also affected by any artificial gradual changes in orientation near very low angle boundaries. For example if a boundary of $\theta = 2°$ is detected as two boundaries of $\theta = 1°$, then the stored energy will be overestimated by approximately 20% according to equation (2). However, if θ falls below θ_{cut} there will be no contribution to the stored energy. The latter has a larger influence on the data in Table 2, as θ_{cut} is relatively high and the stored energy is underestimated compared to the TEM-determined value. In order to obtain the same stored energy of 4.9 MJ m^{-3} [43] as determined from TEM investigations, θ_{cut} has to be as low as 0.35° for the MSMV-filtered map with step size 50 nm. For the segment method therefore the lowest acceptable value of θ_{cut} without obviously introducing artificial misorientations within subgrains should be used for estimation of the stored energy.

4.4. Local estimates of stored energy

The various estimates of stored energy given above are illustrated based on application to a complete EBSD map, i.e. yielding an average value of stored energy for the entire map. Each of these can also be applied to smaller areas to obtain a more local variation in stored energy, and in each case will be subject to all of the same limitations in accuracy described above. In addition, however, there are some specific considerations that should be taken into account for the calculation of stored energy on a local microstructural scale using these measures, particularly with regard to the length scale over which values can be calculated.

Methods A and B rely on the calculation of stored energy as an average over a number of subgrains, and as such these methods are not suitable for providing an estimate of stored energy on the scale of individual features in a deformed microstructure, although they may be used to estimate the stored energy within regions such as the LMRs or localized glide bands if these are sufficiently large. Similarly the individual subgrain-based methods (methods E and F) are limited by definition to providing a value for the stored energy on the scale of each individual subgrain. In one sense this should not be regarded as a problem, because in dislocation-cell forming materials it can be assumed that the majority of the stored energy of deformation is located at the deformation-induced boundaries. As such, a value of the stored energy for an individual subgrain has a clear physical meaning as the average stored energy for all the boundaries enclosing the subgrain. A corresponding limitation, however, is that the stored energy is necessarily associated with the entire subgrain, so that on a map-pixel basis an energy is unrealistically associated with the interior of each subgrain. Such a method though can be useful for example for models of recrystallization where it is necessary to decide the location where nucleation of new grains may take place.

The two remaining methods for estimation of stored energy (the segment and KAM methods) both have the advantage of being able to give a value for local stored energy on the scale of individual map pixels, and therefore with a spatial resolution limited by the EBSD step size. The KAM-based estimate is an appealing parameter, not least in that it provides a value directly for each pixel of a map, and hence in a format that can be easily visualized as a map, or imported into a model. As noted in section 4.3 previously, however, there are some problems with the way this parameter is calculated, particularly in that an average value is calculated for each pixel based on the misorientation to all the neighbour pixels within the kernel, typically including pixels that are not in fact separated from the kernel-centre pixel by real boundaries.

In contrast, the segment method has the advantage for quantification of local variations in stored energy in that the contributions from individual boundaries, which are assumed to exist wherever a pixel-to-pixel misorientation is larger than a pre-defined lower misorientation angle threshold (required to limit the influence from orientation noise in the EBSD data), are counted directly. The main disadvantage of this method stems from the fact that the stored energy is necessarily evaluated *between* map pixels and as such the result for a map of M x N pixels cannot be mapped directly on a pixel grid of similar size. One simple solution, however, to this is to calculate the stored energy based on a resampled grid where the new pixels are centred on the gridlines of the original map. Accordingly it can be concluded that the segment method should be the preferred choice for evaluation of the local variation in stored energy using EBSD data on the scale of the map step size.

5. Summary

Heterogeneity is a general characteristic of the microstructure of metals deformed over a wide range of plastic strains. For samples deformed to large strains a key heterogeneity is in the refinement of the microstructure by high angle boundaries, where even in samples deformed to very large strains a significant volume fraction of material can have a microstructure where locally the boundaries are predominantly only of low misorientation angle. This heterogeneity can be quantified by partitioning the microstructure into low misorientation regions (LMRs) and high misorientation regions (HMRs). The extent of such heterogeneity can be described by the area fraction of the LMRs (f_{LMR}), and a combination of parameters can be used to quantify the length scale of the LMR heterogeneity.

A detailed characterization of deformation microstructure heterogeneity is particularly important for studies of recrystallization, as this process can depend strongly on local variations in the deformed microstructure, and hence stored energy. For this purpose the EBSD technique has a strong advantage over other microstructural-based techniques in that it allows characterization over a wide range of length scale. Several methods have been proposed for estimation of stored energy from EBSD data. All of these methods are, however, sensitive to a lesser or greater extent to the limited angular resolution of the EBSD technique, reflecting the fact that even after a large deformation many boundaries of low angle are still present. It can be noted that even the most accurate microstructure-based stored energy estimates (i.e. from TEM inspections) typically give values lower than found from differential scanning calorimetry (DSC) methods [43,56]. The reason for this difference is still not understood, but may relate to the energy of triple lines and/or the assumptions over the relationship between misorientation angle and boundary energy [56].

In any case the relative local variation in stored energy is important, and of the EBSD methods available, estimation of the stored energy by a summation of the contributions from individual boundary segments is considered to be the best suited at present for characterization of the local variation in stored energy on the scale of the dislocation boundary features. Given the advantage of the EBSD technique for collecting data over large areas with relatively high spatial resolution, the accuracy with which local variations need to be characterized (as limited by the angular resolution of the EBSD technique) is an important question. Improvements in angular resolution through new approaches to EBSD pattern analysis [61,62] may be possible, though the extent to which these methods can be used on samples deformed even to moderate strains is still not known. Similarly, although X-ray synchrotron measurements may in the near-future allow mapping of deformed microstructures with as low as 100nm resolution and an angular resolution of a fraction of a degree [63], only small volumes can be mapped with such a method. Here it is possible that models that capture the effect of stored energy variations on recrystallization can also help to provide guidance on the sensitivity to relative variations in stored energy, and hence the requirements for the accuracy in experimental characterization.

Acknowledgements

The authors gratefully acknowledge the support from the Danish National Research Foundation (Grant No DNRF86-5) and the National Natural Science Foundation of China (Grant No. 51261130091) to the Danish-Chinese Center for Nanometals, within which this work has been performed.

References

[1] Hansen N and Kuhlmann-Wilsdorf D 1981 *Mater. Sci. Eng.* A **81** 141
[2] Hansen N 2001 *Metall. Mater. Trans. A* **32** 2917
[3] Hansen N and Juul Jensen D 2011 *Mater. Sci. Technol.* **27** 1229
[4] Godfrey A and Hughes D A 2004 *Scr. Mater.* **51** 831
[5] Hughes D A and Hansen N 2014 *Phys. Rev. Lett.* **112** 135504
[6] Huang X and Winther G 2007 *Philos. Mag.* **83** 5189
[7] Winther G and Huang X 2007 *Philos. Mag.* **83** 5215
[8] Le G M, Godfrey A, Hong C, Huang X and Winther G 2012 *Scr. Mater.* **66** 359
[9] Wert J A, Liu Q and Hansen N 1995 *Acta Metall. Mater.* **43** 4153
[10] Huang X and Hansen N 1997 *Scr. Mater.* **37** 1
[11] Chen H S, Godfrey A, Hansen N, Xie J X and Liu Q 2008, *Mater. Sci. Eng. A* **483-484** 157
[12] Xing Q, Huang X and Hansen N 2006 *Metall. Mater. Trans. A* **37** 1311
[13] Godfrey A, Juul Jensen D and Hansen N 1998 *Acta Mater.* **46** 835
[14] Hughes D A and Hansen N 1993 *Metall. Trans. A* **24** 2021
[15] Cizek P, Whiteman J A, Rainforth W M and Beynon J H 2004 *J. Microsc.* **213** 285
[16] Duggan B J, Hatherly M, Hutchinson W B and Wakefield P T 1978 *Metal Sci.* **12** 343.
[17] Jia N, Eisenlohr P, Roters F, Raabe D and Zhao X 2012 *Acta Mater.* **60** 3415
[18] Armstrong R W and Walley S M 2008 *Int. Mater. Rev.* **53** 105
[19] Mishin O V, Bay B, Winther G and Juul Jensen D 2004 *Acta Mater.* **52** 5761
[20] Zhang Y H, Juul Jensen D, Zhang Y B, Lin F X, Zhang Z and Liu Q 2012 *Scr. Mater.* **67** 320
[21] Xu W, Ferry M, Cairney J M and Humphreys FJ 2007 *Acta Mater.* **55** 5157
[22] Paul H, Driver J H and Jasieńskia Z 2002 *Acta Mater.* **50** 815
[23] Moelans N, Godfrey A, Zhang Y and Juul Jensen D 2013 *Phys. Rev. B* **88** 054103

[24] Godfrey A, Hansen N and Juul Jensen D 2007 *Metall. Mater. Trans. A* **38** 2329
[25] Godfrey A, Juul Jensen D and Hansen N 2001 *Acta Mater.* **49** 2429
[26] Liu Q 1995 *Ultramicrosc.* **60** 81
[27] Liu Q, Huang X and Yao M 1992 *Ultramicrosc.* **41** 317
[28] Godfrey A, Wu G L and Liu Q 2002 *Mater. Sci. Forum* **408-412** 221
[29] Godfrey A, Mishin O V and Liu Q 2006 *Mater. Sci. Technol.* **22** 1263
[30] Mishin O V, Godfrey A and Juul Jensen D, Analysis of deformation structures in FCC materials using EBSD and TEM techniques *Electron Backscatter Diffraction in Materials Science* ed A J Schwartz et al (New York: Springer Science+Business Media) chapter 19 p263
[31] Segal V M 1995 *Mater. Sci. Eng. A* **197** 157
[32] Tsuji N, Ito Y, Saito Y and Minamino Y 2002 *Scr. Mater.* **46** 893
[33] Mishin O V, Godfrey A, Juul Jensen D and Hansen N 2013 *Acta Mater.* **61** 5354
[34] Mishin O V, Bowen J R and Lathabai S 2010 *Scr. Mater.* **63** 20
[35] Godfrey A, Mishin O V, Bowen J and Liu Q 2009 *Proc. 30th Risø International Symposium on Materials Science* ed J C Grivel et al. (Roskilde: Risø-DTU) p 31
[36] Mishin O V and Godfrey A 2008 *Metall. Mater. Trans. A* **39** 2923
[37] Mishin O V and Bowen J R 2009 *Metall. Mater. Trans. A* **40** 1684
[38] Luo Z P, Mishin O V, Zhang Y B, Zhang H W and Lu K 2010 *Scr. Mater.* **66** 335
[39] Mishin O V, Bowen J R and Godfrey A 2012 *Mater. Sci. Forum* **715-716** 825
[40] Mishin O V, Segal V M and Ferrasse S 2012 *Metall. Mater. Trans. A* **43** 4767
[41] Zhang Y B, Mishin O V and Godfrey A 2014 *J. Mater. Sci.* **49** 287
[42] Zhang Y B, Juul Jensen D and Godfrey A 2014 *Metall. Mater. Trans. A* **45** 2899
[43] Knudsen T, Cao W Q, Godfrey A, Liu Q and Hansen N 2008 *Metall. Mater. Trans. A* **39** 430
[44] Godfrey A, Cao W Q, Hansen N and Liu Q 2005 *Metall. Mater. Trans. A* **36** 2371
[45] Taheri M, Weiland H and Rollett A 2006 *Metall. Mater. Trans. A* **37** 19
[46] Choi S H and Jin Y S 2004 *Mater. Sci. Eng. A* **371** 149
[47] Takayama Y and Szpunar J A 2005 *Mater. Trans.* **45** 2316
[48] Lee K, Reis A C C, Kim G and Kestens L 2005 *J. Appl. Crystallogr.* **38** 668
[49] Humphreys F J 1999 *J. Microsc.* **195** 170
[50] Choi SH and Cho J H 2005 *Mater. Sci. Eng. A* **405** 86
[51] Cao W Q, Godfrey A and Liu Q 2003 *Mater. Sci. Eng. A* **361** 9
[52] Liu Q, Juul Jensen D and Hansen N 1998 *Acta Mater.* **46** 5819
[53] Kobayashi M, Takayama Y and Kato H 2004 *Mater. Trans.* **45** 3247
[54] Hughes D A and Hansen N 2000 *Acta Mater.* **48** 2985
[55] Godfrey A 2004 *Scr. Mater* **50** 1097
[56] Godfrey A and Liu Q 2009 *Scr. Mater.* **60** 1050
[57] Mishin O V, Östensson L and Godfrey A 2006 *Metall. Mater. Trans. A* **37** 489
[58] Mishin O V and Godfrey A 2007 *Mater. Sci. Forum* **550** 295
[59] Bate P S, Knutsen R D, Brough I and Humphreys F J 2005 *J. Microsc.* **220** 36
[60] Barou F, Maurice C, Feppon J and Driver J 2009 *Int. J. Mater. Res.* **100** 516
[61] Wilkinson A J, Britton T B, Jiang J and Karamched P S 2014 *IOP Conf. Ser.: Mater. Sci. Eng.* **55** 012020
[62] Maurice C, Quey R, Fortunier R and Driver J H 2013 High Angular Resolution EBSD and Its Materials Applications *Microstructural Design of Advanced Engineering Materials* ed D A Molodov (Weinheim, Germany: Wiley-VCH Verlag GmbH & Co KGaA) chapter 14 p 341
[63] Ice G E et al. 2009 *Mater. Sci. Eng. A* **524** 3

IOP Conf. Series: Materials Science and Engineering **89** (2015) 012004 doi:10.1088/1757-899X/89/1/012004

The interplay between grain boundary structure and defect sink/annealing behavior

J Han[1], V Vitek[1], and D J Srolovitz[1,2]

[1]Department of Materials Science and Engineering, University of Pennsylvania, 3231 Walnut St., Philadelphia, PA 19104 USA
[2]Department of Mechanical Engineering and Applied Mechanics, University of Pennsylvania, 3231 Walnut St., Philadelphia, PA 19104 USA

E-mail: srol@seas.upenn.edu

Abstract. We present a series of results from atomistic simulations in three different materials (3 crystal structures) that demonstrate that the multiplicity of grain boundary (GB) structures at fixed macroscopic GB degrees of freedom is both extremely large and ubiquitous. The GB energy vs. misorientation curve that is commonly discussed is in fact a wide band, with many GB states very close in energy. The existence of so many GB states suggests that GB configurational entropy S_c is important for GB properties. We demonstrate that the GB S_c consists of two major contributions, one of which is geometric in nature and one that depends on bonding. We then show how this concept can be employed to predict GB relaxation dynamics by analogy with Adam-Gibbs theory, originally derived to predict the properties of glass forming liquids. Finally, we apply these predictions to understand GB denuded zone size during irradiation.

1. Introduction

Grain boundaries are widely thought of as being perfect sinks and/or sources of defects in polycrystals. Two classical examples are grain boundaries as sinks for vacancies or interstitials during irradiation damage or in creep and grain boundaries behaving as sinks for dislocations. As the dislocation example illustrates, absorbing dislocations changes grain boundary structure because net Burgers vector can neither be created nor destroyed. When a dislocation enters a grain boundary it can lower its energy by decomposing into partial (grain boundary) dislocations with Burgers vectors much smaller than are possible in the crystal, consistent with the DSC (displacement shift complete) lattice [1]. Such DSC dislocations change the boundary structure.

There are many examples in the literature (e.g., see [2-6]) of the existence of multiple grain boundary structures corresponding to the same five macroscopic bicrystal parameters. Some more recent examples include dislocation pairing in low angle grain boundaries [7] and the observation of grain boundary structural phase transitions [8]. Structural multiplicity is a general feature of grain boundaries [9]. The existence of multiple grain boundary structures is not determined by the structure of the coincidence site lattice alone, since not all symmetry-allowed structures are stable.

In this paper, we discuss the multiple possible structures of a grain boundary of fixed macroscopic degrees of freedom, elucidate the implications of these for grain boundary thermodynamics and kinetics, including the case of adsorption of point defects as occurs, for example, during radiation damage. We do this via application of atomistic simulation techniques and to demonstrate the generality of the results we consider a set of symmetric tilt grain boundaries over the full angular range for materials with three different crystal structures and two types of atomic bonding. Finally, we compare experimental and predicted point defect adsorption kinetics in a radiation damage environment.

2. Multiplicity of Grain Boundary Structures

Metastable GB structures correspond to local minima in the potential-energy landscape for fixed macroscopic bicrystal geometry. To describe an elemental (or disordered solid solution) bicrystal, five macroscopic geometrical degrees of freedom (DOFs) are required; they can be represented by a rotation axis $\boldsymbol{o}$ (2), a rotation angle θ (1), and the GB plane normal $\boldsymbol{n}$ (2). Any atomistic description of a GB in a bicrystal must also account for three microscopic geometrical DOFs: the rigid-body displacement of two adjoining crystals with respect to each other [1]. Such displacement has two components, t_1 and t_2, parallel to the GB plane and one component, t_3, perpendicular to the plane. The latter represents expansion or contraction of the bicrystal. For a given t_1 and t_2, t_3 is determined uniquely by energy minimization, leaving only $\boldsymbol{t} \equiv (t_1, t_2)$ as the free microscopic DOFs. In a grand-canonical ensemble another microscopic DOF arises. This involves adding/removing atoms to/from the GB, i.e., changing the GB atomic fraction, φ [10]. Adding/removing a monolayer of atoms to/from the GB plane simply shifts one grain relative to the other. Hence, the GB structure repeats as the GB atomic fraction, $\varphi \equiv$ (number of added atoms)/(number of atoms per monolayer), varies from 0 to 1. Changes in φ may lead to additional metastable structures [8,10]. Therefore, metastable GB structures can be found by minimizing the energy of a bicrystal model with respect to the atomic coordinates and the variables $\boldsymbol{t}$ and φ for fixed macroscopic DOFs. Energy minimization starting from different initial translation states $\boldsymbol{t}$ with certain atomic fraction φ may result in different metastable states, characterized by $(\boldsymbol{t}^{(i)},\varphi^{(i)})$, where the superscript (i) represents one of the N accessible metastable GB states for fixed $\{\boldsymbol{o},\theta,\boldsymbol{n}\}$.

For given macroscopic DOFs, different metastable GB structures may be found for different $\boldsymbol{t}$ and

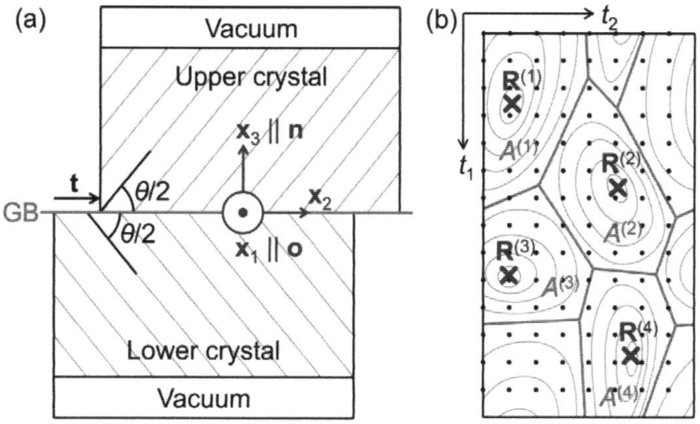

Figure 1. (a) Schematic of the simulation supercell; here, we use $\boldsymbol{o}$ = [001] and focus on symmetric tilt boundaries. (b) Schematic of the sampling of translation GB states (points) and the associated GB energy contours (gray curves) in the periodic GB cell. $\boldsymbol{R}^{(i)}$ denotes the ith metastable structure found for a metabasin of area $A^{(i)}$ that is delimited by dividing surfaces marked by red lines.

φ; this is called GB structural multiplicity. It has been observed in both experiments and atomistic simulations [8-14]. There is evidence that GB structural multiplicity holds the key to understanding such dynamic phenomena as GB sliding [3] and point-defect absorption at GBs [8, 11]. However, at present the idea of structural multiplicity has only been used to explain microscopic mechanisms rather than predict properties. Here we propose a novel approach to predict nonequilibrium GB structures and properties, accounting for the important effect of GB structural multiplicity.

As illustrated in Fig. 1(a), a series of symmetric tilt GBs in cubic materials were constructed by rotating two crystals with respect to each other by ±θ/2 about the axis ***o*** = [001] and joining them at the plane with normal ***n*** = [010]; the indices refer to the unrotated crystals. Using this approach, we constructed 110 coincidence-site-lattice GBs with 0° ≤ θ ≤ 90° and reciprocal coincident site density of Σ ≤ 941. The simulation supercells were created as shown in Fig. 1(a). Periodic boundary conditions were applied along x_1 and x_2 and there was a vacuum above and below the bicrystal (in x_3). The thickness of each crystal along x_3 was 60-80 nm. To determine the generality of the results, we considered three model materials with different crystal structures and bonding: embedded atom-method (EAM) Al [15] (fcc), Finnis-Sinclair (FS) W [16] (bcc), and Tersoff Si [17] (diamond cubic).

To investigate the effect of microscopic DOFs, the initial GB configurations were explored by uniformly sampling the translation $\{\boldsymbol{t}\}$ space for each misorientation θ (see Fig. 1(b)). To do this, one crystal was rigidly displaced by the vector ***t*** with respect to the other until the vector swept the entire periodic GB cell. The density of the sampled configurations in the periodic GB cell area A_{cell} is $94a_o^{-2}$ for Al and W, and 257 a_o^{-2} for Si (a_o is the cubic lattice constant). For each configuration, the energy

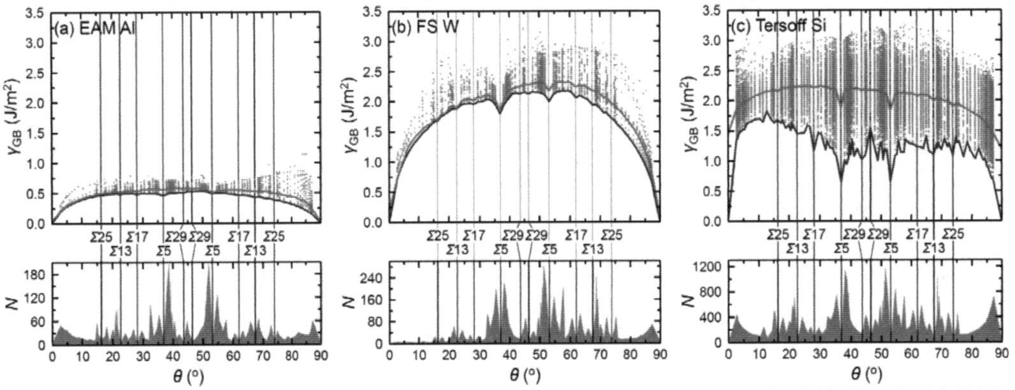

Figure 2. GB energy γ versus misorientation θ for a series of [001] symmetrical tilt GBs in (a) EAM Al (fcc), (b) FS W (bcc), and (c) Tersoff Si (diamond cubic). In the top panel each data point represents the GB energy associated with one metastable structure $\gamma^{(i)}$ (corresponding to the translation $t^{(i)}$); the blue curves denote the lowest energies $\gamma_{\min}(\theta)$ and the red curves denote the basin-ensemble-averaged GB energies $\langle\gamma(\theta)\rangle$, as defined in Section (3). The lower panel shows the number of metastable GB states $N(\theta)$. The vertical lines label the misorientations with Σ < 30.

was minimized with respect to the atomic coordinates and translation vector ***t*** using a steepest descent procedure. GB configurations with different initial ***t*** relaxed into a finite set of translation states corresponding to local minima in the GB energy with respect to $\{\boldsymbol{t}\}$ (stable or metastable structures). Associated with each translation $\boldsymbol{t}^{(i)}$ is a particular metastable GB with energy $\gamma^{(i)}$ and a basin of attraction (metabasin) within which any sampled configuration relaxes into the same structure $\boldsymbol{R}^{(i)}$.

The energy of each metastable structure $\gamma^{(i)}$ is shown in the top panel of Fig. 2 as a function of misorientation θ. The minimum-energy structures possess the energy $\gamma_{\min}$ (blue curves in the top panel of Fig. 2). The $\gamma_{\min}(\theta)$-curves for Al and W feature a series of cusps at misorientations corresponding to small Σ values [18] while the curve for Si is highly serrated with particularly sharp cusps at two Σ5 GBs. For nearly all misorientations, multiple metastable states exist, each with a unique energy. Then, considering all accessible states, we see that the classical GB energy versus misorientation curve broadens into a GB-energy band (γ-band). The existence of such finite energy bands has been neglected in most earlier studies of GB properties.

The thickness/width of a γ-band varies considerably from material to material, as seen in Fig. 2. The band width for a GB with misorientation θ is $\Delta(\theta) \equiv \gamma_{max}(\theta)-\gamma_{min}(\theta)$, where $\gamma_{max/min}(\theta)$ is the maximum/minimum energy amongst all metastable GBs with θ. The average band width $\langle\Delta\rangle$ and the average relative width $\langle\delta\rangle \equiv \langle\Delta\rangle/\gamma_{min}$ over all θ values were found to be $\langle\Delta\rangle$ = 0.27, 0.66 and 1.6 J/m^2, and $\langle\delta\rangle$= 0.80, 0.40 and 1.4 for Al, W and Si, respectively. The GB-energy variation in Si is the largest. Al exhibits smaller $\langle\Delta\rangle$ value but larger $\langle\delta\rangle$ value than W. These differences may stem from differences in bonding amongst these materials; covalent bonds in Si may stabilize more metastable GB structures than metallic bonds. Another measure of a γ-band is the number of metastable GB states $N(\theta)$ (the bottom panels of Fig. 2). Examination of the $N(\theta)$ curves shows a strong similarity amongst the three model materials. Such material-independence is related to the geometric nature of GB configurational entropy (see below).

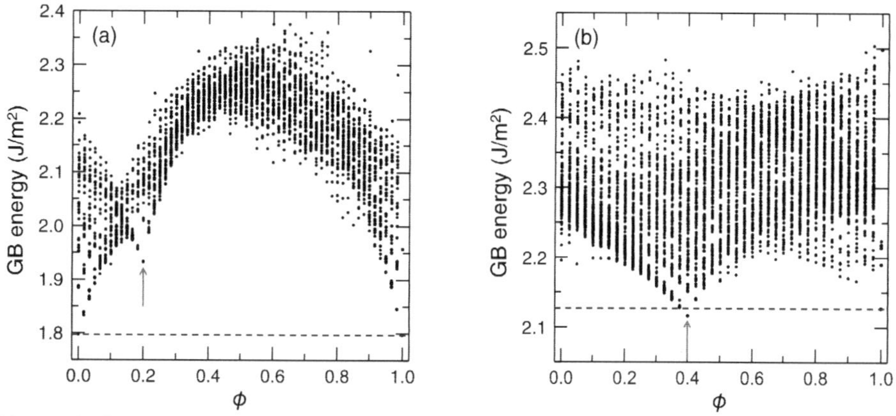

Figure 3. GB energies as a function of atomic fraction φ for (a) Σ5 36.9° (310) and (b) Σ29 46.4° (730) GBs in W. Each data point corresponds to a different metastable translation state t. The new metastable states identified by varying φ are indicated by the red arrows. The horizontal dashed blue lines indicate the lowest energies found by translations alone (φ = 0).

Atoms can be added to or removed from a grain boundary (GB) without changing its macroscopic bicrystallography (misorientation, boundary inclination) or translational state. The GB atomic fraction φ is varied by adding or removing atoms in the GB plane; adding a full atomic plane (parallel to the GB) does not change the GB structure. We investigated this additional degree of freedom (DOF) following the non-conservative sampling method proposed by Frolov *et al.* [8]. We examined Σ5 36.9° (310) and Σ29 46.4° (730) GBs in W in the full $\{t, \varphi\}$ space, where 60 and 40 fundamental GB cells were in the GB simulation cell, respectively. The GB-energy spectrum as a function of φ and $\boldsymbol{t}$ is shown in Fig. 3 for these GBs. Introducing φ as a variable produces, in each case, exactly one additional metastable state (i.e., one for which γ is a minimum with respect to $\boldsymbol{t}$ and φ). The extra minimum for the Σ5 (310) GB corresponds to a metastable GB state, whereas in the Σ29 (730) case the new structure has even lower energy than the structures created by translations alone (i.e., φ = 0). Note, the points in these plots not corresponding to the bottom of the cusps do not represent new GB structures, but rather one of the metastable GBs with a non-periodic array of point defects. Since adding additional atoms to the GB rarely add more than a single additional metastable state, we neglect those found from varying φ in the GB statistics, below.

The metastable states in the grain boundary bands in Fig. 2 are not arbitrary, but are closely related to the GB structure. The structure of grain boundaries can be well described within the structural unit

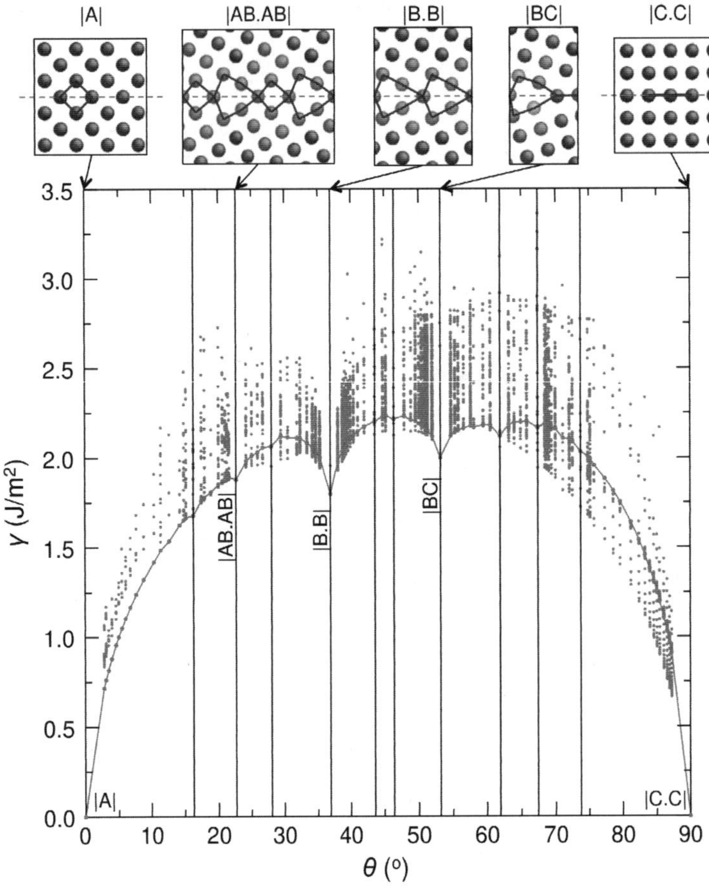

Figure 4. The grain boundary energy band structure from Fig. 2b for W. The red curve shows a path through the individual stable (or metastable) states that correspond to combinations of the structural units A for 0° and the Σ5 boundary (37°) structural units B and between 37° and 90° a combination of C units.

(SU) model (e.g., see [19, 20]), which posits that GBs are composed of a periodic array of a single fundamental structural unit in the case of certain high symmetry boundaries at specific misorientations (corresponding to cusps in a GB energy versus misorientations plot) and for misorientations between these delimiting boundaries the structure is composed of an arrangement of the two types of structural units found in the delimiting boundaries composed of single SUs. Consider the energy band structure of the relatively simple case of the symmetric [001] tilt GBs in W, which is shown in Fig. 2b. Boundaries corresponding to the red curve between 0° and 37° (a Σ5 boundary) consist entirely of the structural units from these two end points as shown in Fig. 4. The 0° and 37° boundaries correspond to A and B SUs, respectively, and for all of the misorientations in between, the boundary structures delineated by the red curve indicate, boundaries that are solely composed of A and B SUs. For misorientations between 37° and 90° the GBs are solely composed of B and C units. Since units A and B correspond to perfect crystals, there is only one "boundary" type at 0° and 90°. For the 37° Σ5 GB, there are three possible GB structures (1 of these is associated with changing φ) and so, in principle, there should be three distinct paths through the metastable states corresponding to combinations of A and B, and A and B' for the misorientations range between 0° and 37°. However, there are clearly more than two possible paths; i.e., there are many metastable GB structures at misorientations between these two limits. This implies that there are other delimiting boundaries, consisting of repeating (complex) unit cells at misorientations between 0° and 37° degrees that are represented by very small cusps in the GB energy versus misorientations plot. All of the metastable states seen in Figs. 2 and 4 can be connected through such a SU model construction.

3. Grain Boundary Statistics

The results presented above demonstrate that for grain boundaries with almost any set of macroscopic degrees of freedom, there are many possible, metastable grain boundary structures and a correspondingly many possible grain boundary energies. This raises the question, which grain boundary structure or which distribution of structures will be observed in experiments. The answer will obviously be different depending on whether or not the grain boundary is in equilibrium. If the grain boundary is not in equilibrium, the structure or distribution of structures will be sensitive to such things as how the polycrystals were synthesized, the thermal history of the polycrystals, and grain boundary dynamical processes operative at the time the structure is observed (or immediately prior to quenching) such as grain boundary migration, grain boundary sliding, etc.

In equilibrium (annealing over infinite time), the distribution of grain boundary structures should be determined by the Boltzmann statistics. The probability of observing a grain boundary in the state i, $p_e^{(i)}(\theta) = \exp(-E^{(i)}(\theta)/kT) / Z$, where $E^{(i)}(\theta) = A_{cell}\,\gamma^{(i)}(\theta)$, kT is the thermal energy, and Z is the partition function. The blue curves in Fig. 2 correspond to the equilibrium internal energies of the grain boundaries as a function of misorientation at zero Kelvin (in this case, the only GB structures observed correspond to the minimum energy structures). At finite temperature, the average grain boundary energy is $\langle \gamma(\theta) \rangle_{\mathrm{e}} = \Sigma_i\, p_e^{(i)}(\theta)\gamma^{(i)}(\theta)$, where the summation is over all $N(\theta)$ GB states and $p_e^{(i)}$ is evaluated at the temperature of interest. We can also consider an extreme non-equilibrium (xne) state, where the probability of a particular GB state is determined by the size of the GB translations phase space it occupies, that is $p_{xne}^{(i)} = A^{(i)}/A_{cell}$, where $A^{(i)}$ is the area of the basin of attraction corresponding to state i and A_{cell} is the area of the entire translation space as seen in Fig. 1b. The red curves in Fig. 2 corresponds to the average grain boundary energy for this ensemble; i.e., $\langle \gamma(\theta) \rangle_{\mathrm{xne}} = \Sigma_i\, p_{xne}^{(i)}(\theta)\gamma^{(i)}(\theta)$. Examination of the results in Fig. 2c for Si, shows that while the equilibrium grain boundary energy (blue curve) shows a rapid oscillation with θ not normally observed in experiments, the non-equilibrium grain boundary energy (red curve) shows a relatively smooth variation with θ except for the cusps observed at low Σ (i.e., $\Sigma 5$), as commonly seen in experiments. This difference between equilibrium and extreme non-equilibrium is not as strong for Al or W. In most non-equilibrium situations, we expect that the average grain boundary energy is $\langle \gamma(\theta) \rangle_{\mathrm{e}} < \langle \gamma(\theta) \rangle < \langle \gamma(\theta) \rangle_{\mathrm{xne}}$ (this is, however, not a rigorous statement).

The presence of a large number of possible grain boundary structures can be described through the configurational entropy, as defined in information theory (i.e., the Shannon entropy [21]). For the extreme non-equilibrium case, we can write this configuration entropy S_c as

$$\frac{S_c}{k_\mathrm{B}} = -\sum_{i=1}^{N(\theta)} p^{(i)}(\theta) \ln p^{(i)}(\theta) = \ln\frac{A_{cell}(\theta)}{a_0^2} - \langle \ln\frac{A^{(i)}(\theta)}{a_0^2} \rangle \qquad (1)$$

where $p^{(i)}$ here corresponds to the xne case. Figure 5 shows this configurational entropy for Al, W and Si from our simulations. Note that except for the case of W at low misorientations (below ~20°), there is a striking similarity in the data amongst the three different materials, except for the overall magnitude, which does appear to be material dependent.

The last expression in Eq. (1) suggests that the entropy can be divided into two parts. The first is associated with A_{cell} and is purely geometric, i.e., it does not depend on material type. We also plot this geometry-dependent term in the entropy $k_B\ln(A_{cell}/a_0^2)$ in Fig. 5 (red curve). The data clearly show that S_c and $k_B\ln(A_{cell}/a_0^2)$ vary with misorientation in a very similar manner (except at small θ in W). This suggests that the second term on the right hand side of the last expression in Eq. (1) is nearly constant (as seen from the blue symbols in Fig. 5) and hence can be written in a θ-independent form as $k_B\ln(\rho a_0^2)$, where ρ is a material parameter we call the "effective density of states". If all $A^{(i)}$ were identical, then $\rho = N/A_{cell}$, i.e., exactly the density of states in $A_{cell}(\theta)$. Here we find that $\rho \approx 7.0a_0^{-2}$, $18.5a_0^{-2}$ and $171a_0^{-2}$

for Al, W and Si, respectively. This again suggests that Si has GBs with the highest structural multiplicity and the degree of structural multiplicity varies from material to material. Based on Eq. (1)

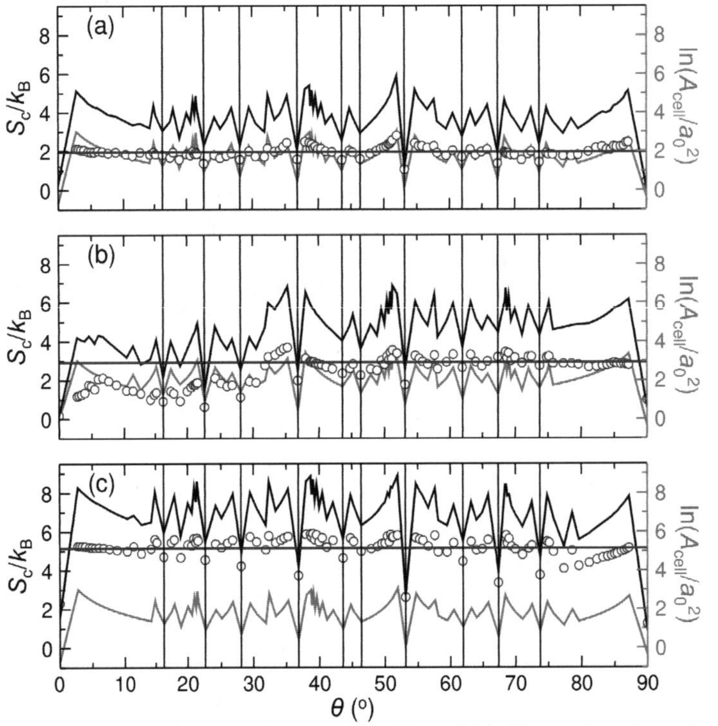

Figure 5. GB configurational entropy of (a) Al, (b) W, and (c) Si as a function of misorientation (black curve) from the data in Fig. 2 and Eq. (1). The dotted red curve and the red axis on the right show the first term in the configurational entropy from Eq. (1). The open blue circles are the result of subtracting the red curve from the black curve, i.e., the second term in Eq. (1).

and employing $A_{cell}/a_0^2 \propto \Sigma^\beta$, where β is a constant (β = 1/2 or 1 for symmetrical tilt or twist GBs, respectively), we find that S_C varies linearly with ln Σ.

4. Glass-Like Grain Boundary Behaviour

The extremely high density of metastable grain boundary structures that are close in energy is analogous to the high density of nearly degenerate metastable states that exist in a glass or glass forming liquid. More than 100 years ago, Rosenhain and Ewen [22] postulated that grains in cast iron were "cemented" together by a thin "amorphous" materials ''identical with or at least closely analogous to the condition of a greatly undercooled liquid.'' While there is no doubt today that grain boundary structure is highly ordered [20], this analogy between GBs and glasses (and glass-forming liquids) is persistent [23-28]. Grain boundary properties may indeed exhibit glass or glass-forming liquid like properties; these include GB sliding [27], GB diffusivity [28] and GB mobility [26,28]. Although highly ordered, GB metastable structures constitute a dense spectrum of local states in which GB structure can be trapped. Just as many glass properties are determined by transitions between nearly degenerate states [29], GBs properties kinetic properties may exhibit similar glass-like properties. For example, the Vogel-Fulcher relation provides a much better fit to the temperature-dependence of grain boundary mobility than does the Arrhenius relation (as for relaxation times in glass forming liquids) [26]. We examine this conjecture via the relationship between the multiplicity of grain boundary states, the configurational entropy, and glass-like relaxation behavior within the framework of Adam-Gibbs theory [30].

Adam-Gibbs theory [30] explains the temperature dependence of the relaxation behavior in glass-forming liquids in terms of the temperature variation of the size of the cooperatively rearranging region in the material. As the temperature is lowered, the system is able to explore smaller and smaller regions of phase space within any fixed time period (e.g., that of a physical measurement). One conclusion from this approach (and many other works since) is that the typical relaxation time τ scales with temperature as $\tau \sim \exp(C/TS_c)$, where C is a constant, T is temperature and S_c is the configurational entropy. For describing grain boundaries, we employ the configurational entropy from Eq. (1). As discussed above, $A_{cell}/a_0^2 \propto \Sigma^\beta$ and Σ varies with θ while the effective density of states ρ does not. This implies that the misorientation dependence of the kinetic relaxation time should scale with misorientation θ as

$$\tau(\theta) = \tau_0 e^{C/TS_c(\theta)} = \tau_0 \exp\left[\frac{C'/k_B T}{\ln(A_{cell}(\theta)/a_0^2)+\ln(\rho a_0^2)}\right] = \tau_0 \exp\left[\frac{C'}{k_B T \ln(D A_{cell}(\theta)/a_0^2)}\right] \quad (2a)$$

$$\tau(\theta) = \tau_0 \exp\left[\frac{C''}{k_B T \ln(D\Sigma(\theta))}\right] \quad (2b)$$

where C, C', C'' and D are all constants (independent of θ).

5. Application: defect absorption at grain boundaries

The above analysis demonstrates the connection between the high density of metastable structures in grain boundaries, the configurational entropy and grain boundary kinetics. As a demonstration of the applicability of these results, we consider the variation in structure that occurs at grain boundaries from thermal fluctuations, from adsorption of point defects, and adsorption of point defects at grain boundaries (e.g., during irradiation).

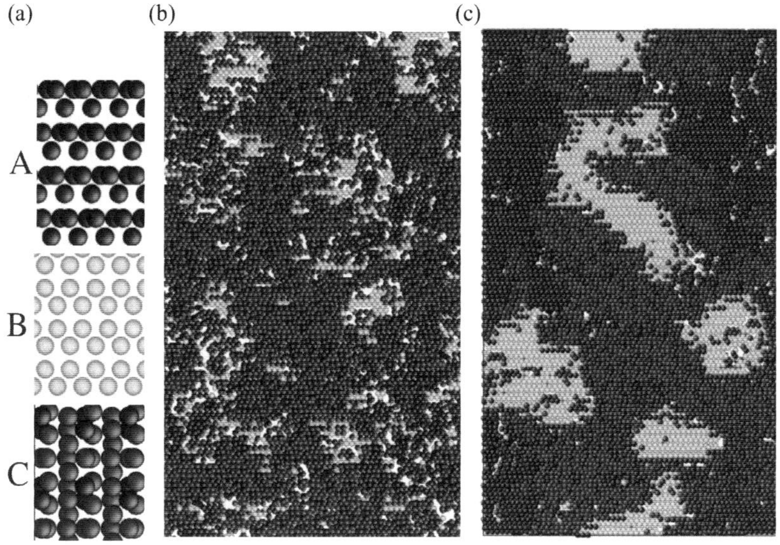

Figure 6. (a) Three different Σ5 (310) [001] symmetric tilt GB structures in W (the labelling A, B, C simply identifies metastable states). The colours label classes of topological type and are arbitrary. (b) An MD simulation cell of a 61,778 Å^2 GB of the same type, equilibrated at 3000 K. (c) Same as (b) but following adsorption of 2400 interstitial atoms into the GB at 1500 K.

In order to distinguish different grain boundary structures, we focus on low Σ boundaries where the number of available states make it possible to distinguish one grain boundary structure from another. We first consider the Σ5 (310) [001] symmetric tilt GB in W, for which there are three distinct metastable structures that we arbitrarily label A, B, and C, as shown in Fig. 6(a). We identify different sites in the bicrystal by finding the Voronoi cell around each atom and labeling each site according to its complete topology [31], and not considering atoms whose Voronoi cell topology corresponds to the BCC crystal lattice. In this way, atoms in different GB structures can be identified [32]. Figure 6(b) shows a 61,778 Å^2 GB of this type, equilibrated at 3000K and quenched to 0K. Examination of the GB shows that the boundary consists of coexisting domains of all three structures. The lowest energy state corresponds to A in Fig. 6(a).

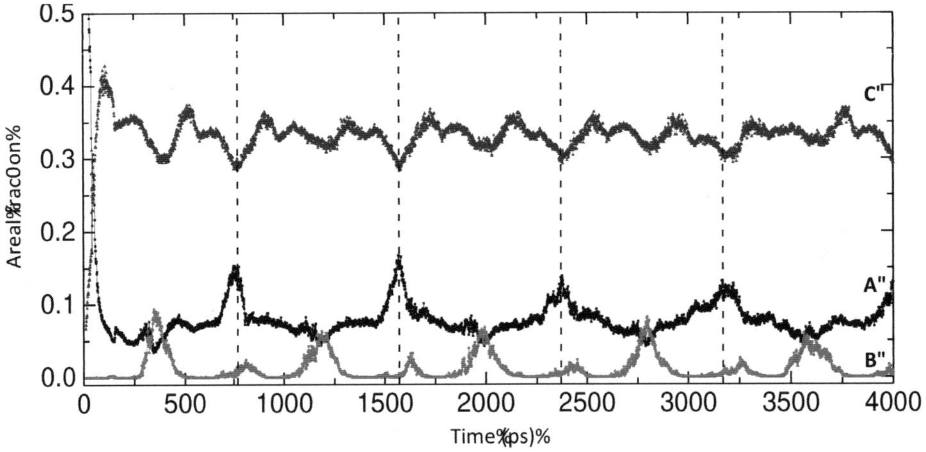

Figure 7. The area fraction of the Σ5 (310) [001] symmetric tilt GB in W of each of the three metastable structure types (A-black, B-red, C-blue curves) shown in Fig. 6(a) as a function of time from a molecular dynamics simulation performed at 1500K where interstitials are introduced into the 61,778 Å^2 boundary at a rate of 10/ps to simulate point defect adsorption at a GB during radiation damage.

Adsorbing 2400 interstitial atoms into the GB in Fig. 6(b) at 1500K transforms this structure into that shown in Fig. 6(c). Compared to the original structure in Fig. 6(a) this grain boundary has larger domains of all three boundary types; likely a result of lowering the temperature. We introduce interstitials into this same grain boundary at a fixed rate during a molecular dynamics simulation at 1500K to simulate point defect annihilation at a grain boundary during irradiation. Figure 7 shows the fraction of the boundary occupied by each of the three structural types shown in Fig. 6(a) as a function of time where interstitials are introduced into the 61,778 Å^2 GB at a rate of 10/ps to simulate point defect adsorption at a GB during radiation damage. We note that the dominant boundary type is C during this dynamic simulation even though type A has the lowest energy. The other interesting feature of these dynamics is that the fraction of the boundary occupied by each type is a nearly periodic function of time. This time period corresponds to that necessary to add the number of interstitials equal to the number of atoms on a single plane parallel to the boundary, since adsorption of this number of interstitials corresponds to the translation of the GB by a single atomic plane (~10^4 W interstitials).

We now turn to the analysis of an experiment in which polycrystalline Cu was irradiated with He ions at elevated temperature [33]. (We note that while the experiments were performed on an FCC metal and the simulations on a BCC metal, the results presented above demonstrate that the main predictions are crystal structure independent in cubic materials.) While dislocation loops and voids formed within grains, regions of near constant width where these defects were nearly absent were found in the vicinity of some grain boundaries but not others. The width of these denuded zones (DZ) varied from GB to GB;

suggesting that some GBs are more efficient point defect sinks than others. We model this process by assuming that the adsorption and incorporation of point defects into GBs leads to the shift of the GB structure between metastable states. To analyze this situation, we return to the Adam-Gibbs theory which suggests that relaxation rates scale with configurational entropy as ~exp(-C/TS_c). From the dependence of the configurational entropy on the metastable state statistics, discussed above, this implies that the adsoprtion rate of point defects should depend on misorientation and through this on Σ. The denuded zone width λ is the compromise between the rate at which point defects diffuse to the grain boundary and the rate at which they can be adsorbed/relaxed into the boundary. During irradiation at a constant rate, a steady-state point defect profile is established such that λ is proportional to the rate of adsorption and hence varies inversely with relaxation time τ. Given the relationship between Σ and τ in Eq. (2b), this implies that $\lambda = \lambda_0 \exp(-F/T\ln\Sigma)$, where F is a constant, or at fixed temperature (as in these experiments), $\lambda = \lambda_0 \exp(-F'/\ln\Sigma)$.

We replot the DZ width λ data from [33] for small Σ versus Σ in Fig. 8. Not surprisingly, the DZ width decreases as Σ decreases, since the fewer metastable states available (typical of low Σ boundaries) the more difficult (longer relaxation times) it is to relax the point defects in the boundaries. We fit the data to the theoretical form from above $\lambda = \lambda_0 \exp(-F'/\ln\Sigma)$ and compare the prediction with the experimental data in Fig. 8. As expected, the model shows that larger-Σ GBs have higher sink efficiency than lower-Σ GBs; this dependence, however, becomes weak for large Σ ($\lambda \rightarrow \lambda_0$ in this limit). The agreement between the GB property prediction and the experimental measurements supports the γ-band model, the underlying GB statistics upon which this prediction rests, and the Adam-Gibbs approach to determining GB relaxation kinetics.

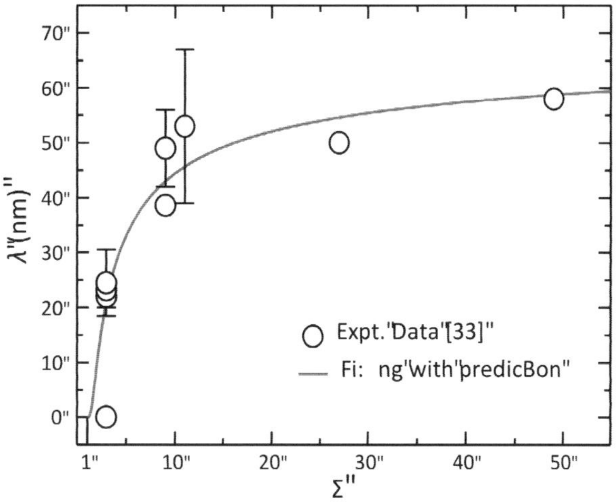

Figure 8. The denuded zone width λ versus the reciprocal coincident site lattice density Σ. The experimental data are from [33] and the red curve is the best fit to those data using, $\lambda = \lambda_0 \exp(-F'/\ln\Sigma)$.

6. Discussion and Conclusions

The results presented above provide compelling evidence from atomistic simulations in three different materials with three different crystal structures that the multiplicity of grain boundary structure is ubiquitous and that the magnitude of this multiplicity is much larger than generally appreciated. While we characterized this multiplicity for a series of [001] symmetric tilt GBs in these materials, we expect that such large multiplicities are the general rule for all crystalline materials and most grain boundaries. Corresponding to this large set of metastable structures for fixed macroscopic degrees of freedom is the broadening of the GB energy versus misorientation curve into a fairly broad and continuous band of energies versus misorientation. Hence, the common picture of grain boundary thermodynamics is rather oversimplified. The GB energy band versus misorientation can be understood in terms of the structural unit model by admitting multiple possible GB structures at delimiting GBs.

The existence of a large multiplicity of GB states at fixed macroscopic degrees of freedom with a near continuous set of energies suggests that GB configurational entropy S_c should be important for GB properties. We demonstrated that the GB S_c consists of two major contributions, the first is purely geometric and depends on the reciprocal coincident site lattice density Σ and the second is sensitive to bonding and, hence, to material type and crystal structure.

When a large set of metastable states of a particular grain boundary is available, grain boundary dynamics should be sensitive to transitions between nearby metastable states. At high temperature, multiple states should be observable, as expected on the basis of equilibrium thermodynamics. In addition, when grain boundaries are out-of-equilibrium, for example, following a quench or during dynamical processes such as GB sliding, GB migration, and during irradiation, the distribution of GB structures is expected to differ significantly from those observed in equilibrium. This is akin to a liquid where it dynamically shifts between nearly degenerate states. However, GB properties are likely more similar to glass forming liquids where the ability to access other states decreases as the temperature is lowered. In this view, the relaxation dynamics/kinetics of a grain boundary, like a glass-forming liquid, can be described using Adam-Gibbs theory that suggests that relaxation time grows with decreasing temperature or S_c as $\tau \sim \exp(C/TS_c)$.

Using the dependence of GB S_c on the reciprocal coincident site lattice density Σ found in atomistic simulations we can make predictions of how grain boundary relaxation kinetics depends on the five macroscopic GB degrees of freedom. We presented some evidence from a radiation damage scenario that this approach captures the dependence of point defect adsorption kinetics on grain boundary type.

While the results and analysis described herein is a combination of simulation results, statistical mechanics ideas and, admittedly, some speculation, it does provide a new direction in which to think of the relationship between defect structure and defect properties, that is particularly appropriate for 2-dimensional defects, such as GBs, heterophase interfaces and surfaces that are not singular. It also suggests that the renewed attention to structural phase transitions in grain boundaries is interesting, even though most of the extant work only considers extremely simple (near singular) grain boundaries. The more general case is particularly fascinating.

Acknowledgments

The authors gratefully acknowledge Emanuel A Lazar for his assistance in analysing the GB structures in Figs. 6 and 7 and for making his unpublished topological analysis codes available to the authors for this work. JH and DJS gratefully acknowledge the support of the Center for the Computational Design of Functional Layered Materials, an Energy Frontier Research Center funded by the U.S. Department of Energy, Office of Science, Basic Energy Sciences under Award # DE-SC0012575.

References

[1] Bollmann W 1970 *Crystal Defects and Crystalline Interfaces* (Berlin: Springer)
[2] Vitek V 1970 *Scripta Metall.* **4** 725
[3] Bristowe P D and Crocker A G 1975 *Phil. Mag.* **31** 503
[4] Smith D A, Vitek V and Pond R C 1977 *Acta Metall.* **25**, 1363
[5] Pond R C, Smith D A and Vitek V 1979 *Acta Metall.* **27**, 235
[6] Bristowe P D and Crocker A G 1978 *Phil. Mag.* **38A** 487
[7] Olmsted D L, Buta D, Adland A, Foiles S M, Asta M, and Karma A 2011 *Phys. Rev. Lett.* **106**, 046101
[8] Frolov T, Olmsted DL, Asta M and Mishin Y 2013 *Nature Comm.* **4** 1899
[9] Vitek V, Sutton A P, Wang G J and Schwartz D 1983 *Scripta Metal.* **17** 183
[10] von Alfthan S, Haynes P D, Kaski K, and Sutton A P 2006 *Phys. Rev. Lett.* **96** 055505
[11] Vitek V, Minonishi Y, and Wang G J 1985 *J. Phys. (Paris)* **46 C4** 171

[12] Tarnow E, Dallot P, Bristowe P D, Joannopoulos J D, Francis G P, and Payne M C 1990 *Phys. Rev. B* **42** 3644
[13] Krakow W 1991 *Philos. Mag. A* **63** 233
[14] Mills M J, Daw M S, Thom G J and Cosandey F 1992 *Ultramicroscopy* **40** 247
[15] Mishin Y, Farkas D, Mehl M J, and Papaconstantopoulos D A 1999 *Phys. Rev. B* **59** 3393)
[16] Ackland G J and Thetford R 1987 *Philos. Mag. A* **56** 15
[17] Tersoff J 1988 *Phys. Rev. B* **37** 699
[18] Wang G J and Vitek V 1986 *Acta Metall.* **34** 951
[19] Sutton A P and Balluffi 1995 *Interfaces in Crystalline Materials* (Oxford: Oxford University Press)
[20] Sutton A P and Vitek V 1983 *Phil. Trans. Royal Soc. (London) A* **309** 1
[21] Shannon C E 1948 Bell *System Technical Journal* **27** 379
[22] Rosenhain W and Ewen D 1913 *J. Inst. Met.* **10** 119
[23] Ashby M F 1972 *Surf. Sci.* **31** 498
[24] Keblinski P, Phillpot S R, Wolf D, and Gleiter H 1996 *Phys. Rev. Lett.* **77** 2965
[25] Wolf D 2001 *Curr. Opin. Solid State Mater. Sci.* **5** 435
[26] Zhang H, Srolovitz D J, Douglas J F, and Warren J A, *PNAS* **106** 7735
[27] Ke T S 1947 Phys. Rev. **71** 533
[28] Schoenfelder B, Keblinski P, Wolf D and Phillpot S R 1999 *Intergranular and Interphase Boundaries in Materials* (Pfaffikon, Switzerland: TransTech)
[29] Berthier L and Biroli G 2011 *Rev. Mod. Phys.* **83** 587
[30] Adam G and Gibbs J H, 1965 *J. Chem. Phys.* **43** 139
[31] Lazar E A, Mason J K, MacPherson R D, and Srolovitz D J 2012 *Phys. Rev. Lett.* **109** 095505
[32] Lazar E A, Han J, and Srolovitz D J, to be published
[33] Han W Z, Demkowicz M J Fu E G, Wang Y Q, and Misra A 2012 *Acta Mater.* **60** 6341

Particle-assisted abnormal grain growth

EA Holm[1,3], TD Hoffmann[2], AD Rollett[3], and CG Roberts[4]
[1]Sandia National Laboratories, Albuquerque, NM 87185-1411 U.S.A.
[2]Arizona State University, Tempe, AZ 85287-6106 U.S.A.
[3]Carnegie Mellon University, Pittsburgh, PA 15213-3890 U.S.A.
[4]Carpenter Technology Corp., Reading, PA 19612-4662 U.S.A.

E-mail: eaholm@cmu.edu, ddffnn@gmail.com, rollett@cmu.edu, croberts@cartech.com

Abstract. Abnormal grain growth is observed in systems that are nominally pinned by static particle dispersions. We used mesoscale simulations to examine grain growth in three-dimensional polycrystals containing stable, inert particles located at grain boundaries. In the absence of pinning particles, only normal grain growth occurs. When particles are present, some normal grain growth occurs, until a Zener-Smith pinned state is achieved. However, after a long incubation time, a few grains can thermally fluctuate away from their particle clouds and grow abnormally. The abnormal events are rare and stochastic. The abnormal grains are always among the largest initial grains, but most of the largest initial grains do not grow abnormally.

1. Introduction

Abnormal grain growth (AGG) occurs in a wide variety of metallurgical systems: textured and nontextured metals, thin films, strained systems, nanocrystalline materials, and systems with second phase precipitates. The proposed mechanisms for AGG are equally varied, including grain boundary energy anisotropy, inhomogeneous strain energy density, initial size advantage, favorable free surface energy, and grain boundary mobility variations. Clearly, AGG is one phenomenon with many causes.

Particularly baffling is the occurrence of AGG in polycrystals that are nominally stable due to the presence of inert, static pinning particles [1]. Dispersions of second phase particles are often used to inhibit normal grain growth in polycrystalline metals [2]. Since the particles exert a pinning force that balances the curvature driving force for grain growth (as first described by Zener and Smith in 1948 [3]), it stands to reason that they should prevent abnormal grain growth as well. However, AGG is observed in systems that are nominally pinned by static particle dispersions, such as nickel-based superalloys [4].

In this paper we describe a new mechanism by which static second phase particles not only fail to inhibit AGG, but actually cause it to happen. We term this phenomenon 'particle-assisted abnormal grain growth'.

2. System and Methods

We were inspired by a recent simulation result by Roberts [5], who observed a few instances of AGG in nominally pinned microstructures that contained particles preferentially located on grain boundaries. We wanted to devise a model system to improve our understanding of the cause and phenomenology of AGG in systems with static pinning particles.

We began with equiaxed, nontextured, 3D grain structures grown via a Monte Carlo Potts model (MCPM) grain growth simulation to various initial grain sizes [6]. We assigned uniform and isotropic grain boundary properties (energy and mobility), and then deposited pinning particles, which were 3x3x3 cubes of inert sites (r = 3), at random on the grain boundaries with volume fractions ranging from 2% to 10%. The polycrystal was then annealed using a standard MCPM simulation [7-10]. Under these conditions, the only driving force for microstructural evolution is the normal, isotropic curvature driving force for grain growth.

As with all particle pinning simulations, the system sizes are required to be rather large [11]. Each simulation used a system of at least 150^3 lattice sites evolved for at least one million Monte Carlo timesteps (MCS) at a simulation temperature of 1.5. Because AGG in particle pinned systems is rare and stochastic, we ran many independent trials. Achieving this throughput required us to utilize an efficient, parallel kinetic Monte Carlo code that can run on both workstations and large cluster computers [12].

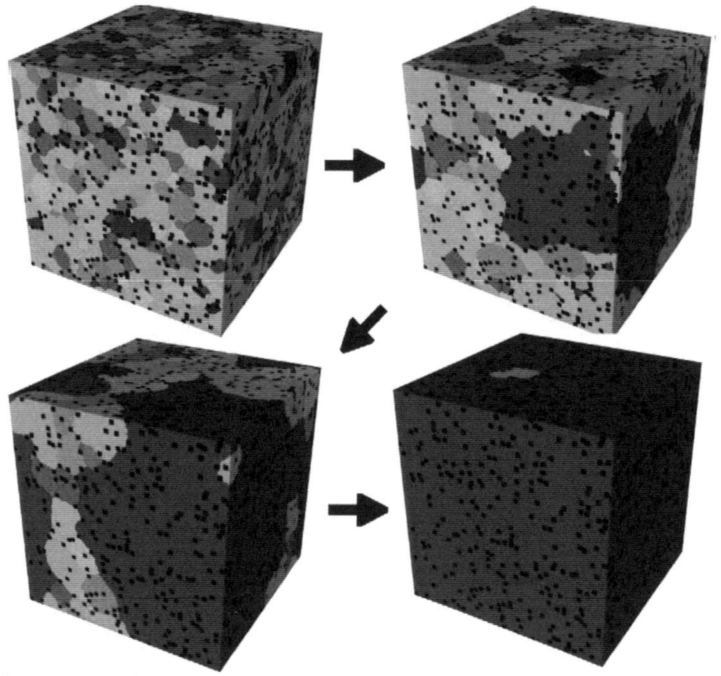

Figure 1. Abnormal grain growth in a particle-pinned system. The initial grain structure (upper left, t=0 MCS) is equiaxed, with particles decorating the grain boundaries. After some incubation time, an abnormal grain appears (light gray grain in upper right structure, t = 10^5 MCS). The abnormal grain grows (lower left, t = $2x10^5$ MCS) and eventually consumes the other pinned grains (lower right, t = $3x10^5$ MCS). Particles are shown as black squares.

3. Results and Discussion

In the absence of pinning particles (particle volume fraction $f = 0$), the isotropic growth conditions ensure that grain growth proceeds normally and uniformly to a single crystal state.

For a grain diameter $R_0 = 10$ and particle fraction $f \leq 8\%$, the grain structure is not initially pinned. Upon annealing, some normal grain growth occurs until a pinned grain size is achieved. In the pinned state, not all the particles are located on grain boundaries. The final pinned grain size is approximately the Zener-Smith pinned size predicted by similar simulations for spatially random particle distributions [11]. This pinned state appears stable for at least 10^6 timesteps.

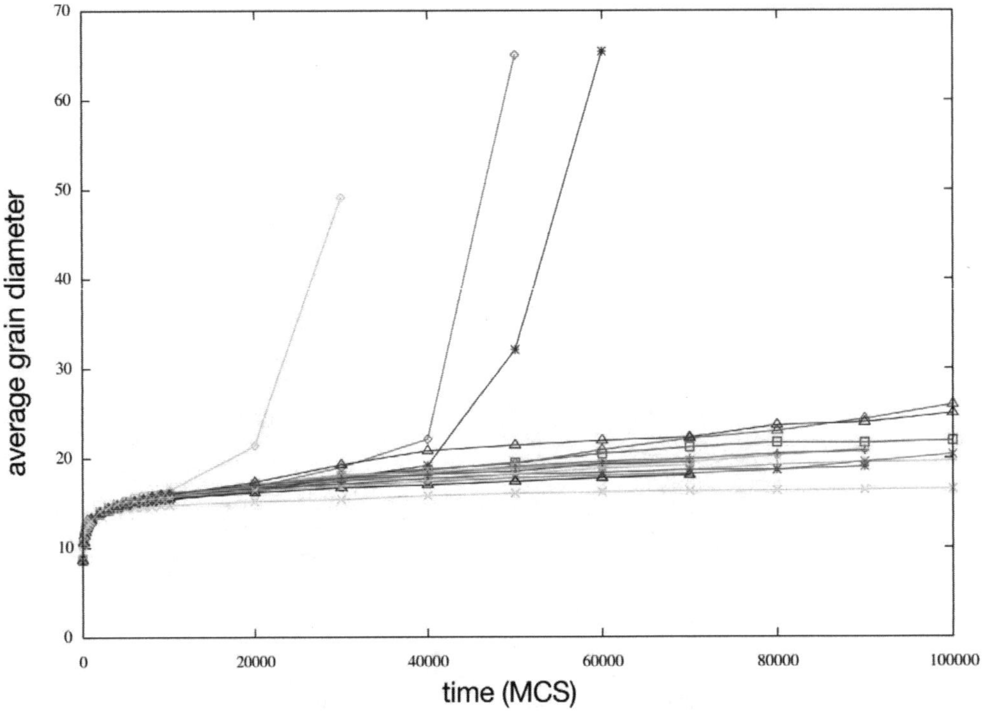

Figure 2. Grain growth kinetics for 15 independent simulations with particle volume fraction of 10 %. All systems initially grow from initial grain diameter of 10 to the Zener-Smith pinned size of approximately 22. Most systems remain stagnant up to 10^5 timesteps, but in three cases, an abrupt increase in grain size indicates an abnormal grain growth event occurs.

In a system with a particle volume fraction f = 10% and the same initial grain diameter, the microstructure also grows to a grain size approximately equal to the Zener-Smith pinned size. However, we found that after long anneals, a few grains grow abnormally, as shown in Fig. 1. The initiating event is a grain boundary that thermally fluctuates off from its pinning particles. If the free boundary belongs to a grain that is inclined to grow, it will grow without hindrance from its neighboring grains, since their boundaries remain immobile. Likewise, the particles do not inhibit the growth of the abnormal grain, since it has more neighbors, thus more driving force for growth, than the average, pinned grains. As the grain grows, its size advantage – thus its ability to grow at the expense of its neighbors – increases. In the right local environment, the result is persistent, abnormal growth of the free grain. Thus, we find the counterintuitive result that pinning particles cause AGG in systems not otherwise inclined to grow abnormally. This is particle-assisted abnormal grain growth.

Figure 2 shows grain growth kinetics for 15 independent simulations with R_0 = 10 and f = 10% over 10^5 timesteps. All the systems grow briefly, then stagnate at a grain size approximately equal to the Smith-Zener pinned grain size of 22 [11]. However, three simulations result in AGG events, indicated by a sudden increase in average grain size. It is interesting that AGG occurs at different times in different systems, indicating that it is a stochastic process. Once AGG takes place, the abnormal grain grows quickly to consume the system, so we typically can observe only one abnormal event per system. Although the other systems in Fig. 2 appear to stagnate, we found that in longer anneals (up to 10^6 MCS) every one eventually underwent an AGG event.

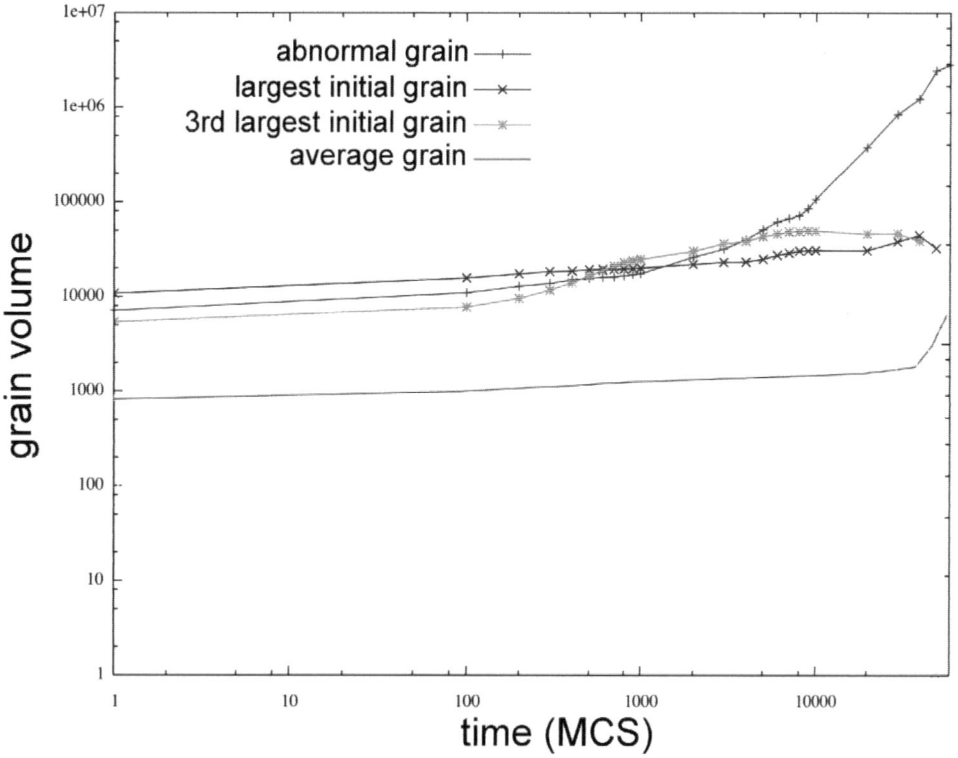

Figure 3. The kinetics of growth of the three largest grains and the average grain in a system that evolves via particle-assisted AGG. The eventual abnormal grain begins as the second largest grain. The largest grain shows no signs of abnormal growth.

In analyzing the abnormal grains, we find that they are always among the largest initial grains, but they are often not the largest. The experimental work of Chen *et al.* [13] also showed this, which suggests there must be some additional advantage besides size that leads to AGG. Figure 3 shows the kinetics of growth for the three largest grains in a microstructure that eventually undergoes AGG. These three large grains are about 14 times bigger by volume than the average grain, which is only a factor of about 2.4 in grain diameter; thus, the largest grains are well within the typical lognormal grain size distribution for an equiaxed microstructure.

The eventual abnormal grain begins as the second largest grain in the system and initially grows at about the same rate as the largest grain. Interestingly, the third largest grain initially grows faster than both of the larger grains and surpasses them in size for a while. However, at around 1000 MCS, the abnormal grain begins growing much faster, presumably due to a critical boundary breakaway event. It quickly becomes the largest grain in the system and ultimately consumes the others. The average grain size evolves very little until the abnormal grain has become very large.

While the abnormal grain always begins as a large grain, most large grains never grow abnormally. We were unable to find a simple marker that indicates which grain in an initial microstructure would become abnormal. For example, in a particular underlying grain structure, different random particle placement causes different grains to become abnormal. In addition, in a given grain/particle structure, different grains may become abnormal depending on simulation parameters (i.e. simulation temperature, choice of computer code, etc.). Since experiments have shown that particle-boundary correlations are far from random [14], it may, for example, be useful to measure the local density of particles per unit grain boundary area. This might reveal local variations in the pinning force on the boundaries of a given grain and thus provide the additional selection criterion for individual grains to become abnormal. In any case, these observations reinforce the conclusion that particle-assisted AGG is a stochastic phenomenon.

4. Conclusions

Abnormal grain growth is a single phenomenon with many causes. The observation of AGG in polycrystals that are nominally stable due to the presence of inert, static pinning particles is particularly perplexing. Mesoscale grain growth simulations of 3D polycrystals containing static particle dispersions exhibit AGG in systems that do not evolve abnormally in the absence of particles. The AGG mechanism is the thermal fluctuation of a grain boundary away from its particle cloud. Because AGG does not occur in these systems without a particle dispersion, we term this AGG mechanism particle-assisted abnormal grain growth. Particle-assisted AGG adds a new explanation for the AGG phenomenon in a previously baffling class of materials.

Acknowledgements

Sandia is a multi-program laboratory operated by Sandia Corporation, a Lockheed Martin Company, for the United States Department of Energy's National Nuclear Security Administration under contract DE-AC0494AL85000. This work was supported in part by the DOE Office of Basic Energy Sciences and Sandia's Laboratory Directed Research and Development program, as well as by National Science Foundation awards DMR-1307138 and DMR-1107986.

References

[1] May JE and Turnbull D 1958 *Trans. AIME* **212** 769-781
[2] Manohar PA, Ferry M, Chandra T 1998 *ISIJ Int.* **38** 913-924
[3] Smith CS 1948 *Trans. AIME* **175** 15-51
[4] Mino K, Nakagawa YG, Ohtomo A 1987 *Metal. Trans. A* **18** 777-784
[5] Roberts CG 2007 Ph.D. Thesis, Carnegie Mellon University
[6] Holm EA, Battaile CC 2001 *JOM* **53** 20-23
[7] Anderson MP, Grest G, Srolovitz DJ 1989 *Phil. Mag. B* **59** 293-329
[8] Anderson MP, Srolovitz DJ, Grest G and Sahni P 1984 *Acta Metal.* **32** 783-791
[9] Miodownik MA 2007 Monte Carlo Potts Model, in: Janssens KGF et al. (Eds.), *Computational Materials Engineering: An introduction to microstruture evolution*, Elsevier Academic Press, London, pp 47-108
[10] Rollett AD and Manohar PA 2004 The Monte Carlo Method, in: D. Raabe, et al. (Eds.), *Continuum Scale Simulation of Engineering Materials*, Wiley-VCH, Weinheim, Germany, pp 77-114
[11] Miodownik MA, Holm EA and Hassold GN 2000 *Scripta Mater.* **42** 1173-1177
[12] Plimpton SJ, Thompson A and Slepoy A 2009 *SPPARKS Kinetic Monte Carlo Simulator*, Sandia National Laboratories
[13] Chen N, Zaefferer S, Lahn L, Gunther K and Raabe D 2003 *Acta Mater.* **51** 1755-1765
[14] Roberts CG, Semiatin SL, A. Rollett AD 2007 *Scripta Mater.* **56** 899-902

Investigating the mechanisms of grain boundary migration during recrystallization using molecular dynamics

E R Homer

Department of Mechanical Engineering, Brigham Young University, 435 CTB, Provo, UT 84602, USA

E-mail: eric.homer@byu.edu

Abstract. Resolving the mechanisms associated with grain boundary migration is a difficult task. This work is focused on detailing techniques to utilize atomistic simulations to better understand the energy, structure and mobility of grain boundaries (GBs). Various techniques are detailed on how to construct different simulation cells, select boundary plane normals, and measure simulated mobility. A limited number of GBs, inspired by experimental observations, are selected for the present work. The GBs studied indicate that the structure of the theoretical set of GBs has a structure consistent with an overall minimization of energy. The mobility simulations indicate that shear coupling is the preferred migration mechanism. However, when the same GBs are constrained, as might occur in polycrystalline networks, the relative mobilities of the set of GBs are significantly altered. Finally, the influence of deformation structures, as they pertain to GB migration during recrystallization, are discussed.

1. Introduction

Thermomechanical processes are essential in the production and manufacturing of a large portion of the world's materials. These thermomechanical treatments subject the microstructure within the material to deformation, recovery, recrystallization, or grain growth [1]. Each of these steps is critical to achieving a material with a given defect density, texture, grain size and structure, which in turn play an important role in defining a material's strength, ductility, and property anisotropy. For example, defect density correlates with strength and ductility, grain size affects both creep resistance and strength (as in the Hall-Petch relationship [2,3]) and texture is the basis for property anisotropy.

The three stages in process annealing are: (1) recovery, which is marked by the annihilation of and rearrangement of dislocations; (2) recrystallization, which marks the nucleation of new defect-free crystals that grow to consume a previously deformed material; and (3) grain growth, which involves the subsequent growth of these grains and overall reduction of energy by the reduction of grain boundary area. The latter two processes both involve the migration of GBs, although the driving force for migration is different for the two processes. In addition, recent work has demonstrated that recovery can even involve the migration of grain boundary triple junctions [4]. As a result, understanding the mechanisms and influences of boundary migration is essential to fully resolving the different annealing processes.

As noted above, GB migration during the different stages of annealing is influenced by various factors. Complete decoupling of these factors, such as the inherent mobility of a GB from the driving force facilitating its motion, the defect structure being consumed, and the GB network constraining that motion, is at present an impractical procedure. One can, however, gain insight into the relative importance of the various factors affecting migration by examining the factors individually and comparing their effect with those directly observed.

GB migration during recrystallization, in particular, is primarily influenced by the stored strain energy due to prior strain hardening [1]. While this stored energy provides the primary driving force for migration, recent work has shown that the alignment of dislocation boundaries along particular crystal planes can lead to faster grain boundary migration [5]. This leads to the question of whether the structure and orientation of dislocations that build up during deformation play a more important role on the migration of a given GB than its inherent mobility. This is especially pertinent since it is known that deformation prior to recrystallization can have a significant influence on the resulting texture following recrystallization [1]. Finally, it is known that GB plane normal plays an important role in defining the properties and structure of different GBs, even when the GBs have the same misorientation.

The general aim of this research is to detail methods that could be used to examine the various factors influencing the migration of recrystallizing GBs in deformed aluminum. Recrystallization GB migration in aluminum is selected because of the availability of data from nondestructive techniques such as 3D X-ray Diffraction (3DXRD) [6,7] and 3D X-ray crystal microscopy (3DXRM) [8]. These synchrotron techniques provide a full 3-D voxelized map of the microstructure. The orientation is recorded for each 3D voxel, with a spatial resolution in the micron to submicron scale. By using X-ray techniques with the highest spatial resolution and comparing the orientations in these voxels, one can even characterize the local deformation microstructure.

These works often examine GBs that are nominally of a single misorientation with the surrounding deformed structure. In the experiments of interest, the recrystallized sample is examined at various times during the recrystallization. In the region examined in these snapshots, a recrystallizing GB is advancing to consume the deformed aluminum. The GBs exhibit interesting migration patterns: stop and go motion, regions that are fairly planar that migrate slowly, curved regions that migrate at higher rates, etc. [6,7]. The planar and curved regions show a particularly strong dependence on GB plane normal.

Unfortunately, the 3DXRD and 3DXRM methods do not provide mobility measurements and cannot resolve the nature of the boundary migration or the reasons that some portions of the GB remain stagnant while other portions migrate relatively quickly. One could analyze individual boundaries and their migration by more standard experimental techniques [9-14], though we seek instead to simulate these behaviors at an atomic level.

The present work utilizes atomistic simulations to investigate the structure, energy and mobility of various GB plane orientations in a high-angle GB. The nature of the study utilizes methods similar to a previous survey of 388 GBs [15-19], but the focus is on creating GBs to match conditions in experiments rather than the construction of random GBs. While the conditions cannot be identical to those in experiments, the approach does allow one to study various influencing factors and gain insight into their relative importance as they might influence experimental observations.

2. Methodology

2.1. Bicrystallography to match experimental and atomistic GBs

The GB of interest would separate a recrystallized grain from a deformed microstructure, and therefore have a range of misorientations. As a result, one must either choose to study a range of misorientations and boundary plane orientations or one can study the average misorientation, as we choose to do here. We select a high-angle GB about an arbitrary axis, not one of the commonly studied 100, 110 or 111 axes. We identify the coincidence site lattice (CSL) Σ45c GB, which is given by 53.1301° rotation about the <221> axis. The Σ45c CSL GB can be studied for various boundary plane orientations to gain an understanding of how different regions of a GB, curved or flat, might respond during migration. To eventually provide optimal comparison between experiments and the simulated

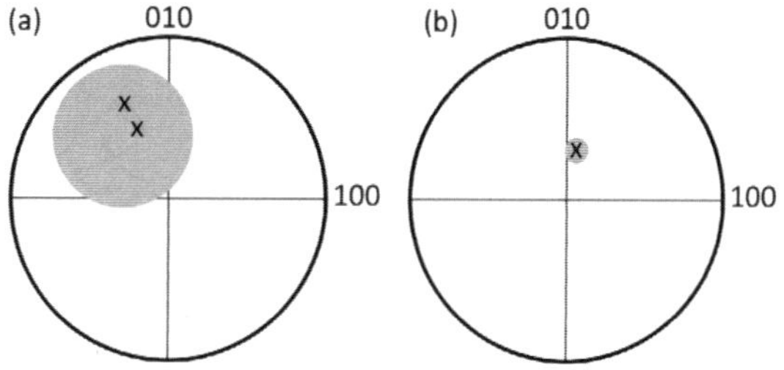

Figure 1. Equal angle projection of boundary plane normal distributions for (a) a diffuse distribution representative of curved GB normals and (b) a sharp distribution representative of flat GB normals. The reference frame for the projection are the crystallographic coordinate system of the recrystallized grain. Normals of particular interest are marked by an 'x'.

atomistic GBs, the crystallography obtained from experimental work must be examined, however here we have chosen an example GB which are inspired by the experimental observations.

The boundary plane orientations for the various GBs studied in this work are selected from a hypothetical distribution of plane normals, as defined in the crystal frame of the recrystallized grain. These normals are separated into two populations: one population of normals will have a diffuse distribution over a large range of normals, corresponding to a 'curved' region of a GB, and the second population will be a sharp distribution over a small range of normals, corresponding to a 'flat' region of the same GB. These two distributions of normals are presented in Figures 1a and 1b. The two populations overlap slightly but the curved GBs are obviously more diffuse. The normal for the 'flat' portion is identified as [0.05 0.55 0.83]. Since the 'curved' distribution of normals is diffuse, we select two normals in the region, one with a normal equal to [-0.39 0.83 0.41], and the other to be selected for reasons described later. A total of 3 GBs will be studied, one representing the 'flat' GB and two representing the 'curved' portions of the same GB.

To obtain the atomic structures for these GBs, an adaptation of the methodology used by Olmsted et al. is used [15]. These techniques employ CSL lattice theory to determine boundary structures so that periodicity in the boundary plane can be employed to avoid the effects of free surfaces. In the Olmsted methodology, one seeks CSL bicrystals that fit within a given set of orthogonal boundary plane dimensions, thereby limiting what boundary plane normals can be studied. Since we are instead interested in creating bicrystals for specific normals, we adapt Olmsted's methodology in the following way. First, we define the CSL lattice for the misorientation using Grimmer's method [20]. This CSL lattice is defined in the crystal frame of one of the two bicrystals. A small portion of the CSL lattice for the Σ45c misorientation is plotted in Figure 2a, where each CSL lattice site is given by an 'x'. Using a set of CSL lattice points, one can define a set of vectors that begin at the origin and end at each of the defined lattice points, as shown in Figure 2a. Vectors that have the same direction as other vectors are discarded, and for efficiency, the smallest magnitude vector for each direction is retained. This results in a set of vectors which all have unique directions.

Given this set of vectors with unique direction, one can begin to define planes by the combination of any two vectors. Since these two vectors correspond to CSL lattice points, the boundary plane they define will capture the natural periodicity of both crystals (one on each side of the GB), thus ensuring that periodicity for the GB is maintained so free surface effects can be avoided. These two vectors become the basis vectors, $\vec{a}_1$ and $\vec{a}_2$, and form the boundary repeat unit which is typically a parallelogram as shown in Figure 2b, since most combinations of $\vec{a}_1$ and $\vec{a}_2$ are not orthogonal. The normal, $\hat{n}$, for each boundary plane is defined by the normalized cross product of $\vec{a}_1$ and $\vec{a}_2$. While

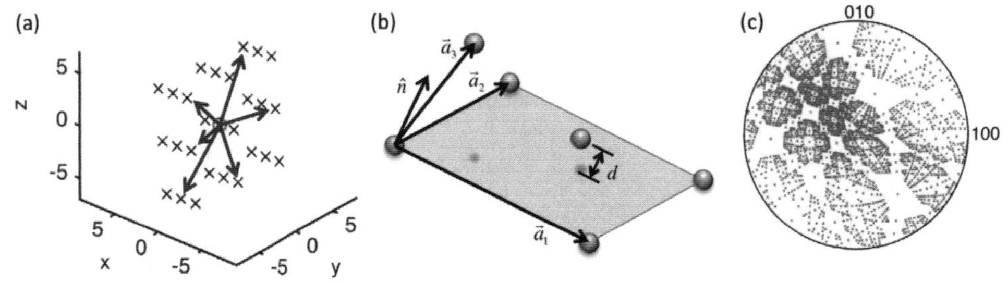

Figure 2. Methodology used to construct bicrystals with boundary planes that match those of experimental interest. (a) Partial Σ45c CSL lattice, marked by 'x', in the crystal coordinate frame. Also includes a small sampling of possible vectors that initiate at the origin, which can be used as potential basis vectors for boundary planes. (b) A set of two basis vectors, $\vec{a}_1$ and $\vec{a}_2$, define a unique boundary plane with normal $\hat{n}$. The third basis vector $\vec{a}_3$ based on a CSL lattice point ensures periodicity of the simulation cell. The distance d of the nearest CSL lattice point to the boundary plane defines the maximum position for unique placement of the boundary plane. (c) Stereographic project of boundary plane normals derived from the combination of basis vectors obtained in a block of ±3 CSL lattice points away from the origin.

the boundary plane repeat unit may not be orthogonal, this method has the advantage of producing significantly smaller GBs than if the constraint of orthogonality was enforced as in the work of Olmsted et al. [15].

In order to match the boundary plane normal to those desired (cf. Fig 1), as we seek to do here, we must find a boundary plane that is sufficiently close to the normal of interest. Using the CSL points and their corresponding vectors, one can obtain a family of possible normals from the set of selected CSL lattice points. For example, using a set of unique vectors defined by ±3 CSL lattice points away from the origin, one can obtain all boundary plane normals defined in Figure 2c. Since there may be multiple combinations of vectors that define the same plane, the combination that defines the smallest repeat unit area is kept. It is noted that in Figure 2c, certain regions have a high density of boundary plane normals and other regions have a low density of boundary plane normals.

One can select the boundary plane which is nearest to the desired normal, or if there is no boundary plane sufficiently close, one can use larger blocks of CSL lattice points (±4 or more CSL lattice points away from the origin) to improve the overall density of boundary planes near a desired normal. Finally, it is noted that for the Σ45c misorientation, a full hemisphere of unique boundary plane normals exists, or in other words, the Σ45c boundary plane orientation fundamental zone is a full hemisphere [21].

In the present work, once the planar boundary repeat unit is determined, a third vector, $\vec{a}_3$, is identified to determine the periodicity of the CSL lattice in a direction out of the boundary plane. In the present work, the selection of this third vector is another CSL lattice point is based on a combination of orthogonality to the boundary plane and magnitude of the vector. This results in a periodic triclinic simulation cell [22], which can be easily accommodated by software such as LAMMPS. The balance between the two factors is critical for efficient computations, since orthogonality maximizes parallel computing efficiency and a small magnitude minimizes the number of atoms contained in each CSL unit.

If one simulates a fully periodic bicrystal (containing 2 GBs), the basis vectors for the simulation cell must follow all three vectors identified above to maintain periodicity. On the other hand, it was recently determined that if one simulates a bicrystal with only 1 GB, this constraint of using a non-orthogonal $\vec{a}_3$ is not necessary. Since periodicity will not be enforced in this dimension, this does enable the possibility of defining a basis vector that is orthogonal to the boundary plane, resulting in a monoclinic simulation cell that can have improved computational efficiency. The magnitude of this orthogonal basis vector would simply be the vertical distance of the next CSL lattice point that is still contained within the dimensions of the two basis vectors defining the boundary plane. Despite the possibility to define the third basis vector of the simulation cell orthogonal to the first two, the present simulation utilizes the triclinic non-orthogonal basis vectors that lie at CSL lattice points.

With the three basis vectors defined, one can evaluate the microscopic degrees of freedom that enable a particular GB to achieve its minimum energy structure. These microscopic degrees of freedom are identified as: three degrees of freedom defining the relative shift vector between the two crystals, one degree of freedom defining the placement of the actual GB plane relative the boundary plane defined by the basis vectors, one degree of freedom for the allowed proximity between neighboring atoms, and one degree of freedom determining which overlapping atoms should be discarded [15]. These microscopic degrees of freedom are essential to defining a GB because they affect how the atoms line up and coordinate at the interface of the two crystals. The shift vector adjusts the position of the two crystals relative to each other, the GB plane placement determines which two planes of the each of the crystals will line up next to each other, proximity cutoff determines how close atoms can get to one another before they must be removed, and the selection of which atoms to be deleted determines where a given atom may be placed.

The present work uses the methods established by Olmsted et al. to identify the unique configurations and combinations of these variables [15]. Unique combinations of the relative shift vector between the two crystals are determined from the displacement shift complete (DSC) lattice. Since these represent continuous degrees of freedom, which would be difficult to fully evaluate, fractions of the three basis vectors of the DSC lattice are typically considered. In the present work, each vector is divided into two increments for a total of 8 relative shift vectors. The placement of the GB plane must be somewhere between the original boundary plane and the minimum distance, *d*, to the next CSL lattice point contained within the simulation cell (illustrated in Figure 2c), since this CSL lattice point reflects the next periodic image of the same atomic configuration. Since the lattice spacing of atoms in the two crystals is discrete, one can identify the exact placement of this plane that will result in unique configurations. Finally, following Olmsted, we consider three cases of discarding overlapping atoms that are within the proximity cutoff: deleting overlapping atoms from the first crystal, deleting overlapping atoms from the second crystal, and removing overlapping atoms from both crystals and replacing the both with an atom at a location halfway between the two removed atoms. The proximity cutoff is evaluated from 1/3 to 85% of the nearest neighbor distance evaluated in 0.1 Å steps. These combinations can result in 100s to 10,000s of configurations that may give the minimum energy configuration.

The construction of the full simulation cells also follows the method of Olmsted et al. [15]. The bicrystals are periodic in the two dimensions within the GB plane and non-periodic in the direction normal to the GB, resulting in a single GB within the simulation cell. The size of the simulation cell is set to a minimum of $17a_0/2$, where a_0 is the lattice parameter, or at least two repeats of the CSL lattice for both directions in the plane of the GB, whichever is larger. In the direction normal to the plane, each grain on either side of the GB is set to a minimum of 40 $a_0/2$ or at least two repeats of the CSL lattice, which ever is larger. An additional CSL unit cell is added in between the two grains for the purposes of the GB construction and additional CSL unit cells are added to each end of the bicrystal, which are constrained to move as a block to ensure that free surface effects do not influence the GB energy calculation during minimization. The aluminum embedded-atom method (EAM) interatomic potential developed by Mishin et al. is utilized for atomistic simulations analyzed here [23].

The minimum energy structure for a given GB is obtained by energy minimization of each of the possible configurations over the microscopic degrees of freedom. In the present work, the conjugate gradient minimization technique in LAMMPS is used. We use tolerance cutoff values of 10^{-10} eV/Å for the force and a relative error of 10^{-8} % for the energy. These values balance achieving

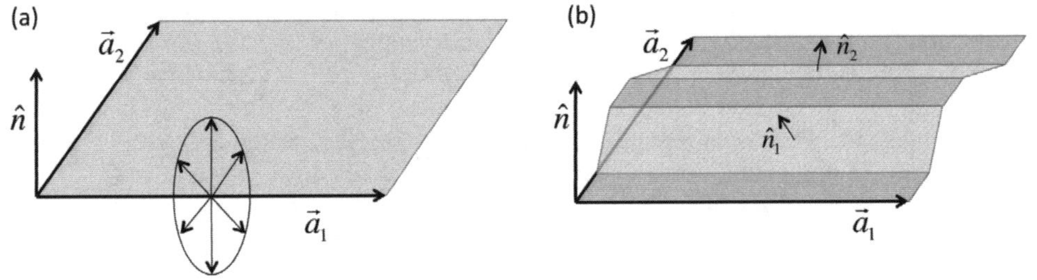

Figure 3. Methodology for creating multi-normal GBs. (a) To ensure that periodicity of the simulation cell is maintained, one can consider sets of alternate boundary planes, whose normals are all perpendicular to either the $\vec{a}_1$ and $\vec{a}_2$ basis vector. Only the $\vec{a}_1$ set of possible normals is indicated here. (b) A multi-normal GB constructed from the original boundary plane in (a), with additional normals $\hat{n}_1$ and $\hat{n}_2$, which preserve the periodicity of the simulation cell.

minimization convergence of sufficient accuracy with the time required to achieve convergence since the evaluation of up to 10,000s of configurations can be time consuming.

All these steps can be automated for the creation of GBs to match those desired for different experiments. Access to high performance computing resources makes these calculations tractable. In an effort to make these approaches accessible to the research community, another group recently published a similar algorithm to this present work. While the algorithms are not the same, their algorithm is an open source Python utility called GBpy, which performs many of the same functions as those discussed here to determine the minimum energy structure for a GB of interest [24].

2.2. Multi-normal GBs.

One minor drawback to studying GBs solely in bicrystal form is that one can miss the effects of other boundary normals or network constraints. We do not study the effects of network constraints here because the experiments of interest frequently only examine a single GB [6,7]. However, the GBs are typically curved and therefore have a range of normals as illustrated in Figure 1. To better understand how neighboring regions of a different normal affect the overall GB motion, we also create a multi-normal GB. This is similar to a faceted GB, but the facets are large compared to the atomic level facets that may exist in a GB of a single normal. The GBs here will be composed of multiple 'flat' segments but will be connected to other 'flat' segments with different boundary plane orientations. The periodic boundary conditions utilized to create the GBs requires that all facets also obey the periodic boundary conditions. Thus, along the two periodic dimensions of the plane, $\vec{a}_1$ and $\vec{a}_2$, an additional facet must either be non-existent or more than one facet must exist to allow the boundary to return to its original position before it repeats. Thus facets that will be repeated across a periodic boundary can only exist for a circle of directions, all of which must be normal to either $\vec{a}_1$ and $\vec{a}_2$, an example of which is shown in Figure 3a for circle of normal about the $\vec{a}_1$ axis. By selecting facets of these normals, one can see in Figure 3b how a boundary of multiple facets can return a GB to the original normal before repeating along the $\vec{a}_2$ direction. These facets are continuous along the $\vec{a}_1$ direction, thereby preserving the periodicity of the simulation cell. In this manner, the influence of multiple GB normals on each other can be studied.

2.3. Mobility.

The mobility of the GBs is examined using the synthetic driving force method [25]. In short, the method utilizes an order parameter to distinguish between the atoms in two grains of differing orientation. Additional potential energy is added to atoms of one grain selected to be the unfavored orientation, while atoms in the other grain have no added energy, making it the favored orientation.

The gradient in energy across the GB, leads to a driving force, *f*, inducing GB migration and growth of the grain with the favored orientation. Using the magnitude of the applied synthetic driving force, *f*, and the measured velocity of the GB migration, *v*, one can calculate the mobility of the GB, *M*, according to $v = M \cdot f$ [9].

While the synthetic driving force does not have a physical equivalent method to add energy to just one orientation of crystals, GB mobility measurements by this method have been demonstrated to be independent of the type of driving force [26]. Futhermore, this method has been compared to other boundary migration techniques and found to provide nominally identical results [27]. More recently, an improved synthetic driving force method was introduced, called the energy conserving orientational driving force [28]. While the newer method does overcome minor issues of the original method, the two give nominally similar results, indicating that the results contained herein are valid [28].

Similar to Olmsted et al. [15], the GBs studied in this work are tested under three driving forces in the set of (0.005, 0.010, 0.025) eV/atom. To mimic temperatures near those indicated in the experimental work, the mobility measurements have been carried out at 600 K using an NVT ensemble; additional simulations were conducted for a few boundaries at 800 K for purposes that will be explained below. The simulation cell is equilibrated over 50 picoseconds prior to the application of the synthetic driving force. The GB position is then tracked over a period of time and the velocity is extracted from the slope of the GB position-time data. The end groups of atoms are set to be free in some cases, enabling shear coupling to occur if the GB motion selects that, or constrained in other cases, to prevent shear coupled GB motion in cases it is desirable to be avoided.

3. Results

3.1. Energy and structure of bicrystal GBs

Using the techniques described above, it was determined that the (5 47 76) crystal plane coincides well with the 'flat' boundary plane normal; the angular deviation is 1.6°. For the 'curved' boundary plane normal noted above, the $(\bar{1}\,2\,1)$ crystal plane was found to be within 1.4° degrees of the desired normal. Finally, a second migrating boundary plane normal, with a $(\overline{55}\,113\,109)$crystal plane, was selected so it could be tested independently and in conjunction with a (5 47 76) plane as part of a multi-normal GB. The (5 47 76) and $(\overline{55}\,113\,109)$ bicrystals each have triclinic simulation cells while the higher symmetry of the $(\bar{1}\,2\,1)$ bicrystal basis vectors enables a monoclinic simulation cell.

The (5 47 76), $(\bar{1}\,2\,1)$, and $(\overline{55}\,113\,109)$ bicrystals had 32,310, 4,275, and 585 microscopic degrees of freedom, respectively, of which each configuration was minimized. A plot of the GB energy distributions for the configurations is provided in Figure 4a. The minimum GB energy for the (5 47 76), $(\bar{1}\,2\,1)$, and $(\overline{55}\,113\,109)$ GBs are 397 mJ/m^2, 466 mJ/m^2, and 445 mJ/m^2, respectively. These are higher than the textbook "high angle" GB energy listed of 324 mJ/m^2 [1], though the 'flat' boundary normal has energy lower than the 'curved' boundary normals and thus closer to the textbook value.

In Figure 4a it can be seen that minimum energy structures are far more sensitive to the starting configuration than those shown by Olmsted et al. [15]. The structures for these three GBs are shown in Figure 4b-d, where only those atoms with a centrosymmetry parameter greater than 1 are shown. Thus only those atoms in the GB that are not ideally coordinated are shown and the ideally coordinated surrounding atoms are not included in the image.

For the (5 47 76) GB shown in Figure 4b, it can be seen that there are very clear facets to the GB. For example, the overall plane normals are [5 47 76] and [47 5 76] in the lower and upper grain respectively. The long facet in the GB is close to a (3 8 10) and a (8 3 10) plane in the lower and upper crystals respectively, while the short facet is close to a $(\bar{4}\,5\,14)$ and $(5\,\bar{4}\,14)$ plane in the lower

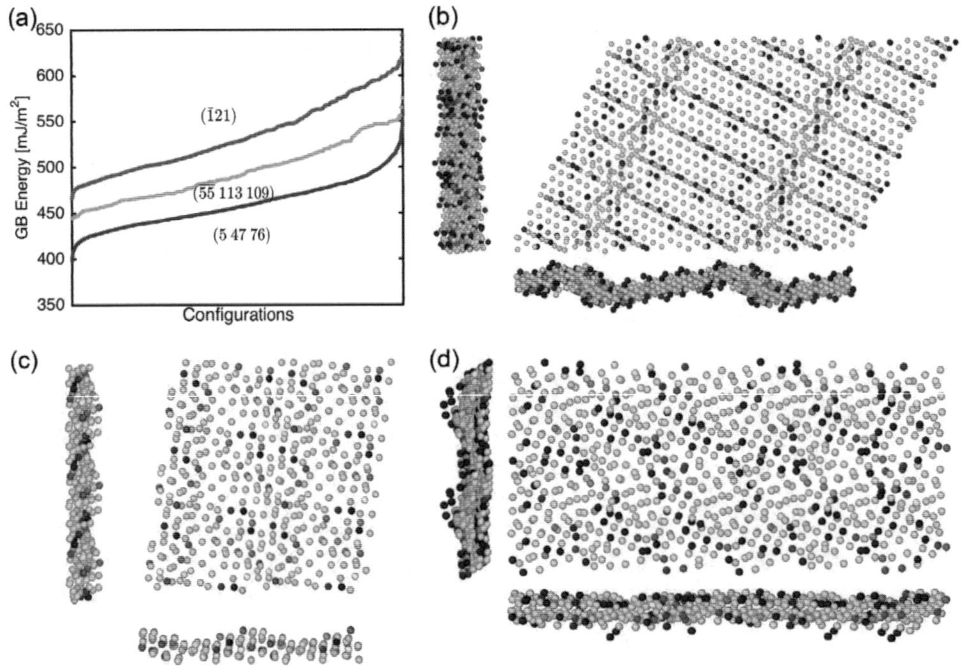

Figure 4. Energy and structure for the three GBs examined in this work. (a) Distribution of GB energy for all possible configurations examined to find the minimum energy structure. GB structures plotted using atoms whose centrosymmetry parameter is greater than 1 for the (b) (5 47 76), (c) $(\bar{1}\,2\,1)$, and (d) $(\overline{55}\,113\,109)$ boundary plane GBs.

and upper crystals respectively. Since the GB configuration was tested in so many other configurations (Figure 4b is the low energy configuration), these facets clearly play a role in achieving this low-energy structure.

In contrast, the $(\bar{1}\,2\,1)$ GB shown in Figure 4c appears to have facets in the upper grain as indicated by the lower side view, but the lower grain only has slight serrations that differ by an atomic plane or so. Furthermore, the upper grain also shows some minor facets in the left side view. Upon 3D inspection, an array of peaks and valleys in the atoms of the upper crystal can be seen, while the lower crystal is more or less flat. This difference in the two grains likely results from the lack of symmetry in GB plane normal, where the lower crystal has normal of $[\bar{1}\,2\,1]$ and the upper crystal has a normal of $[\bar{1}\,2\,7]$.

Finally, the $(\overline{55}\,113\,109)$ GB shown in Figure 4d is relatively flat in the upper grain but the lower grain has some minor peaks and valleys as indicated in the side view, though these are minor This GB is also asymmetric in that the lower and upper crystals have normals of $[\overline{55}\,113\,109]$ and [15 15 165], respectively.

3.2. GB Mobility
The three GBs were each subjected to the synthetic driving force at various magnitudes of the applied driving force. Unexpectedly, all three GBs demonstrated shear coupled GB motion [29,30]. The mobility values and shear coupling factors are listed in Table 1. The (5 47 76) GB had a higher mobility than the $(\bar{1}\,2\,1)$ and $(\overline{55}\,113\,109)$ GBs by almost an order of magnitude, except at the highest driving force where it is still higher but by a smaller magnitude. It is noted that the mobility of

Table 1. Mobility and shear coupling factor for three Σ45c GBs under different driving forces.

Driving Force		Mobility [m/(s GPa)]			Shear coupling factor $[v_{\parallel} / v_{\perp}]$		
Nominal [eV/atom]	Actual [GPa]	(5 47 76)	$(\bar{1}\,2\,1)$	$(\overline{55}\,113\,109)$	(5 47 76)	$(\bar{1}\,2\,1)$	$(\overline{55}\,113\,109)$
0.005	0.044	242	37	21	0.31	0.28	0.33
0.010	0.089	347	44	22	0.31	0.24	0.26
0.025	0.223	440	78	117	0.31	0.33	0.26

Table 2. Mobility for three Σ45c GBs under shear constrained conditions

Driving Force		Shear Constrained Mobility [m/(s GPa)]		
Nominal [eV/atom]	Actual [GPa]	(5 47 76)	$(\bar{1}\,2\,1)$	$(\overline{55}\,113\,109)$
0.005	0.044	34	67	17
0.01	0.089	45	66	30
0.025	0.223	87	77	71

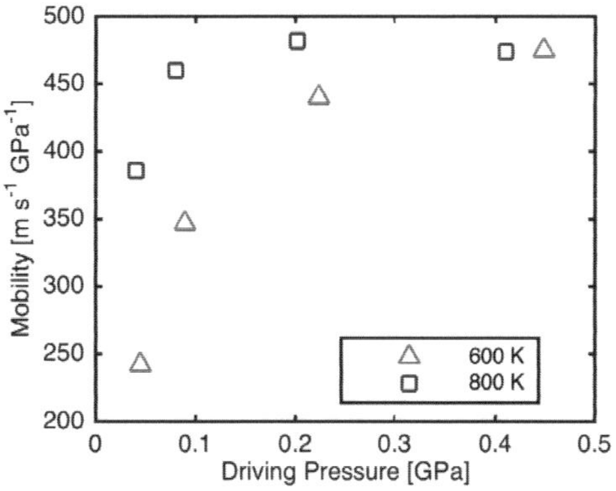

Figure 5. Driving force (pressure) dependence of mobility for the (5 47 76) GB at two different temperatures.

these GBs depends on the magnitude of the applied driving force. As demonstrated by Olmsted et al. [27], this phenomenon appears to be associated with the roughening transition temperature. Below the GB roughening transition temperature, boundaries exhibit a larger dependence on the driving force, and above the temperature, boundaries show very little dependence on the magnitude of the driving force [27]. This decrease in dependence with increasing temperature is demonstrated in Figure 5, where it can be seen that the mobility of the (5 47 76) GB shows a smaller dependence on the driving force magnitude at 800 K than it does at 600 K. Since the melting point of Al is 933K, it is likely that this boundary will not roughen before it melts.

As shear coupled GB motion requires lateral motion of the surrounding grains or deformed matrix material, it is possible that this type of motion would be constrained under experimental circumstances. As a result, all three GBs were examined for their mobility under constraints that inhibited shear coupling. To do this, blocks of atoms at the two ends of the bicrystal were constrained to not move. The resulting mobilities are presented in Table 2. In this case it can be seen that all three GBs have similar magnitudes of mobility at the highest driving force but at lower driving forces, the $(\bar{1}\,2\,1)$ GB has the highest mobility.

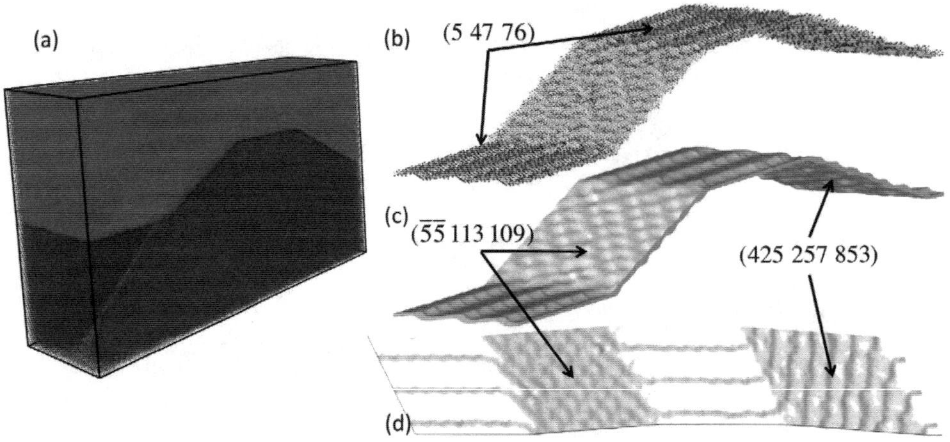

Figure 6. Multi-normal GB created for this work. (a) Entire simulation cell showing the two different crystals. (b) Atomic structure at the GB plotted using the centrosymmetry parameter. (c-d) Two different views of an Ovito surface construction [31] showing the faceting of the (5 47 76) GB and its influence in faceting on the adjacent regions with different normals.

3.3. Multi-normal bicrystals

To investigate the effect of a GB with multiple normals, as is the case with curved boundaries observed in many experiments, a bicrystal with multiple normals was created. The creation of this bicrystal follows the methodology discussed in section 2.2. The simulation cell for this multi-normal bicrystal starts with the (5 47 76) GB, since it is associated with a 'flat' GB configuration. The (5 47 76) simulation cell is enlarged by several times and then the boundary plane is allowed deviate according to normals that follow the outline of section 2.2. For the multi-normal bicrystal shown in Figure 6a, it can be seen that there are three normals present in the simulation cell. The 'flat' (5 47 76) crystal plane is allowed to exist on both a lower and higher plane. One additional plane is selected as the $(\overline{55}\ 113\ 109)$ crystal plane because it is contained within the distribution of 'curved' normals. To maintain periodicity, an additional normal is required; this additional normal is selected to be the mirror image of $(\overline{55}\ 113\ 109)$ crystal plane and results in a (425 257 853) plane, which is not studied individually in this work. The multi-normal bicrystal is triclinic, has volumetric dimensions of 128 x 566 x 295 $Å^3$ and contains 1.28 million atoms.

Figures 6b, 6c, and 6d illustrate the structure of the multi-normal GB. It can be seen that the (5 47 76) regions contain the structure characteristic for this boundary plane normal, as illustrated in Figure 4b. However, it is difficult to achieve exactly the same structure in the $(\overline{55}\ 113\ 109)$ region as is illustrated in Figure 4d because the lattice shifts between the crystals for the minimum energy structures are not identical between the (5 47 76) and $(\overline{55}\ 113\ 109)$ GBs. As a result, the bicrystal was constructed using the microscopic degrees of freedom that enabled the minimization of the (5 47 76) structure. It can be seen that the $(\overline{55}\ 113\ 109)$ region has a grid of minor facets while the (425 257 853) region has facets that are continuous with the facets of the (5 47 76) structure. The GB energy of this structure is 485 mJ/m^2, which is higher than three GBs studied individually in this work and much higher than textbook values.

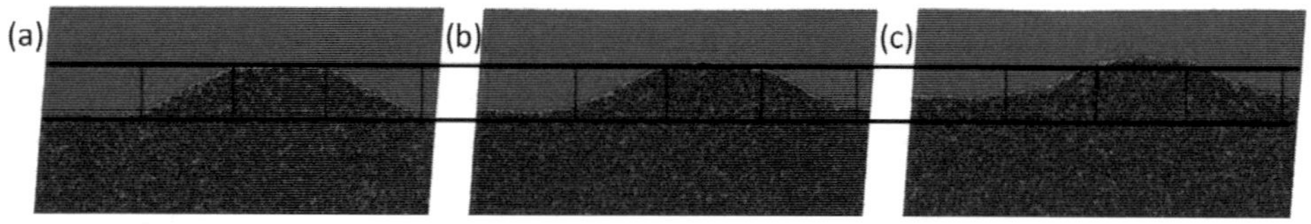

Figure 7. Examination of boundary migration in the multi-normal GB at three snapshots (a-c). Black lines are overlayed in order to view the change in boundary position. Note the faster migration on the lower (5 47 76) GB than in any other region.

When subjected to the synthetic driving force, the lower (5 47 76) regions grow the quickest of all the different normal regions, as shown in Figure 7. This includes faster growth than the upper (5 47 76) region. It is reasonable to expect that the curvature driving force has an effect here. Though it is noted that these results are consistent with the individual examinations of (5 47 76) and ($\overline{55}$ 113 109), which would suggest that whether shear coupling is constrained or not, the (5 47 76) GB should move more quickly. It is noted that this GB, even with its multiple normals does demonstrate some magnitude of shear coupling, though the magnitude is difficult to quantify since the GB does not move uniformly.

4. Discussion

4.1. GB Energy and Structure

The three individual GBs and one multi-normal GB indicate very interesting correlations between energy and structure. From these simulations it appears that the lowest energy structures for a given boundary plane are most frequently associated with a faceted structure. This is in line with previous work and theory [32-35] that indicate that the lowest energy structure can often be comprised of two or more crystal planes with differing orientation and energy that give the same net boundary plane but with lower energy. This is very apparent in the case of the (5 47 76) boundary plane, where the facets are the same in both crystals due to the higher symmetry across the GB. In contrast, the ($\overline{1}$ 2 1) and ($\overline{55}$ 113 109) boundary planes show more complex faceting, which faceting is different depending upon which crystal is examined. As a result, it is not as surprising that these GBs have higher energy.

It is also possible that other boundary plane could have lower energy than the (5 47 76) crystal plane. For example, since the (5 47 76) plane is dominated by the long facet with a (3 8 10) crystal plane, it is likely that this boundary plane would have even lower energy. Nevertheless, this low energy crystal plane is not in the distribution of 'curved' boundary normals shown in Figure 1a. Thus, from the set of GBs examined here, a theoretical structure comprised only of the examined GBs would be consistent with a GB that has been minimized for energy.

4.2. GB Mobility

The mobility values of the GBs examined in this work indicate a number of possibilities for how the actual migration in a recrystallization experiment could occur. While experimental conditions cannot be replicated exactly in atomistic simulations, there is information that can be gained about how different factors may be influencing the actual GB migration in the experiment.

First, it is surprising that of the 3 GBs examined, all exhibit shear coupling when not constrained. In Olmsted's set of 388 GBs, six Σ45c CSL nickel GBs were examined, of which four exhibited shear coupling [16,30]. In other words, the Σ45c GBs examined here show a high propensity for shear coupling. This is significant because shear coupling has been speculated as a possible mechanism underlying stress-driven grain growth [36]. Whether or not shear coupling actually occurs may be subject to various factors. A curved boundary with a distribution of boundary planes, as is often seen

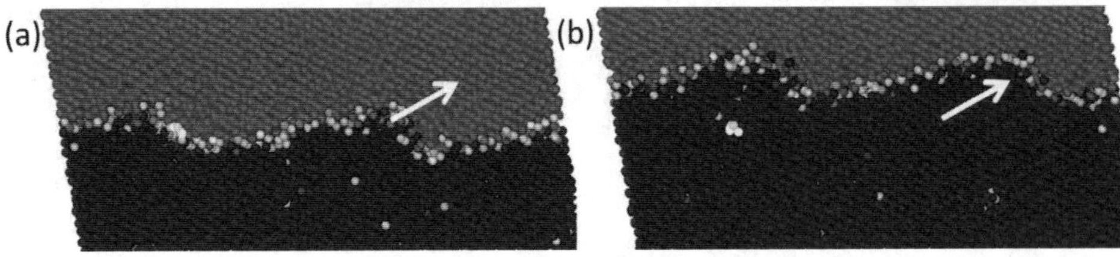

Figure 8. Examination of the boundary migration of the (5 47 76) GB at two different snapshots (a-b). Not the lateral motion of the small facet dominating the migration, though it is noted that the boundary as a whole simultaneously migrates as the facets move.

in experiments, may or may not enable shear coupling. Though the multi-normal GB examined here did exhibit some magnitude of shear coupling. One complicating factor with this shear-coupling scenario though is that if 'flat' GBs typically migrate slowly, this is inconsistent since the (5 47 76) crystal plane has the highest mobility under shear-coupled conditions.

Second, when the three GBs are constrained in a manner to inhibit their shear coupling, the mobility values change significantly. The 'flat' (5 47 76) GB that had been the fastest when not shear constrained now becomes slower than the 'curved' $(\bar{1}\,2\,1)$ GB at low driving forces. However, the 'curved' $(\overline{55}\,113\,109)$ GB has a mobility lower than the 'flat' (5 47 76) GB. Thus, one 'curved' GB is faster than the 'flat' GB while the other 'curved' GB is slower. One would then need to resolve the distribution of different boundary plane orientations with the relative mobility of each to determine whether the 'curved' GB has higher inherent mobility than the 'flat' GB.

Third, in further examination of the shear-constrained motion of the (5 47 76) GB, it can be seen that the $(\bar{4}\,5\,14)$ facet is critical to the boundary migration. Even in this shear-constrained state these small facets migrate laterally as the GB migrates in its normal direction, as indicated in Figure 8. While most of the movement is lateral, this does not entirely account for the vertical motion of the GB. The simulation indicates that for every angstrom the short facet moves laterally, the long facet moves half an angstrom vertically. Since the $(\bar{4}\,5\,14)$ facet falls in the middle of the distribution of 'curved' normals in Figure 1a, its high mobility may be further indicative of high mobility among the 'curved' boundary normals.

Fourth, it is noted once again though, that despite the fact that the (5 47 76) GB has a slower mobility than the $(\bar{1}\,2\,1)$ GB or the $(\bar{4}\,5\,14)$ facet within the (5 47 76) GB, the (5 47 76) GB's mobility is non-zero, indicating that under the proper driving force it could move, and in experiments, 'flat' portions of the recrystallized GB remain fairly stagnant or migrate very slowly.

Finally, the present work does not take into account the inhomogeneous distribution of stored energy in the deformed microstructure, nor to any possible preference for migration along specific dislocation boundaries [5]. Also, the presence of medium to high angle dislocation boundaries in the microstructure (e.g. [37]) causes sufficient lattice curvature that there is not a single orientation within the deformed material. As a result, the misorientation across the entire experimental GB is not constant. Furthermore, since dislocation structures can impact GB migration, as has been indicated by Godiksen et al. [38], it is possible that the mobilities simulated in this work would be significantly altered by the presence of dislocation structures and even enable or constrain shear coupled GB migration.

In the end, any one of these various factors influencing the migration of recrystallized GBs in experiments could dominate over the others. It is difficult to know exactly which effect is then controlling or exerting the largest influence on the actual migration. Nevertheless, based on the data and discussion above, it is reasonable to deduce a number of facts about what may likely happen. First, it is unlikely that many materials experience sustained shear coupling in a single direction over large distances. Depending on the distance migrated by the GB, this would produce significant offsets; based on shear coupling factors in Table 1, offsets would be equal to ~1/3 the distance migrated by the GB. Furthermore, the boundaries do not always migrate in a uniform fashion, so varying degrees of shear coupling would have to be accommodated and this could be difficult. In addition, not all GB plane orientations shear couple in the same direction or by the same magnitude, further complicating shear coupling over a large area. Second, the fact that the shear constrained mobilities of the GBs show some level of correlation between high/low mobility and 'curved'/'flat' boundary planes, means that the mobility may be a good indicator of when a given boundary will migrate even in the presence of additional factors. Third, there may be some factor that would cause a 'flat' (5 47 76) GB to be less mobile, despite the fact that the atomistic simulations predict finite mobility values. Perhaps the deformation structure somehow facilitates faster motion of the 'curved' portions of the boundary and also manages to decrease the mobility or completely immobilize boundaries that might otherwise exhibit significant migration.

5. Conclusions

The present work seeks to provide insight into possible factors affecting the migration of recrystallizing GBs, by comparing to atomistic simulations of GB energy, structure, and mobility. The work details methods for selecting specific CSL bicrystals that can be chosen to match experimental GBs. Additionally, methods for constructing multi-normal bicrystals are presented.

As a case study, three GBs are selected to match a theoretical GB with both 'curved' and 'flat' boundary plane orientations. The structure and energy of the three GBs indicate that 'flat' portions of the recrystallization GB have lower energy than the 'curved' portions of the GB. This structure is consistent with the principle of a minimum energy GB structure, meaning that a GB will maximize the area of low energy boundary plane orientations while minimizing the area of high energy boundary plane orientations that may be required to accommodate geometry.

The three selected GBs prefers to migrate by shear coupled mechanisms when not constrained. However, since many polycrystalline materials would likely restrict shear coupled GB motion, these same three GBs are also studied using boundary conditions that restrict lateral migration of the GBs. Under these boundary conditions, the GBs are still mobile, but the 'flat' boundary is faster than one 'curved' GB, while slower than the other. As a result, one can reasonably conclude that inherent GB mobility is an important factor in the motion of recrystallization boundary migration. However, this fact alone is not sufficient to accurately describe or predict recrystallization GB migration in experiments. The deformation microstructure in the non-recrystallized regions also likely plays a roll due to the fact that the simulated 'flat' GB is mobile while experiments typically show slow or negligible migration of 'flat' GBs over long periods of time. The GB migration mechanisms (moving facets in the case of the 'flat' GB) may also interact with deformation microstructures in such a way as to accelerate or decelerate the inherent mobility of a GB.

Together, the atomistic methods and case study presented here are intended to demonstrate how atomistic simulations provide valuable insight into the physical factors of GB migration that are not easy to resolve using experimental techniques alone.

Acknowledgements

This work was supported by the U.S. Department of Energy, Office of Science, Basic Energy Sciences under Award # DE-SC0012587. Thoughtful discussion with D. Juul Jensen and Y. Zhang on experimental measurements of recrystallization GB migration is gratefully acknowledged.

References

[1] Humphreys F J and Hatherly M 2004 *Recrystallization and Related Annealing Phenomena* (Boston: Elsevier)
[2] Hall E O 1951 *Proc. Phys. Soc. B* **64** 747–53
[3] Petch N J 1953 *J Iron Steel Inst* **174** 25–8

[4] Yu T, Hansen N and Huang X 2013 *Acta Mater.* **61** 6577–86
[5] Fan G H, Zhang Y B, Driver J H and Juul Jensen D 2014 *Scripta Mater* **72-73** 9–12
[6] Schmidt S, Nielsen S F, Gundlach C, Margulies L, Huang X and Juul Jensen D 2004 *Science* **305** 229–32
[7] Van Boxel S, Schmidt S, Ludwig W, Zhang Y, Juul Jensen D and Pantleon W 2014 *Mater Trans* **55** 128–36
[8] Ice G E and Larson B C 2000 3D X-ray crystal microscope *Advanced Engineering Materials* **2** 643–6
[9] Gottstein G and Shvindlerman L S 2010 *Grain Boundary Migration in Metals* (Boca Raton: CRC Press)
[10] Molodov D A, Gottstein G, Heringhaus F and Shvindlerman L S 1998 *Acta Mater.* **46** 5627–32
[11] Molodov D A, Ivanov V A and Gottstein G 2007 *Acta Mater.* **55** 1843–8
[12] Molodov D A, Gorkaya T and Gottstein G 2011 *J Mater Sci* **46** 4318–26
[13] Gorkaya T, Burlet T, Molodov D A and Gottstein G 2010 *Scripta Mater* **63** 633–6
[14] Gorkaya T, Molodov D A and Gottstein G 2009 *Acta Mater.* **57** 5396–405
[15] Olmsted D L, Foiles S M and Holm E A 2009 *Acta Mater.* **57** 3694–703
[16] Olmsted D L, Holm E A and Foiles S M 2009 *Acta Mater.* **57** 3704–13
[17] Holm E A, Olmsted D L and Foiles S M 2010 *Scripta Mater* **63** 905–8
[18] Holm E A, Rohrer G S, Foiles S M, Rollett A D, Miller H M and Olmsted D L 2011 *Acta Materialia* **59** 5250–6
[19] Rohrer G S, Holm E A, Rollett A D, Foiles S M, Li J and Olmsted D L 2010 *Acta Materialia* **58** 5063–9
[20] Grimmer H 1984 *Acta Crystallogr A* **40** 108–12
[21] Patala S and Schuh C A 2013 *Philos Mag* **93** 524–73
[22] Tadmor E B and Miller R E 2011 *Modeling Materials* (New York: Cambridge University Press)
[23] Mishin Y, Farkas D, Mehl M J and Papaconstantopoulos D A 1999 *Phys Rev B* **59** 3393–407
[24] Banadaki A D and Patala S 2015 *Journal of Applied Crystallography* **48** 585–8
[25] Janssens K G, Olmsted D L, Holm E A, Foiles S M, Plimpton S J and Derlet P M 2006 *Nat Mater* **5** 124–7
[26] Mendelev M I, Deng C, Schuh C A and Srolovitz D J 2013 *Model Simul Mater Sc* **21** 045017
[27] Olmsted D L, Foiles S M and Holm E A 2007 *Scripta Mater* **57** 1161–4
[28] Ulomek F, O'Brien C J, Foiles S M and Mohles V 2015 *Model Simul Mater Sc* **23** 025007
[29] Cahn J W, Mishin Y and Suzuki A 2006 *Acta Mater.* **54** 4953–75
[30] Homer E R, Foiles S M, Holm E A and Olmsted D L 2013 *Acta Mater.* **61** 1048–60
[31] Stukowski A 2010 *Model Simul Mater Sc* **18** 015012
[32] Hoffman D W and Cahn J W 1972 Vector *Surface Science* **31** 368–88
[33] Cahn J W and Hoffman D W 1974 *Acta Metall Mater* **22** 1205–14
[34] Tschopp M A and McDowell D L 2007 *J Mater Sci* **42** 7806–11
[35] Merkle K and Thompson L 1999 *Phys Rev Lett* **83** 556–9
[36] Rupert T J, Gianola D S, Gan Y and Hemker K J 2009 *Science* **326** 1686–90
[37] Hansen N and Jensen D J 2011 *Mater Sci Tech-Lond* **27** 1229–40
[38] Godiksen R B, Trautt Z T, Upmanyu M, Schiotz J, Juul Jensen D and Schmidt S 2007 *Acta Mater.* **55** 6383–91

Dynamic recrystallization behaviour at grain boundaries and triple junctions

H Miura

Department of Mechanical Engineering, Toyohashi University of Technology, Tempaku-cho, Toyohashi, Aichi, 441-8580, Japan

Email: miura@me.tut.ac.jp

Abstract. Dynamic recrystallization (DRX) behaviour and nucleation mechanisms were investigated using copper and copper alloy bicrystals, tricrystals and polycrystals. New grains were preferentially formed along grain boundaries in the bicrystals. After grain-boundary migration and bulging, nuclei appeared behind the deeply bulged grain boundary regions. The critical strain for nucleation was about one-quarter to one-half of the peak strain. The characteristics of nucleation at a grain boundary depended sensitively on grain boundary character. In copper alloy bicrystals, nucleation was much delayed due to solute drag of migrating grain boundaries. The nucleation at triple junctions, in contrast, took place at a much lower strain. New grain formation at triple junction was stimulated by development of folds. All the new grains were twin-related (Σ3) to the matrix and were formed behind the migrating grain boundaries. Therefore, it was revealed that the DRX mechanism in copper and copper alloys was essentially controlled by annealing twin formation. Variant selection of the twinning plane depended sensitively on the direction of the grain-boundary migration and on the geometry, however, was not affected by activated slip plane or dislocation glide. The DRX nucleation mechanisms at grain boundaries and at triple junctions are discussed with respect to grain-boundary migration and annealing twin formation.

1. Introduction

Nucleation of dynamic recrystallization (DRX) is known to occur preferentially near regions of high strain gradient and high dislocation density, e.g., deformation bands [1], particle/matrix interfaces [2, 3], grain boundaries [4, 5], and triple junctions (TJs) [6]. All these are characteristic sites where non-homogeneous deformation takes place. For example, grain boundaries and TJs are locations of inhomogeneous deformation because of the plastic strain compatibility that is necessary to accommodate a dissimilar shape change of the component grains or when grain boundary sliding (GBS) is blocked. This induces preferential nucleation of DRX even at a much lower strain, i.e., at 1/4 to 1/2 of the peak strain. There are a plenty of reports about nucleation mechanisms at grain boundaries. Bailey and Hirsch [7] as well as Luton and Sellars [8] proposed a simple bulging mechanism. This process is known to occur in low-to-medium stacking fault energy metals and alloys. Gottstein et al. [9] proposed that DRX concerns annealing twin formation in low stacking fault energy alloy. Miura et al. [5] have revealed from the experiments using copper bicrystals that nucleation takes place at the largely and deeply bulged areas induced by grain-boundary migration (GBM) and that all the new grains are of annealing twins. The bulging occurs towards the directions where high strain energy and high dislocation density were derived by inhomogeneous plastic deformation. While nucleation at grain boundaries takes place at rather lower strain region than the peak strain. Nucleation at TJ is even possible at strains below yielding, i.e., without macroscopic plastic deformation. Andiarwanto et al. [10] reported that nucleation at TJs onsets at 1/10 to 1/20 of the peak strain in a Fe-Ni alloy polycrystal. Such nucleation is accelerated and enhanced by the extensive occurrence of GBS. When the sliding is blocked at TJs, the required strain concentrations for GBM that triggers annealing twinning are thus provided. In contrast, TJs play reduced roles during static recrystallization because TJs are thermally stable [11]. TJs can, however, migrate along folds which are formed at TJs by impediment of GBS during high temperature deformation [6]. This enables nucleations at lower strains before macroscopic yielding.

When GBS takes place, grain-boundary steps induced by plastic deformation can impede the sliding. The impediment causes inhomogeneous strain around the bearing steps, which sequentially causes GBM and therefore, formation of new grains. Even while GBS can occur extensively below the yield strain, formation of steps by plastic deformation is required for this mechanism. Therefore, nucleation at grain boundary steps starts at a much higher strain region compare with that at TJs while both are induced by GBS [6,12].

When the most essential DRX mechanism in copper and copper alloys is annealing twin formation, the variant selection would strongly affect the texture formation after DRX. Miura et al. revealed that the twinning-plane variant is selected on the basis of the direction of GBM [5, 12]. In case of nucleation at TJs, however, the twinning-plane selection seemed not definitely controlled by the migration direction of TJs [13]. The mechanisms of the variant selection of twinning plane, therefore, still remain obscure.

In the present paper, DRX and nucleation behaviour in copper and copper alloys are introduced from our basic experiments using bicrystals, tricrystals and polycrystals. And the essential nucleation mechanisms are precisely discussed. The use of bicrystals and tricrystals provided ideal experimental conditions under which the presence of other grain boundaries or TJs could not interfere with the analysis.

2. Experimental

2.1 Procedure of preparation of bicrystal samples and microstructural observations

Orientation controlled bicrystals of 99.99% purity copper containing straight symmetrical [001] tilt boundaries were grown from seed crystals by the Bridgman method. The tilt angles, θ, were 20 ° and 64 ° according to the definition of the symmetrical [001] tilt boundary as described in Fig. 1. Because for the grain boundaries were aligned parallel to the tensile axis, GBS is geometrically difficult except some cases such as GBS for accommodations of the component crystals deformed. For comparison, a θ = 20 ° bicrystal with the boundary inclined 45° against the loading axis was also prepared, in which GBS takes place easily. These bicrystals are referred to as θ = 20 ° (//) and θ = 20 ° (<) bicrystals, respectively. Furthermore, a single crystal, which was cut out of a component single crystal of the θ = 20 ° (//) bicrystal, and polycrystal samples (initial grain size: D_0 = 23.3 μm) were also prepared. The tensile tests

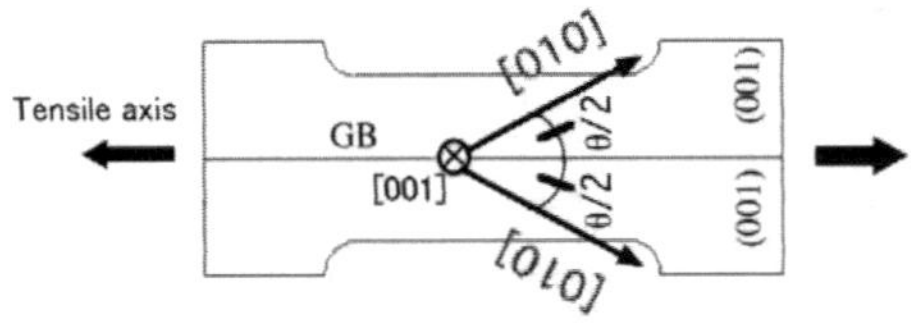

Figure 1 Geometry and dimensions of the bicrystal sample with [001] symmetrical tilt θ boundary aligned parallel to the tensile axis. Samples with grain boundaries aligned 45° to the loading axis were also prepared for comparisons. These bicrystals are referred to as θ (//) and θ (<) bicrystals, respectively [5].

were carried out in vacuum at 923 K and at a strain rate of 4.2×10^{-4} s^{-1} on an Instron-type mechanical testing machine. The specimens were deformed to a strain (ε = 0.1) lower than the peak strain and quenched using H_2-gas to freeze the microstructure for metallographic observations. Metallography was performed on the [001] plane, using optical microscopy after polishing and electrolytic etching. For a crystallographic analysis of the DRXed grains, orientation-imaging microscopy (OIM (Hitachi S-3500/TexSEM Inc.)) was employed.

2.2 Procedure of sample preparation of Cu-Sn-P alloy bicrystals and compression tests
Cu-0.19mass%Sn and Cu-0. 66mass%Sn alloy bicrystals having [001] twist 17° grain boundaries were grown by the Bridgman method using seed crystals. After discharge machining to rectangular shaped samples with dimensions of 12 x 9 x 6 mm^3, they were deformed in compression in the direction parallel to twist [001] axis, i.e., the direction normal to the grain boundary plane. High temperature compression tests were carried out in vacuum at 1073K and at a strain rate of $\dot{\varepsilon} = 2 \times 10^{-2} s^{-1}$ to various strains of ε = 0.2, 0.5, 1.0 followed by water quenching. Microstructural observations were carried out on the plane parallel to the compression axis. A polycrystalline Cu-0.66mass%Sn sample with an initial grain size of about 200 μm was also prepared and hot deformed at the same conditions for comparison.

2.3 *Procedure of preparation of pure copper tricrystals and compression tests*
A 99.9999%high purity copper tricrystal with dimensions of 8 x 8 x 32 mm^3 was cut out of an ingot unidirectionally solidified, which was composed of a columnar-grained structure. A schematic illustration of the tricrystal is exhibited in Fig. 2. The details of the four cube compression samples cut out from the large tricrystal as well as the geometries of the grain boundaries with respect to the compression axis are also indicated. A straight TJ line runs through the whole tricrystal. Although the orientations of the component three single crystals remain fixed, the grain boundary geometry changed depending on its position within the tricrystal. The component grains and grain boundaries will be referred to as GI, GII and GIII, and GBI, GBII and GBIII, respectively. GBI, GBII and GBIII are all random type boundaries with misorientation angles of 24.6°, 31.8° and 45.0°, respectively. Such random boundaries are known to slide easily at high temperatures [4]. The cube tricrystal samples were compressed in an Instron-type mechanical testing machine at 673K in a vacuum at a true strain rate of $2.9 \times 10^{-4} s^{-1}$. After compression tests to a true strain of 0.05, they were water quenched. Microstructural observations were carried out from the direction parallel to the TJ lines.

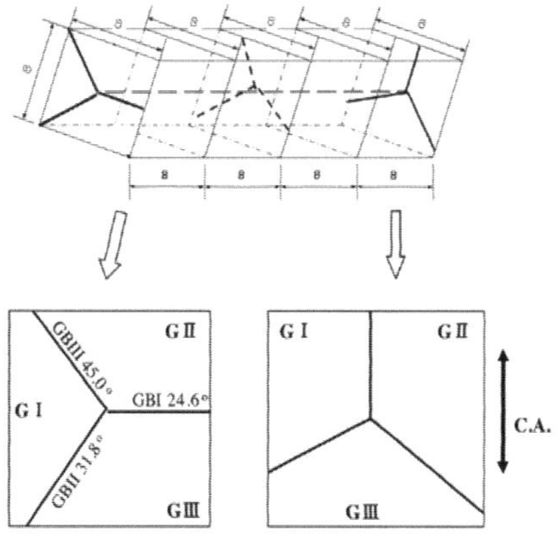

Figure 2 Schematic illustrations of the tested tricrystals cut out from a large columnar polycrystal. The large original tricrystal is composed of a twisted arrangement of grain boundaries. The component single crystals in the small four tricrystals cut out of the large original tricrystal possess completely the same crystallographical orientation [12].

2.4. Procedure of pure copper polycrystal sample preparation and tensile tests
A 99.99% purity polycrystalline copper sheet having an initial grain size of about 460 μm prepared by cold rolling and annealing, was discharge machined to tensile specimens with gauge dimensions of 12

x 4 x 1 mm^3. The coarse initial grain size was for easy distinction of newly developed grains. Tensile tests were carried out in vacuum on an Instron-type testing machine at 873 K and various initial strain rates from 4.2 x 10^{-5} to 4.2 x 10^{-3} s^{-1} followed by water quenching. Observations of folds were carried out on the plane normal, and then, samples were mechanically polished and etched for microstructural observations of DRXed grains.

3. Results and discussion

3.1 DRX behaviour in copper bicrystals

3.1.1Typical flow curves

The true stress vs. true strain curves obtained by tensile tests are shown in Fig. 3. After yielding, both bicrystals show gradual work hardening up to a peak stress followed by work softening due to the extensive occurrence of DRX and, then, eventually the flow stress drops to rupture (Fig. 3 (a)). The different hardening behaviour of the two samples in Fig. 3 (a) is attributed to the orientation geometry in addition to the changes in the activated slip systems. In both geometries, the individual component single crystals ideally deform by non-coplanar double slip. The large work-hardening region compared with polycrystalline materials is, therefore, due to widely appeared stage II deformation, which would be not so sufficient to rapidly achieve critical dislocation density for DRX nucleation. The tensile axis in the 20°(//) bicrystal was close to [016] with an equal Schmid factor in both component crystals. And the tensile axis in the 20°(<) bicrystal was close to [023] again with the same Schmid factor. This is evident from the work-hardening behaviour which reveals a stage I characteristic in the early stages of deformation for the 20°(<) sample in contrast to the 20°(//) bicrystal that reveals a stage II soon after yielding. Therefore, the peak strain in the 20°(<) bicrystal is higher than in the 20°(//) bicrystal but the peak stresses are comparable. When compared with the flow curves of the single crystal, bicrystal and polycrystal, the peak strain appears earliest in the polycrystal sample (Fig. 3 (b)). These peak strains in the bicrystal and single crystal look almost the same and delayed compared with that of the polycrystal. This result indicates that appearance of the peak strain should be strongly affected by the density of grain boundaries and TJs as well as dislocations. This tendency is well known that DRX onsets more easily and earlier with a decreasing grain size [14]. The higher flow stress in the bicrystal than that in the single crystal even while the same component single crystals and tensile direction would be due to

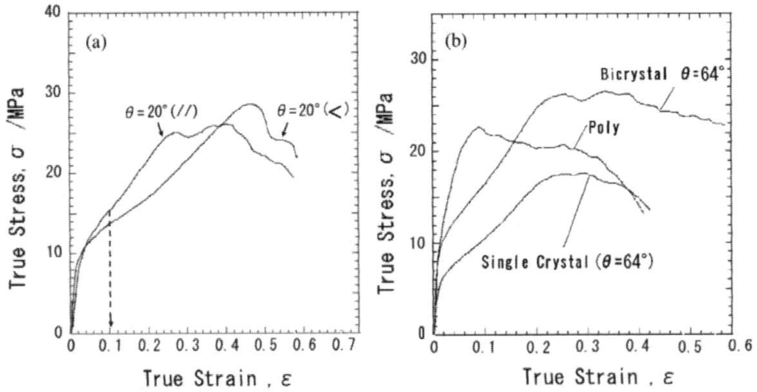

Figure 3 True stress vs. true strain curves for (a) θ = 20° bicrystals with boundary parallel (//) to and 45° (<) inclined to the loading direction and (b) θ = 64° bicrystal, the component single crystal and polycrystal. The broken line in (a) indicates the strain of water quenching for structural observations [5, 13].

the formation of a grain-boundary affected zone (GBAZ) to induce larger strain hardening. This will be discussed in detail later.

3.1.2 Evolved microstructure obtained by tensile tests

The surface morphology of the 20° bicrystal (//) after straining to $\varepsilon = 0.1$ is shown in Fig. 4. The conjugated slip traces of the primary and secondary slip planes can be observed. The appeared stage II deformations in Fig. 3, therefore, can be attributed to the conjugated slips. Mild GBM can be also seen to take place in Fig. 4.

The microstructure developed in the $\theta = 20°$ (//), $\theta = 20°$ (<) and $\theta = 64°$ bicrystals at a strain of $\varepsilon = 0.1$ was investigated using optical microscopy (Fig. 5). It is evident that the evolved microstructure differs sensitively depending on the misorientation angle and geometry of the bicrystal. Relatively rough GBM took place when the tensile axis is parallel to the grain boundary (Figs. 5 (a), (c)), while that is quite smooth and uniform in the bicrystal in which extensive GBS was possible (Fig. 5 (b)). In the θ =

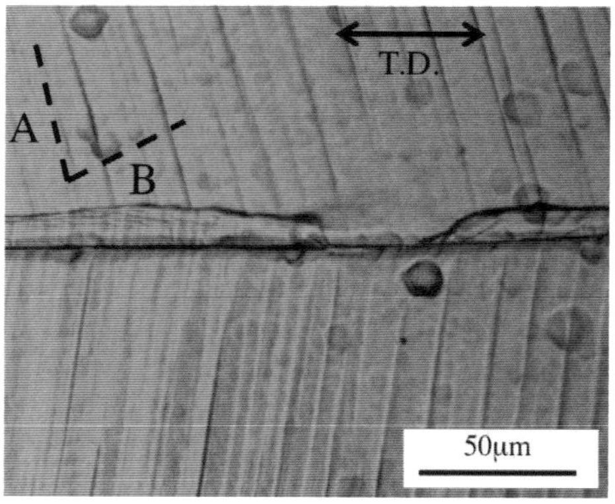

Figure 4 Surface morphology formed on the 20° bicrystal (//) after straining to $\varepsilon = 0.1$. A and B indicate the conjugated slip traces of the primary and 2ndary, and 3rd and 4th slip planes, respectively. T.D. is the tensile direction [5].

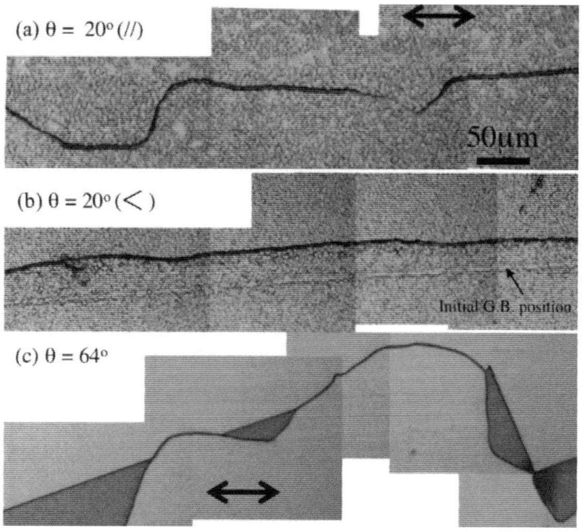

Figure 5 Evolution of substructure along the grain boundary after straining to $\varepsilon = 0.1$ of bicrystals of (a) $\theta = 20°$ (//), (b) $\theta = 20°$ (<) and (c) $\theta = 64°$. Arrow marks indicate the tensile direction [5,13].

64 ° bicrystal, new grains were formed along the deeply migrated grain boundary. The strain ε = 0.1 is 2/5 of the peak strain. It is evident, therefore, DRX nucleation preferentially onsets at the grain boundaries. GBS is assumed to promote stress concentration to trigger GBM and DRX nucleation [15, 16]. On the other hand, it is reported that deep GBM, which is related with faster GBM, induces more rapid formation of annealing twins [13]. Figure 5 indicates that the latter should be more effective on the nucleation. In short, grain-boundary serration and GBM should play the most important role for nucleation right from the beginning of deformation.

When slips take place in the component single crystals, GBAZs are soon formed along grain boundaries to accommodate the asymmetrical plastic deformation. It causes complicated plastic deformation as well as accumulation of dislocations along the grain boundary region and, then, should induce deep and severe GBM. Mahajan et al. [17] and Miura et al. [6,13] have reported such annealing twin formation behind migrating grain boundaries. This coincides with the hypothesis proposed by Gleiter that annealing twins are formed by the development of stacking faults during GBM [18]. Such stacking faults should occur more frequently when GBM occurs more rapidly and inhomogeneously as observed in the θ = 64° bicrystal (Fig. 5 (c)). The twinning process sets off massive DRX along the grain boundary. Thus, the faster GBM causes, the earlier DRX nucleation at grain boundaries takes place. Other influences include the grain-boundary character [19] and the inhomogeneous formation of GBAZs [6, 20] to cause severe serration and deep GBM.

The evolved DRX grains were investigated using OIM and the result is exhibited in Fig. 6. It is clear that all the new grains are of 1st order Σ3 annealing twins. It is assumed, therefore, that twinning must be the most essential mechanism of discontinuous DRX. While the new grains were formed behind the deeply migrated grain boundary, the twinning plane is normal to the GBM direction rather than the activated slip plane. This would suggest that twinning-plane variant selection is not related with deformation twinning but with stacking faults during GBM. The mechanism of nucleation and twinning variant selection is illustrated in Fig. 7. Twinning due to staking faults preferentially takes place behind

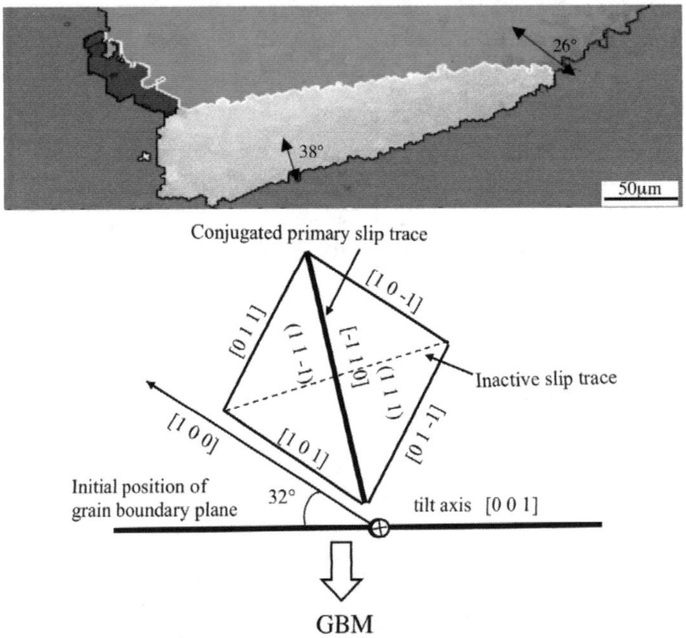

Figure 6 OIM map of the 64° bicrystal deformed to ε = 0.1. Σ3 boundaries are indicated by white lines. The crystallographical orientation is schematically represented below the OIM map [13].

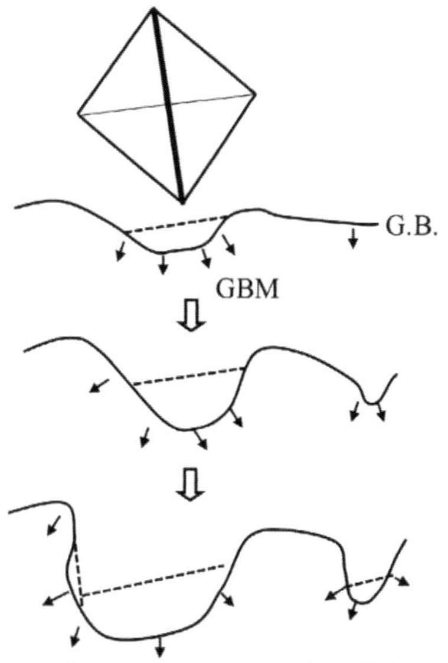

Figure 7 Schematic illustration representing the nucleation mechanism and twin-variant selection. All the straight dotted lines are twin boundaries. The tetrahedron at the top illustrates the crystallographic orientation of the upper grain.

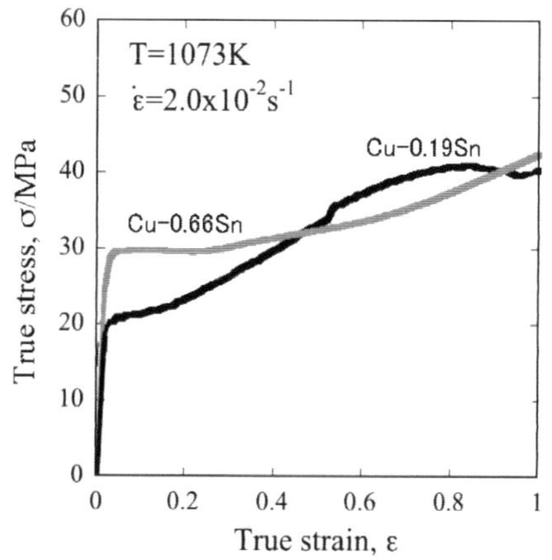

Figure 8 True stress vs. true strain curves of Cu-0.19mass%Sn and Cu-0.66mass%Sn alloy bicrystals with [001] twist 17° boundaries. The compression axis was parallel to [001].

the deeply and rapidly migrating grain boundary and the twinning-plane variant is selected by the direction of GBM, i.e., the closest normal plane to the GBM direction.

3.2 Effect of element addition on DRX nucleation behaviour

It is revealed from the above experiments using bicrystals that GBM and annealing twin formation should the most essential mechanism for DRX nucleation in pure copper. However, a question of what can be the mechanisms for the nucleation when GBM is suppressed would arise. To reveal the DRX behaviour and the mechanisms of nucleation in copper alloys in which GBM is suppressed by the solute drag effect, Cu-Sn alloy bicrystals were grown and tested in compression at 1073 K. The obtained flow curves are shown in Fig. 8. Although deformation was carried out at a sufficiently high temperature, the

IOP Conf. Series: Materials Science and Engineering **89** (2015) 012007 doi:10.1088/1757-899X/89/1/012007

peak stress appeared only in Cu-0.19mass%Sn alloy bicrystal at around $\varepsilon = 0.8$. Work hardening due to stage II deformation is still continued even at a strain of 1.0 in the Cu-0.66mass%Sn alloy bicrystal. In both samples, the peak strain became much larger compared with that in pure copper (Fig. 3). This result clearly indicates that the onset of DRX is notably delayed by Sn addition, even though the effect of strain rate is considered.

The change in the microstructure during compression tests was investigated using OIM. The summarized results are displayed in Fig. 9. In the Cu-0.66mass%Sn alloy bicrystal, new grain formation associated with an annealing twin was strongly suppressed. Instead of twins, substructure and low angle boundaries are developed along grain boundaries and they become more evident with increasing strain. That is, new grains were formed along a grain boundary accompanied by gradual increase in misorientation. This gradual increment in the misorientation should be induced by formation of GBAZ rather than the mechanisms of continuous DRX. Contrary to that, in the Cu-0.19mass%Sn alloy, new grains associated with twins are formed at $\varepsilon = 0.5$ along a grain boundary and it spreads into the whole grain interior at $\varepsilon = 1.0$. The strain where extensive DRX is achieved is much larger compared with that in pure copper. However, the microstructure developed at $\varepsilon = 0.5$ (Fig. 9 (e)) should be noticed that a few twins were formed. Most of the new grains were mainly composed of sub and low angle boundaries. It is particular to see that the substructure was significantly developed in the grain interiors. The shape and the geometry of the substructure look closely related with slip lines in the component grains. Therefore, it is assumed that the substructure and GBAZ were developed at low to medium strain regions, and then, extensive GBM took place accompanied by annealing twin formation at a high strain region due to highly stored strain energy and dislocation density in the Cu-0.19mass%Sn alloy (see Fig. 9 (e)). Any way, in both bicrystals, a mechanism similar to continuous DRX can also possibly play an important role in grain refinement especially in the grain interiors. The microstructure of Cu-0.66mass%Sn alloy polycrystal deformed to $\varepsilon = 1.0$ is shown in Fig. 10. Plenty of DRX grains associated with annealing twins were formed along the initial grain boundaries. This result indicates that DRX is

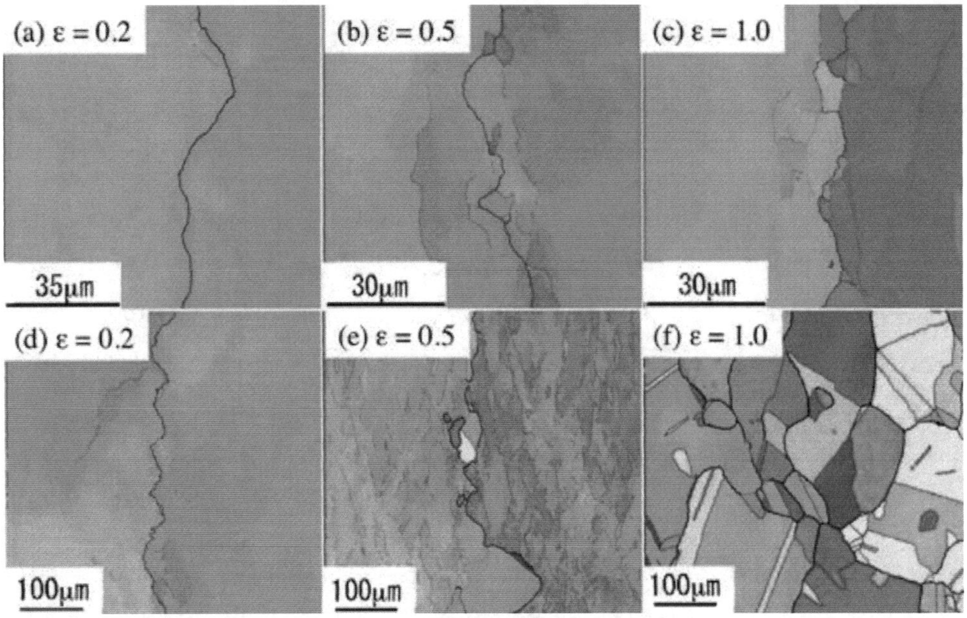

Figure 9 OIM maps of Cu-0.66mass%Sn, (a), (b), (c), and Cu-0.19mass%Sn, (d), (e), (f), alloy bicrystals with [001] twist 17°grain boundaries tested in compression at 1023 K and at a strain rate $\dot{\varepsilon} = 2 \times 10^{-2}\ s^{-1}$ to various strains. White, thin black, bold black and brawn lines indicate sub, low angle, high angle and twin boundaries, respectively.

Figure 10 OIM map of Cu-0.66mass%Sn alloy polycrystal deformed in compression at 1073K to $\varepsilon = 1.0$ and at a true strain rate of $\dot{\varepsilon} = 2x10^{-1}\ s^{-1}$. White, thin black, bold black and brawn lines indicate sub, low angle, high angle and twin boundaries, respectively.

not itself difficult in Cu-0.66mass%Sn alloy and is assisted by presence of a high density of grain boundaries and TJs. This view is in good agreement with the report that the onset of DRX becomes easier with decreasing grain size [14] and with the result of much delayed DRX nucleation in the bicrystals (Fig. 9). The role of TJs on DRX nucleation will be discussed in the next session.

3.3 DRX nucleation at TJs

Even while TJs are one of the most important components in polycrystalline materials, studies about the effect of TJs on DRX nucleation are still not so popular. Observed results of DRX nucleation at TJs in tricrystals and polycrystals are introduced below.

3.3.1. Nucleation at TJs in tricrystals

Typical OIM maps of the grains nucleated at the TJs in tricrystals at a strain of $\varepsilon = 0.05$, which is shortly

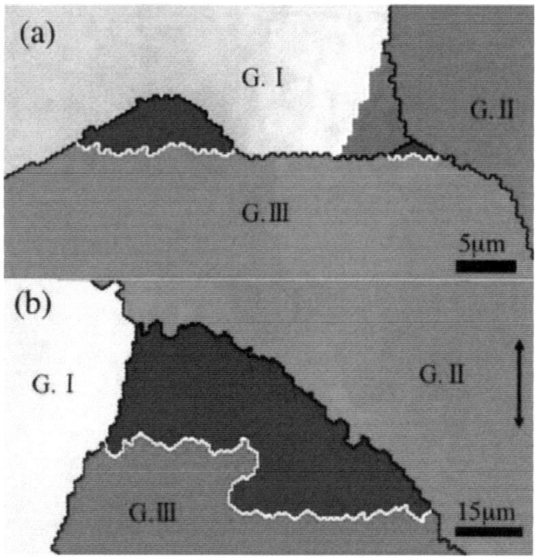

Figure 11 Nucleated grains at triple junctions in tricrystals at a strain of $\varepsilon = 0.05$. In both samples, only the grain boundary geometry is different [12]. White line indicates twin boundary. Arrow mark is the compression axis.

after yielding, are displayed in Fig. 11. Because of the large Schmid factors roughly between 0.49 and

0.47 in the component grains, double slips began to take place soon after the yielding. It is evident in Fig. 11 that DRX grains have most preferentially nucleated at TJs. It should be noted that all DRX grains were accompanied by deep migration grain boundaries and TJs. In total 13 nuclei were observed in whole tricrystal samples. In the case of Fig. 11 (b), TJ migrated towards GI and G II and a twin was formed in GIII behind the migrating TJ. Therefore, the most important DRX mechanism at TJs in copper appears also the formation of an annealing twin. It is interesting to see that TJs can easily migrate during high temperature deformation. Such behaviour is in contrast with the high thermal stability of TJs during static annealing [11]. The observed migration of TJs is brought by the assistance of GBS that takes place and formation of folds during high temperature deformation [6].

Miura et al. have shown from the experiments using copper tricrystals that TJs are the most preferred sites for nucleation under conditions where extensive GBS takes place [12]. In such a case, DRX nucleation began to appear only at around $\varepsilon = 0.01$ strain, which is about 1/10 to 1/20 of the peak strain. This was explained in terms of the sequential occurrence of GBS, folding in the GBS blocking grain, strain-induced migration of a TJ along the fold and twinning behind the migrating TJ. Though GBS is sensitively influenced by the geometry and grain-boundary character [21], the geometry independent nucleation in the present tricrystal should be attributed to the random type grain boundaries that can thus slide easily. A few DRX grains were nucleated also at grain boundaries (Fig. 11(a)), while this was always accompanied by nucleation at TJs.

A typical photograph of folding is displayed in Fig. 12. The fold was formed at a TJ because of blocking of GBS. The direction of the fold is not related with slip planes in the blocking grain. The fold is, however, more readily formed when the directions of GBS and of slip in the grain impeding GBS are close [6]. In Fig. 12, TJs and grain boundaries have already slightly migrated which can be found from the slip traces evolved on the surface.

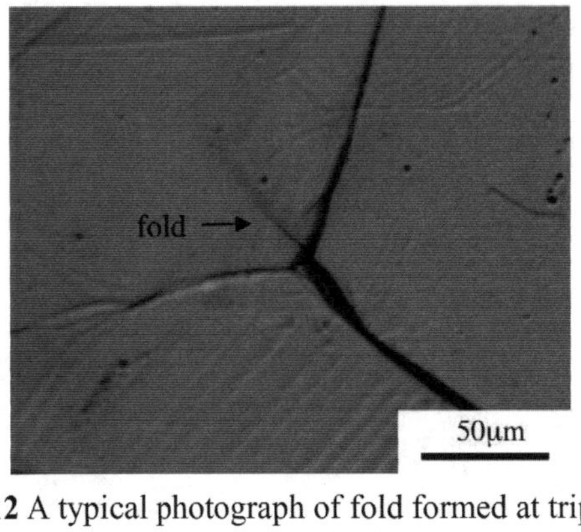

Figure 12 A typical photograph of fold formed at triple junction.

3.3.2. Nucleation at TJs in polycrystals

Pure copper polycrystal samples were tensile tested up to a strain $\varepsilon = 0.06$ at maximum, which was 2/3 and 1/3 of the peak strains when tested at 4.2×10^{-5} to 4.2×10^{-3} s^{-1} respectively. The change in the probability of nucleated TJs and the number of formed folds are summarized as a function of strain in Fig. 13. It is marvelous that DRX nucleation at TJs could be recognized even at the strain of $\varepsilon = 0.01$ which was at around yielding. And the amount of DRX nucleation at TJs increased almost linearly with strain. Furthermore, the probability increased as the strain rate decreased. Because GBS takes place more easily and more extensively with a decreasing strain rate [16], the deformation concentration and therefore folding at TJs becomes more easily provided. This tendency is clearly shown by the observation exhibited in Fig. 13 (b), in which the number of the detected folding increased with strain. However, the probability of nucleation at TJs reduced to half when the TJs were composed of twin ($\Sigma 3$) boundaries. This would be because the twin boundary hardly slides [21]. Hence, the deformation concentration and folding at TJs are difficult. Actually, nucleation at TJs appears less significant under the conditions where GBS is difficult [13].

Figure 14 shows the analysed results of crystallographical orientation distribution of the nucleated DRX grains at TJs. Here, only the smallest □ value or the smallest misorientation angle among the orientation relationships between the mother grains and the new DRX grains were plotted. It is revealed from Fig. 14 that more than 80% of the nucleated DRX grains at TJs were □3 twins. This result is quite similar to those in copper tricrystals tested in compression [22, 23]. These results suggest that the most essential and operative DRX mechanism at TJs in copper, in other words, metallic materials of low-medium stacking fault energy, is twinning. The result that the nucleated grains excepting □3 orientation had quite large misorientations would also well explain the nature of discontinuous DRX behaviour in copper.

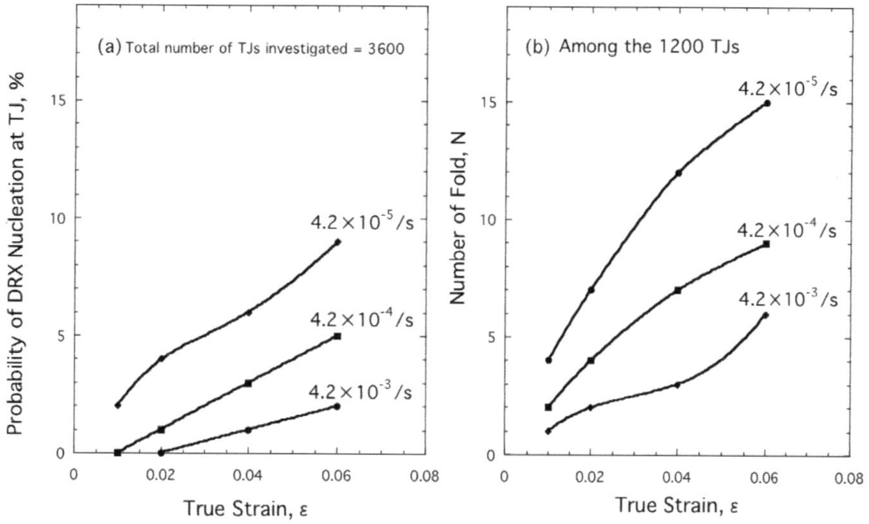

Figure 13 Summarized results of changes in (a) probability of DRX nucleation and (b) number of folds formed at triple junctions with increasing strain.

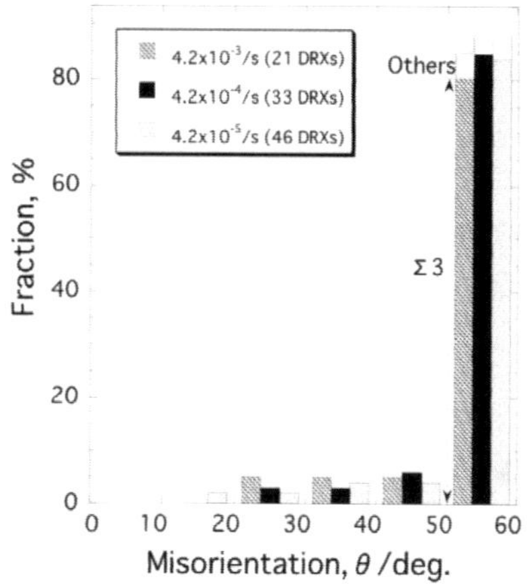

Figure 14 Strain-rate dependence of misorientation and Σ value distributions of nucleated DRX grains at triple junctions.

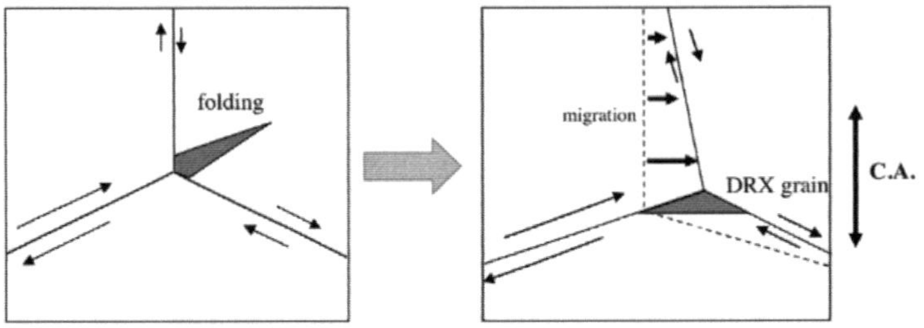

Figure 15 Schematic illustrations representing mechanisms of DRX nucleation at a triple junction accompanied by migration of grain boundaries and the triple junction. Strip and bold arrows in the grain interiors indicate grain boundary sliding and grain boundary migration. C.A is the compression axis.

A mechanism of DRX nucleation at TJ is schematically drawn in Fig. 15. Folding takes place in the grain blocking the extensive GBS. Grain boundaries migrate along the fold. A twin is formed behind the fastest migrating grain boundary. Because TJ can migrate fastest, twins are most preferentially formed at the TJ. GBS towards the TJ becomes more notable due to softening after nucleation at the TJ, which would accelerate the TJ migration. This process can operate even before yielding.

It is known that DRX becomes easier with a decreasing grain size [14]. It is assumed that a high density of grain boundaries, which are the one of the most preferential nucleation sites, stimulates DRX. Serration and bulging necessary for the nucleation mechanism at grain boundaries becomes, however, more difficult with a decreasing the grain size due to the limited space. Nucleation at TJs, therefore, becomes more important when the grain size becomes finer. Nucleation at quadruple junctions must be most operative in a fine-grained structure in such conditions.

4. Summary

The nucleation of dynamic recrystallization (DRX) at grain boundaries and triple junctions (TJs) were systematically investigated using copper and copper alloy bicrystals, tricrystals and polycrystals. DRX nucleation started at a grain boundary at strains much lower than the peak strain. This proves that nucleation takes place preferentially at a grain boundary. However, nucleation at TJs can onset in a further a lower strain region around yielding. All these nucleations were always accompanied by migration of grain boundaries and TJs along a grain-boundary affected zone or fold. Because grain-boundary sliding causes stress concentration and the formation of fold, DRX is much affected by its occurrence. All nucleated DRX grains were annealing twins to one of the adjoining grains in the pure copper bicrystals and the tricrystals, which substantiates the idea that the DRX nucleation mechanism and thus the origin of discontinuous DRX in metals of low stacking fault energy is annealing twin formation. Annealing twins appeared behind the migrating boundaries, preferentially at locations that are the most deeply bulged. However, annealing twin formation becomes difficult by element addition to retard grain boundary migration. In this case, mechanisms of continuous DRX can contribute to nucleation. The twinning-plane variant of the nucleated twins behind the migrating grain boundary was not selected at random but was related to the direction of grain boundary migration. Activated slip planes or dislocation motion did not affect this valiant selection.

Acknowledgments
This research was supported by Japan Science and Technology Agency (JST) under Industry-Academia Collaborative R&D Program "Heterogeneous Structure Control: Towards Innovative Development of Metallic Structural Materials" and Japan Copper and Brass Association.

References
[1] Blaz L, Sakai T and Jonas JJ 1983 *Metal Sci.* **17** 609.

[2] Humphreys FJ 1979 *Acta Mater.* **27** 1801.
[3] Ferry M and Humphreys FJ 1996 *Acta Mater.* **44** 1293.
[4] Miura H, Aoyama H and Sakai T 1994 *J. Japan Inst. Metals* **58** 267.
[5] Miura H, Sakai T, Mogawa R and Gottstein G 2004 *Scripta Mater.* **51** 671.
[6] Miura H, Sakai T, Andiarwanto S and Jonas JJ 2005 *Phil Mag.* **85** 2653.
[7] Bailey JE and Hirsch PB 1962 *Proc. Roy. Soc.* (London) **A267** 11.
[8] Luton MJ and Sellars CM 1969 *Acta Metall.* **17** 1033.
[9] Gottstein G, Zabardjadi D and Mecking H 1979 *Metal Sci.* **13** 223.
[10] Andiarwanto S, Miura H and Sakai T 2004 *Tetsu-To-Hagane* **90** 21.
[11] Gottstein G, Ma Y and Schvindlerman LS 2005 *Acta Mater.* **53** 1535.
[12] Miura H, Sakai T and Jonas JJ 2006 *Scripta Mater.* **55** 167.
[13] Miura H, Sakai T, Mogawa R and Jonas JJ 2007 *Phil. Mag.* **87** 4197.
[14] Sakai T and Jonas JJ 1984 *Acta Metall.* **32** 189.
[15] Miura H, Aoyama H and Sakai T 1994 *J. Japan Inst. Metals* **58** 267.
[16] Miura H, Ozama M, Mogawa R and Sakai T 2003 *Scripta Mater.* **48** 1501.
[17] Mahajan S, Pande SC, Iman AM and Rath BB 1977 *Acta Mater.* **48** 2633.
[18] Gleiter H 1969 *Acta Met.* **17** 1421.
[19] Winning M, Gottstein G and Shvindlerman LS 2001 *Acta Mater.* **49** 211.
[20] Peralta P and Laird C 1997 *Acta Mater.* **45** 3029.
[21] Monzen R, Sumi Y, Kitagawa K and Mori T 1990 *Acta Metall Mater.* **38** 2553.
[22] Miura H, Andiarwanto S and Sakai T 2003 *Mater. Sci. Forum* **426-432** 4387.
[23] Miura H, Andiarwanto S and Sakai T 2002 *Mater. Trans.* **43** 494.

Grain boundary motion and grain rotation in aluminum bicrystals: recent experiments and simulations

DA Molodov, LA Barrales-Mora and J-E Brandenburg
Institute of Physical Metallurgy and Metal Physics, RWTH Aachen University, D-52056 Aachen, Germany

E-mail: molodov@imm.rwth-aachen.de

Abstract. The results of experimental and computational efforts over recent years to study the motion of geometrically different grain boundaries and grain rotation under various driving forces are briefly reviewed. Novel in-situ measuring techniques based on orientation contrast imaging and applied simulation techniques are described. The experimental results obtained on specially grown aluminum bicrystals are presented and discussed. Particularly, the faceting and migration behavior of low angle grain boundaries under the curvature force is addressed. In contrast to the pure tilt boundaries, which remained flat/faceted and immobile during annealing at elevated temperatures, mixed tilt-twist boundaries readily assumed a curved shape and steadily moved under the capillary force. Computational analysis revealed that this behavior is due to the inclinational anisotropy of grain boundary energy, which in turn depends on boundary geometry. The shape evolution and shrinkage kinetics of cylindrical grains with different tilt and mixed boundaries were studied by molecular dynamics simulations. The mobility of low angle <100> boundaries with misorientation angles higher than 10°, obtained by both the experiments and simulations, was found not to differ from that of the high angle boundaries, but decreases essentially with further decrease of misorientation. The shape evolution of the embedded grains in simulations was found to relate directly to results of the energy computations. Further simulation results revealed that the shrinkage of grains with pure tilt boundaries is accompanied by grain rotation. In contrast, grains with the tilt-twist boundaries composed of dislocations with the mixed edge-screw character do not rotate during their shrinkage. Stress driven boundary migration in aluminium bicrystals was observed to be coupled to a tangential translation of the grains. The activation enthalpy of high angle boundary migration was found to vary non-monotonically with misorientation angle, whereas for low angle boundaries the migration activation enthalpy was virtually the same. The motion of the mixed tilt-twist boundaries under stress was observed to be accompanied by both the translation of adjacent grains parallel to the boundary plane and their rotation around the boundary plane normal.

1. Introduction

The grain microstructure of crystalline solids decisively determines their physical and mechanical properties. Grain boundaries, i.e. interfaces between crystallites of the same phase with different crystallographic orientations, have the unique property to react to applied forces by their displacement. Grain boundary migration, by which the crystalline solid is reconstructed atom by atom, as well as grain rotation are essential parts of grain microstructure evolution in polycrystals during plastic deformation and in the course of recrystallization and grain growth during annealing at elevated temperatures.

It is commonly accepted that grains rotate towards a low energy grain boundary configuration associated with the low Σ CSL orientation relationships [1-3]. This was confirmed by atomistic computer simulations by several groups [4-6]. According to the theoretical model by Cahn and Taylor [7], grain rotation can be induced by grain boundary motion due to the coupling between boundary motion and tangential translation of the adjacent grains, which is proportional to the normal boundary displacement. Since Cahn and Taylor have proposed their approach to describe grain boundary motion, grain translation and grain rotation [7], significant progress was achieved with respect to the theoretical analysis of the shear-migration coupling [8-15]. Also, a substantial body of experimental evidence of this phenomenon was obtained for planar boundaries [16-23]. Further, it has been experimentally shown that stress driven boundary migration can be accompanied by grain rotation [24].

The motion of grain boundaries is controlled by their mobility m and the applied driving force p. As was predicted theoretically [25,26] and shown by numerous experiments [16,27-31], the rate of boundary migration v is proportional to the acting driving force

$$v = mp \tag{1}$$

The grain boundary mobility can be extracted from the temporal change of the grain size during annealing, since polycrystals are always liable to grain growth owing to a driving force arising from the curvature of the grain boundaries. The large part of the grain boundary mobility data was derived just from such kind of experiments. However, measurements of boundary mobility based on the change of the mean grain size with time in polycrystals give only a mean value, averaged over many different types of grain boundaries. Specific data on grain boundary mobility can only be obtained from the behavior of individual boundaries with well-defined geometry. This can be accomplished in experiments on specially grown bicrystal.

It is well established that both grain boundary energy and mobility to a great extent depend on grain boundary character, which is commonly reduced to misorientation between neighboring grains and inclination of the boundary plane [32]. A dependence of grain boundary energy on inclination can result in the formation of low energy grain boundary facets [33-40], which in turn can substantially affect grain boundary motion [41-45] and, therefore, be crucial for processes of microstructural evolution in polycrystals [46-50].

It is widely accepted that grain boundary faceting relates to high angle boundaries with misorientations close to low Σ CSL orientation relationships [33-36]. It has been, however, experimentally proved that the same behavior also applies to low angle tilt grain boundaries with low index rotation axis [51-54]. In experiments on aluminum bicrystals with <100> and <111> tilt grain boundaries with misorientation angles θ in the transition range from low to high angles it was found that, in contrast to expectations for the used bicrystal geometry (Fig. 1), only boundaries with misorientation angles larger than 15° could assume a curved shape and steadily move under a capillary driving force. Low angle boundaries ($\theta < 15°$) did not assume a continuously curved shape [55] and, correspondingly, did not move under a capillary driving force [51-54]. In the entire investigated temperature range they remained flat or formed facets which met the initial boundary at a sharp edge (e.g. Fig. 5 in Ref. [52]). The observed faceting behavior of tilt grain boundaries and its change with increasing misorientation was attributed to the dependence of grain boundary energy on inclination [51,52].

It is important to note, however, that a geometrically pure tilt grain boundary with a low index rotation axis is a specific model case. Actually, most previous experimental and modeling studies of grain boundary dynamics were performed on tilt boundaries [32], since they are easier to fabricate in both experiment and simulations. In a real polycrystal most boundaries are not pure tilt or pure twist but rather of general type. Their structure is more complex than that of pure tilt boundaries, which are composed of edge dislocations. Therefore, the migration and faceting behavior of random grain boundaries can be expected to differ substantially from that of pure tilt boundaries. Also, the dynamic/rotation behavior of grains with random boundaries can be very different from that of grains with pure tilt boundaries. In the current paper the results of experimental and computational efforts over recent years to study the motion of geometrically different grain boundaries and grain rotation under various driving forces are briefly reviewed.

2. Applied driving forces and measuring techniques

The experiments were conducted on specimens (Fig. 1) fabricated by electro-discharge machining from high purity Al bicrystals (99.9995%) grown by the vertical Bridgman technique. Details of the crystal growth, bicrystal characterization and sample preparation are given elsewhere [16-18,51,52].

2.1. Boundary migration under a curvature driving force

The well approved experimental technique [51-63] for measuring grain boundary migration under a constant curvature driving force p_c provided by the boundary energy γ, $p_c = \gamma / a$ (Fig. 1) was applied. Grain boundaries with various rotation angles around a <100> axis and different orientations of this axis with respect to the boundary plane and sheet normal were examined. The investigated mixed tilt-twist grain boundaries were composed of a rotation angle θ around a common [001] axis and grain boundary plane rotated by an angle $\xi \approx 20°$ with respect to the tilt boundary plane (Fig. 1a). The angle ξ, therefore, specifies a twist component of a mixed grain boundary. In particular, <100> mixed boundaries with misorientations 4.9°, 9.1°, 12.3° and 20.9° were investigated. For comparison the behavior of a 9.1° [001] pure tilt boundary, i.e. with ξ = 0°, was also examined. In two further investigated bicrystals with 7.0° and 11.4° $[\bar{1}00]$ boundaries a common rotation axis of the adjacent grains was parallel to the bicrystal surface as shown in Fig. 1b. The initial inclination of a 7.0° $[\bar{1}00]$ boundary was parallel to the growth (x -) axis of the bicrystal, i.e. this boundary had tilt geometry. The plane of the 11.4° $[\bar{1}00]$ boundary was initially inclined from the pure tilt position by about 6° (rotated around z-axis), i.e. it was of a mixed geometry with a twist component ξ = 6°.

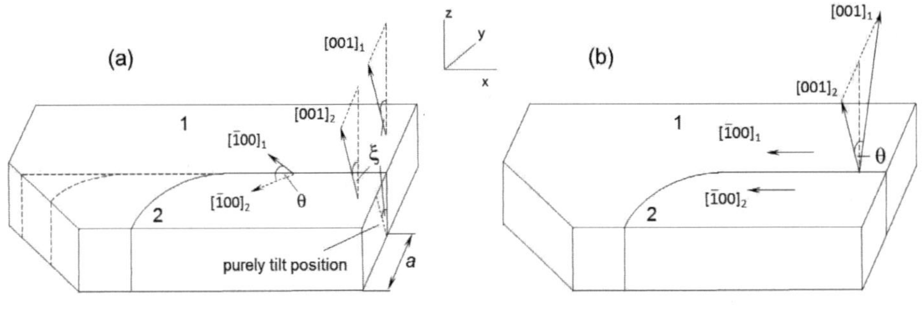

Figure 1 Bicrystal geometries for measuring the grain boundary migration. (a) Bicrystal with a θ <100> tilt-twist grain boundary (ξ is a twist component); (b) bicrystal with a θ <100> tilt boundary with a rotation axis parallel to the specimen surface [54].

The measurements of grain boundary shape and motion in the temperature range between 390°C and 640°C were performed by an in-situ technique in a SEM equipped with a specially designed heating stage [64]. The grain boundary shape and location were determined utilizing the orientation contrast revealed by a secondary electron detector (Fig. 5). The measuring procedure is described in Refs. [51,52].

2.2. Stress-induced grain boundary migration

Grain boundaries can also be driven by an applied mechanical stress. For a tilt grain boundary with misorientation angle θ subjected to shear stress τ the migration driving force p_s can be expressed as [65]

$$p_s = 2\tau cos\varphi \cdot tan(\varphi/2) \quad (2)$$

where $\psi/2$ is the angle between the grain boundary normal and the effective Burgers vector $\boldsymbol{b}$ per unit boundary length as determined from the Frank-Bilby equation [8,9]. For example, for <100> tilt boundaries $\varphi = \theta$ for the <100> mode of migration-shear coupling and $\varphi = \theta - 90°$ for the <110> coupling mode [8,9]. For low angle <100> tilt boundaries, for which φ is small or close to 90° Eq. (2) can be approximated by [16]

$$p_s \approx \tau sin\varphi \qquad (3)$$

The recent measurements of stress induced grain boundary migration [19-21,24,66,67] were performed by in-situ observations and recording of grain boundary migration using a hot deformation stage integrated in a SEM [19,66].

For an application of a shear stress to the grain boundaries two sets of grips were designed and fabricated (Fig. 2). Each pair of the grips is applicable to specimens with a particular geometry. Grain boundary behaviour under an applied stress can be observed and measured in-situ utilizing the orientation contrast on the specimen surface revealed by a secondary electron detector (Fig. 3). An analysis of the recorded orientation image sequences with a special software reveals the normal boundary displacement d and the lateral grain translation s.

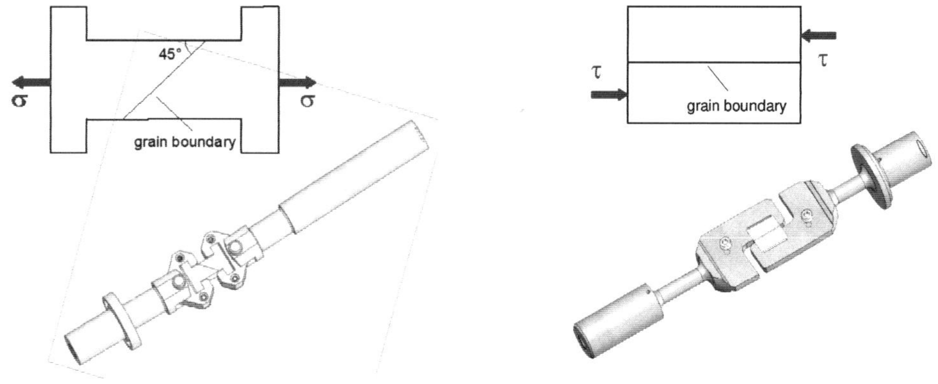

Figure 2 Bicrystal geometry and corresponding grips for the application of a shear stress to the grain boundary [19].

Figure 3 Motion of a 76.4°<100> tilt grain boundary in an Al bicrystal under an applied tensile stress of 0.26 MPa during annealing at 380°C.

3. Applied modeling techniques

In order to determine the grain boundary energy of both the <100> tilt and mixed grain boundaries the computation procedure proposed by Lee and Choi [68] was applied. The method does not require periodic boundary conditions, typically assigned to the simulation sample in the two directions parallel to the boundary and, thus, can be utilized for computing grain boundary energy for any misorientation and inclination angles. The second nearest-neighbor modified embedded atom method (2NN MEAM) potential for Al developed by Lee et al. [68-71] was utilized to describe the interaction between atoms. The energy of the created bicrystal sample was subsequently minimized by application of 0K molecular statics. The grain boundary energy was determined as the difference between the computed potential energies of the bicrystal and the original single crystal sphere [54].

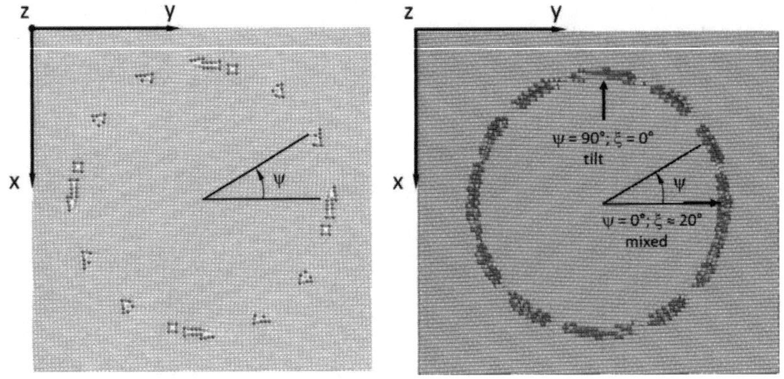

(a) (b)

Figure 4 Cross-sections of the simulation box with the embedded cylindrical grain with the 5.5°<100> (a) tilt and (b) mixed tilt-twist boundary. The grain and the matrix are rotated by $\pm\, \theta/2$ (here θ_0=5.45°) around the common [100] axis, which (a) for grain with the pure tilt boundary is aligned parallel to the z axis and for grain with the mixed boundary is tilted by an angle $\xi \approx 20°$ around the x axis.

In order to analyze the effect of the inclinational anisotropy of grain boundary energy on the grain shape evolution and grain growth kinetics molecular dynamics (MD) simulations were employed. The large-scale atomic/molecular massively parallel simulator (LAMMPS) code [72] was utilized. The atomic interactions were described by the same 2NN MEAM potential as employed in the molecular statics (MS) simulations. The configuration used in these MD simulations was composed of a cylindrical grain embedded in a differently oriented crystalline matrix. The capillary driven shrinkage of such a grain misoriented with respect to the matrix by different angles θ around a common <100> axis was studied. Four pairs of grains with pure tilt and mixed tilt-twist boundaries (Fig. 4) with the misorientation angles θ_0= 5.45°, 10.38°, 16.26°, and 22.61° were examined. The grains with tilt boundaries were misoriented around the [100] axis aligned parallel to the z axis of the simulation box (Fig. 4a). The boundaries of these grains retain their pure tilt character over the entire circle/length. For computing the grains with the mixed boundaries the common [100] rotation axis was additionally rotated counterclockwise by an angle of about $\xi \approx 20°$ ($\xi \equiv \mathrm{acos}([100] \cdot \bar{c}/||c||)$) around the crystallographic direction parallel to the x axis of the simulation box. At the inclination $\psi = 0°$, therefore, these boundaries had a tilt-twist character with a 20° twist component. With increasing inclination ψ the twist component ξ decreases down to zero at $\psi = 90°$ (Fig. 4b). All simulations were performed at 400°C. Grain rotation during shrinkage was not prohibited to allow the natural development of the system. The change of the grain volume with time was tracked by using an orientation dependent order parameter for the atoms, as defined in [73].

4. Faceting and migration behavior of low angle boundaries of different character

The experiments revealed that all investigated mixed <100> tilt-twist grain boundaries readily assumed a continuously curved shape (Fig. 5) and while migrating retained their shape self-similar in the entire investigated temperature range. All boundaries moved steadily at constant temperature (Fig. 6a), such that the velocity v for each temperature could be calculated. From the obtained boundary velocity the reduced mobility A was determined as

$$A \equiv v \cdot a = A_0 exp(-H/kT) = m\gamma \tag{4}$$

where a is the width of the shrinking grain (Fig. 1), H is the activation energy and A_0 the pre-exponential factor. The temperature dependence of the reduced mobility for all investigated mixed boundaries is shown in Fig. 6b.

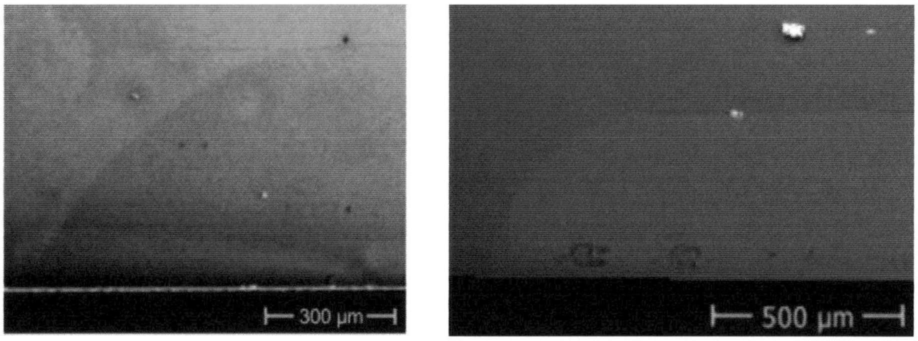

(a) **(b)**

Figure 5 Images of the moving mixed tilt-twist grain boundaries during annealing at elevated temperatures. (a) 4.9°<100> (twist component $\xi \approx 20°$) (Fig. 1a); (b) 11.4°<100> boundary, which in the original straight configuration (Fig. 1b) had a twist component of $\xi \approx 6°$.

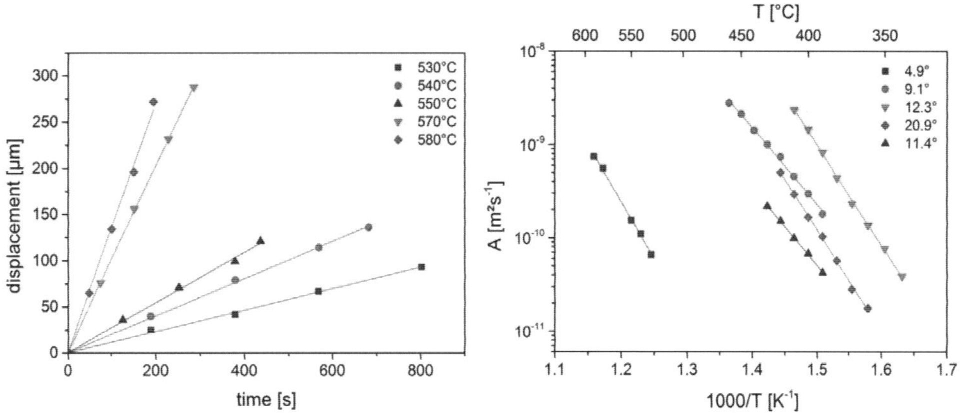

(a) **(b)**

Figure 6 (a) Displacement vs. time at various temperatures for migration of a 4.9°<100> mixed boundary; (b) Reduced mobility, i.e. boundary velocity v, normalized by the reciprocal width of the shrinking grain, a^{-1} (see Eq. (4)), vs. temperature for the migration of the investigated <100> mixed boundaries.

In contrast to the behavior of the mixed <100> boundaries, the 9.1° and 7.0° <100> pure tilt boundaries (with geometry shown in Fig 1a and 1b, respectively) did not assume a curved shape. They remained straight and, therefore, did not move during annealing at elevated temperatures up to 640°C. These observations are consistent with the results of previous experiments on low angle <100> and <111> tilt boundaries [51,52]: the boundary remained flat and did not move under the driving force provided by its energy. Obviously, this is due to the anisotropy of its energy with respect to boundary inclination (Fig. 7). In the original position both the 9.1° and 7.0° boundaries have a local minimum energy at the symmetrical configuration with $\psi = 0°$ (as 10.4° <100> pure tilt boundary in Fig. 7). Accordingly, due to their minimal energy the symmetrical 9.1° and 7.0° tilt boundaries retained the initial flat/faceted configuration and, therefore, did not move under the driving force provided by their energy.

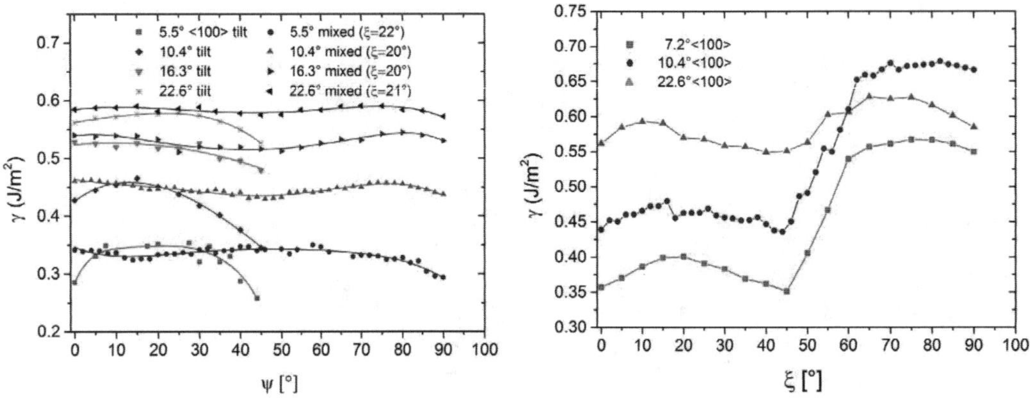

(a) **(b)**

Figure 7 Computed dependence of the grain boundary energy (a) on boundary inclination for <100> tilt and mixed (twist component $\xi \approx 20°$) boundaries and (b) upon change of the boundary geometry from pure <100> tilt ($\xi = 0°$) to pure <100> twist ($\xi = 90°$) at 0 K.

As revealed by MS simulations, the change of the boundary geometry from pure tilt to mixed tilt-twist resulted in an essential change of the inclination dependence of the boundary energy. As seen in Fig. 7a, the energy of the mixed grain boundaries is essentially constant throughout the entire inclination range. As a result, according to the tendency of the system to minimize the total boundary area, the initially straight boundary configuration can easily be replaced by a continuously curved boundary shape that allows the boundary to move under a capillary driving force during heating at elevated temperature, as observed in the experiments (Fig. 5).

The computed inclination dependence of grain boundary energy $\gamma(\xi)$ in Fig. 7b provides the explanation of the behavior of 11.4° mixed boundary (see Fig. 1b), which in contrast to 7.0° tilt boundary became curved and mobile (Fig. 5b). The energy of the investigated 11.4° <100> mixed boundary in its original position with $\xi = 6°$ is already outside the energetic minimum at the pure tilt position with $\xi = 0°$. This is obviously high enough to avoid the faceting (retaining the origin straight configuration) during annealing at elevated temperatures.

5. Grain shape evolution during shrinkage of initially cylindrical grains with boundaries of different character

The computed temporal evolution of the shape and size of the shrinking initially circular grains with different boundaries was found to depend on both the misorientation angle and character of the examined boundaries [54]. Very soon after the beginning of the respective simulation runs the grains with low angle 5.5° and 10.4° tilt boundaries assumed a faceted shape with well pronounced flat sides with inclinations close to $\psi = 0°/90°$ and $\psi = 45°$ (Fig. 8). The grain with the 16.3° tilt boundary was much more rounded but observed to form non-stable facets appearing and disappearing during the grain shrinkage. In contrast, the shapes of the grains with the same misorientations with respect to the matrix but mixed tilt-twist boundaries were observed to be essentially rounded. The grains with the high angle 22.6° boundaries both tilt and mixed were well rounded close to a circle during shrinkage [54].

Apparently, the results of MD simulations are in direct relation to the results of MS simulations. The formation of a faceted octagon-like shape of grains with low angle 5.5° and 10.4° tilt boundaries can be rationalized in terms of the essential inclinational anisotropy of their energy revealed by MS simulations. The flat sides of these grains represent symmetrical boundary configurations with inclinations ψ = 0°/90° and ψ = 45°, where the energy attains distinct minima (Fig. 7a). A closer inspection of the shape of grains with low angle tilt boundaries (after 1 ns computational annealing) revealed that they comprise eight symmetrical boundary facets, which in turn are composed of edge dislocations with Burgers vectors b = a<100> and a/2<110> [54]. The formation of a faceted shape of grains with low angle tilt boundaries in MD simulations also conforms to the behavior of a 9.1° pure tilt boundary observed in the experiment, where this boundary remained straight in the whole temperature range investigated.

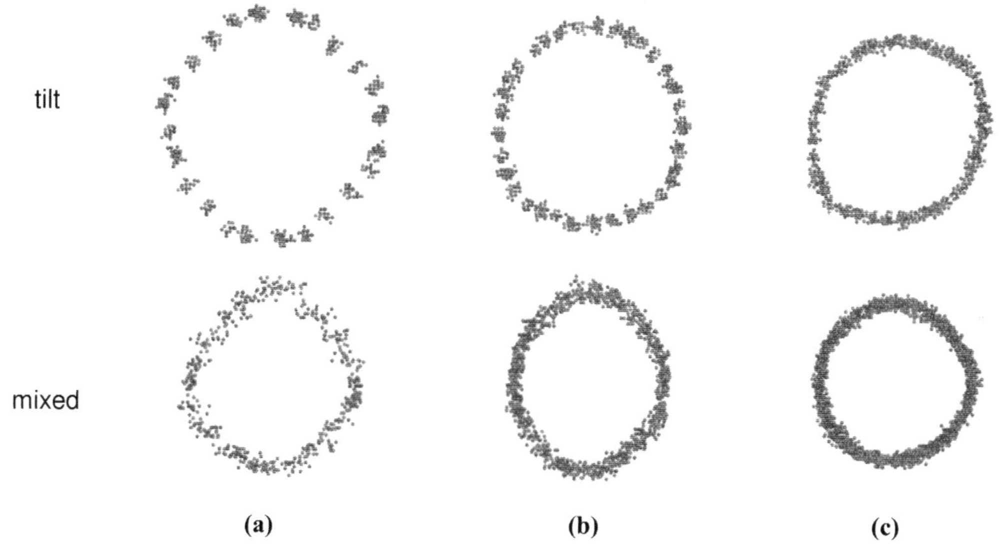

Figure 8 Snapshots of MD simulations of shrinking initially cylindrical grains with tilt and tilt-twist character at equal time of shrinkage t = 1.01 ns. (a) 10.4°; (b) 16.3°; (c) 22.6°. The grain boundaries are revealed by displaying only atoms with a coordination number z = 10 and z = 13.

As revealed by MD simulations, in contrast to grains with low angle tilt boundaries, the grains with 16.3° and 22.6° tilt boundaries shrank keeping their round shape (Fig. 8). This is obviously due to a reduction of inclinational anisotropy of grain boundary energy with increasing misorientation, which is well seen in Fig. 7 by comparing the $\gamma(\psi)$ dependency for the investigated boundaries. Furthermore, the almost isotropic energy of mixed low angle boundaries revealed by MS simulations (Fig. 7a) seems to account for the formation of a rounded shape of grains with 5.5° and 10.4° mixed boundaries as observed in MD simulations. As discussed above, for the same reason the experimentally examined 12.3°, 9.1° and 4.9° mixed grain boundaries, in contrast to the 9.1° pure tilt boundary, could readily assume the curved shape and move under the capillary driving force.

6. Mobility of <100> low angle boundaries

The results of the measurements of low angle <100> mixed boundary motion [54] revealed that the mobilities of boundaries with relatively high misorientations (9.1°, 11.4° and 12.3°) essentially did not differ from that of the 20.9° boundary and those of high angle tilt boundaries, which due to their sufficiently isotropic energy are capable to move having pure tilt geometry. This is particularly apparent in Fig. 9a, where the mobilities of the investigated boundaries at 410°C are shown together with the data for high angle 18.3°, 20.0° and 23.5° <100> tilt boundaries in Al bicrystals of the same purity obtained previously [52]. As seen, the mobilities obtained for low angle boundaries with $\theta > 9°$ are in the range of values, which are typical for high angle <100> boundaries.

In contrast, the 4.9° boundary in our experiments moved with similar velocities as the other boundaries only at substantially higher temperatures between 525°C and 590°C (Fig. 6b). Consequently, at a given temperature it is comparatively much less mobile. As seen in Fig. 9a, its mobility interpolated to the temperature of 410°C is about three orders of magnitude lower than that of the other (low angle mixed and high angle tilt) boundaries. This result is in a good agreement with the data obtained in experiments on Al polycrystals [74,75] as well as measurements on Cu bicrystals [76].

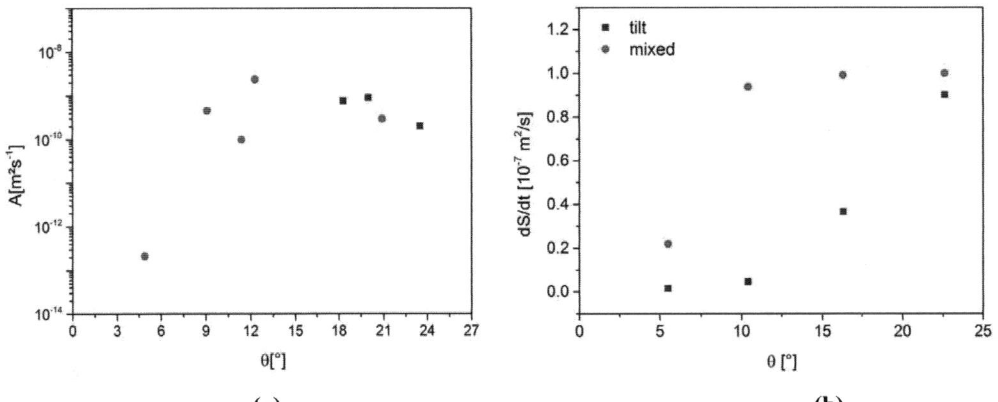

Figure 9 (a) Reduced mobility $A = v \cdot a$ of investigated <100> mixed tilt-twist boundaries (circle symbols) at 410°C. The square symbols represent data obtained for the migration of high angle <100> pure tilt boundaries with misorientations 18.3°, 20.0°, 23.5° in Al bicrystals of the same purity at the same temperature [52]. (b) Misorientation dependence of the computed rate of grain area change dS/dt as obtained in MD simulations.

Figure 9b shows the rate of grain area change dS/dt with growing misorientation θ for all tilt and tilt-twist boundaries as obtained in MD simulations. For both types of boundaries the rate of grain area change was found to increase with misorientation, although in different ways. For low angle (5.5° and 10.3°) tilt boundaries it did not differ significantly, but rose much faster with a further increase of θ. It can be understood, taking into account that both the lower energy γ of low angle boundaries and the smaller local curvature κ of most boundary segments of these grains (due to the faceting) caused a reduction of the capillary driving force $p_c = \gamma\kappa$. The grain with the 5.5° mixed boundary shrank faster, but not substantially faster, than that with the 5.5° tilt boundary. With an increase of misorientation from 5.5° to 10.3°, however, the rate of area change for a grain with the mixed boundary increased very rapidly up to the level of dS/dt values for grains with the medium angle 16.3° and high angle 22.6° mixed boundaries (Fig. 9b).

It is remarkable that the both computed misorientation dependence of dS/dt and dependence $A(\theta)$ obtained in experiments (Fig. 9) are qualitatively similar. In both the experiments and MD simulations the low angle <100> mixed boundaries with a misorientation angle around 10° were found to be quite mobile with the mobility similar to that of <100> high angle boundaries (no matter tilt or mixed). Also, akin to experiments, the simulations revealed that with decreasing misorientation down to about 5° the boundary mobility decreases substantially.

7. Rotation behavior of grains with pure tilt and mixed <100> boundaries

The remarkable result of the MD simulations performed is that they revealed the crucial impact of grain boundary character on a grain rotation during shrinkage [54,77]. The simulated grains with pure tilt boundaries rotated towards higher θ, while grains with the mixed tilt-twist boundaries practically did not rotate during their shrinkage. This is illustrated in Fig. 10, where snapshots of the grains with misorientation angles θ_0= 16.26° and 22.61° after their shrinkage down to the normalized grain volume of V/V_0= 0.25, where V_0 is the initial grain volume, are depicted.

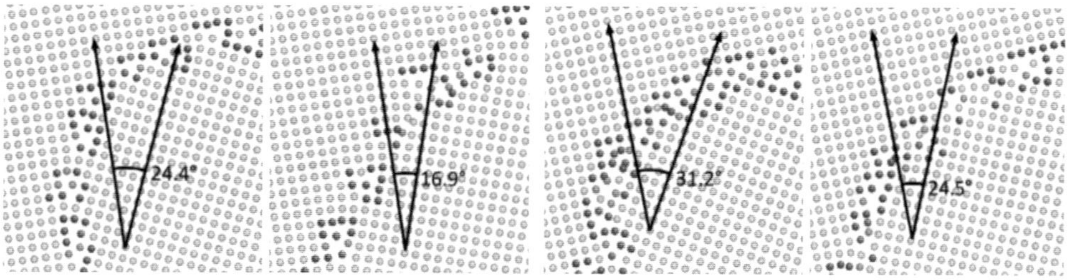

(a) (b)

Figure 10 Grain misorientation and grain boundary structure as observed on (100) crystallographic planes for grains with the (a) θ_0= 16.26° and (b) θ_0= 22.61° tilt and mixed boundaries (left and right, respectively) after grain shrinkage to V/V_0= 0.25 (V and V_0 are a shrinking and initial grain volume, respectively).

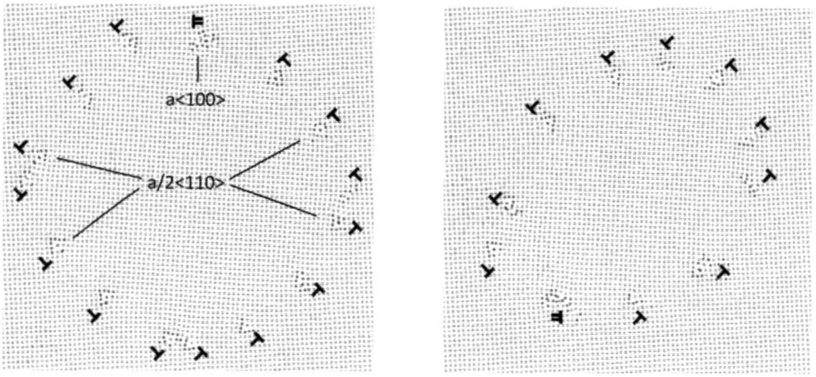

(a) (b)

Figure 11 Dislocation structure of the grain boundary with θ_0= 5.45° as it appears on a {100} crystallographic plane after annealing for (a) 0.11 ns and (b) 0.42 ns. Atoms are colored according to their coordination number [77].

The observed rotation of grains with tilt boundaries can be understood in terms of the grain boundary motion coupled to shear [7,8]. The grain rotation induced by grain boundary motion was analyzed theoretically [7] and observed in computer simulations [78-80]. According to the model [7], the grain rotation caused by the boundary motion perfectly coupled to shear is characterized by the conservation of the dislocation content of the boundary and, therefore, rotation of the shrinking grain towards higher θ is associated with the substantial increase of the dislocation density in the grain boundary. This case is obviously applied to the results of our MD simulations performed for grains with pure tilt boundaries composed of edge dislocations.

On the other hand, the grains with mixed boundaries were observed not to rotate during their shrinkage [54,77]. The dislocation content of these boundaries, therefore, was not characterized by the invariant dislocation number, but rather by the constant dislocation density that preserved essentially the grain boundary structure. This behavior requires an effective annihilation of grain boundary dislocations. Indeed, a closer inspection of the atomic configurations in the boundary revealed that dislocations with the mixed edge-screw character, which compose the respective boundaries, can easily interact with each other and annihilate as a result of the chains of dissociation and recombination reactions (Fig. 11) [77].

Therefore, the results of the performed MD simulations distinctly demonstrated a sensitivity of grain rotation during curvature driven grain shrinkage on grain boundary structure. The availability of the effectively operating mechanisms of dislocation annihilation in the mixed grain boundaries, for instance as identified in Ref. [77], was the most likely reason for the missing rotation during shrinkage of the island grains in thin films of Al [81] and Au [82,83], where grain boundaries were not of ideally pure tilt character with low index rotation axis.

Obviously, based on the simulation results obtained for the three investigated pairs of orientation relationships, it is impossible to conclude about the possibility of rotation of grains with all various boundary structures. However, it can be suggested that if the further studies will confirm the confinement of grain rotation during capillarity driven shrinkage to grains with pure tilt boundaries only, this phenomenon can be considered as playing a minor role for microstructure development during grain growth in conventional polycrystals, but can be probably appreciable in the development of sharply textured columnar structures.

8. Mechanically induced grain boundary migration and grain rotation

Stress driven grain boundary motion was predicted theoretically by Read and Shockley for low angle symmetrical tilt boundaries [84] and first experimentally observed for low angle boundaries in Zn [85-87]. It has been found, also in experiments with some high angle boundaries with low Σ CSL relationships in bicrystals of Al, Zn and ZrO_2 [88-95], that boundary migration under an applied shear stress is coupled to a strain, which in bicrystals is observed as a translation of the grains parallel to the boundary plane. According to the common understanding based on these observations, the coupled boundary migration is confined to symmetrical low angle boundaries composed of edge dislocations and special low Σ high angle boundaries comprising secondary grain boundary dislocations: the dislocation glide under stress leads to a shear deformation of the crystal region swept by the moving boundary. According to the dislocation model of low angle tilt grain boundaries by Read and Shockley [84], for symmetrical low angle tilt boundaries the relation between normal boundary displacement d and grain translation s, also referred to in literature as coupling factor $\beta = s/d$ (Fig. 12), changes with misorientation angle θ according to the tangent rule $\beta = 2tan(\theta/2)$. More recent theoretical models [8-12] predict that the boundary migration-shear coupling can occur for high angle boundaries as well. This has been also confirmed by computer simulation studies [10,96-99] and observed for random boundaries by TEM investigations in nanocrystalline and ultrafine-grained polycrystals [100-102].

8.1. Experimental proof of boundary migration-shear coupling

In a series of experiments [16-18] the stress-driven migration of planar symmetrical <100> tilt boundaries with misorientations in the entire angular range (Fig. 13) was measured. It has been proved that stress driven grain boundary motion is coupled to a shear strain of the crystal region behind the moving boundary. The important result of these experiments with various <100> tilt boundaries is that contrary to the common understanding the boundary migration – shear coupling is not confined to low angle and some low Σ high angle boundaries, but is a typical observation for any symmetrical <100> tilt boundaries in the entire range of misorientations between 0° and 90°, no matter how close to low Σ CSL orientation relationships. Moreover, the experimental results have verified the prediction of the model by Cahn et al. [8,9] that the tangent rule for calculation of the coupling factor β retains its validity over the entire misorientation range including the high angle regime: as depicted in Fig. 13a, the values of the coupling factor calculated according to the model are in excellent agreement with the values obtained in the experiment. Two branches in the misorientation dependence of the coupling factor in Fig. 13a confirm that for <100> tilt boundaries there are indeed two geometrically different mechanisms of shear-coupled boundary migration [8], which can formally be represented by slip of dislocations with Burgers vector b=a<010> on a {001} plane or b=a/2<110> on {110}, respectively.

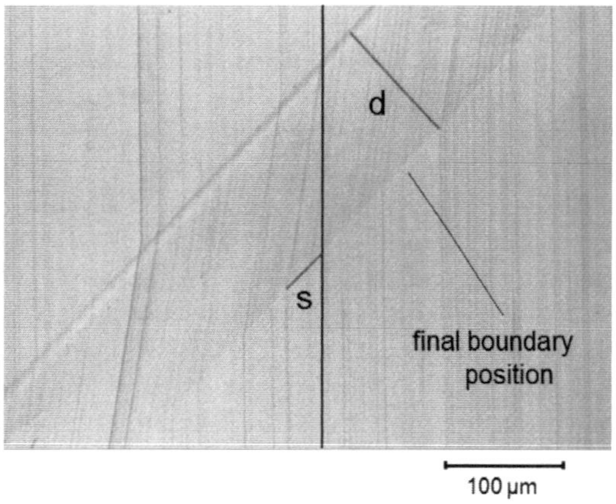

Figure 12 Migration-shear coupling of a 23.6°<111> tilt boundary moving under a tensile stress of 0.34 MPa (after 30 minutes annealing at 340°C) [18]. Coupling between grain boundary migration and produced shear is characterized by the factor determined as $\beta = \gamma = s/d$. The measured β for the 23.6°<111> tilt boundary is β = 0.42, which is equal to the theoretical value β_{theor}.

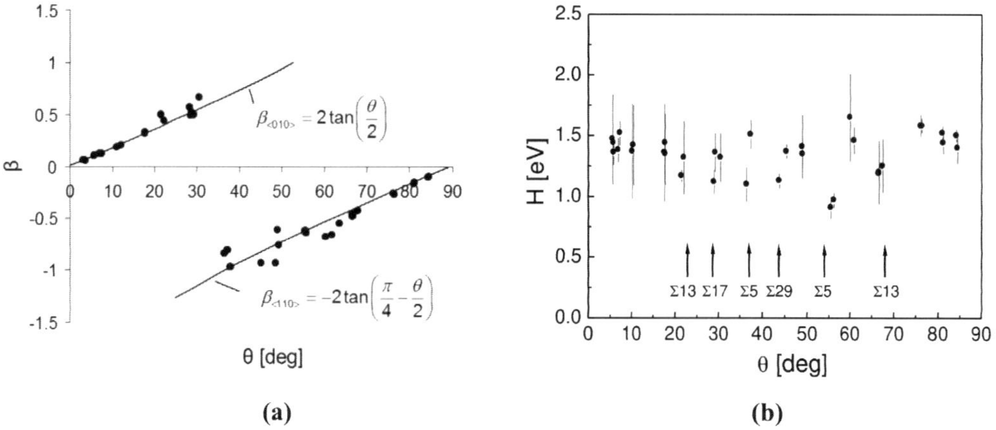

Figure 13 (a) Measured (points) and calculated (lines) coupling factors β and (b) activation enthalpy H of stress driven migration vs. misorientation angle θ for symmetrical <100> tilt grain boundaries [18].

The boundary migration – shear coupling was observed not only for symmetrical but also for asymmetrical tilt grain boundaries. The structure of asymmetrical low angle boundaries is represented by a mixed arrangement of dislocations with different Burgers vectors. Particularly, asymmetrical <100> tilt boundaries consist of dislocations with both a<010> and a/2<110> Burgers vectors. Under a stress these dislocations should move on different planes. According to Read and Shockley [84] this renders their concurrent motion impossible without hindering and locking each other. Consequently, asymmetrical tilt boundaries in the dislocation model cannot be driven by a stress and, therefore, cannot cause a shape change of a bicrystal. Contrary to this prediction, the experiments with <100> tilt boundaries [20] as well as measurements on bicrystals with <111> tilt boundaries [67], showed that the asymmetrical tilt boundaries can move under an applied stress and produce a shear, although the obtained coupling factors distinctly deviated from the coupling factor calculated according to the tangent rule.

8.2. Mobility of stress driven boundaries, its dependence on temperature and misorientation
Measurements on bicrystals with various grain boundaries performed at different elevated temperatures revealed that stress driven shear-coupled boundary migration is thermally activated, and the boundary migration rate v depends on temperature according to an Arrhenius relation $v = v_0 exp(-H/kT)$, where H is the activation enthalpy of grain boundary migration. Figure 13b depicts the misorientation dependence of the migration activation enthalpy for <100> tilt boundaries [18]. As seen, in both low angle regimes ($\theta < 18°$ and $\theta > 76°$) the measured activation enthalpy remains virtually the same with a value of about H=1.45 eV. In the high angle regime, however, the migration activation enthalpy changes with the angle in a non-monotonic fashion, such that the lower H values correspond to boundaries with misorientations close to low Σ CSL orientation relationships (Fig. 13b). This suggests that the migration mechanism, which is reflected in the migration activation enthalpy, remains essentially unaltered for all low angle boundaries, whereas it varies substantially with the character of structurally different high angle boundaries.

8.3. Stress driven migration of Σ7 tilt boundary in Al
An unexpected boundary behavior under an applied stress was observed in experiments with Σ7 tilt boundaries [21,67]. Specifically, three symmetrical tilt grain boundaries with misorientations close to 38.2°<111>, 73.4°<201> and 135.6°<112>, i. e. the same special Σ7{132} CSL boundary, were examined. The experiments showed that the boundaries in the investigated bicrystals moved under an applied stress, but their migration was not accompanied by a shear strain, i.e. tangential displacement of the adjacent grains. This result provides evidence that the migration behavior of high angle low Σ grain boundaries under stress can differ from that of random boundaries, and their stress driven migration must not necessarily be coupled to shear, as predicted by the geometrical models of stress induced boundary migration.

Crystallographically equivalent boundaries in bicrystals of different geometric description behaved differently with respect to the migration rate and its temperature dependence [21]. The migration activation enthalpy ranged from 0.7 to 1.3 eV. This reveals that the kinetics of stress driven boundary migration and its operative mechanism depend on the direction of the applied stress.

8.4. Concurrent grain boundary motion and grain rotation
The remarkable boundary response to an applied stress – boundary migration accompanied by both a translation of the adjacent grains parallel to the boundary plane and their rotation around the boundary plane normal was observed in experiments with a 18.2°<100> mixed tilt-twist grain boundary [24]. This boundary was composed of a rotation angle θ = 18.2° around a common [001] axis and grain boundary plane rotated by 20° with respect to the tilt boundary plane, i.e. it had a twist component ξ = 20°.

The observed behavior was quantitatively characterized by measuring the surface topography of the bicrystal by means of atomic force microscopy after annealing under stress (Fig. 14). The topographic profiles across the boundary measured on the opposite sample surfaces clearly evidenced that the surface of the shrinking crystal (grain 2) after annealing was rotated around the boundary normal. Also, from the measured topography it is apparent that this rotation proceeded simultaneously with the normal boundary displacement (Fig. 14), providing evidence that the two processes were related.

The results of this experiment can be rationalized in terms of the dislocation structure of the mixed boundary, which is assumed to contain a set of edge dislocations accommodating the tilt component of the misorientation and a set of screw dislocations accommodating the twist component. Under the applied stress edge dislocations move in the direction normal to the boundary plane and produce shear deformation of the lattice and, therefore, the coupled boundary motion. At the same time, the applied stress imposes Peach-Kohler forces on the screw dislocations causing their glide in the boundary plane. This process induces grain rotation, the sign of which depends on the direction of the shear stress. It is noteworthy that this grain rotation changes the twist component of the grain misorientation (increases or decreases), demonstrating the important effect of changing the grain boundary character by applied stresses.

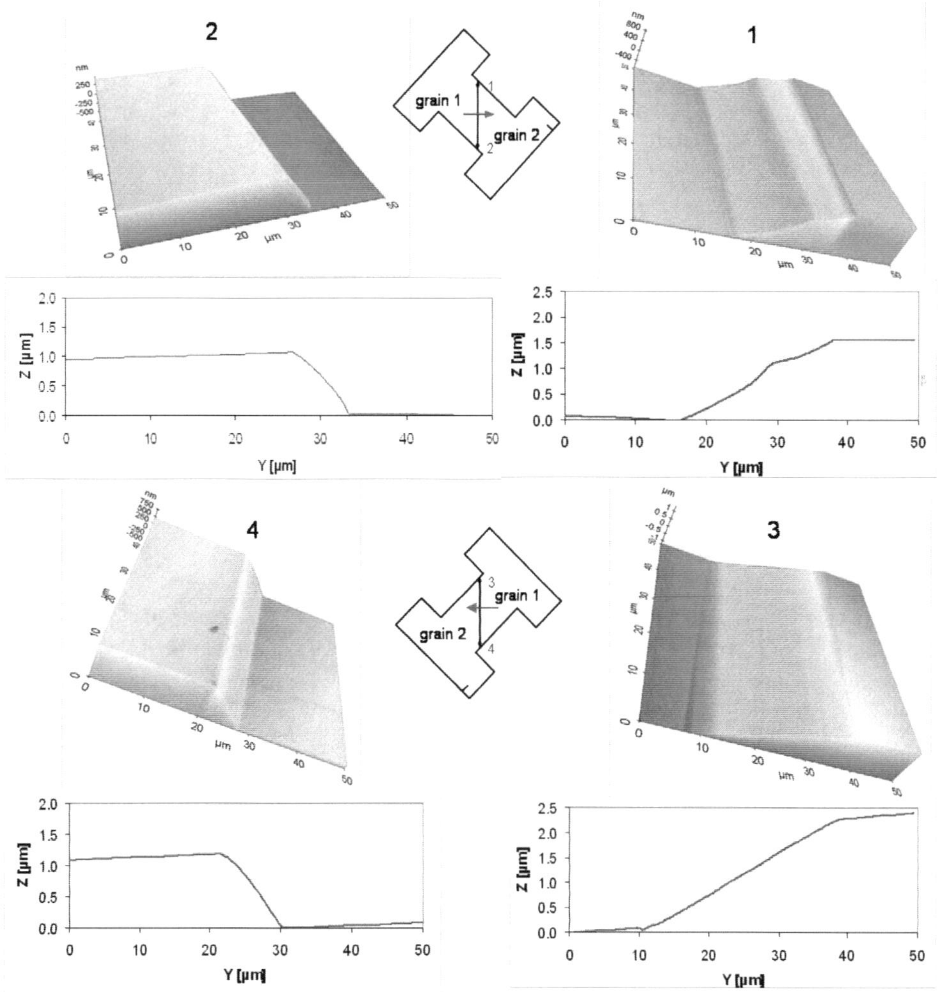

Figure 14 Surface topography of the bicrystal with 18.2° <100> mixed tilt-twist (ξ = 20°) boundary after annealing at 392°C under applied tensile stress of 0.23 MPa [24].

Acknowledgements
The authors express their gratitude to the Deutsche Forschungsgemeinschaft (DFG) for financial support (Grants MO 848/10, MO 848/14).

References
[1] Harris KE, Singh VV and King AH 1998 *Acta Mater.* **46** 2623
[2] Klinger L and Rabkin E 2011 *Acta Mater.* **59** 6691
[3] Bachurin DV, Nazarov AA and Weissmüller J 2012 *Acta Mater.* **60** 7064
[4] Ashkenazy Y, Averback RS and Albe K 2001 *Phys. Rev. B* **64** 205409
[5] Haslam AJ, Phillpot SR, Wolf D, Moldovan D and Gleiter H 2001 *Mater. Sci. Eng.* A **318** 293
[6] Upmanyu M, Srolovitz DJ, Lobkovsky AE, Warren JA and Carter WC 2006 *Acta Mater.* 54 1707
[7] Cahn JW and Taylor JE 2004 *Acta Mater.* **52** 4887
[8] Cahn JW, Mishin Y and Suzuki A 2006 *Acta Mater.* **54** 4953
[9] Cahn JW, Mishin Y and Suzuki A 2006 *Philos. Mag.* **86** 3965
[10] Ivanov VA and Mishin Y 2008 *Phys. Rev. B* **78** 064106
[11] Caillard D, Mompiou F and Legros M 2009 *Acta Mater.* **57** 2390
[12] Mompiou F, Legros M and Caillard D 2009 *Acta Mater.* **58** 3676
[13] Khater HA, Serra A, Pond RC and Hirth JP 2012 *Acta Mater.* **60** 2007
[14] Rajabzadeh A, Mompiou F, Legros M and Combe N 2013 *Phys. Rev. Letters* **110** 265507
[15] Taupin V, Capolungo L and Fressengeas C 2014 *Int. J. Plast.* **53** 179
[16] Molodov DA, Ivanov VA and Gottstein G 2007 *Acta Mater.* **55** 1843.
[17] Molodov DA, Gorkaya T and Gottstein G 2007 *Mater. Sci. Forum* **558-559** 927
[18] Gorkaya T, Molodov DA and Gottstein G 2009 *Acta Mater.* **57** 5396
[19] Gorkaya T, Burlet T, Molodov DA and Gottstein G 2010 *Scripta Mater* **63** 633
[20] Molodov DA, Gorkaya T and Gottstein G 2011 *J. Mat. Sci.* **46** 4318
[21] Molodov DA, Gorkaya T and Gottstein G 2011 *Scripta Mater.* **65** 990
[22] Rajabzadeh A, Legros M, Combe N, Mompiou F and Molodov DA 2013 *Philos. Magazine* **93** 1299
[23] Rajabzadeh A, Mompiou F, Lartigue-Korinek S, Combe N, Legros M and Molodov DA 2014 *Acta Mater.* 77 223
[24] Gorkaya T, Molodov KD, Molodov DA and Gottstein G 2011 *Acta Mater.* **59** 5674
[25] Smoluchowski R 1951 *Phys. Rev.* **83** 69
[26] Turnbull D 1951 *Journal of Metals* **3** 661
[27] Vandermeer RA 1965 *Transactions AIME* **233** 266
[28] Vandermeer RA, Juul-Jensen D and Woldt E 1997 *Metall. Mater. Trans. A* **28** 749
[29] Aristov V Y, Fradkov VE and Shvindlerman LS 1979 *Phys. Met. Metall.* **45** 83
[30] Molodov DA, Gottstein G, Heringhaus F and Shvindlerman LS 1998 *Acta Mater.* **46** 5627
[31] Günster Ch, Molodov DA and Gottstein G 2013 *Acta Mater.* **61** 2363
[32] Gottstein G, Molodov DA and Shvindlerman LS 2010 *ASM Handbook* vol 22B, ed DU Furrer and SL Siemiatin (Materials Park: ASM International) p 67
[33] Sutton AP and Balluffi RW 1995 *Interfaces in Crystalline Materials* (Oxford: Clarendon Press)
[34] Hsieh TE and Balluffi RW 1989 **37** 2133
[35] Straumal BB, Polyakov SA, Bischoff E, Gust W and Mittemeijer EJ 2001 *Interface Science* **9** 287
[36] Lee SB, Sigle W and Rühle M 2003 *Acta Mater.* **51** 4583
[37] Rohrer GS 2005 *Annu. Rev. Mater. Res.* **35** 99
[38] Rohrer GS 2011 *J. Mater. Sci.* **46** 5881
[39] D. Y. Yoon, and Y. K. Cho 2005 *J Mater Sci*, **40**, 861-870.
[40] Molodov DA, Günster Ch, Gottstein G and Shvindlerman LS 2012 *Philos. Magazine* **92** 4588
[41] Straumal BB, Rabkin E, Sursaeva VG and Gornakova AS 2005 *Zt. Metallkd.* **96** 161

[42] Straumal BB, Sursaeva VG and Gornakova AS 2005 *Zt. Metallkd.* **96** 1147
[43] Sursaeva VG, Straumal BB, Gornakova, Shvindlerman LS and Gottstein G 2008 *Acta Mater.* **56** 2728
[44] Sursaeva VG 2010 *Materials Letters* **64** 105
[45] Brandenburg J-E, Barrales-Mora LA, Molodov DA and Gottstein G 2013 *Acta Mater.* **61** 5518.
[46] Lee SB, Yoon DY and Henry MF 2000 *Acta Mater.* **48** 3071
[47] Lee SB, Hwang NM, Yoon DY and Henry MF 2000 *Metall. Mater. Trans. A* **31** 985
[48] Kazaryan A, Wang Y, Dregia SA and Patton BR 2002 *Acta Mater.* **50** 2491
[49] Rabkin E 2005 *J. Mater. Sci.* **40** 875
[50] Kang S-JL, Lee MG, and An SM 2009 *J. Am. Ceram. Soc.* **92** 1464
[51] Kirch DM, Zhao B, Molodov DA and Gottstein G 2007 *Scripta Mater.* **56** 939
[52] Kirch DM, Jannot E, Barrales-Mora LA, Molodov DA, Gottstein G 2008 *Acta Mater.* **56** 4998
[53] Brandenburg J-E, Barrales-Mora LA, Molodov DA and Gottstein G 2013 *Scripta Mater.* **68** 980
[54] Brandenburg J-E, Barrales-Mora LA and Molodov DA 2014 *Acta Mater.* **77** 294
[55] Verhasselt JCh, Gottstein G, Molodov DA and Shvindlerman LS 1999 *Acta Mater.* **47** 887
[56] Surholt T, Molodov DA and Herzig Chr 1998 *Acta Mater.* **46** 5345
[57] Molodov DA, Czubayko U, Gottstein G and Shvindlerman LS 1995 *Scripta Metall. Mater.* **32** 529
[58] Gottstein G, Molodov DA, Czubayko U and Shvindlerman LS 1995 *J. de Phys. IV* **5** 89
[59] Molodov DA, Czubayko U, Gottstein G and Shvindlerman LS 1998 *Acta Mater.* **46** 553
[60] Gottstein G, Molodov DA and Shvindlerman LS and Srolovitz DJ 2001 *Curr. Opin. Solid State Mater. Sci.* **5** 9
[61] Ivanov VA, Molodov DA, Shvindlerman LS and Gottstein 2004 *Mater. Sci. Forum* **467-470** 751
[62] Ivanov VA, Molodov DA, Shvindlerman LS and Gottstein G 2004 *Acta Mater.* **52** 969
[63] Furtkamp M, Gottstein G, Molodov DA, Semenov VN and Shvindlerman LS 1998 *Acta Mater.* **46** 4103
[64] Kirch DM, Ziemons A, Burlet T, Lischewski I, Molodova X, Molodov DA and Gottstein G 2008 *Rev. Sci. Instrum.* **79** 043902
[65] Molodov DA and Mishin Y 2013 *Microstructural Design of Advanced Engineering Materials*, ed DA Molodov (Weinheim: Wiley-VCH) p 201
[66] Gorkaya T, Burlet T, Molodov DA and Gottstein G 2012 *Mater. Sci. Forum* **715-716** 819
[67] Molodov DA, Gorkaya T and Gottstein G 2012 *Mater. Sci. Forum* **706-709** 2886
[68] Lee B-J and Choi S-H 2004 *Modelling Simul. Mater. Sci. Eng.* **12** 621
[69] Lee B-J and Baskes MI 2000 *Phys. Rev. B* **62** 8564
[70] Lee B-J, Baskes MI, Kim H and Cho YK 2001 *Phys. Rev. B* **64**:184102
[71] Lee B-J, Shim J-H and Baskes MI 2003 *Phys. Rev. B* **68** 144112
[72] Plimpton SJ 1995 *Comput. Phys.* **117** 1
[73] Janssens KGF, Olmsted D, Holm EA, Foiles SM, Plimpton SJ and Derlet PM 2006 *Nature Materials* **5** 124
[74] Huang Y and Humpreys FJ 2000 *Acta Mater.* **48** 2017
[75] Yang CC, Rollett AD and Mullins WW 2001 *Scripta Mater.* **24** 2735
[76] Viswanathan R and Bauer CL 1973 *Acta Metall.* **21** 1099
[77] Barrales-Mora LA, Brandenburg J-E and Molodov DA 2014 *Acta Mater.* **80** 141
[78] Srinivasan SG and Cahn JW 2002 *Science and technology of interfaces,* ed S Ankem, CS Pande, I Ovidko and R Ranganathan (Seattle: TMS) p 3
[79] Trautt ZT and Y Mishin 2012 *Acta Mater.* **60** 2407
[80] Wu KA and Voorhees PW 2012 *Acta Mater.* **60** 407
[81] Mompiou F, Legros M, Radetic T, Dahmen U, Gianola DS and Hemker KJ 2012 *Acta Mater.* **60** 2209
[82] Babcock SE and Balluffi RW 1989 *Acta Metall.***37** 2367

[83] Radetic T, Ophus C, Olmsted DL, Asta M and Dahmen U 2012 *Acta Mater.* **60** 7051
[84] Read WT and Shockley W 1950 *Phys. Rev.* **78** 275
[85] Washburn J and Parker ER 1952 *Trans AIME* **194** 1076
[86] Li CH, Edwards EH, Washburn J and Parker ER *Acta Metall.* 1953 **1** 223
[87] Bainbridge DW, Li CH and Edwards EH 1954 *Acta Metall.* **2** 322
[88] Watanabe T, Kimura SI and Karashima S 1984 *Philos. Mag. A* **49** 845
[89] Horiuchi R, Fukutomi H and Takahashi T 1987 *Fundamentals of Diffusion Bonding*, ed Y Ishida (Amsterdam: Elsevier) p 347
[90] Fukutomi H and Kamijo T 1985 *Scripta Metall.* **19** 195
[91] Fukutomi H, Iseki T, Endo T and Kamijo T 1991 *Acta Metall. Mater.* **39** 1445
[92] Sheikh-Ali AD and Valiev RZ 1990 *Phys Status Solidi (a)* **117** 429
[93] Sheikh-Ali AD, Lavrentyev FF and Kazarov YuG 1997 *Acta Mater.* **45** 4505
[94] Sheikh-Ali AD and Szpunar JA 1998 *Mater. Sci. Eng.* **A245** 49
[95] Yoshida H, Yokoyama K, Shibata N, Ikuhara Y and Sakuma T 2004 *Acta Mater.* **52** 2349
[96] Suzuki A and Mishin Y 2005 *Mater. Sci. Forum* **502** 157
[97] Mishin Y, Suzuki A, Uberuaga BP and Voter AF 2007 *Phys. Rev. B* **75** 224101
[98] Zhang H, Du D and Srolovitz DJ 2008 *Philos. Magazine* **88** 243
[99] Lim AT, Haataja M, Cai W and Srolovitz DJ 2012 *Acta Mater.* **60** 1395
[100] Legros M, Gianola GS and Hemker KJ 2008 *Acta Mater.* **56** 3380
[101] Rupert TJ, Gianola DS, Gan Y and Hemker KJ 2009 *Science* **326** 1686
[102] Mompiou F, Caillard D and Legros M 2009 *Acta Mater.* **57** 2198

Deformation, recovery and recrystallization from heterogeneous shear bands in steel sheets

K Murakami[1], **M Sugiyama**[2] **and K Ushioda**[3]
[1] Yawata R&D Lab., Technical Research & Development Bureau, Nippon Steel & Sumitomo Metal Corp., 1-1, Tobihatacho, Tobata-ku, Kitakyusyu City, Fukuoka Prefecture, 804-8501 Japan
[2] Graduate School of Engineering, Osaka University, 2-1 Yamada, Suita, Osaka, 565-0871, Japan
[3] Technical Research & Development Bureau,Nippon Steel & Sumitomo Metal Corp., 20-1, Shintomi, Futtsu, Chiba, 293-8511, Japan

Email: [1]murakami.4fz.kenichi@jp.nssmc.com; [2]sugiyama@mat.eng.osaka-u.ac.jp; [3]ushioda.cd8.kohsaku@jp.nssmc.com

Abstract. In order to clarify the mechanism of Goss texture formation along shear bands, the effect of cold rolling reduction on shear band formation and crystal orientation within shear bands and annealing texture were investigated in Fe-3%Si {111}<112> single crystals. Shear bands were formed intermittently during cold rolling; shear bands with smaller angles were formed earlier and those with larger angles were formed later. This result supports geometrical softening as a mechanism for shear band formation. Crystal rotation along shear bands became large in proportion to reduction. In calculation crystal rotation in ideal matrix, (111)[11-2], did not reach Goss, however, with consideration of orientation fluctuation the rotation angle exceeded Goss orientation and it coincided with the experimental data. After annealing recrystallized grains along shear bands were mainly Goss grains regardless of cold rolling reduction. The speculated reason for the dominance of Goss during annealing is that Goss cells have a comparatively lower density of dislocations and are surrounded by largely deformed areas.

1. Introduction

Properties of steel sheets, for instance deep drawing and magnetic flux density, largely depend on their textures and hence it is of great importance to control texture as desired. In the process, hot rolling, if necessary hot band annealing, cold rolling followed by final annealing are critical for controlling the texture of final products. In general, deformation structures after rolling are heterogeneous and recrystallization progresses heterogeneously through the process of nucleation and growth.

In poly-crystals which steel industries normally manufacture, vicinities of grain boundaries before deformation are one of the strongly preferable nucleation sites. Therefore grain sizes prior to cold rolling can be an essential parameter to determine texture after annealing following cold rolling [1]. Furthermore, in detail, boundaries which comprise {111} grains can be preferential nucleation sites because of local multiple slips [2].

For another preferential nucleation site, shear bands, which may occur after heavy planar strain in {111}<112> grains, are remarkably effective. Several studies about recrystallization texture after deformation of {111}<112> single crystals have argued that heterogeneous shear bands act as preferential nucleation sites of recrystallized grains with Goss orientation, {110}<001> [3-10]. However, the mechanism of shear band formation and Goss texture development during annealing is not clearly understood although the Goss orientation is extremely important for steel sheets products.

Our final aim is to clarify the mechanism of Goss texture formation along shear bands. Studies about shear band formation and crystal orientation within them depending on cold rolling reduction have not been investigated so far though annealing texture are affected strongly by cold rolling reduction. Therefore, the purpose of study is to investigate the effect of cold rolling reduction on shear band formation, crystal orientation within shear bands and annealing texture in Fe-3%Si {111}<112> single crystals, which can help to achieve our final aim.

2. Experimental procedure

2.1 Sample preparation

An ingot of Fe-3%Si was produced in a vacuum melting furnace in the laboratory. After reheating it for 48 hours at 1573K in a nitrogen atmosphere, the ingot was cut into pieces to measure the crystal orientation using X-ray back-reflection Laue diffraction method. Then, specimens of (111)[11-2] orientation single crystal were cut into sizes of 20mm in width, 20-30mm in length and 2.85mm in thickness. The chemical composition of samples was Si 3.0, C 0.002, P 0.001, N 0.001wt% and Fe for the balance.

Regarding cold rolling, in order to deform the above small specimens effectively, a single crystal sample was inserted in a rectangular die in a polycrystalline hot band of 90mm width x 100mm length x 2.85mm thickness in size. Specimens were then sandwiched between two sheets of other hot bands of 90mm width x 100mm length x 2.0mm thickness in size, and the four sides of the three hot bands were welded together.

The sandwiched specimens were cold rolled by 28, 40, 50 and 68% reduction with multi-pass rolling as denoted in Table 1. In some specimens annealing for 10 seconds at 823K was performed to obtain recovered structures. As for annealing texture, samples were annealed for 30 minutes at 923K in a salt bath then quenched into water.

Table 1. Samples for measurements

Samples	Initial thickness(mm)	Reduction(%)	Final thickness(mm)
a	2.85	28	2.05
b	2.85	40	1.71
c	2.85	50	1.43
d	2.85	68	0.92

2.2 Measurement and analysis of crystal orientations

In order to obtain microstructures mechanical polishing was performed in ND-RD (rolling and normal direction) sections followed by chemical etching with nitric acid and an ethanol solution. Orientation measurements and analyses were carried out in the ND-RD sections after chemical polished specimens using an EBSD (electron back-scattered diffraction pattern) system by TSL Solutions equipped in a SEM, JSM6500F(JEOL).

Quantitative data about crystal orientation within shear bands were taken from EBSD mapping data. In each specimen, three areas were taken and an area of a map was 30x40μm^2 in ND-RD sections with 0.1μm as the measuring step. The total points of the three maps in each specimen were, therefore, 417 thousand. Among these total points, less reliable values were eliminated; that is, if the CI values (confidence index values) in the analysis software in TSL Solutions were less than 0.1, the measurements points were eliminated. Furthermore, in the analysis of crystal orientation, rotation around TD (transverse direction) axis is important so that orientations within 10 degrees from the initial TD axis [-110] were analyzed and evaluated in this study.

2.3 Calculation of crystal rotation rates
In order to estimate crystal rotation within shear bands, Taylor factors and crystal orientation rates were calculated with a viscoplastic Taylor model [11]. Within shear bands simple shear was considered and, as for crystal rotation, crystal orientation rates were calculated as dividend by strains.

In relation to calculations considering orientation fluctuations, initial orientations were slightly changed around ND and RD axis of ideal matrix, (111)[11-2]

3. Results

3.1 Formation angles of shear bands
Optical micrographs of cold rolled specimens are shown in Fig.1. As the signature of each direction is very important because of single crystals, the orientations and signature of directions were corresponded in the schematic figure in the right side of Fig. 1. All specimens were observed in ND-RD sections and SBs (shear bands) were observed from lower left sides to upper right sides, which corresponded to previous studies [7,8,10]. And it was recognized that these SBs included the TD axis.

In the specimen with 28% reduction, only one type of SBs with the angle of 23 degrees (A1) to the RD (rolling direction) were formed as shown in Fig.1(a). In the 40% reduction in Fig. 1(b), two types of SBs were observed and this micrograph was very similar to the previous result by Ushioda and Hutchinson [8]. From Fig. 1(b) one type had a larger angle, 31 degrees (B1), and the other had a smaller angle, 15 degrees (B2). In the 50% reduction as shown in Fig. 1(c), several types of SBs could be observed and it was difficult to specify types of SBs, but the largest angles was 26 degrees (C1). In Fig. 1(c) black lines are deformation twins as denoted by arrows. In the 68% reduction in Fig. 1(d), it was the most difficult to specify SBs and the largest angles of SBs to RD were 30 degrees (D1).

3.2 Crystal rotation within shear bands
Crystal orientations within SBs were analyzed from EBSD data. In each specimen, the EBSD mapping data in the vicinity of the largest angle of SBs were taken into account. The important orientation fiber, which has a common TD axis, [-110], includes the initial orientation, (111)[11-2] and Goss, (110)[001], and, they have common Euler angle with 90degrees of $\phi1$ and 45degrees $\phi2$ (Bunge). Therefore, the orientation distribution was described as a function of Φ, in which the initial (111)[11-2] is 55degrees and Goss 90degrees, as shown in Fig.2.

In Fig.2, orientation distribution from the initial became large in accordance with rolling reduction. The orientation end towards Goss (positive direction of Φ) also became large. For example, the positive end in (a) 28% reduction did not reach Goss, however, those in (b),(c) and (d) with a larger reduction than 40% exceeded Goss to 120degrees of Φ. It can be stated that the crystal rotation within SBs progressed in proportion to the cold rolling reduction.

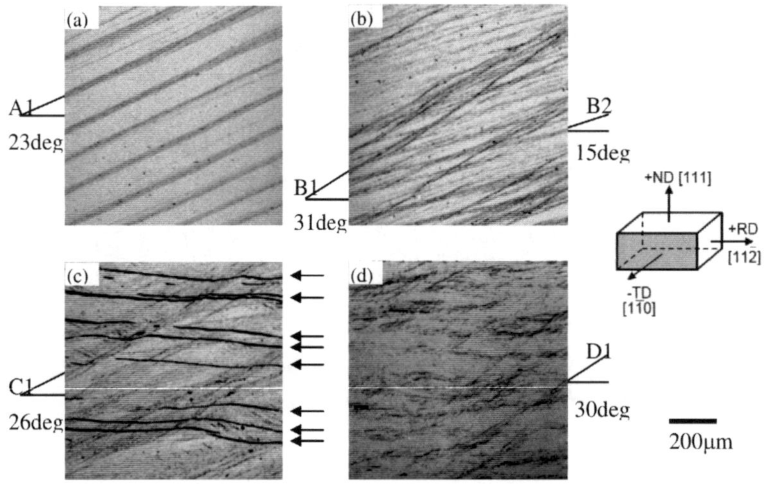

Figure 1. Optical micrographs of cold rolled specimens in ND-RD sections; Cold rolling reductions are (a) 28%, (b) 40%, (c) 50% and (d) 68%. Arrows in (c) show deformation twins.

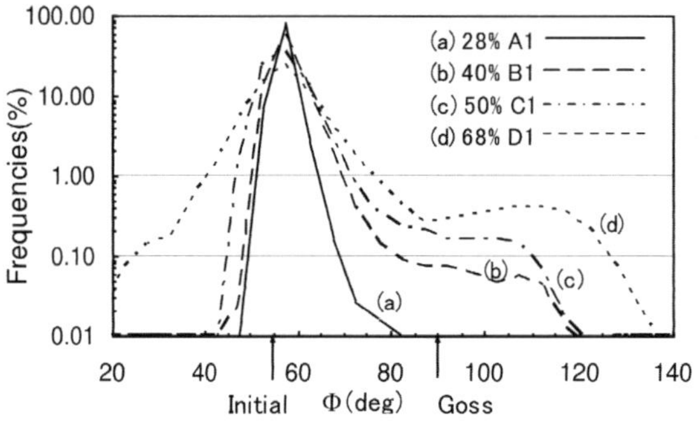

Figure 2. Orientation distribution around TD axis within 10 degrees. (a) A1 in 28%, (b) B1 in 40%, (c) C1 in 50% and (d) D1 in 68% of reduction.

3.3 Annealing texture

Fig.3 shows annealing microstructure and texture taken by EBSD. The microstructures are quality images of Kikuchi patterns [12]. Darker images indicate the area with higher density of dislocations. In 28% of reduction (Fig.3(a)), elongated recrystallized grains along SBs were observed and almost half of them were Goss grains. However, not all of SBs were occupied by recrystallized grains even after 30 minutes of annealing, and therefore, the annealing texture remained the initial orientation, (111)[11-2]. In 40% (Fig.3(b)), almost all SBs were occupied by recrystallized grains, and most of which were Goss grains. As the SBs did not cover all the area and the initial matrix still remained, the texture had two strong orientations, that is, the initial and Goss orientations

In 50% (Fig.3(c)), recrystallization occurred in the whole area and previous SB regions before annealing could be observed as inclined areas with small grains. In this case the main texture was the Goss orientation. The situation of 68% (Fig.3(d)) was similar to that of 50%. Recrystallized grains with a homogenous size were observed in the whole area and the main texture was Goss orientation. On the whole, it is recognized that recrystallized grains along SBs were mainly Goss regardless of cold rolling reduction.

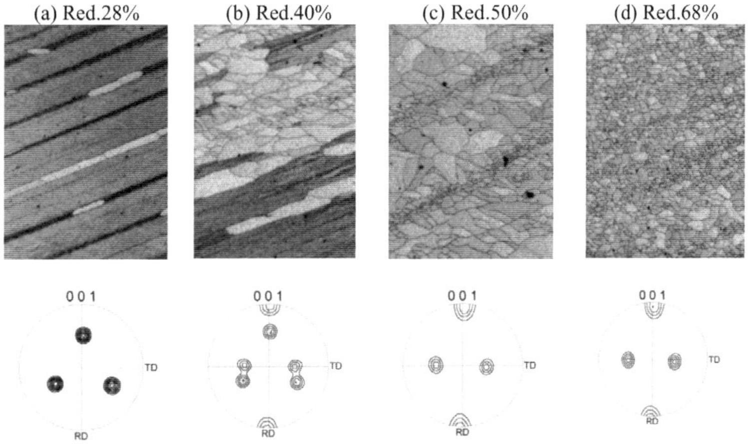

Figure 3. Annealing microstructure described by Kikuchi pattern quality and {100} pole figures taken by EBSD. Annealing was performed for 30 minutes at 923K. Size of each area is 550μm x 800μm. The contour of pole figures is 4-8-12-16-20-24-28-32.

4. Discussion

4.1 Formation of shear bands

According to a previous study [8], two types of SBs were reported after 50% reduction; the one was 35degrees and the other 17degrees to RD in the ND-RD section. The situation is similar to the case of 40% reduction in the present study, in which two types of SBs were observed. However, this is not the general case, since only one type of SB after 28% reduction was observed, and the order of SB formation will be discussed below.

Fig.4 shows the transition of SB angles before and after rolling reduction. If SBs with a thickness of h_1 are formed to h_2 after reduction, the angle of θ_1 becomes θ_2. The angle θ_2 can be easily evaluated; $\tan\theta_2 = \tan\theta_1 \times h_2^2/h_1^2$. According to this formula, the A1 with the angle of 23degrees in Fig.1(a)28% changes to 16.4degrees, which was very close to the B2 with the angle of 15 degrees in Fig.1(b)40%. The origin of B2, therefore, is speculated to be A1.

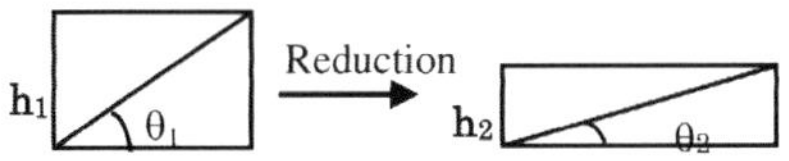

Figure 4. Change of SB angles during reduction. The angle of θ_1 becomes θ_2, in which $\tan\theta_2 = \tan\theta_1 \times h_2^2/h_1^2$.

Similarly, B1 with the angle of 31degrees in Fig.1(b) changes to 22.6 degrees at a reduction of 50%. In Fig.1(c) several kinds of SBs were observed, and one of them had an angle close to 22.6 degrees (not denoted in Fig.1(c)), which was smaller than that of C1, 26 degrees. Therefore, the order of SB formation is considered to be the first A1 (=B2), then B1, and the last C1. From this discussion, if we have both a smaller and a larger angle of SBs at the same reduction, for example B2 and B1 in Fig.1(b), the former(B2) is considered to be formed earlier than the latter(B1). Furthermore, it can be mentioned that the first type of SBs is formed at a certain reduction, and then after some reduction, the second type of SBs is formed at one time, and this process can be continued during further reduction. Therefore, it is speculated that, the SB formation continues intermittently during deformation.

Concerning a mechanism for SB formation, geometrical softening has been suggested [13,14], in which crystal rotation and Taylor factors or M values were calculated considering simple shear within SBs dependent on shear strains and SB angle to RD, as shown in Fig.5. This mechanism has been discussed also in other papers [7,8,13]. In one of the papers, however, a SB with a smaller angle (17degrees) could not be explained by geometrical softening while that with larger angles (35degrees) could be explained [8]. From the present discussion above, it is thought that the SBs with smaller angles were formed earlier and their initial angle just after formation should have been larger than 17degrees. And, if the initial angle is larger than 19.5degrees, geometrical softening can be one of the possible mechanisms for the SB formation because, in the angle from 19.5 to 45 degrees (the region of arrows in Fig.5(b)), Taylor factors became smaller as deformation progressed in Fig.5. Therefore, it can be mentioned that the present results support geometrical softening for the SB formation.

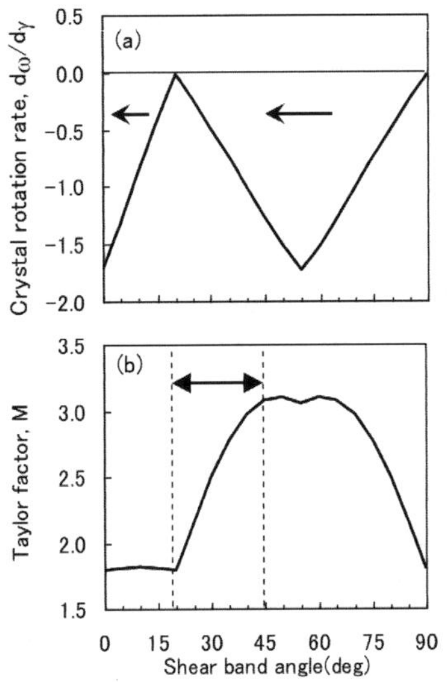

Figure 5. Crystal rotation and Taylor factors along shear bands considering simple shears. (a) The arrows denote the direction of the rotation, and (b) the region between arrows shows $dM/d\gamma<0$ (from 19.5 to 45 degrees).

4.2 Crystal rotation within shear bands after cold rolling

From Fig.2, the orientation distribution within SBs introduced in higher rolling reduction became larger than that of SBs in a lower rolling reduction. First, crystal rotation within a SB was calculated with a consideration of simple shear in SBs. In this calculation, the angle of a SB was set to be 35 degrees to RD, since the initial angle of SBs was not clearly known and the maximum angle is 35 degrees among previous studies and this study, which is the study of Ushioda and Hutchinson [8].

Calculation results are shown in Fig.6, in which rotation angles in a SB is dependent on strains of simple shear. Strains of simple shear are estimated easily if we have a straight maker before SB formation like deformation twins in Fig. 1(c). The estimated strain was 2.3 in this study and 2.9 in a previous study [8]. The rotation angle of a SB, however, was 15.5 degrees and did not reach the Goss orientation after strains of 2.3/2.9 or more.

The reason why calculated crystal rotation in a SB was 15.5 degrees is as follows; From Fig.5 (a), if our starting point is 35 degrees, it rotates in the left direction. However the rotation stops at 19.5 degrees, since the crystal rotation rate is zero at that point. Subtraction of 19.5 from 35 is just 15.5. When geometrical slip system are considered, after rotation of 15.5 degrees one of {211} slip planes is parallel to the SB angle, and therefore, crystal rotation within the SB stops.

The calculation result above was not consistent with the experimental result as shown in Fig. 2. In order to consider the discrepancy, effects of orientation fluctuation just before SB formation were estimated. Even if the matrix {111}<112> is very stable in planar strains, after reduction more than 20%, there can be some orientation fluctuations around RD and ND. Calculations of crystal rotation in SBs from initial orientations slightly different from ideal {111}<112> were performed. In these calculation results crystal rotation rate at 15.5 degrees in Fig. 5 (a) was not zero, and this means that crystal rotations within SBs continue in accordance with shear strains.

Results of rotation angle in SBs are shown in Fig. 7. With consideration of orientation fluctuations around ND axis, rotation angles increased with strains and it surpassed the Goss orientation in more than strains from 2 to 3, which are actual values of experiments. And from Fig. 7, if orientation fluctuation is larger, rotation angle progresses more largely, which means that, crystal rotations in SBs which formed at a later stage of rolling deformation can rotate in larger angles than those in SBs at an earlier stage of rolling.

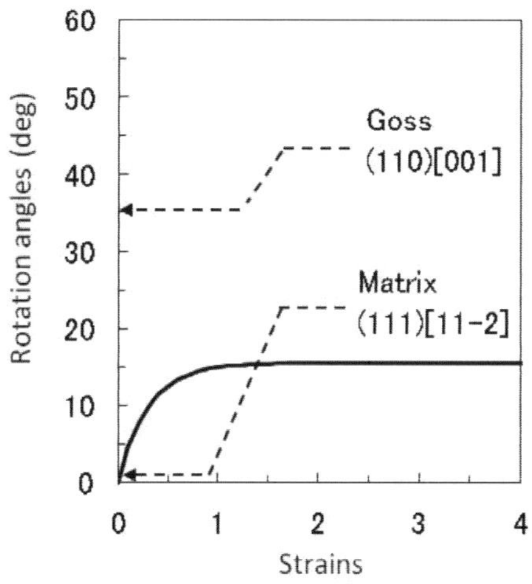

Figure 6. Calculation result of rotation angles within a shear band with 35degrees of RD dependent on strains of simple shear. The arrows denote the matrix (0deg. ; (111)[11-2]) and Goss (35deg. ; (110)[001]).

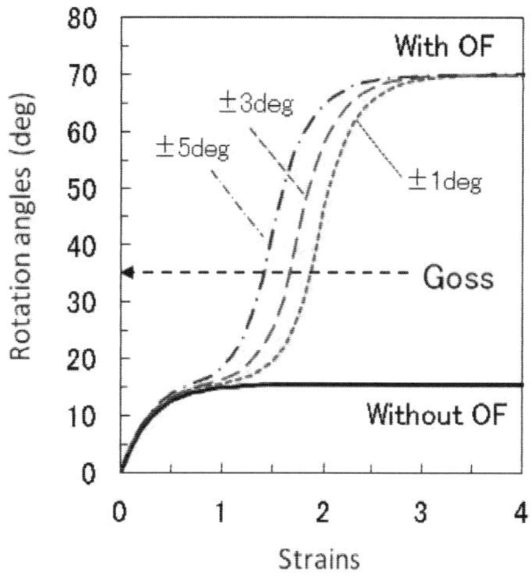

Figure 7. Calculation result of rotation angles within a shear band with and without consideration of orientation fluctuation (OF). Numbers denotes fluctuation angles from initial matrix ((111)[11-2]) around normal direction.

4.3 Formation of Goss texture during annealing

From Fig.3 recrystallized grains along SBs were Goss orientation while orientation distributions within SBs after cold rolling were widely spread and Goss was not the special orientation. In order to obtain the information of dominance of Goss grains during annealing, recovered orientation maps along SBs were obtained with EBSD.

Euler angles Φ of subgrains dependent on diameters are shown in Fig. 8. From this result, it is considered that Goss orientation becomes gradually dominant during subgrain growth while initial small subgrains do not have strong preferential orientations. Similar results were obtained by Furubayashi [6] and Tsuzaki et al. [10] though grain sizes in which Goss becomes dominant are different. In this study the width of SBs are from 2 to 5 micrometers and Goss becomes dominant during subgrain growth within SBs.

In the present study, recrystallized grains along SBs had mainly the Goss orientation and this result was similar to previous results along SBs [7,8,10] and along band structures [5,6,9]. The reason why Goss grains are dominant along SBs or band structures has been argued in these studies [5-10], however, any explanations that can describe the Goss formation well has not been obtained yet. Among the previous studies, Tuzaki et al. argued several hypotheses, and consequently, the most possible one is that Goss locates at the edge of the orientation distribution along SBs [10]. According to the hypothesis, if Goss locates at the edge of the orientation distribution, Goss subgrains can recrystallize earlier than others because of high angle boundaries to the matrix. In Fig. 2, however, orientation distribution changed dependent on rolling reduction while the annealing orientations along SBs were almost Goss texture independent of reductions, and this result contradicted the hypothesis of "Goss location in the edge of the orientation distribution along SBs".

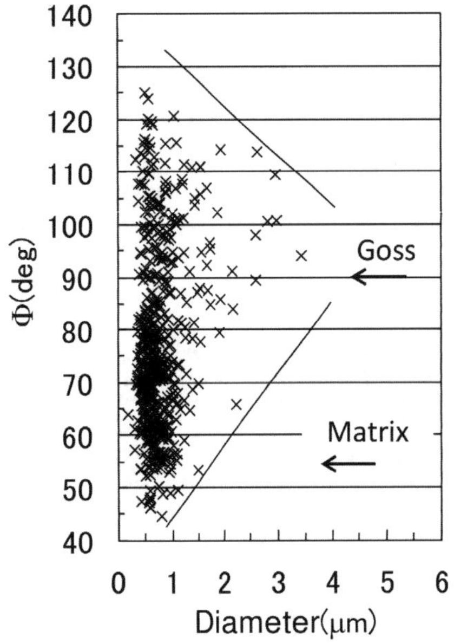

Figure 8. Orientation distributions of subgrains within shear bands C1(reduction 50%) after annealing for 10s at 823K. Transverse direction is [10-1] within 10 degrees.

In order to consider another hypothesis for the dominance of Goss, careful observation has been done in EBSD maps. Figure 9 shows the orientation and quality maps in the vicinity of Goss along a SB in a sample of 40% reduction. The green color in Fig.9(a) denotes Goss orientation and white circles in Fig.9(b) show areas including Goss which are associated with Fig.9(a). From Fig.9 (b), the circled dark areas indicate comparatively large dislocation densities [12], but Goss areas (or Goss cells) inside circles are less dark than the surrounding areas. This means that, Goss cells exist in the region with larger amounts of dislocations, but the Goss themselves have smaller amounts of dislocations than just neighboring areas, presumably due to dynamic recovery. Therefore, it is speculated that Goss cells grow earlier than other cells and they consume surrounding areas with higher stored energies. However, the above reason, why Goss cells are formed as less deformed sites than other areas, has not been elucidated yet. The reason should be investigated further in terms of orientation stabilities of Goss grains during rolling and direct observation of dislocations around Goss cells.

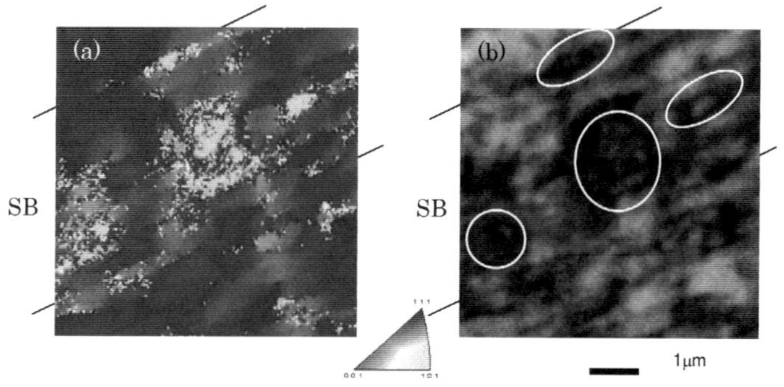

Figure 9. The orientation and quality maps of EBSD. (a) The ND orientation and (b) the quality map of sample after 40% of reduction (white circles shows Goss orientations).

5. Summary

In order to clarify the mechanism of Goss texture formation along shear bands, the effect of cold rolling reduction on shear band formation and crystal orientation within shear bands and annealing texture were investigated in Fe-3%Si {111}<112> single crystals.

Shear bands were formed differently depending on rolling reduction; In 28% reduction, only one type of shear bands were observed, but in 40% two types and in more than 50% several types were formed. From estimation of the formation angle to the rolling direction, shear bands with smaller angles were formed earlier and those with larger angles were formed later. Thus, shear bands were formed intermittently during cold rolling. This result supports geometrical softening as a mechanism for shear band formation.

Crystal rotation along shear bands around TD//[1-10] progressed in accordance with rolling reduction; the orientation distribution from the initial matrix became large in proportion to the reduction and even exceeded Goss orientation when the rolling reduction became larger than 40%. Crystal rotation was calculated on the basis of simple shear deformation and in the initial matrix, (111)[11-2], rotation angle within shear bands did not reach the Goss orientation. With consideration of orientation fluctuation, however, the rotation angle progressed more and exceeded the Goss orientation. It means that crystal rotations in shear bands which formed at a later stage of deformation can rotate to larger angles than those in SBs at an earlier stage.

After annealing recrystallized grains along shear bands were mainly Goss grains regardless of cold rolling reduction. From analysis of recovered specimens Goss grains became dominant during subgrain growth within shear bands. The speculated reason for the dominance of Goss during annealing is that Goss cells have a comparatively lower density of dislocations and are surrounded by largely deformed areas. The reason why Goss cells are formed in less deformed than other areas along shear bands should be investigated further.

References

[1] Abe M, Kokabu Y, Hayashi Y and Hayami S 1982 *Trans. Jap. Inst. Met.* **23** 718
[2] Ushioda K, Nakanishi S, Morikawa T, Higashida K, Suwa Y and Murakami K 2013 *Mat. Sci. Forum* **753** 58
[3] Dunn C G 1954 *Acta Met. 2* 173
[4] Hu H 1960 *Acta Met.* **8** 124
[5] Taoka E, Furubayashi E and Takeuchi S 1968 *Trans. Nat. Res. Ins. Met.* **9** 155
[6] Furubayashi E 1969 *Trans. Iron Steel Inst. Jpn.* **9** 222
[7] Haratani T, Huchinson W B, Dillamore I L and Bate P 1984 *Met. Sci.* **18** 57
[8] Ushioda K and Hutchinson W B 1989 *Iron Steel Inst. Jpn. inter.* **29** 862

[9] Furubayashi E 1998 *Iron Steel Inst. Jpn. Inter.* **38** 510
[10] Tsuzaki K, Kobayashi H, Tanaka H and Maki T 1999 *Iron Steel Inst. Jpn.* 40
[11] Bacroix B, Hu Z 1995 *Met. Mater. Trans. A* **26A** 601
[12] Wright S I and Nowell M M 2006 *Microscopy and microanalysis* **12** 72
[13] Dillamore I L, Roberts J G and Bushi A C 1979 *Met. Sci.*, **13** 73
[14] Huchinson W B 2007 *Mat. Sci. Forum* **558-559** 13

IOP Conf. Series: Materials Science and Engineering **89** (2015) 012010 doi:10.1088/1757-899X/89/1/012010

Crystallographic aspects of nucleation and growth during primary recrystallization in stable single crystals of Al and Al-1%Mn alloy

H Paul[1], M M Miszczyk[1] and J H Driver[2]

[1] Polish Academy of Sciences, Krakow, PL

[2] Ecole des Mines de Saint-Etienne, Saint Etienne, FR

Email: h.paul@imim.pl

Abstract. The early stages of recrystallization have been characterized in stable, single crystals of Al and an Al-1%wt. Mn alloy to rigorously quantify the orientation relations between nuclei and simple deformation structures. Goss{110}<001> and brass{110}<112> oriented samples were deformed in a channel-die to develop a homogeneous structure composed of two sets of symmetrical microbands and then lightly annealed. SEM/EBSD analyses demonstrate a strong relation between as-deformed orientations and the limited number of recrystallized grain orientations. The misorientation angles across the recrystallization front are mostly grouped in the ranges of 25-55° around axes located near, but not at, the normals of all four {111} planes. In some cases the orientation of the growing grain transforms through the formation of a first generation twin with the twinning plane normal lying near the rotation axis. The intensity of new grain nucleation in near-surface areas is significantly greater than those observed in central parts of the sample. Many new grains grow from the surface into the sample centre mostly along the {111} planes. However, irrespective of the section new grains possess the same type of orientation relationship with respect to the deformed state.

1. Introduction

The orientations of recrystallized grains are considered to 'originate' from the orientations of the deformed matrix regions from which they grow, e.g. [1, 2]. Particles [3-5], strain heterogeneities [6-10], original grain boundaries (including deformation twin boundaries) [8], etc. are the privileged places of nucleation. However, in the absence of these well-known nucleation sites, recrystallized grains have been shown to appear extensively in the deformed crystals of stable orientations, where high angle boundaries occur only accidentally [11-13]. In these simple cases the conventional mechanism of primary nucleation, based on the presence of pre-existing nuclei in the deformed structure possessing a high angle boundary, cannot explain texture change during recrystallization.

According to the oriented growth theory the growth potential of a given nucleus depends on the orientation difference at the migrating recrystallization front [14] and the fastest growing nuclei contribute to the final recrystallization texture. (However, it is still an open issue to define the most advantageous disorientation across the migrating recrystallization front that is favourable to rapid growth). This assumption leads to the conclusion that in a uniform deformation microstructure with only one misorientation relation, almost equiaxed grains should be formed during recrystallization. However, in the early stages of recrystallization of grains without (significant) deformation gradients a considerable number of recrystallized grains exhibit a strongly elongated shape [11]. This anisotropic growth suggests that some directions or planes are more privileged for growth, despite a near-uniform disorientation relation in all directions. This leads to the important conclusion *that it may not be the disorientation across the recrystallization front but probably the type of migrating grain boundary, e.g. their twist or tilt character, which is the most important factor controlling migration velocity,* as discussed several times in the past, e.g. [1, 2, 4]. Since the movement of pure screw or edge dislocation groups along specific planes leads to a twist or tilt boundary [4], respectively, identification of the boundary type across the recrystallization front is crucial for a description of the dislocation mechanisms responsible for grain growth. Identifying the boundary character requires an exact knowledge of the 3D spatial situation of the boundaries between the new grain and as-deformed neighbourhood, together with the disorientation axis with respect to the boundary plane. This is possible by detailed inspection of three perpendicular sections, e.g. ND/ED and ND/TD planes, where ND, ED and TD are the normal, extension and transverse directions, respectively.

Moreover, there are also other aspects of the phenomenon such as the *influence of the free surface* on the morphological and crystallographic aspects of recrystallization and the role of recrystallization twinning in texture transformation even in high SFE metals. The intensity of recrystallization in near-surface areas is significantly greater than that observed in central parts of the sample and new grains grow from the surface to the sample centre. But it is not clear if the orientation of grains nucleating near the surface and in the centre are similar or not. This leads to the question – is it possible to compare the textures obtained in recrystallization experiments of bulk samples with those obtained during in-situ heating in SEM or TEM, where the influence of free surface is evident. As regards recrystallization twinning in fcc metals an important question is whether all boundaries showing twin-oriented relationship (60°<111>) are in fact twin type boundaries, i.e. does the boundary between the twin-related areas coincide with a {111} plane?

In order to clarify those problems, it is advisable to perform experimental research under conditions where the number of 'free parameters' affecting this transformation is minimized. Therefore, the current experimental investigations have focused on an analysis of transformations which occur in model single crystals, with stable orientations, of a face centered cubic (fcc) metal (Al and Al-1%Mn). This approach to the problem makes its description simpler and clearer since the stability of the orientation during deformation enables one to precisely define the orientation relationship which appears when the 'primary nucleus' is formed. Another important aspect is that the deformation of stable orientation crystallites (up to large strains) practically does not lead to the appearance of high-angle boundaries. In the absence of high-angle boundaries, the conventional mechanism of primary nucleation, based on the presence of pre-existing nuclei in the deformed structure, cannot be accepted as valid for explaining a texture change during recrystallization. The basic research technique employed local orientation measurements based on scanning electron microscopy (SEM/EBSD). Since the 'tilt or twist' nature of a grain boundary can only be precisely defined by measurements in three sections the disorientation characterization across a recrystallization front is complemented by a spatial description of boundaries of grains showing anisotropic growth.

2. Experimental

The single crystal bars of high purity Al and Al-1%wt.Mn alloys were grown by directional solidification. The Goss (110)[00-1] and brass (110)[1-1-2] oriented samples were carefully cut from the bars using a wire saw to dimensions of 10 x 10 x 10 mm^3 (height x width x length). The deformation was carried out by plane strain compression (PSC) in a channel-die. Figure 1 shows the configuration of the most active slip systems for ideal Goss and brass orientations. The multi-stage test was performed at 293 K at the initial strain rate of 0.02 s^{-1} to a logarithmic strain of 0.51 ($\varepsilon = ln(h_{initial}/h_{final})$). In order to limit friction between the sample, the punch and the walls of the channel-die, each sample was wrapped in 0.2 mm thick PTFE films. To investigate the interrelation between as-deformed and recrystallization textures, the deformed samples were annealed to the first stages of primary recrystallization, typically in the range of 400 - 700 K for less than 60 s. Annealing was carried out in an air furnace. All specimens were water quenched post-annealing. The new grain orientations were mostly analysed on the ND/ED plane in the centre section of the annealed sample. However, for a 3D analysis of the grain boundaries additional orientation maps were made on two other sections, i.e. ND/TD and ED/TD. The influence of the section depth was investigated on the free surface as well as on ¼ and ½ sample thickness. Prior to annealing all the samples were mechanically ground on all faces to 4000 grid SiC paper and electropolished. The samples were characterized by scanning electron microscopy (SEM) using a JEOL 6500F or Zeiss Supra 55VP. The SEM's were equipped with an electron backscattered diffraction (EBSD) system, working in automatic beam scanning mode.

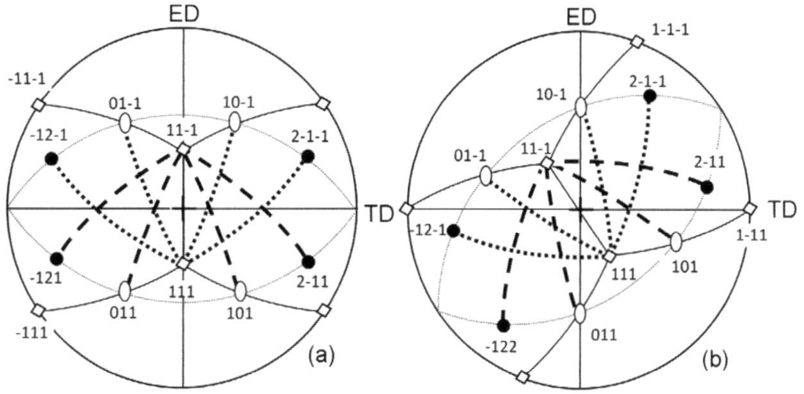

Figure 1. Stereographic projections showing configuration of the most active slip systems in ideal (a) Goss (110)[00-1] and (b) brass (110)[1-1-2] orientations [12].

3. Results and discussion

3.1. Deformed state

3.1.1. Microband structure and orientation stability. At a logarithmic strain of 0.51, the microstructures of Goss and brass oriented crystals consist of two complementary sets of elongated microbands delineated by dislocation walls, as presented in Fig. 2 for Al and Al-1%Mn alloy of Goss orientation and also discussed many times in the past, e.g. [15-19]. Based on SEM/EBSD measurements, rectangular cells were formed as a result of microband intersections. Both sets of dislocation walls were inclined very close to the expected traces of the active {111} slip planes; they were always within the scatter limit of normals to the {111} planes. The orientations of areas corresponding to the above microstructures confirm the stability of the initial orientations at logarithmic strains of 0.51. The samples revealed a relatively small orientation spread, as shown for Al-1%Mn alloy single crystals of brass orientation (Fig. 3), even if the initial orientation deviated a few degrees from the ideal position. However, in the case of initial samples showing 'larger deviation' from ideal positions cyclical orientation changes were superimposed on a strong, continuous lattice rotation. The majority of adjacent microbands displayed opposite rotation senses, as reported earlier for an Al-0.3%Mn alloy, e.g. [20].

3.1.2. The as-deformed rotation axes and angles. The distribution of disorientation axes between adjacent pixels of mapped areas of the deformed single crystals is visualized by stereographic projections for the ED/TD reference system, as presented in Fig. 4a and b for Al single crystals of Goss and brass orientations, respectively. The maximum density of disorientation axes occurred in the vicinity of TD but with some systematic variations. In the case of *Goss-oriented crystals* (for which the intersection of the active (111) and (11-1) planes coincides with TD), there is a small spread away from TD so that the most frequent axes were near the closest <112> axes to TD. Moreover, the spread is sufficiently wide to observe disorientation axes near [-221]. In the case of *brass-oriented crystals*, the maximum disorientation axis is located between TD and the intersection trace of the active (111) and (11-1) planes. The disorientation axes display a systematic departure from TD with a wide spectrum of disorientation axes between [-11-1] (||TD), [-22-1] (location of maximum density), [-110] (intersection trace of slip planes). For both orientations the distributions of the disorientation angles between particular pixels of the orientation maps were significantly below 10° (see Fig. 3).

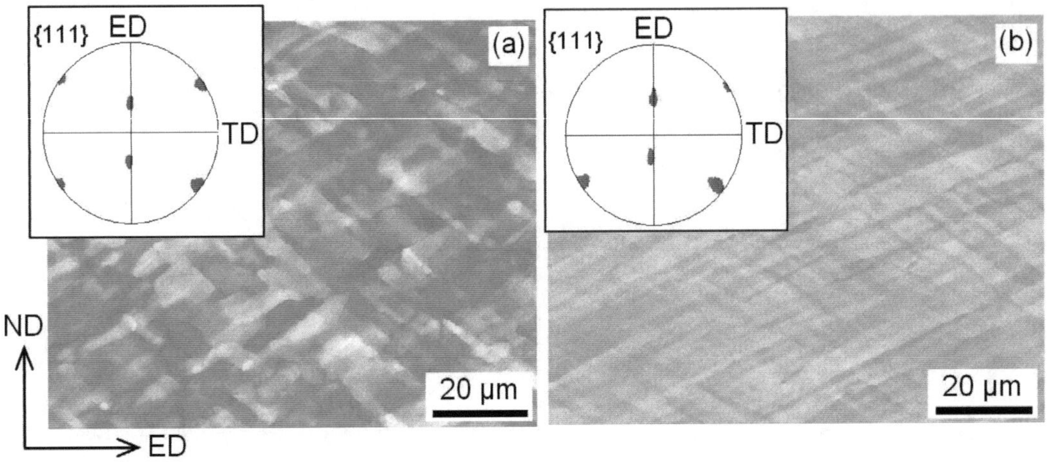

Figure 2. Microstructures and corresponding {111} pole figures observed in Goss oriented single crystals of (a) Al and (b) Al-1%Mn alloy. SEM/EBSD local orientation measurements in ND/ED section with a step size of 100nm. IPF colour code was applied [12].

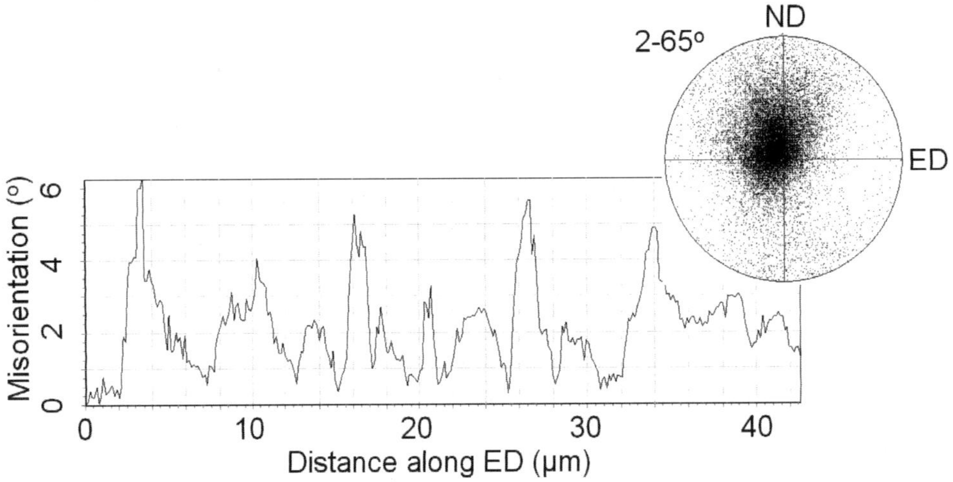

Figure 3. Misorientation line scan along ED and misorientation axes distribution in misorientation relation between neighbouring pixels of the deformed state. Al-1%wt.Mn alloy single crystal of brass orientation deformed to a strain of 0.51.

3.2. Early stages of recrystallization

After short annealing only discontinuous recrystallization was observed and these partially recrystallized structures give rise to new orientations against a background of well-defined microtexture components of the as-deformed state. The nucleation characteristics observed at the SEM scale (in the middle of the bulk sample) turn out to be the same for both metals and orientations. The new grains were uniformly distributed in the deformed matrix. Only a small fraction of the grains was twinned. The microstructures were composed predominantly of single, isolated grains that show an elongated shape and some quantity of grains forming compact chains. The longer axes of some elongated grains in the longitudinal section were closely parallel to the traces of the active {111} slip planes.

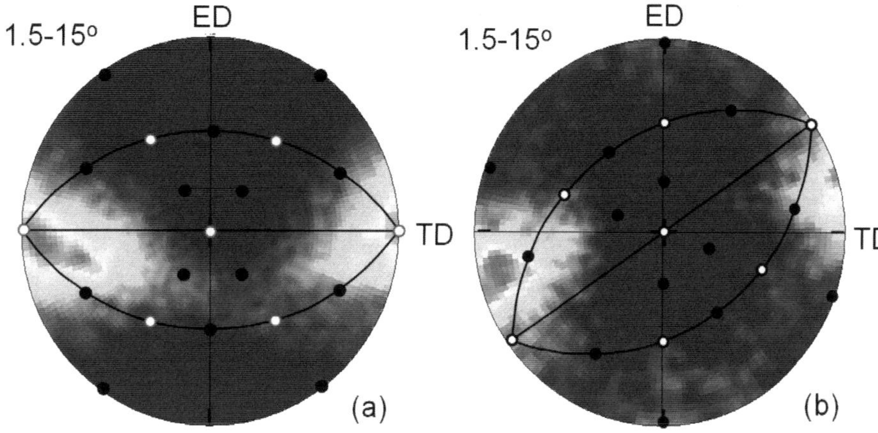

Figure 4. The distributions of misorientations in the as-deformed state. Stereographic projections showing the <112> and <110> directions for Al of (a) Goss and (b) brass orientations. White and black circles marked positions of <011> and <112> directions, respectively.

In the following sections the different aspects of the early stages of recrystallization will be discussed. They include: (i) a description of preferential grouping of new grain orientations, (ii) a determination of the grain boundary type between the as-deformed areas and new grains (analysis of anisotropic grain growth on three perpendicular sections), (iii) recrystallization twinning during the early stages of recrystallization, and, (iv) the influence of the free surface on the intensity and crystallography of recrystallization.

3.2.1. The orientations of recrystallized grains. Orientations of single isolated nuclei/grain. As can be seen in Fig. 5 the orientations of recrystallized grains are different from those of the deformed state. A general rule is that the <111> poles of the deformed crystal do not coincide with the poles of the recrystallized grains, or more precisely coincide very rarely [11]. Despite the scatter of the recrystallized grain orientations, it can be claimed with certainty that they are not random. Only a finite number of groups of recrystallized grain orientations were observed, as presented in Fig. 5 for Al-1%Mn single crystals of Goss and brass orientations.

In the case of both orientations the 'texture image' of the first recrystallized grains is quite symmetrical with respect to the external coordinate directions. A primary nucleus of homogeneous orientation is related to the adjacent deformation orientation by a rotation around an axis located near the pole of one of the four {111} planes of its deformed environment. It was noticed that if the formation of a grain orientation occurs by a positive rotation around a certain axis in a given region, grains characterized by a negative rotation appear in neighbouring regions. (Note that if those grains meet a pseudo-twin boundary is formed, i.e. the grains possess a twin orientation relationship but the common {111} plane does not coincide with the boundary plane). This phenomenon of opposite local rotations is observed clearly when recrystallization twins occur and the twinning plane normal is placed near the disorientation axes between each part of the new grain and its nearest as-deformed neighbourhood (Fig. 6).

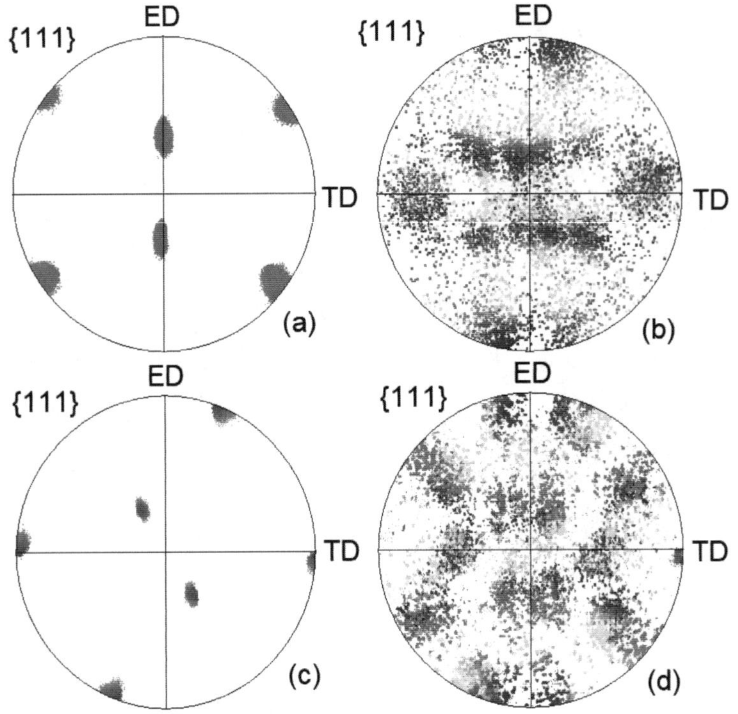

Figure 5. {111} pole figures showing the as-deformed (a,c) and recrystallized (b,d) grain orientations in samples of Goss (a,b) and brass (c,d) orientations for the Al-1%Mn alloy. Recrystallization anneals: Goss – 618 K/25 s, brass – 696 K/25 s [12].

First recrystallization twins. Once a nucleus appears, it can transform through the formation of a first generation recrystallization twin. Annealing twins are frequently observed in fcc metals of medium-low SFE. Little significance was attached to annealing twins in Al and its alloys until the publication of results in Göttingen based on the application of high voltage TEM to recrystallization [21] produced evidence of extensive twin formation during the early stages of recrystallization of several fcc metals including pure Al. The orientation maps made for both metals and orientations indicate that some of the new grains undergo twinning. However, the quantity of recrystallization twins strongly depends on the analysed surface; the intensity is the highest on the free sample surface and significantly decreases as the sample centre approaches [22]. The most frequent situation occurs when the twinning plane normal is situated near the rotation axis (near of all the four {111} planes).

It was also observed that the normals of the twinning planes lie among the rotation axes of the disorientation relations between the recrystallized grains and their as-deformed neighbourhood. This situation is shown in Fig. 6. The area described as R1 is rotated by an angle of about 30-35°, whereas the region of R2 represents a 25-30° rotation of opposite sense around axes lying near the [11-1] pole of the deformed state. Since both twin-related regions represent nearly the same disorientations with respect to the as-deformed neighbourhood it is difficult to indicate which region was formed first. Another case of twinning occurs if the normal to the twinning plane is situated well away from the rotation axis, i.e. the twinning in the grain occurs on a plane other than that of the <111> pole lying near the rotation axis, as observed earlier for medium to low SFE metals [11]. This usually leads to different disorientations of both recrystallized areas with respect to the deformed state. Nevertheless it is not clear at the moment if recrystallization twining starts when normal grain growth is stagnated or if twinned areas appear simultaneously and they grow together, or whether both these mechanisms are valid.

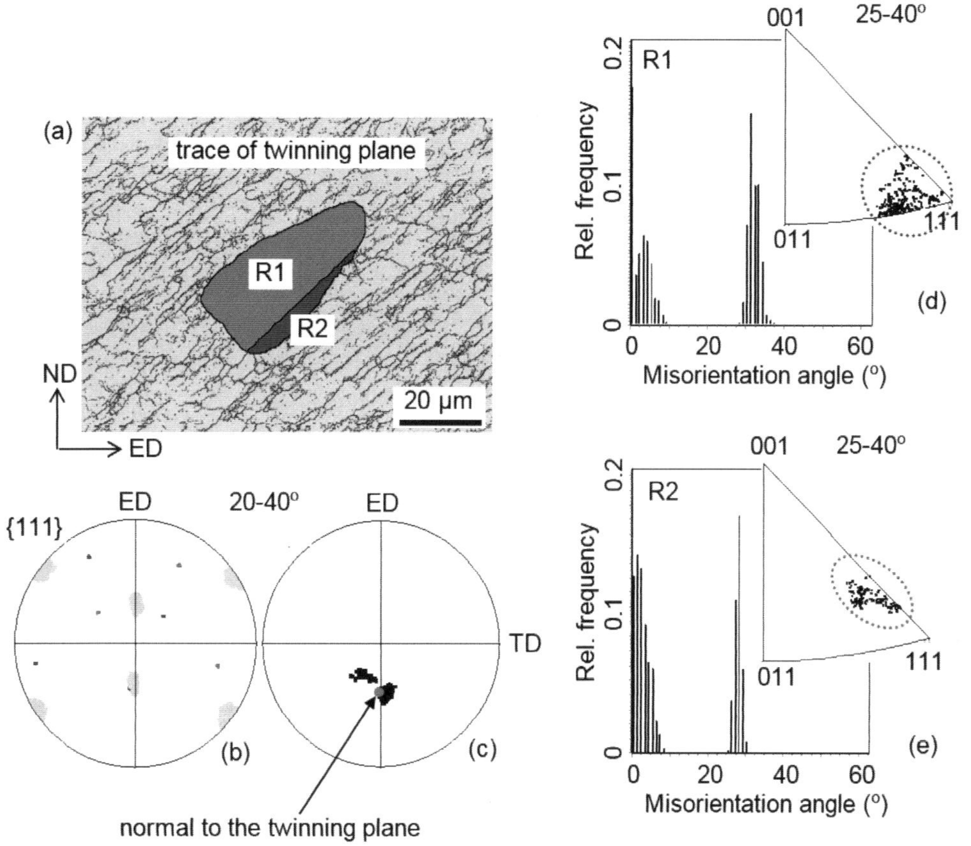

Figure 6. Recrystallization twinning in pure Al. (a) SEM/EBSD orientation map and (b) corresponding {111} pole figure showing new grain and as-deformed state orientations. (c) Misorientation axis (in sample coordinates) distribution across the recrystallization front. Misorientation angle and axis distribution (presented in crystallite coordinates) for (d) R1 and (e) R2 areas. Sample recrystallized at 618 K for 25 s.

'Pseudo-twin' boundaries between growing grains. The two cases described above concern the twin boundaries that represent the 'ideal twin disorientation' (~60°<111>) or the boundaries that were only slightly deviated (<2°) from this ideal relation. However, the SEM/EBSD system detects boundaries, as twin boundaries, if they satisfy the Brandon criterion [23]. It is in fact possible to find boundaries with disorientation angles and axes more deviated from the ideal position, but within the limit of the Brandon criterion. Most of these boundaries are 'pseudo-twin boundaries'; they appear between the growing grains. Very often it was observed that if, within a given area, a new grain forms as a result of a positive ~25-35° rotation around a near <111> axis, in the neighbouring area a new grain appears as result of a negative ~25-35° rotation around an axis close to the previous one. The combination of these rotations leads to the formation of a grain boundary between impinging grains

that fulfils Brandon's criterion for twin boundaries. But their boundary is not coherent since their common {111} plane does not coincide with the boundary plane. For a precise identification of coherent twin boundaries the position of the common the {111} plane is needed.

A new algorithm to identify (coherent) twin boundaries was applied. In the first step the grain boundaries with the twin orientation relationship, i.e. 60°<111>, were found on the orientation maps using Brandon's criterion (for the maximum deviation from the ideal axis and angle positions) [23]. In the second step the boundary direction was determined. Then for each boundary (line) with a twin orientation relationship the common {111} plane was identified; their normal (in fact their projection on the analysed plane) was marked by black arrows. Using this method, it is possible to reject boundaries which are clearly incoherent. But, it is important to note that (on the basis of two dimensional orientation maps) this procedure cannot absolutely guarantee a twin boundary. Ambiguous cases occur if the intersection of the common {111} plane and the boundary plane lies on the analysed section. Figure 7a shows a grain boundary map with marked boundaries between areas showing twin type orientation relationship (red lines) and coherent twin boundaries (yellow lines). The number of neighbouring grains showing twin-type orientation relationships was higher in Goss-oriented samples then in brass ones (Fig. 7b). Moreover, in both crystal orientations the intensity of twinning was stronger in near-surface volumes then in the sample centre.

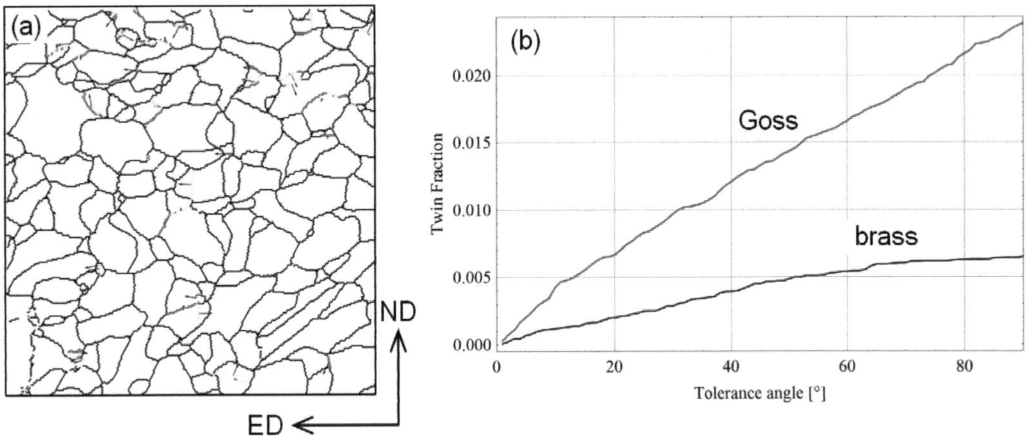

Figure 7. (a) Grain boundary map with marked boundaries between areas showing a twin orientation relationship (in red) and coherent twin boundaries (in yellow), where the normal to a common {111} plane is perpendicular to the boundary, with the 20° tolerance of the orthogonality (b). Fraction of twin boundaries on the orientation map in relation to tolerance angle of the orthogonality grain boundary direction and the direction of projection of <111> direction. Al-1%Mn alloy recrystallized at 680 K for 70 s and at 723 K for 565 s of Goss and brass orientations, respectively.

3.2.2. Disorientation axes distributions. Figures 8 show the disorientation axes distributions in the ED/TD sample coordinate system. The disorientations were calculated separately for the neighbouring pixels of the deformed/recovered areas (Fig. 8a) and across the recrystallization front, i.e. between the neighbouring pixels of new grains and the nearest deformed/recovered neighbourhood (Fig. 8b). A comparison of the preferred disorientation axes situation shows that there is a lack of correspondence between the distributions for the deformed crystal and the one identified across the recrystallization front. This leads to the important conclusion that the mechanisms responsible for crystal lattice rotations during deformation do not exactly match those controlling the formation of the recrystallized grains.

Recovered state. The disorientation axes between particular pixels of the recovered state were generally similar to those observed in the deformed state. Most of the disorientation axes coincide with <111> and <011> directions for the brass and Goss oriented samples, respectively. However, a discernible scattering towards some other axes (also near TD) was also observed, as discussed earlier for the deformed state.

Disorientation across the recrystallization front. For the orientation relationships of boundaries between isolated single grains of homogeneous orientation (surrounded by high-angle boundaries) and deformed/recovered regions the numbers of disorientation angles are presented in Fig. 9. The range of potential disorientation angles was binned by intervals of 1.25° and the number of disorientations across the recrystallization front corresponding to a given angular range was counted. The most populated disorientation angles in the disorientation relation between the new grain and the deformed areas varied between 25° and 55° with a wide maximum between 30° and 45°. Most of the single, isolated, nuclei are misoriented by 25-30° with respect to the deformed state, so that the special role of near-40° disorientations is not confirmed, in contradiction to much previous literature, e.g. [1, 2, 24, 25].

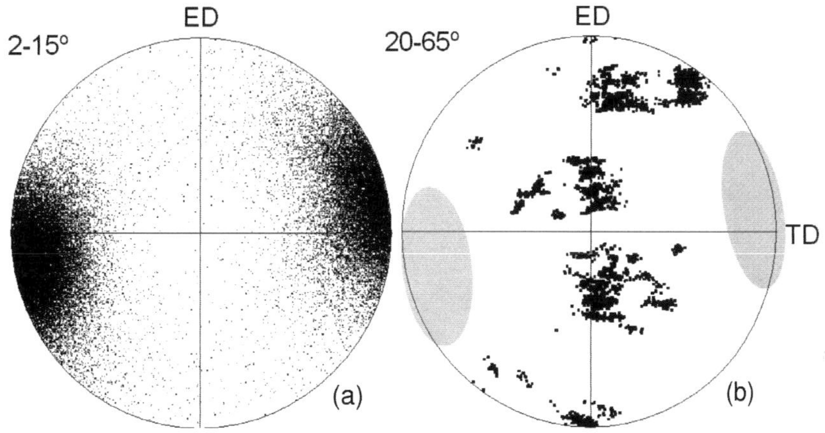

Figure 8. Interrelation between the distribution of misorientation axes at early stages of recrystallization observed for the Al-1%Mn alloy with the brass (110)[1-1-2] orientation. (a) The distribution of misorientation axes for the deformed/recovered state between particular pixels of the orientation map and (b) the misorientation axes between deformed/recovered and recrystallized areas. Sample recrystallized at 680 K for 25 s.

It is evident that the formation of new grain orientations, which is not random, leads to some preferences in the distribution of the disorientation axes between the deformed/recovered and recrystallized areas. For both the Goss and brass orientations (Fig. 9), a definite majority of disorientations axes are located around the poles of the (111) and (1-1-1) planes, i.e. those of the active slip systems during deformation. Nevertheless, the misorientation axes do not coincide exactly with the poles of these planes. Distributions of misorientation axes show that only a relatively small fraction of the disorientation axes coincide with poles of the {111} planes (Fig. 10). This means that the orientation relationship, usually described in the literature by *α<111>* relations, is in fact very rarely observed. The most populated disorientation axes were usually shifted by about 20°-25° towards the <221>, <012>, <123> or <112> directions, lying close to selected normals of the {111} slip planes. This shift was significantly greater than the orientation spread typically observed inside the deformed state (usually below 10°). It should be mentioned that in some rare cases, disorientation axes were found quite distant from the {111} plane poles of the active systems (including the vicinity of the two remaining <111>poles).

3.2.3. The directionality of grain growth. Analysis in the ND/ED section. The orientation maps measured on the ND-ED plane for both metals and orientations show that a considerable number of recrystallized grains are of elongated shape. The straight high angle grain boundaries were observed along traces of {111} planes. These grains exhibit a well-defined lattice rotation with respect to the deformed/recovered areas. An example of the crystal lattice rotation, related to the appearance of heavily elongated grains, was carried out on Al-1%Mn alloy single crystals of Goss{110}<001> orientation (Fig. 11). To a first approximation, this aspect of nucleation and grain growth leads to the following conclusions:

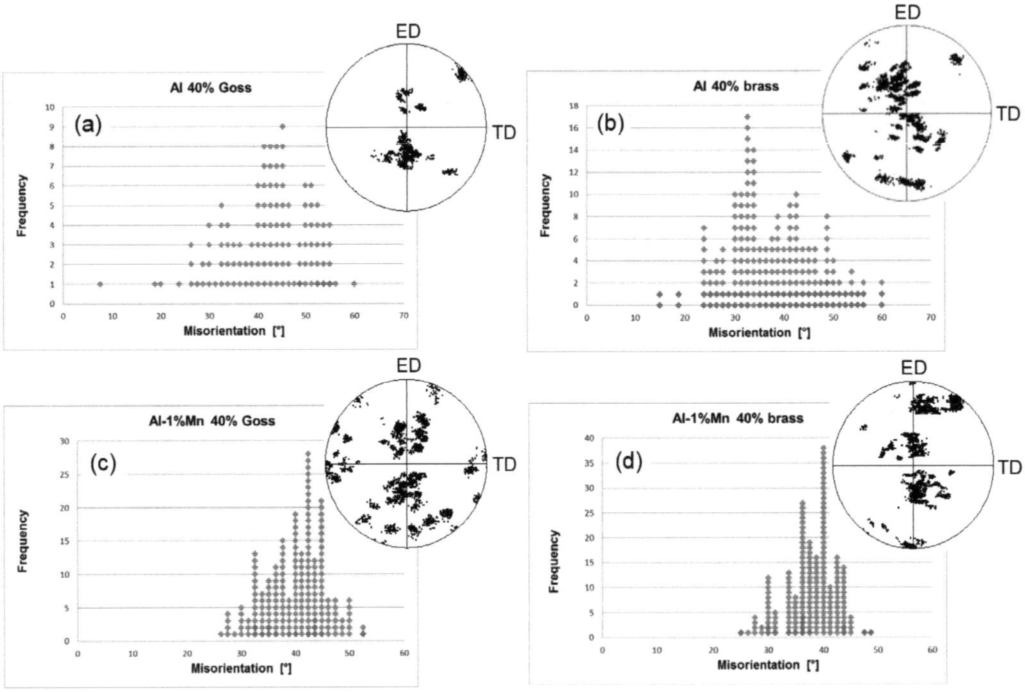

Figure 9. Misorientation relation at early stages of recrystallization between recrystallized area and nearest as-deformed neighbourhood. Distribution of misorientation angles (plots uses 1.25° bins) and axes for (a, b) Al and (c, d) Al-1%Mn with (a, c) Goss) and (b, d) brass orientations. Red points marked misorientation relation for twinned areas. Recrystallization anneals: (a) - 618 K/25 s, (b) – 696 K/25 s, (c) - 686 K/25 s and (d) - 696 K/30 s.

- The longest grain direction runs mostly along the traces of (111) and (11-1) planes, i.e. the most active planes during strain, and which are symmetrically inclined to the ND-TD plane.
- The orientations of the elongated grains are related to those of the group of the deformed crystals. The crystal lattice of each of the grain groups is rotated (positively or negatively) around diverse groups of disorientation axes.
- The rotation axes are situated near the poles of (111) and (11-1) planes. This may suggest a mechanism for the formation of a new grain by thermally activation movement of dislocation families, 'travelling' in pairs on one of the above-mentioned planes, as proposed in [11]. Although the rotation axes approach the normal vector of the preferred slip planes they only rarely coincide with the exact location of the <111> directions.

- The crystal lattice rotation of elongated grains takes place around axes near the <111> pole, corresponding to the {111} plane across which preferred grain growth occurs. This is shown by stereographic projections of the rotation axis distributions between deformed (recovered) and recrystallized regions. The crystal lattice of the grains elongated along a (111) plane undergoes rotation around axes near the [111] direction. By analogy, the crystal lattice of the grains elongated along the (11-1) plane undergoes a rotation around the near [11-1] direction.

3.2.4. 'Tilt' or 'twist' grain boundaries. Analysis on three perpendicular sections. The anisotropic growth of new grains in a relatively homogeneous structure of the deformed state, as observed in the middle of the longitudinal section of Goss{110}<001> oriented samples, enables one to elucidate the mechanisms responsible for rapid grain growth. However, it is crucial to determine the grain boundary plane with respect to the rotation axis vector [1]. This opens up the possibility of determining the 'twist' or 'tilt' character of the grain boundary. But the tilt or twist nature of a grain boundary formed between new grains and the as-deformed state can only be precisely defined by SEM/EBSD measurements in three perpendicular sections.

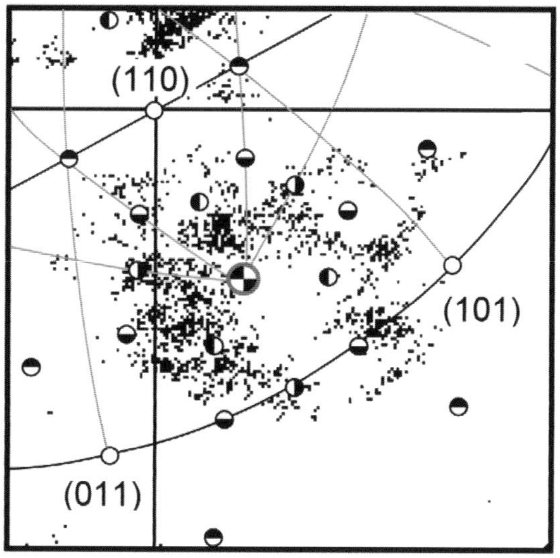

Figure 10. The scattering of misorientation axes around the [111] pole for Al-1%Mn of brass orientation. Detail from Fig. 9d.

The orientation maps in the ND/TD and ED/TD sections confirm that the majority of grains are characterized by a strongly elongated shape. The longer axes of grains run horizontally or vertically, along the traces of the {111} planes in both sections. In the case of the ND/TD section the traces of the highly active slip planes are situated horizontally, whereas the less active are vertical (Fig. 12a). In the case of the ED-TD section the traces of the highly active slip planes are situated vertically, whereas the less active are inclined at ~35° to ED (Fig. 12b). In all sections the elongated grains of both families are growing from the surface into the deformed structure. It is clear that the traces of elongated grain boundaries are not random and coincide with traces of the {111} planes. Therefore, it is concluded that the majority of flat grains are situated along {111} planes (Fig. 12c).

Based on the results of 3D analysis it is concluded that the long straight boundaries perpendicular to the <111> axis (which is also the rotation axis in the disorientation relation between as-deformed state and new grains) were twist boundaries. This idea, based on the coordinated movement of dislocations at the head of the recrystallization front leading to twist boundary formation with a disorientation axis parallel to the [111] direction, is linked to the explanation of Ridha and Hutchinson

[26] for the rotations around <111> axes. Consequently, the boundaries at the head of the growing grain were tilt boundaries. Since grain growth is due to rapid motion of the tilt boundary at the head of a growing grain this description is consistent with the results of Kohara et al. [27] and Parthasarathi and Beck [28] who observed that tilt boundaries move faster than twist boundaries in fcc metals.

3.3. The influence of the free surface on the morphology and crystallography of new grains

The fraction of recrystallization strongly depends on the sample section location. It is clear that for the applied recrystallization anneal the surface is almost completely covered by new grains, whereas in the ¼ and ½ sample sections the density of new grains significantly decreases. The shape of the recrystallized grains on the surface is nearly equiaxed and the average grain size is smaller than that observed on the other two sections. Moreover, the large grains in the ¼ and ½ sample sections reveal a strong tendency to grow within volumes of slightly larger lattice disturbances.

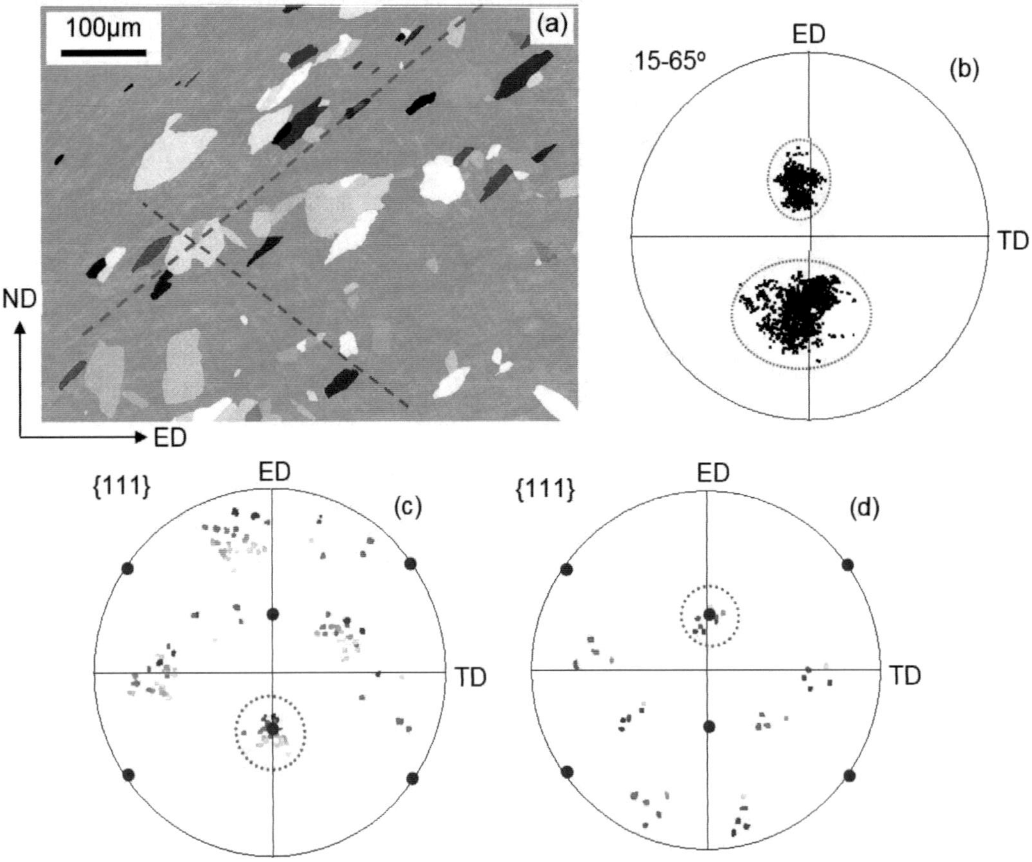

Figure 11. 'Directionality' of grain growth at the early stages of recrystallization observed in Al single crystal of Goss orientation. (a) Orientation map. (b) The distribution of the misorientation axes between deformed/recovered and recrystallized areas. (c) and (d) {111} pole figures showing the orientations of recrystallized grains growing along traces of the {111} planes marked as red and blue, respectively.

The observed positive or negative ~30-35° rotations of a new grain orientation around axes lying close to the normal of the most active {111} planes lead to four variants of near S{123}<634> components. Irrespective of the surface the pole figures of new grains revealed a quite symmetrical character. Since nearly the same groups of new grain orientations were detected on each section, (Fig. 13), this led to the same type of disorientation with respect to the components of the as-deformed state. The positive and negative rotations around axes situated near the normals of the most active {111} planes are able to explain the strong cube {100}<001> texture component formation during recrystallization of many Al alloys. The <111> poles of four variants of S orientation 'marked' the position of poles characteristic of the cube orientation. (Four variants of S orientation are formed as a result of positive and negative ~ 30° rotations of the Goss orientation around axes near the <112> direction). The orientations maps do not show cube-oriented grains (in the 20° scattering from ideal position) or more precisely they are detected very rarely. However, a strong cube texture is observed on pole figures especially those presented in 'continuous' form. This phenomenon is discussed in more detail in another work [29].

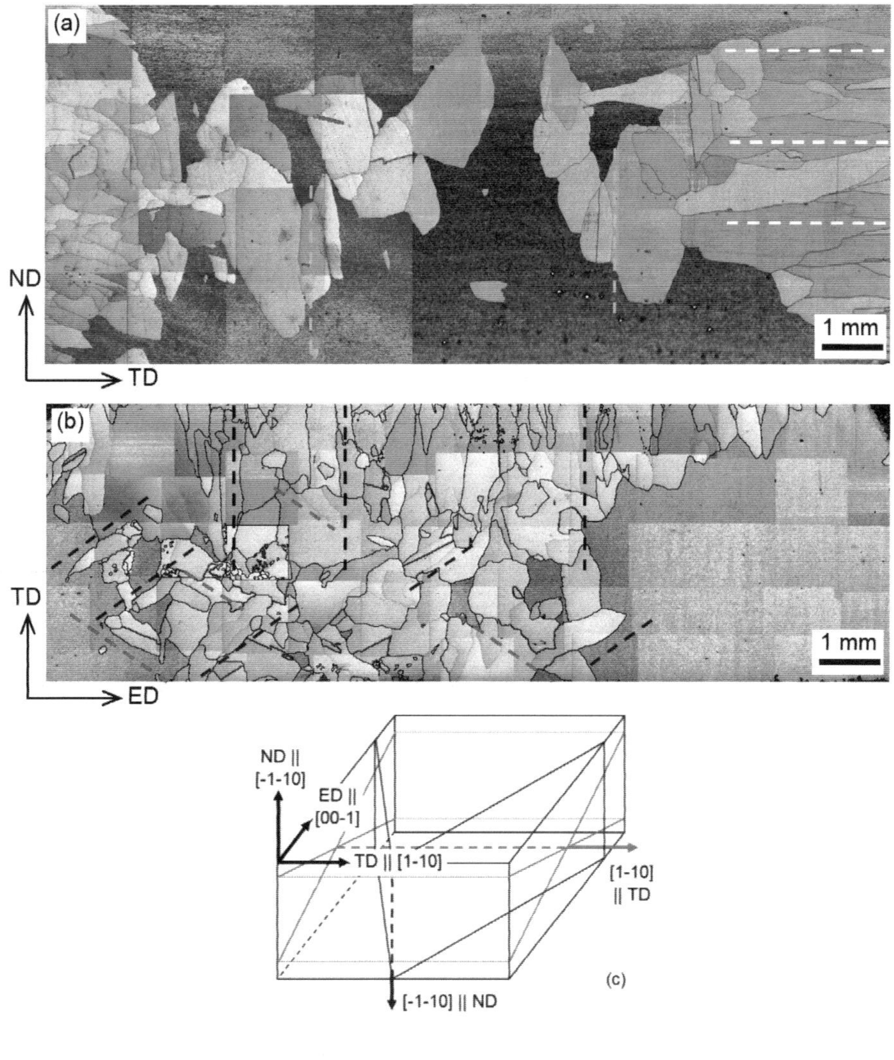

Figure 12. Directionality of grain growth observed on (a) ND/TD and (b) ED/TD planes. Dashed lines marked the directions of the privileged growth of new grains; they corresponds to the position of traces of the {111} planes. (c) Schematic presentation of the situation of the {111} planes.

4. Conclusions

Based on local EBSD orientation measurement the orientation relations which appear at the initial stages of recrystallization between a new grain and the as-deformed areas of stable single crystals have been analyzed. The results allows us to draw the following conclusions:

- The orientations of new grains during primary recrystallization are not random. For Al and Al-1%Mn and both Goss and Brass orientations at the early stages of recrystallization, the appearance of a limited number of new grain orientation groups was demonstrated. The orientation relation across the recrystallization front defines the final rotation by angles in the range 25-55° around axes mostly grouped about the <122>, <012>, <112> and <111> directions. For a single isolated nucleus of uniform orientation, the most preferred rotation angles were near 25-30°, whereas the rotation axes are usually grouped around one of the normal of all four {111} planes but only rarely coincide with them.

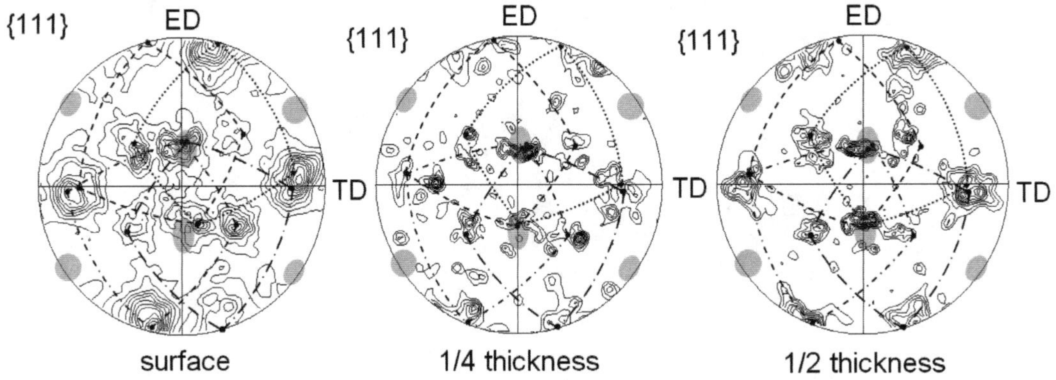

Figure 13. The texture image observed in different sections of the Goss-oriented single crystal of Al-1%Mn alloy recrystallized at 696 K for 70 s.

- The growth direction of new grain indicates the privileged role of the {111} planes in the initial stages of recrystallization. In most cases, the new grains were strongly elongated with planar facets close to those of dislocation slip planes.
- The orientation of the growing new grain in some cases transforms through the formation of a first generation twin. The most frequent situation occurs when the normal of the twinning face plane is situated near the rotation axis, around which the crystal lattice of the 'primary nuclei' rotates.
- A large fraction of new grains, regardless of the metal, develop a strongly elongated shape. For both initial orientations, preferred grain growth occurred mostly along {111}-type planes. However, the most active planes during strain, located symmetrically to the ND-TD plane, play a decisive role in this process. The orientations of the elongated grains were directly related to those of the group of as-deformed orientations. The crystal lattice of the grains elongated along a {111} plane undergoes a general rotation around an axis near to their normal.
- It is proposed that the thermally activated, coordinated movement of dislocation groups can lead to the creation of a twist or tilt grain boundary according to their screw or edge character.

Acknowledgments

The authors are grateful to Dr P. Drzymała for the numerical program enabling a statistical analysis of twin boundaries. This work was supported in part by the Polish Ministry of Science and Higher Education, projects: 2014/13/B/ST8/04291 and 'Iuventus Plus' IP2011 036471. M.M. acknowledges the Region Rhone Alpes for a graduate student scholarship.

References

[1] Humphreys F J and Hatherly M 2004 *Recrystallization and related annealing phenomena* (Pergamon) 2nd Edition
[2] Doherty R D, Hughes D A, Humphreys F J, Jonas J J, Juul Jensen D, Kassner M E, King W E, McNelly T R, McQueen H J and Rollett AD 1997 *Mater Sci Engn A* **238** 219
[3] Ardakani M G and Humphreys F J 1977 *Acta Metall* **42** 763
[4] Rios P R and Sandim H R Z 2005 *Mater Res* **8/3** 225
[5] Ardakani M G and Humphreys F J 1994 *Acta Metall Mater* **42/3** 763
[6] Engler O, Kong X W and Lücke K 2001*Acta Mater* **49** 1701
[7] Paul H, Morawiec A, Baudin T. 2012 *Metall Mater Trans A* **43/12** 4777
[8] Doherty R D 1978 in *Recrystallization of Metallic Materials*, ed. F. Haessner, Dr. Rieder Verlag, Berlin **23**
[9] Paul H, Driver J H and Jasieński Z 2002 *Acta Mater* **50** 815
[10] West S S, Winther G and Juul Jensen D 2011 *Metall Mater Trans A* **42** 1400
[11] Miszczyk M M, Paul H. Driver J H, Maurice C 2015 *Acta Mater* **83** 120
[12] Miszczyk M, Paul H, Driver J H and Maurice C 2011 *Arch Metall Mater* **56** 933
[13] Paul H, Driver J H and Lens A 2005 *Mater Sci Forum* **495-497** 1249
[14] Barrett C S 1940 *Trans Metall Soc AIME* **37** 128
[15] Borbély A, Maurice C, Piot D and Driver J H 2007 *Acta Mater* **55** 487
[16] Albou A, Driver J H and Maurice C 2010 *Acta Mater* **58** 3022
[17] Paul H, Maurice C and Driver J H 2010 *Acta Mater* **58** 2799
[18] Godfrey A, Juul Jensen D and Hansen N 1998 *Acta Mater* **46** 823
[19] Godfrey A, Juul Jensen D and Hansen N 2001 *Acta Mater* **49** 2429
[20] Paul H, Maurice C, Driver J H and Miszczyk M M 2010 *IOP Conf. Series, Mater Sci Engn A* **7** 1
[21] Berger A, Wilbrandt P J, Ernst F, Klement U and Haasen P 1988 *Prog Mater Sci* **32** 1
[22] Humphreys F J and Ferry M 1996 *Scripta Mater* **35** 99
[23] Brandon D P 1966 *Acta Metal* **14** 1479
[24] Engler O, Kong X W and Yang P 1997 *Scripta Mater* **37/11** 1665
[25] Fan G H, Zhang Y B, Driver J H and Juul Jensen D 2014 *Scripta Mater* **72-73** 9
[26] Ridha A A and Hutchinson W B 1984 *Acta Metall* **30** 1929
[27] Kohara S, Parthasarathi M N and Beck P A 1958 *J Appl Phys* **29** 1125
[28] Parthasarathi M N and Beck P A 1961 *Trans Metall Soc AIME* **221** 1831
[29] Miszczyk M M and Paul H *this issue*

Preferred lattice misorientations in rolled aluminium: tracking experiments and crystal plasticity simulations

R Quey[1], P Dawson[2] and J H Driver[1]
[1] Ecole des Mines de Saint Etienne, CNRS, France
[2] Sibley School of Mechanical and Aerospace Engineering, Corenell University, USA

Email: driver@emse.fr

Abstract. The microstructural heterogeneities developed in the grains of hot deformed Al have been quantified by applying the microtexture tracking technique to hot plane strain compressed Al-0.1%Mn. By successive EBSD measurements over the same (internal) surface at different strains, a large data set of lattice orientation and disorientation development has been obtained in over 150 grains up to strains of 1.2. Simultaneously, high resolution finite element simulations have been carried out to large strains with the same grain orientations; both experiments and simulations focus on the orientation distributions developed within individual grains. It is shown that 15-20% of the grains undergo orientation splitting, usually when the orientation is both symmetrical with respect to the loading and divergent in terms of potential lattice rotations. An analysis of the reorientation velocity field predicted for hot PSC provides a first order indication of the particular grain orientations expected to undergo orientation splitting together with their splitting modes. In the majority of grains which simply undergo orientation spreading, a detailed analysis of the disorientation axes within grains has been carried out. At low strains, there is a very high density of near TD disorientation axes which progressively evolve with plastic strain towards RD. A similar, although faster, evolution to both RD and ND is predicted by the FECP simulations. Original explanations for these low and high strain disorientation axes are proposed, based first on the influence of random local stress variations on lattice rotations and then on the reorientation velocity field.

1. Introduction

Plastic deformation is inherently inhomogeneous over many scales from the macroscopic to the micrometer level and, as is well known, this heterogeneity strongly affects microstructural evolution during subsequent annealing. Much work has been carried out in this area by cold deformation, but hot deformation can offer some advantages for fundamental studies. Although it is generally considered that hot deformation, by virtue of dynamic recovery processes, leads to a more homogeneous microstructure, many heterogeneities still occur, albeit on a coarser scale than at low temperatures. This turns out to have the advantage of facilitating their characterization by EBSD and also their modelling by finite element crystal plasticity (FECP) simulations. In particular, strain-induced orientation gradients across grains can be characterized and, in some cases, modelled. The present work is mostly focused on the hot deformation microstructures, microtextures and orientation distributions developed in hot, plane strain compressed, Al-Mn polycrystals.

Since FECP simulations can model the evolution of grain orientations during plastic straining it follows that experiments should be able to do the same. In the present case, we have chosen a tracking method by EBSD mapping of the inside surface of a split sample deformed in channel-die compression.

It was possible to follow over a hundred grains during successive plastic strains to a final strain of 1.2. Our previous publications have shown that the average behavior of the grains is reasonably consistent with the Taylor model of lattice rotations [1, 2], but that about 15% of the grains break up into two or more distinct orientations. This work looks in more detail at the local behavior within the grains, particularly grain orientation break-up and also the sub-grain orientation spreads.

2. Experimental and modelling procedures

The "microtexture tracking" technique consists of taking orientation maps of the same grains of a polycrystal on the internal surface of a split sample submitted to successive deformations in plane strain compression (PSC) [1]. A high-purity Al–0.1 wt%Mn alloy with an average grain size of about 300 μm and a weak initial crystallographic texture was used. The split sample was made of two identical parts of 8x3.5x10 mm each, assembled along the transverse direction (TD). Prior to deformation, an orientation map of the grain structure was carried out by EBSD on a 4x4 mm^2 region located at the centre of one of the two sample parts, using a step size of 5 μm. The sample was then deformed in plane strain compression at 400 °C (0.72 Tm) in a channel die rig [3] to successive strains of 0.19, 0.42, 0.77 and 1.20. After each deformation step, the sample was quenched to retain the "hot deformation" microstructure. Without any additional polishing, an orientation map was carried out on the same area of the sample, using the same spatial resolution. At $\varepsilon = 0.41$, 0.77 and 1.20, some finer orientation maps were also acquired on a region of the sample covering a few grains, using a step size of 0.5–0.6 μm. This enables one to directly analyze the evolution of the subgrains, which are 5–10 μm in size [4]. It was previously shown in Refs. [1, 4] that following the grain rotations on the internal surface of a split sample does not significantly affect the sample texture nor the local orientations. As a consequence, in the present work, the grains are considered to deform as grains in a real polycrystal.

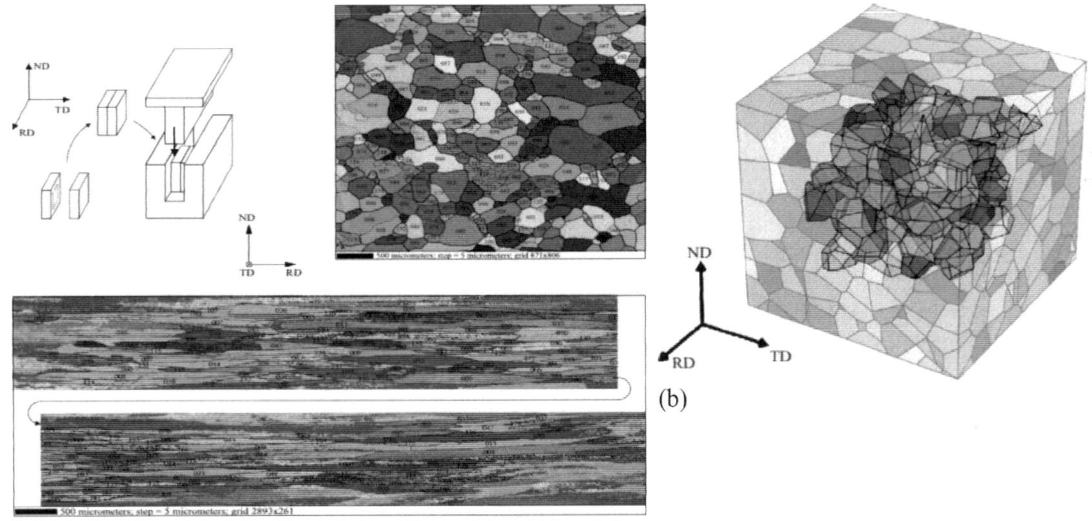

Figure 1. (a) Experimental set-up and grain structures on the ND-RD plane before and after straining to 1.2, (b) 3-D Voronoi tessellation of same grain orientations used for FECP simulation. Rodrigues vector orientation colouring.

To model the local plastic strains and microtextural evolutions, a finite element crystal plasticity analysis was performed to simulate hot deformation of an equivalent Al polycrystal using an FECP model [5] and the initial grain orientations [6]. The microstructure was represented by a 1000-grain three-dimensional Voronoi tessellation of which 182 central grains were assigned the experimental orientations, as illustrated in Fig. 1b. The surrounding grains were assigned orientations taken randomly from a uniform distribution. This means that the initial polycrystal reproduces the experimental orientation distribution, but not the grain morphology and topology, due to the fact that the (3-D) experimental polycrystal morphology is not completely known. To follow these large plastic strains, remeshing was applied to correct the mesh for distorted elements, and the microstructural variables were transported from the old mesh to the new mesh, using the Neper polycrystal generator [7, 8]. The FECP simulation simulated the experimental deformation sequence, including the unloadings at $\varepsilon = 0.19$, 0.42 and 0.77. To enable the in-grain orientation spreads to develop properly, the grains were finely discretized into tetrahedral elements. The initial average mesh density was 600 elements per grain, and the final one ($\varepsilon = 1.2$) was 1250 elements per grain, so that element sizes are roughly 30-40 µm (compared to the grain size of about 300 µm).

Following previous studies on aluminium single crystals of similar compositions deformed at the same temperature [9, 10], we consider a mixed slip model, where different system families are potentially active: {111}<110>, {100}<110> and {110}<110> (24 systems). Slip is assumed to be viscoplastic with the slip rate $\dot{\gamma}^{\alpha}$ on a system α related to the resolved shear stress τα through the power law:

$$\dot{\gamma}^{\alpha} = \dot{\gamma}_0 \left|\frac{\tau^{\alpha}}{g^{\propto}}\right|^{1/m} sgn(\tau^{\alpha}) \tag{1}$$

$$\dot{g}^{\alpha} = h_0 \left(\frac{g_s - g^{\propto}}{g_s - g_0}\right)\dot{\gamma}, \qquad where\ \dot{\gamma} = \sum_{\propto} |\dot{\gamma}^{\alpha}| \tag{2}$$

By definition, sgn(x) = 1 if $x \geq 0$ and -1 if $x < 0$. The values of the hardening parameters have been determined from experimental measurements, and are: $\dot{g}_0 = 1$, $m = 0.12$, $h_0 = 4$ MPa, $g_0 = 8g_r$ MPa, and $g_s = 12g_r$ MPa, where $g_r = 1$ for {111}<110>, 1.4 for {100}<110> and 0.9 for {110}<110> (the g_r values are from Ref. [10]). A complete description of the constitutive model and the FECP implementation can be found in Refs. [11, 12].

3. Results

3.1 Orientation distributions.

Orientation distributions in deformed grains are readily visualized by standard pole figures as shown in Fig. 2 where a) shows orientation spreading of a deformed grain and b) orientation splitting of the same grain at higher strain. However, their quantitative analysis is better performed using the quaternion and/or Rodrigues vector representations as originally described by Glez and Driver [13] and Barton and Dawson [14].

The fundamental Rodrigues region of a cubic crystal has the shape of a 3D polyhedron (a truncated cube) and provides a low distortion picture of orientations, or disorientations, without the degeneracy problems of many other representations. As an example, Fig. 2c plots out the Rodrigues vector distributions of orientations in the same, fragmented, grain at a strain of 1.2. The quaternion method provides the average orientations [15, 16] and also the size and shape of the disorientation spread around this average. Following Refs. [13, 14], the shape of this distribution is characterized by an eigen decomposition of the set of quaternion disorientations into eigenvectors vi (principal spread directions) and eigenvalues λi (spread amplitudes). They are arranged so that $\lambda 1 > \lambda 2 > \lambda 3$ and v1 is taken as the principal disorientation axis.

Given that the disorientations are small, the principal angular spreads are obtained as:

$$\dot{\theta}^{i} = 2\arctan(\sqrt{\lambda_i})$$

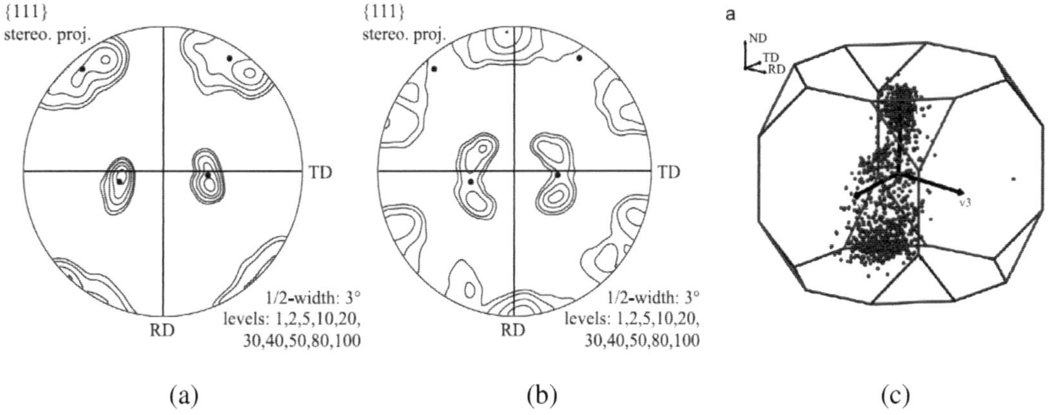

Figure 2. Example of a grain undergoing fragmentation; a) {111} pole Fig. at strain 0.42 (spreading), b) pole Fig. at strain 0.77 (fragmentation) and c) Rodrigues vector representation of the disorientation distribution at strain 1.2. Original grain orientation given as points in a) and b).

3.2 Orientation Splitting.

From the set of 176 grain orientations that were followed experimentally during hot PSC (and simulated numerically using the FECP method), about 10% were observed to undergo clear orientation splitting at the final strain of 1.2. Their proportion increased monotonically, and almost linearly, with plastic strain showing that large strain experimental techniques are essential to characterize this phenomenon. Splitting is a phenomenon which can occur throughout a grain, i.e. within any of the RD-TD, TD-ND or RD-ND planes, but as a result of the 2-D EBSD measurements on the TD-ND plane, no splitting could be detected on this plane. The true fractions of orientation splitting should therefore be increased by 3/2 to about 15%. The 3-D FECP simulations, which do not have this limitation, predicted that about 20% of the grains should split at the final strain.

Fig. 3 gives an example of the orientation splitting of a grain initially near 20° ND rotated Cube which breaks up into 2 distinct orientations by opposite ND rotations – in both the experiments and the simulations. The upper {111} pole figures illustrate the orientation distributions determined by EBSD (a) and calculated by FECP (b), which are in good agreement.

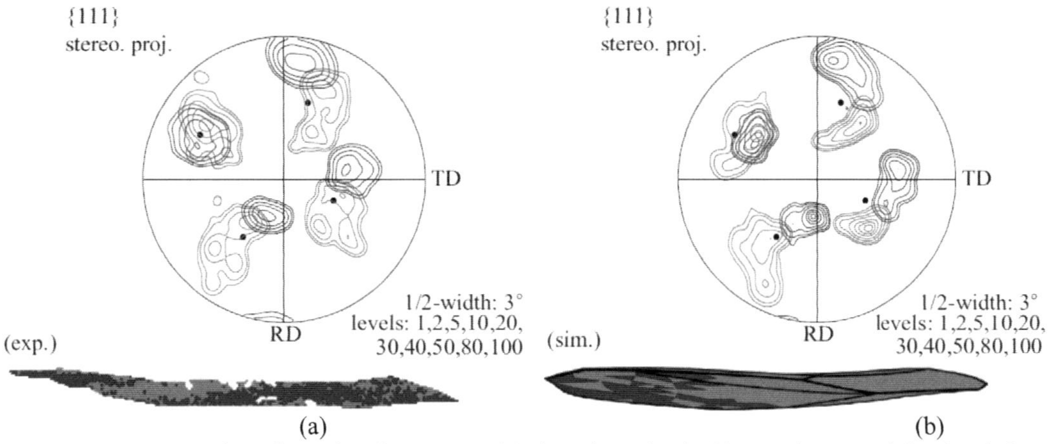

Figure 3. Fragmentation of a grain of ND-rotated Cube orientation in (a) experiment and (b) simulation. The two modes are shown in red and blue on the pole Figs. and on the maps.

The lower orientation maps show the scale of splitting and their spatial distributions; perfect agreement between these EBSD and FECP maps is not really expected since the latter has an element size almost one of magnitude greater than the EBSD pixels. Nevertheless, similar tendencies can be observed.

As could be expected, the grains that split, generally into two orientations, possess symmetrical orientations with respect to the loading directions (Figs. 2 and 3). They were principally located near Cube {100}<001>, 45°/ND rotated Cube {110}<001>, U {110}<011> and Goss {110}<001> together with two grains near {233}<311>, an orientation on the TD <110> fiber. The reason for an increasing number of fragmented grains with strain is that fragmentation occurs more efficiently for orientations which are initially very close to the symmetry fibres. For orientations which are further away, the instability conditions discussed below are not so strong, and they develop significant spreading before they actually split. A more detailed analysis of orientation splitting requires a knowledge of the lattice stability with respect to possible rotations during plastic straining. This has been developed assuming a Taylor type model for hot, plane strain compression of Al and mixed slip as described above. The results of this stability analysis are illustrated on Figs. 4 and 5.

Fragmentation can be understood by examining the properties of the reorientation velocity field. This field is shown in Fig. 4 for hot plane strain compression of Al at the surface of the Rodrigues fundamental region, following previous work by Kumar and Dawson [17], but here with the 24 mixed slip systems. If an orientation is a stable equilibrium point, it constitutes a sink of the flow field. For an orientation to be asymptotically stable, all reorientation trajectories in the vicinity of an orientation must converge on the orientation. Using linearized stability analyses, the character of the equilibrium can be determined to be stable or unstable from the signs of the eigenvalues of the reorientation velocity gradient [6].

The regions surrounding the sinks are basins. A characteristic of most reorientation velocity fields associated with slip is that the orientations move quickly toward one of the fibres and then move more slowly along the fibre toward an equilibrium point. We can interpret the fragmentation results with respect to these characteristics of the reorientation velocity field, keeping in mind that neither the experimental observations nor the simulation adhere strictly to the Taylor assumption. The sink and basin features of the velocity field shown in Fig. 4 are evident, as is the close connection of these features to the symmetry fibres.

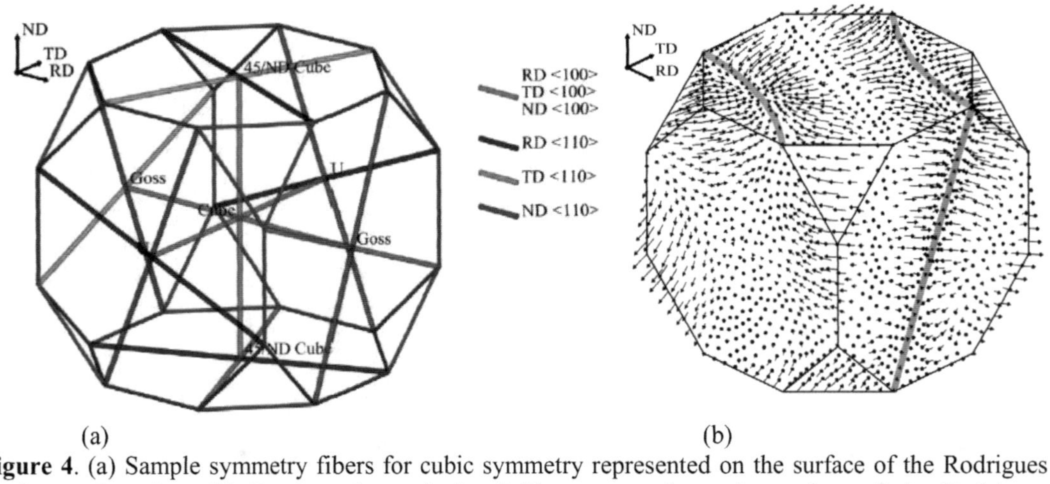

(a) (b)

Figure 4. (a) Sample symmetry fibers for cubic symmetry represented on the surface of the Rodrigues fundamental region. (b) Reorientation velocity field represented on the surface of the Rodrigues fundamental region showing both divergence and convergence towards the α (blue) and β (red) fibres.

Thus, fragmentation is more likely with orientations that lie at the boundary of a basin, because they tend to be drawn equally strongly into two different basins and toward two different points of the predominant texture fibres. So, the boundaries of the basins can be considered as equivalent to ''watershed'' regions. An example is the case of the grains of U orientation. As can be seen in Fig. 4, U is an equilibrium orientation. Moreover, it is stable about RD, but unstable about ND. This favors fragmentation by opposite reorientations about ND. Such a fragmentation actually takes place for the experimental grain of orientation close to U as illustrated in Fig. 2.

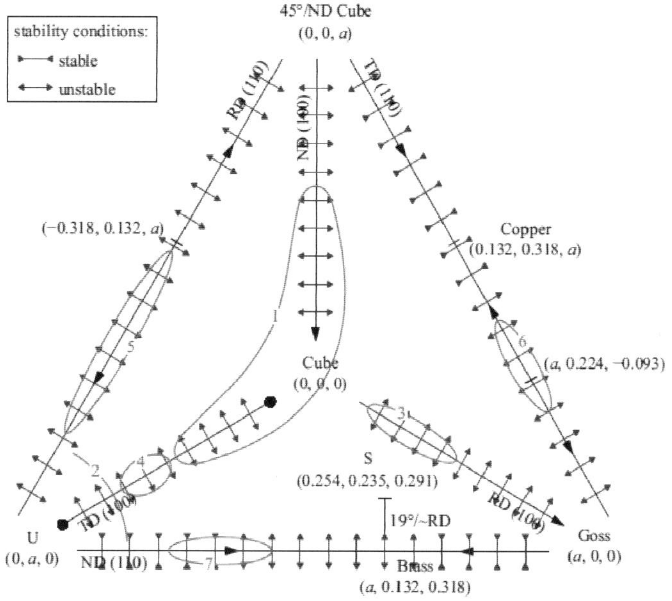

Figure 5. Fragmentation diagram for hot PSC Al showing potential reorientation velocity fields for sample symmetry orientations.

The second diagram, Fig. 5, depicts the reorientation velocity field for the principal grain orientations undergoing hot PSC. As above, instability is associated with potential divergence as seen for example for the 45°/ND Cube with respect to all rotations and also the Goss orientation with respect to TD and ND rotations. It is worth recalling that compared with other deformation conditions, hot PSC is not expected to generate large fractions of grains that undergo orientation splitting. Fragmentation is generally favoured by strain paths such as shear (e.g. torsion or ECAP) which do not develop stable grain orientations nor stable deformation textures. Cold work also strongly favours grain orientation splitting. The present study on hot, plane-strain compressed Al actually corresponds to a relatively low level of splitting, as occurs experimentally (~15%) and numerically (~20%) at strains of order unity. In this context, the agreement between FECP simulations and experimental results can be considered as satisfactory, at least in terms of lattice reorientations. In many cases the orientation splitting occurs in the predicted grain orientations about the predicted rotation axes. An analysis of the reorientation velocity field predicted for hot PSC by a Taylor type model provides a first order indication of the particular grain orientations expected to undergo orientation splitting together with their splitting modes. However, the agreement between experiments and simulations is not found for all orientations. In some cases, the discrepancies may be due to the influence of the neighbouring grains which are not identical in the present experiments and numerical simulations.

In conclusion, for this part of the work concerning grain orientation fragmentation, it appears possible to model grain orientation splitting (or its absence) in most, but not all grains undergoing hot PSC to high strains, i.e. the equivalent of industrial rolling.

3.3 Orientation spreading.

As pointed out above in the same set of experiments, the majority of the grains undergo orientation spreading, a general phenomenon which also merits further analysis. Of course there have been many studies of disorientation development in deformed grains, particularly during cold plastic straining and Risø has taken a major role in this field, e.g. [18, 19]. Contrary to cold deformation, hot plastic straining by conventional shaping methods has usually been found to lead to an average disorientation that stabilizes to a near constant value of a few degrees at strains of order 0.5 to 1 [20-23].

In the present study, we shall focus on an aspect of the problem that, until recently, has not received much attention: the disorientation axes of the intra granular orientation spread. The same methods as above are employed, but the set of grains whose behaviour is examined in detail is taken from the complementary set which did not fragment and which is sufficiently large to provide valid spread data, i.e. at least 1000 orientation measurements. A total of 92 grains met this requirement. The same set of 92 grains was considered in the simulation metrics, with a slight difference due to a limited number of grains that underwent orientation fragmentation in the experiment, but not in the simulation (or vice versa). In the following, we first analyze the evolution of the disorientation distributions over this set of grains. We then look for correlations between the disorientation distribution properties and the grain average orientations.

The distributions of the average disorientation angles are presented in Fig. 6 for the experiment (a) and simulation (b). Average values at successive strains are listed in Table 1. It should be recalled that these are point-to-average disorientations which usually tend to be significantly higher than point-to-point (or subgrain-to-subgrain) disorientations since they include orientation gradients across entire grains. As a consequence the axes associated with point-to-average disorientations are usually defined with greater accuracy than those of point-to point disorientations. This method was also adopted by Pantleon et al [24] in a similar study of cold rolled aluminum. For both experiment and simulation, the average disorientation angle increases rapidly early in the deformation ($\varepsilon < 0.5$) to comparable values between 5 and 6°. At strains above 0.5, the rate of increase drops off more rapidly in the experiment than in the simulation, such that the experimental average tends towards a saturation value of 7–8°. It is difficult to estimate a saturation value in the simulations. In both simulation and experiment, there is a high degree of variability in the average disorientation, ranging in the experiment from 3.6 to 19.7°.

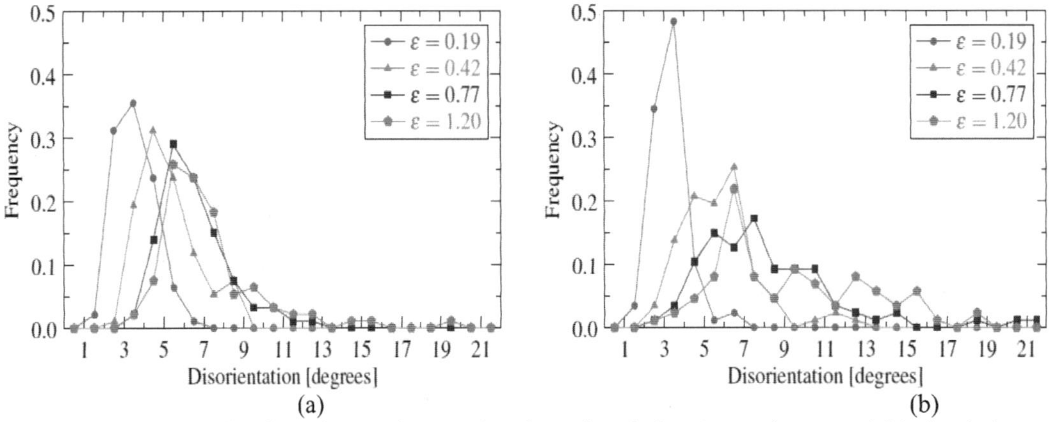

Figure 6. Average disorientation angles as a function of strain in (a) experiment and (b) simulation.

Fig. 7 gives an example of the disorientation distribution of a deformed grain in Rodrigues space and which is clearly anisotropic. Using the eigenvalues of the covariant matrix of the Rodrigues orientation distribution as described above, this anisotropy can be conveniently quantified by the factor $\lambda_a=\lambda_1/\sqrt[3]{\lambda_1 \lambda_2 \lambda_3}$ which takes a value of 1 for an isotropic distribution. The average values of this anisotropy factor are given in Table 1 along with the average disorientation angles. The λ_a values are typically of order 1.5-2, but there is an interesting difference between the experiments where λ_a decreases with strain and the simulations where they tend to increase with strain.

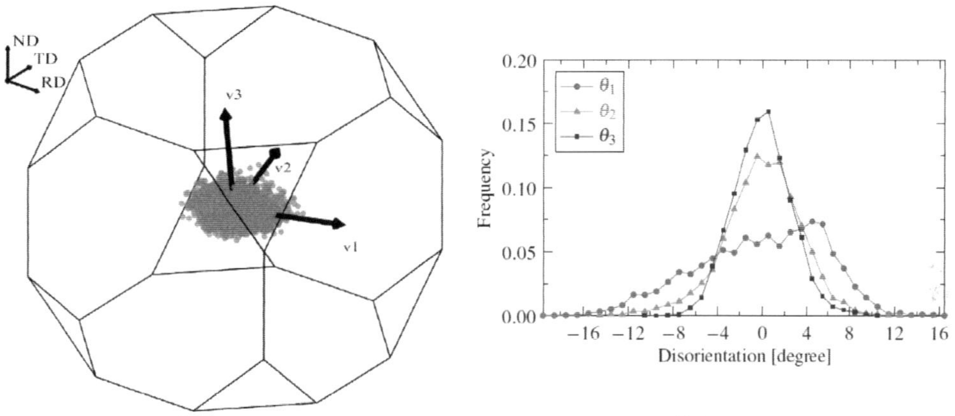

(a) (b)

Figure 7 (a) Anisotropic disorientation distribution seen in the Rodrigues fundamental region and principal axes. (b) Angular distributions along the three principal axes.

Table 1. Average disorientation angles and anisotropy factors as a function of strain in the experiment and simulation.

Strain	**Average disorientation angle (°)**		**Anisotropy factor**	
	experiment	simulation	experiment	simulation
0.19	3.5	3.3	1.91	1.62
0.42	5.2	5.7	1.75	1.64
0.77	6.5	8.0	1.55	1.84
1.20	7.2	9.9	1.49	2.13

The distributions of the preferential disorientation axes at successive strains are depicted in Fig. 8, as equal-area projections onto the sample RD-ND plane, for both experiment and simulation. The projections are reduced to one quarter using orthotropic sample symmetry. In addition, the simulated distribution at $\varepsilon = 0.02$ is also provided to analyze the early stage of plastic deformation. Fig. 8 shows that the experimental axes are strongly aligned with TD up to $\varepsilon = 0.42$. This initial TD disorientation axis is consistent with some earlier results on hot deformed Al single crystals by Glez and Driver [23] and work by Pantleon et al. on 38% cold rolled Al [24] and tensile tested Cu polycrystals [25]. However, the tendency for preferential TD axis is much stronger here than reported for cold deformation; typically the axis densities around TD are about twice those after cold rolling.

IOP Conf. Series: Materials Science and Engineering **89** (2015) 012011 doi:10.1088/1757-899X/89/1/012011

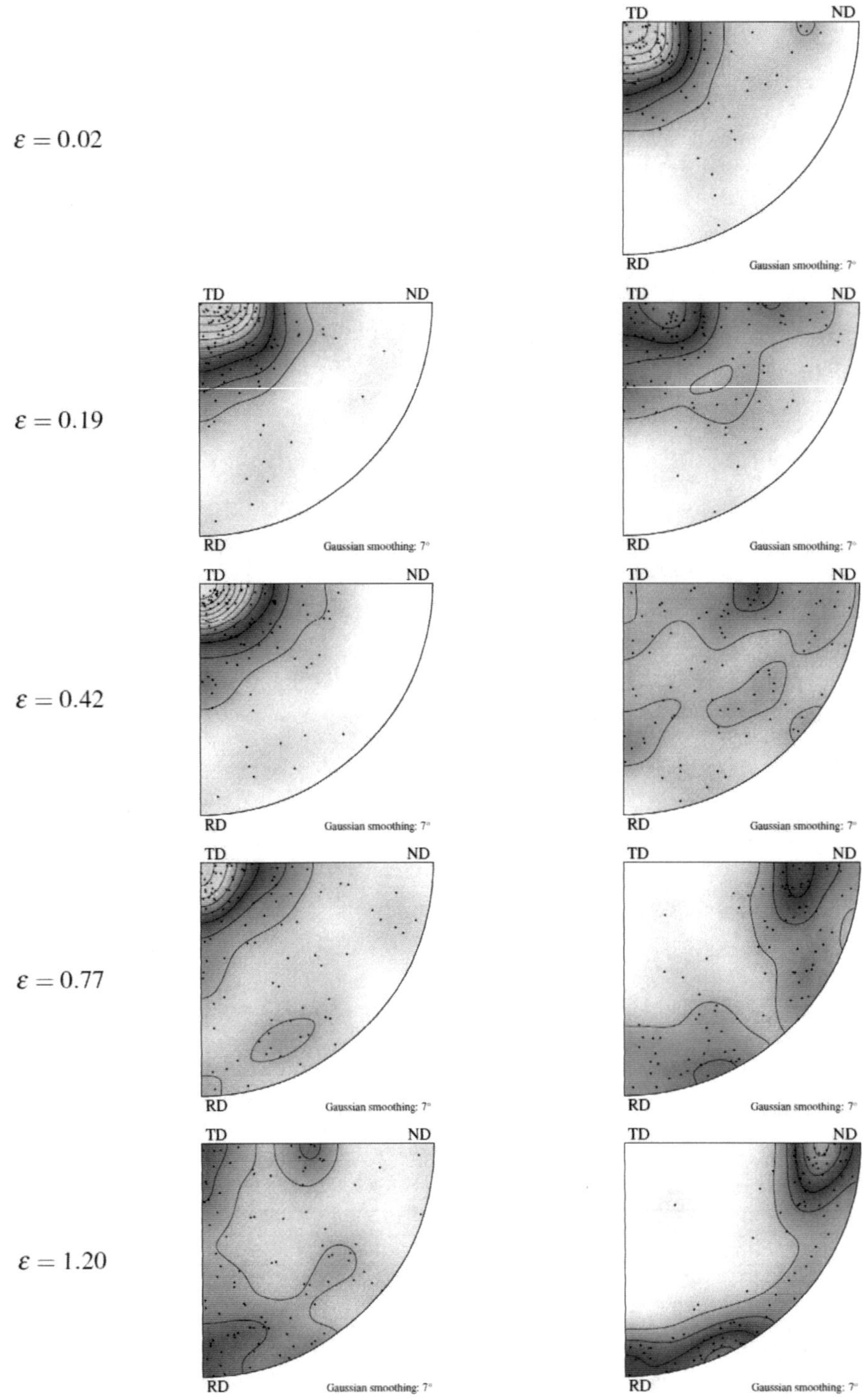

Figure 8. Preferential disorientation axis distributions as a function of strain, for (a) experiment and (b) simulation, shown as equal-area projections onto the RD-ND plane.

A second important result from Fig. 8 is the change in disorientation axis with strain. Looking at the experimental results, the initial strong preference toward the TD axis ($\varepsilon = 0.19$) weakens significantly at a strain of 0.77 and evolves to a preference toward mixed TD+RD at a strain of 1.2 (with some axes located halfway between RD and ND). In the simulation, the preferred axes are also close to TD at small strains. A rapid transition then occurs, between $\varepsilon = 0.19$ and 0.42, so that the preferred axes exhibited in the large-strain distribution is shared between RD and ND, without any TD. This change in disorientation axis with plastic strain does not appear to have been reported in previous works, which were carried out at smaller strains, but could have significant implications for the nature of the sub-grain boundaries. We shall examine below the possible mechanisms that lead to TD disorientation axes and then their evolution at high strains.

First, we have checked that these disorientation axes are essentially related to the sample axes. We have plotted the disorientation axes in the crystal coordinates and found them to be nearly uniformly distributed in the crystal stereographic triangle. At small strains, this was expected from the TD distribution of the preferential disorientation axes and the uniform distribution of the average grain orientations. Second, it is known that the grain rotation axes during PSC also tend to align with TD, so one could ask whether there is a stronger correlation with sample TD or with grain rotation axis? The frequencies of these two distributions have been determined and it was found that the sample TD frequencies (both experiments and simulations) are about twice those of the grain rotation axis [26].

So what favours these disorientation axes? Clearly, they are related to local variations in slip rates, which generate local variations of lattice rotations, but which must possess specific properties. An elementary analysis of spatial variations in slip rates under the conditions of PSC where the predominant slip systems have planes inclined near 45° between RD and ND and orthogonal directions in this plane would suggest local rotations about the direction perpendicular to the slip plane normals and directions, i.e. TD. A more general analysis can be made by looking at the effect of variations in stress state. The latter would occur as a consequence of microstructure heterogeneities principally between dislocation cells or sub-grains and their interior. One expects them to vary in a near random way about the average stress state of the grain which would be near the Taylor, Bishop and Hill stress. By virtue of Equation 1 stress variations give rise to variations of slip rates. Using the viscoplastic relation (1) between stress and slip rates, it can be shown [26] that the rotation velocity field varies with the stress vector according to

$$\frac{\partial \dot{\mathbf{r}}}{\partial \sigma_v} = -\sum_{\alpha} \frac{\partial \dot{\gamma}^{\alpha}}{\partial \tau^{\alpha}} (\mathbf{t}^{\alpha} \otimes \mathbf{p}^{\alpha}) \tag{3}$$

where $\mathbf{p}^{\alpha}$ stands for the vector form of the symmetrical Schmid tensor, and $\mathbf{t}^{\alpha}$ the spin vector. This relation has been applied to the initial grain orientations to determine the rotations expected of small random stress variations.

The distribution of the associated preferential directions is provided in Fig. 9a as a projection on the RD-ND plane. The distribution exhibits an intense focus about TD that is in remarkable quantitative agreement with observations of Fig. 8 (at $\varepsilon = 0.02$–0.19). The distribution can also be represented over the Rodrigues fundamental region (Fig. 9b) by batons aligned with their associated principal axes and coloured according to their components along sample directions: green for TD, red RD or blue ND. This figure confirms that a TD preferential disorientation axis is expected independently of grain orientation.

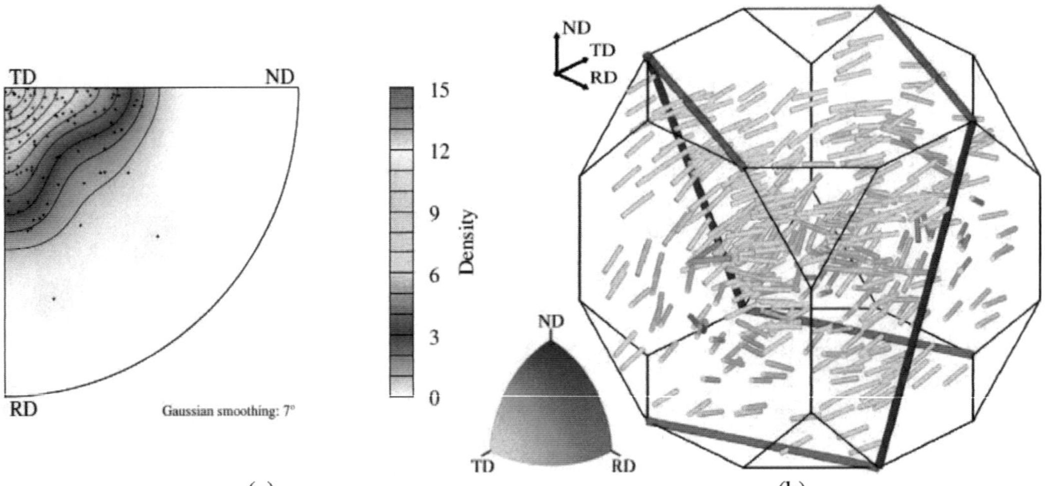

(a) (b)

Figure 9. Preferential disorientation axes simulated by random stress variations shown (a) as an equal-area projection onto the RD-ND plane and (b) at the initial average grain orientations in the Rodrigues fundamental region. On (b), the blue and red lines stand for the α and β fibers, respectively.

The general trend to develop TD preferential disorientation axes can therefore be explained from the expression of $\partial\dot{r}/\partial v$ (Equation 3), which involves the slip geometry and the slip behaviour. Under the effect of a random, isotropic stress distribution about the nominal value, all slip systems are subject to equivalent resolved shear stress distributions. On a given slip system, the resolved shear stress perturbations lead to slip rate perturbations proportional to $\partial\dot{r}/\partial v$. These values are significantly higher for the active systems, which have high τ^{α} values and spin vectors close to TD. This ultimately leads to a reorientation velocity perturbation which has a higher component along TD.

This analysis applies well to small strains, where the orientation spreads are relatively small. When an orientation distribution is wide enough, its orientations are subject to different reorientations even under the same stress or strain. This influences the orientation distribution anisotropy and can be analyzed from the reorientation velocity field, basically as done for orientation fragmentation above. The Eigen decomposition of the velocity gradient indicates how an orientation distribution narrows or broadens about specific directions during deformation. The directions (and the associated intensities) change widely over orientation space, but as shown in Fig. 10, the values of greatest interest are those in the vicinity of stable fibres. In this figure, the same colouring scheme is applied for the components of the reorientation velocities. Most are reddish (RD) with some blueish (ND) and the corresponding stereographic projections indicate near RD (+ some ND) axes.

The predicted directions evolve smoothly along the β and α fibres from RD about the Brass orientation, to RD+ND about the S and Copper orientations, and to ND about the Goss orientation.

In summary, these two mechanisms favor different disorientation axes and compete to provide an evolution of the preferential disorientation axis distributions with strain that is dominated at low strain by one mechanism and at large strain by the other. Assuming a constant stress variability in the grains, the stress-variability mechanism predicts a linear growth of the average disorientation angle with stable, intense anisotropy (anisotropy factor of 3 in average) and preferential disorientation axis distribution about TD. In contrast, the reorientation-velocity mechanism influences the anisotropy properties only when the orientation distributions are large enough and most grains have reached the stable texture fibres. By continuously broadening the orientation distributions along a direction between RD and ND (the first eigenvector of the reorientation velocity gradient, which depends on the orientation) rather than along the original TD, this second mechanism leads to a transition to RD or RD+ND.

It should be noted that orientation distributions would necessarily align with the first eigen vector of the reorientation velocity gradient only at very large strains. Because a grain orientation distribution which reaches the stable orientation fibre is not uniformly distributed about an average orientation, but rather with a strong TD component, it needs a significant amount of additional strain for the disorientation axes to migrate to RD or RD+ND. In the experiment, this movement occurred for only half of the grains at 1.2, but is expected to be complete at higher strain.

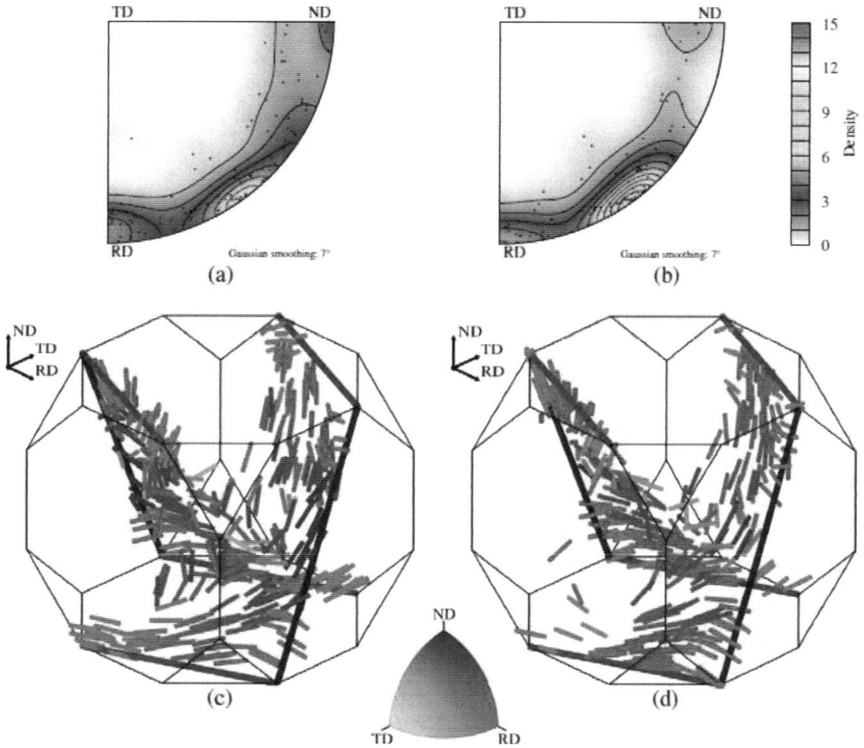

Figure 10. Preferential disorientation axes resulting from the reorientation velocity conditions shown (a, b) as an equal-area projection onto the RD-ND plane and (c, d) at the final average grain orientations in the Rodrigues fundamental region. (a, c) refers to the experiment and (b, d) to the simulation. On (c, d), the blue and red lines stand for the α and β fibers, respectively.

Interestingly, the transition of the preferential disorientation axes from TD to RD or RD+ND also indicates a change of the dislocation structures, i.e. the character of the dislocation subgrains. Specifically, knowing that the slip system spin vectors are preferably distributed about TD at all strains, a TD preferential disorientation axis at small strain indicates a higher content of edge dislocations (Burgers vector perpendicular to the rotation axis) while an RD or RD+ND preferential disorientation axis at large strain indicates a higher content of screw dislocations (Burgers vector parallel to the rotation axis).

4. Conclusions

The orientation distributions developed within individual grains from small to large strains at 0.72 Tm have been quantified and compared to high resolution FECP simulations. By means of the microtexture tracking technique applied to hot, plane strain compression of an Al-0.1 wt.% Mn alloy, a large data set of orientation and disorientation development has been obtained in over 150 grains up to strains of 1.2. Simultaneously, FECP simulations have been carried out at the scale of about 10^3 elements per grain to compare with the same experimental grain orientations to large strains. A mixed slip, viscoplastic model is used for these high temperature deformation conditions.

Apart from the fact that most grain reorientations are close to those of the Taylor model, it is shown that a significant proportion undergo orientation splitting (15-20%), usually when the orientation is both symmetrical with respect to the loading and divergent in terms of potential lattice rotations (termed watershed orientations). An analysis of the reorientation velocity field predicted for hot PSC provides a first order indication of the particular grain orientations expected to undergo orientation splitting together with their splitting modes. In the majority of grains, which simply undergo orientation spreading, a detailed analysis of the disorientation axes within grains has been carried out.

At low strains, there is a very high density of near TD disorientation axes which progressively evolve with plastic strain towards RD. A similar, although faster, evolution to RD+ND is predicted by the FECP simulations. Original models for these low and high strain disorientation axes are proposed: i) at low strains random local stress variations create slip rate perturbations which are significantly higher for the active systems characterized by high τ^{α} values and spin vectors close to TD; ii) at high strains most grain orientations are close to the rolling texture components whose reorientation velocity fields favour RD+ND axes.

Acknowledgements
Partial support has been provided by ONR under grant number N00014-09-1-0447.

References
[1] Quey R, Piot D and Driver J H 2010 *Acta Mater.* **58** 1629–1642
[2] Quey R, Piot D and Driver J H 2010 *Acta Mater.* **58** 2271–2281
[3] Maurice C, Piot D, Klocker H and Driver J H 2005 *Metall. Mater. Trans.* **36A** 1039-1047
[4] Quey R and Driver J H 2011 *Mater. Charac.* **62** 1222–1227
[5] Dawson P R 2000 *Int. J. Solids and Structures* **37** 115-130
[6] Quey R, Dawson P and Driver J 2012 *J. Mech. Phys. Solids* **60** 509–524
[7] Quey R, 2009 Neper: a 3-D random polycrystal generator for the finite element method, *http://neper.sourceforge.net*
[8] Quey R, Dawson P R and Barbe F 2011 *Comput. Methods Appl. Mech. Engrg.* **200** 1729–1745
[9] Maurice C, Driver J H 1997 *Acta Mater.* **45 (11)** 4627–4638
[10] Perocheau F and Driver J H 2002 *Int. J. Plasticity* **18 (2)** 185–202
[11] Marin E B and Dawson P R 1998 *Comput. Methods Appl. Mech. Engrg.* **165** 1–21
[12] Marin E B and Dawson P R 1998 *Comput. Methods Appl. Mech. Engrg.* **165** 23–41
[13] Glez J C and Driver J2001 *J. Appl. Crystallogr.* **34** 280–288
[14] Barton N R and Dawson P R 2001 *Modell. Simul. Mater. Sci. Eng.* 9 433–463
[15] Krieger Lassen N C, Juul Jensen D and Conradsen K 1994 *Acta Crystall.* **A50** 741–748
[16] Humbert M, Gey N. Muller J and Esling C 1996 *J. Appl. Crystallogr.* **26** 662–666
[17] Kumar A and Dawson P 2009 *J. Mech. Phys. Solids* **57** 422–445
[18] Hughes D and Hansen N 2000 *Acta Mater.* **48** 2985–3004
[19] Liu Q, Huang X, Lloyd D and Hansen N 2002 *Acta Mater.* **50** 3789–3802
[20] McQueen H and Hockett J 2000 *Metall. Trans.* **1** 2997–3004
[21] Marthinsen K and Nes E 2001 *Mater. Sci and Techn.* **17** 376–388
[22] Humphreys F, Bate P 2007 *Acta Mater.* **55** 5630–5645
[23] Glez J C and Driver J 2003 *Acta Mater.* **51 (10)** 2989–3003
[24] Pantleon W, He W, Johansson T and Gundlach C 2008 *Mater. Sci. Eng. A* **483-484** 668–671
[25] Krog Petersen S, Bowen J, Pantleon W 2009 *Int. J. Mater. Research* **100** 433–438
[26] Quey R, Driver J H and Dawson P R, submitted to *J. Mech. Phys. Solids.*

The kinematics of deformation and the development of substructure in the particle deformation zone

J Quinta da Fonseca and L Ko

School of Materials, The University of Manchester, Manchester M13 9PL, UK

E-mail: joao.fonseca@manchester.ac.uk

Abstract.
The details of the deformation near undeformable particles influence the recrystallization behaviour of commercial aluminium alloys. It has been shown that the size of the particle deformation zone (PDZ) and the local lattice rotation decrease with particle size and that a uniform distribution of very small closely spaced particles makes the deformation more homogeneous. Although a number of models have been proposed to describe the development of the PDZ, an explanation for this particle size effect remains elusive. In this paper, high resolution digital image correlation was used to map the deformation around particles of different sizes in a model Al-Si alloy. Deformation maps were compared to maps of local lattice rotation obtained using electron back scattered diffraction (EBSD) to elucidate the link between the kinematics and the resultant substructure. These results were also compared to crystal plasticity finite element predictions. The deformation maps revealed that deformation is concentrated in intense slip bands that have a characteristic spacing and which are strongly correlated to the deformation substructure. Whereas particles larger than this characteristic spacing interact strongly with the slip bands, causing large local rotations, smaller particles either interact more weakly or not at all. Since the strain between the bands is only a fraction of that in the bands, the rotations associated with the smaller particles are invariably smaller. Very small, closely spaced particles change the slip band pattern, decreasing the band spacing and decreasing the amount of shear within the bands. CFEM modelling was able to reproduce general features of the particle deformation zone but because it does not predict the development of slip bands and their characteristic spacing it fails to predict an effect of size on the local rotation.

1. Introduction

The microstructure of aluminium alloy products is determined during thermomechanical processing. The processing parameters are controlled to produce the grain size and texture required by the application. Deformation breaks down the cast structure and provides the driving force for subsequent recrystallization during annealing. Almost all commercial alloys contain second phases of varying sizes and origins, which are well known to affect the development of the deformed state and the recrystallisation behaviour during annealing [1]. Small , closely spaced particles <1 μm are known to retard and hinder recrystallization [2], whereas larger particles can stimulate the nucleation of new grains during recrystallization, which helps to refine the grain size and also randomise the texture, which is important to reduce the anisotropy of the final product [1]. The stimulating effect of large particles originates from the formation of a particle deformation zone (PDZ) in the matrix near the second phase particles, where the deformation is different from that in the bulk. This difference gives rise to highly misoriented regions within the PDZ, which act as preferential nucleation sites during recrystallization. The local misorientation was studied extensively by Humphreys and collaborators using both TEM and X-ray diffraction on single crystals [3, 4] and later using electron backscattered diffraction (EBSD) on single crystal [5] and polycrystalline samples [6, 7]. They found that the amount of rotation in the PDZ in a single crystal increased with particle size with the rotation near a 4 µm being almost twice that near a 1 µm particle. Beyond a particle size of 3-4 µm the amount of rotation remains essentially unchanged. In polycrystalline material, the measured local rotations are smaller [4] and show a considerable amount of scatter [6]. When the particles are very small, of the order of 100 nm, evenly dispersed and closely spaced , not only are the PDZ and associated rotations very small, the deformation is generally more homogeneous and intergranular misorientations are reduced throughout the material [7, 8].

A number of models have been proposed to explain the formation of the PDZ and associated lattice rotations. Early models focused on how the misfit between the matrix and the deformable particles could be accommodated by dislocation generation and dynamic recovery [9, 3] but this treatment is only applicable to small strains. Later models attempted to explain how the presence of the particle affects local plasticity using Taylor multislip modelling [4, 5] and more recently crystal plasticity finite element (CPFE) modelling [10, 7]. CPFE models successfully predict the general shape of the PDZ around large particles, with good agreement with measured local lattice rotations for single crystals. They also predict that the shape and extent of the deformation zones around particles depends on the initial grain orientation, which implies that the PSN effect will depend on the orientation of the grain in which it develops. However these models cannot capture the effect of particle size on the lattice rotation in the PDZ , even when using strain gradient plasticity to introduce a natural lengthscale [10], and fail to explain the large scatter in local lattice rotation measured in polycrystals [6], which could be related to the low efficiency of PSN. These models also cannot explain why very small closely spaced particles homogenise the deformation structure [8, 7].

One of the difficulties in understanding the limitations of current models is that comparisons with experimental data are limited to lattice rotation data. This is problematic for two reasons. Firstly, modelling predictions are different for different starting grain orientations but experimentally, the starting orientation of a grain is usually not known. Secondly, the lattice curvature measured using EBSD or TEM does not uniquely determine the kinematics of the deformation because there is more than one possible strain distribution that is compatible with a given lattice curvature and therefore agreement between the lattice rotation predicted by a crystal plasticity model does not fully validate the model. One way to address these difficulties is to measure directly the strain distribution around the particle as well as the lattice rotation in a grain with known starting orientation.

In this work, high resolution digital image correlation (HRDIC) [11] was used to map the kinematics of deformation around particles of different sizes in a model Al-Si alloy during deformation in plane strain compression. EBSD was used to determine the orientation of the matrix before deformation and to map the local lattice rotation after deformation. The experimental results were also compared with predictions from a CPFE simulation using an image based mesh. The aim was to measure how the kinematics of deformation is affected by the particle size and to explain why and how continuum crystal plasticity models fail to predict these size effects.

2. Material and experimental procedure

2.1. Material

The material studied was a model Al-Si alloy which can be heat treated to precipitate particles of varying sizes. The chemical compositions of the alloy is given in table 1. The material was first solution heat treated at 580 °C for 6 hours and then water quenched. Two particle distributions

Table 1. Chemical composition of the model Al-Si alloy studied (wt%).

Si	Fe	Cu	Ti	B	Al
0.81	0.002	0.001	10.0003	0.0005	Bal.

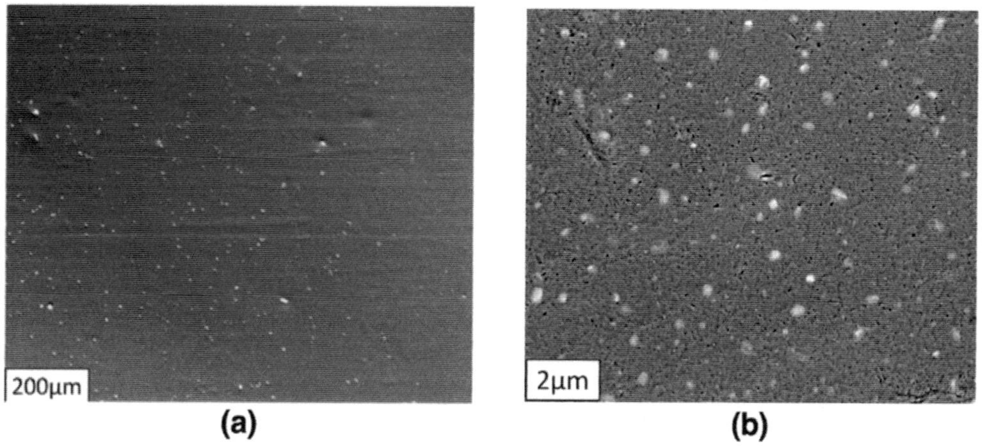

Figure 1. SEM images of the starting microstructure of the alloy in the two conditions (a) following a prolonged heat treatment to give large, widely spaced particles (b) following solution heat treatment and low temperature heat treatment to precipitate small sub micron sized particles.

were then produced by ageing: large, widely spaced particles were generated by heat treating at 490 °C for 120 hours, followed by slowly cooling at 1 °C/h to room temperature, and very small, closely spaced particles by heat treating the material at 200 °C for 90 minutes, which is part of the gold remodelling procedure required for the strain mapping. The two microstructures produced are shown in figure 1. The material containing coarser particles shows range of particle sizes at a range of distances with an average particle size of 10 µm, which at volume fraction of 0.8% corresponds to an average distance between particles of 23 µm. However, there is a preference for precipitation at the grain boundaries and therefore inside the grains the interparticle spacing is larger. The second heat treatment produces evenly spaced small particles ranging in size from 100 to 200 nm with an average spacing of 1-2 µm spacing. The grain size was about 1 mm in both materials.

After heat treatment the samples were machined into 5×5×5 mm cubes for testing using a diamond saw. The surfaces to be studied were then polished using diamond paste of size 6 , 3 and 1 µm, followed by 30 minutes of OPS polishishing to achieve a shiny, deformation free surface ready for EBSD.

2.2. EBSD

EBSD was used to map the starting orientations of the grains before deformation and to map the lattice rotations after deformation. For the initial maps, 2 × 2 mm areas were mapped with a 2 µm step size in the centres of the compression samples, using a CamScan Maxim FEGSEM operating at an accelerating voltage of 20 kV controlled via the HKL Channel5 software from Oxford Instruments. These maps were also used to select areas of interest for the deformation

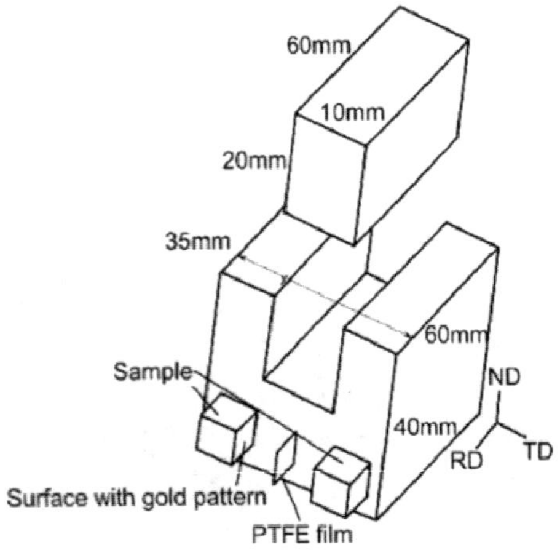

Figure 2. Channel die used to deform the sample. The sample is made up of two 5×5× 5 mm cubes, which are coated with a gold remodelled surface and are separated by a thin PTFE film.

mapping.

The maps after deformation were made with a much smaller step size of 0.1 µm, which is similar to the spatial resolution of the deformation mapping measurements, using a Magellan XHR FEGSEM operating at an accelerating voltage of 20 kV and also controlled via the HKL Channel5 software from Oxford Instruments. These final EBSD maps had indexing rates of around 95%. Non-indexed pixels were extrapolated using the neighbour orientation using the Channel 5 software. Then the data was filtered with two passes of a Kuwahara filter with 5×5 pixel block as implemented in VMAP, which allows misorientations as small as 0.5° to be detected [12]. The region of interested was also imaged using electron channelling contrast imaging (ECCI) after deformation to reveal the deformation substructure using a Magellan XHR FEGSEM at 15 kV with a working distance of 5 mm.

2.3. Deformation mapping

After EBSD, the samples were prepared for the HRDIC measurements using the procedure described by Di Gioacchino and Fonseca [11]. This involves coating the polished surfaces with a thin layer of gold (∼ 50 nm) using an Edwards S150B sputter coater. This surface layer is then remodelled by placing it in a water vapour environment at 200 °C for 90 minutes. This produces a fine, uniform, characteristic distribution of gold speckles on the surface, which makes it possible to map local strains with 0.1 µm resolution using digital image correlation.

The samples were compressed at room temperature using an Instron 5569 testing machine using a channel die to simulate plane strain compression (PSC) as shown in figure 2. Two cubes, each with a surface prepared using the gold remodelling procedure, were placed in the die with the surface of interest normal to the transverse direction. A thin (40 µm) PTFE film was placed between the cubes for lubrication. Each loading step had a target strain of 5%. Since the required displacement was very small (≤0.25 mm), the crosshead movement was not a good indicator of the strain in the sample and, in some cases, 2-3 attempts were needed to obtain a 5% reduction. After each 5% increment, the cubes were removed from the die, separated and imaged. These experiments were limited to a reduction of 30% because further compression damaged the gold pattern beyond use.

A high resolution Scanning Electron Microscope (Sirion FEGSEM) was used for backscattered electron imaging of the regions of interest. The imaging conditions were optimised for Z-contrast, so the gold particles appear bright and the aluminium background appears dark, in low contrast with the silicon particles. The working distance was kept low at 5 mm to maximise the BSE signal. An 8 bit depth was used during imaging. After each strain increment, the sample was placed in the microscope and the image was focused ensuring that the working distance was maintained constant at exactly 5 mm. A magnification of 2000× was chosen so that each gold spot covers at least 3×3 pixels. The sample is always aligned with the bottom edge, which is parallel to the rolling direction during deformation.

After acquisition, the images were processed using the software Davis, from LAVision, Gottingen. The image correlation uses an adaptive multipass FFT implementation, where the search window is shifted and deformed iteratively by displacements of the neighbouring windows as described in [13]. The initial window size used is 512×512 pixels with an overlap of 50%. This is incrementally halved until a final window size of 8×8 pixels, at which 3 iterations are performed. The final overlap between windows is 50%. A circular Gaussian weighting window is used at the final size, which slightly improves the signal to noise ratio of the image correlation. Digital image correlation gives maps of displacement from which the in-plane compressive axial strain, ϵ_{yy}, and the in-plane rotation, ω_z, can be calculated:

$$\epsilon_{yy} = ln(1 + \frac{dv}{dy}) \tag{1}$$

$$\omega_z = \frac{1}{2}(\frac{du}{dy} - \frac{dv}{dx}) \tag{2}$$

where du and dv are the displacements along x (RD) and y (ND) respectively.

3. Results

Sequential images were acquired at a large number of different locations, sampling different grain orientations, however due to space constraints only the results of three representative regions are shown here for the material containing coarse particles: a region containing no particles, a region containing a large (>10 µm) particle and another containing a smaller (<2.5 µm) particle. For the material containing very fine (100 nm to 200 nm) particles, the results from only one representative region are presented. All deformation maps are presented in the deformed configuration.

3.1. Deformation away from particles

In the material with coarse particles, the average distance between particles is about 50 µm and therefore in much of the material deformation occurs away from the influence of particles and is representative of the deformation of the pure metal [3]. Images of the gold remodelled surface in a region without particles are shown, at different deformation levels, in figure 3. The figure shows a region within one grain where the gold remodelling process produced, fortuitously, a straight boundary separating slightly different speckle morphologies. The deformation is concentrated on thin bands of slip, separated by regions with much less strain. As can be seen by the relative movement of the circled gold speckles, the distance between the top and middle circles decreases much more quickly than that between the middle and bottom one, which remains almost constant during deformation. This difference is caused by the larger number of slip bands on the top half of the region imaged. Before deformation starts, the boundary separating the two speckle regions is parallel to the compression axis. During compression, the material lattice starts to deform and the speckle boundary is sheared and broken up by the appearance of slip bands. As the boundary is broken up, it starts to rotate anti-clockwise as do the slip bands. This rotation is clearly not uniform and appears to be related to the different amounts of shear in the different bands.

This deformation heterogeneity can be quantified using HRDIC. The axial strain (ϵ_{yy}) maps for this region are shown in figure 4 There are two sets of maps, each showing the strain after

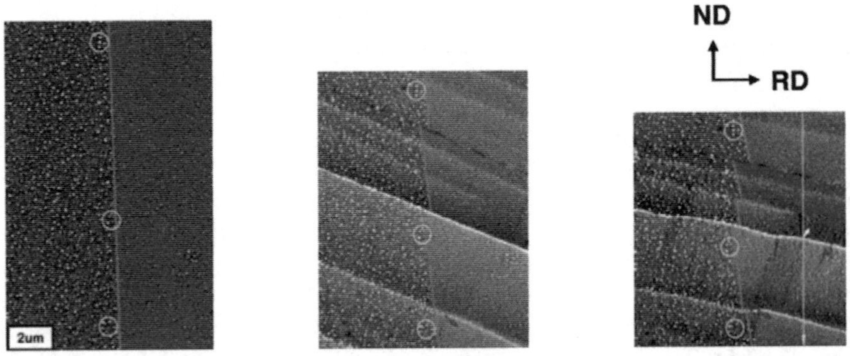

Figure 3. A small region within one grain far from any particles or grain boundaries imaged at different strain levels (a) 0% - a straight boundary between two regions of the surface where the speckle pattern in slightly different (b) after 16% strain the boundary is broken by a series of intense slip bands with the segments rotating to different extents (c) after 30% compression the boundary is broken down even further and becomes more diffuse. The top of the region has clearly deformed more that the bottom as the relative movements of the circles indicate.

11%, 16% and 30%. In the first one, shown in figures 4 (a) - (c) the maps show the increments of strain between successive images whereas in the second, shown in figures 4 (d) - (f) they show the total strain.

The strain maps show clearly that the number of slip bands increases slightly with strain, with a secondary set of bands, which are perpendicular to the primary set, appearing at the 16% strain increment. When new primary band A appears at 16% strain, it seems to inhibit slip in the neighbouring band B. The increase in the number of bands makes the bands in the total strain map (figure 4) appear to broaden. The primary slip bands are aligned at 30° to the rolling direction after a reduction of 11% and rotate towards RD with increasing strain, approaching 15° after a reduction of 30%. Although the slip bands are approximately parallel, this alignment is clearly not perfect: the bands are not straight, and new bands that appear later in the deformation are at a higher angle to the existent bands. This is particularly clear in figure 4 (c). The plot in figure 5 shows the strain along the line marked in figures 3 (c) and 4 f. The plot confirms that deformation occurs mainly in the slip bands with the matrix between the bands deforming much less. The strain in the regions between the slip bands appears to increase at 30% applied strain but it is clear from figure 4 (c) that this increase in strain is caused by an increase in the number of bands parallel to the primary bands and the appearance of complementary secondary bands. It is not possible to generalise from observations of a single, small region, but by repeating this analysis over several different regions the following generalisations can be made. The deformation of the matrix away from the particles and grain boundaries is very heterogeneous, with most of the strain being born by almost discrete slip bands which are about 100 nm thick and which effectively broaden with deformation by the activation of new parallel bands. The spacing of the slip bands ranges from 3 to 4 μm irrespective of grain orientation and decreases only slightly with deformation. Although the primary bands in the region presented here start off at 30°, the initial alignment of slip bands is usually closer to 45° but soon develops a different preferential alignment, which sometimes coincides with {111} slip traces but usually does not. At very low strains ($\leq$ 5%) there are usually two sets of weak slip bands with opposite shears (not shown here) and one of these soon dominates to become the primary set, as shown in figure 4, up until a strain of 0.2 to 0.25 when a second complementary set reappears, not usually

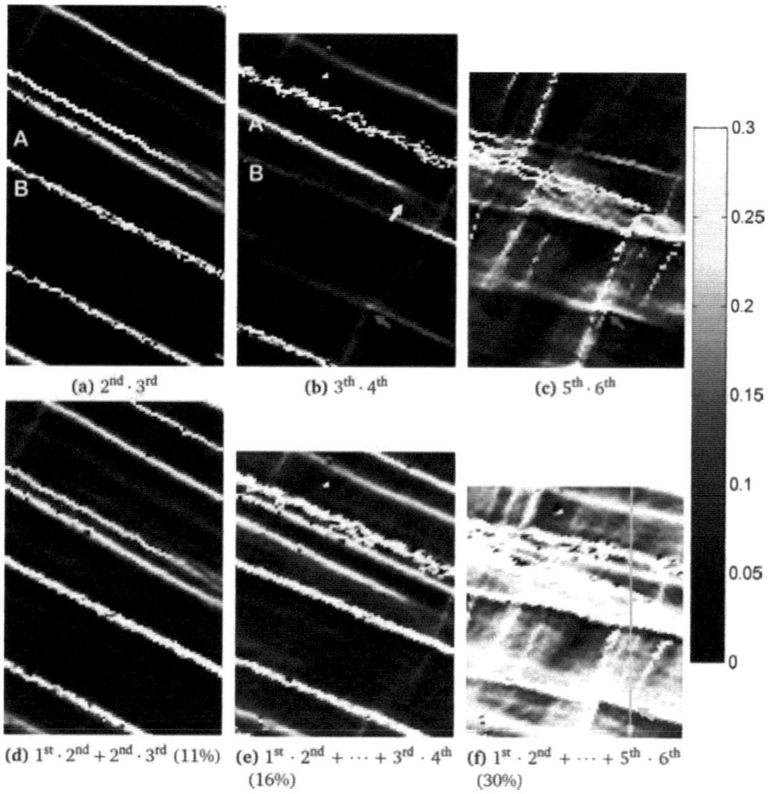

Figure 4. Axial compressive strain maps for a region away from any particles and away from a grain boundary after overall compression by 11%, 16% and 30%. (a)-(c) show the increment in strain between successive deformation steps, which are indicated below the images. (d)-(f) show the total strain at the 3 increments.

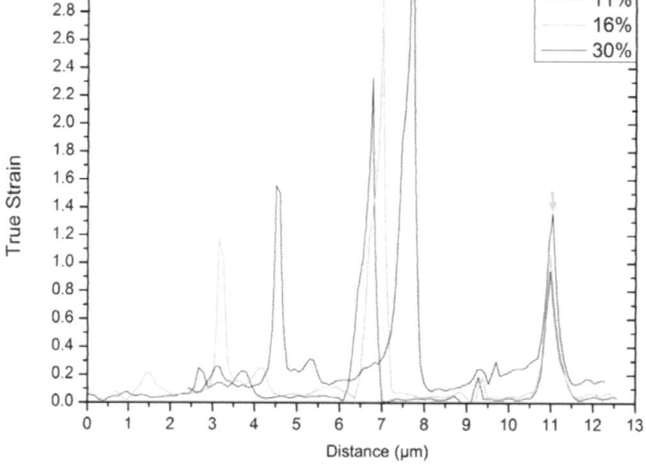

Figure 5. Compressive axial strain profiles along the line marked in figure 4 (f). The profiles are aligned with the bottom slip band, with the corresponding peaks marked with an arrow.

aligned with slip traces. After 30% reduction the average strain accumulated between the slip bands is about 0.2, whereas the strain in the bands is about 1.5 and can be as high as 3. Thus, although the slip bands are much narrower that the distance between them, they still account for the majority of the deformation at 30% applied strain. The alignment of the dominant bands is clearly dependent on crystal orientation, although orientation is not the defining criterion. In fact, some grains show different slip band alignment in different parts of the grains. For near stable orientations, slip bands are fairly straight and tend to align between ±30° and ±40°. For more unstable orientations, the slip bands are less straight and therefore show a greater range of alignment angles.

3.2. Deformation near an isolated 10 μm particle

3.2.1. Deformation mapping An example of the development of strain and material rotation during plastic deformation in the matrix around a large particle is shown in figure 6 for a grain with near cube orientation (ϕ_1 =80°, Φ =14°, ϕ_2 =9°). As can be seen, both strain and rotation are mainly concentrated along primary slip bands. At the beginning of deformation (average strain in the region of 0.08) the local strain is around 0.5-0.8 in slip bands, up to 10× the applied strain. In contrast, the strain in areas between slip bands is only around 0.05, slightly less than the applied strain. When the strain increases to 0.2 and 0.3, the strain in the slip bands increases to almost 1 and new slip bands appear, including a second set at around 0.25 strain. This shows that the pattern of deformation near a particle is, in general terms, very similar to the deformation away from the particle. However the presence of the particle changes the details of the pattern in the regions near the particle.

The first obvious effect is that the particle interrupts intersecting primary slip bands from the start of deformation. Interrupted bands are indicated by arrows in figure 6 (a), (c) and (e). The bands are not interrupted at the particle but at some distance away from it, approximately equal to the particle size. The presence of the particle also appears to cause some of these primary slip bands to bend. A good example is the band that propagates from the interrupted band above and to the left of the particle. This bending occurs at very low strains and therefore is not caused by local lattice distortions. This also implies that the bands are not crystallographically aligned. The third effect is an increase in the density of slip bands in the regions near the sides of the particle which are parallel to the slip bands. In this case, the particle increases the density of both the primary bands, especially below the particle and at larger deformations that of the secondary bands, to the left and right of the particle. Interestingly one secondary band seems to be nearly continuous across the particle and is not interrupted like the primary bands were.

The HRDIC maps in figure 6 (b), (d) and (f) show that rotation is also concentrated in the bands, which implies that deformation in the bands occurs by unidirectional shear. However, it is also seen that the rotation in the regions between the bands, where there is much less shear, is not homogeneous. At ϵ_T = 0.08, a −5° rotation (a negative rotation is anti-clockwise here) can be seen at the top left and bottom right corner of the particle (b). With an increase in strain to 0.2, there is a global rotation of −5° of the entire area and the rotation around the particle increases to −10° in parts of (b). When the strain further increases, both positive and negative rotations around different sides of the particle can be seen. At an applied strain of 0.3, the regions near the particle have rotated to 5° and −15°.

3.2.2. Lattice rotation mapping Figure 7 shows the EBSD derived local lattice rotations of the region mapped in 6 relative to a point near the particle, marked with a ∗, where the material has remained undeformed. The maximum misorientation measured is lower than that measured in single crystals [3] but is in better agreement with more recent EBSD measurements in polycrystals [7]. The maps show the amount of lattice rotation around the RD (x), TD (y) and ND (z) axes, where a positive angle represents a clockwise rotation. The rotations are

IOP Conf. Series: Materials Science and Engineering **89** (2015) 012012 doi:10.1088/1757-899X/89/1/012012

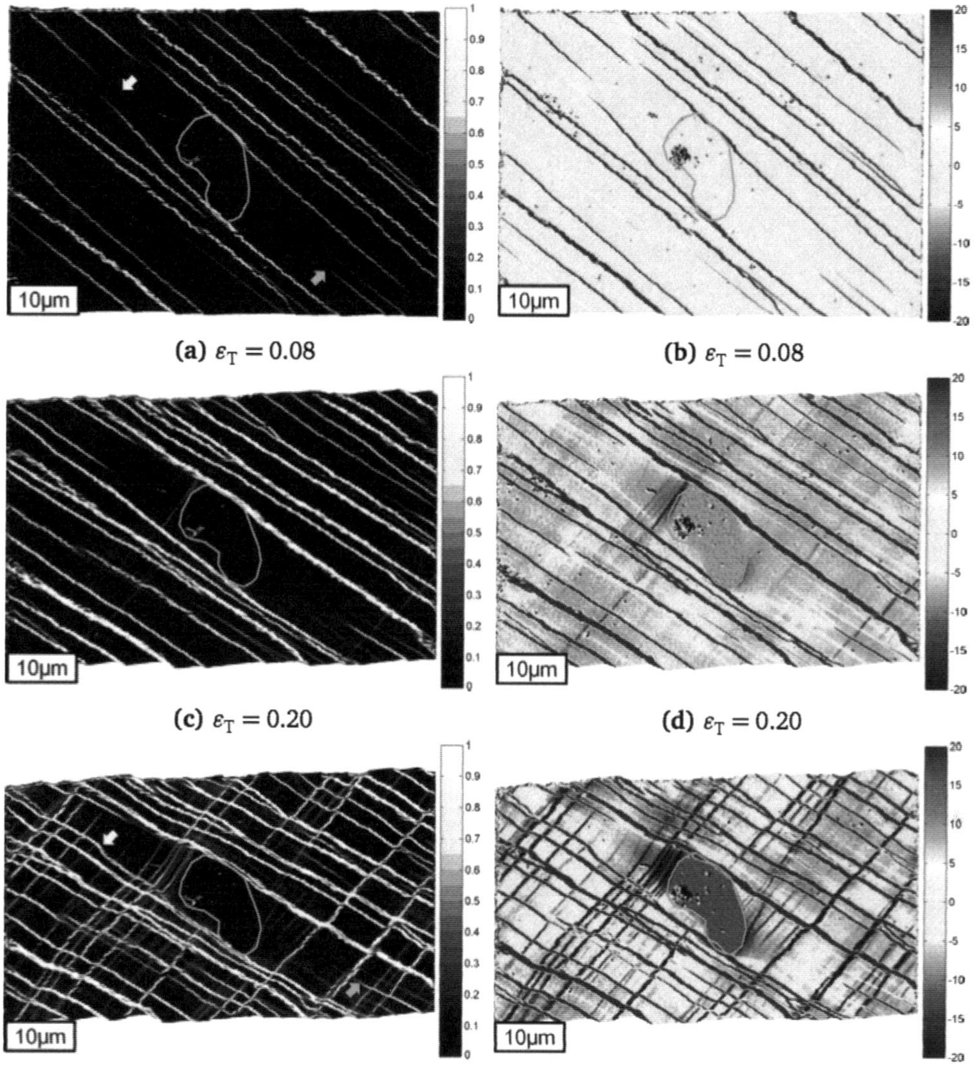

(a) $\varepsilon_{\mathrm{T}} = 0.08$ (b) $\varepsilon_{\mathrm{T}} = 0.08$

(c) $\varepsilon_{\mathrm{T}} = 0.20$ (d) $\varepsilon_{\mathrm{T}} = 0.20$

(e) $\varepsilon_{\mathrm{T}} = 0.31$ (f) $\varepsilon_{\mathrm{T}} = 0.31$

Figure 6. Maps of the local deformation around a large particle measured using HRDIC. Axial strain, ϵ_{yy}, is mapped in (a), (c) and (e), and the rotation, ω_z is mapped in (b),(d) and (f). Note that ϵ_T is the total strain in the area mapped.

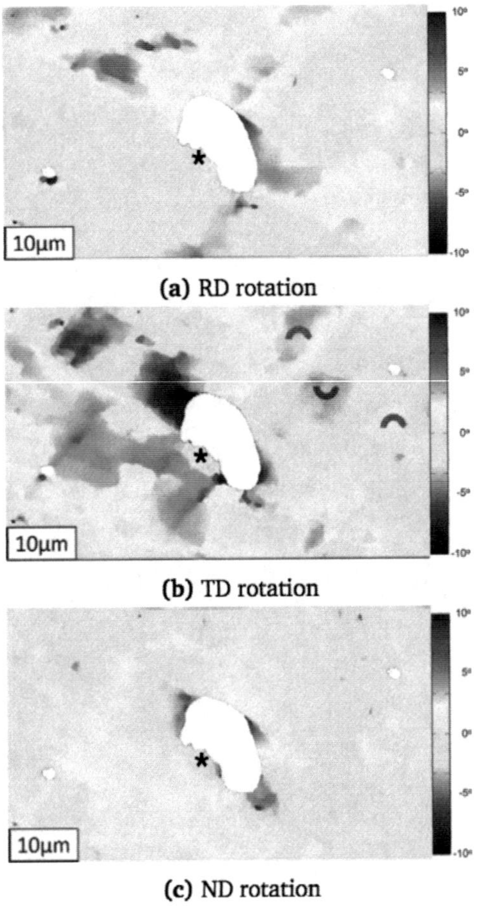

Figure 7. Lattice misorientation maps around a large particle. Misorientation is relative to position $*$ and resolved along RD (a), TD (b) and ND (c).

largest around TD but are non-zero around RD and ND. This implies that although the imposed deformation is plane strain, the resultant lattice rotations are 3 dimensional. The particle clearly affects the local lattice rotation, forming the typical pattern of rotations associated with the particle deformation zones observed in aluminium alloys. A region of negative TD rotation of about $-10°$ can be seen at the upper left and lower right corners whilst a region of positive rotation exists in the upper right and lower left corners of the particle, of around 3° and 10° respectively. These regions have sizes comparable to that of the particle and are remarkably similar both in shape and magnitude to the material rotations in the regions between the bands in figure 6 (f). In addition, there are some very localised 1-2 μm regions of high positive local lattice rotation at the bottom of the particle. These seem to be directly related to very local changes in the thickness and direction of the slip bands.

3.3. Deformation near an isolated 2μm particle

3.3.1. Deformation mapping The deformation maps for a region containing a smaller 2 μm particle are shown in figure 8. The orientation of the grain containing this particle is different ($\phi_1 = 4°, \Phi = 40°, \phi_2 = 26°$) and, as a consequence, the evolution of deformation heterogeneity is also slightly different demonstrating that although the alignment of slip bands is not crystallographic, it is orientation dependent. Although these maps were obtained at the

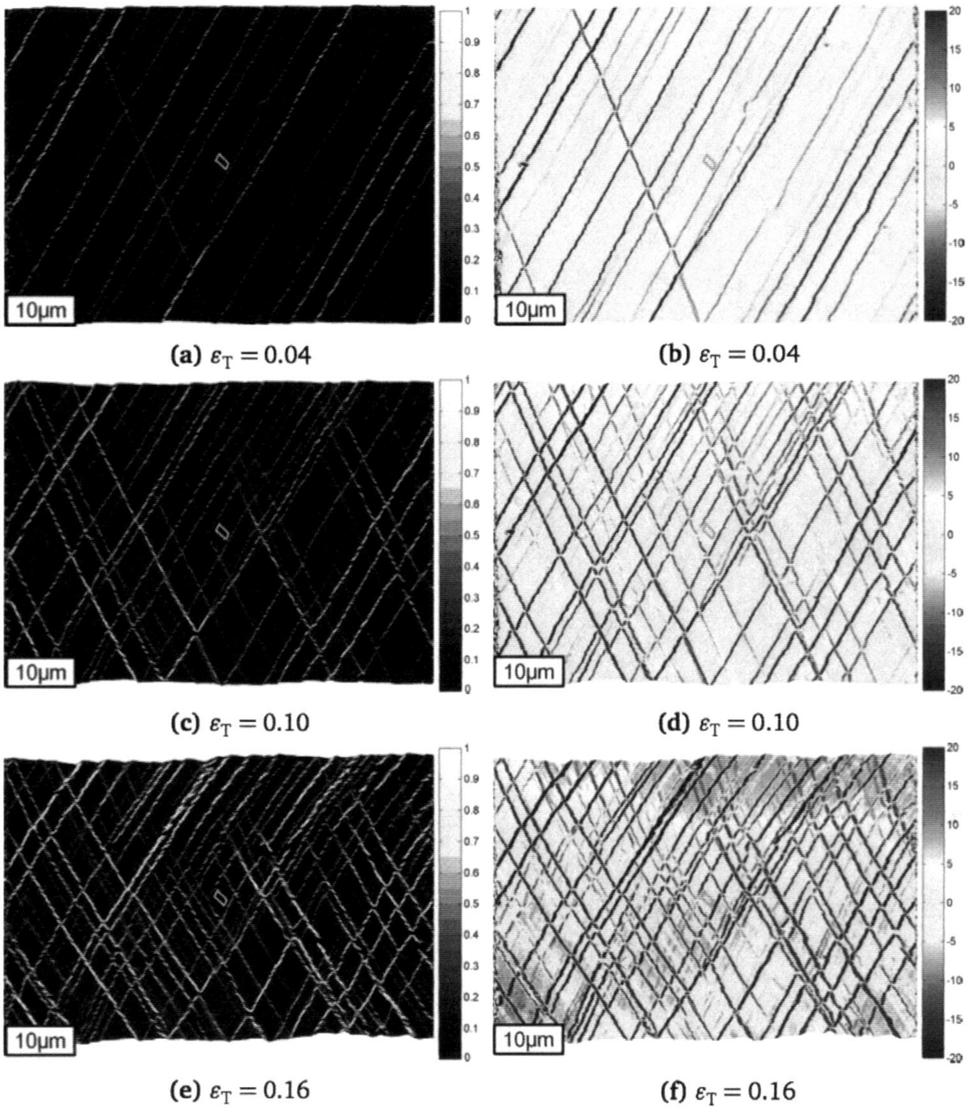

Figure 8. Maps of the local deformation around a small particle measured by HRDIC. Axial strain is mapped in (a), (c) and (e), and the rotation is mapped in (b),(d) and (f). Note that ϵ_T is the total strain in the area mapped. The actual applied strain is about twice ϵ_T.

same strain increments as for the large particle, the local strain in this region is about half, demonstrating the large variations in local strain possible in a polycrystalline material. As before, most of the deformation occurs by shear in discrete slip bands. However, in this case, the second set of slip bands is already visible after an average strain (for the region) of just 0.04 and well established at 0.10. It is worth noting, however, that presumably because this is a harder grain, the macroscopic strain is about twice as high and therefore these secondary bands appear at the same macroscopic strain increment as they did near the larger particle. The number of slip bands increases progressively with deformation and the distance between them

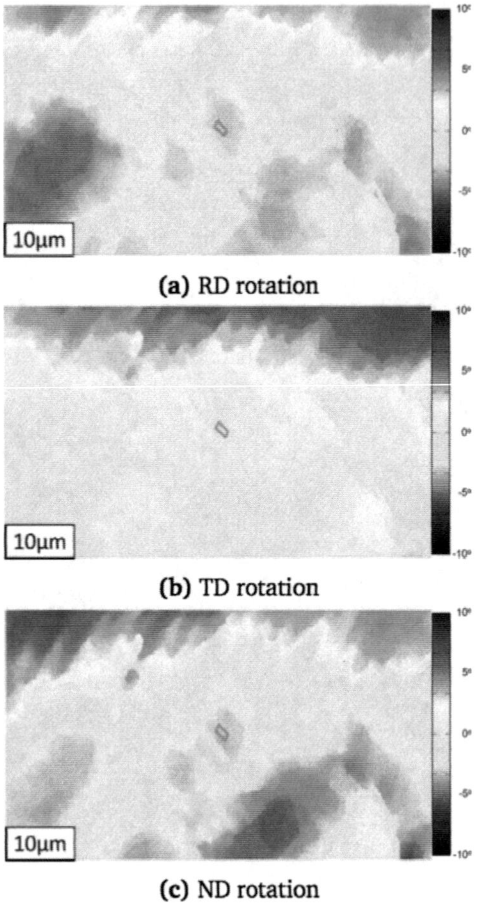

Figure 9. Lattice misorientation maps around a small particle. Misorientation is resolved along RD (a), TD (b) and ND (c).

decreases accordingly. However, the spacing of slip bands is always larger than the particle size. Unlike in the case of the large particle there is no obvious effect of the presence of the particle on the deformation around it. However the distribution of slip bands varies across the region, with a slightly higher density of slip bands just to the right of the particle, for example. At $\epsilon_T = 0.16$ there are two clear sets of complementary bands everywhere apart from the top right region, where the primary bands dominate.

The rotation maps show that the region first rotates homogeneously clockwise by up to 2°, and then splits up, with the central region rotating to about $-6°$ and the top right region, where the primary bands dominate, rotates to 7°. As in the strain maps, the presence of the particle does not seem to affect the local rotation. Rather, it seems to be affected by the intensity, density and spacing of the slip bands which varies with position independently of the presence of the particle.

3.3.2. Lattice rotation mapping Figure 9 shows the local lattice rotation maps obtained by EBSD for the same region. As with the large particle, the change in lattice rotation is highest around TD, but in this case both the ND and RD show considerable orientation gradients of over 10°. However, the highest gradients are not associated with the particle. Instead they appear to delimitate regions far from the particle where the slip band characteristics are different. Around

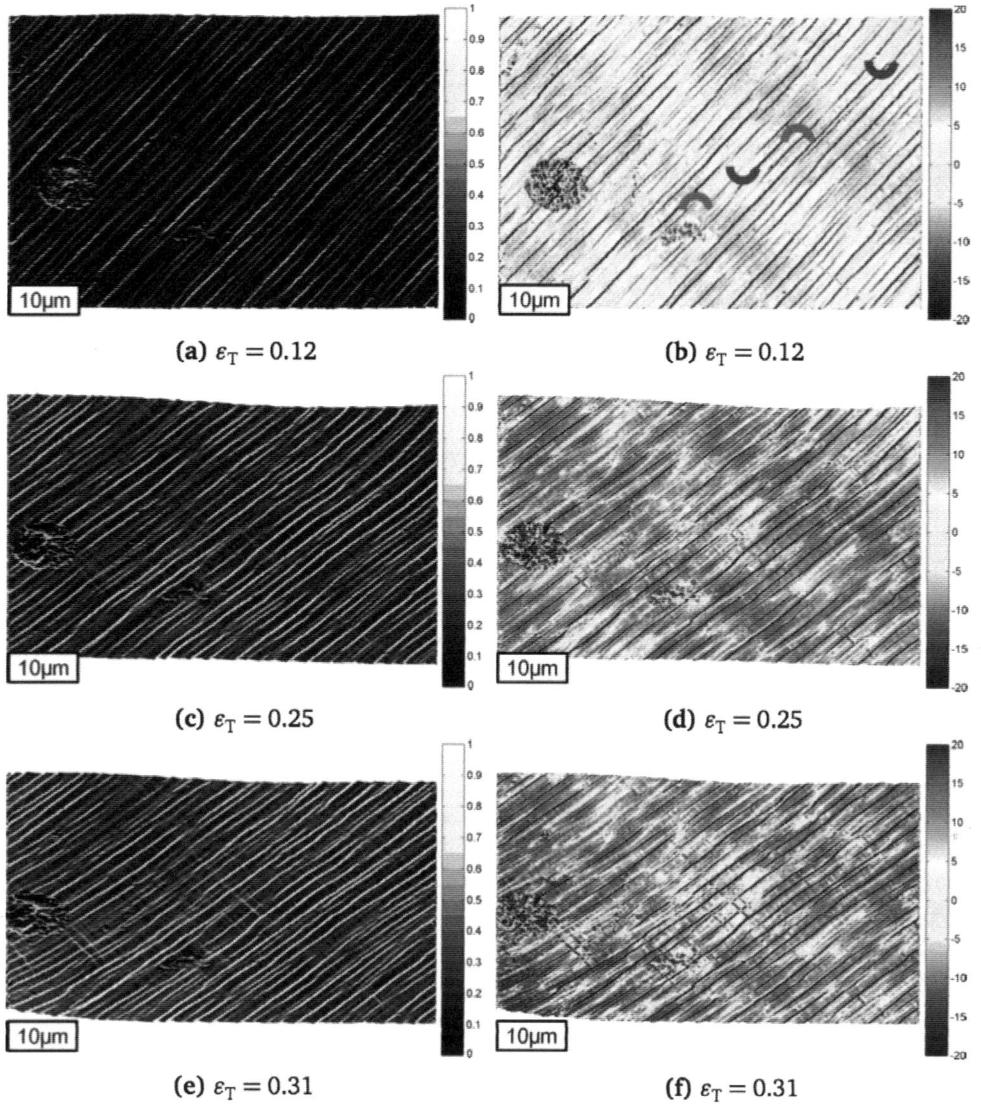

Figure 10. Maps of the local deformation in a material containing very small, evenly spaced particles measured by HRDIC. Axial strain is mapped in (a), (c) and (e), and the rotation is mapped in (b),(d) and (f). Note that ϵ_T is the total strain in the area mapped.

TD, the biggest gradient is at the top of the image, towards the right, which corresponds to the region where only primary bands can be seen in the strain maps in figure 8 (e). Again the lattice rotations are very similar to the material rotations in the regions between the slip bands although in this case the changes in material rotation cannot be related to the particle. In fact there is a strong correspondence between the slip bands and the lattice rotation maps, with features at the top well aligned with the primary deformation bands and those at the bottom aligned with the secondary bands.

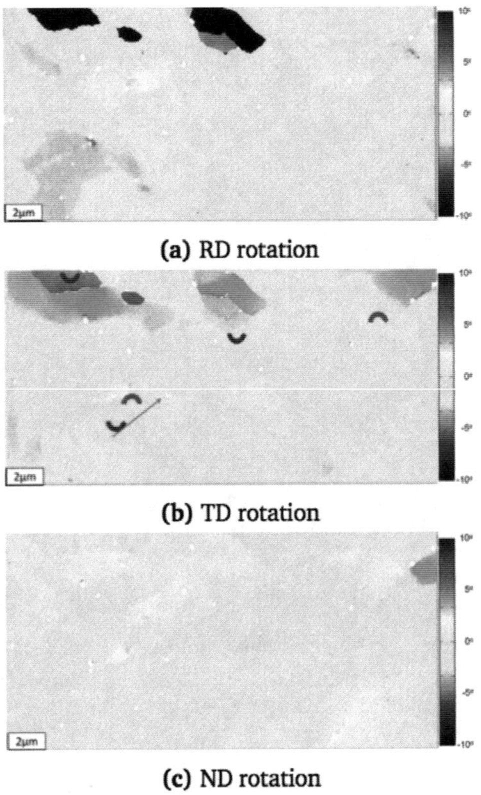

Figure 11. Misorientation maps for a material containing very small, evenly spaced particles. Misorientation is resolved along RD (a), TD (b) and ND (c).

3.4. Deformation in the presence of closely spaced, small (< 200nm) particles

3.4.1. Deformation mapping Forgoing the ageing heat treatment causes the gold remodelling procedure to precipitate very small particles about 100 to 200 nm in size with 1 to 2 µm spacing. Figure 10 shows the deformation in this material, in a grain with orientation $\phi_1 = 31°, \Phi = 1°, \phi_2 = 72°$. Deformation is clearly still dominated by the appearance of slip bands but these have a significantly smaller spacing of about 1 to 2 um, identical to the interparticle spacing. Unlike in the material with coarse particles, the amount of strain within the bands varies noticeably along their length. In fact the bands often appear to be interrupted and slightly offset. Careful examination of the strain maps and the source images reveals that these interruptions and the associated variations in strain are associated with the presence of particles. At a mean strain of 0.12, the strain in slip bands varies from 0.1 to 0.8, with an average of 0.4. At a mean strain of 0.3, this increases to 0.653. In between the bands, the strain varies from 0.03 - 0.07 at $\epsilon_T = 0.12$ and between 0.12 - 0.17 at $\epsilon_T = 0.3$. As was the case in the aged material, secondary slip bands appear at a strain of 0.25 but they are much less intense and much more closely spaced. Clearly the presence of closely spaced small particles makes the deformation much more homogeneous.

The material rotation (ω_z) maps in figures 10 (b), (d) and (f) show that the the rotation of the regions between the primary slip bands starts out very homogeneously. However, with deformation, a patchwork of regions several µm in size with rotations ranging from 0 to 15° develop. However, careful observation reveals that these regions do not correspond to rigid body rotations like those found in the material with coarse particles. Instead they originate from the shear in closely spaced secondary slip bands.

3.4.2. Lattice rotation mapping Figure 11 shows the EBSD result in the centre of the region mapped in figure 10. A smaller subregion was mapped at a higher magnification to reveal the subtle substructure that emerges in this case. The lattice rotation maps reveal patches of alternate lattice rotations 4 - 6 µm in size that correspond to the patches of material rotation identified in the DIC rotation maps. However the misorientations between patches are very small, with a maximum of 2.5°. Some patches are separated by segments of low angle grain boundary, as shown by the misorientation profile in 11 (e). This means that although the shape of the patches is similar in both maps, the amount of lattice rotation between the bands is much smaller than the amount of material rotation measured, unlike in the material containing coarse particles where the two were very similar.

4. Discussion

4.1. The kinematics of deformation and local lattice rotation

The HRDIC measurements have revealed that the deformation in the absence of particles is heterogeneous, with the formation of one set of primary slip bands at the start of deformation and the activation of complementary bands at higher strains. The occurrence of slip bands during deformation of aluminium and its alloys has long been recognised [14] but because these bands are narrow and sparse, their contribution to the overall deformation was assumed to be minimal. However, these new measurements demonstrate, unambiguously, that most of the deformation is in fact accommodated by shear in these bands. The bands have a characteristic spacing of 3-4 µm, which is comparable to the size of the cells formed in this material, suggesting there is a direct relationship between the slip bands and the low angle grain boundaries observed after deformation. The bands are not aligned with slip traces, implying multiple slip within bands, but their alignment does depend on the crystal orientation.

Although the imposed macroscopic deformation is plane strain compression, the local deformation in all the regions studied show significant shear and rigid body rotation. This implies that within a grain inside a polycrystal, the deformation constraints are somewhat relaxed with overall compatibility being achieved by heterogeneous deformation elsewhere in the grain or even in neighbouring grains. Paradoxically, this suggests that the deformation of a grain embedded in a polycrystal deforming in plane strain might be less constrained than a single crystal deformed in plane strain. The consequence is that for much of the deformation most of the shear occurs on one set of primary bands which causes large rotations of the grain, with a second set of complementary bands appearing only at a strain of about 0.25.

The presence of large particles affects the deformation by interacting locally with the slip band pattern. The stresses associated with the particle cause changes in the direction, width, intensity and number of slip bands in the particle deformation zone. This is consistent with the idea that the particle modifies the local plasticity by changing the local slip system activity, as proposed by Humphreys and Kalu [5]. However the details of the kinematics are quite different from those usually assumed. Whereas the in their model the deformation was assumed to be homogeneous with the particle shadowing slip in the different regions around the particle, these results show that in practice this is instead achieved by modifying the discrete slip band pattern that characterises deformation at this scale.

One remarkable observation is that the EBSD lattice rotations are almost identical to the HRDIC material rotations of the regions between the slip bands. This is unexpected, since in crystal plasticity material rotation and lattice rotation should not coincide. This is why the slip bands appear in the material rotation maps and not in the lattice rotation maps. However, because the strain between slip bands is much less that that in the bands, their rotation is determined by the shear in the slip bands and not by the local plastic strain. The stresses around the particle affect the intensity, width, spacing and direction of the slip bands changing the magnitude of the average local shear, and giving rise to the rotations in the PDZ around

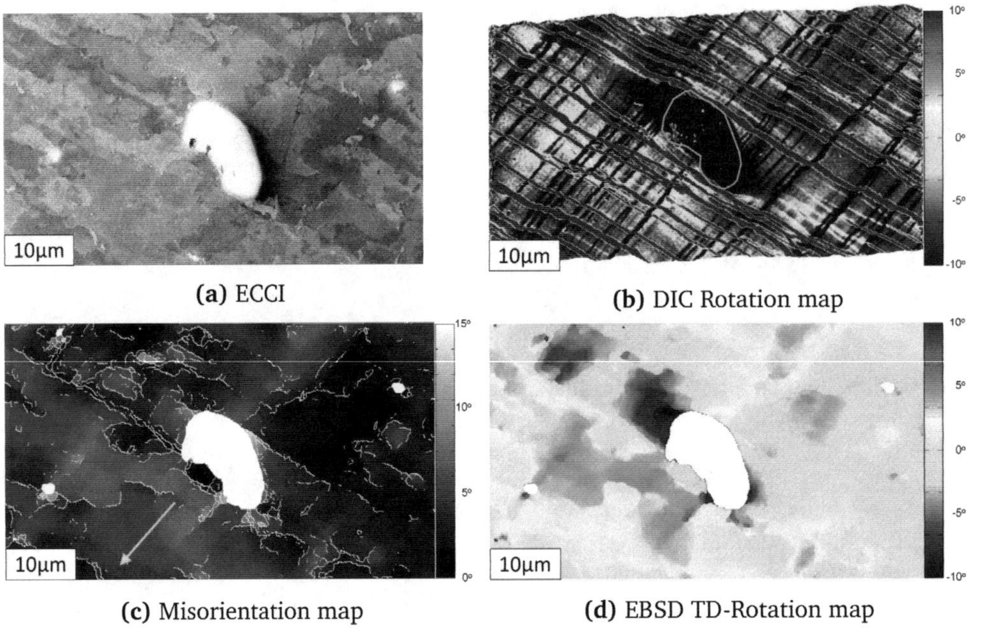

Figure 12. Comparison between images of the substructure following deformation around a large particle and the rotation measured using HRDIC (a) Electron channeling contrast imaging (b)

the particle, which extend out to about 1 particle diameter away from the interface. In addition to these rotation zones, there are also more localised lattice rotations associated directly with changes in the width and direction of the slip bands, which give rise to much smaller $\sim$1 μm misoriented regions near the particle. This is a consequence of the multislip nature of the deformation in these bands, which are actually made up of a fine structure of nonaligned slip, as shown in figure 4 (f).

All these observations suggest that the development of the substructure is directly related to the heterogeneity of deformation. This relationship is confirmed in figure 12 where the region around the largest particle is shown in back scattered contrast alongside the rotation map from HRDIC, a misorientation map from EBSD and the lattice rotation around ND. The electron channelling contrast image shows substructural features aligned with the slip bands in the DIC map. This alignment is even clearer in the misorientation map shown in figure 12 (c) where low angle grain boundaries are outlined. There is direct correspondence between slip bands and low angle grain boundaries in many instances. Although not all slip bands have a corresponding low angle grain boundary, the thicker, more intense ones invariably do, especially those which are forced to bend by the presence of the particle. The regions close to the bottom right corner of the particle, where the slip bands are much closer are offset and also bifurcate, correspond to small subgrains, which were probably formed by the complex local deformation and the very high levels of shear strain associated with the slip bands. Not all bands are associated with subgrains. One such region is that under the green arrow in figure 12 (c), where there is an absence of low angle grain boundaries, even though there are clearly several slip bands. The most likely reason for the absence of low angle grain boundaries here is the lower magnitude of shear in these bands.

If the particle deformation zones are formed by the interaction of the particles with the slip band pattern, this offers an explanation for why the size of the PDZ and lattice rotation is smaller for smaller particles. As the deformation maps near the smaller particle exemplify (figure 9), when the particle size drops below the slip band spacing, the probability of a particle interacting with the slip band decreases and when it does, the effects will be smaller than with large particles. If the particle does not interact with a slip band then the effects will be even smaller because, although the deformation between bands is not zero, it is much less that the strain in the slip bands. The region near the small particle also shows how the lattice rotation generated away from particles by changes in the deformation pattern can be as large as that in a PDZ. The change in slip band alignment gives rise to misorientations of 15°, almost as large as those seen near the large particle. This behaviour is similar to that observed on cube oriented single crystals deformed in uniaxial compression by Liu and Hansen [15], and is therefore not limited to polycrystalline materials. The interaction of the particle with the deformation pattern will depend not only the particle size but also its shape and orientation with respect to the slip pattern and the characteristics of the deformation pattern will depend on grain orientation and local stresses within the grain, which in a polycrystalline sample is affected by neighbouring grain interactions. Taken together, these observations also imply that, in practice, there should be significant variability in the size of the PDZ and on the local misorientations generated, particularly for smaller particles. This is in agreement with previous measurements [6] and consistent with the randomization of texture associated with PSN.

The idea that heterogeneous deformation in the form of intense slip bands is essential to the formation of the deformation substructure can also explain the change in behaviour observed in the presence of very fine, closely spaced particles [8]. The results here show that small, closely spaced particles modify the deformation pattern, causing it to be more homogeneous. Discrete slip bands still appear but they are more closely spaced and less intense. Although as particles become smaller the mechanism of interaction between the particles and the deforming particle will change [16], another important consideration must be the distance between the particles. According to our results, the interaction between the deformation pattern and the particle should decrease as the particle size decreases. However, if as the particles become smaller they also become more numerous and more closely spaced, as often is the case, then interaction with the deformation pattern becomes unavoidable. In the material studied here, the particles are 1-2 µm apart and are about 100 nm in size. The end result is that every particle interacts and interrupts a slip band after only a few micrometres, forcing the slip band spacing to decrease. Deformation continues but at larger stresses, which, in combination with the much more homogenous slip allows local stress variations to be accommodated without significant large accompanying lattice rotations.

4.2. Comparisons with crystal plasticity finite element modelling

Crystal plasticity finite element modelling (CPFEM) was used used to model the deformation around the large particle. CPFEM is used for a variety of applications [17] including the modelling of deformation at the microstructural scale. The implementation used in this study was developed by Bate [18] and has been used to predict the formation of deformation bands and PDZ in both 2D [10] and 3D [7]. Although the same implementation is used the details of the model used here are quite different. In an attempt to capture any geometric effects of particle geometry, the mesh was created by sampling the EBSD map before deformation and using quadratic hexahedral elements to make up a volume which was 1 element thick, with the area mapped represented by an array of 70×95 elements. The boundary conditions were applied to simulate plane strain compression to a compressive strain of 0.3 in increments of $1{\times}10^{-4}$. This model is clearly unrealistic in several ways. In practice, deformation will be 3 dimensional and the material around the particle is therefore over-constrained in the model.

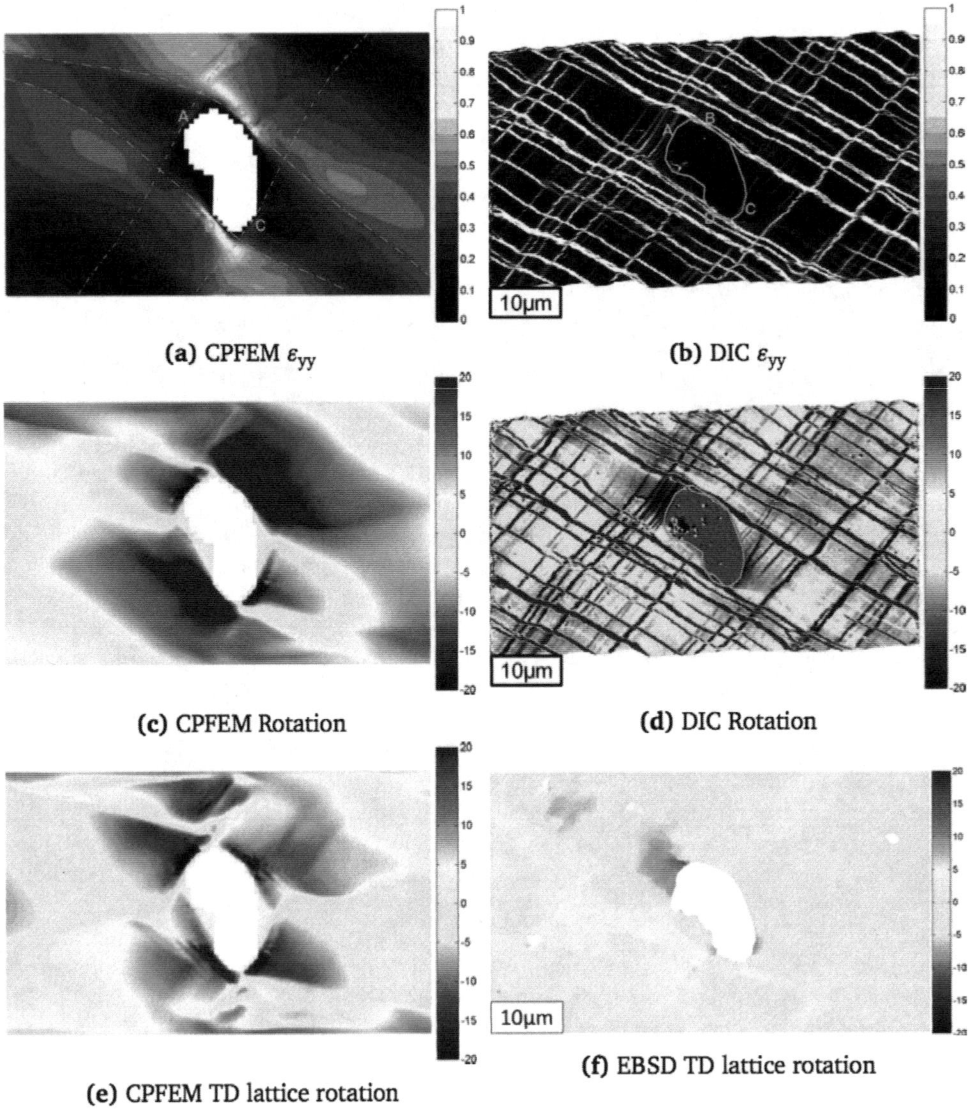

Figure 13. Comparison between the predictions of crystal plasticity modelling (a), (c) and (e) and deformation mapping results using EBSD (b), (d) and (f).

This is particularly true for modelling the plane strain compression experiment because although the particle is constrained to some extent, a small amount of out plane deformation can and does occur. However it is difficult to do much better without knowledge of the full shape of the particle and grain below the surface. This means that quantitative comparisons must be made with care and the impact of these shortcomings should be considered when comparing the results of the simulation with those of the experiment. The material model, the details of which can be found in [18] ,[10] and [7], was calibrated using both the stress-strain data from compression testing and plane-strain compression testing of the material containing coarse particles, with a strain rate sensitivity of 0.02, making the behaviour effectively rate insensitive.

The predicted strain distributions are shown in figure 13(a), alongside the HRDIC measurements. Although the predicted strain distribution is clearly very different from that measured, it has similar characteristics. The model is clearly unable to capture the slip band patterning and therefore the maximum value of strain predicted ($\sim$0.8) is much less than that measured, which in some bands exceeds 2. There is however a relationship between the slip band density and the predicted strain intensity, that is, where the predicted strain is high, above and below the particle, the slip band density is higher. The direction of the dominant shear is predicted well, which suggests it is determined by the crystal orientation. Unlike the bands obtained experimentally, however, the bands of strain in the model are curved, which is a consequence of the plane strain boundary constraints imposed. The experimental strain map shows that the entire region is sheared and rotated during compression, which does not happen in the simulation. This rotation also causes the particle to rotate, which also does not happen in the model. Despite this, there are similarities in the way that the particle affects the deformation in the surrounding matrix, and the shape of the PDZ that develops. The change in slip activity is captured to some extent, with the positions of regions with low strain coinciding with those in the DIC measurements. From the rotation (both rigid-body and lattice) results, it can be seen that the direction of rotation predicted by CPFEM in the matrix around the particle generally matched with the results measured in DIC and EBSD, however the magnitudes of the rotations predicted are significantly larger than those measured.

There are two main reasons for the discrepancies between the modelling predictions and experimental results. The first lies in the unrealistic boundary conditions used. As discussed previously, the region studied shows a significant amount of shear indicating that the local deformation is less constrained than that imposed in the model. This added constraint can help explain the higher values of local misorientation predicted. This could also help explain why measurements on single crystals show larger rotations in the PDZ than those in polycrystals, since it is possible that in a polycrystal neighbouring grains can help relax the constraints locally. The second reason for the differences observed is the inability of the continuum CPFE model to simulate the slip pattern development in these materials. The model predicts a smooth, continuous variation in strain which is clearly not representative. In face of these limitations, it is perhaps surprising that such a simple model predicts many aspects of the local deformation well, suggesting it can be useful to understand effects of particle shape and grain orientation. However, it is clear that its inability to predict slip patterning makes it unable to predict the effect of particle size on the local rotation. This also explains why simple strain gradient plasticity theories where length scale considerations affect the work hardening but not the the distribution of strain are also unable to predict this effect [18].

5. Summary and conclusions

High resolution digital image correlation was used to map the deformation around different size particles in a polycrystalline Al-Si alloy during plane strain compression. These measurements revealed that in the material containing coarse particles deformation is inhomogeneous at the scale of the particle, forming a pattern of slip bands with a characteristic spacing. This pattern correlates well with the cell structure found in the material after deformation. The slip pattern is not crystallographically aligned, but its alignment depends on the crystal orientation. The presence of particles modifies the pattern locally, causing slip bands to intensify, broaden, bend and increase in number. These changes in the slip pattern, which can be interpreted as changes in local slip activity, cause the local lattice rotations usually observed in the particle deformation zone. However, when the particle size drops below the slip band spacing, so does the probability of particle slip band interaction. Since the strain between slip bands is significantly less than the strain in the slip bands, the deformation zone of a particle smaller than the slip band spacing is smaller and produces smaller rotations. The presence of very small, closely spaced particles changes the deformation pattern everywhere, because slip cannot occur without interacting with the particles. This has the effect of increasing the flow stress and of homogenising deformation. Limited comparisons with crystal plasticity finite element modelling predictions show that these models can capture important features of the PDZ like its shape and orientation, confirming the idea that it depends on crystal orientation. However, because the model does not predict the heterogeneous nature of deformation and its characteristic spacing it cannot predict the effect of particle size on the development of the PDZ. One interesting implication is that in a material containing both large particles and a distribution of closely spaced, fine particles, the effect of size on the magnitude of the lattice rotation in the PDZ should disappear and show good agreement with CPFE predictions.

Acknowledgements
This research was funded by the LATEST2 EPSRC programme grant EP/H020047/1. The authors would like to thank Pete Bate for starting this work and for providing the crystal plasticity finite element code used.

References
[1] Humphreys F J and Hatherly M 2004 *Recrystallization and Related Annealing Phenomena (Second Edition)* ed Hatherly F J H (Oxford: Elsevier)
[2] Nes E, Ryum N and Hunderi O 1985 *Acta Metallurgica* **33** 11–22
[3] Humphreys F J 1979 *Acta Metallurgica* **27** 1801–1814
[4] Humphreys F J and Kalu P N 1990 *Acta Metallurgica et Materialia* **38** 917–930
[5] Humphreys F J and Ardakani M G 1994 *Acta Metallurgica et Materialia* **42** 749–761
[6] Clarke A, Humphreys F J and Bate P S 2003 *Materials Science Forum* **426-432** 399–404
[7] Humphreys J F, Bate P S, Gholinia A and Brough I 2012 *Materials Science Forum* **715-716** 23–32
[8] Apps P J, Berta M and Prangnell P B 2005 *Acta Materialia* **53** 499–511
[9] Humphreys F J 1977 *Acta Metallurgica* **25** 1323–1344
[10] Humphreys J and Bate P 2003 *Scripta Materialia* **48** 173–178
[11] Di Gioacchino F and da Fonseca J Q 2013 *Experimental Mechanics* **53** 743–754
[12] Humphreys F J, Bate P S and Hurley P J 2001 *Journal of Microscopy* **201** 50–58
[13] Quinta da Fonseca J, Mummery P and Withers P 2005 *Journal of Microscopy* **218** 9–21
[14] Bay B, Hansen N, Hughes D A and Kuhlmann-Wilsdorf D 1992 *Acta Metallurgica et Materialia* **40** 205–219
[15] Liu Q and Hansen N 1995 *Scripta Metallurgica et Materialia* **32** 1289–1295
[16] Brown L M 1997 *Philosophical Transactions of the Royal Society of London A: Mathematical, Physical and Engineering Sciences* **355** 1979–1990
[17] Roters F, Eisenlohr P, Hantcherli L, Tjahjanto D D, Bieler T R and Raabe D 2010 *Acta Materialia* **58** 1152–1211
[18] Bate P 1999 *Philosophical Transactions of the Royal Society of London A: Mathematical, Physical and Engineering Sciences* **357** 1589–1601

Precipitation, strength and work hardening of age hardened aluminium alloys

Ø Ryen[1], B Holmedal[2], K Marthinsen[2] and T Furu[3]

[1]Oceaneering Asset Integrity, Havnegata 7, 7010 Trondheim, Norway
[2]Norwegian Department of Material Science and Engineering, University of Science and Technology, Alfred Getz veg 2, 7491 Trondheim, Norway
[3]Norsk Hydro ASA, Corporate Technology office, 6600 Sunndalsøra, Norway

E-mail: bjorn.holmedal@ntnu.no

Abstract. The strength and work hardening of age hardened AA6063 and AA6082 alloys have been investigated in terms of a detailed characterization of precipitate and dislocation structures obtained by TEM and SEM. Tensile and compression tests were performed at as quenched, peak aged and severely aged conditions. A strong work hardening in the as quenched condition was found, similar to AlMg alloys with twice as much alloying elements in solid solution. It was found that the initial work hardening rate and the critical failure strain are both smallest at the peak aged condition. During large deformations the needle-shaped precipitates are sheared uniformly by dislocations altering their <001> orientations, which indicates extensive cross slip. In the overaged condition the early initial work hardening is larger than at the peak aged condition, but followed by a weak linear work hardening, apparently directly entering stage IV at a low strain. Cracked, needle-shaped precipitates were seen at larger strains.

1. Introduction

Precipitation of thermodynamically metastable or stable phases in age-hardenable aluminium alloys enables design of a variety of strong and light alloys. Due to a good ductility many of these alloys are applicable for energy absorption applications in safety constructions for military or transportation purposes. In terms of high volume production the AA6xxx alloys are very important with less expensive alloying elements. With recent improvements of analysing techniques, such as high-resolution TEM and atom probe microscopy, it has become possible to identify nearly all the phases present, and to visualise very small details as atomic clusters. The precipitation sequence may be written as:

$$\alpha(\text{ssss}) \rightarrow Mg/Si\text{-clusters} \rightarrow \text{GP-zones} \rightarrow \beta'' \rightarrow (\beta' + B' + U1 + U2) \rightarrow \beta(Mg_2Si) \; + \; Si \quad (1)$$

From the solid solution, several types of atomic clusters with various amounts of Mg and Si are formed [1, 2]. The clusters gradually transform into GP-zones, which are more related to the crystallography of the matrix [3], and further into β"-precipitates. β" has been established as a monoclinic Mg_5Si_6-phase [4], and its occurrence is often observed to coincide with the maximum strength of the alloy (T6). Upon overageing several types of particles are formed, such as B', β' and β, the latter one being the equilibrium phase Mg_2Si. Si-particles may also be formed. Other phases have been found as well, and even more will probably come. In Al-Mg-Si-Cu alloys, for example, variants of a Q-phase are identified [5, 6].

At room temperature the sequence will normally be limited to clustering, while commercial annealing temperatures (150-190°C) will give GP-zones and β", as well as several phases if the annealing time is long enough. Even higher temperatures are required to form the equilibrium phase.

Deformation structures are directly linked to the interaction between dislocations and obstacles in the material. Several authors have treated this topic in detail, e.g. Ashby [7] and Brown and Ham [8] and Ardell [9]. Humphreys and Hatherly [10] give a thorough description of the evolution of deformation structures in various alloys. The large number of publications from the Risø-group [11-13] provide rich evidence for the development of substructure in pure aluminium and Al-Mg alloys. Pure aluminium and low solute alloys generally exhibit a cell- or subgrain structure, i.e. the dislocations organise themselves into boundaries surrounding volumes of a low dislocation density. This self-organisation is limited by the presence of solute atoms, giving a lower rate of dynamic recovery and less well-defined subgrain structure. When the Mg content reaches 4-5% the alloy is often termed non-cell forming. A work by Korbel et al. [14] describes such a microstructure and strain localisation in Al-Mg.

A few articles exist where the deformation structure of aged Al-Mg-Si alloys are studied. It is generally observed that underaged samples get a uniform distribution of dislocations, while the dislocations localise into bands in peak aged samples [15, 16]. This banding is often assumed to be related to the shearing of precipitates, creating softer traces for the dislocation slip. The phenomenon is reported for other age-hardenable alloys as well, i.e. Al-Cu, Al-Li and Al-Zn-Mg [17-19]. Also Dowling and Martin [20] observed these dislocation bands in Al-Mg-Si, and found that if strong particles (dispersoids) were present along with the precipitates the bands disappeared after a few percent strain.

In addition to the development of a dislocation structure, it is of great interest to understand the performance of the precipitates during deformation. As mentioned, the precipitates present at peak hardness are shearable, which is evidenced by the in-situ TEM work of Vivas et al. [21] on an AA6056 alloy. After heavy deformation of peak aged alloys the shearable precipitates dissolve, i.e. they are cut into so small pieces that they become thermodynamically unstable [22-24].

In overaged conditions the precipitates are to a larger extent non-shearable, i.e. they remain undeformed while the matrix around them is deformed. Poole et al. [25] have observed that this is the case for small strain deformation of an Al-Mg-Si-Cu alloy, and that dislocations tangle around the precipitates. When the strain exceeds 20% (in cold rolling) the long rod/lath-shaped precpitates are seen to be fragmented into shorter segments. Nourbakhsh and Nutting [23] observed the same after cold rolling of an overaged Al-Cu alloy. They related the low, and partially negative, work hardening rate of the overaged condition to such a fragmentation of the θ'-plates.

An important and promising application of AA6xxx alloys is in the crash-components of vehicles. Therefore, the properties defining the crashworthiness of such alloys are of great interest. The yield strength is important in order to resist deformation, while the ability to work harden has a significant impact on the energy absorption capacity. In order to avoid cracking and failure the ductility (uniform strain, fracture strain) should not be too low. A parameter that has gained attention is the critical failure strain (CFS), [26] defined from a tensile test by:

$$CFS = -\ln\left(1-\left(\tfrac{1}{2}e_u+\left(1-\tfrac{1}{2}e_u\right)\left(1-R_f/R_m\right)\right)\right) \qquad (2)$$

where e_u is the nominal uniform elongation and R_f and R_m are the nominal fracture stress and maximum stress, respectively. This parameter basically reflects the reduction in area in the neck of a tensile test specimen.

The variation in yield strength of Al-Mg-Si alloys, and other age-hardenable alloys, as a function of ageing time and temperature has been studied extensively, and several groups have provided good models for this important mechanical property [27-29]. Much less attention has been paid to the influence of a precipitate structure on the work hardening behaviour, especially when it comes to large strains. Cheng et al. [30-33] studied the work hardening of several tempers of AAxxx and AA7xxx alloys based on tensile tests, spanning from the supersaturated solid solution to overaged conditions. The initial work hardening rate was seen to decrease with ageing up to the peak-aged condition. Upon overageing it increased again as the particles grew in size and became non- shearable. The transition from shearable to non-shearable precipitates was correlated to a very low hardening rate at the end of the tensile test, a behaviour also observed by Barlat and Vasudevan [34] in AA7075. One limitation of tensile tests is the fact that peak- and overaged samples have low ductility, only a few percent. In order to fully understand the work hardening and deformation mechanisms, it is necessary to study the behaviour up to larger strains and methods other than tensile tests should be used. A few results from large strain experiments on various alloy systems may be found in the literature, and they essentially report the same trend, namely a decreasing work hardening rate with increasing ageing. Gracio et al. [35] performed shear tests on a number of conditions of AA6022, and Lloyd and Kenny [18] torsion tested the alloys AA2036 and AA7010 in their T4 and T6 tempers. Westermann et al. [31-33] looked at 7xxx alloys, and Campbell et al. [36] performed hydraulic bulge tests on an AA6010 alloy and found a very low work hardening rate in the strain range ε=0.1-0.5 in the overaged specimen.

In this paper the initial microstructures of several tempers of AA6063 and AA6082, are studied. The microstructure after plastic deformation is investigated for three different tempers of AA6082 in terms of precipitate structures, dislocation structures and stress-strain behaviour. An attempt to model the work hardening behaviour of the same two alloys is reported in [37].

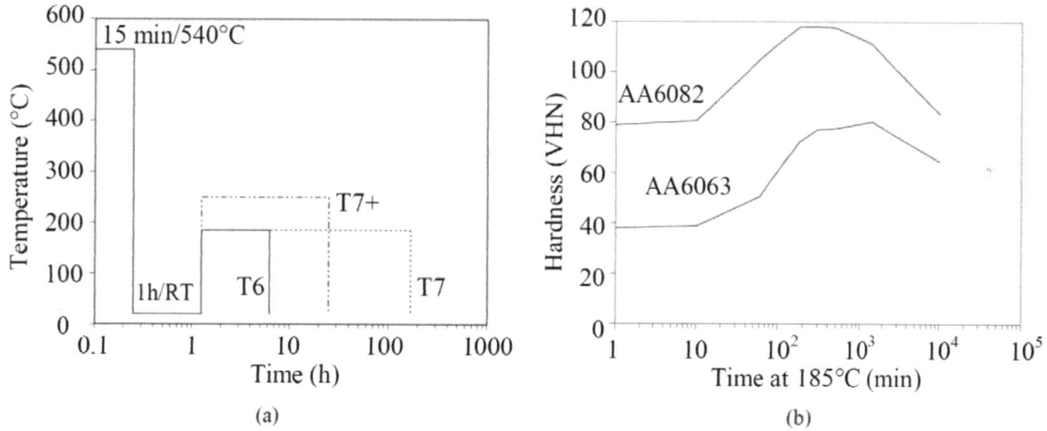

Figure 1: (a) Heat-treatment schedule for the conditions investigated in this report. (b) Hardness as a function of isothermal annealing time at 185°C after 1h/RT.

Table 1. Nominal (and effective) alloy composition of AA6063 and AA6082 (wt%/at%).

Alloy	Mg	Si	Fe	Mn	Cu	eff. Si
AA6063	0.46/ 0.51	0.44/ 0.42	0.19/ 0.09	0.03/ 0.015	0.006/ 0.003	0.38/ 0.36
AA6082	0.67/ 0.74	1.04/ 1.00	0.20/ 0.10	0.54/ 0.27	0.003/ 0.0015	0.79/ 0.76

2. Material and experimental procedures

DC-cast billets of the commercial Al-Mg-Si alloys AA6063 and AA6082 were chosen in favour of extruded profiles in order to avoid effects from texture and deformation structure, but also because of the possibility to make compression specimens of the materials. The chemical composition of the alloys is given in Table 1. It is known that due to the presence of Fe and Mn, a fraction of the Si will be consumed by primary particles and dispersoids. A rule of thumb says that the effective Si-concentration in solid solution in the matrix after casting is reduced to:

$$[Si]_{eff} = [Si]_{nom} - \tfrac{1}{3}\left([Fe]+[Mn]\right) \tag{3}$$

where [X] is the concentration of element X in weight percent. The effective Si-concentration in solid solution is also given in Table 1. The alloys had coarse, equiaxed grain structure (grain size of the order 80-100 μm) and were expected to be texture-free. The billets were homogenised by slow heating (200°C/h) to 580°C, where they were held for 3 h before slow cooling (350°C/h).

The heat-treatment procedure following homogenisation is illustrated in figure 1(a). The alloys were solution heat-treated at 540°C for 15 minutes, quenched, stored at room temperature for one hour and isothermally annealed at 185°C. In figure 1(b) the variation in hardness with ageing time at 185°C is shown. The precipitate structure was studied in the tempers T6, which is the peak strength condition (5h/185°C) and T7, which is overaged (1 week/185°C). In addition, a severely overaged condition of AA6082, called T7+ (24h/250°C), was processed and investigated.

Investigations of the microstructure were performed using a Philips CM-30 transmission electron microscope operating at 150 kV. A Gatan EELS spectrometer was used to measure the foil thickness with an estimated accuracy of 9%. Precipitates were studied with the incident beam parallel to an <001>-direction in the matrix, while the dispersoids were best viewed when tilted out of any zone axis. The size and number of particles were measured on photographic plates. The number density of dispersoids, N_d, was calculated by counting the number, N, over a given area, A, of a specimen of thickness, t:

$$N_d = N/A\left(t+\bar{d}\right) \tag{4}$$

where $\bar{d}$ is the average diameter of the dispersoids. The term in the parantheses is a correction for the fact that some of the particles were larger than the foil thickness. As the dispersoids generally appeared to be somewhat elongated, a maximum and minimum diameter were measured on each particle. By approximating them to spheres, a volume fraction, f_d, could be calculated:

$$f_d = N_d \tfrac{4}{3}\pi\left(\tfrac{1}{2}\bar{d}\right)^3 \tag{5}$$

A similar procedure was used to determine the number density of precipitates, N_p:

$$N_p = 3N/A\left(t+\bar{l}\right) \tag{6}$$

Since the precipitates were needles or rods oriented along the three <100>-directions in the Al- matrix, they appeared as streaks and dots in the TEM. Only the precipitates imaged end-on (dots) were counted (N), as these were the easiest to count. By assuming that they were randomly distributed in three directions, N was multiplied by 3. In order to correct for particles extending out of the foil, the average length, $\bar{l}$, of the precipitates was added to the foil thickness. The volume fraction, f_p, could then be calculated:

$$f_p = N_p \cdot \bar{l} \cdot \bar{A}_p \tag{7}$$

where $\bar{A}_p$ is the average cross-sectional area of the needles/rods.

Tensile testing was performed on flat specimens (thickness 3mm) in an Instron testing machine, or round specimens (diameter 6 mm) on an MTS machine. Some tests were interrupted at strain levels of 1, 2 and 5%. Uniaxial compression tests of circular specimens (diameter 10 mm) were carried out on a Gleeble machine. A strain rate of $0.1s^{-1}$ was used in the Instron and Gleeble machines, while the strain rate in the MTS was $0.01s^{-1}$. Additionally, the T6 and T7 conditions of AA6082 were cold rolled to large strains. Tensile specimens were cut from the rolled sheets and tested at a strain rate of $0.01s^{-1}$.

The T7-curves had an inflection point around ε=0.5, from which it is suspected that some friction effects arise at the highest strains. Therefore, only results up to a strain of ε=0.4 are considered in this report. A second problem with measuring the strain based on tool movements in the compression tests is the low apparent E-modulus as compared to the tensile tests. This obvious error, which is caused by elastic compression of the tools, is corrected for by applying the following expressions:

$$\varepsilon = \ln\left(\exp(\varepsilon') - \frac{\sigma'}{E'\exp(\varepsilon')}\right) \tag{8}$$

$$\sigma = \sigma'\exp(\varepsilon - \varepsilon') \tag{9}$$

where ε', σ' and E' are the strain, stress and E-modulus measured in the compression test.

In order to study the evolution in deformation structure with strain, three tempers of the AA6082 alloy were examined: W60: Solution heat-treated at 540°C for 15 minutes, quenched and stored at room temperature for one hour before deformation. T6: as W60, but annealed at 185°C for 5h before deformation. T7: as W60, but annealed at 185°C for 1 week before deformation. The strain levels at which the microstructural examination could be carried out, were determined by the tests performed. Fractured tensile specimens were studied, assuming that the strain outside of the neck equalled the uniform elongation of the specimen. For W60 this strain was $\varepsilon=0.17$, while it was only $\varepsilon=0.05$ for T6 and T7. The compression tests were stopped at a strain of about $\varepsilon = 0.8$. Since the deformation of compression specimens generally is not homogeneous, they were examined in the middle, i.e. where the effective strain is highest (somewhat higher than the global strain). Both SEM and TEM were used to investigate the microstructure of the deformed materials. Bulk specimens for SEM were prepared by mechanical grinding and polishing, followed by a careful electro polishing (90s/12V) in a methanol-nitric acid solution at -30°C. A Hitachi S-4300SE field emission SEM was used.

Thin foils for TEM were prepared by electrolytical jet-polishing in a methanol-nitric acid solution at -30°C. A Philips CM-30 instrument operating at 150 kV, equipped with a Gatan EELS spectrometer, was used in most of the microstructure investigations. In addition a JEOL JEM-2010 (160 kV) was used. Estimates of the dislocation density were made by counting dislocations (N) intersecting a specified line length (l) on micrographs, taken at positions of known foil thickness (t):

$$\rho = C_1 C_2 N / t\, l \tag{10}$$

where C_1 is a visibility factor (1 for low-index zone axes, except 1.1 for <110>) and C_2 is a stereological factor, in this case 2 [38].

Table 2. Analysis of the dispersoid structure in AA6082.

Mean radius	80 nm
Aspect ratio	1.9
Number density (N_d)	3.3x1018 /m3
Volume fraction (f_d)	1.2 %

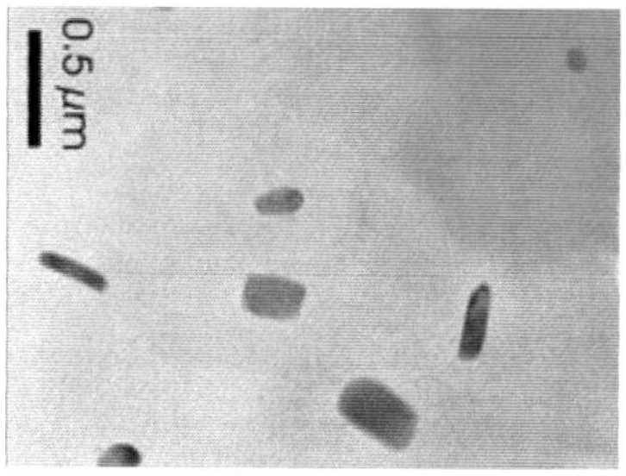

Figure 2. Dispersoids in AA6082.

IOP Conf. Series: Materials Science and Engineering **89** (2015) 012013 doi:10.1088/1757-899X/89/1/012013

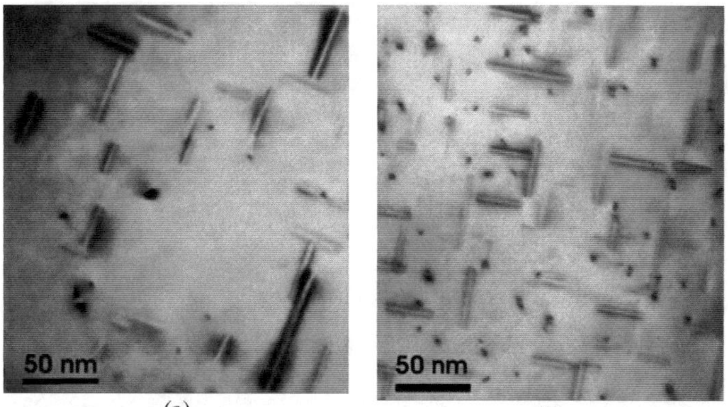

(a) (b)

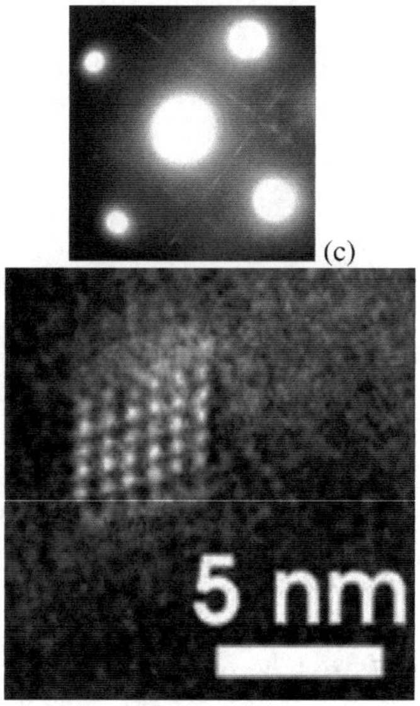

(c)

(d)

Figure 3. β''-precipitates oriented along three orthogonal directions in the T6 condition of (a) AA6063 (foil thickness 56 nm) and (b) AA6082 (foil thickness 43 nm). (c) SA diffraction pattern with characteristic streaks between the (200) matrix reflections. (d) Precipitate viewed end-on, revealing the unit cell of the β''-precipitate. The angle between a and c (105°) is easily seen.

Table 3. Analysis of the precipitate structures in AA6063.

	As quenched	T6
Number density (N_p)	-	7.91 x10^{21} /m^3
Volume fraction (f_p)	-	0.30 %
Length	-	32.1 nm
Cross-section area	-	11.9 nm^2
Cross-section aspect ratio	-	1.5
Equivalent cylinder radius	-	1.95 nm
Equivalent sphere radius	-	4.5 nm
Mg in solid solution	0.51 at%[1]	0.40 at%
Si in solid solution	0.36 at%[1]	0.21 t%

[1]Not measured, but taken from the nominal and effective composition, see above

Table 4. Analysis of the precipitate structures in AA6082.

	As quenched	T6	T7	T7+
Number density (N_p)	-	4.26 x10^{22} /m^3	1.04 x10^{21} /m^3	-
Volume fraction (f_p)	-	0.83 %	1.71 %	-
Length	-	28.8 nm	296 nm	0.2-3μm
Cross-section area	-	6.8 nm^2	55.6 nm^2	-
Cross-section aspect ratio	-	1.5	2.1	-
Equivalent cylinder radius	-	1.1 nm	4.2 nm	7-75 nm
Equivalent sphere radius	-	3.6 nm	15.8 nm	-
Mg in solid solution	0.51 at%[1]	0.44 at%	-	-
Si in solid solution	0.36 at%[1]	0.34 at%	-	-

[1]Not measured, but taken from the nominal and effective composition, see above

3. Microstructure

3.1 Dispersoids

Due to the presence of Mn in AA6082, this alloy contains small dispersoids after homogenisation that are stable during later processing. According to the phase diagram, Al6Mn particles are formed in a binary system, but as both Fe and Si are present the particles have more complicated stoichiometries with various amounts of Al, Mn, Fe and Si.

A micrograph of dispersoids is shown in figure 2. In total 140 dispersoids were counted and measured, and the number density and volume fraction determined using equations (4)-(5). The results are summarised in Table 2. No distinction was made between primary particles and dispersoids, as the density of primary particles was expected to be much lower than the dispersoid density and barely contributes to the average values.

3.2 Precipitate structure in peak aged (T6) condition

TEM micrographs of the peak aged condition (5h/ 185°C) of AA6063 and AA6082 are shown in figure 3. As expected, the alloys contain a high density of small needle-shaped precipitates. A difference in number density between the two alloys is also obvious. The appearance of double-lines confirms that the precipitates are surrounded by a coherency strain field. At least 90% of the precipitates, as the one enlarged in figure 3(d), were identified as β"-precipitates having a unit cell of a=1.516 nm, b=0.405, c=0.674 nm and an angle between a and c of 105.3° [4]. By assuming that all the precipitates are β" with composition Mg_5Si_6, an estimate is given of the amount of Mg and Si that is left in solid solution. In Table 3 and Table 4 measured and calculated average values according to (6)-(7) are given. It is seen that the number density and volume fraction is much less in AA6063 than in AA6082, but at the same time the cross-section area is significantly larger. The precipitates are not exactly cylindrical, as seen by the cross-section aspect ratio of 1.5 in both alloys. However, the area is converted into an equivalent cylinder radius, of which distributions are shown in figure 4 along with the length distributions (normalised by the average value).

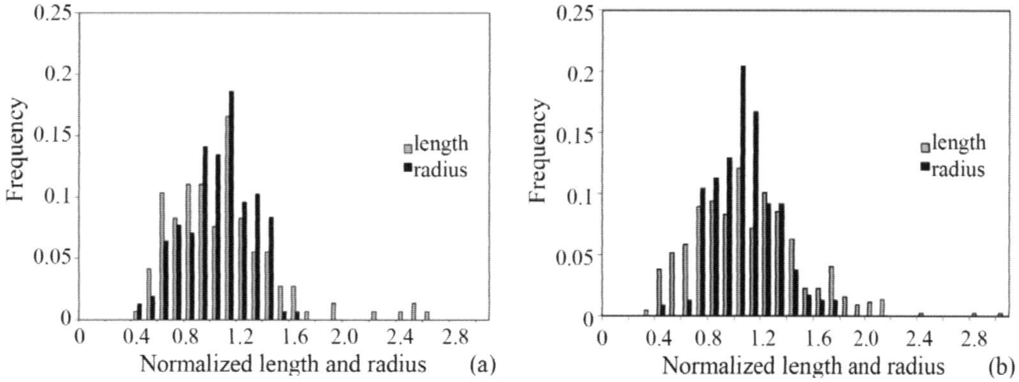

Figure 4. Frequency distributions of length and equivalent cylinder radius, normalised by dividing by the average values, for the T6-condition of (a) AA6063 and (b) AA6082.

3.3 Overaged (T7)

In figure 5 the precipitates in the T7 condition (1 week/ 185°C) of AA6082 are shown. In addition to the obvious increase in cross-section and length, see also Table 4, the size uniformity is much lower than in T6. A larger scatter is seen in the distribution of normalised length and cylinder radius in figure 6, where especially the length distribution has a long tail. The large variation is reflected by the fact that a number of different phases are present. In figure 5(b) the precipitates are identified as β", B', U1 and U2, but some coarse β'-phases were also observed (not seen on this fine scale). About 60% of the particles are identified as B', 14% as β", 10% as U2, while the rest is occupied by U1, β' and some unidentifiable crystal structures. Due to this complexity, and the fact that the composition of B' is not known, it is not possible to estimate the rest of Mg and Si in solid solution. However, because the volume fraction of precipitates is larger than the initial volume fraction of Mg and Si, it is assumed that the rest in solution is close to zero. Another interesting observation is that double-lines, indicating coherency strains, are still present around some of the particles, even though the alloy has been aged for a long time. Note also that obviously some precipitates are formed on dislocations where they appear as pearls on a string. A few dislocations are also seen.

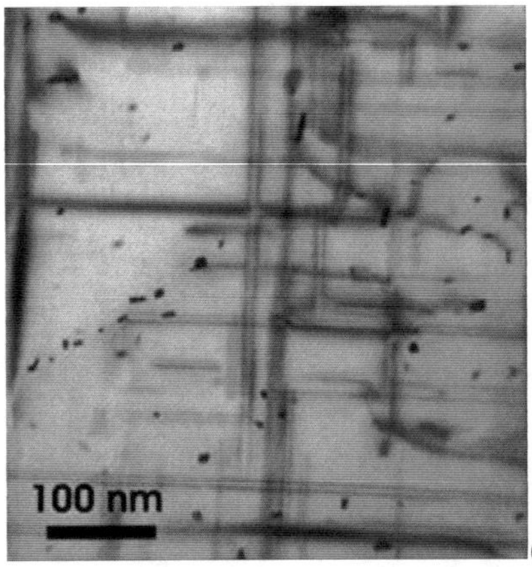

(a)

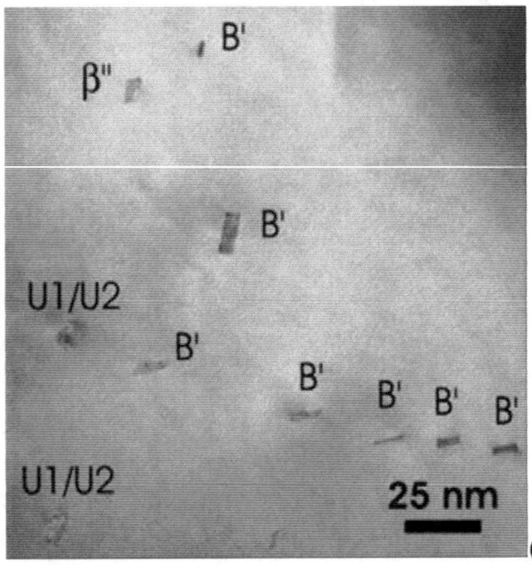

(b)

Figure 5. (a) Precipitates in AA6082 in the T7 condition, oriented along three orthogonal directions in the matrix. (b) Identified phases.

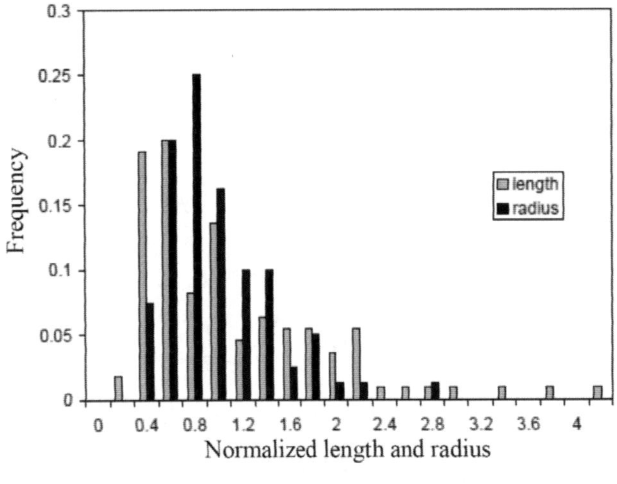

Figure 6. Frequency distributions of length and equivalent cylinder radius, normalised by dividing by the average values, for the T7-condition of AA6082.

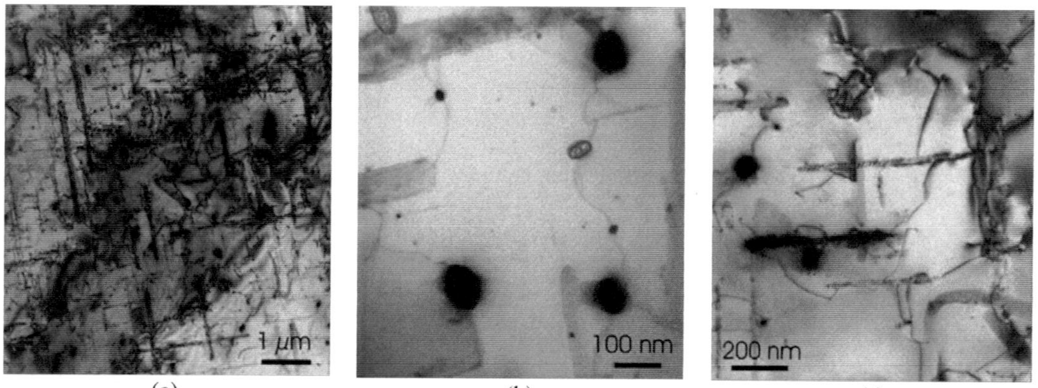

(a) (b) (c)

d)

Figure 7. (a) Precipitates and dislocations in AA6082 in the T7+ condition, oriented along three orthogonal directions in the matrix. (b) Several types of particles with different size. (c) Dislocations around and between particles. (d) Precipitate-free zones on each side of a grain boundary.

3.4 Severely overaged (T7+)

In figure 7 micrographs of the severely overaged condition (24h/250°C) of AA6082 are shown. The precipitates seen in figure 7(a), with a close-up in figure 7(b), are very large, with lengths of about 1-3 μm and diameters of 50-150 nm, which suggests that they are U1-precipitates. However, as seen in figure 7(b), there exist two or three particle populations of different size. A set of small particles are measured to have a typical diameter of about 15 nm. These are estimated to be some hundred nm's long. A striking observation in this specimen is the high density of dislocations, figure 7(a, c), despite the fact that it has not been subjected to any external deformation. These dislocations are absent in the precipitate-free zones, see figure 7(d).

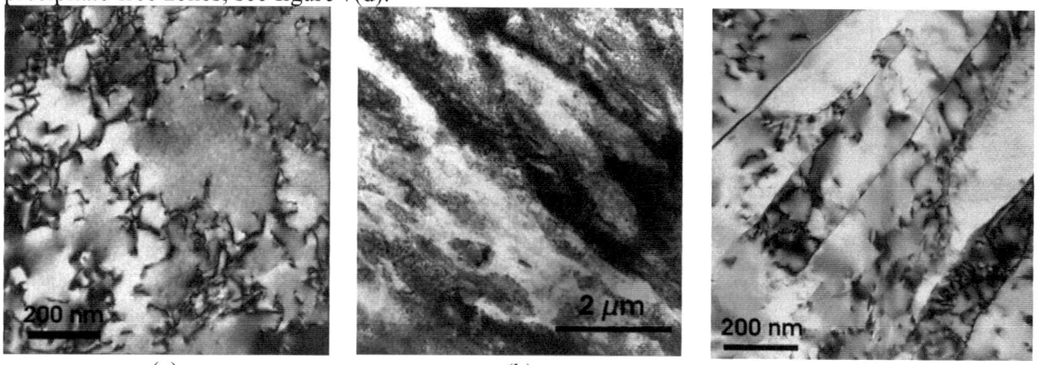

(a) (b) (c)

Figure 8. AA6082. Deformation structures in the W60 condition. (a) ε=0.17 (tensile test specimen) (b-c) ε=0.8 (compression test specimen)

IOP Conf. Series: Materials Science and Engineering **89** (2015) 012013 doi:10.1088/1757-899X/89/1/012013

3.5 Naturally aged (W60) and deformed AA6082

The W60 condition contains approximately 1.5 at% solute (Mg+Si) which is partially clustered. In general the dislocation structure is seen to be rather diffuse, figure 8. The dislocation density is relatively homogeneous, without any clear cell or subgrain structure. After large deformation the structure is dominated by ragged bands that are probably gradually misorientated to each other, figure 8(b), although some sharper well-defined subgrain boundaries are found as well, figure 8(c).

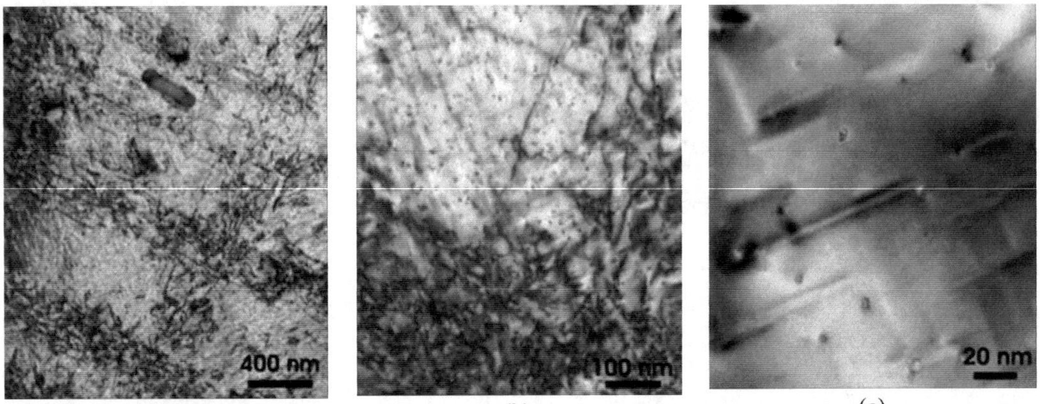

(a) (b) (c)

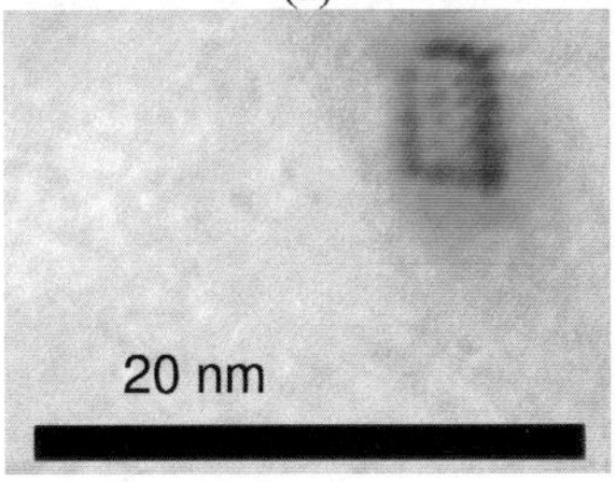

(d)

Figure 9. AA6082. Deformation structure in the T6 condition, ε=0.02 (tensile test specimen). (a) Dislocation bands. (b) Enlarged section of (a). (c) Precipitates. (d) Enlarged section of (c).

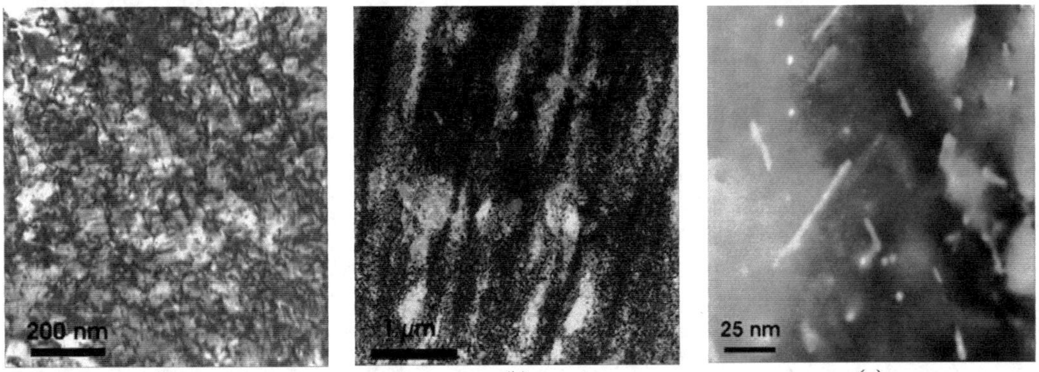

(a) (b) (c)

Figure 10. AA6082. Deformation structure in the T6 condition. (a) ε=0.05 (tensile test specimen) (b-c) ε=0.8 (compression test specimen).

3.6 Peak aged (T6) and deformed AA6082

The peak aged condition has a fine dispersion of needle-shaped β" and a solute content of about 0.8 at% (Mg+Si). After 2% plastic straining the dislocations are seen to be localised into bands surrounding cells of low dislocation density and a size of 0.5-1 µm, figure 9(a-b). The bands are crystallographic in nature, in such a way that when they are viewed along an <001> direction, they form orthogonal bands in the <011> directions. It is therefore most likely that the dislocations lie on parallel {111} planes in the matrix. Such a localisation is often associated with the cutting of precipitates, which is also indicated by the observation of disrupted coherency strain fields, figure 9(c). However, the atomic structure of the precipitates is still clearly visible in figure 9(d).

After 5% straining, these bands have more or less disappeared, and a uniform dislocation structure of a very high density dominates the material, figure 10(a).

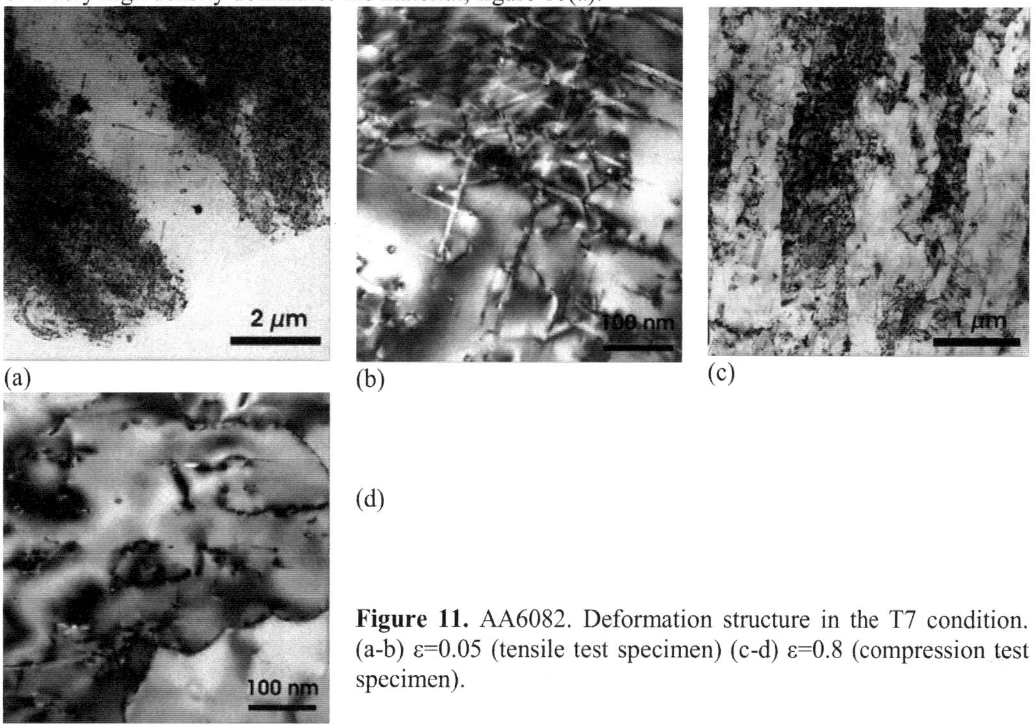

Figure 11. AA6082. Deformation structure in the T7 condition. (a-b) ε=0.05 (tensile test specimen) (c-d) ε=0.8 (compression test specimen).

In the highly deformed material the microstructure resembles that of the W60 condition, i.e. diffuse bands of gradual misorientations, figure 10(b). The precipitates are still visible, but now they create an angle to each other that is different from the initial 90° (in figure 10(c): 60°). Even though some irregularities may be seen, the precipitates are relatively straight. This must imply that the dislocations that cut the precipitates are very homogeneously distributed along one particle.

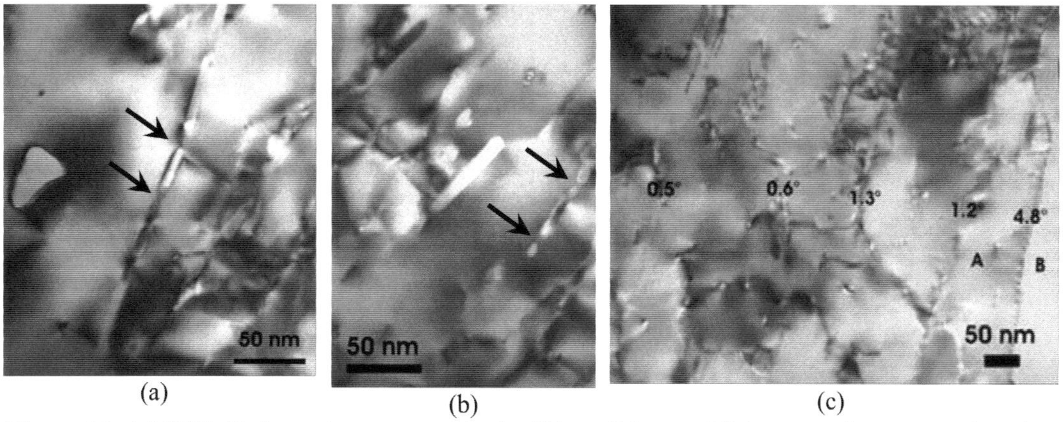

Figure 12. AA6082. Deformation structure in the T7 condition, ε=0.8 (compression test specimen). (a-b) Fragmented precipitates marked with arrows. (c) Measured misorientations across boundaries.

3.7 Overaged (T7) and deformed AA6082

The overaged condition contains a lower density of coarser precipitates than T6, and there is almost no solute left in the matrix. In the early stages of deformation, dislocations are tangled between the precipitates in a rather uniform way, although some dislocation bands were found here as well, figure 11(a-b). At large strains the structure consists of sharp boundaries, figure 11(c), quite different from the diffuse bands in W60 and T6. Obviously, the boundaries have been pinned by the precipitates, as they are sometimes curved into small arcs, figure 11(d). The boundaries separate material volumes of different orientation, similar to the cell block structure that is observed in pure aluminium [21]. Orientation measurements in figure 12(c) show that the misorientation across these boundaries is of the order 3-5°. Within these blocks smaller volume elements can be observed, although they are not as clear as in single-phase materials. Small misorientations (0.5-2°) can be detected across the boundaries.

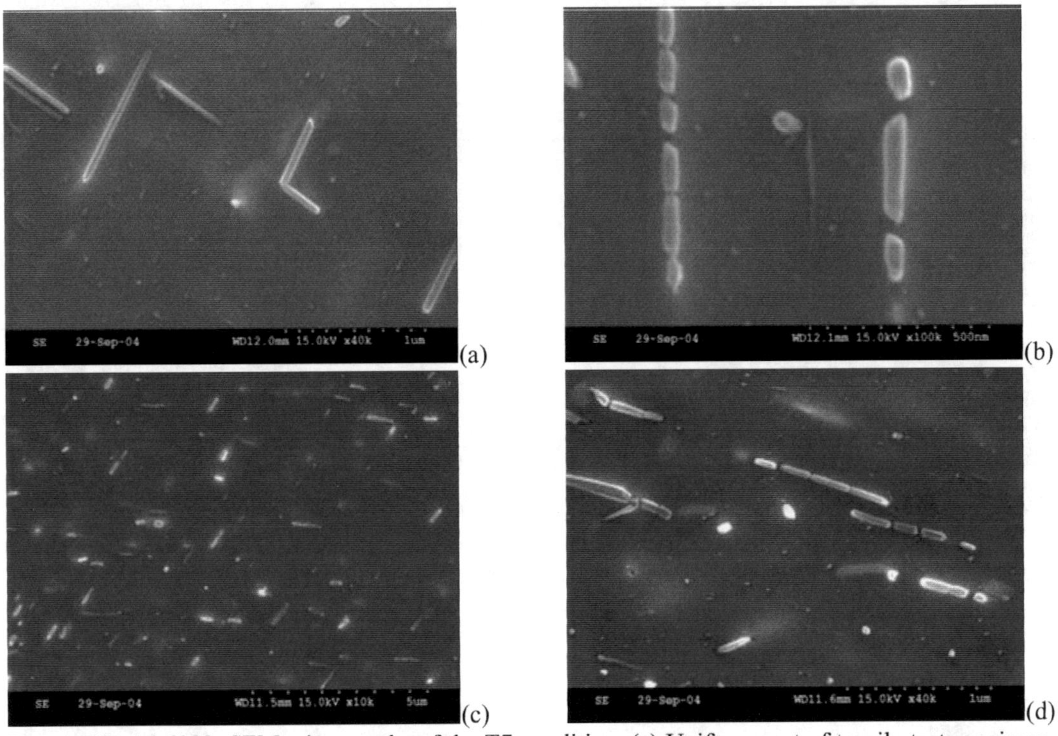

Figure 13. AA6082. SEM micrographs of the T7 condition. (a) Uniform part of tensile test specimen (ε=0.05). (b) Near the fracture surface of the tensile test specimen. (c-d) Compression test specimen (ε=0.8).

When it comes to the precipitates it is seen that they to a large extent are fragmented into smaller pieces, figure 12(a-b). Because of the interface energy between particles and matrix, the fragments are rounded at their ends and in some cases have become spherical. SEM micrographs give information on the population of coarser precipitates in T7. In the uniform region of a tensile test specimen (ε=0.05) the particles are intact, figure 13(a). However, when we move into the neck, figure 13(b), where the local strains have been higher, we see frequent particle fragmentation. Similarly, in the compression tested specimen the particles are broken, figure 13(c-d).

Table 5. Mechanical properties for AA6063 and AA6082

	AA6063				*AA6082*				
	W10	T6x	T6	T7	W5	T6x	T6	T7	T7+
$R_{p0.2}$ (MPa)	36.6	95.6	190.4	165.4	65.0	222.8	321.5	213.8	98.0
R_m (MPa)	121	164.2	222.1	198.7	191.0	299.8	339.5	259.3	172.0
e_u	0.26	0.15	0.082	0.064	0.21	0.13	0.059	0.062	0.072
e_f	0.37	0.25	0.16	0.14	0.30	0.16	0.072	0.11	0.15

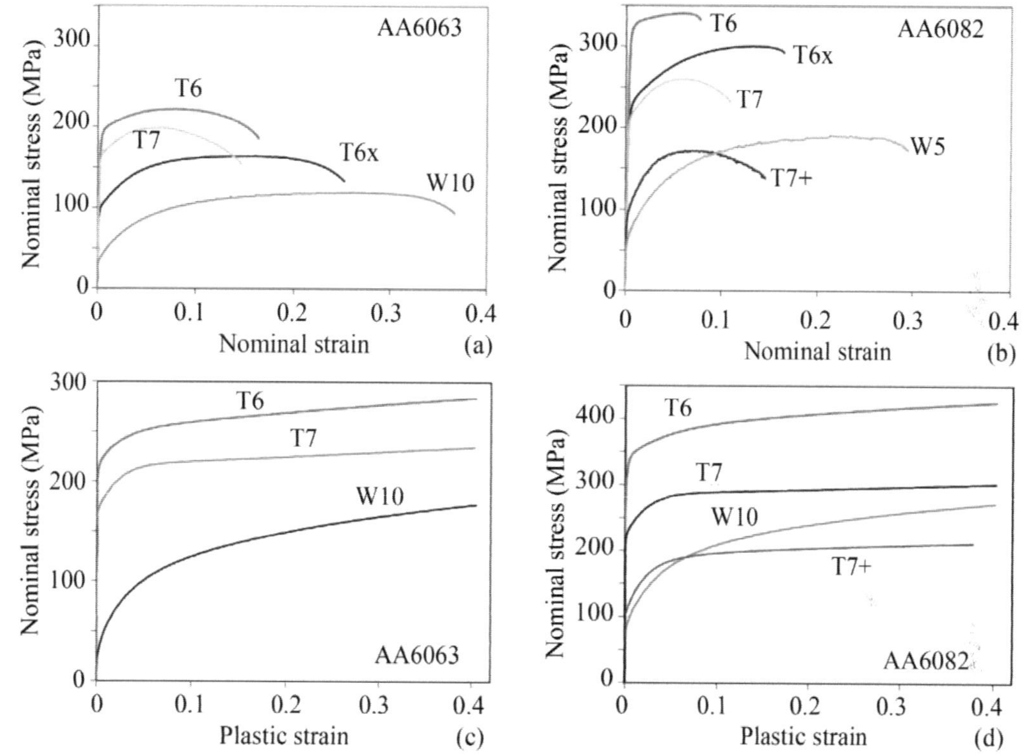

Figure 14. Nominal stress-strain curves. (a-b) by tensile tests, (c-d) by compression tests.

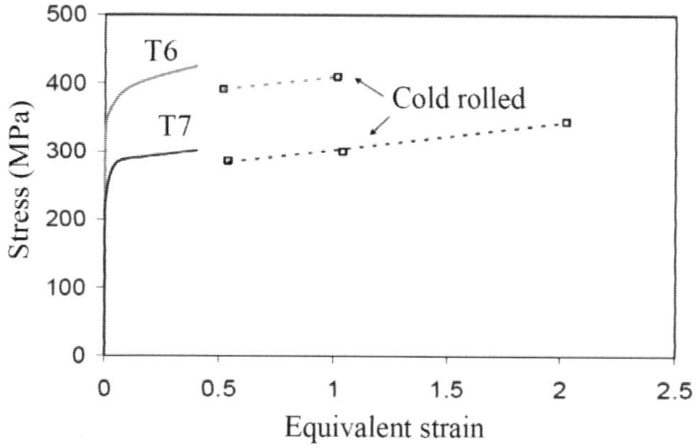

Figure 15. Stress-strain behaviour of AA6082 T6 and T7 obtained from compression tests and cold rolling + tension tests.

4. Mechanical properties

Stress-strain curves by tensile and compression tests of the artificially aged conditions of AA6063 and AA6082 are shown in figure 14 and some properties are listed in Table 5. As expected, the strength ($R_{p0.2}$, R_m) increases and the ductility (e_u, e_f) decreases with ageing from the as-quenched state to T6. During overageing the strength decreases while the uniform elongation remains almost constant. The amount of work hardening (R_m-$R_{p0.2}$) decreases with ageing up to T6, but increases again with overageing. The flow stress from compression tests is plotted according to the strain correction described in (8) and (9). The differences in work hardening behaviour between the tempers are obvious from both the tensile and the compression tests. The compression results essentially give a continuation of the trends from the tensile results. An estimate of the further work hardening, obtained by cold rolling followed by tensile tests, is shown for the conditions T6 and T7 of AA6082 in figure 15.

The post- uniform elongation, as well as the load reduction from maximum load to fracture, are significant in the overaged conditions. This increases the critical failure strain, CFS, (2), but also the higher ductility of the AA6063 alloy has a positive effect on the CFS, see figure 16.

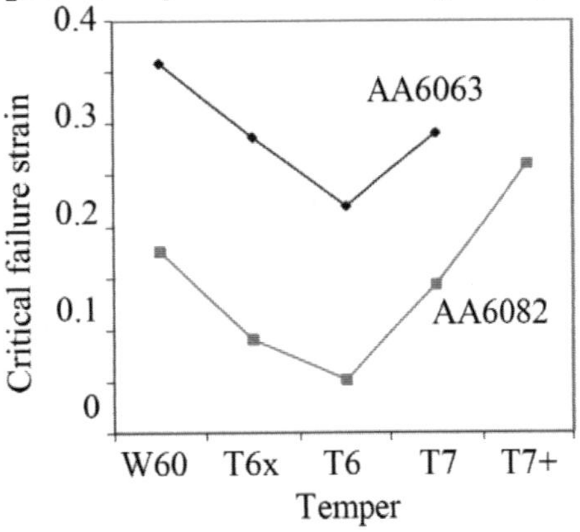

Figure 16. Critical failure strain (CFS) of various tempers of the two alloys.

The rate of work hardening, θ, measured in both tension and compression testing, is plotted against the reduced flow stress for each temper in figure 17 and figure 18. The initial work hardening rate is seen to decrease with ageing up to T6, and the subsequent hardening rate decreases gradually for these conditions. The overaged conditions, however, have a higher initial θ than T6, but decrease rapidly to a low and stable θ.

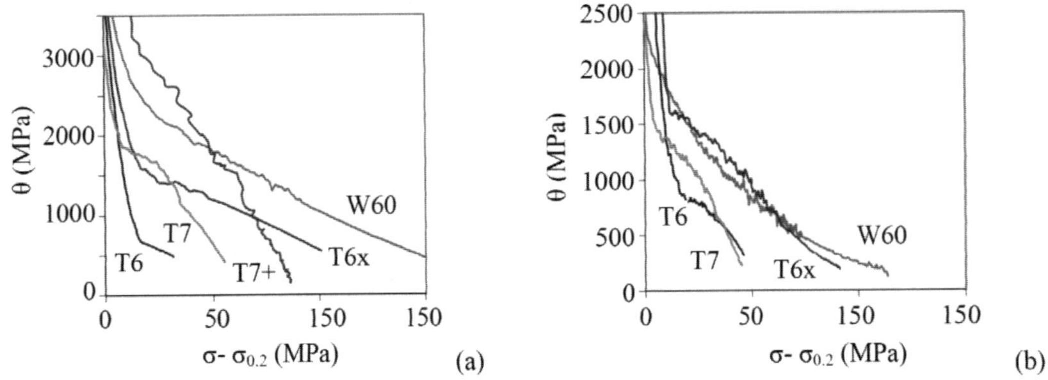

Figure 17. Work hardening rate (θ) as a function of reduced flow stress from tension tests for (a) AA6063 and (b) AA6082.

5. Discussion

The two alloys investigated in this work have a significant difference in composition, as AA6082 contains almost twice as much Mg+Si as AA6063. This alone results in a lower number density and volume fraction of precipitates in AA6063 T6, but it is also responsible for a slower precipitation reaction at room temperature [39]. Therefore, it is reasonable to assume that the number of various Mg and Si containing clusters formed after one hour at room temperature is very different. During ageing at 185°C these clusters act as nucleii for precipitates. Since the number of nuclei is low in AA6063, fewer precipitates will be formed. However, because the spacing of precipitates is larger they can grow faster than in AA6082. This may explain the differences seen in figure 3, Table 3 and Table 4. According to the measurements the amount of Mg and Si in solid solution is reduced by about 50% from the as-quenched state to T6 in AA6082. In AA6063 the reduction is only 30%.

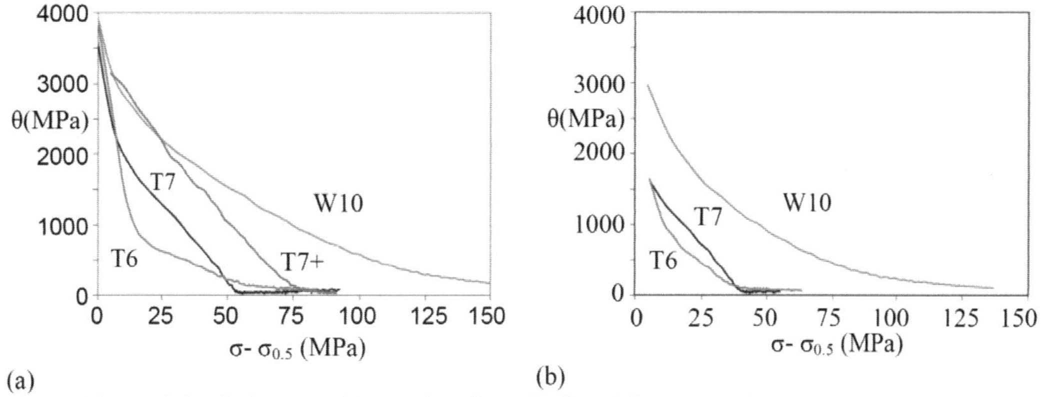

Figure 18. Work hardening rate (Θ) as a function of reduced flow stress from compression tests for (a) AA6063 and (b) AA6082.

As expected, the T6 condition contains a high density of β"-precipitates. No underaged conditions have been investigated in this work, but based on the work by e.g. Marioara et al. [3] it is a reasonable assumption that GP-zones would be found after shorter ageing times.

With further annealing the matrix becomes essentially depleted of solutes, giving a volume fraction of particles that is in fact higher than the original volume fraction of solutes. This is possible, since some of the precipitates contain Al in addition to Mg and Si. Giving a statistical analysis of the particle size, number density and volume fraction becomes increasingly difficult with increased overageing, as the uniformity of the particles is worsened. The high dislocation density in T7+ makes it even more difficult to measure the particles. However, several metastable phases are seen, while no evidence of the equilibrium phase β or Si-particles is found. An ageing temperature of 185°, and even 250°C, has probably been too low for these phases to be formed. It can be concluded that the overageing is quite slow at 185°C, and the presence of coherency strains around the particles in T7 confirms this.

The presence of dislocations arranged around and between the particles in the overaged conditions is worth an extra comment. First of all, it is realized that they must originate from some late stage of the sample preparation, as a majority of the dislocations would have been annealed out during ageing. Therefore the final quench as well as the thin foil preparation and handling are the only possibilities. But since all the dislocations are arranged around and between the particles and none of them in the precipitate-free zones, we can eliminate artefacts from bending of the foil or similar deformation. In the severely overaged sample (T7+) the precipitates are assumed to be incoherent with the matrix. This implies that during the precipitation reaction dislocations must be formed on the interface already before quenching. During cooling the different coefficients of thermal expansion between particles and matrix lead to a difference in volume contraction, and the forces arising may punch dislocations away from the interface. Especially in the T7+ sample, which is quenched from 250°C, large thermally-induced strains can be expected. In the T7 sample this occurs to a much lesser extent, as many of the precipitates are still coherent. The presence of thermally-induced plastic zones is a well-known phenomenon in metal-matrix composites, and local strains of the order 1%, evidenced by slip bands, have for example been observed in the neighborhood of cylinder-shaped SiC-particles [7].

In the peak aged condition, T6, a localization of dislocations into bands at 2% strain indicates that the precipitates are shearable. The disappearance of the bands at higher strain is in accordance with Dowling and Martin [20], who interpreted this as an effect of dispersoids spreading the slip. The bands seen in the compressed specimen, i.e. $\varepsilon=0.8$, are not crystallographic in the same way as at low strains. At $\varepsilon=0.8$ the needle-shaped precipitates are sheared uniformly by dislocations so that they remain straight needles but not in the <001> directions anymore. This indicates extensive cross slip.

In the overaged specimen, T7, it is seen that the low solute content in the matrix gives a "cleaner" dislocation structure. The dislocations have been able to recover and reorganize into sharp boundaries. Above some critical strain the precipitates are fragmented. The SEM micrographs indicate that for the coarsest precipitates this critical strain is somewhere above 5% strain. The smallest precipitates can probably accommodate larger elastic strains, and have therefore undergone some bending before they also fracture, as in the compressed sample.

Generally, it is seen from the results above that AA6063 is softer and more ductile than AA6082, and that the ageing response is faster and stronger in AA6082. This is reasonable, as the AA6082 alloy has a considerably higher amount of Mg and Si, giving more solution hardening and a higher precipitation rate. But it is also seen that the ageing results in a number of effects on the mechanical properties, and in the following a more detailed discussion is given on the behaviour of the different conditions.

A comparison of the as-quenched states of the current alloys, i.e. AA6063 W10 and AA6082 W5, with other solid solution alloys [10] reveals something interesting, see figure 19. The AA6063 alloy, having 0.51 at% Mg and 0.36 at% Si, in total 0.87 at% in solid solution, has a stress-strain curve which is practically identical to an alloy containing 1.1 at% Mg. The AA6082 alloy with 1.5 at% Mg+Si has an identical behaviour as an alloy with 3.2 at% Mg. AA6082 contains a considerable amount of dispersoids that must account for a fraction of the strength. Anyway, from the high hardening rate it is apparent that the effect of having Mg and Si in combination is larger than having only Mg. Such a synergistic effect is also observed by others [40, 41].

However, this must also be viewed in light of what happens to the alloys after only short storage times at room temperature caused by atomic clusters [1], and it is easy to imagine that the formation of such clusters may be accelerated by deformation, i.e. dynamic precipitation, a well-known phenomenon in age-hardenable alloys [42, 43]. Hence, it is possible that the synergistic effect of Mg and Si may be related to the enhanced clustering due to deformation.

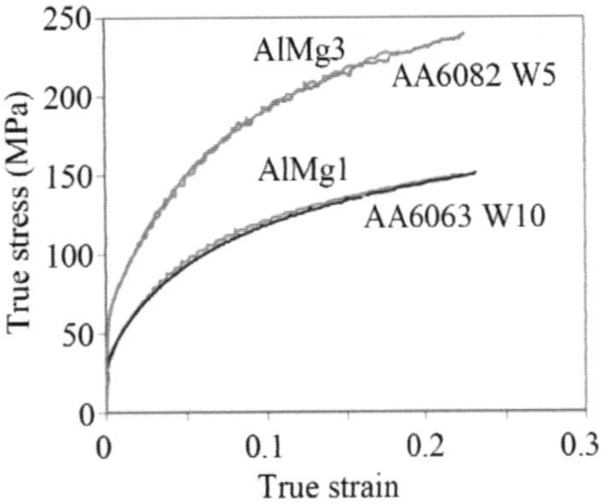

Figure 19. Comparison of tensile tests of AA6063 W10 with AlMg1, and AA6082 W5 with AlMg3.

The increase in yield strength that occurs during ageing up to 5h/185°C is caused by the high density of small, coherent β"-precipitates. Such precipitates are generally considered as shearable, and it has been claimed that they do not contribute to the work hardening, which then is defined solely by the matrix properties [44]. This idea is complicated by the observation of a gradual elastic-plastic transition in the T6-samples of both AA6063 and AA6082, giving a very high work hardening rate immediately after the yield point. Lloyd and Kenny [18] explained this by suggesting that the precipitates are non-shearable at the early stage of deformation, leaving Orowan loops around them. The back stress arising from the dislocation loops will be relieved upon further deformation when the dislocations start cutting through the precipitates. The barrier strength is then reduced and the work hardening rate decreases.

The reduction in number density of precipitates during overageing as well as the loss of coherency, are probably responsible for the reduction in yield strength in T7 and T7+. Additionally, the strain field around precipitates in Al-Mg-Si alloys have been observed to be much stronger if they are rod-shaped than if they are laths [45]. It was seen that the T7-condition of AA6082 contained a large number of lath-shaped B'-precipitates, while the β" in T6 had a more circular cross-section. Hence, the B'-precipitates will contribute less to the strength than the β"-precipitates do, even if they are of similar size. This may cause an extra reduction in yield strength of T7.

A factor of significant importance when deforming two-phase materials is the accumulation of internal back stresses arising from elastic deformation of a second-phase surrounded by a plastically deformed matrix. The back stresses may be evidenced by observations of aided reverse flow, e.g. non-linear unloading or the Bauschinger effect. The theoretical framework was established by Brown and Stobbs [46], and later it has been argued that the stresses can be responsible for large strengths [47]. Since the back stress saturates it does not contribute to the work hardening rate. The cracking of precipitates in the overaged conditions reduces the back stress and softens the material. In this work, however, reverse tests have not been performed to quantify this phenomenon.

The shape of the work hardening curves obtained from tensile tests in figure 17 is analysed using the method described by Cheng et al. [30]. Two parameters, the initial work hardening rate θ_{max} and the rate of decrease of the hardening rate $d\theta/d\sigma$, are read from the curves and plotted as a function of the yield strength in figure 20. Firstly, it is recognised that θ_{max} decreases linearly with the yield strength up to T6, but increases with $\sigma_{0.2}$ at a faster rate in the overaged conditions, figure 20(a,c). The highest initial work hardening rate is found in the condition with the lowest yield strength, and vice versa. Secondly, the rate of decrease of the hardening rate is almost unaffected by the ageing up to T6 (except AA6063 T6, see below), while overageing gives a much more rapid decrease, figure 20(b,d). The observations are in accordance with the ones made by Cheng et al. [30]. They associated θ_{max} to the storage rate of dislocations. The parameter $d\theta/d\sigma$ they linked to the dynamic recovery rate. The dramatic difference between underaged and overaged conditions could then be interpreted as an effect of the transition from shearable to non-shearable particles.

A comparison of the overaged conditions T7 of AA6063 and AA6082 with a commercially pure aluminium alloy is shown in figure 21. The flow stress of the AA1050 alloy is obtained by compression testing to a strain of $\varepsilon=0.5$, and by cold rolling followed by tensile testing at larger strains [48]. It can be seen that the work hardening rate at $\varepsilon>0.1$ in the overaged materials, see also figure 14 and figure 15, is of the same order as that observed in stage IV of work hardening observed in AA1050. This may indicate that the overaged conditions of the 6xxx alloys have reached a stage IV at a much smaller strain than AA1050. The onset of stage IV of an AA1050-type alloy is at a strain of about $\varepsilon=0.7$. The stage is characterised by continued hardening due to decreasing boundary spacing, and a dislocation density in the subgrain interior that, because of dynamic recovery, has saturated at the end of stage III [47].

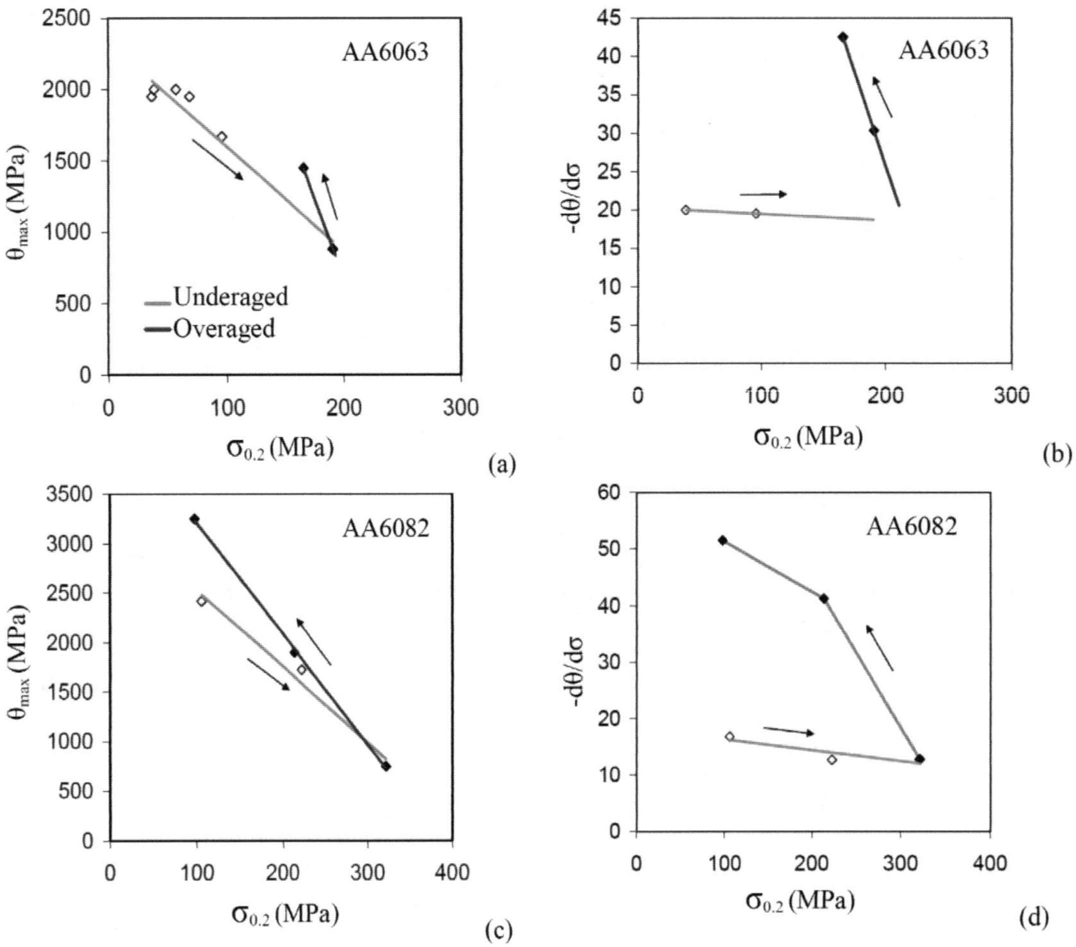

Figure 20. The initial work hardening rate, θ_{max}, and the rate of decrease of the hardening rate, $d\theta/d\sigma$, plotted as a function of the yield strength. Open symbols: underaged conditions, closed symbols: peak aged and overaged conditions. In (a) also the naturally aged conditions are plotted.

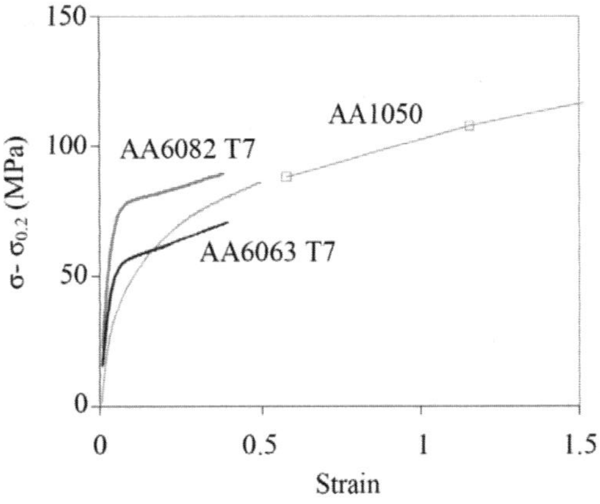

Figure 21. A comparison of the reduced flow stress between AA1050 and the overaged conditions of AA6063 and AA6082.

6. Conclusions

The precipitate structures and corresponding work hardening behaviour in AA6063 and AA6082 have been characterized by TEM and investigated by tensile and compression testing. In the as quenched condition the dislocation structure of the AA6082 is diffuse at small strains. At larger strains ragged bands and subgrains are gradually misoriented to each other. In the T6 condition a dislocation cell structure is recognised at 2% strain but has been replaced by a more homogenous distribution of dislocations at 5% strain. During 80% deformation the needle-shaped precipitates are sheared uniformly by dislocations so that they remain "straight needles" but not with <001>-orientations anymore. This indicates extensive cross slip. At larger strains a subgrain structure is observed. In the overaged conditions dislocations are present surrounding the coarse precipitates in the initial material being formed during the ageing treatment, most likely during quenching. More dislocations are tangled between the coarse precipitates at low strains, whereas a subgrain structure is formed at the larger strains. Cracking of the largest precipitates are observed at large strains in the T7 and T7+ conditions.

The Mg and Si interact and cause a strong work hardening in the as quenched condition, similar to AlMg alloys with about twice as much alloying elements (Mg+Si) in solid solution. It is found that the initial rate of work hardening is lowest at T6, whereas the work hardening rate at slightly larger strains keeps decreasing with increased aging times. In T7 the work hardening rate is similar as if the alloy went straight into stage IV subsequent to the initial. At long aging time large precipitates break and reduce the work hardening rate further. The critical failure strain reaches a minimum in the peak aged conditions.

References

[1] Edwards G A, Stiller K, Dunlop G L and Couper M J 1998 *Acta Mater.* **46** 3893-3904
[2] Murayama M and Hono K 1999 *Acta Mater.* 47 1537-1548
[3] Marioara C D, Andersen S J, Jansen J and Zandbergen H W 2001 *Acta Mater.* **49** 321-328
[4] Andersen S J, Zandbergen H W, Jansen J, Traeholt C, Tundal U and Reiso O 1998 *Acta Mater.* **46** 3283-3298
[5] Marioara C D, Andersen S J, Royset J, Reiso O, Gulbrandsen-Dahl S, Nicolaisen T E, Opheim I E, Helgaker J F and Holmestad R 2014 *Metall. Mater. Trans. A* **45A** 2938-2949
[6] Perovic A, Perovic D D, Weatherly G C and Lloyd D J 1999 *Scripta Mater.* **41** 703-708
[7] Ashby M F 1970 *Philos. Mag.* **21** 399
[8] Brown L M and Ham R K 1971 *Dislocation-particle interactions*, in: A. Kelly, R.B. Nicholson (Eds.) Strengthening Methods in Crystals, Elsevier, Amsterdam, 1971, pp. 9-135
[9] Ardell A J 1985 *Metall. Trans. A* **16** 2131-2165
[10] Humphreys F J and Hatherley M 1996 *Recrystallization and related annealing phenomena*, Pergamon, Oxford.
[11] Bay B, Hansen N, Hughes D A and Kuhlmannwilsdorf D 1992 *Acta Metall. Mater.* **40** 205-219
[12] Bay B, Hansen N and Kuhlmannwilsdorf D 1992 *Mat. Sci. Eng.* **158** 139-146
[13] Hughes D A 1993 *Acta Metall. Mater.* **41** 1421-1430
[14] Korbel A, Embury J D, Hatherly M, Martin P L and Erbsloh H W 1986 *Acta Metal.* **34** 1999-2009
[15] Banizs K 1979 *Mater. Sci. Eng.* **41** 17-24
[16] Zhen L and Kang S B 1997 *Metall. Mater. Trans. A* **28** 1489-1497
[17] Blankenship C P, Hornbogen E and Starke E A 1993 *Mat. Sci. Eng.* **169** 33-41
[18] Lloyd D J and Kenny D 1982 *Metall. Trans. A* **13** 1445-1452
[19] Ryum N, Haegland B and Lindtvei.T 1967 *Z Metallkd* **58** 28-31
[20] Dowling J M and Martin J W 1976 *Acta Metal.* **24** 1147-1153
[21] Vivas M, Lours P, Levaillant C, Couret A, Casanove M J and Coujou A 1997 *Philos. Mag. A* **76** 921-931
[22] Gutierrez-Urrutia I, Munoz-Morris M A and Morris D G 2005 *Mat. Sci. Eng.* **394** 399-410

[23] Nourbakhsh S and Nutting J 1980 *Acta Metal.* **28** 357-365
[24] Sargent C M and Purdy G R 1974 *Scripta Metall. Mater.* **8** 569-572
[25] Poole W J, Wang X, Lloyd D J and Embury J D 2005 *Philos. Mag.* **85** 3113-3135
[26] Yeh J R, Summe T L and Seksaria D C 1999 in: H.F. Mahmood, S.D. Barbat, M.R. Baccouche (Eds.) Craschworthiness, occupant protection and biomechanics in transportation systems (AMD 237/BED 45), ASME, New York, pp. 97-105
[27] Deschamps A and Brechet Y 1998 *Acta Mater.* **47** 293-305
[28] Esmaeili S, Lloyd D J and Poole W J 2003 *Acta Mater.* **51** 2243-2257
[29] Myhr O R, Grong O and Andersen S J 2001 *Acta Mater.* **49** 65-75
[30] Cheng L M, Poole W J, Embury J D and Lloyd D J 2003 *Metall. Mater. Trans. A* **34A** 2473-2481
[31] Westermann I, Hopperstad O S, Marthinsen K and Holmedal B 2009 *Mat. Sci. Eng.* **524** 151-157
[32] Westermann I, Hopperstad O S, Marthinsen K and Holmedal B 2010 *J. Mater. Sci.* **45** 5323-5331
[33] Westermann I, Hopperstad O S, Marthinsen K and Holmedal B 2012 *Int. J. Mater. Res.* **103** 603-608
[34] Barlat F and Vasudevan A K 1991 *Acta Metall. Mater.* **39** 391-400
[35] Gracio J J, Barlat F, Rauch E F, Jones P T, Neto V F and Lopes A B 2004 *Int. J. Plasticity* **20** 427-445
[36] Campbell K, Dover I, Ramachandran T R and Embury J D 1979 *Met. Forum* **2** 229-235
[37] Friis J, Holmedal B, Ryen O, Nes E, Myhr O R, Grong O, Furu T and Marthinsen K 2006 Aluminium Alloys 2006, Pts 1 and 2 519-521 1901-1906
[38] Nord-Varhaug K, Forbord B, Benestad J, Pettersen T, Ronning B, Bardal A, Benum S, Marthinsen K and Nes E 2000 *Mater. Sci. Forum* **331-3** 1387-1392
[39] Ryen Ø, Holmedal B, Li S, van Houtte P, Roven H J and Nes E 2004 in: J.F. Nie, A.J. Morton, B.C. Muddle (Eds.) *Proceedings of the 9th International Conference on Aluminium Alloys,* Institute of Materials Engineering Australasia, Brisbane, Australia, pp. 1004-1009.
[40] Doherty R D and Mcbride J 1993 *Aluminum Alloys for Packaging* 347-368
[41] Suni J P, Shuey R T and Doherty R D 1994 in: T.H. Sanders, E.A. Starke (Eds.) *Aluminum Alloys - Their Physical and Mechanical Properties*, The Georgia Institute of Technology Atlanta, pp. 521-528
[42] Deschamps A, Esmaeili S, Poole W J and Militzer M 2000 *J. Phys. Iv.* **10** 151-156.
[43] Deschamps A, Niewczas M, Bley F, Brechet Y, Embury J D, Le Sinq L, Livet F and Simon J P 1999 *Philos. Mag. A* **79** 2485-2504
[44] Roters F, Raabe D and Gottstein G 2000 *Acta Mater.* **48** 4181-4189
[45] Donnadieu P, Dirras G F and Douin J 2002 *Aluminum Alloys 2002: Their Physical and Mechanical Properties* Pts 1-3 396-4 1019-1024
[46] Brown L M and Stobbs W M 1971 *Philos. Mag.* **23** 1185-1233
[47] Sinclair C W, Embury J D, Weatherly G C, Conlon K T and Engler O 2003 *Mater. Sci. Tech-Lond.* **19** 1321-1329
[48] Ryen O, Laukli H I, Holmedal B and Nes E 2006 *Metall. Mater. Trans. A* **37A** 2007-2013

IOP Conf. Series: Materials Science and Engineering **89** (2015) 012014 doi:10.1088/1757-899X/89/1/012014

Recovery by triple junction motion in heavily deformed metals

T Yu, N Hansen and X Huang

Danish–Chinese Centre for Nanometals, Section for Materials Science and Advanced Characterization, Department of Wind Energy, Risø Campus, Technical University of Denmark, DK-4000, Roskilde, Denmark

E-mail: tiyu@dtu.dk

Abstract. A number of processes may occur during recovery, which reduce the stored energy and coarsen the structure, paving the way for the nucleation of recrystallization. In this review, recovery is discussed based on the initial deformed structure. Recovery in samples deformed to low-to-medium strains is briefly considered, whereas recovery in samples deformed to much higher strains is discussed in more detail. An example is Al, where a new and important recovery mechanism is triple junction motion, which can remove thin lamellae and consequently increase the thickness of neighbouring lamellae. This recovery mechanism therefore coarsens the structure and causes a gradual transition from a lamellar to a near equiaxed structure preceding recrystallization. This mechanism is discussed in conjunction with structural pinning, which stabilizes the coarsening microstructure, thereby balancing the forces controlling the rate of recovery by triple junction motion. Operation of triple junction motion in other heavily deformed metals, such as Ni and Cu, is also briefly discussed.

1. Introduction

Recovery is traditionally defined as all annealing processes occurring in deformed materials without migration of high angle boundaries [1], whose misorientation angles are typically larger than 15°, and it is usually considered as a precursor of recrystallization [2]. Recently, we have documented that coarsening by triple junction motion is an important restoration mechanism in heavily deformed Al subdivided by lamellar boundaries, containing a substantial fraction of high angle boundaries [3]. In that work, triple junction motion has been classified as a recovery mechanism since it occurs before primary recrystallization, although migration of high angle boundaries is heavily involved.

The current paper is focused on new findings on recovery of heavily deformed Al, including mechanisms, kinetics, structural pinning and effects on nucleation of recrystallization. Both deformation structures at different strain levels and recovery behaviours at low-to-medium strains are briefly considered as background. This background is based on polycrystalline metals of medium-to-high stacking fault energies, processed under conditions where plastic deformation takes place by glide of dislocations. Finally, recovery in heavily deformed metals other than Al is also discussed.

2. Deformation microstructure

Deformation of metals of medium-to-high stacking fault energies (e.g. Al, Ni, Cu and Fe) to a low strain typically produces cell-block structures, where a cell block is a group of cells where the same sets of slip systems operate [4]. Most of the dislocations are stored in cell boundaries and cell-block boundaries, whereas regions between these boundaries have a relatively low dislocation density. The structural morphology changes as the strain increases. After deformation to a high strain, for example by cold rolling, the structure is characterized by extended lamellae evolved from cell blocks [5,6]. The lamellar boundaries are finite, forming triple junctions where they meet, and there are interconnecting boundaries linking lamellar boundaries. The two types of boundaries and three types of triple junctions are illustrated in figure 1a in the longitudinal section, containing the rolling direction (RD) and the normal direction (ND), and in figure 1b in three dimensions. Each Y-junction is formed by three lamellar boundaries, each pair of H-junctions is formed by two lamellar boundaries and an interconnecting boundary between them, and each r-junction is formed by three interconnecting boundaries [3].

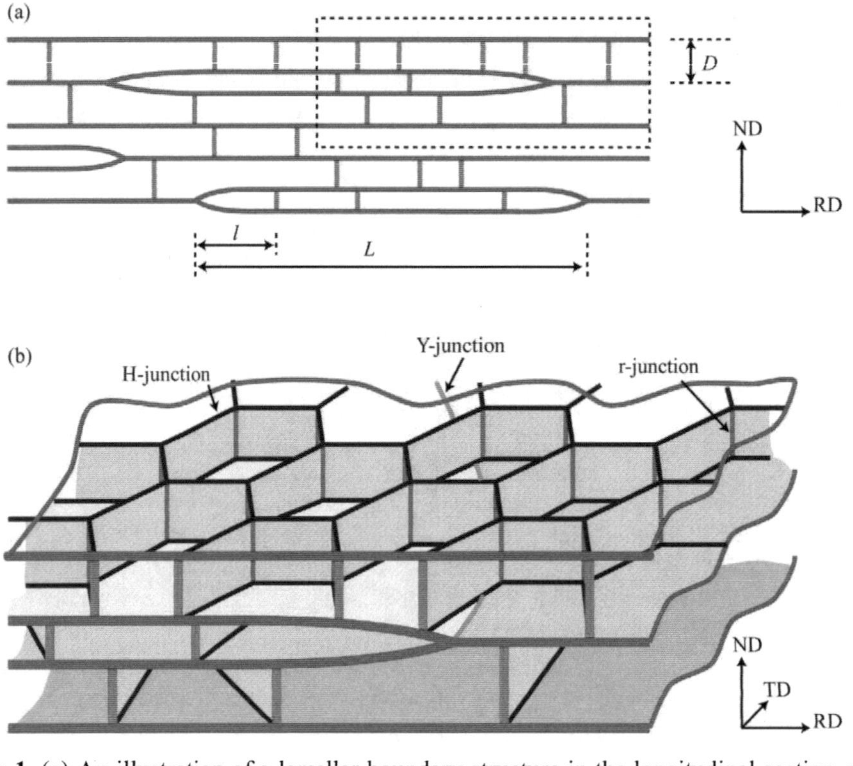

Figure 1. (a) An illustration of a lamellar boundary structure in the longitudinal section, where the lamellar boundary (red) spacing is D, the interconnecting boundary (cyan) spacing is l, and the lamella length is L. (b) An illustration of the three-dimensional boundary structure of the marked region in (a). Lamellar boundaries are shown in different shades of red, with the top layer transparent, and interconnecting boundaries are shown in cyan. One Y-junction is shown in orange; all H-junctions are shown in blue; and all r-junctions are shown in green. The TD is the transverse direction [3].

Both cell-block boundaries at low strains and lamellar boundaries at high strains are geometrically necessary boundaries (GNBs), whereas both cell boundaries at low strains and interconnecting boundaries at high strains are incidental dislocation boundaries (IDBs) since they are assumed to form by mutual trapping of glide dislocations [7,8]. For both GNBs and IDBs, the average boundary spacing decreases and the average misorientation angle across the boundary increases when the strain is increased. However, the rate of change is very different for the two types of boundaries, indicating different mechanisms controlling their formation and evolution. In parallel, the density of three types of triple junctions all increases with increasing strain at different rates [3].

In addition to the change in the structural morphology and structural parameters, the stored energy of deformation also changes when the strain is increased. It increases monotonically with increasing strain over a large strain range although at a decreasing rate [9,10].

3. Recovery in metals deformed to low-to-medium strains

The recovery of metals deformed to low-to-medium strains has been studied extensively more than half a century ago, as reviewed by Bever [11] and Perryman [12]. The early recovery studies were primarily focused on the release of the stored energy and the changes in the physical and mechanical properties, and recovery processes were discussed in terms of annihilation of excess vacancies, mutual annihilation of dislocations with opposite Burgers vectors and rearrangement of dislocations into low angle subgrain boundaries [11,13,14]. Later experimental observations by transmission electron microscopy (TEM) and theoretical analyses showed that subgrain growth is also an important recovery process, either by subgrain coalescence (e.g. [15,16]) or by subgrain boundary migration (e.g. [17,18]). While dislocation annihilation and reconfiguration might be the dominating recovery mechanism in metals of low stacking fault energies, subgrain growth by subgrain boundary migration seemed to be more important in metals of high stacking fault energies (e.g. [19-21]). The growth of subgrains is of particular interest since large subgrains partly or entirely surrounded by high angle boundaries may develop into nuclei for recrystallization. Figure 2 shows an example, where the spacing of boundary dislocations in a low angle twist boundary (A-B) connected to a high angle boundary (C-D) is not uniform. The spacing near the high angle boundary is larger than that far away from this boundary. This is considered as evidence of dislocation absorption in a high angle boundary and a precursor of subgrain coalescence, which may lead to nucleation of recrystallization [22]. For reviews in recovery, see also [23] and [24].

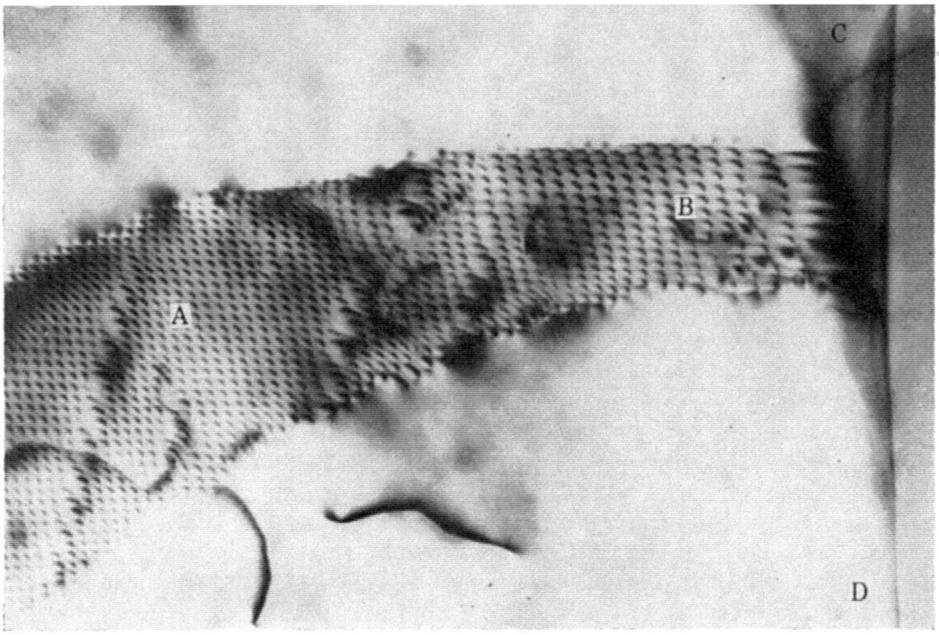

Figure 2. A bright-field micrograph of a low angle twist boundary (A-B) which is linked to a high angle boundary (C-D). The dislocation spacing within the net of the low angle boundary is found to increase from about 27 nm near A to about 33 nm near B [22].

Specific recovery mechanisms have been analysed based on kinetic equations where parameters are dislocation density, subgrain size and flow stress [16,20,25-28]. However, recovery consists of a number of complex processes, involving a number of parameters, and in order to describe the global recovery kinetics, empirical rate equations [11] are often used.

If P is a material property, decreasing during annealing, then a logarithmic decay of the property corresponds to the following rate equation:

$$\frac{dP}{dt} = -\frac{c_1}{t}, \tag{1}$$

where c_1 is a constant and t is the annealing time. However, the relationship between the property and the time is indirect, and equation (1) leads to an infinite rate at the beginning of recovery. To overcome the above two problems, the first-order reaction rate equation may be used:

$$\frac{dP}{dt} = -c_2 P \exp(-\frac{Q_0}{RT}), \tag{2}$$

where c_2 is a constant, Q_0 is the activation energy independent of temperature, R is the gas constant, and T is the absolute temperature. Equation (2) assumes a constant activation energy. However, it has been shown that the apparent activation energy for recovery increases as recovery proceeds (e.g. [20,29,30]), and therefore the recovery rate equation first proposed by Kuhlmann [31] (later as Kuhlmann-Wilsdorf) may give a better description:

$$\frac{dP}{dt} = -c_3 \exp(-\frac{Q_0 - \beta P}{RT}), \tag{3}$$

where c_3 and β are constants. However, equation (3) has a problem at the end of recovery, i.e. the rate never goes to zero. A more flexible equation has been proposed by Borelius and co-workers [32]:

$$\frac{dP}{dt} = -c_4 P \exp(-\frac{Q_0 - \beta' P}{RT}), \tag{4}$$

where c_4 and β' are constant. Equation (4) may be recognized as a first-order reaction rate equation with an apparent activation energy depending on the extent of recovery which already has taken place. It may describe the whole range of recovery as suggested recently [30,33].

4. Recovery in Al deformed to high strains

In recent years there has been an increasing interest in metals and alloys deformed to large strains with a very high strength. The structure is on a fine scale and the stored energy is high. Consequently their thermal stability is of significant interest. Al has a high stacking fault energy, so cross slip is easy and recovery plays an important role during annealing. For highly strained Al, more than half of the stored energy can be released by recovery processes before the first recrystallization nucleus is observed [30]. Due to its large extent of recovery, Al has been favoured in many studies (e.g. [20,34,35]) and some detailed work even demonstrated the dependence of the recovery behaviour on local texture [34,35]. During plastic deformation excess vacancies are produced, but they annihilate rapidly below room temperature in Al almost independent of the strain and the impurity level [36,37]. Therefore in this section Al is chosen to demonstrate recovery after deformation to high strains, and recovery is discussed in terms of dislocations, interconnecting boundaries and lamellar boundaries, whereas excess vacancies are ignored.

4.1. Initial stage of recovery

In a recent study of recovery in commercial purity aluminium AA1050 (99.5% purity) cold rolled to a true strain of 5.5 (99.6% thickness reduction), the initial stage of recovery was investigated by isothermal annealing in the temperature range 5-100 °C [38]. Post-mortem microstructural characterization by electron backscatter diffraction (EBSD) and TEM showed that the spacings of lamellar boundaries and interconnecting boundaries were almost unchanged during annealing, pointing to a negligible effect of subgrain coalescence and migration of low and high angle boundaries. However, the density of dislocations in regions between these boundaries was found to decrease significantly during annealing, indicating dislocation annihilation as an important recovery mechanism in the initial stage of recovery. In addition, extensive TEM observations of thin foils at the same areas before and after annealing not only confirmed annihilation of loose dislocations and dislocation bundles during low temperature annealing, but also revealed sharpening of interconnecting boundaries, indicating reorganization and possibly annihilation of dislocations in the boundary. An example of such TEM observations is shown in figure 3.

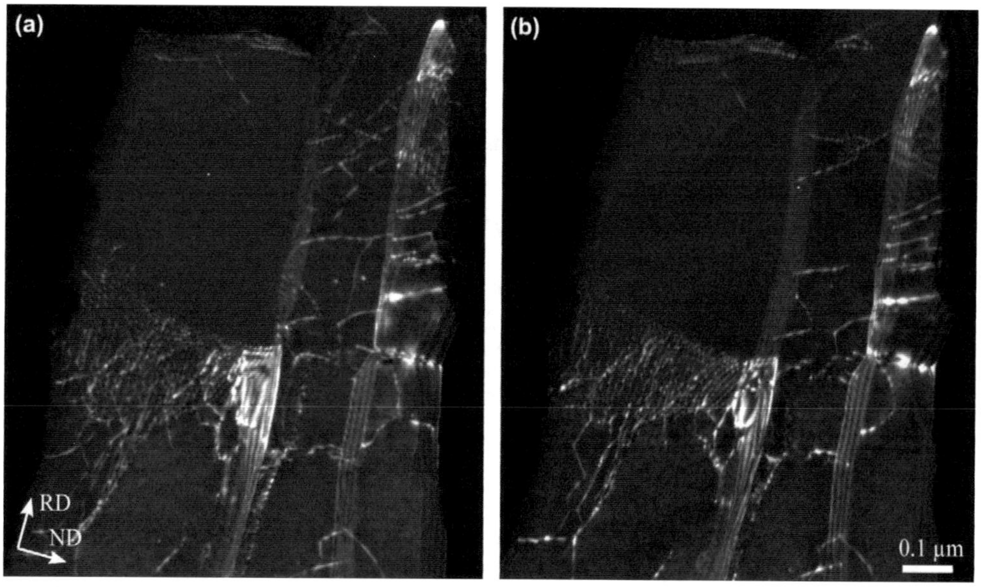

Figure 3. Weak-beam dark-field TEM micrographs showing annealing of dislocations and reorganization of interconnecting boundaries during storage of a TEM thin foil at room temperature for 9 days. (a) Before storage; (b) after storage.

In the same study [38], the recovery kinetics was followed by hardness tests. Due to the small recovery window examined, a constant activation energy was assumed and thus equation (2) was applied. The analysis led to a small apparent activation energy, 60-86 kJ/mol. However, this apparent activation energy may not directly correspond to an ideal diffusion mechanism since different diffusion species may be involved, different diffusion paths may contribute and ideal diffusion mechanisms may be affected by the deformation structure. Nevertheless, the low apparent activation energy is consistent with annihilation of dislocations between boundaries and within interconnecting boundaries, and the energy barriers for diffusion processes may be lowered by the presence of defects, e.g. a high density of dislocations and finely spaced boundaries.

4.2. Recovery by triple junction motion

While the initial stage of recovery only had a limited influence on the structure and the strength of the deformed Al, the following stage of recovery led to both significant coarsening of the microstructure and a marked decrease in strength [33]. In a recent study of recovery in heavily cold-rolled commercial purity aluminium AA1050 (true strains 2, 4 and 5.5), thermally activated migration of triple junctions formed by lamellar boundaries (i.e. Y-junctions) was identified as a key recovery mechanism, leading to removal of thin lamellae and uniform coarsening of lamellar structures [3].

During annealing of deformed samples in the temperature range 120-220 °C, general post-mortem TEM and EBSD investigations showed that the lamellar boundary spacing measured along the ND was increased but the lamellar morphology was largely maintained. After longer annealing periods, a gradual decrease of the structural aspect ratio led to a gradual transition from the lamellar morphology into a near equiaxed one [3]. Extensive TEM observations were carried out for a number of thin foils at the same areas before and after annealing. It was found that the process responsible for the removal of lamellar boundaries (and thus the increase of lamellar boundary spacing) was Y-junction motion. One example is shown in figure 4 where the thin foil was annealed at 120 °C. The Y-junction migrated up, removing the middle lamella in black contrast and increasing the lamellar boundary spacing of two neighbouring lamellae while keeping the lamellar morphology, in agreement with the general post-mortem observations. At higher annealing temperatures, Y-junction motion was found to be much more frequent, and an example is shown in figure 5 for annealing at 220 °C. Significant Y-junction motion took place in an initially finely spaced region, leading to doubling of the lamellar boundary spacing locally after annealing for 1 h. The microstructure after annealing evolved into being more equiaxed, but it kept a clear lamellar morphology.

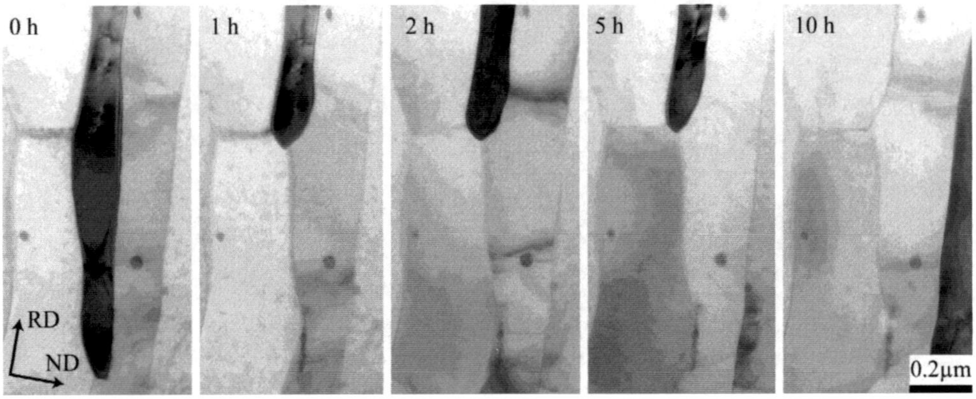

Figure 4. An example of Y-junction motion observed in the longitudinal section of Al cold rolled to a true strain of 5.5 and annealed at 120 °C for different total periods of time [3].

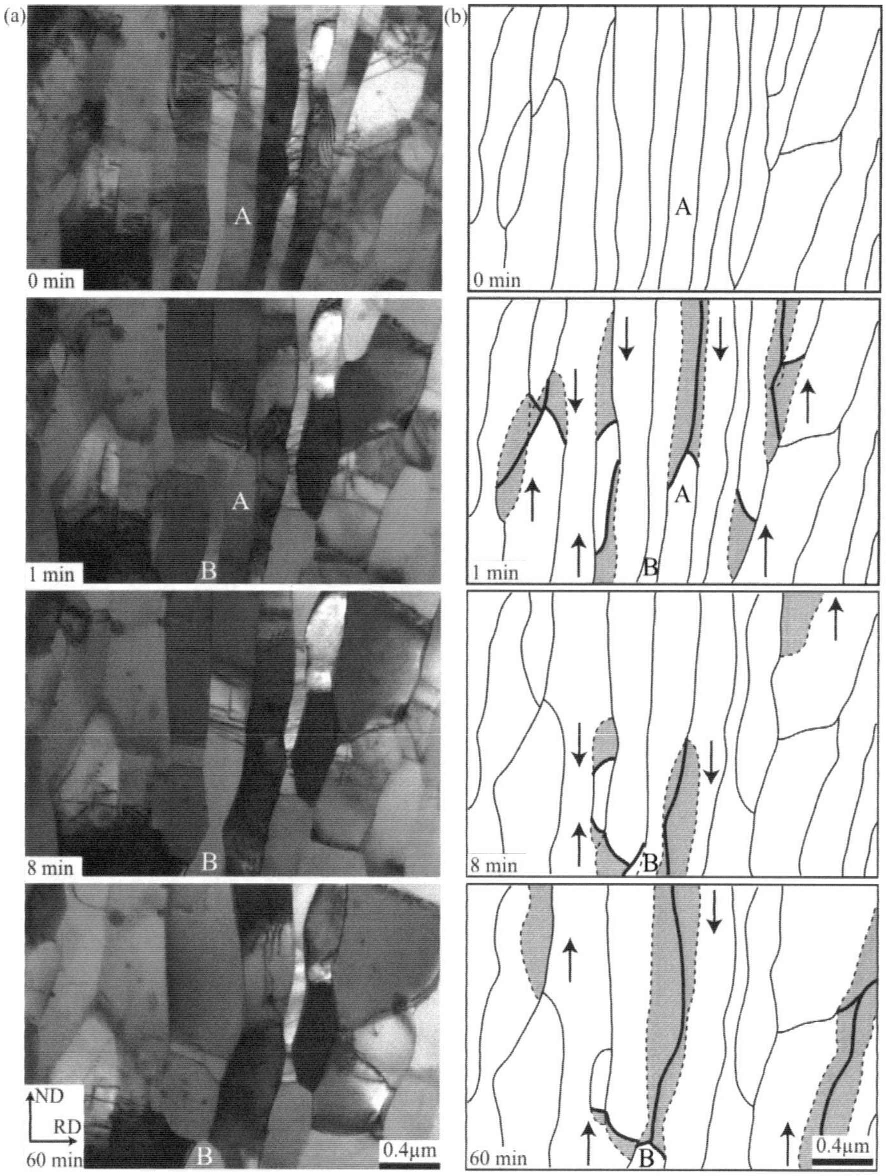

Figure 5. TEM observations of Y-junction motion during annealing at 220 °C. (a) Microstructure of the same area before and after annealing for 1, 8, and 60 minutes. (b) Sketches of the same area at different states corresponding to (a). For the states after annealing, each sketch combines the current state with the previous state to show the structural evolution. Solid thin lines represent lamellar boundaries which were unchanged; dashed lines represent boundaries which had migrated with new positions represented by solid bold lines. The areas swept by migrating boundaries are shown in grey, and the directions of the Y-junction motion are indicated by arrows. Lamella "A" was completely removed after 8 minutes and lamella "B" coarsened at the beginning due to removal of lamella "A" but then was shortened greatly by Y-junction motion [3].

Besides observations by TEM on thin foils, Y-junction motion was also observed by EBSD on the surface of bulk samples deformed to different strains [3]. Figure 6 shows an example, where the differences in the microstructure of the same area at different cumulative annealing periods are indicated. As shown in this case, Y-junction motion still played an important role although the lamellar boundary spacing had increased significantly (from 0.24 to 0.57 μm) after annealing for 2 h. Further coarsening by Y-junction motion and general boundary migration led to nucleation of recrystallization, which was observed after annealing for 96 h at this temperature [39].

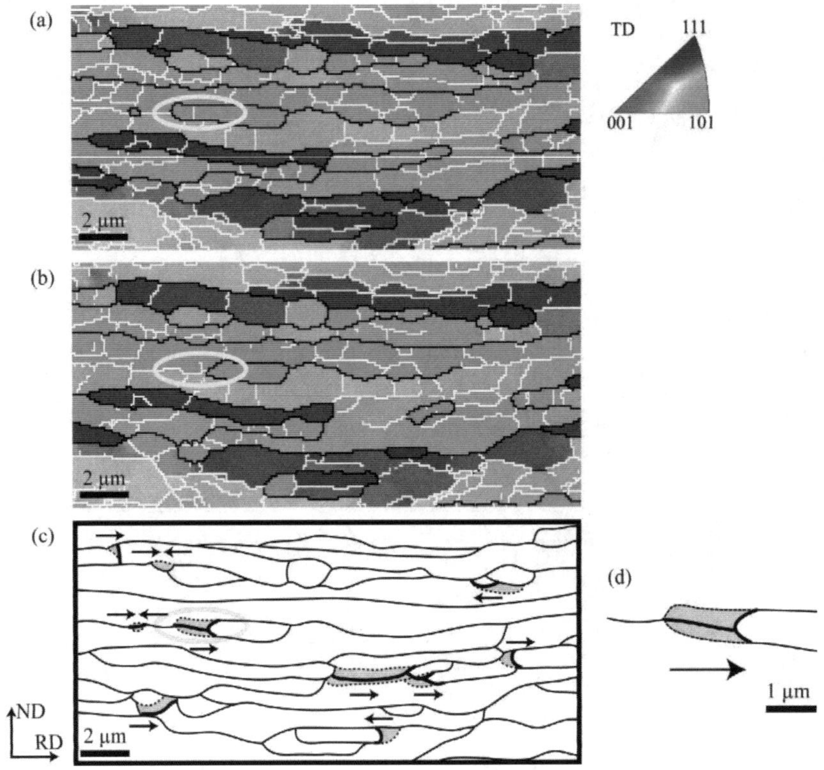

Figure 6. Examples of Y-junction motion in the longitudinal section of Al deformed to a strain of 5.5 and annealed at 220 °C. (a) Annealed for 2 hours; (b) annealed for 4 hours accumulatively. The colours are coded according to the orientation of each pixel after mild noise reduction using Channel 5. High angle boundaries (>15°) are shown in black and low angle boundaries (1.5~15°) in white. (c) A sketch of the lamellar boundaries combining the two states, following the same markings as in figure 5. The marked region is magnified in (d) [40].

Based on observations and analyses, Y-junction motion was proposed as a dominant recovery mechanism in a finely spaced lamellar structure [3]. The sketch shown in figure 7 illustrates how such a structure coarsens via Y-junction motion and evolves into a more equiaxed structure gradually, and after further annealing some coarsened subgrains may develop into nuclei of recrystallization.

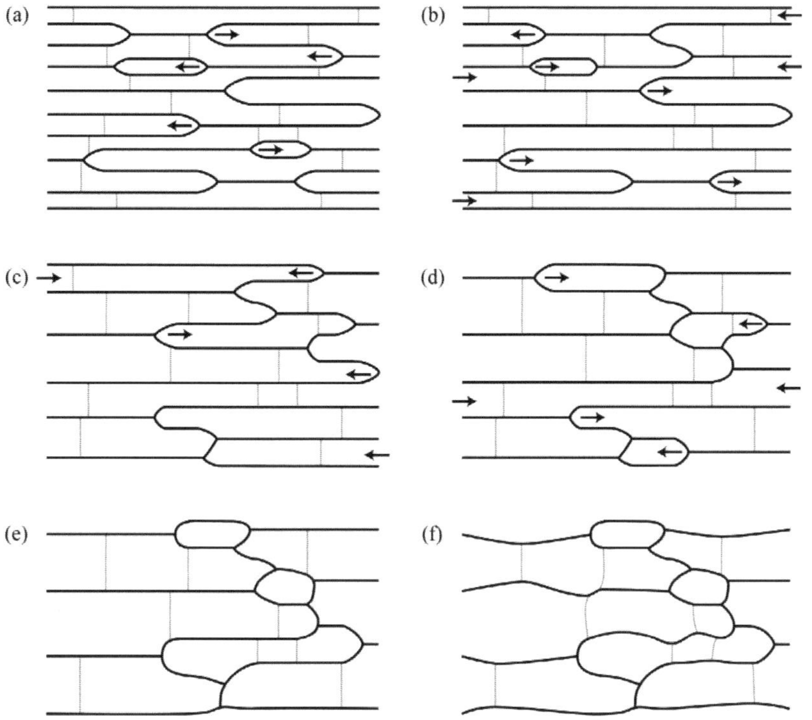

Figure 7. Schematic diagrams illustrating structural coarsening via Y-junction motion. (a) The original deformation structure with typical lamellar boundaries (bold black lines) and interconnecting boundaries (thin grey lines). This lamellar structure coarsens via Y-junction motion with arrows showing the directions of motion. (b)-(d) Intermediate coarsening steps. (e) A more equiaxed structure has evolved after coarsening via Y-junction motion. (f) A modified equiaxed structure considering also the drag of interconnecting boundaries at H-junctions (i.e. interconnecting boundaries resist widening) [3].

4.3. Triple junction motion and structural parameters

The geometry of a Y-junction is important in determining its stability, in particular the spacing D and the dihedral angle 2θ associated with the middle lamella. The dihedral angle and the lamellar boundary spacing shown in figure 8 were measured from TEM micrographs of deformed samples, and the categorization into two groups was based on the annealing results – circles for not-migrated Y-junctions and dots for migrated ones. The plot shows that: (i) the estimated dihedral angle appeared to increase with increasing lamella spacing although the exact value depended on the local configuration (e.g. affected by neighbouring H-junctions); (ii) a migrating Y-junction generally had a small dihedral angle and the corresponding shortening lamella generally had a small spacing; and (iii) neither a lamella spacing below ~50 nm nor a dihedral angle below ~30° was observed. However, the large dispersion of data shown in figure 8 indicates that other factors may also play a role in determining the stability of a Y-junction [3].

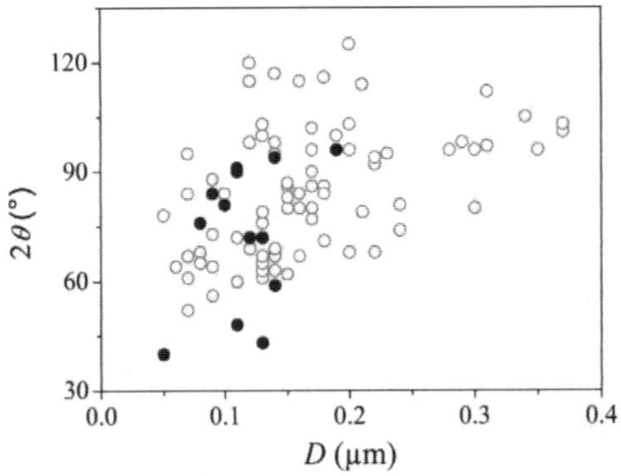

Figure 8. Distribution of the dihedral angle and the lamella spacing of Y-junctions in Al deformed to a strain of 5.5. The pairs of dihedral angle and lamella spacing were estimated from TEM micrographs taken from the longitudinal section. Open circles represent random measurements, whereas solid points represent measurements of Y-junctions observed to have migrated after annealing at 120 °C for 5 hours [3].

Most of the migrating Y-junctions were composed of three high angle boundaries or a combination of high angle boundaries and low-to-medium angle boundaries, but cases of migrating Y-junctions formed by three low-to-medium angle boundaries were also found although much less frequently. The deformation structures were subdivided into typical rolling texture components, and Y-junction motion was examined for each rolling texture component but no significant orientation preference was found. Y-junction motion was more frequently found at higher temperatures during recovery annealing, and also in samples deformed to higher strains, which have smaller boundary spacings and higher stored energy [3].

The driving force for Y-junction motion comes from many sources, e.g. Y-junction line energy and strain energy from dislocations, but the principal source was considered to be from the boundary energy (grain boundary surface tension). However current theories of boundary mobility [41,42] cannot explain the rapid Y-junction motion observed at low annealing temperatures, nor the absence of dihedral angles below 30° or lamellae thinner than 50 nm. This has led to the tentative suggestions that (i) the triple junction mobility may depend on its dihedral angle, being very large for a sharp triple junction, and (ii) an internal stress may build up at sharp triple junctions during deformation, leading to mechanically assisted triple junction motion and removal of very thin lamellae [3].

4.4. Activation energy

The coarsening kinetics of the same heavily cold-rolled Al was studied recently [39]. In that study, deformed Al was subjected to a series of isothermal annealing treatments in the temperature range 140-220 °C. Both EBSD and electron channel contrast (ECC) imaging were used to characterize the microstructural evolution, and the measured lamellar boundary spacings are shown in figure 9 for samples that had undergone microstructural coarsening before recrystallization. The increase of the lamellar boundary spacing approximately followed a linear relationship in the log-log plot and the slope ($1/n$) increased with increasing annealing temperature. Based on equation (2), the annealing time t and the annealing temperature T (K) required for achieving the same degree of recovery were related by the following equation:

$$\ln t = c + \frac{Q_0}{RT}, \qquad (5)$$

where c is a constant. Since the activation energy for recovery increases significantly in the course of recovery (e.g. [20,29,30]), equation (5) may only apply to each small recovery window, where the variation of the activation energy is small. Based on equation (5), the activation energy for each small recovery window was determined from (ln *t*) vs 1/*T* plots. The analysis led to a low activation energy (110 kJ/mol) at the beginning of coarsening and a high one (240 kJ/mol) at the end of uniform coarsening, i.e. when the lamellar boundary spacing had increased to 0.9 μm (figure 10). In this interval, the microstructural coarsening was dominated by Y-junction motion.

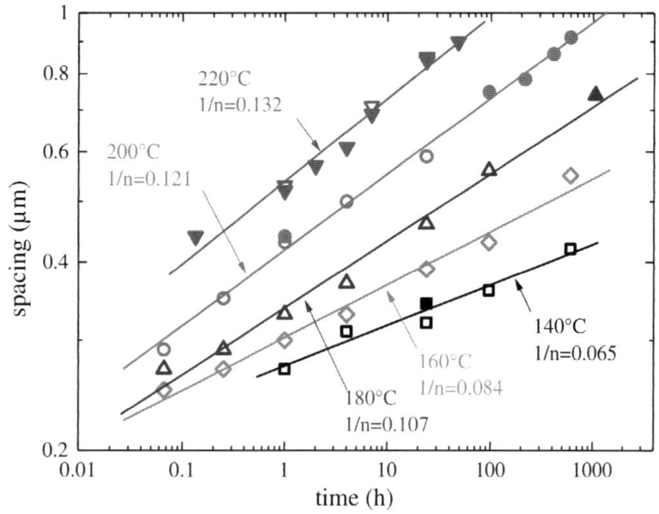

Figure 9. EBSD (solid symbols) and ECC (open symbols) data of the average lamellar boundary spacing during recovery of cold-rolled Al. The spacing was measured by the line intercept method along the ND. Different slopes (1/n) of linear least squares fittings in log–log plots are indicated for different temperatures [39].

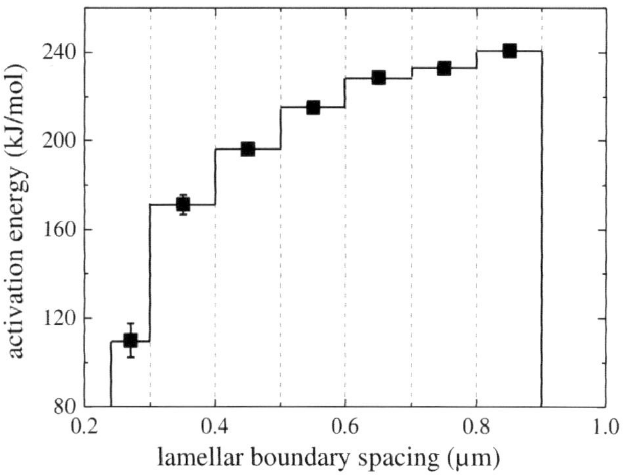

Figure 10. Variation of the apparent activation energy during microstructural coarsening of cold-rolled Al [39].

The estimated apparent activation energy at the beginning of coarsening was close to that for dislocation annihilation in the initial stage of recovery, and the activation energy at the end of uniform coarsening was close to that for the subsequent recrystallization, which was previously reported to be 230-235 kJ/mol for AA1050 [43,44]. The agreement in these numbers suggests a continuous increase of the apparent activation energy during annealing, including both recovery and recrystallization.

It should also be noted that, based on the results shown in figures 9 and 10, the microstructural coarsening of the same material at other annealing temperatures may be predicted by a simple extrapolation. For example, an extrapolation to 300 °C predicts that about 1 minute is needed for the lamellar boundary spacing to increase to 0.9 μm, in good agreement with experimental results [45]. It was indeed reported [45,46] that Y-junction motion still played an important role in Al during annealing at 300 °C, where the recrystallization kinetics was accelerated, and that recovery by Y-junction motion affected boundary populations in recovered microstructures and thus subsequent recrystallization.

In another study [40], the recovery kinetics was studied based on hardness measurements. Instead of being divided into small recovery ranges, the whole kinetics was fitted to equation (4), which assumes that the apparent activation energy increases linearly with decreasing stored energy. The results were generally consistent with the analysis based on boundary spacings presented above.

4.5. Structural pinning

In a recent study, Y-junction motion in heavily cold-rolled Al was studied by in situ TEM [47]. The in situ observations showed that the migrating Y-junctions were pinned by encountered interconnecting and lamellar boundaries, which slowed down the recovery process and led to a stop-go migration pattern. The dynamic pinning force for Y-junction motion was directly observed by the bowing of encountered boundaries and their component dislocations as illustrated in figures 11 and 12. This pinning mechanism stabilizes the recovered microstructure, i.e. the structure is stabilized by balancing the driving and pinning forces controlling the rate of Y-junction motion.

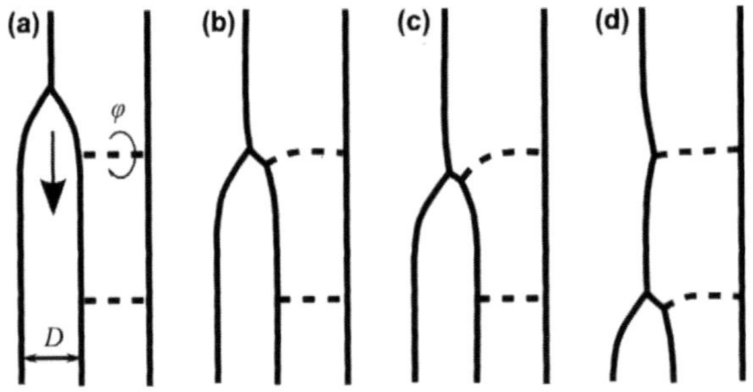

Figure 11. Schematic drawing that shows the pinning and depinning interactions between a moving Y-junction and interconnecting boundaries (dashed lines), including the bowing and slight extension of the interconnecting boundary during annealing. The arrow indicates the direction of Y-junction motion, D is the thickness of the shortening lamella and φ is the misorientation angle of the interconnecting boundary [47].

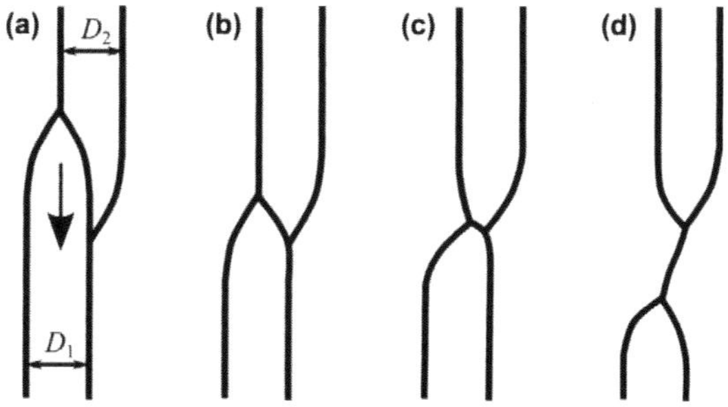

Figure 12. A sketch showing pinning and depinning interactions between a moving Y-junction and an attached lamellar boundary. The arrow indicates the direction of Y-junction motion, and D_1 and D_2 are the thickness of lamellae [47].

It was also found in TEM foils cut from bulk annealed samples that Y-junctions were typically in the vicinity of attached interconnecting or lamellar boundaries (see figures 11b and 12b). These attached boundaries increased the dihedral angles at the Y-junctions, thereby reducing the driving force for Y-junction motion and stabilizing the partially recovered structure. Quantitative measurement showed that the average projected distance from a Y-junction to the nearest attached position was only 94 nm, which was about 1/3 of the distance for a random distribution of Y-junctions [47]. Such a big difference indicated that Y-junctions were not randomly distributed, but formed stable configurations, in agreement with in situ observations. Therefore, the effect of these deformation-induced boundaries is paradoxical: on one hand they store deformation energy and provide the driving force for recovery and recrystallization, but on the other hand they stabilize the microstructure by retarding or pinning Y-junction motion and boundary migration.

5. Recovery in other metals deformed to high strains

In other metals deformed by dislocation glide, e.g. Ni and Cu, significant structural differences exist, which may affect the recovery mechanisms and kinetics. For example, both the spacings between lamellar boundaries and the dihedral angles are significantly smaller in Ni and Cu, and as a consequence recovery by triple junction motion may be facilitated in these metals. Besides, the stored energy in Ni and Cu is higher than that in Al in the high strain region. However, there are also factors against recovery in Ni and Cu, for example the competition between recovery and recrystallization. In Al there is a significant recovery range with a large decrease in mechanical properties before the onset of recrystallization. On the other hand Ni and Cu usually show only a small decrease in hardness before recrystallization starts, and structural changes related to recovery processes are expected to be small. Based on the above points, it is difficult to speculate whether recovery by triple junction is more or less pronounced in Ni and Cu, which also represent a lower stacking fault energy and a higher melting point. Preliminary experiments have been carried out on high purity Ni and Cu, showing less pronounced triple junction motion. This observation suggests that other factors have to be taken into account, for example structural pinning, which based on structural parameters may be different in other metals. This is to be explored in future research together with changes in other parameters.

Three important factors are the stacking fault energy, solutes and particles. It is well known that the stacking fault energy plays a key role in the recovery of dislocation networks, especially in pure metals, and therefore compared to Al smaller stacking fault energies in Ni and Cu lead to less recovery of dislocation networks, which may provide a stronger structural pinning effect. Solutes are also very important since they can strongly interact with dislocations, impede dislocation motion and thereby increase structural pinning. The mechanisms associated with solute dislocation interactions are diverse and include elastic size effects, as well as changes to shear modulus and stacking fault energy. Solutes may also segregate at boundaries, resulting in higher activation energies for boundary migration [20,39] and thereby delay recrystallization [48]. This delay gives a larger recovery window for triple junction motion, which has been shown to operate at lower activation energies (figure 10) and may be less affected by solutes due to its special geometry, i.e. solutes may be deposited at the tailing boundary. Moreover, when there is a dispersion of small particles, Zener drag will decrease the driving pressure for recrystallization and thereby delaying recrystallization, whereas triple junction motion may be less affected due to its high driving force when the lamellar boundary spacing is fine and the dihedral angle is small.

Therefore, future research may be carried out to extend the window of recovery by delaying recrystallization kinetics through alloying [48,49] or by replacing cold deformation with warm deformation. This is to reduce the driving pressure for recrystallization while maintaining a large fraction of deformation induced high angle boundaries at high strains. Based on such structures, it will also be possible to optimize annealing treatments of metals and alloys deformed to large strains to balance their mechanical properties, e.g. strength and ductility.

6. Concluding remarks

The recovery of heavily deformed metals have been demonstrated with a focus on Al, where triple junction motion has been identified as an important mechanism, resulting in structural coarsening and therefore being a precursor of recrystallization. In situ observations showed that migration of Y-junctions was a stop-go migration process, where a rate reduction was caused by the interaction of migrating junctions with deformation induced boundaries. This pinning mechanism stabilized the structure by balancing the driving forces, thereby increasing the thermal stability of the highly strained material. This is of importance when developing ultra-strong metals and alloys by plastic deformation to very high strains.

Acknowledgement

The authors gratefully acknowledge the support from the Danish National Research Foundation (Grant No. DNRF86-5) and the National Natural Science Foundation of China (Grant No. 51261130091) to the Danish-Chinese Center for Nanometals, within which this work has been performed. Dr. D. A. Hughes is thanked for helpful discussions.

References

[1] Humphreys F J and Hatherly M 2004 *Recrystallization and Related Annealing Phenomena* (Oxford: Pergamon)
[2] Doherty R D 1980 *Proc. 1st Risø International Symposium on Metallurgy and Materials Science,* ed N Hansen et al. (Roskilde, Denmark: Risø National Laboratory) p 57
[3] Yu T, Hansen N and Huang X 2011 *Proc. R. Soc.* A **467** 3039
[4] Hughes D A and Hansen N 2004 *ASM Handbook* vol 9, ed G F Vander Voort (Materials park, OH: ASM International) p 292
[5] Hughes D A and Hansen N 2000 *Acta Mater.* **48** 2985
[6] Liu Q, Huang X, Lloyd D J and Hansen N 2002 *Acta Mater.* **50** 3789
[7] Kuhlmann-Wilsdorf D and Hansen N 1991 *Scr. Metall. Mater.* **25** 1557
[8] Hansen N 2001 *Metall. Mater. Trans.* A **32** 2917
[9] Bever M B, Holt D L and Titchener A L 1973 *Progress in Materials Science* **17** 5
[10] Godfrey A and Liu Q 2009 *Scr. Mater.* **60** 1050
[11] Bever M 1957 *Creep and Recovery,* ed R Maddin (Cleveland, OH: Am. Soc. Met.) p 14
[12] Perryman E 1957 *Creep and Recovery,* ed R Maddin (Cleveland, OH: Am. Soc. Met.) p 111
[13] Friedel J 1964 *Dislocations* (Oxford: Pergamon)
[14] Lytton J, Westmacott K H and Potter L 1965 *Transactions of the Metallurgical Society of AIME* **233** 1757
[15] Hu H 1962 *Transactions of the Metallurgical Society of AIME* **224** 75
[16] Li J C M 1962 *J. Appl. Phys.* **33** 2958
[17] Gleiter H 1969 *Philosophical Magazine* **20** 821

[18] Smith C and Dillamore I 1970 *Metal Science* **4** 161
[19] Hasegawa T and Kocks U 1979 *Acta Metall.* **27** 1705
[20] Furu T, Orsund R and Nes E 1995 *Acta Metall. Mater.* **43** 2209
[21] Nes E 1995 *Acta Metall. Mater.* **43** 2189
[22] Jones A R, Ralph B and Hansen N 1979 *Proc. R. Soc.* A **368** 345
[23] Cahn R W 2004 *Phsical Metallurgy* vol 3, ed R W Cahn and P Haasen (Amsterdam: Elsevier) p 2400
[24] Mankins W L 2004 *ASM Handbook* vol 9, ed G F Vander Voort (Materials park, OH: ASM International) p 207
[25] Sandstrom R 1977 *Acta Metall.* **25** 897
[26] Sandstrom R 1977 *Acta Metall.* **25** 905
[27] Prinz F, Argon A and Moffatt W 1982 *Acta Metall.* **30** 821
[28] Doherty R and Szpunar J 1984 *Acta Metall.* **32** 1789
[29] Michalak J T and Paxton H W 1961 *Transactions of the Metallurgical Society of AIME* **221** 850
[30] Vandermeer R A and Hansen N 2008 *Acta Mater.* **56** 5719
[31] Kuhlmann D 1948 *Zeitschrift Fur Physik* **124** 468
[32] Borelius G, Berglund S and Sjoberg S 1953 *Arkiv for Fysik* **6** 143
[33] Yu T, Hansen N, Huang X and Winther G 2009 *Proc. 30th Risø International Symposium on Materials Science,* ed J Grivel et al. (Roskilde, Denmark: Risø DTU) p 393
[34] Albou A, Borbely A, Maurice C and Driver J H 2011 *Philosophical Magazine* **91** 3981
[35] Xing Q, Huang X and Hansen N 2006 *Metall. Mater. Trans.* A **37** 1311
[36] Haessner F and Schmidt J 1988 *Scripta Metall.* **22** 1917
[37] Schmidt J and Haessner F 1990 *Z. Phys.* B **81** 215
[38] Yu T, Hansen N and Huang X 2012 *Philos. Mag.* **92** 4056
[39] Yu T, Hansen N and Huang X 2013 *Acta Mater.* **61** 6577
[40] Yu T 2011 *Recovery and recrystallization of nanostructured metals – Mechanisms and kinetics* PhD thesis (Roskilde, Denmark: Risø DTU)
[41] Galina A V, Fradkov V Y and Schvindlerman L S 1987 *Fiz. Metallov Metalloved.* **63** 1220
[42] Gottstein G and Shvindlerman L S 2010 *Grain Boundary Migration in Metals: Thermodynamics, Kinetics, Applications* (Boca Raton, FL: CRC Press)
[43] Chen S, Zuidwijk T and van der Zwaag S 2001 *Proc. 1st Joint Int. Conf. on Recrystallization and Grain Growth* ed G Gottstein and D A Molodov (Berlin: Springer-Verlag) p 821
[44] Verlinden B, Ratchev P, Van den Broeck R and van der Winden M 2001 *Proc. 1st Joint Int. Conf. on Recrystallization and Grain Growth* ed G Gottstein and D A Molodov (Berlin: Springer-Verlag) p 1329
[45] Mishin O V, Godrey A, Juul Jensen D and Hansen N 2013 *Acta Mater.* **61** 5354
[46] Mishin O V, Godrey A, Yu T, Hansen N and Juul Jensen D 2015 *IOP Conf. Series: Materials Science and Engineering* **82** 012083
[47] Yu T, Hughes D A, Hansen N and Huang X 2015 *Acta Mater.* **86** 269
[48] Zhang H W, Lu K, Pippan R, Huang X and Hansen N 2012 *Proc. 33rd Risø International Symposium on Materials Science,* ed S Fæster et al. (Roskilde, Denmark: Risø DTU) p 399
[49] Tian H, Suo H L, Mishin O V, Zhang Y B, Juul Jensen D and Grivel J-C 2013 *J. Mater. Sci.* **48** 4183

Boundary migration during recrystallization: experimental observations

Y B Zhang, D Juul Jensen

Section for Materials Science and Advanced Characterization, Department of Wind Energy, Technical University of Denmark, Risø Campus, Roskilde 4000, Denmark

E-mail: yubz@dtu.dk

Abstract. Quantitative analysis of boundary migration during recrystallization is a key task to understand the recrystallization process and to improve recrystallization models. In the last 25-30 years, quantification of boundary migration has mostly been conducted in term of average growth rates in many materials. This type of analysis has largely been based on the assumption that all or groups of recrystallizing grains grow in the same uniform manner, therefore the results represent average growth behaviors on a *macro* scale. Recently, significant efforts have been made to quantify the boundary migration during recrystallization on *local* grain scales, using different advanced experimental characterization and computer simulation techniques. This paper aims at summarizing these recent achievements with focus on the potentials of the various advanced experimental characterization techniques. Suggestions for new experimental and simulation work important for advancing the current understanding of local boundary migration are finally discussed.

1. Introduction

During recrystallization, nearly perfect, dislocation-free nuclei develop at preferential sites in the deformed microstructure and grow by boundary migration through the deformed matrix, until the whole deformed matrix is replaced by recrystallized grains. The boundary migration rate, v, is generally expressed as:

$$v = MF \tag{1}$$

where M is the boundary mobility and F is the driving force for migration. For recrystallization F is typically considered as the energy stored in the deformed matrix [1]. Experimentally, this equation has been frequently validated for boundary migration during the grain growth process (typically occurring after recrystallization is complete) [2], but more rarely for boundary migration during recrystallization. The few recrystallization studies quantifying equation (1) have been conducted by static stereological 2D characterizations, and the analysis is thus only valid for the average behavior of groups of grains (e.g. [3]). Also along the same lines, many studies of recrystallization have focused on recrystallization kinetics and estimated average boundary migration rates using e.g. the Cahn-Hagel [4] and the extended Cahn-Hagel method [5] based on stereological analysis or using maximum grain sizes [6].

In the last 25-30 years, lots of works on characterization of deformation microstructures have shown that deformation microstructures in most metals are highly heterogeneous (e.g. [7]) and depend on the crystallographic orientations of the deformed grains (e.g. [8]). At the same time, several new experimental characterization techniques have been developed, such as in-situ measurements using electron backscattered diffraction (EBSD) techniques (e.g. [9]) and 3D X-ray diffraction (3DXRD) techniques (e.g. [10]), which allow detailed quantification of boundary migration and in some cases of the deformed microstructure in front of the migrating boundary on a *local* grain scale in 2D or 3D. These techniques have been used to study recrystallizing boundaries and their migration into different deformed grains on the local scale. The results have shown that locally the migration is much more complex than hitherto assumed: the migration of individual boundary segments occurs in a jerky stop-go fashion, and fairly large protrusions and retrusions often form and evolve on many boundaries (e.g. [11]). It has therefore been suggested that the local heterogeneities in the deformed microstructures have to be taken into account for a comprehensive understanding of the local recrystallization boundary migration [12].

The aim of this paper is to summarize experimental achievements in this relatively new research field of local recrystallization boundary migrations. The main focus will be on the quantification of local protrusions/retrusions and of their effects on the boundary migration as well as on the effects of the local deformed microstructures in front of the recrystallizing boundaries. The paper is organized in sections according to the experimental techniques used for the studies: i.e. 2D (x, y), 3D (x, y, z) and (x, y, t) as well as 4D (x, y, z, t). Potentials of the experimental techniques and suggestions for future work will be discussed at the end.

2. 2D (x, y) static characterizations

Partially recrystallized microstructures can be revealed by many microscopical techniques such as optical microscopy (OM), electron channeling contrast (ECC), electron backscattered diffraction (EBSD), and transmission electron microscopy (TEM). Among these techniques, OM is certainly the easiest to use, and used to play an important role in the early days for recrystallization studies (e.g. [13]). Nowadays, due to the wide availability of scanning electron microscopes (SEMs), OM is generally only used as a simple tool to get a quick characterization of the recrystallization state on a macro scale.

The ECC and EBSD within SEMs are now more commonly used to visualize recrystallizing structures. The ECC techniques can easily provide spatial resolution of ~1 nm [14], whereas EBSD can typically provide a spatial resolution of 20-30 nm. This lower resolution is because of the large interaction volume of the electron beam and sample surface at a high tilting angle [15]. EBSD techniques however provide unique possibilities for mapping microstructures with full information of crystallographic orientations (angular resolution of 1.5-2°), which allows quantitative analysis of local microstructures and textures of both the deformed matrix and the recrystallized grains [9]. EBSD is therefore nowadays the most frequently used technique for recrystallization studies, e.g. in the latest recrystallization conference proceeding ~76% of the papers on experimental studies of recrystallization were based on measurements conducted using EBSD techniques [16].

TEM (incl. high voltage TEM) techniques can characterize microstructures with the highest spatial resolution down to the atomic level. Free dislocations and the dislocation content in dislocation boundaries (e.g. [17]) can be characterized with TEM but not with any of the above mentioned techniques. Before the appearance and commercializing of the EBSD techniques, TEM was the most important tool for detailed recrystallization studies (e.g. [18]). One of the drawbacks with TEM is that it is not so easy to determine crystallographic orientations over large area as with EBSD. Even using semi-automatic methods [19] by which the crystallographic orientations can be determined at a reasonable speed, the efficiency can still not be comparable with that of EBSD. A relatively high angular resolution (~0.5°) can, however, be obtained. Besides that, the sample preparation is difficult for TEM and to prepare a good large TEM sample for statistical analysis is even more challenging.

Examples of partially recrystallized structures visualized by OM, ECC, EBSD and TEM are given in figure 1. By looking at all these partially recrystallized micrographs, a common feature can be noticed namely that recrystallizing boundaries are rough, locally consisting of many protrusions and retrusions with sizes of few to tens micrometres, which reflects that the migration of recrystallizing boundaries is inhomogeneous. Although this feature was reported in the literature even in 1950's (see figure 1a), not much attention has been paid to these local 'anomalies' until recently. In order to understand the formation of protrusions and retrusions as well as their importance for the boundary migration, quantification methods of individual protrusions/retrusions and correlations between protrusions/retrusions and the neighboring deformation matrix are needed.

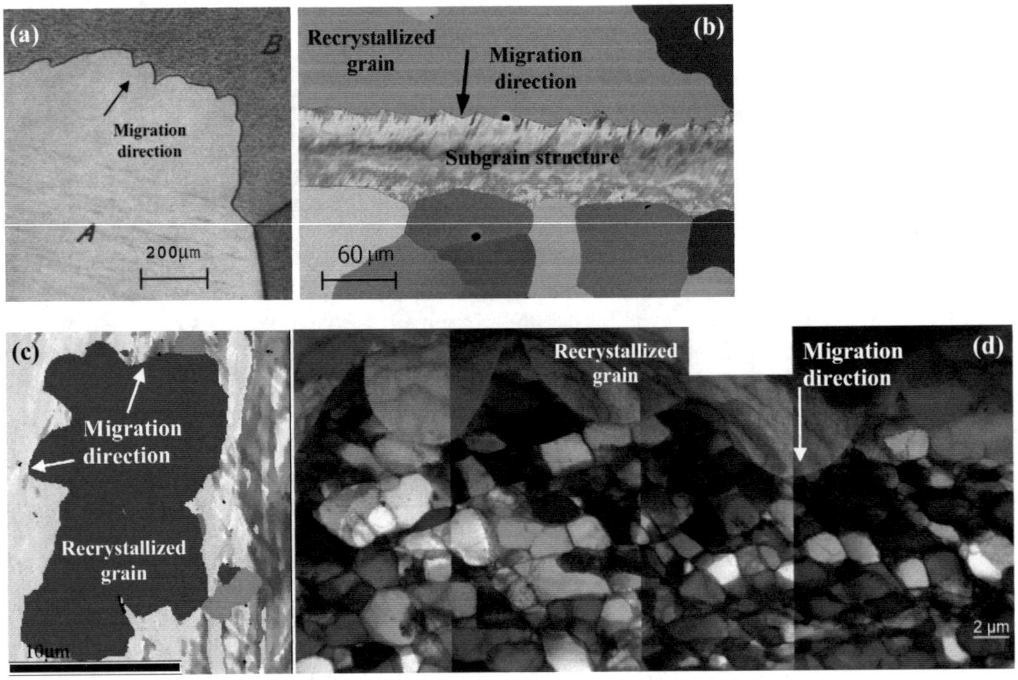

Figure 1. Examples showing rough recrystallizing boundaries in different partially recrystallized materials using different characterizing techniques: (a) optical micrograph showing strain induced boundary migration in high purity aluminium (Reproduced with permission from [13], Copyright 1950, AIP Publishing LLC.), (b) ECC image showing a grain boundary with protrusions migrating into a recovered region during recrystallization of high purity tantalum (body-centered cubic) (Reprinted from [20] with permission from Elsevier), (c) EBSP map of partially recrystallized microstructure of 96% cold rolled pure (99.996%) nickel [21], (d) TEM image showing in detail the rough features on a recrystallizing boundary in a partially recrystallized pure aluminium sample.

2.1. Quantification of protrusions/retrusions

To describe quantitatively protrusions/retrusions, Martorano et al. [22] used a sinusoidal function to simulate a variation in driving force and consequently boundary shape in a numerical analysis of growth during recrystallization. Considering the experimental fact that most protrusions/retrusions do not have a perfect sinusoidal shape, Zhang et al. developed another method, using polynomial equations to fit the shape of protrusions/retrusions, to obtain a more precise quantification of the individual protrusions/retrusions [12, 21]. The method is illustrated in figure 2. Individual protrusions/retrusions are first selected manually from the 2D experimental images, and then the boundary segments are fitted using a 3rd or 4th order polynomial equation. The radius of the curvature of the boundary can then be calculated by:

$$r = (1+\dot{y}^2)^{3/2}/\ddot{y}, \qquad (2)$$

where $\dot{y}$ and $\ddot{y}$ represent the first and second derivatives of the selected polynomial equation, respectively. The curvature-based driving force for each protrusion/retrusion is then calculated using equation (3) [1]:

$$F_\sigma = 2\sigma/r, \tag{3}$$

where σ is the grain boundary energy and r is the radius of the curvature of the boundary.

This method can be used to quantify the protrusions/retrusions on recrystallizing boundaries observed with all the characterization techniques mentioned above and as shown in figure 1. Among these micrographs, ECC images are generally the most appropriate ones, considering their high spatial resolution, short acquisition time and easy sample preparation. Only when really small protrusions/retrusions have to be characterized, TEM is needed. In term of automatic grain boundary detection for quantification, however, EBSD maps are obviously the best choice because of the additional orientation information, which helps separating recrystallized grains from the matrix.

As the curvature varies along a protrusion/retrusion on the recrystallization boundaries, the maximum curvature-based driving force, $F_{\sigma,max}$, at the position where the radius of the curvature is minimum is suggested to be used to quantify each protrusion/retrusion [12]. By measuring all the protrusions and retrusions on a boundary, the distribution of $F_{\sigma,max}$ can be obtained. An example of such a distribution is shown in figure 3b, where $F_{\sigma,max}$ values are calculated for all protrusions and retrusions on the boundary in figure 3a. Large spreads of the distribution of $F_{\sigma,max}$ are seen for both protrusions and retrusions, implying large curvature differences between individual protrusions/retrusions. The tail in the distribution for retrusions is much longer than that for protrusions, revealing that retrusions on this boundary are much sharper than protrusions. This type of distribution can be used directly to compare characteristics of different boundaries in the same or different materials, and thereby provide a basis for understanding the correlation of boundary structures and local deformed microstructure in front of the moving boundary, as will be discussed in the next section 2.2.

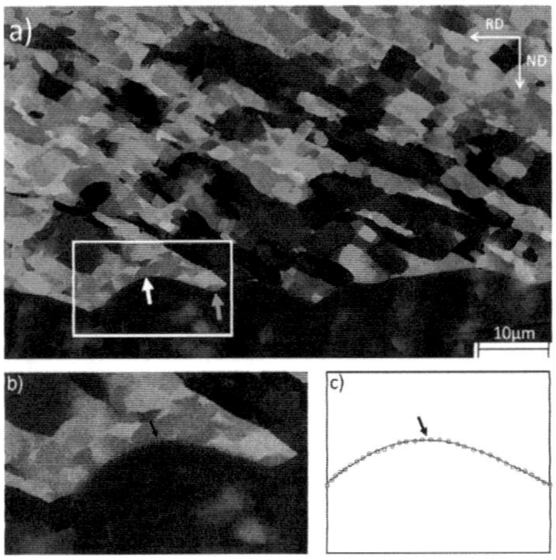

Figure 2. Illustration of the procedure for quantification of protrusions/retrusions. (a) ECC image showing a partially recrystallized microstructure of 50% cold-rolled aluminium annealed at 250 °C for 10 minutes. A white and a grey arrow are used to mark a protrusion and a retrusion, respectively. (b) Enlarged image of the microstructure within the white rectangle in (a), with the boundary segments traced by red circles. (c) The fitting result, as shown by the thick black line. Black arrows in (b) and (c) are used to mark the place where the boundary curvature is minimum and thus the dragging force is maximum. [12]

A drawback of this method is that it is a time consuming process to obtain the distributions of $F_{\sigma,max}$, even for one boundary, as the boundary is determined manually. Recently, Sun et al. have developed a fully automatic method, where a parameter called area integral invariant is used to quantify protrusions/retrusions on a boundary and thereby to quantify the boundary roughness. A detailed description of this method can be found in this proceeding [23].

2.2. Correlation to the deformed microstructure

To correlate the local protrusions/retrusions to the local deformed microstructure in front of the boundary, quantitative crystallographic information over a large area are required. For this purpose,

EBSD generally is the best choice. But to optimize the data collection process, EBSD can be used together with ECC: ECC for quick, high-resolution characterization of protrusions/retrusions and EBSD for characterization of deformation microstructures on a coarser scale.

A simple calculation has been made to correlate the average maximum curvature dragging/driving forces from protrusions/retrusions (i.e. $<F_{\sigma,max}>$ for protrusions/retrusions) with the average stored energies in the deformed microstructure in front of the boundaries [21]. The stored energy is estimated based on EBSD maps. In that work, $<F_{\sigma,max}>$ for protrusions was calculated based on sinusoidal equations. A more precise calculation of $<F_{\sigma,max}>$ for both protrusions and retrusions using the method described in equations (2) and (3) has been conducted in the present work, and the correlation is shown in figure 4. For protrusions, an overall trend can be seen: higher average stored energy leads to higher $<F_{\sigma,max}>$, although with a large scatter. For retrusions, the magnitude of $<F_{\sigma,max}>$ is much larger (note the scale difference in figure 4a and 4b), and the spread is larger.

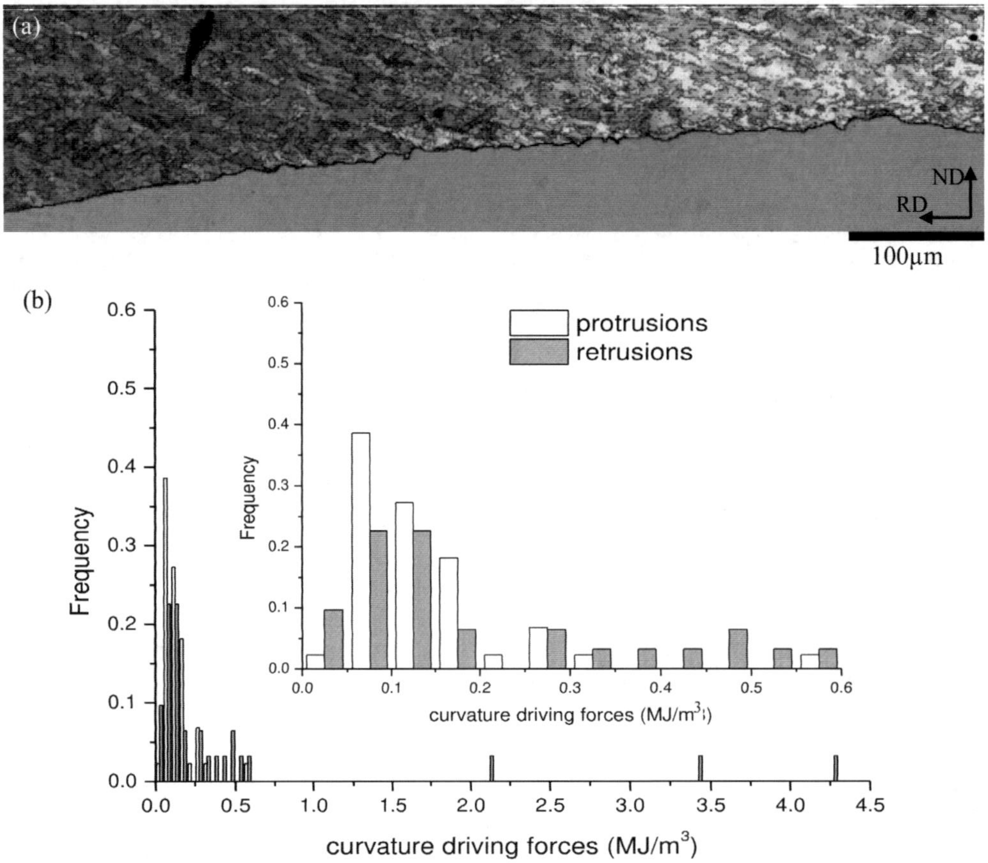

Figure 3. (a) EBSD map showing a long rough recrystallization boundary in 50% cold-rolled aluminium annealed at 250 °C for 10 minutes. (b) Distribution of $F_{\sigma,max}$ for both protrusions and retrusions along the boundary in (a). Insert in (b) showing the details of the distribution in the range of 0-0.6 MJ/m^3.

It has to be noted that the curvature of a protrusion provides a dragging force for the boundary migration. The size of protrusions can therefore not be so large, as the curvature dragging force cannot exceed the local driving force from the stored energy in the deformed matrix, i.e. the protrusion size is limited by the magnitude of the local stored energy. On the other hand, the curvature of a retrusion provides an extra driving force for the boundary migration to be added to that of the stored energy in the deformed matrix. The retrusion size is therefore not limited by the stored energy. However, why the retrusions on some boundaries can become so large is not trivial to understand, considering the fact that the curvatures of retrusions can provide much larger driving forces than the stored energy in the deformed matrix, and that neighboring stagnated segments of the boundary may be pulled forward. Simple energy balances as that suggested for the protrusions may therefore not be valid.

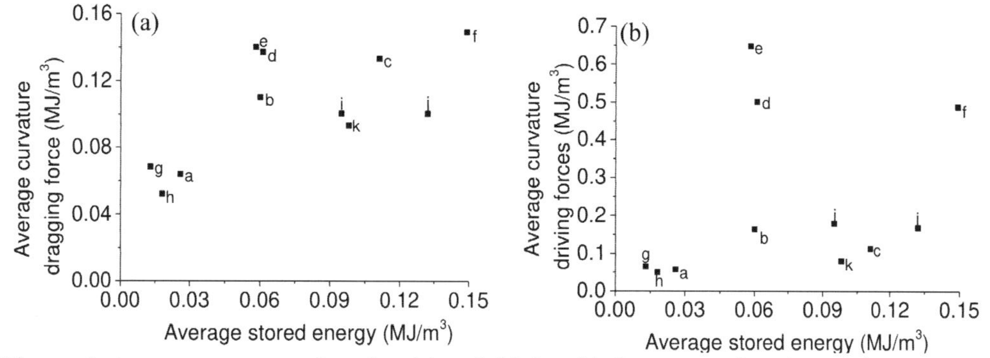

Figure 4. Average curvature dragging (a) and driving (b) forces as a function of the average stored energy in the deformed matrix for 11 different recrystallizing boundaries (marked with letters a-k) in partially recrystallized 50% cold-rolled aluminium annealed at 250 °C for 10 minutes.

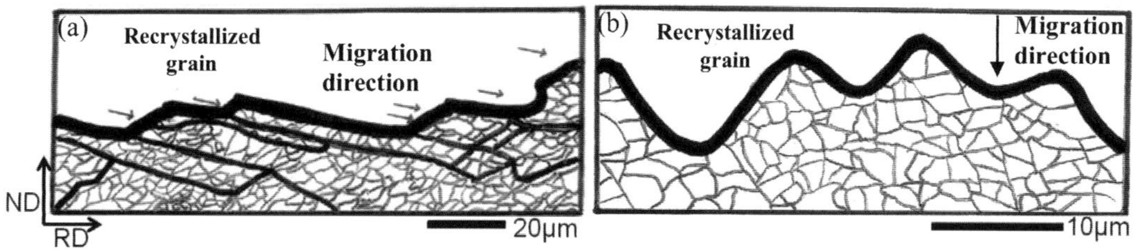

Figure 5. Sketches of protrusions and retrusions on recrystallization boundaries in 4N pure Al (cr 50% and annealed at 250 °C for 10 minutes). (a) Sketch based on ECC image showing saw-tooth shaped protrusion. The fat lines in the deformation microstructure represent extended dislocation boundaries across which a significant contrast variation is observed. The arrows indicate the expected direction of boundary migration. (b) Sketch based on a TEM image showing rounded protrusions. [24]

Several reasons may explain the large scatter in figure 4. (i) Both the sizes of protrusions/retrusions and the local stored energy vary from place to place, even along the same boundary. (ii) The deformed microstructures are typically heterogeneous, depending strongly on the grain orientations. Therefore, even for two deformed grains with similar stored energy, the deformed microstructures may be completely different, and hence the resulting protrusions/retrusions may be of different types. For example, round and saw-tooth types of protrusions have been observed along different recrystallizing boundaries in partially recrystallized pure aluminium depending on the local microstructure in front of the recrystallizing grains (see figure 5). (iii) The protrusions/retrusions and stored energies are characterized based on 2D images. (iv) According to equation (1), variations in boundary mobility may also be a reason. Lower mobilities for some segments may lead to large retrusions. (v) As the analysis in figure 4 is based on statistical characterizations, the real deformation microstructures that result in the observed protrusions/retrusions have at least partly been consumed. The stored energy calculated in figure 4 is therefore not the actual driving force that led to the formation of protrusions/retrusions.

To clarify these possibilities, future work may include: correlating each individual protrusions/retrusions with *local* stored energy in the deformed microstructure to clarify point (i) above; morphological analysis of the recrystallizing boundaries and the deformed microstructures to investigate possible correlations and thus clarify point (ii); the unknown variations in the third dimension have to be investigated by 3D (x, y, z) characterizations (point (iii) above); points (iv) and (v) can be analyzed by dynamic studies.

3. 3D (x, y, z) static characterizations

3D (x, y, z) characterization techniques can generally be divided into two groups: destructive ways using serial sectioning, and non-destructive ways using e.g. high energy X-rays. The destructive ones leading to static characterization are discussed in this section, while non-destructive ones will be discussed in section 5 in connection with dynamic 3D studies, because the non-destructive techniques offer this possibility.

Characterizations of 3D volumes using destructive serial sectioning are normally achieved by repeating two processes: sectioning and microstructural characterization. For the microstructural characterization, microscopy, e.g. OM, ECC and EBSD can be used (e.g. [25, 26]). Earlier, OM was the main characterization tool, but nowadays EBSD is more and more used. The sectioning process can be conducted using focused ion beam (FIB) inside an SEM (e.g. [27]) or using electro- (e.g. [28]), chemical polishing (e.g. [29]) or laser ablation (e.g. [30]). One of the advantages with FIB and laser ablation is a relatively high resolution (down to a few nanometres at the best) for each milling layer, thanks to the high precision of the beam. With commercial software the whole characterization process can be automated, which reduces significantly the amount of work for later alignment between layers. The advantage with manual mechanical, electro- or chemical polishing is that there is no limit in the size of the characterization area at each layer. The minimum sectioning thickness is, however, much coarser than with FIB and laser. Even with the most advanced automatic characterization system, such as Robo-Met. 3D [31], the best resolution can only be 200 nm, while with a human-based polishing process 2 μm at the best is more realistic.

Among all the 3D studies published, deformation microstructures are more often characterized than recrystallization (e.g. [32, 33]), and few papers relevant for local grain boundary migration are published. One example is given in the following.

3.1. 3D characterization of recrystallization boundary, using FIB

To visualize the 3D shape of a recrystallizing boundary, a 3D microstructure containing a recrystallizing boundary separating a recrystallized grain and deformed matrix has been characterized using FIB as a serial sectioning milling tool and EBSD for microstructural mapping. In total, 40 layers with 0.5 μm resolution along all three directions have been obtained (see figure 6a) [34]. The characterized volume has a size of 50 × 60 × 20 μm^3. The 3D morphology of the recrystallizing boundary can be easily revealed because of the difference in orientations across the boundary. The 3D rendering of the recrystallizing boundary in figure 6b shows clearly the heterogeneity of protrusions/retrusions in different sample directions. This data showed that in at least this sample the 3D protrusions/retrusions are not isolated peaks but elongated ridges along TD. This result indicates that an estimation of protrusion/retrusion sizes and curvatures based on 2D microstructural investigations in the RD-ND section for cold rolled samples may not deviate too much from the true 3D results (at least for the present example).

By having the fine spatial resolution, the deformed microstructures can also be reconstructed in 3D, and the 3D influence of deformation microstructure on the formation of protrusions/retrusions can be examined. The 3D result clearly reveals that the dislocation boundaries are aligned almost parallel to the TD direction (see figure 7), which indicates that the actual 3D bands in the deformed microstructure can be reasonably represented by 2D bands seen on the RD-ND section. The protrusions/retrusions appear to be flatter in the RD-TD plane compared to the RD-ND plane, which supports the suggestion that the extended banded deformation structure along TD affects the evolution of protrusions/retrusions.

With this data set, the characteristics of the recrystallization boundary can be obtained. Misorientation analysis shows no systematic difference between the boundary segments of the protrusions and retrusions. The boundary plane normal of the recrystallizing boundary can also be obtained. Detailed correlation of the boundary plane normal with the local deformed microstructure can therefore also be made. One drawback of this result is the volume characterized is relatively small and thus not good for statistical analysis as that conducted based on the 2D images. Furthermore, the destructive sectioning does not allow following the boundary migration dynamically.

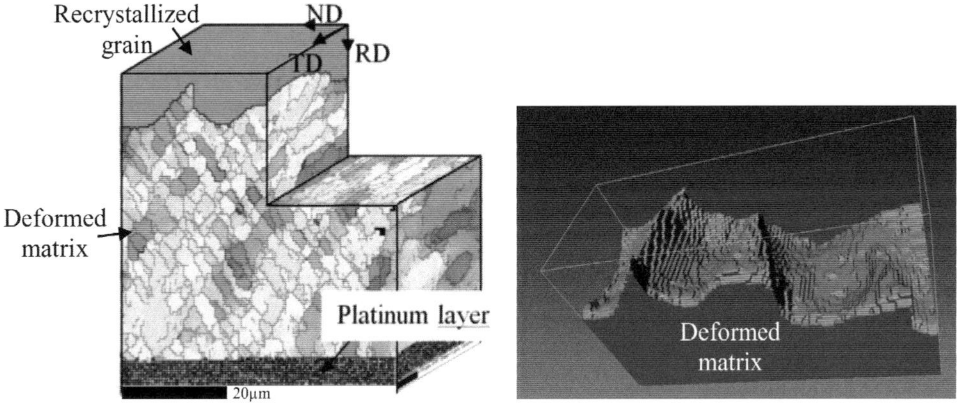

Figure 6. Example showing the 3D characterization using FIB as milling tool inside a SEM. (a) 3D volume showing the reconstructed microstructure around a recrystallization boundary in partially recrystallized pure aluminium (cr 50% and annealed at 250 °C for 10 minutes). (b) Top view of the recrystallizing boundary surface, the volume visualized as a solid is the deformed matrix. The bottom random colored layers in (a) are caused by platinum deposition. [34]

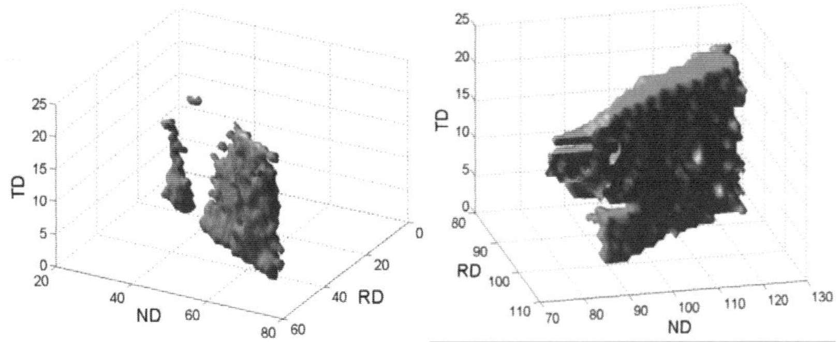

Figure 7. Examples showing two 3D alignment of bands of dislocation boundaries subtracted from the deformed matrix in figure 6a, where 2° is used to detect dislocation boundaries. [34]

4. In-situ/ex-situ 2D (x, y, t) characterizations

Dynamic studies following in situ or ex situ the boundary migration process are crucial for understanding the migration mechanisms. The direct observations from such studies can avoid the so-called 'destroy-evidence' problem [35], and thus be used to test mechanisms or theories as those obtained based on the 2D post-mortem analysis e.g. those given in section 2. Moreover, in-situ/ex-situ studies can provide new possibilities for quantification of *local* microstructural changes during recrystallization, e.g. *local* boundary migration rate, local stored energies for the migration, *local* boundary misorienations, and recovery in the deformation matrix.

To perform in-situ/ex-situ characterization, all 2D characterization techniques can be used. TEM and high voltage TEM were used intensively in the 1970-80's (e.g. [36, 37]). Because only relatively small areas can be observed using TEM, most of these investigations are focusing on the early stages of nucleation. As only thin foils can be probed by TEM, the investigations may be affected by the two free surfaces and the wedge shape of the thin sample area. The phenomena observed may therefore not represent the true bulk behavior.

Since EBSD became a common characterization technique, more and more in-situ/ex-situ recrystallization studies are conducted with EBSD instead of TEM. Although the spatial resolution does not compare with that of TEM, there are several advantages in using EBSD for in-situ/ex-situ studies. Besides those mentioned in section 2, a large characterization area, only one free surface, easy sample handling, and resistance to sample buckling are additional advantages with EBSD compared to TEM for in-situ studies. As the sample preparation for ECC characterization is the same as for EBSD, ECC has also been used for in-situ/ex-situ studies, though less frequently than EBSD because it does not provide enough crystallographic information. The time resolution for in-situ/ex-situ ECC characterizations can be much higher than that for EBSD. It may therefore be beneficial to combine EBSD with ECC for in-situ/ex-situ studies: using the EBSD mapping with a sufficient angular resolution for calculation of boundary misorientations and local stored energies, while using ECC images to have both a higher spatial resolution and a faster data acquisition of the microstructural evolution.

One of the difficulties to combine EBSD mapping with ECC imaging is the drift problem, which typically results in non-linear spatial distortion in the EBSD maps [38]. Due to the distortion, features in the EBSD maps may not directly be registered with those in the ECC images. To solve this problem, a correction method, thin plate spline, has been developed to correct the spatial distortion in the EBSD maps [38]. After correction, EBSD maps can easily be fused with ECC images for quantitative analysis.

In recent years, both EBSD and ECC have been used to follow recrystallization process in situ/ex situ (e.g. [40-43]). Many of those studies are, however, of demonstration types focusing on methodology and qualitative description of recrystallization process but not so much on quantification. Even in the latter case, most of the works are still based on the average properties of the whole samples or over the whole characterization areas [41, 42, 44]. A few examples, in which quantitative analyses of boundary migration have been conducted on the local scale, are summarized in the following.

4.1. Ex-situ ECC studies of boundary migration

The migration of a recrystallizing boundary in partially recrystallized high purity aluminium has been followed using the ECC technique, with the aim of quantifying local protrusions/retrusions and understanding their effects on boundary migration [12]. The boundary migration process can be seen in figure 8, where numbers are used to identify protrusions and retrusions. The results reveal that the migration of recrystallization boundaries is quite inhomogeneous and differs from place to place: some boundary segments move faster while others move slower. Also some segments migrate in a stop-go type of fashion, while others migrate during all annealing steps (see figure 8d).

Detailed quantification of protrusion and retrusion sizes using the methods described in section 2.1 has shown that during the whole investigated migration period, the values of $F_{\sigma,max}$ for protrusions are mainly in the range of -0.05~-0.2 MJ/m^3, whereas for retrusions $F_{\sigma,max}$ can be much higher, with most of them in the range of 0.2~1.0 MJ/m^3. This suggests that during the migration process the magnitude of both the dragging and driving forces from the evolving protrusions and retrusions, respectively, can be of comparable magnitude to that from the energy stored in the deformed matrix. Equation (1) is therefore suggested to be modified as:

$$v=M\cdot(F_S + F_\sigma), \tag{4}$$

where F_S and F_σ are the driving force contributions from the stored energy of the deformed structure and boundary curvature, respectively. The ex-situ results have also confirmed that during boundary migration the protrusions are relatively small, and the protrusions can become stationary at some stage (e.g. protrusion no. 1 in figure 8) once the dragging force from protrusions, F_σ, counterbalanced the driving force from the stored energy, F_S according to equation (4). These observations have further shown the importance of F_σ for understanding local boundary migration.

It is however rather difficult to understand the local migration behavior of retrusions based on equation (4), as both driving forces from stored energy and from boundary curvature of retrusions have positive signs, and the sum of both is much higher than the driving force for the migration of protrusions. It is clear that if the local driving force in front of a retrusion is low, the retrusion may not move forward and thus develop to become very large [45, 46]. This is however not the situation here. For example, it is found that during the whole annealing process followed here, segments at the tip of retrusion no. 2 only migrate when $F_{\sigma,max}$ is higher than 0.55~0.60 MJ/m^3 [12]. It is possible that the mobility for segments at retrusion no. 2 is smaller than their neighboring, as the contrast of the deformed cells in front of these segments (as marked by the ellipse in figure 8a) is different from their neighboring. But how different the mobility can be, and how the mobility differences can lead to stagnation of the boundary segments cannot be investigated by this experiment, as quantification of boundary characteristics and stored energy in the deformed matrix cannot be obtained based on ECC images.

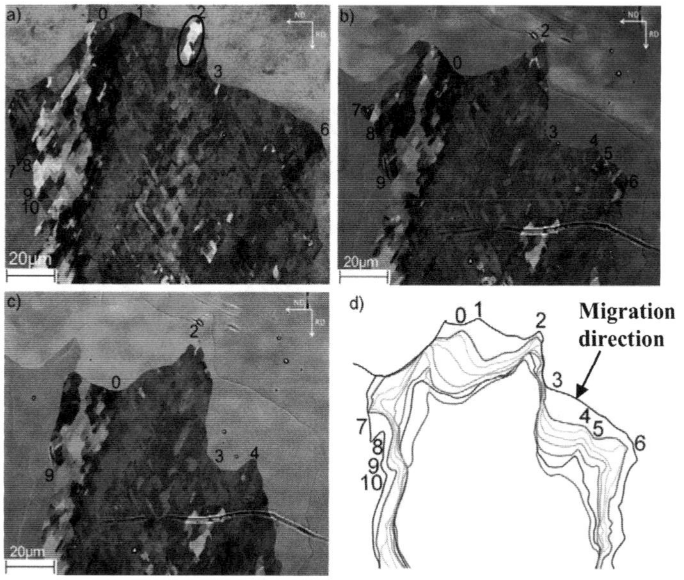

Figure 8. ECC images showing the microstructural evolution of 50% cold rolled aluminium during annealing at 250 °C for: (a) 10 minutes; (b) 10+3x15 minutes; (c) 10+6x15 minutes. (d) Locations of the recrystallization boundary during the entire annealing sequence. Specific protrusions and retrusions are identified using numbers. The ellipse in (a) marks deformation cells with a different contrast from their neighbors. [12]

4.2. Combining EBSD and in-situ ECC to follow boundary migration

Ex-situ EBSD studies have been carried out to follow the growth of individual grains during recrystallization in heavily cold rolled nickel [47]. The local microstructures that were consumed during annealing were extracted and detailed misorientation analyses of moving boundaries and non-moving boundaries were carried out. The results have, however, shown that even neighboring boundary segments, with very similar misorientations to the neighboring deformed microstructure and with similar driving force (F) from the deformed matrix, can behave quite differently [47]. The EBSD mapping step size for this study was coarse (2 μm), and effects of local deformation microstructures on the growth could therefore not be revealed.

Very recently, new experiments have been conducted to follow the local boundary migration: EBSD was used to characterize the starting microstructure and the boundary migration was followed using ECC during in situ annealing inside the SEM [48]. Figure 9a and 9b show a subset of the starting microstructures, while figure 9d shows 7 boundary traces that have been analyzed in detailed. The general boundary migration pattern is similar to that found in the above mentioned ex-situ studies. It is seen, that after the second time step the two ends of the boundary remain fixed, where long, flat boundary segments develop, which do not migrate further during the entire duration of the annealing experiment. In between these fixed points, the migration of the recrystallizing boundary is quite heterogeneous: many small protrusions/retrusions form during the migration, and the migration velocity varies in space and time, with fast migration observed in particular during time interval #2. The misorientation analysis has shown that not much difference is observed between the migrating and non-migrating boundary segments [48].

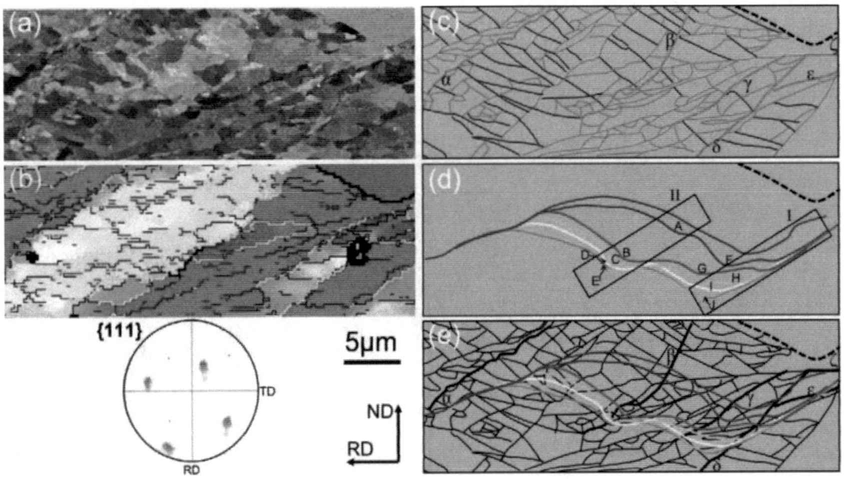

Figure 9. EBSD and ECC techniques used in combination to characterize the microstructure and to follow in situ the boundary migration in partially recrystallized pure aluminium (cr 50% followed by rolling at liquid-nitrogen temperature to a total reduction of 86%) during annealing at 50 °C. (a) ECC image, (b) EBSD map and (c) sketch based on ECC image in (a) showing the microstructure in the area where detailed boundary migration are analyzed. (d) Traces of the positions of the recrystallizing boundary during the first seven 100 s time steps. Box I and II mark local areas where detailed analysis is carried out. Letters A-J are used to identify protrusions. (e) Non-colored image of the deformed microstructure (from c) superimposed with recrystallizing boundary traces shown color. In (c) and (e) boundaries containing segments with misorientation >10° are highlighted by thicker lines and marked by α-ζ in (c). [48]

By correlating the boundary traces with the deformed microstructure (figure 9e), it is evident that the coarsely spaced dislocation boundaries, marked by Greek letters in figure 9c, which have higher misorientations, have dominated the migration process and stimulated the formation of protrusions (see figure 10a). The higher local stored energies induce larger protrusions, and the maximum curvature-based dragging forces from protrusions are typically smaller than the local stored energies, with one exception for protrusion C. It should be noted that only the maximum curvature dragging force is used for the protrusions, while the stored energy is calculated as an average energy density over the small area under each complete protrusion profile. According to equation (4), the protrusions B, C, D should have a reduced migration rate or stop moving at next annealing step, and protrusions A,

F, G can continue migrate during next annealing step. This agrees well with the experimental observations, i.e. the migration is understood qualitatively.

However, quantitative understanding of the migration is more challenging. Even when the local net curvature dragging/driving forces from protrusions/retrusions are included in a detailed quantification of boundary migration, a large scatter between the migration velocity, v and the modified driving force, $F_S + F_\sigma$ is found (see figure 10b), where a significant deviation from a simple straight line relationship as predicted by equation (4) is observed. Several reasons have been suggested to understand this result [48]. Among those, lack of knowledge about the appropriate value of mobility, M, which may be complex and may depend on many parameters, is suggested to be the main factor to be examined in future experiments. Also the fact that individual boundary segments are strongly affected by the motion of neighboring segments has also to be considered. Both advanced modeling and 3D experimental tools are required to clarify these issues.

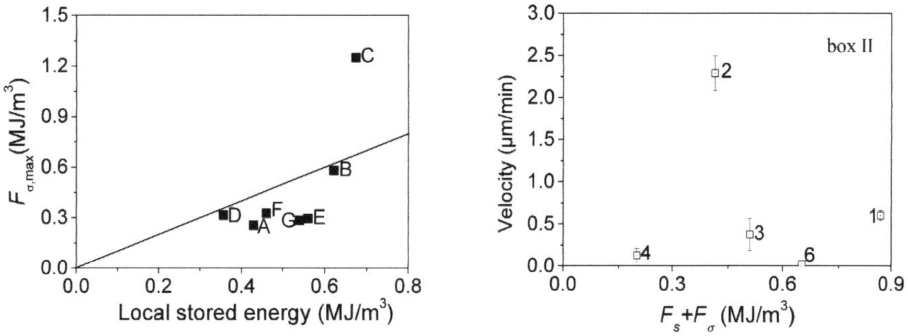

Figure 10. (a) Maximum curvature dragging forces ($F_{\sigma,max}$) from the protrusions A-G (marked in figure 9d) against the local stored energies calculated from the deformed microstructure consumed by the protrusions. (b) Migration velocity of the recrystallizing boundary as a function of total driving force, $F = F_s + F_\sigma$, for boundary segment in box II in figure 9d for each time interval. Numbers show the time intervals. [48]

5. In-situ/ex-situ 3D (*x*, *y*, *z*, *t*) characterizations

In the last 15-20 years, two major groups of non-destructive 3D characterization techniques have been developed and used for recrystallization studies, namely 3DXRD techniques (e.g. [49, 50]) and differential aperture X-ray microscopy techniques (also called 3D X-ray crystal microscopy (3DXRM)) (e.g. [51]). With these techniques, the boundary migration can be traced in situ/ex situ in 3D, so that full boundary information described by 5 parameters (3 for misorientation and 2 for boundary plane normal) can be obtained, and the influence of 3D deformed microstructure on boundary migration can be directly visualized.

The 3DXRD techniques utilize typically monochromatic, high energy, high brilliance, and hard synchrotron X-rays at large third generation synchrotron facilities. For reconstruction of the 3D volume, 'tomographic' reconstruction algorithms are used for analysis of the diffraction data recorded using detectors at different distances to the sample. Several operation modes can be chosen to optimize spatial, angular or time resolution in a 3DXRD experiment. Typically, due to limitation of the detector resolution to ~1 μm, the spatial resolution can be fine-tuned in the range of ~1-5 μm depending on the experimental requirements. Time resolution of a few minutes to a few hours allows in-situ or interrupted in-situ studies. Angular resolution is ~0.1°. However, the 3DXRD techniques are currently still not capable of resolving deformed microstructure, especially for deformation >20% [10, 49, 50].

Very recently, based on the 3DXRD techniques, a new concept — dark-field X-ray microscopy (DFXRM) — has been developed and implemented [52]. This technique utilizes an objective (typically consisting of many compound refractive lenses (CRLs)) to magnify a diffracted beam by a factor of typically 10-20. With this technique, 3D maps of orientations within both individual deformed and recrystallized grains with a spatial resolution of ~300 nm have been achieved [52].

The 3DXRM techniques utilize a focused, polychromatic synchrotron X-rays with energy of 5-30 keV [51, 53]. A Pt wire is used as a differential aperture to differentiate diffraction signal from different sample depth along the beam, and crystallographic information is resolved by indexing the Laue diffraction patterns at each depth [51]. Practically, due to limitations from the imperfection of the focusing mirror, the spatial resolution is ~500 nm. As the 3D mapping is in a scanning mode, the time resolution of this technique is normally poor. Nowadays, to map a reasonable volume can take up to 2-3 days. Therefore only interrupted in-situ or ex-situ studies are possible. The angular resolution is ~0.01°. Using these techniques, 3D deformed microstructures in deformed samples (<50% deformation) have been resolved with submicrometre resolution [54].

Both groups of techniques have been used for studies of local boundary migration. The achievements in understanding of recrystallization boundary migration as well as future potentials are summarized in the following.

5.1. Watching growth of individual grains into a lightly deformed single crystal

With 3DXRD techniques, the first state-of-the-art experiment to follow the growth of a recrystallized grain into a lightly deformed Al1050 single crystal during in situ annealing was reported in 2004 [11]. Rough boundaries and very heterogeneous migration patterns of individual boundary segments were observed in 3D, through the whole growth period. For example, the formation and evolution of protrusions/retrusions on the growing boundary and stop-go type of motion were for the first time demonstrated in that experiment to be a bulk behavior and not artifacts from 2D observations. The spatial resolution of this 3D experiment was about 5 μm along the *x*, *z* direction and 20 μm along the *y* direction (beam direction). [11]

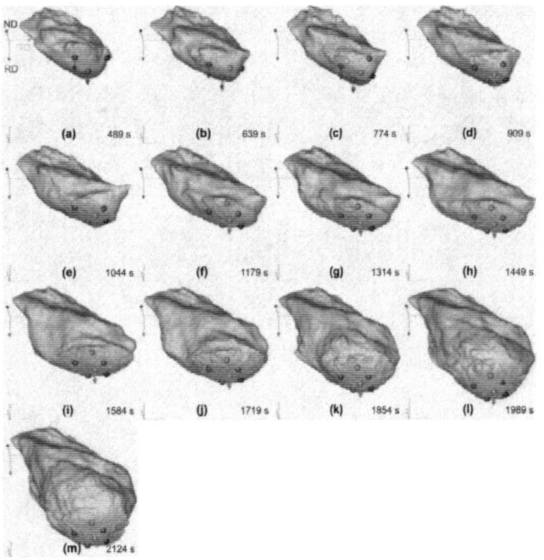

Figure 11. Snapshots of a grain at different times during its growth through the deformed matrix of a near cube oriented single Al crystal. Protrusions/retrusions are seen on many boundary segments and the dots and arrows show the formation and migration of a coarse scale facet. The dimension is given by the tripod, with arms 42 μm long. (a)-(e) are annealed at 290 °C, while (f)-(m) are annealed at 300 °C. [55]

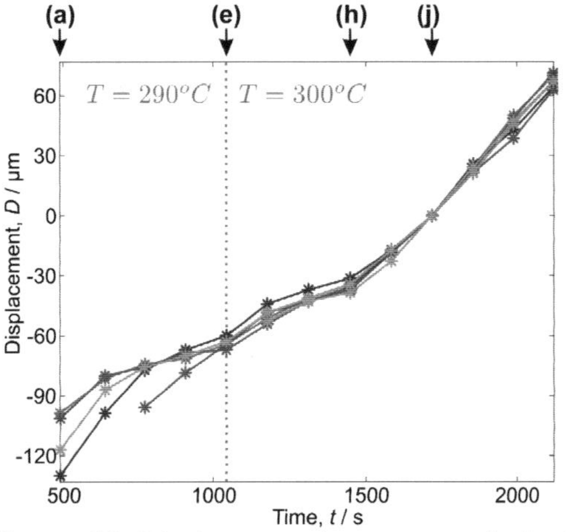

Figure 12. Displacement-time curves obtained from the 3D reconstructions along the set of parallel lines indicated in figure 11. The displacements of local grain boundary segments are characterized by the distances of the successive intersection points along the parallel lines to their reference positions (which are the positions of the intersection points at 1719 s). [55]

To overcome this limited spatial resolution, the topo-tomography technique, which is a kind of 3DXRD, has recently been used to follow the growth of a recrystallized grain in a deformed single crystal of near cube orientation [55]. Figure 11 shows the grain after different annealing times [55]. As this technique gives a better spatial resolution of 1.4 μm along all directions, details of the local protrusions/retrusions and facets on the boundary are clearly visualized, and quantifications of boundary morphology are possible. For example, the evolution of the facet, marked by color dots in figure 11, has been analyzed (see figure 12). It is seen that the facet develops only after a certain annealing time and afterwards grows as a unity. An interesting observation is that the facet is not parallel to any of the crystallographic {111} or {100} planes, but parallel to the macro sample RD direction. To understand how this larger than 100μm-size facet develops and why it grows preferentially along RD require detailed characterization of the deformed microstructure, which is still not possible with this technique.

5.2. Ex-situ studies of boundary migration with 3DXRM.

As the 3DXRM techniques enable resolving deformed microstructures at the submicrometre scale, the 3D microstructure shown in figure 6 can be obtained non-destructively, and the migration of a 3D boundary and its correlation to the 3D deformed microstructure can also be obtained. The authors have recently conducted successfully such type of experiments. Both the deformed microstructure and recrystallized grains separated by a recrystallizing boundary have been indexed in 3D with a spatial resolution of 1μm along all directions, and the migration of the boundary has been followed by ex situ annealing. With this 4D data set, the larger goal of studying effects of both the misorientations (angle/axis pair) and the boundary plane normals for the 3D boundary can be achieved. Also possible effects of the 3D deformed microstructure on the migration can be examined. The data processing is in progress.

5.3. Recrystallization boundaries studied by DFXRM

With the ability to resolve both deformed microstructure and recrystallized grains, the DFXRM technique is also a powerful tool for 4D recrystallization studies. Very recently a layer (200 nm thick) of a recrystallized grain in a partially recrystallized Al1050 sample has been characterized using this technique [56]. The diffraction signal from the investigated recrystallized grain is magnified 18 times using Be CRLs as objective lens. The spatial resolution, which is limited by the optical performance of the lens, is ~200 nm. The ultra-high angular resolution of ~0.005° allows detailed characterization of e.g. very low angle boundaries. The full 3D shape of the grain can be reconstructed layer by layer using this technique, and the growth of the grain can be followed by repeating the 3D reconstruction after different annealing times. As the technique can also resolve the deformed microstructure with high spatial resolution, the effects of the deformed microstructure on the growth can in theory be investigated [56].

6. Conclusions and outlook

6.1. Evaluation of the characterization techniques

The present paper has demonstrated essential potentials of advanced characterization techniques for understanding the recrystallization boundary migration process. All the 2D characterization techniques mentioned in the paper have become standard tools in most research laboratories. Combined with the fact that it is relatively easy to learn to operate these techniques, they therefore are important 'work horses' for studies of boundary migration during recrystallization. The drawback with 2D characterization is it only reveals a 2D section of the whole bulk sample. Although statistical quantification of average parameters can be obtained using stereological methods, local parameters for individual grains or boundaries are always suffering from uncertainties in the third dimension. Similarly, kinetic information can be obtained by analyzing statistical results from a series of samples with e.g. different recrystallization fractions. But these analyses are post-mortem, and sometimes suffer from the so-called 'destroy-evidence' problems.

Destructive 3D (x, y, z) characterization techniques have had a renaissance and have become more frequently used for materials science research in recent years. In many scientific studies, the need for 3D (x, y, z) techniques has been demonstrated, to obtain a complete picture of the microstructure, and to remove the bias, uncertainties, and sometimes misinterpretations from 2D characterizations. In some cases 3D data sets provide completely new findings. The efforts that are needed to collect 3D data sets are however large compared to those for a 2D data set. First the access to equipment that enables 3D characterizations is still relatively limited. Furthermore significant experience is required to obtain good 3D data sets. Also the data acquisition time is long. To have a reasonable volume characterized using FIB milling may take days or weeks, while with manual milling it may take weeks to months. Finally, the destructive methods inherently do not allow any dynamic studies. Data post processing of 3D data sets (e.g. data alignment and segmentation) and visualization are still typically challenging due to the large data set and lack of suitable software. Currently, several open source software, e.g. Image J [57], Dream3D [58] and MTEX [59], have been developed for handling 3D data sets, which may ease the situation, but more functionalities have to be developed.

Compared to the 3D (x, y, z) characterizations, the efforts required for conducting good in-situ/ex-situ 2D (x, y, t) studies are less, especially for ex-situ studies, for which no heating device inside e.g. the SEM is required. With in-situ/ex-situ studies, so-called 'destroy-evidence' problems are solved. Similar to the 3D (x, y, z) characterizations, there is, however, still a lack of in-house or commercial software that allow quantifications of local microstructural changes. Additionally, this type of studies always suffers from surface problems [60], which in some cases may be critical for interpreting the results.

The synchrotron 3D techniques are no doubt the most powerful tools for recrystallization studies. They provide unique possibilities for non-destructive in-situ 3D studies, from which the microstructural evolution can be revealed in full 4D. The drawback of these techniques is related to the very limited access to synchrotron facilities. In order to get beam time, a high scientific level proposal and cutting edge type of experiments are typically required. Even if beamtime is allocated (typically maximum a week), the success rate of synchrotron experiments may be low, due to complex instrumentation, and other unexpected problems. Therefore very careful preparation of the experiment is highly recommended, which typically can take half a year or more. Another major drawback of synchrotron experiments is that the limited beamtime does generally not allow for series of experiments and thus not for very systematic studies of effects of metallurgical parameters on, e.g. boundary migration.

Currently, the spatial resolutions of the non-destructive 3D synchrotron techniques are all lower than those of the 2D electron microscopy techniques. For the 3DXRM techniques, the spatial resolution is limited by the focusing mirror, i.e. by the perfection of the mirror. Nowadays, the best ultra-precise K-B mirror fabricated by JTEC Corporation (a spin-off company from Osaka University) can focus hard X-rays at the sub-100 nm level [61]. With this type of mirror installed in the near future, the 3DXRM techniques will allow research of boundary migration at the nanometre scale. Similarly, with better objective lenses fabricated in the future, a spatial resolution of 10-30 nm can in theory be achieved with the new 3DXRD technique, DFXRM. These further advancements of synchrotron techniques will in the future allow recrystallization studies for highly strained nanostructured materials.

6.2. Recrystallization boundary migration

The present paper has demonstrated that local boundary migration is a complex process. It is evident that the local variations within the deformation microstructures are very important for boundary migration during recrystallization. A better understanding of mobilities for a recrystallizing boundary is essential but poorly understood. It remains as open question if one can really operate with simple mobilities in the investigations of local boundary migration during recrystallization. More work has to be done to solve the problems we are facing and to advance the understanding of local boundary migration. Suggestions include:

- More 2D (x, y) and 3D (x, y, z) experimental characterizations of recrystallization boundaries and the local deformation microstructures in front have to be conducted, in order to correlate the local microstructural parameters, e.g. misorientation angles and spacings of the dislocation boundaries, with the local protrusions/retrusions along the boundaries. In this respect, deformed grains oriented differently, different deformation strains, different annealing temperatures and different materials have to be included.

- More in-situ/ex-situ 2D (x, y, t) and 3D (x, y, z, t) experiments have to be conducted. From such experiments one could try to appropriately quantify boundary mobilities using local boundary migration velocities and local stored energy.
- Computer modelling should be included in the interpretations of experimental results and in defining the most critical experiments to be done. For example using phase-filed modelling to investigate the effects of local microstructural parameters in idealized deformed structures on the boundary migration [45, 46]. Currently, molecular dynamics has been used to determine boundary mobilities (e.g. [62, 63]). Direct comparison between experimental and simulation results will be of importance.
- Currently, the effects of residual stresses in the deformed microstructures are to the authors' knowledge not considered in recrystallization studies. Whether the formation of protrusions/retrusions at different length scales are affected by residual stresses has to be quantified. Here one may choose to focus first on materials with graded structures, in which the residual stresses may be pronounced.

Acknowledgements
The authors gratefully acknowledge the support from the Danish National Research Foundation (Grant No DNRF86-5) and the National Natural Science Foundation of China (Grant No. 51261130091) to the Danish-Chinese Center for Nanometals, within which this work has been performed.

References

[1] Haessner F 1978 *Recrystallization of metallic materials.* (Stuttgart: Dr. Riederer Verlag GmbH.)
[2] Gottstein G and Shvindlerman L S 2010 *Grain boundary migration in metals: Thermodynamics, Kinetics, Applications*, (London: CRC Press Taylor & Francis Group)
[3] Vandermeer R A, Juul Jensen D and Woldt E 1997 *Metall. Mater. Trans.* **28A** 749-54
[4] Cahn J W and Hagel W C 1963 *Acta Metal.* **11** 561
[5] Juul Jensen D 1992 *Scripta Mater.* **27** 533
[6] Lu Y, Molodov D A and Gottstein G 2011 *Acta Mater.* **59** 3229-43
[7] Hansen N and Juul Jensen D 2011 *Mater. Sci. Techno.* **27** 1229-40
[8] Huang X and Winther G 2007 *Phil. Mag.* **87** 5189-214
[9] Schwartz A J, Kumar M, Adams B L and Field D P 2009 *Electron Backscatter Diffraction in Materials Science*, (New York: Springer)
[10] Poulsen H F 2004 *Three-Dimensional X-Ray Diffraction Microscopy, Mapping polycrystals and their dynamics* (Berlin: Springer)
[11] Schmidt S, Nielsen S F, Gundlach C, Margulies L, Huang X and Juul Jensen D 2004 *Science* **305** 229-32
[12] Zhang Y B, Godfrey A and Juul Jensen D 2011 *Scripta Mater.* **64** 331-4
[13] Beck P A, Sperry P R and Hu H 1950 *J. Appl. Phys.* **21** 420-5 doi: 10.1063/1.1699676
[14] Joy D C, Newbury D E and Davidson D L 1982 *J. Appl. Phys.* **53** 81-122
[15] Humphreys F J 2001 *J. Mater. Sci.* **36** 3833-54
[16] Barnett M 2013 *Recrystallization and grain growth V* (Switzerland: Trans Tech Pulications Ltd)
[17] Hong C, Huang X and Winther G 2013 *Phil. Mag.* **93** 3118–3141
[18] Jones A R, Ralph B and Hansen N 1979 *Proc R Soc London A* **368** 345
[19] Liu Q 1995 *Ultramicroscopy* **60** 81-9
[20] Martorano M A, Sandim H R Z, Fortes M A and Padliha A F 2007 *Scripta Mater.* **56** 903-6
[21] Zhang Y B, Godfrey A and Juul Jensen D 2009 *Computers, Materials & Continua* **14** 197-207
[22] Martorano M A, Fortes M A and Padilha A F 2006 *Acta Mater.* **54** 2769-76
[23] Sun J, Zhang Y B, Dahl A B, Conradsen K and Juul Jensen D 2015 *Proc. 36th Risø Int. Symp. on Mater. Sci* (*Roskilde, DK, 2015*), ed S Fæster et al. (Roskilde: Risø)
[24] Juul Jensen D, Lin F X, Zhang Y B and Zhang Y H 2013 *Mater. Sci. Forum* **753** 37-41

[25] Spanos G, Rowenhorst D J, Chan S and Olson G B 2010 *Proc. 31st Risø Int. Symp. on Mater. Sci (Roskilde, DK, 2010)*, ed N Hansen et al (Roskilde: Risø) pp. 159-70
[26] Uchic M D, Holzer L, Inkson B J, Principe E L and Munroe P 2007 *MRS Bull* **32** 408-16
[27] Zaefferer S, Wright S I and Raabe D 2008 *Metall. Mater. Trans.* **39A** 374-89
[28] Zhang Y, Juul Jensen D, Zhang Y B, Lin F X, Zhang Z and Liu Q. 2012 *Scr. Mater.* **67** 320-3
[29] Rowenhorst D J, Gupta A, Feng C R and Spanos G 2006 *Scr. Mater.* **55** 11-6
[30] Echlin M P, Husseini N S, Nees J A and Pollock T M 2011 *Adv. Mater.* **23** 2339-42
[31] Spowart J E 2006 *Scr. Mater.* **55** 5-10
[32] Fan G H, Zhang Y B, Driver J H and Juul Jensen D 2014 *Scr. Mater.* **72-73** 9-12
[33] Xu W, Quadir M Z and Ferry M 2009 *Metall. Mater. Trans.* **40A** 1547-56
[34] Zhang Y B, Godfrey A, MacDonald N and Juul Jensen D 2012 *Proc. 1st Int. Conf. on 3D Material Science* ed M De Graef et al. (Wiley) pp 31-6
[35] Duggan B J, Lucke K, Kohlhoff G and Lee C S 1993 *Acta Metall. Mater.* **41** 1921
[36] Faivre P and Doherty R D 1979 *J. Mater. Sci.* **14** 879-919
[37] Berger A, Wilbrandt P-J, Ernst F, Klement U and Haasen P 1988 *Prog. Mater. Sci.* **32** 1-95
[38] Zhang Y B, Elbrønd A and Lin F X 2014 *Materials Characterization* **96** 158-65
[39] Zhang Y B, Godfrey A and Juul Jensen D 2013 *Mater. Sci. Forum* **753** 117-20
[40] Zhang Y B, Godfrey A and Juul Jensen D 2010 *Proc. 31st Risø Int. Symp. on Mater. Sci (Roskilde, DK, 2010)*, ed N Hansen et al. (Roskilde: Risø) pp. 497-503
[41] Lens A, Maurice C and Driver J H 2005 *Mater Sci Eng A* **403** 144-53
[42] Le Gall R, Liao G and Saindrenan G 1999 *Scr. Mater.* **41** 427-32
[43] van der Zwaag S, Anselmino E, Miroux A and Prior D J 2006 *Mater. Sci. Forum* **519-521** 1341-8
[44] Hurley P J and Humphreys F J 2004 *J. Microsc* **213** 225-34
[45] Moelans N, Godfrey A, Zhang Y B and Juul Jensen D 2013 *Phys. Rev. B* **88** 054103
[46] Moelans N, Zhang Y B, Godfrey A and Juul Jensen D 2015 *Proc. 36th Risø Int. Symp. on Mater. Sci (Roskilde, DK, 2015)*, ed S Fæster et al. (Roskilde: Risø)
[47] Zhang Y B, Godfrey A, Liu Q, Liu W and Juul Jensen D 2009 *Acta Mater.* **57** 2632-39
[48] Zhang Y B, Godfrey A and Juul Jensen D 2014 *Metal. Mater. Trans.* **45A** 2899-905
[49] Juul Jensen D and Poulsen H F 2012 *Materials Characterization* **72** 1–7
[50] Poulsen H F, Ludwig W, Lauridsen E M, Schmidt S, Pantleon W, Olsen U L, Oddershede J, Reischig P, Lyckegåd A, Wright J and Vaughan G 2010 *Proc. 31st Risø Int. Symp. on Mater. Sci (Roskilde, DK, 2010)*, ed N Hansen et al. (Roskilde: Risø) pp. 101-120
[51] Larson B, Yang W, Ice G, Budai J and Tischler J 2002 *Nature* **415** 887–890
[52] Simons H, King A, Ludwig W, Detlefs C, Pantleon W, Schmidt S, Snigireva I, Snigirev A and Poulsen H F 2014 *Nature Communications* **6** 6098
[53] Ice G E, Pang J W L, Barabash R I and Puzyrev Y 2006 *Scr. Mater.* **55** 57-62
[54] Pang J W L, Ice G E and Liu W 2010 *Mater. Sci. Eng.* **528** 28-31
[55] van Boxel S, Schimdt S, Ludwig W, Zhang Y B, Juul Jensen D and Pantleon W 2014 *Materials Transactions* **55** 128–136
[56] Ahl S R, Simons H, Jakobsen A C, Zhang Y B, Juul Jensen D and Poulsen H F 2015 *Proc. 36th Risø Int. Symp. on Mater. Sci (Roskilde, DK, 201)*, ed S Fæster et al. (Roskilde: Risø)
[57] http://imagej.nih.gov/ij/index.html
[58] Groeber M A and Jackson M A 2014 *Intergrating Materials and Manufacturing Innovation.* **3** 5
[59] Bachmann F, Hielscher R and Schaeben H 2010 *Solid State Phenomena* **160** 63-8
[60] Gottstein G and Shvindlerman L S 1992 *Scripta Metall. Mater.* **27** 1521-6
[61] http://www.j-tec.co.jp/english/focusing/index.html
[62] Olmsted D L, Holm E A and Foiles S M 2009 *Acta Mater.* **57** 3704-13
[63] Homer E 2015 *Proc. 36th Risø Int. Symp. on Mater. Sci (Roskilde, DK, 2015)*, ed S Fæster et al. (Roskilde: Risø)

Dark field X-ray microscopy for studies of recrystallization

S R Ahl[1], H Simons[1,2], A C Jakobsen[1], Y B Zhang[3], F Stöhr[4], D Juul Jensen[3] and H F Poulsen[1]

[1] Department of Physics, Technical University of Denmark, 2800 Kgs. Lyngby, Denmark
[2] European Synchrotron Radiation Facility, CS 40220, 38043 Grenoble Cedex 9, France
[3] Department of Wind Energy, Technical University of Denmark, 4000 Roskilde, Denmark
[4] DTU Danchip, Technical University of Denmark, 2800 Kgs. Lyngby, Denmark

E-mail: sroh@fysik.dtu.dk

Abstract. We present the recently developed technique of Dark Field X-Ray Microscopy that utilizes the diffraction of hard X-rays from individual grains or subgrains at the (sub)micrometre-scale embedded within mm-sized samples. By magnifying the diffracted signal, 3D mapping of orientations and strains inside the selected grain is performed with an angular resolution of 0.005° and a spatial resolution of 200 nm. Furthermore, the speed of the measurements at high-intensity synchrotron facilities allows for fast non-destructive *in situ* determination of structural changes induced by annealing or other external influences. The capabilities of Dark Field X-Ray Microscopy are illustrated by examples from an ongoing study of recrystallization of 50% cold-rolled Al1050 specimens.

1. Introduction

Current methods for microstructural mapping such as Electron Back-Scatter Diffraction (EBSD [1]) and standard Transmission Electron Microscopy (TEM [2]) are inherently two-dimensional (2D) and can therefore not directly map the microstructure of bulk materials. X-ray techniques such as 3D X-Ray Diffraction (3DXRD [3]), Diffraction Contrast Tomography (DCT [4]), and 3D Polychromatic X-Ray Microscopy [5] can provide 3D maps of polycrystals, but cannot resolve nano-scale features. Recently, Dark Field X-Ray Microscopy (DFXRM) demonstrated the ability to measure 3D maps of orientations within individual deformed grains with 300 nm spatial resolution [6]. Importantly, this non-destructive technique offers the unique possibility of *in situ* studies of 3D structural changes during recovery and recrystallization processes in polycrystalline metals. In addition to sub-micrometre spatial resolution, the remarkable angular resolution, better than 0.005°, allows for mapping of internal structures with very small angular differences - orders of magnitude less than those resolvable by other means.

With this new tool it will potentially be possible to map not only the recrystallized grains but also the deformed structures surrounding the grain. Thus we will now be able to study individual new grains, their possible internal structure of very low angle boundaries, and their growth into the deformed material, resolving individual dislocations and dislocation boundaries in the deformed material. Strain fields at grain boundaries and around single dislocations may be mapped, and we may be able to obtain new insight into the internal structure of partly recrystallized metals as well as the processes occurring during recovery. For instance, it has

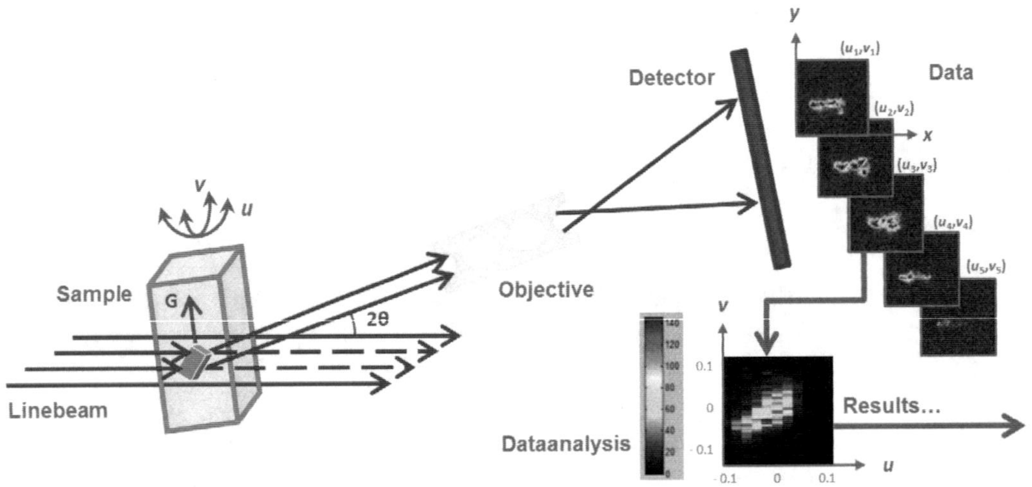

Figure 1. Sketch of the experimental setup of DFXRM and data acquisition. A line beam illuminates a plane of the polycrystalline sample. The diffracted signal is magnified by the objective and direct space images are collected. The orientation of the scattering vector **G** is varied by tilting the sample slightly in the two perpendicular angles u and v. From the raw images a 2D rocking curve for each pixel is constructed and further analysis of the internal structure of the sample carried out.

been hypothesized that when subgrains in deformed material merge during recovery, their lattices rotate towards a common orientation [7], yet this rotation has not been directly observed. With DFXRM, we are able to resolve very small angular changes in lattice orientation, potentially enabling direct observations of this phenomenon. Furthermore, grain boundaries may be resolved with sub-micrometre resolution through a series of annealing steps, allowing us to study grain coarsening in a fully recrystallized sample.

In this paper we outline the DFXRM technique and describe our current instrumentation, which is an *ad hoc* setup being developed in collaboration between the Technical University of Denmark (DTU) and the European Synchrotron Radiation Facility (ESRF) in France. To illustrate the capabilities of DFXRM we present preliminary data from an ongoing study of the recrystallization of aluminium. The spatial resolution of the data presented is better than 200 nm and angular differences of 0.005° are resolved. We demonstrate that inside micrometre-sized recrystallizing grains in the present sample there exist very low angle boundaries with angular differences of less than 0.1°.

2. Instrument description

The defining feature of DFXRM is that it magnifies the X-rays scattered by a crystalline material through Bragg diffraction. For crystal lattice planes separated by a distance d, X-rays of wavelength λ will be diffracted through the angle 2θ according to Braggs law: $\lambda = 2d\sin\theta$. DFXRM, see Figure 1, utilizes hard X-rays with energies of at least 15 keV, which are necessary

Figure 2. Beamline ID06. The instrument is set up for the experiment described in the case study below. A gas blower is mounted for annealing studies.

to penetrate into the bulk of large (typically millimeter-sized) samples. To ensure that sampling times are commensurate with the material dynamics, DFXRM uses intense x-rays from the high-brilliance x-ray source of the ESRF. Specifically, the instrument is located at beamline ID066, see Figure 2.

When the sample is properly aligned in the beam, a grain inside the sample with a clear diffraction spot is selected on a near-field detector positioned 1-5 cm away from the sample. The diffracted signal is then magnified through an objective onto a far-field detector several meters away, giving a direct space image of the diffracted signal from the illuminated part of the sample at this particular sample orientation.

While the concept of imaging the diffracted intensity is common with Diffraction Topography [8], it is the additional inclusion of the imaging objective that makes DFXRM unique in its angular and spatial resolution. Furthermore, we are currently working towards a setup with variable magnification, to give the opportunity of zooming in and out. With this a spatial resolution of 50 nm should be within reach. The objective is a compound refractive lens [9] that consists of a series of individual refractive X-ray lenses that combine to give a practically manageable focal length, typically 20-40 cm. Thus, the objective can be positioned along the diffracted beam such that it forms a spatial image on the far-field detector. The contrast in this image arises from the diffraction process, meaning that local changes in d-spacing and orientation in the sample are visible as intensity gradients in the image. Due to the absorbing nature of the X-ray objective, the numerical aperture (i.e. angular acceptance) of the system is very small (0.1°). This implies that we observe the spatial distribution of regions with a specific orientation and d-spacing of the lattice planes.

Since the instrument only images a very small range of orientations and lattice spacings, mapping samples with any orientation and strain distributions requires taking multiple exposures while scanning the sample and/or the axis of the diffracted beam and recombining these images into a composite. For example, by tilting the sample slightly through the two perpendicular angles (u, v), see Figure 1, the internal deviation of the lattice plane orientation is mapped for the illuminated region of the sample. Specifically, raw images for a 2D-set of perpendicular tilt-angles are collected and from this set the measured small section of the full pole figure is extracted as a 2D rocking curve for each direct space coordinate, see Figure 3 and 5. The (u, v)-values for the maximum intensity is determined from a Gaussian moment fit and this information is collected in a colour-coded direct space image, showing the distribution of lattice orientations, see Figure 4(c). Similarly, elastic strain (i.e. the variations in lattice spacing) may

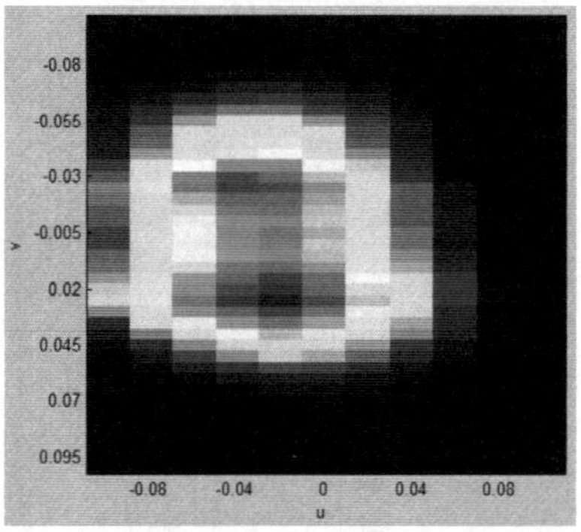

Figure 3. Summed 2D rocking curve. The summed intensity of the image for each value of the tilts (u, v) is mapped in arbitrary units where red indicates large intensity and blue is small.

be mapped through consecutive scans in which the sample is kept stationary and the objective and detector are moved together in a narrow 2θ-range around the nominal diffraction angle for the unstrained lattice.

The image visible in the detector constitutes a 2D projection through the illuminated volume of the sample. Therefore, 3D mapping with high resolution must be performed layer-by-layer and a small vertical dimension of the beam is required for the detector image to represent a single slice of the sample. This is achieved by condensing the incoming beam to a line beam. Then a 3D volume can be mapped by repeating the above procedure while moving the sample vertically in steps. The slices are then combined layer-by-layer to a 3D image during postprocessing.

Annealing, deformation, and other transformative processes may be introduced and measurements may be performed stop-go, to map the structural changes during the processes. Repeating the measurement after external stimuli provides information on the induced structural changes in 3D.

3. Case study: Partly recrystallized Al1050

Here we present preliminary results from an ongoing study of recrystallization of aluminium to illustrate the capabilities of DFXRM.

3.1. Sample

A sample of aluminium Al1050 cold-rolled to 50% reduction in thickness was cut into a rod of 300×300 μm^2 cross-section perpendicular to the rolling direction (RD) and heat-treated for 50 minutes at 325°C (almost 50% recrystallized). The rod was aligned in the line beam and tilted to $\theta = 11.8^\circ$ to look at the (200)-reflection ($d = 4.04$ Å) with the lattice plane normal along the RD. A recrystallizing grain of $15 \times 20 \times 20$ μm^3 in size was located within the bulk of the sample. It is presumed, but not investigated, that the grain is surrounded by deformed material.

3.2. Experimental

The DFXRM instrument is located at the beamline ID06 at the ESRF, see Figure 2. A monochromatic X-ray beam at an energy of 15 keV ($\lambda = 0.83$ Å) was generated by an undulator and selected by a Si (111) double bounce monochromator. The condensing system comprised 5 200 μm-radius Be-lenses located 20 m from the source followed by a set of slits to restrict the beam size to 200×400 μm^2 and a 1D silicon compound refractive lens with 40 50 μm-radius lenslets located 55 m from the source. This produced a line beam of 200 nm in height and 200

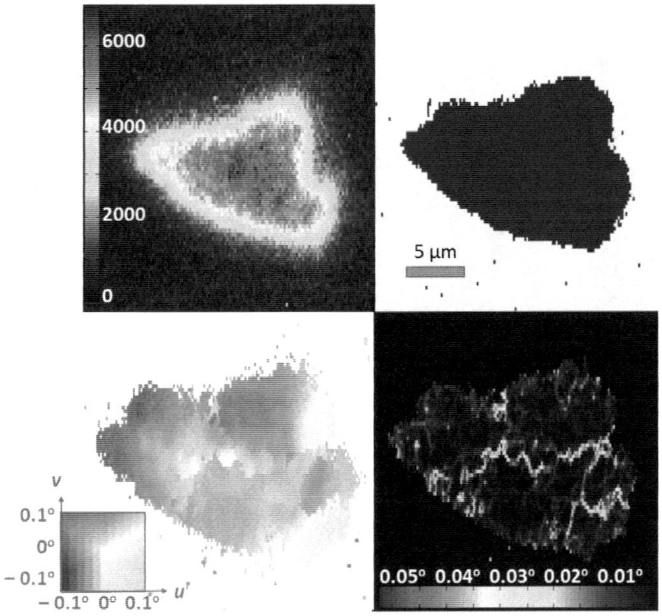

Figure 4. Preliminary data. The data shown is for a single layer of an approximately $15 \times 20 \times 20$ μm^3 recrystallizing grain. (*a*) Adding images for all measured tilts (u, v) gives a measure of the spatial distribution of total intensity, shown here in arbitrary units. (*b*) Setting a threshold for the total intensity gives an estimate of the grain shape on the micrometre-scale. (*c*) Colour-coding each pixel by the center position of the peak in the 2D rocking curve gives a map of the distribution of orientations. (*d*) The difference in orientation of neighbouring pixels gives an impression of very low angle boundaries within the grain.

μm wide, allowing us to illuminate one slice of the sample at a time with sufficient sampling statistics to capture images in 3-5 seconds. The objective consists of 72 50 μm-radius Be-lenses, which ultimately provided a magnification of 18 times over the 5.5 m sample-to-detector distance. The far-field detector was a FReLoN CCD detector with 2048 × 2048-pixels with an effective pixel size of 1.2 μm. The spatial resolution was thus limited by the optical performance of the objective lens, which is still better than 200 nm.

Images were collected for small tilts in the two perpendicular angles (u, v), see Figure 1. Both tilt-series covered $\pm 0.1^{\circ}$, in steps of 0.005° for angle u along the θ-direction and in steps of 0.02° for the perpendicular angle v.

3.3. Results

From the total intensity of each image with different values of the tilts (u, v) a summed 2D rocking curve was obtained, see Figure 3. The two distinct peaks observed here indicate that at least two different orientations are present inside the grain. Adding all images gives an estimate of the grain morphology, see Figure 4(*a*) and setting an intensity threshold as in Figure 4(*b*) gives a measure of the grain shape, allowing us to compare grain shapes for example before and after annealing. 2D rocking curves for each pixel were constructed from the intensity of a single pixel in each of the 451 images. The lattice orientation associated with each pixel was determined as (u, v) of the maximum intensity in the 2D rocking curve. This is summarized

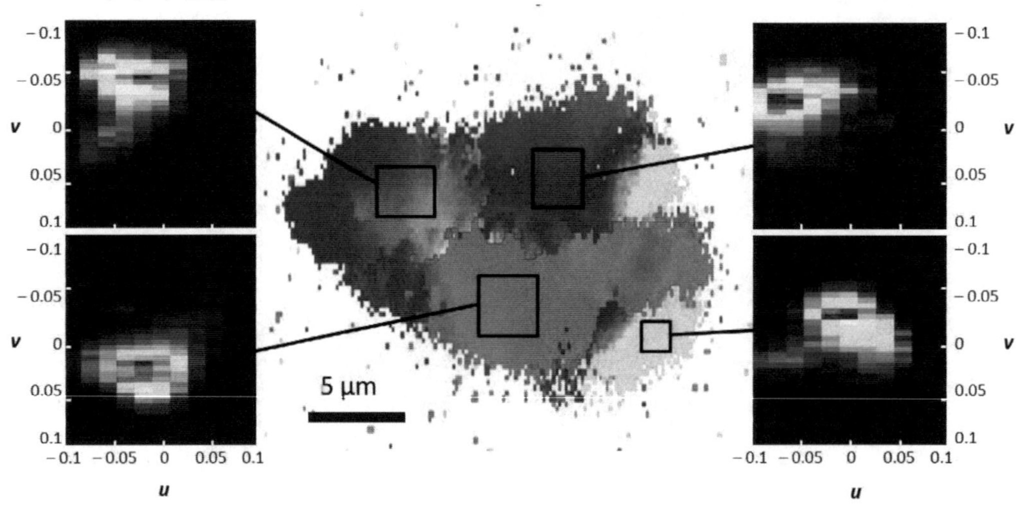

Figure 5. 2D Rocking curves of regions with slightly different orientation. Central image shows the recrystallizing grain colour-coded according to the lattice orientation. Insets show the 2D rocking curves in (u, v) of the marked regions.

in the colour-coded image in Figure 4(c). The difference in orientation between neighbouring pixels is shown in Figure 4(d).

Interestingly, the grain is observed to be subdivided by several very low angle boundaries separating regions of slightly different lattice plane orientations. These regions are $2 - 5$ μm across and their orientational differences are of the order of 0.1^o or less as seen in the 2D rocking curves for different regions in Figure 5. The very low angle boundaries of 0.1^o angular difference correspond to about $4 \cdot 10^6$ dislocations/m or a distance of more than 0.2 μm between dislocations. The internal structure of the recrystallizing grain may be due to the impurities present in Al1050 or be reminiscent of the deformed structure since the features are of similar size [10]. Note that the apparent sharpness of the very low angle boundaries in Figure 4(d) may be an artefact of the reconstruction process. Therefore, a thorough analysis of the image registration is necessary before further interpretation and characterization of these boundaries can be conducted.

The conclusion holds, however, that within a recrystallizing grain in the present sample there exist very low angle boundaries with 0.2 μm or longer between dislocations. This is at the limit of the current instrumental resolution, but in the near future we hope to increase the spatial resolution to about 50 nm. This will enable us to investigate in detail the very low angle subdivision of the recrystallizing grain and the strain fields associated with these internal boundaries. Further potential studies include *in situ* annealing and characterization of such recrystallizing grains to make films of the growth of recrystallizing grains and the possible changes within the grains, as well as characterization of the surroundings of the grains, deformed material or other grains.

4. Conclusion

We have demonstrated that Dark Field X-Ray Microscopy is a promising technique for studying polycrystalline materials in 3D. The technique offers relatively fast data acquisition and involves a hard X-ray line beam 200 nm in height and a set of X-ray lenses that magnifies the diffracted signal. This setup allows us to non-destructively construct 3D maps of orientation and strain of micrometre-sized structures embedded in mm-sized samples of the same material with an angular resolution of about 0.005° and spatial resolution of 0.2 μm, thereby providing a significant step forward for 3D materials characterization techniques.

Acknowledgments

The authors are grateful to W. Pantleon, S. Schmidt and C. Detlefs for scientific advice, ESRF for providing beamtime and the Danish Instrumentation Centre, Danscatt. The work is part of an ERC grant "diffraction based x-ray transmission microscopy".

References

[1] Schwartz A J, Kumar M, Adams B L and Field D P 2009 *Electron Backscatter Diffraction in Materials Science* (Berlin: Springer)
[2] Williams D B and Carter C B 2009 *Transmission Electron Microscopy: A Textbook for Materials Science* (Berlin: Springer)
[3] Poulsen H F 2004 *Three-Dimensional X-Ray Diffraction Microscopy: Mapping Polycrystals and their Dynamics* (Berlin: Springer)
[4] Ludwig W, Reischig P, King A, Herbig M, Lauridsen E M, Johnson G, Marrow T J and Buffire J-Y 2009 *Rev. Sci. Instrum.* **80** 033905
[5] Larson B C, Yang W, Ice G E, Budai J D and Tischler J Z 2002 *Nature* **415** 887
[6] Simons H, King A, Ludwig W, Detlefs C, Pantleon W, Schmidt S, Stöhr F, Snigireva I, Snigirev A and Poulsen H F 2015 *Nature Comm.* **6** 6098
[7] Hu H 1962 *Trans. Met. Soc. AIME* **224** 75
[8] Tanner B K 1976 *X-Ray Diffraction Topography* (Oxford: Pergamon Press)
[9] Snigirev A, Kohn V, Snigireva I and Lengeler B 1996 *Nature* **384** 49
[10] Hansen N and Juul Jensen D 2011 *Materials Science and Technology* **27** 1229

Phase decomposition and nano structure evolution of metastable nanocrystalline Cu-Co solid solutions during thermal treatment

A Bachmaier[1], M Stolpe[2], T Müller[1,3] and C Motz[1]

[1] Chair of Materials Science and Methods, Campus D2.2, Saarland University, Saarbrücken, Germany
[2] Metallic Materials, Campus C6.3, Saarland University, Saarbrücken, Germany
[3] Erich Schmid Institute of Materials Science, Austrian Academy of Science, Jahnstraße 12, Leoben, Austria

E-mail: a.bachmaier@matsci.uni-sb.de

Abstract. Nanocrystalline and ultrafine-grained $Cu_{100-x}Co_x$ (x =26 and 76) solid solutions have been prepared by severe plastic deformation (SPD) of elemental powder mixtures. For both concentrations a supersaturated solid solution fcc phase was identified after the deformation process with grain sizes of less than 50 nm for Co rich solutions and around 100 nm for Cu rich solutions. Additionally, synthesis of nanocrystalline materials in the Cu-Co alloy system by electrodeposition has been conducted. Microstructural characterization by scanning and transmission electron microscopy, differential scanning calorimetry, and microhardness measurements are used to investigate the structural evolution, the thermal stability and mechanical properties of the different nanocrystalline Cu-Co alloy materials during isothermal and non-isothermal annealing. In this study it is shown that the phase decomposition of the metastable Cu-Co solid solutions has a significant influence on their thermal stability, which can be linked to the underlying microstructure that forms during annealing.

1. Introduction

Nanocrystalline (nc) and ultrafine-grained (ufg) materials have been a subject of extensive research over the past couple of decades due to their extraordinary mechanical and physical properties like high strength [1-3]. Nc and ufg materials are, however, thermally and mechanically unstable due to the large amount of enthalpy stored in their large grain boundary area, and structural coarsening affects material properties considerably even for annealing at low temperatures for short times [4]. Hence, the stability of nc and ufg materials is a fundamental issue, but strategies to enhance grain size stability are still under research.

In this study, synthesis of bulk nc and ufg materials is conducted by severe plastic deformation (SPD) and electrodeposition (ED) in the Cu-Co model system. The peritectic Co-Cu system has a positive heat of mixing extending over the whole composition range and no room temperature solubility at all [5]. Due to strain-induced mechanical mixing, nc alloys with solubilities far beyond equilibrium have been already obtained in different immiscible Cu-based alloy systems by SPD [6-9]. In [10] it is reported, that Cu-Co solid solutions can be synthesized by ED as well. The metastable Cu-Co solid solutions are not in an equilibrium state after processing. During subsequent annealing treatments, phase decomposition is expected. Tailoring material properties for potential technological applications requires an understanding of the structural changes which occur during annealing and an evaluation of a possible impact of phase decomposition on thermal stability and mechanical-physical properties. In this study, microstructural evolution during annealing and the thermal stability of electrodeposited and severe plastically deformed nc and ufg Cu-Co solid solutions are investigated. It is shown that the phase decomposition process has a major impact on the thermal stability with a direct relation of structure to stability.

2. Experimental

The initial two-phase $Cu_{100-x}Co_x$ materials (x=26 and 76 at. %) were hot-isostatically pressed at 900°C under vacuum by RHP-Technology (Seibersdorf, Austria) from Co powders (purity 99.9%) and Cu powders (purity 99.99%) into samples with a cylindrical shape and subsequently processed by SPD (high-pressure torsion (HPT) deformation at room temperature with a pressure of 5 GPa and a rotation speed of 0.2 rpm). To study the influence of the impurity content, Co and Cu powders with a lower purity (99.8 and 99.9%) were additionally prepared in the same composition by applying the same HPT deformation parameters. The strain value applied (equivalent strain ≥ 210 for a distance ≥ 1.5 mm from the sample centre) is high enough to obtain the steady state region of the microstructural refinement (i.e. further HPT deformation does not induce further grain refinement) in the samples [11,12].

Synthesis of $Cu_{24}Co_{76}$ alloys was further conducted by ED in reducing simultaneously the Cu^{2+} and Co^{2+} ions onto a Cu substrate from an electrolyte solution containing 112 g/l $CoSO_4{\cdot}7\ H_20$, 6 g/l $CuSO_4$, 142 g/l Na_2SO_4, 59 g/l $C_6H_5Na_3O_7{\cdot}2\ H_20$, 19 g/l H_3BO_3, 2g/l $C_7H_5NO_3S$ and 0.1 g/l $C_{12}H_{25}NaO_4S$ at 40°C. The deposition parameters were adjusted to obtain a Co content similar to the HPT deformed materials resulting in a current density j= − 30 mA/cm² at a pH= 4.5.

All samples were further subjected to isothermal heat treatments at 300°C, 400°C, 600°C for 1 h in air. Selected HPT deformed samples are also annealed at the same temperatures for 7 h and 100 h to investigate the long term stability of the processed materials. Vickers microhardness was measured on all samples in as-deformed/deposited and annealed conditions. Microstructural characterization was carried out by scanning electron microscopy (SEM) using a Zeiss SIGMA™-VP field emission SEM device. Electron-back scatter diffraction (EBSD) is conducted using the EBSD system (Oxford Nordlys EBSD detector) attached to the SEM and Oxford HKL Channel 5 analysis software. Transmission electron microscopy (TEM) investigations of selected samples were performed on a JEOL JEM 2011 instrument at an acceleration voltage of 200 kV.

Differential scanning calorimetry (DSC) and differential thermal analysis (DTA) measurements on selected samples were performed using a Perkin Elmer DSC 8000 and a Netsch STA 446/C/6/MFG/G Jupiter device. Subsequent re-runs served as the reference measurement and baseline for the analysis for all measurements. For the determination of the activation energies with the method of Kissinger, different linear heating rates of 10, 20, 60 and 120 °C/min in a temperature range from 25°C to 720°C were applied.

3. Results and Discussion

3.1. As-deformed state

Both Cu-Co alloy materials are deformed until the steady state is reached. During the HPT process, the microstructure of the initial two-phase Cu-Co materials is strongly refined. The microstructures of the alloys in this steady state are displayed in Figure 1, which shows SEM images recorded with back scattered electrons. Both samples are homogeneously refined. From the SEM images, it is not possible to distinguish between the Cu and Co phases. With increasing Co content in the alloy, the grain size is reduced. Similar types of nanostructures are obtained for the samples from the low-purity Cu-Co materials. Vickers hardness measurements reveal for both compositions the typical high hardness of HPT deformed materials. The hardness in the as-deformed state is 314±11 HV and 444±12 HV for the $Cu_{74}Co_{26}$ and $Cu_{24}Co_{76}$ alloy obtained from the high-purity material, respectively. The hardness of the samples made of the lower-purity material is 281±4 HV ($Cu_{74}Co_{26}$) and 437±4 HV ($Cu_{24}Co_{76}$). All alloys exhibit a higher hardness compared to pure HPT deformed Cu and Co [13,14], which is due to the smaller grain size in the alloys compared to pure HPT deformed materials.

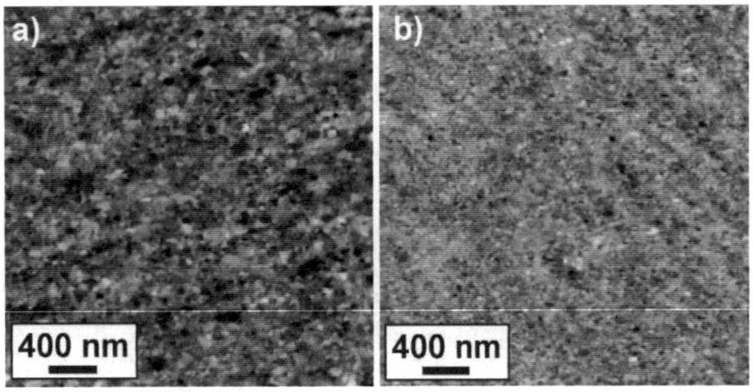

Figure 1: SEM images recorded with back-scattered electrons of the microstructure of the samples in the as-deformed condition: a) $Cu_{74}Co_{26}$ and b) $Cu_{24}Co_{76}$.

Figure 2 shows bright-field TEM images of the as-deformed Co- and Cu-rich samples. From these images, a grain size less than 50 nm for Co rich alloy (Figure 2a) and around 100 nm for Cu rich alloy (Figure 2b) can be confirmed. In both selected area diffraction (SAD) patterns, a single set of diffraction rings is visible, which are either fcc Cu or fcc Co Debye-Scherrer rings of {111}, {200}, {220}, {311} and {222} planes. The fact that the fcc Cu and fcc Co Debye-Scherrer rings merge into each other might be explained by the formation of a supersaturated solid solution in the Cu-Co alloys. Due to the small difference in the lattice parameters of fcc Cu and fcc Co, a clear discrimination between the reflections from the SAD patterns is challenging. However, large supersaturations up to 26 at. % Co in fcc Cu have been recently confirmed by atom probe tomography (APT) in the $Cu_{74}Co_{26}$ alloy [15]. To prove the formation of supersaturated solid solutions in the alloys with a higher Co-content, APT investigations are currently conducted.

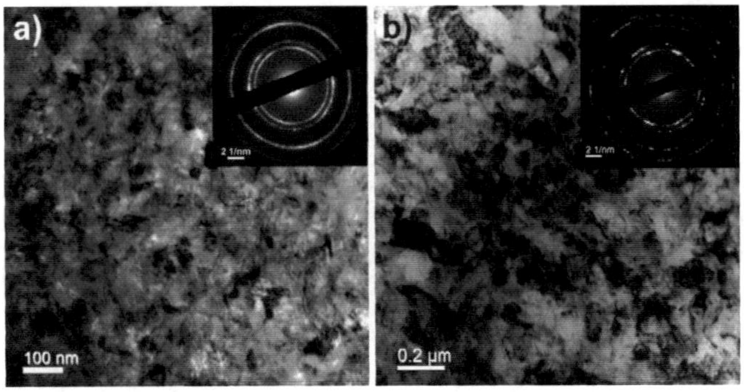

Figure 2: TEM bright field images with corresponding SAD patterns (insets) of the microstructure of the samples in the as-deformed condition: $Cu_{24}Co_{76}$ (a) and $Cu_{74}Co_{26}$ (b).

3.2. *As-deposited state*

Synthesis of bulk nc Cu-Co alloys is additionally conducted by ED. By adjusting the deposition parameters, a Cu-Co alloy with the same Co content as the HPT deformed Co-rich alloy ($Cu_{24}Co_{76}$) is obtained. Smaller Co contents similar to the composition of the HPT deformed Cu-rich alloy could not be achieved up to now. Figure 3a shows the microstructure of the as-deposited alloy recorded with back scattered electrons. Similar to the HPT deformed Co-rich alloys the grain size is nc and hardly resolvable by SEM. In Figure 3b, a bright-field TEM image with corresponding SAD pattern of the as-deposited Co-Cu alloy sample is shown. Only one set of fcc Debye-Scherrer rings (either fcc Cu or fcc Co) is observable in the SAD pattern, which also suggests a mixture of the two phases in the as-deposited condition. The ED Cu-Co alloy sample exhibits a quite high hardness in the as-deposited state, which exceeds the hardness of the HPT deformed sample. The microhardness of the $Cu_{24}Co_{76}$ alloy is 471±9 HV.

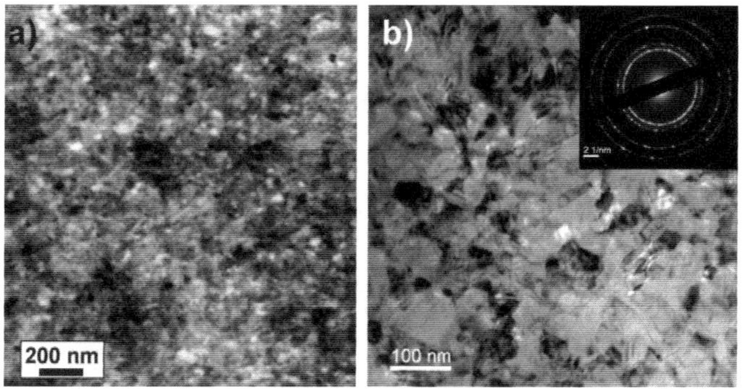

Figure 3: SEM image (a) recorded with back-scattered electrons and TEM bright field image (b) with corresponding SAD patterns (inset) of the microstructure of the $Cu_{24}Co_{76}$ alloy sample in the as-deposited condition.

3.3. Non-isothermal and isothermal annealing

To investigate the thermal stability of the Cu-Co alloys, the samples are initially annealed for 1h in air at three different temperatures (300°C, 400°C, 600°C), which corresponds to ~35, 40 and 52 % T_m with T_m being the melting temperature of the corresponding $Cu_{24}Co_{76}$ or $Cu_{74}Co_{26}$ alloys, respectively. The microhardness as a function of the annealing temperature after annealing for 1h in air is plotted in Figure 4. The microhardness of the HPT deformed $Cu_{24}Co_{76}$ alloy decreases only slightly for annealing temperatures up to 400°C. The hardness of the Cu rich alloy ($Cu_{74}Co_{26}$) even increases during annealing at the same temperatures. At the highest annealing temperature (600°C), a drop in hardness can be observed for both HPT deformed samples, which is less pronounced for the Co rich alloy. The microhardness of the ED $Cu_{24}Co_{76}$ alloy increases for the lowest annealing temperature (300°C) as well. However, at an annealing temperature of 400°C, a strong hardness decrease is observed (Δ HV = 133).

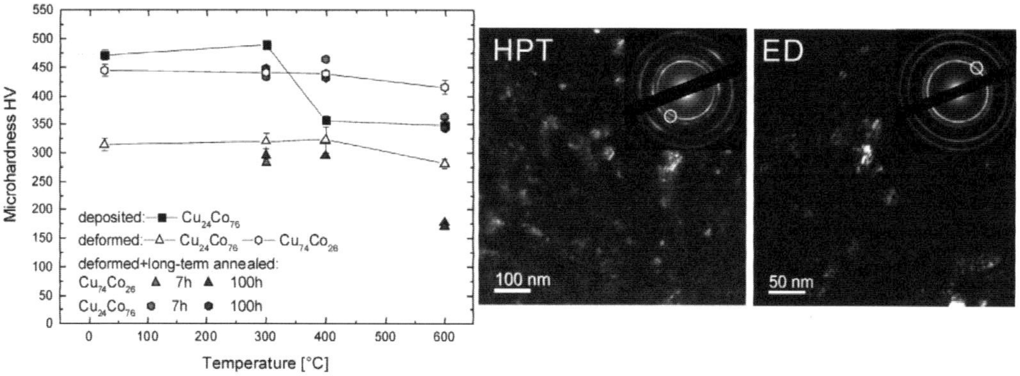

Figure 4: Microhardness of HPT deformed and ED Cu-Co alloy samples as a function of annealing temperature (300°C, 400°C, 600°C) after 1h annealing time. The microhardness of HPT deformed samples annealed at 300°C, 400°C and 600°C for 7 h and 100 h is also shown in the plot. Dark-field TEM images of the HPT deformed and ED $Cu_{24}Co_{76}$ alloy samples annealed for 1h at 300°C obtained by selecting the lattice reflections as indicated in the SAD patterns.

In Figure 4, dark-field TEM images of selected annealed microstructures (HPT deformed and ED $Cu_{24}Co_{76}$ alloy annealed for 1h at 300°C) are additionally shown. In both samples a nc structure is still visible. The corresponding SAD patterns (see insets) reflect the nc structure as well. Evaluation of the annealed microstructures by SEM after annealing at 400°C for 1h shows that the HPT deformed

Cu-Co alloys exhibit no distinct change compared to the microstructure in the as-deformed state. By contrast, massive grain growth occurs in the ED $Cu_{24}Co_{76}$ alloy (Figure 5a). Inside the micrometer-sized grains, a fine scaled substructure is visible (Figure 5b).

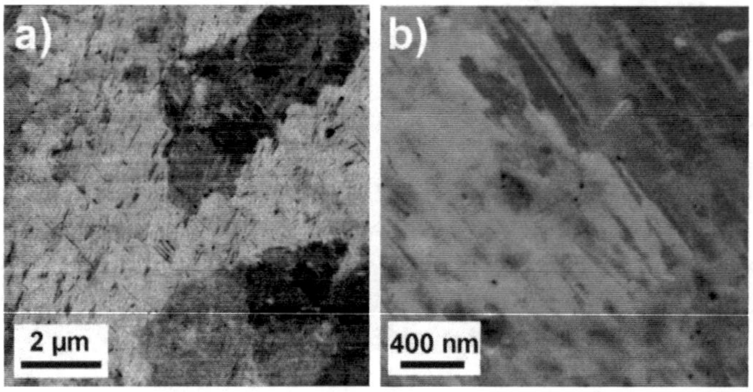

Figure 5: SEM images recorded with back-scattered electrons of the microstructure of the ED $Cu_{24}Co_{76}$ alloy sample after annealing for 1h at 400°C: Overview (a) and detailed view with a high magnification (b).

Thermal analysis was further performed to determine the decomposition temperature and annealing kinetics of the HPT deformed Cu-Co alloy materials. The $Cu_{74}Co_{26}$ and $Cu_{24}Co_{76}$ alloy samples show broad and overlapping exothermic peaks in the DSC scans (Figure 6a and b). Overlapping peaks are reported to occur frequently during annealing of supersaturated solid solutions [16]. The thermal processes of the Cu-Co alloys are further not completed at the maximum temperature (720°C) which can be applied during the DSC measurements. Therefore, additional DTA measurements were carried out with a linear heating rate of 20 °C/minute in a temperature range from 25°C to 1000°C (Figure 6c). In these scans, additional broad exothermic peaks at higher temperatures are visible.

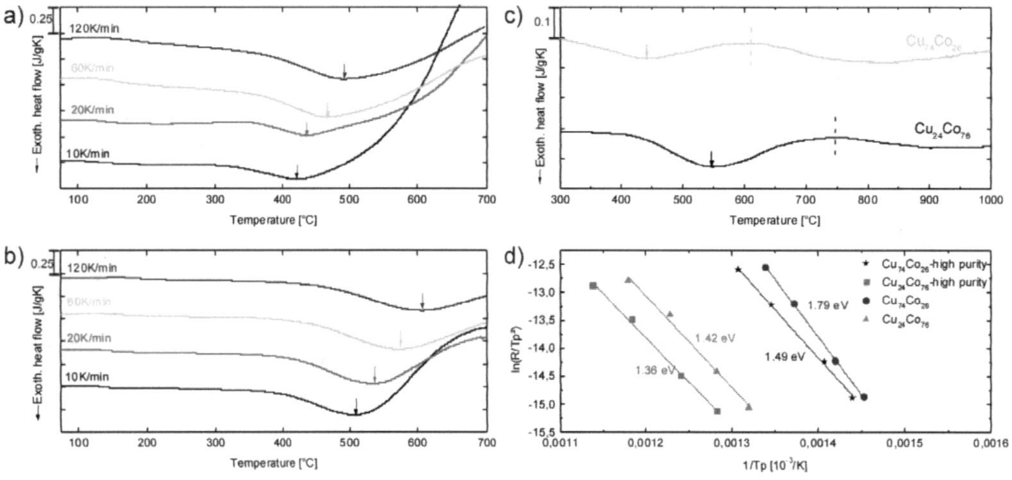

Figure 6: DSC scans with different heating rates for the Cu-Co alloys after HPT deformation: a) $Cu_{74}Co_{26}$ and b) $Cu_{24}Co_{76}$. DTA scans of both Cu-Co alloy samples with a heating rate of 20°C/minute (c). Kissinger plot for the Cu-Co alloy samples with different initial purities (d).

SEM microstructural characterization of the DSC samples heated up to 700°C shows no significant grain growth although distinct heat releases are clearly observable in the DSC scans (Figure 6a and b). On the contrary, SEM analysis shows that grain growth and phase decomposition take place in the Cu-Co alloy samples after the DTA scans (maximum temperature 1000°C). As an example, an EBSD phase map of the $Cu_{74}Co_{26}$ alloy sample after the DTA scan is displayed in Figure 7. Areas appearing red and blue in the map represent the fcc Cu phase and the hcp Co phase, respectively.

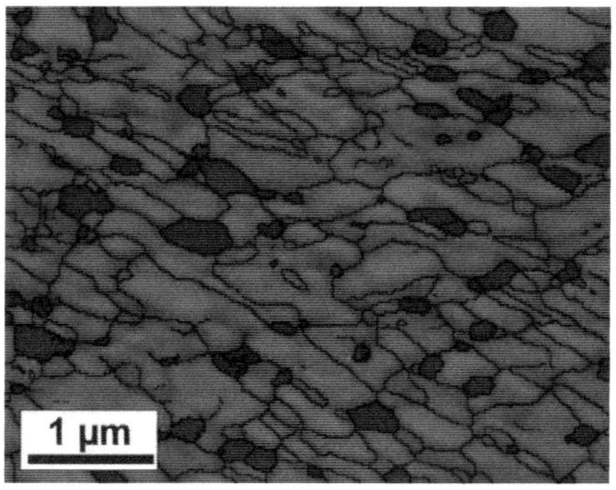

Figure 7: EBSD scan of the $Cu_{74}Co_{26}$ alloy sample after the DTA scan, with all grains belonging to the hcp Co and fcc Cu are marked as blue and red, respectively. The phase map is superimposed by an image quality map.

Hence, the exothermic peaks at intermediate temperatures (indicated by the arrows in the DSC and DTA scans) might be attributed to the decomposition process of the metastable Cu-Co solid solutions. With increasing Co content, the peak temperature of the decomposition process increases from ~440°C ($Cu_{74}Co_{26}$) to ~545°C ($Cu_{24}Co_{76}$). Additionally, the starting temperature of the second exothermic process (i.e. grain growth) is shifted to higher temperatures with increasing Co content (marked with dashed line in Figure 6c).

For the $Cu_{74}Co_{26}$ and $Cu_{24}Co_{76}$ alloy samples (low and high purity initial materials), the peak temperatures of the decomposition process at various heating rates are also evaluated by the method of Kissinger (Figure 6d). Due to the occurrence of broad and poorly separated peaks, the validity of the Kissinger analysis in terms of a single process is not strictly fulfilled and the obtained activation energies provide only a qualitative hint for the underlying processes occurring during phase decomposition. For the $Cu_{24}Co_{76}$ alloy samples, similar activation energies of 1.42±0.05eV (lower purity material) and 1.36±0.09eV (higher purity material) are determined. For the $Cu_{74}Co_{26}$ alloy samples, slightly different activation energies of 1.79±0.03eV (lower purity material) and 1.49±0.04eV (higher purity material) are observed. The value of activation energies for phase decomposition are considerably lower than values typical found for Co tracer diffusion in bulk Cu, 2.2 eV [17], Cu tracer diffusion in bulk Co, 2.85 eV, [18] and Co-Cu interdiffusion, 2.5 eV [19] and Cu and Co self-diffusion activation energies in the fcc lattice [20,21].

3.4. Long term stability

Selected samples are also annealed at 300°C, 400°C and 600°C for 7 h and 100 h to investigate the long term stability of the HPT processed materials. The measured microhardness after annealing is summarized in Table 1 and additionally plotted in Figure 4. At low annealing temperatures (300°C and 400°C), nearly no change in microhardness independent on annealing time is measured. In contrast, the microhardness decreases with increasing annealing time at the highest annealing temperature (600°C). After an annealing time of 7 h, a microhardness of 178±4 HV and 363±20 HV is measured for the $Cu_{74}Co_{26}$ and $Cu_{24}Co_{76}$ alloy, respectively. After 100h annealing time, the microhardness is further reduced to 171±3HV ($Cu_{74}Co_{26}$) and 343±13HV ($Cu_{24}Co_{76}$).

Long-term annealing at 400°C leads only to minor microstructural coarsening. Annealing at 600°C for different times results in notable coarsening instead. To give an example, TEM images of the $Cu_{74}Co_{26}$ alloy after annealing for 7h and 100h at 400°C and after annealing for 100h at 600°C are displayed in Figure 8. A grain size on the average around 100 nm is visible in the TEM bright field image of the microstructure annealed at 400°C for 7h (Figure 8a). Inside the grains, fine scaled contrast variations are additionally visible. Tilting the electron beam reveals that nearly all of the large grains show this kind of contrast modulations in the regions explored with TEM. In the SAD pattern, only fcc Debye-Scherrer rings of {111}, {200}, {220}, {311} and {222} planes are observed. Annealing for longer times at the same annealing temperature does not lead to further coarsening or to a significant change in the resulting microstructure (Figure 8b).

Table 1. Microhardness of annealed HPT processed samples.

microhardness	HPT $Cu_{74}Co_{26}$		HPT $Cu_{24}Co_{76}$	
annealed	7h	100h	7h	100h
300°C	283±10 HV	296±6 HV	433±7HV	448±8HV
400°C	320±12 HV	295±11 HV	464±13 HV	432±12 HV
600°C	178±4 HV	171±3HV	363±20 HV	343±13HV

On the contrary, the TEM bright field image of the $Cu_{74}Co_{26}$ alloy annealed for 100h at 600°C displays a microstructure in which significant grain growth occurred (Figure 8c). Similar modulated contrast variations as displayed in Figure 8a-b are visible in some of the grains. Some grains, however, fail to show any of these contrast variations. The SAD pattern shows two sets of spotty diffraction rings (fcc Cu and fcc Co) further confirming the observed microstructural coarsening.

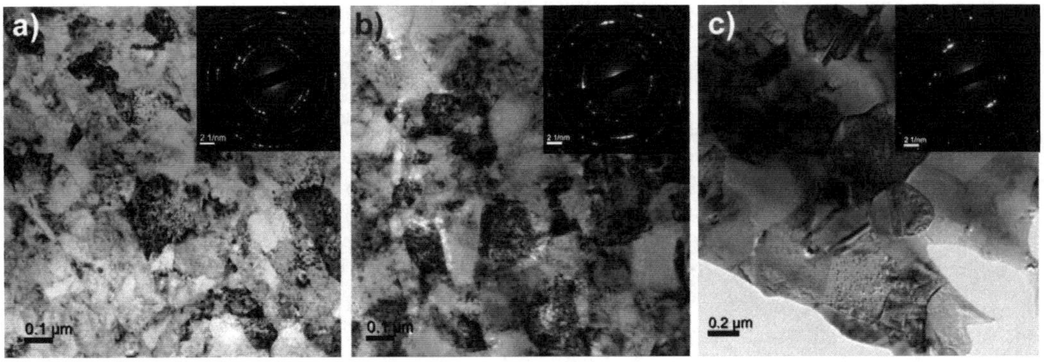

Figure 8: TEM bright field images with corresponding SAD patterns (insets) of the microstructure of the $Cu_{74}Co_{26}$ alloys after annealing for 7h (a) and 100 h (b) at 400°C and 100h at 600°C (c).

4. Summary

Nc and ufg bulk Cu-Co alloys have been prepared by HPT and ED with similar Co contents and their thermal stability during isothermal and non-isothermal annealing is investigated. The grain size and hardness in the as-deposited or as-deformed condition depends on the Co content. Samples with a higher Co content exhibit a higher hardness and smaller grain sizes. For both alloy compositions, the formation of fcc solid solutions is observed.

No significant grain growth is observed before completion of phase separation in the SPD-processed Cu-Co solid solutions. At intermediate annealing temperatures (≤400°C), the Cu-Co alloys are thermally stable even after annealing for several days. At the same time, a high hardness is maintained in the HPT deformed alloy materials. The onset temperature of phase separation is a function of the Co content and increases with increasing Co content. TEM investigations reveal that the vast majority of the grains at these temperature show modulated contrast variations which might indicate spinodal decomposition inside the grains.

For isothermal annealing at 600°C, phase separation and clearly visible microstructural coarsening is observed in the HPT deformed Cu-Co alloys. Although decomposition and microstructural coarsening cannot be hindered, an ufg structure is still maintained in the Cu-Co alloys even after annealing at 600°C (~50% Tm) for 100h. In principle, such phase decomposition mechanisms can be used to produce tailored nano structures to optimize e.g. thermal or mechanical properties.

Similarly, no significant change in the microstructure and microhardness is observed in the ED $Cu_{24}Co_{76}$ alloy sample annealed at 300°C. On the contrary, significant coarsening is observed in the ED alloy at an annealing temperature of 400°C. Inside the micrometer-sized grains, a fine-scaled substructure is observable, which might be responsible for the still quite high hardness of the ED Cu-Co alloy in the annealed state. The reasons for the differing behavior of electrodeposited and severe plastically deformed alloys is the scope of future work.

Acknowledgments
We gratefully acknowledge the financial support by the Austrian Science Fund (FWF): J3468-N20.

References

[1] Gleiter H 1989 *Prog. Mater. Sci.* **33** 223-315
[2] Gleiter H 2000 *Acta Mater.* **48** 1-29
[3] Meyers M A, Mishra A and Benson DJ 2006 *Prog. Mater. Sci.* **51(4)** 427-556
[4] Hibbard G, Aust KT, Palumbo G and Erb U 2001 *Scripta Mater.* **44** 513
[5] Nishizawa T and Ishida K 1984 *Bull. Alloy Phase Diagrams* **5** 161
[6] Sauvage X, Jessner P, Vurpillot F and Pippan R 2008 *Scripta Mater.* **58** 1125-28
[7] Bachmaier A, Rathmayr G B, Bartosik M, Apel D, Zhang Z and Pippan R 2014 *Acta Mater.* **69** 301-13
[8] Bachmaier A, Kerber M, Setman D and Pippan R 2012 *Acta Mater.* **60** 860
[9] Edwards D, Sabirov I, Sigle W and Pippan R 2012 *Phil. Mag.* **92** 4151–66
[10] Bran J, Jean M, Lardé R, Sauvage X, Le Breton JM, Morin-Grognet S and Pautrat A 2014 *J. Alloy Comp.* **596** 118-24
[11] Pippan R, Scheriau S, Taylor A, Hafok M, Hohenwarter A and Bachmaier A 2010 *Annu. Rev. Mater. Res.* **40** 319-43
[12] Hohenwarter A, Bachmaier A, Gludovatz B P, Scheriau S G and Pippan R 2009 *Int. J. Mater. Res.* **100** 1653-61
[13] Edalati K, Fujioka T and Horita Z 2008 *Mater. Sci. Eng. A* **497** 168–73
[14] Edalati K, Toh S, Arita M, Watanabe M and Horita Z 2013 *Appl. Phys. Letters* **102** 181902-1
[15] Bachmaier A, Aboulfadl H, Pfaff M, Mücklich F and Motz C 2015 *Mater. Charact.* **100** 178-91
[16] Gente C, Oehring M and Bormann R 1993 *Phys. Rev. B* **48** 13244–52
[17] Mackliet C A 1958 *Phys. Rev.* **109** 1964-70
[18] Arita M, Nakamura M, Goto K S and Ichinose Y 1984 *Pans. Jap. Inst. Met.* **25** 703
[19] Bruni F I and Christian J W 1973 *Acta Met.* **21** 385
[20] Neumann G and Tolle V 1986 *Phil. Mag. A* **54** 619-29
[21] Bussman W, Herzig C, Rempp W, Maier K and Mehrer H 1979 *Phys. Stat. Sol. (a)* **56** 87-98

IOP Conf. Series: Materials Science and Engineering **89** (2015) 012018 doi:10.1088/1757-899X/89/1/012018

Influence of precipitate size and morphology on grain refinement in nickel aluminium bronze

C J Barr[1,2] and K Xia[1,2]
1 Department of Mechanical Engineering, University of Melbourne, Victoria 3010, Australia

[2] Defence Materials Technology Center, Hawthorn, Victoria, 3122, Australia

E-mail: k.xia@unimelb.edu.au

Abstract. Nickel aluminium bronze (NAB) was subjected to equal channel angular pressing (ECAP) using routes B_A and C at 400°C to investigate the effect of precipitate size and morphology on grain refinement in low stacking fault energy alloys. Both routes produced dynamically recrystallised grains of ~550 nm in size although only route B_A was able to create a uniform distribution of the refined grains. The large unrefined regions in NAB processed via route C was thought to arise from its inability to redistribute the various precipitate phases, as recrystallisation was enhanced around the coarse κ_{II} rosettes and refined κ_{III} lamellae but reduced in the areas containing fine κ_{IV} precipitates.

1. Introduction

The interactions between dislocations and precipitates are well known for their ability to increase strength and work hardening rates [1, 2]. The dislocation structures resulting from these interactions are highly dependent on the size and morphology of the precipitate phase [3-8] and can have either a beneficial [3, 4] or detrimental [5] effect on recrystallisation and grain refinement. Understanding how best to utilise these effects is of great importance to ultrafine grained (UFG) materials produced through severe plastic deformation (SPD), as grain refinement could be completed at lower strains compared to their single phase counterparts. Previous studies have analysed the effect of precipitate size in aluminium following equal channel angular pressing (ECAP) [3-5]. Being a high stacking fault energy (SFE) material, aluminium undergoes grain refinement through the recovery of tangled dislocations into low angle grain boundary (LAGB) cells and the subdivision of the original grains by high angle grain boundaries (HAGBs) originating from dense dislocation walls and microbands [9-15]. Introducing very fine precipitates (~20 nm) was seen to inhibit recovery and the formation of LAGB cells as well as to suppress microband formation through homogenising the slip [5]. In contrast, matrix rotations around coarse precipitates (~2 μm) led to the rapid production of HAGBs which subdivided the original grains to a greater extent than in the single phase material for the same strain [3, 4], thus accelerating the transformation from a coarse-grained to a UFG microstructure.

While these results are useful for high SFE materials, the limited cross-slip and recovery in low SFE materials result in new grain refinement processes based on twinning [16-18], and potential changes to the influence of precipitate size and morphology on grain refinement. However, little attention has been paid to the effect of precipitates on low SFE materials processed by SPD. Nickel aluminium bronze (NAB) is an ideal material to investigate such changes, as its matrix is similar to the well researched single phase low SFE Al-bronzes [17, 18] while also containing coarse, fine and lamellar precipitates [19-21]. The influence of the lamellar phase is of particular interest as it has been shown to break down into fine fragments following ECAP in route B_A [22] whereas a near lamellar structure is kept in route C [22, 23]. This paper aims to use electron backscatter diffraction (EBSD) to examine the final NAB microstructures following ECAP and highlight the role of routes and precipitate size and morphology on grain refinement.

2. Experimental materials and procedures

A cast NAB plate with a composition of Cu–8.8 Al–4.4 Fe–5.2 Ni–1.1 Mn (wt.%) was supplied by VEEM Australia, after heat treatment at 675°C for 6 h to transform any retained β′ phase into $\alpha + \kappa_{III}$ [21]. The plate was machined into bars of 8 × 8 × 100 mm, which were lubricated using graphite before being inserted into a 90° ECAP die with a cross section of 9 × 9 mm. The die was heated to 400°C and after the temperature stabilised, pressing was conducted at 5 mm/min. In addition to a forward plunger in the entrance channel, a back plunger was placed in the exit channel to provide a constant back pressure of 50 MPa during ECAP. Samples were pressed for 4 passes via route B_A and 8 passes via route C, with samples quenched upon retrieval from the die. Microstructures were examined using scanning electron microscopy (SEM, backscattered electron imaging at an accelerating voltage of 15 kV) on samples prepared to 0.5 μm diamond polish before finishing with colloidal silica. Grain structures were then analysed using electron backscatter diffraction (EBSD), using an accelerating voltage of 12 kV and step sizes of 1500, 100 and 50 nm for the as-received, route C and route B_A samples, respectively, prepared using electropolishing at 30 V in a solution of 30% nitric acid in methanol at –32°C.

3. Results

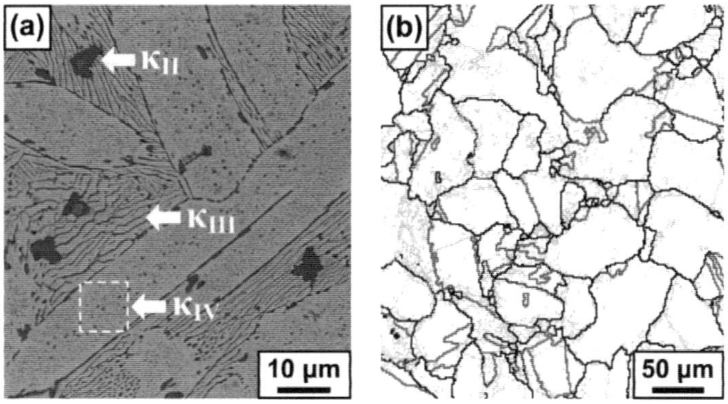

Figure 1. (a) SEM and (b) EBSD microstructures of the as-received NAB, with different colours denoting different boundary types including HAGBs (black), LAGBs (grey), twin boundaries (red) and various coincidence site lattice boundaries (other colours).

The SEM and EBSD microstructures for the as-received NAB are shown in figures 1a and 1b, respectively. The κ_{II} rosettes, lamellar κ_{III} and κ_{IV} precipitates [19, 20] were all present in the α matrix (arrowed, figure 1a), although the larger κ_I rosettes were not seen. Due to the evolution of the various phases during cooling [21], the lamellar regions can be easily identified, and the κ_{II} phase is only found in the areas where κ_{III} is present. From the EBSD mapping (figure 1b), the grains were largely equiaxed with an average size of ~40 μm, with many containing a single twin and very few LAGBs.

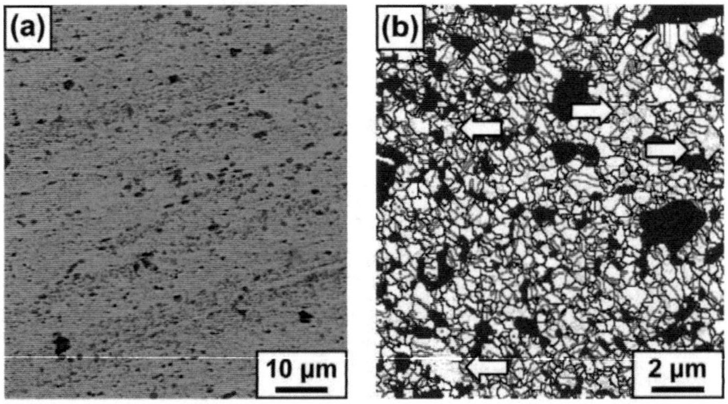

Figure 2. (a) SEM and (b) EBSD microstructures following 4 passes of ECAP at 400°C via route B_A, showing uniform precipitate and grain structures with a few small pockets dense with LAGBs (arrowed). The different coloured lines have the same meanings as in figure 1, with the dark areas indicating the presence of precipitates.

The microstructures following 4 passes via route B_A are shown in figure 2. In contrast to the as-received one, the route B_A microstructure is homogeneous, with little distinction between the κ_{IV} and the refined κ_{III} (figure 2a). While most of the κ_{III} had broken down into fine fragments similar in size to the κ_{IV} (~350 nm), some had transformed into coarser spheroids (~1 μm). The EBSD mapping (figure 2b) also shows a uniform grain structure consisting mostly of recrystallised grains (~550 nm in size) with annealing twins, as well as a few small pockets dense with LAGBs.

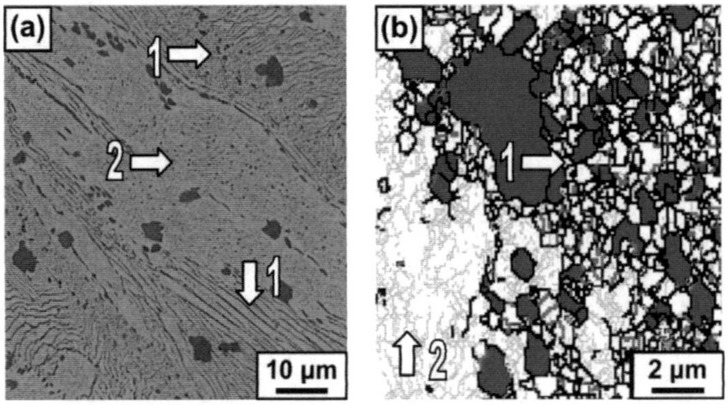

Figure 3. (a) SEM and (b) EBSD microstructures following 8 passes of ECAP at 400°C via route C. Both grain and precipitate structures are inhomogeneous, with fine grains found only in the areas containing κ_{III} (arrowed 1) while the areas containing κ_{IV} remain unrefined (arrow 2). The different coloured lines have the same meanings as in figure 1, with the dark areas indicating the presence of precipitates.

The microstructures following 8 passes via route C are shown in figure 3. Unlike route B_A, route C is unable to fully break down the lamellae (figure 3a) even after twice the number of passes. This is due to the restorative action of redundant strain on each even numbered pass, which prevents the redistribution of the κ_{III} phase and maintains its lamellar morphology [22]. The grain structure shown by EBSD mapping (figure 3b) is also vastly different, with distinct areas of refined and unrefined grains. While the recrystallised grains were of similar size to those via route B_A, they could only be found in the vicinity of the κ_{II} rosettes and refined κ_{III}. In contrast, the areas of $\alpha + \kappa_{IV}$ remained unrefined and were found to have a cellular structure of LAGBs. Such cells are uncommon for low SFE materials in which linear arrays of stacking faults or twins along a preferred crystal orientation usually form [16].

4. Discussion

4.1. Influence of precipitates on grain refinement
The above results indicate that grain refinement in NAB at 400°C is governed by dynamic recrystallisation and any enhancement of grain refinement must stem from an increase in the recrystallisation sites. Precipitates larger than 1 μm, such as the κ_{II} rosettes, are particularly well suited to this task as they rapidly produce 'deformation zones' with locally high dislocation densities and lattice rotations necessary for recrystallisation. This process is known as particle stimulated nucleation (PSN) [8] and is not seen for finer precipitates in which deformation is more readily accommodated through the punching of prismatic dislocation loops [6, 7]. This may explain the distinct lack of recrystallised grains in the vicinity of the κ_{IV} particles for route C, as the precipitates are too small to cause PSN. In the absence of precipitates, a critical boundary misorientation is required before recrystallisation can take place [8]. While a high concentration of LAGBs exists around the κ_{IV} particles, these boundaries are insufficient to act as nucleation sites. It remains to be seen whether these LAGB structures are an improvement over those in the single phase material.

The unchanging nature of the κ_{II} and κ_{IV} makes their effect on recrystallisation easier to interpret, unlike the κ_{III} which is seen to evolve from the original lamellae to either fragments or spheroids. Regardless of the final morphology, recrystallisation is promoted around all the refined κ_{III}. This is to be expected around the spheroidised κ_{III}, as the transformation coarsens them to ~1 μm when PSN becomes active. The same process, however, cannot apply to the fragmented κ_{III} as it remains the size of the κ_{IV}. It is therefore believed that the enhanced nucleation originates at an earlier stage where the lamellae are still intact. Further EBSD studies are required to analyse the developing grain structures during the initial few passes of ECAP to pinpoint this mechanism.

4.2 Effect of ECAP routes on grain refinement
Although PSN is seen as the main contributor to the enhanced grain refinement following ECAP at 400°C, recrystallisation through PSN is limited to the deformation zones around each precipitate [8]. This can be clearly seen in route C with recrystallised grains solely in the vicinity of the κ_{II} or κ_{III} particles, despite the high LAGB density in the neighbouring κ_{IV} regions. At the same time, the microstructure via route B_A consists almost entirely of recrystallised grains, complementing its homogeneous distribution of precipitate phases. The homogeneity in precipitate distribution is attained through the non-redundant strain applied in route B_A, which spreads the various phases throughout the microstructure. As the κ_{III} and κ_{IV} are forced into increasingly thin layers, the deformation zones around the κ_{III} are more likely to overlap with the κ_{IV} regions, thus allowing the whole microstructure to become recrystallised. In route C, strain is reversed on even numbered passes and this limits the extent of redistribution and the interior of the κ_{IV} regions remain unaffected.

Another contributing factor leading to more efficient grain refinement via route B_A may lie in its four unique shear planes compared to the single shear plane in route C [24] upon which the shear direction reverses on each pass. It has been shown that processes with strain reversals were only 65%

as effective in promoting recrystallisation as those in which strain is applied in a single direction [8]. The effect of this redundant strain must be significant, as the route C sample has had twice the total strain as route B_A yet the κ_{IV} areas show no sign of recrystallisation.

5. Conclusions

(1) Grain refinement in NAB following ECAP at 400°C takes place through dynamic recrystallisation, producing equiaxed grains of ~550 nm in size.

(2) ECAP routes have a significant influence on the extent of grain refinement, with route B_A producing a microstructure consisting mostly of refined grains but route C giving rise to large unrefined regions.

(3) The nucleation of recrystallised grains is dependent on the size and morphology of the surrounding precipitate phases. The coarse κ_{II} rosettes enhance the nucleation of new grains through particle stimulated nucleation. Nucleation is also enhanced around the refined κ_{III}, although the changing morphology makes the precise mechanism difficult to identify.

(4) The fine κ_{IV} precipitates limit nucleation, leading to the associated unrefined areas via route C.

Acknowledgements
The authors would like to acknowledge the support of the Defence Materials Technology Centre (DMTC). The DMTC was established and is supported under the Australian Government's Defence Future Capability Technology Centres Programme. Dr. Daniel McDonald, the Monash Centre for Electron Microscopy (MCEM) and the Australian Centre for Microscopy and Microanalysis (ACMM) are acknowledged for assistance with SEM and EBSD analysis.

References

[1] Gladman T 1999 *Mate. Sci. Tech.* **15** 30-6
[2] Ebeling R and Ashby M 1966 *Phil. Mag.* **13** 805-34
[3] Apps P J, Bowen J R and Prangnell P B 2003 *Acta Mater.* **51** 2811-22
[4] Berta M, Apps P J and Prangnell P B 2005 *Mater. Sci. Eng. A* **410** 381-5
[5] Apps P J, Bowen J R and Prangnell P B 2005 *Acta Mater.* **53** 499-511
[6] Ashby M F 1966 *Phil. Mag.* **14**, 1157-78
[7] Humphreys F and Stewart A 1972 *Surf. Sci.* **31** 389-421
[8] Humphreys F and Hatherly M 2004 *Recrystallization and related annealing phenomena 2nd ed.* (Oxford: *Elsevier)*
[9] Gil Sevillano J and Aernoudt E 1987 *Mater. Sci. Eng* **86** 35-51
[10] Bay B, Hansen N and Kuhlmann-Wilsdorf D 1989 *Mater. Sci. Eng. A* **113** 385-97
[11] Hansen N and Hughes D 1995 P*hysica Status Solidi A* **149** 155-72
[12] Hansen N 1990 *Mater. Sci. Tech.* **6** 1039-47
[13] Bay B, Hansen N, Hughes D A and Kuhlmann-Wilsdorf D 1992 *Acta Metall. Mater.* **40** 205-19
[14] Kuhlmann-Wilsdorf D and Hansen N 1991 *Scr. Metall. Mater.* **25**, 1557-62.
[15] Nes E 1997 *Prog. Mater. Sci.* **41** 129-93.
[16] Asgari S, El-Danaf E, Kalidindi S R and Doherty R D 1997 *Metall. Mater. Trans. A* **28** 1781-95

[17] Qu S, An X H, Yang H J, Huang C X, Yang G, Zang Q S, Wang Z G, Wu S D and Zhang Z F 2009 *Acta Mater.* **57** 1586-601
[18] An X, Lin Q and Qu S 2009 *J. Mater. Res.* **24** 3636-46
[19] Hasan F, Jahanafrooz A, Lorimer G W and Ridley N 1982 *Metall. Mater. Trans. A* **13** 1337-45
[20] Culpan E and Rose G 1978 *J. Mater. Sci.* **13** 1647-57
[21] Jahanafrooz A, Hasan F. Lorimer G W and Ridley N 1983 *Metall. Mater. Trans. A* **14** 1951-6
[22] Barr C J, McDonald D T and Xia K 2013 *J. Mater. Sci.* ***48*** 4749-57
[23] McDonald D T, Barr C J and Xia K 2013 *Metall. Mater. Trans. A* **44** 5556-66
[24] Valiev R Z and Langdon T G 2006 *Prog. Mater. Sci.* **51** 881-981

SEM observations of grain boundary mobility under thermal and plasticity effects

B Béucia[1], P Franciosi[2], S Queyreau[1,2], D Chaubet[1] and B Bacroix[2]

[1]Université Paris 13, Sorbonne Paris Cité, LSPM, CNRS UPR3407, F-93430, Villetaneuse, France
[2]CNRS UPR3407, LSPM, Université Paris 13, Sorbonne Paris Cité, F-93430, Villetaneuse, France

E-mail: bermane.beucia@lspm.cnrs.fr

Abstract. Grain growth in high purity aluminum was studied by in situ experiments in a SEM in order to follow grain boundary mobility under thermal and plasticity effects. After a first cold rolling stage followed by a first heat treatment, samples are submitted to one or two sequences of 3% plastic strain followed by an additional heat treatment. SEM observations allow following the resulting mobility for representative sets of grain boundaries with either low or large misorientations between differently oriented grain pairs. This paper reports on the quantitative investigation of the driving force contributions due, on the one hand, to grain boundary curvatures as measured from experimental observations and, on the other hand, to differences of dislocation densities between the adjacent grains of each examined boundary, as estimated from numerical simulations of polycrystalline plasticity. A "migration diagram" representing these two contributions is presented and discussed with regard to existing velocity laws and models.

1. Introduction

Grain boundary straining and mobility features are studied in polycrystals because they play an important role in the thermo-mechanical properties of metallic materials that is still far from being understood. In order to study grain boundaries experimentally, it is necessary to overcome the difficulty of accessing within a polycrystal several key parameters for a great number of boundaries, such as the plane of the boundary, its curvature and structure. Up to now, among the few well assessed characteristics, it has been shown that boundary migration was depending on the locally stored energy within the material [1], according to a velocity law that can be simply written as $V = MP$. In this formula, V is the grain boundary velocity, M is the boundary mobility which depends on temperature and boundary misorientation, and P is the driving force necessary for the boundary mobility, which can be decomposed into two contributions: one is due to the stored energy during plastic strain and is simply proportional to the differential of dislocation densities $\Delta\rho_{ij}$ between the adjacent (i, j) grains of a i/j boundary, that we here denote $E_{\Delta\rho}$ and the other one, associated with curvature, , is proportional to the grain boundary energy and will be denoted E_c.

If the mobility of the grain boundaries is known to furthermore depend on the grain orientations and misorientations [2],[3], the way all these dependencies combine witch each other is still a challenging point. Scanning Electron Microscope (SEM) techniques combined with electron backscattering diffraction (EBSD) [4] provides an efficient tool to study the migration of grain boundaries, although the 3D information (about curvature for instance) remains out of reach. In this study, also based on SEM+EBSD experimental examinations, polishing techniques have been used in order to at least verify that the grain dimensions in the sample depth were similar to the surface one. The estimates of the dislocation densities stored in the individual grains after a 3% plastic strain have been obtained from now classical plasticity models for polycrystals.

2. Experimental method

From a pure aluminum ingot characterized by an average grain size about 2 centimeters, a first sample was cold rolled to 85% reduction (from 15 millimeters to 2.25 millimeters) and a second one to 90% reduction (from 15 millimeters to 1.5 millimeters). For in situ annealing and further mechanical and microstructural characterization, both samples were then spark cut to the dimensions shown in figure 1. The final useful surface for observation has thus a length of 5 mm and a width of 2 mm. In order to obtain good images from the SEM, each sample was subjected to mechanical polishing with several silicon carbide papers from 1200 to 4000 mesh and to a final electrolytic polishing within a solution composed of 5% of perchloric acid, 65% of ethanol, 10 % of 2-butoxyethanol and 15% of water.

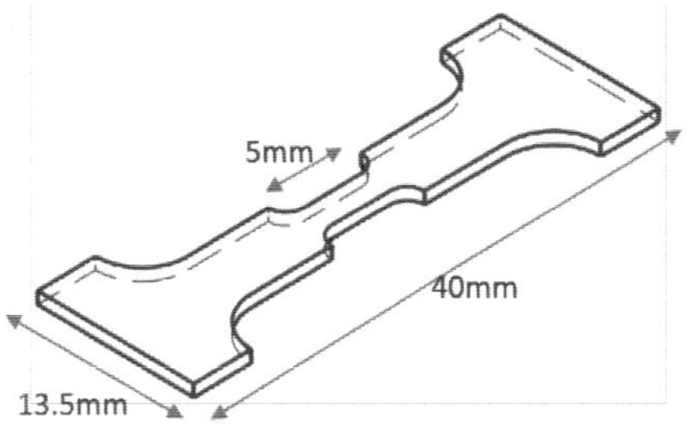

Figure 1. Sample geometry and dimension.

Figure 2. Device for in situ hot tensile tests

After polishing, the 85% reduced sample (to be denoted as S85) was annealed at 400°C for 3 minutes and the 90% reduced sample (to be denoted as S90) at 350°C for 30 minutes [5]. Both were then characterized by electron backscattering diffraction (EBSD). Then, these samples were deformed in tension up to 3% plastic deformation, in order to introduce sufficient energy into the material for triggering grain boundary mobility during subsequent annealing. The S85 sample was annealed 5h at 400°C and the S90 sample 1h at 350°C. For the S85 sample, a second sequence of a 3% plastic straining followed by the same thermal treatment (5 hours at 400°C) was applied in situ inside the SEM. For the S90 sample, the first sequence of straining + heating (1 hour at 350°C) was performed in situ. All these treatments are presented in Table 1.

Table 1: Thermomechanical treatments imposed to the two investigated samples (in bold the in situ treatments).

Sample	Rolling step	T_1 treatment (recrystallization)	1st critical strain step	T_2 treatment (grain growth)	2nd critical strain step	T_3 treatment (grain growth)
S85	85%	400°C 3minutes	3%	400°C 5hours	3%	400°C 5h
S90	90%	350°C 30minutes	3%	350°C 1hour	-	-

For in situ testing, a tension micro machine system was used (see figure 2) [6]. A resistive furnace is placed below the sample for heating and the heating rate is assigned to be 10°C/s. Images of the sample surface were taken every five minutes in order to follow grain growth, and grain boundary displacement.

At the end of the test, the microstructure was analyzed by EBSD to quantify the grain growth and the change of the crystallographic texture. Additionally, the EBSD software was used to calculate the Schmidt factors of the possible slip systems. Then, it could be checked that the observed traces did correspond to the system(s) associated with maximal Schmidt factor.

The EBSD analyzes were performed using a scanning electron microscope with a field emission gun Zeiss Supra 40VP. Kikuchi diagrams were obtained by tilting the samples at 70° with respect to the incident electron beam. Data acquisition was performed using the software Nordif TSL-OIM data collection, with a step of 1 micrometer. Data processing was then done using the OIM data analysis software.

3. Data analysis

In order to examine the correlation between the driving force and the grain boundary motion, this driving force is simply estimated from a relation of the type $P = \alpha E_c + \beta E_{\Delta\rho}$. E_c is the boundary energy resulting from the curvatures of the boundaries and is taken of the form $2\gamma/R$, where $\gamma = \gamma_m \,\theta/\theta_m(1 - ln\,\theta/\theta_m)$ for a Low Angle Grain Boundary (LAGB) and $\gamma = \gamma_m$ for a High Angle Grain Boundary (HAGB) and R is the curvature radius of the boundary. It is estimated locally through image processing. It most often corresponds to the maximal value observed for a given grain boundary. γ_m is taken as $0.324J/m^2$ [7]. $E_{\Delta\rho}$ is the energy part which depends on the dislocation densities in the grain pair adjacent to the considered boundary. This energy part is given by the formulae $E_{\Delta\rho} = 0.5\mu b^2 \Delta\rho$ where μ is the shear modulus, equal to $25.4GPa$ for aluminium, and b the Burgers vector, equal to $2.76 * 10^{-10}m$. ρ is the grain dislocation density and $\Delta\rho = \rho_i - \rho_j$ is the difference for a (i,j) grain boundary.

The dislocation density within the individual grains is calculated by using a viscoplastic "Relaxed Constraint" Taylor model which includes the single crystal constitutive law developed by Tabourot and al. [8] and further modified by Bacroix and Brenner to account for the observed anisotropy of the recovery processes [9]. This model allows the determination of the reorientation and the hardening of every plastically deformed crystal. The evolution of the dislocation density during straining is given by the equation $\frac{d\rho^s}{dt} = \frac{1}{b}.\left\{\frac{\sqrt{\sum a^{sl}\rho^l}}{K} - 2\sum\left(\frac{b^{sl}\rho^l}{\rho^s}\right)\rho^s\right\}.|\dot{\gamma}|^s$, in which ρ^s stands for the dislocation density and $\dot{\gamma}^s$ for the slip rate on the system s. The hardening a^{sl} and recovery b^{sl} coefficients, as well as the parameter K are obtained through an identification procedure on available stress – strain tensile curves for aluminum single crystals [10]. These tests, as well as the experimental ones described earlier, are simulated by prescribing a uniaxial stress state to each crystal. The identified coefficients are gathered in Table 1. The two hardening and recovery matrices are characterized by only 5 distinct coefficients, which describe the interaction types between systems (from 1 to 5: self-hardening, collinear, coplanar, orthogonal and others) [8]. The initial dislocation density ρ_0^s is taken equal to 10^9 m^{-2} for all systems. It is worth noting that the parameters are not strictly the same for all single crystals, but two main classes can be distinguished: the <111> and <100> orientations (these indices represent the Miller indices of the tensile direction) which correspond to the simultaneous activation of at least 6 systems on one hand and the <123> and <112> orientations for which one or two systems only are simultaneously activated on the other hand.

Table 2: Typical parameters values of the single crystal hardening law

Orientation	K	a^1	a^2	a^3	a^4	a^5	b^1	b^2	b^3	b^4	b^5
111	90	0.4	0.5	0.6	0.6	0.5	5.0	0.5	1.0	4.0	0.5
100	95	0.3	0.5	0.6	0.6	0.5	5.0	0.5	0.5	5.0	0.5
112	105	0.2	0.4	0.6	0.6	0.35	1.5	0.0	0.0	3.0	0.0
123	105	0.25	0.4	0.6	0.6	0.35	1.5	0.0	0.0	3.0	0.0

4. Results

We first examine and report on the S85 sample. Figures 3, 4 and 5 represent respectively the sample surface after 85% rolling and T1 treatment, after the 2nd critical strain step, and after T3 treatment (Table 1). They show (according to the color convention of the reported reference triangle) three main orientations: the <111> orientation, and the <112> and <102> orientations which are close to <100>. Grain sizes of 200 to 500 μm are obtained. The grains are still elongated in the rolling direction.
Figures 3, 4 and 5 summarize the grain boundary evolution due to the assigned strain + heat treatment sequences. In figure 5, the new shapes of the same grains after T_3 treatment characterized by EBSD are superimposed to the grain contours prior to annealing (in red). We clearly observe the Strain Induced Boundary Migration (SIBM) of a part of the GBs present in the material. Furthermore, in most cases, we clearly see red grains growing at the expense of blue ones.

However, other orientations exist below the observed surface of the sample, which present smaller grain sizes and which can affect the growing sequence observed on the extreme surface. The microstructures characterized at 200 and 400 μm below the surface are thus presented in figures 6 and 7. It is observed that the dimension of the grains through the thickness of the sample is of same order of magnitude as the grain size on the surface with a larger number of small grains.

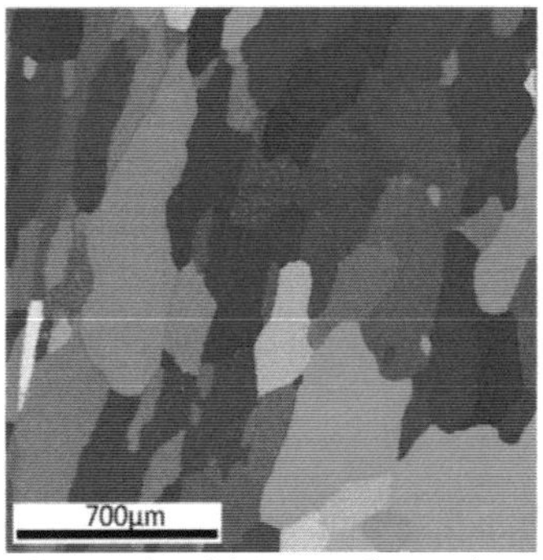

Figure 3. S85 sample surface after 85% rolling and T_1 treatment.

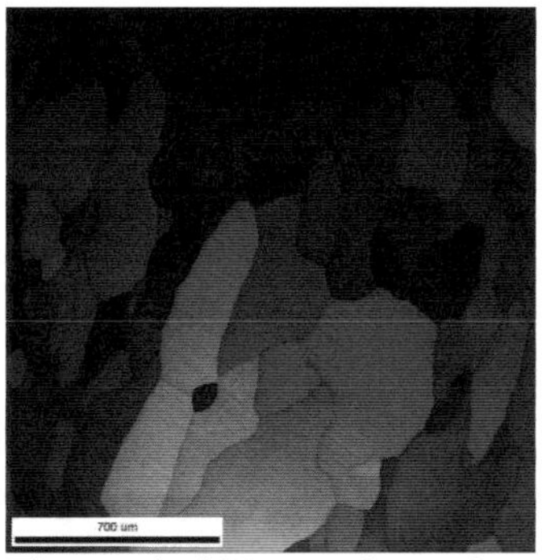

Figure 4. S85 sample surface after T_2 treatment + 3% strain.

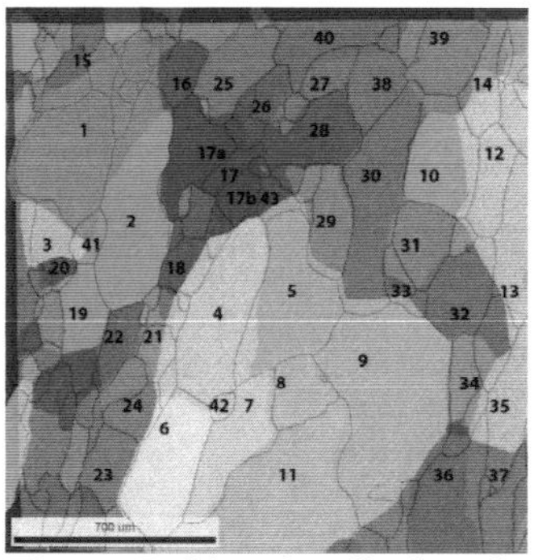

Figure 5. S85 after final T_3 in situ annealing with grain contours of figure 4 (red).

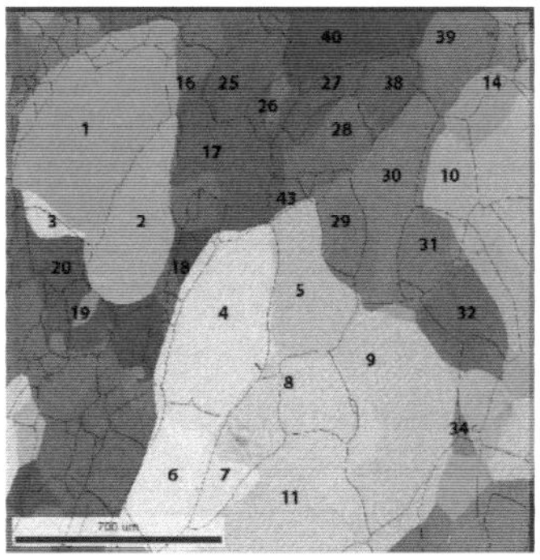

Figure 6. Microstructure at a depth of 200 μm with grain contours on the surface (red).

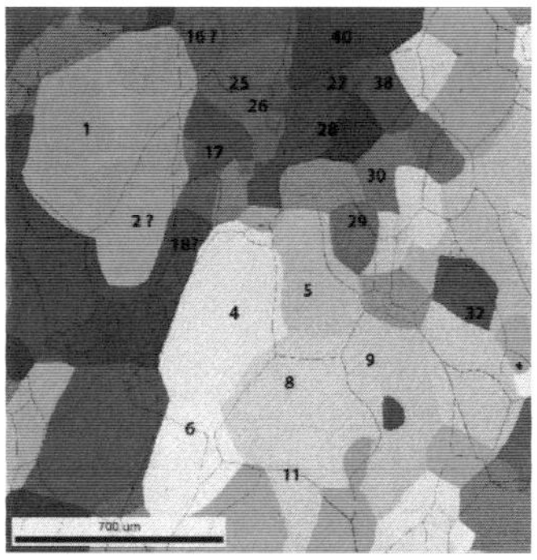

Figure 7. Microstructure at a depth of 400 μm with grain contours at depth of 200μm.

In all these figures, the grains have been numbered to follow the boundaries between adjacent pairs: the characteristics of these boundaries are collected in Table 2. For the S85 sample, 30 such boundaries have been followed, for which the misorientation (θ), boundary displacement (L) and curvature have been extracted from the EBSD maps, whereas the dislocation density gradient has been calculated with the identified constitutive law. These calculations show a higher stored energy ($E_{\Delta\rho}$) for the <111> grains in which multiple slip occurs. The superposition of the inverse pole figure mapping before and after grain growth shows that these most deformed grains tend to shrink while the surrounding grains expand.

A similar examination is performed on the S90 sample. The number of examined boundaries in this case is 50. It is observed in Figures 8 and 9 that (i) the grains are globally smaller than in the S85 sample; (ii) the orientations are more on the <100> side of the reference triangle (only a few <111> oriented grains are present); it is also seen in Table 2 that the boundary curvatures are more marked and the difference of dislocation densities are in contrast much weaker than for the S85 sample. The grains with the largest growth are those (of white color as grain 70 or 116) oriented in the center of the reference triangle. About half of the collected data are reported in Table 2 to save room.

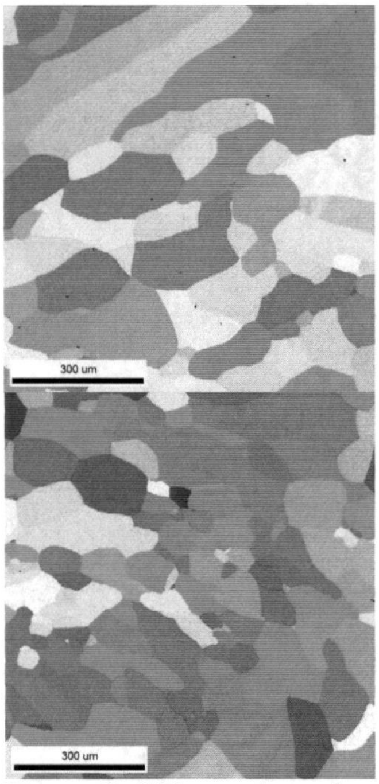

Figure 8. S90 sample surface after 90% rolling + annealing.

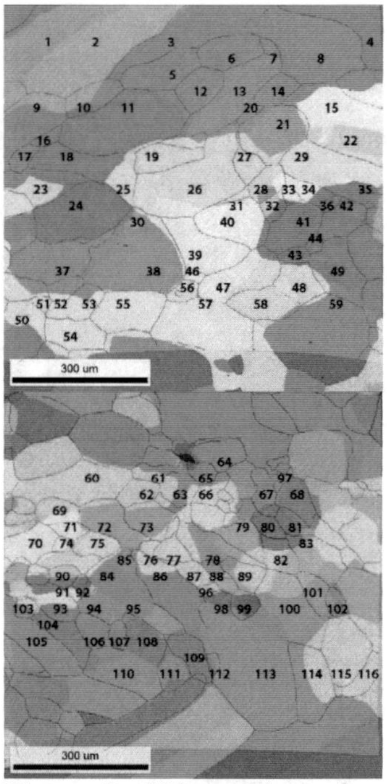

Figure 9. S90 sample surface after 3% strain + annealing with grain contour of figure 8.

5. Discussion

By convention, a grain boundary displacement from a grain i to a grain j is always counted positive. Both the related curvature (the E_c energy contribution) as well as the related differential of the dislocation density (the $E_{\Delta\rho}$ energy contribution) can either be positive or negative which yields four possibilities. Since the grain boundary motion is expected in the direction where the energy $P = \alpha E_c + \beta E_{\Delta\rho}$ is positive, only three of the four possibilities can a priori be associated with GB motion. If the migration rate (or distance) is proportional to this energy, isovalues $\alpha E_c + \beta E_{\Delta\rho} = cst$ are straight lines parallel to a line $\alpha E_c + \beta E_{\Delta\rho} = 0$ passing through the origin of a $(E_c, E_{\Delta\rho})$ plot. It is also expected that the slope of these lines $E_{\Delta\rho}/E_c = -\alpha/\beta$, is negative.

The « migration diagram » (an approximate of a velocity map) shown in figure 10 collects all the examined grain boundaries for the two analyzed samples on a $(E_c, E_{\Delta\rho})$ plot: each point in this diagram has been given a dimension proportional to the measured boundary displacement ΔL after a same time interval Δt.

This displacement would be an estimate of the velocity V if the displacement rate was constant, which is not the case. Nevertheless, it is clearly observed at first when plotting this displacement in terms of the two energy contributions, that there are no points in the negative "forbidden sector" ($E_c < 0, E_{\Delta\rho} < 0$) as expected. It is further observed in the limits of the data uncertainty and dispersion, that: (i) most of the points are is the positive sector; (ii) the points with a large radius (most mobile boundaries) are farther from the diagram origin than the points with a small radius (weakly mobile boundaries); (iii) the dislocation density effect seems less marked than the curvature effect, at least for the S90 material for which the more isotropic distribution of grain orientations reduces the dislocation density gradient and the grain size is much smaller leading thus to higher curvatures.

We can also note, that if we try to plot a straight line which would represent the P = 0 limit and place all measured points on the side of a positive driving force, only a few points remain below, and are thus associated with an overall negative driving force. These few cases need to be further examined in order to see if this is due only to data dispersion.

Table 3: Characteristic of the examined GBs for both samples.

S85					S90				
Boundary	θ	L(μm)	Δρ/ρ0	Ec(J/m³)	Boundary	θ	L(μm)	Δρ/ρ0	Ec(J/m³)
$\mathbf{B}_{16,17}$	4	0	7.74E+02	9.43E+02	$\mathbf{B}_{97,67}$	12	5	1.85E+02	-1.10E+04
$\mathbf{B}_{10,31}$	43	0	1.94E+02	3.13E+03	$\mathbf{B}_{100,101}$	6	6	1.64E+02	8.29E+03
$\mathbf{B}_{19,22}$	9	3	9.38E+02	5.89E+02	$\mathbf{B}_{97,68}$	9	6	4.58E+01	9.07E+03
$\mathbf{B}_{2,17}$	59	7	1.18E+03	2.59E+03	$\mathbf{B}_{13,6}$	47	10	1.26E+01	7.56E+02
$\mathbf{B}_{28,30}$	10	8	1.23E+03	1.51E+03	$\mathbf{B}_{15,21}$	53	10.5	3.07E+01	1.26E+04
$\mathbf{B}_{36,9}$	50	17	-1.74E+03	-1.01E+03	$\mathbf{B}_{62,73}$	51	11	-2.87E+02	6.21E+03
$\mathbf{B}_{5,30}$	53	18	3.28E+02	2.60E+03	$\mathbf{B}_{109,112}$	14	12	-2.93E+02	1.33E+04
$\mathbf{B}_{30,39}$	12.5	18	6.82E+01	6.72E+03	$\mathbf{B}_{13,14}$	50	43	4.21E+01	1.26E+04
$\mathbf{B}_{8,5}$	7	26	5.66E+01	-2.69E+03	$\mathbf{B}_{116,114}$	43	44	6.38E+01	2.78E+03
$\mathbf{B}_{5,43}$	55	35	1.49E+03	-4.32E+03	$\mathbf{B}_{64,65}$	54	45	-3.48E+01	1.89E+03
$\mathbf{B}_{9,33}$	56	36	5.72E+02	-2.82E+03	$\mathbf{B}_{31,26}$	45	46	-4.21E+01	3.70E+03
$\mathbf{B}_{4,42}$	45	39	1.76E+03	5.18E+03	$\mathbf{B}_{54,52}$	41	50	-7.61E+01	1.06E+04
$\mathbf{B}_{5,29}$	46	45	3.32E+02	4.97E+03	$\mathbf{B}_{3,5}$	59	60	-8.40E+01	4.57E+03
$\mathbf{B}_{9,30}$	55	50	3.27E+02	3.59E+03	$\mathbf{B}_{70,71}$	46	64	-2.11E+02	6.87E+03
$\mathbf{B}_{4,18}$	46	60	2.32E+03	2.25E+03	$\mathbf{B}_{3,10}$	35	67	3.81E+01	1.70E+04
$\mathbf{B}_{2,41}$	51	60	5.82E+02	6.22E+03	$\mathbf{B}_{19,26}$	40	67	-1.92E+02	4.86E+03
$\mathbf{B}_{39,10}$	47	60	-1.20E+03	1.80E+03	$\mathbf{B}_{76,77}$	15	73	3.35E+02	1.79E+04
$\mathbf{B}_{1,15}$	60	65	6.54E+02	3.29E+03	$\mathbf{B}_{116,115}$	52	112	1.16E+02	2.13E+04

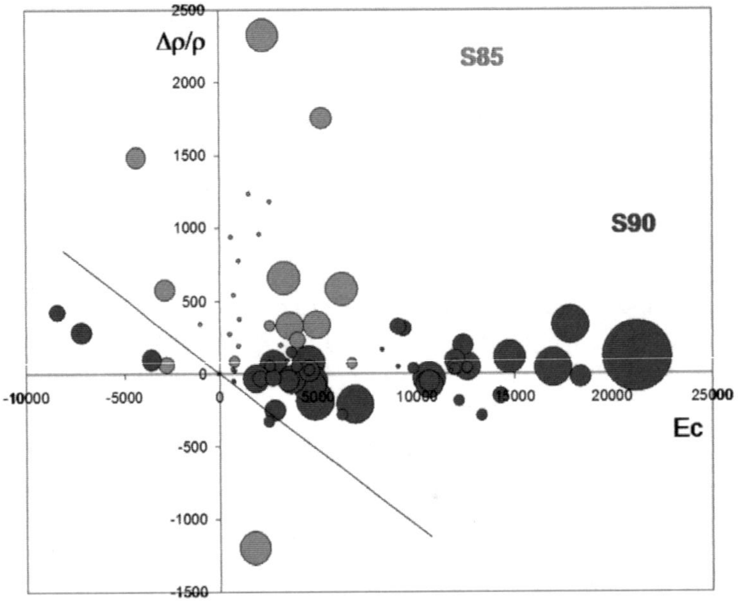

Figure 10. Migration diagram for the grain boundaries examined in the aluminium samples S85 and S90.

6. Conclusions

The performed SEM in situ examination of grain boundary migrations in aluminium polycrystals has been combined with numerical simulation of plastic straining in order to obtain boundary curvature measurements and dislocation density estimates. The method has been shown capable to yield a "migration diagram" from which the relative contributions to the boundary migration driving force, i-e the curvature and density contributions can be compared. For more precision, it appears form these results that with an increased number of examined boundaries a quantitative analysis of the data is at hand. This work is on progress.

References

[1] Humphreys F J and Hatherly M 2004(Elsevier)
[2] Lücke K, Rixen R and Senna M 1976 *Acta Metallurgica* **24** 103–10
[3] Gottstein G and Shvindlerman L S 1992 *Scripta Metallurgica et Materialia* **27** 1515–20
[4] Humphreys F J and Ferry M 1997 *Materials Science and Technology* **13** 85–90
[5] Huang Y and Humphreys F J 1999 *Acta Materialia* **47** 2259–68
[6] Ritter R 1999 (Elsevier)
[7] Murr L E 1975 (United States: Addison-Wesley Publishing Company)
[8] Tabourot L, Fivel M and Rauch E 1997 *Materials Science and Engineering: A* **234-236** 639–42
[9] Bacroix B and Brenner R 2012 *Computational Materials Science* **54** 97–100
[10] Hosford W, Fleischer R and Backofen W 1960 *Acta Metallurgica* **8** 187–99

Recovery and recrystallisation in mechanically alloyed and annealed, legacy, FeCrAlY ODS alloy precursor powders

K Dawson, A Rao, G J Tatlock and A R Jones
Centre for Materials and Structures, School of Engineering, University of Liverpool, Liverpool, L69 3GH, UK

E-mail: k.dawson@liverpool.ac.uk

Abstract. This study presents findings related to the recrystallisation behaviour in Mechanically Alloyed (MA) and annealed powders of legacy commercial Oxide Dispersion Strengthened (ODS) FeCrAl alloys PM2000, MA956 and ODM751. Annealing of as-MA ODS alloy powders at temperatures ≥ 800 °C induced primary recrystallisation. The volume fraction (V_f) recrystallised increased with higher annealing temperatures in the range studied (~800-1050 °C). However, low temperature (650 °C) recovery reduced the subsequent kinetics of recrystallisation in PM2000 alloy. Transmission Electron Microscopy (TEM) analysis of annealed PM2000 and MA956 alloy powders indicates that precipitation of nano-particulate Y-Al-O phases begins at temperatures as low as 650 °C and microstructural changes during annealing of ODS powders involved interactions between nano-particle formation and recovery/recrystallisation processes. High number densities ($N_V > 10^{23}$ m^{-3}) of coherent nano-precipitates were identified in both recovered and recrystallised regions of powder particles. These formed over a range of temperatures used in the consolidation processing of ODS alloys. The orientation relationship between nano-particles and the matrix was identical in both recovered and recrystallised grains, indicating that particles were dissolved at recrystallising interfaces and subsequently reprecipitated. Examination and comparison of as-MA and annealed powder specimens suggests that nuances in the manufacturing of these three, nominally similar, alloys leads to differences in recovery/recrystallisation behaviour, which may influence microstructure and, ultimately, properties in the final product form.

1. Introduction

The production of oxide dispersion strengthened alloys manufactured by mechanical alloying of metallic powders to incorporate powdered yttria was first demonstrated by Benjamin in 1970 [1]. The MA process creates an intimate mixing of constituents typically comprising either blended elemental powders or pre-alloyed metal powders and oxide powders (usually Y_2O_3). The alloy and oxide powders are loaded into milling containers along with milling media (usually steel balls) and during the MA process powders trapped between colliding balls are repeatedly flattening, cold welded, fractured and re-welded. When the MA process is optimised elemental homogenisation can be approached and each MA powder particle should have an identical composition [2]. During the MA process true strains in the region of 9 may be imparted to the powder [3] and high dislocation densities and nano-crystalline structures are produced [4]. Once milled, the MA powders are further processed to produce a consolidated product. During these subsequent steps the powders, which are heavily deformed, undergo thermal degassing treatments, then heating cycles in preparation for densification, typically by extrusion or hot isostatic pressing (HIP). In consolidated form the alloys, depending on composition, contain high number densities (10^{20} m^{-3}<N_V<10^{24} m^{-3}) of oxide particles and a fine grained but, usually, highly dislocated microstructure.

These materials were designed for use at high temperatures where good creep strength and oxidation resistance are required. The creep performance of this group of alloys can in part be attributed to the dispersion of nano-oxides but is also further enhanced by the generation of high aspect ratio grains formed during secondary recrystallisation annealing treatments which are performed at high homologous temperatures (approximately 0.9 T_M). Oxidation resistance is afforded by an adherent alumina scale which forms due to the necessary addition of approximately 10 at% aluminium. By making such aluminium additions, the oxide dispersion type and number density is significantly affected. In alloys which do not contain aluminium but are made with additions of yttria and titanium, high number densities (>10^{23} m^{-3}) of, typically, 2 nm yttrium-titanium oxides precipitate during consolidation. However, in the presence of Al, Ti is displaced from the oxide system and relatively coarse (ca. 20 nm) yttrium-aluminium oxides are formed in number densities ca. 10^{21} m^{-3}.

A recent study [5] has shown that during annealing treatments of PM2000 MA powders, at 650 °C, nano-particles approximately 2 nm in diameter are precipitated within a recovered, fine grained, matrix. It was also shown that, during annealing treatments made at higher temperatures (850 °C and 950 °C), the microstructure evolves from the wrought micro-forged structure to a mixed recovered/recrystallised grain structure. It was demonstrated that high number densities of nano-particulates were present in both the recovered grains and recrystallised grains and that an identical orientation relationship existed between the nano-particles and the recovered and recrystallised matrices. In addition, a particle free zone formed in recrystallised grains adjacent to the migrating recrystallising boundary. Based on these findings it was evident that the nano-particles formed during recovery were first dissolved by mobile recrystallising interfaces and then were re-precipitated in the recrystallised grains.

This study builds on our previous work and compares microstructural evolution in annealed, legacy MA precursor powders of ODM751, MA956 and PM2000 ODS alloys.

2. Experimental

The nominal compositions of the three alloy powders studied are presented in Table 1. Powders were annealed for 1 hour over a range of temperatures between 800 °C and 1050 °C. In a separate experiment, powders were first subjected to recovery anneals, for various durations at 650 °C, prior to 1 hour anneals over the temperature range 800 to 1050 °C.

As-MA and annealed powders were mounted in conductive resin and prepared, by standard metallographic techniques, for observations in a scanning electron microscope (SEM). A final polish was performed using a 40 nm colloidal silica suspension; this preparation step enabled backscattered electron channelling contrast images to be recorded. The volume fraction of recrystallised material was measured from electron micrographs using the mean linear line intercept method.

Table 1. Nominal compositions (wt%) of ODM751, MA956 and PM2000.

Alloy	Manufacturer	Cr	Al	Ti	Mo	Y_2O_3	Fe
ODM751	Dour Metal S.A.	16	4.5	0.6	1.5	0.5	Bal
MA956	Special Metals Corp.	20	4.5	0.4	_	0.5	Bal
PM2000	Plansee GmbH	20	5.5	0.3	_	0.5	Bal

Mechanically alloyed and annealed MA956 and PM2000 powder particles were prepared for TEM/STEM analysis using the method described by Dawson and Tatlock [6]. Further specimens were prepared from ODM751, by the FIB lift-out method [7], in an FEI Helios 600i FIB instrument. TEM/STEM analysis was performed in a probe-side aberration-corrected JEOL 2100FC_S microscope operated at 200kV fitted with an EDAX windowless energy dispersive x-ray spectrometer (EDS).

3. Results

3.1. As-Mechanically Alloyed Powder

By far the most significant differences in as-MA microstructures could be observed between the ODM751 powder (figure 1a) and the MA956 (figure 1b) and PM2000 (figure 1c) powders. The ODM751 powder was inhomogeneous, containing large (<5 μm diameter) aluminium-rich particles, chromium-rich particles and molybdenum rich areas. It is likely that these un-mixed alloy components were present due to a processing route that employed elemental powders mixed with pre-alloyed powders to achieve target alloy compositions [8]. In addition to the heterogeneous distribution of elements, the scale of the as-MA wrought microstructure in ODM751 was not as refined as that displayed in either PM2000 or MA956 powders. Areas as large as 10 μm in diameter composed of micrometre sized grains were not uncommon within ODM751 powder particles. Also apparent was the variation in microstructure and homogeneity of mixing from particle to particle within ODM751 powder; this was, again, far more obvious in ODM751 powder than in either MA956 (figure 1b) or PM2000 (figure 1c).

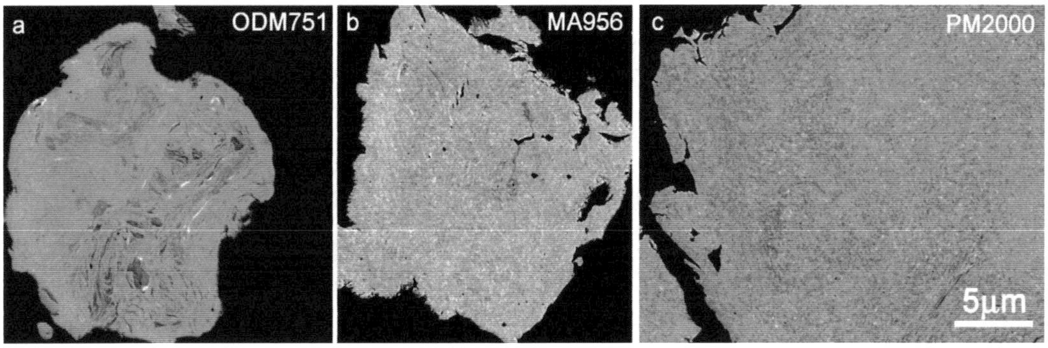

Figure 1. Backscattered electron images showing the wrought microstructures of as-MA (a) ODM751. (b) MA956 and (c) PM2000

SEM analysis comparing the wrought as-MA microstructures of MA956 and PM2000 powders revealed only subtle differences. Occasional bands of Cr-Al and Fe enriched materials, exhibited as swathes of unmixed alloy, were entrained along the cold welded fold lines of the micro-wrought MA microstructure in the MA956 powder. In addition to the swathe-like structures, which measured micrometres in length, numerous contaminant particles were also seen. The most common particles were titanium carbonitrides (Ti(C,N)), measured in the size range 100 to 500 nm; but silica and alumina particles were also observed. Occasional yttrium-rich particles were also identified but such particles were only a few 10s of nm in diameter. By comparison, PM2000 powder displayed a relatively homogeneous deformed grain structure, revealing grain diameters within the range 20 to 100 nm surrounded by matrix that displayed no definable structure (figure 2a). SEM observations, combined with EDS results, indicated there were fewer contaminant particles such as Ti(C,N) in the PM2000 powder. Figures 2 b and c show different microstructures observed in as-MA ODM751. Many areas displayed well defined grains, often distorted or elongated (figure 2b) whereas highly disordered regions (figure 2c) were similar in appearance to the milled structures of MA956 and PM2000 powders.

Transmission electron microscopy studies, which included large area selected area diffraction analysis, did not reveal evidence of unmixed yttria in as-MA PM2000; all patterns were consistent with the BCC ferrite structure.

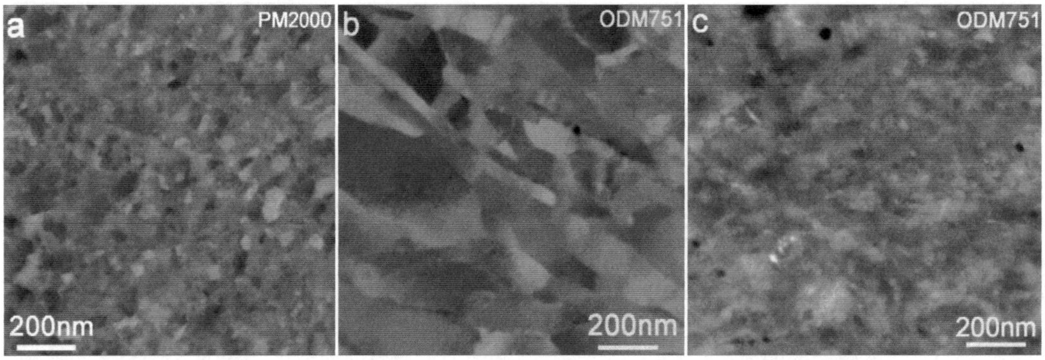

Figure 2. High resolution backscattered electron channelling contrast image showing the wrought nanostructure of (a) as-MA PM2000. Figures b and c show the varied microstructure observed in as-MA ODM751powders.

3.2. Evolution of recovered and annealed microstructures

The annealing of MA956 and ODM751 alloy powders, at 950 °C for 1 hour, produced both general recrystallisation and bands of finer recrystallised grains (figures 3a and 3b). PM2000 powders, on the other hand, annealed under identical conditions, showed a propensity to form large recrystallised grains whilst retaining areas of fine-grained recovered alloy (figure 3c). It was evident that each of the nominally similar legacy commercial ODS alloy powders studied exhibited a specific pattern of recrystallisation behaviour with differences apparent in both the morphology of microstructural evolution and process kinetics.

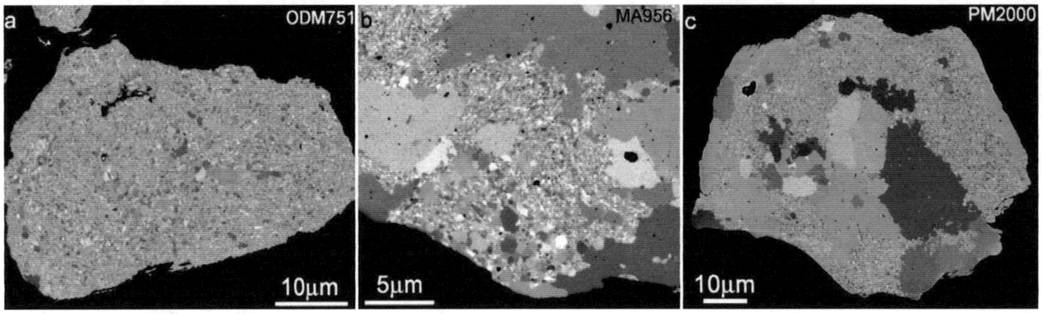

Figure 3. BSE images of annealed (1 hour at 950 °C) ODS alloy precursor powders (a) ODM751, (b) MA956 and (c) PM2000

In the annealed condition, ODM751 displayed regions of recovered fine grained material and areas which had recrystallised. Calculating the volume fraction of recrystallised material in ODM751 proved difficult as there was an overlap in the grain size distributions of the two regions. TEM analysis of 850 °C annealed ODM751 powder also revealed occasional small diameter recrystallised grains within fine grained recovered regions.

In addition to difficulties associated with differentiation between fine recrystallised and recovered ODM751 grain structures, large areas of coarse grained material were apparent in the as-MA powder. Moreover, considerable variations in the V_f of coarse and fine-grained material were observed from one particle to another. This is consistent with the inhomogeneous nature of the as-MA ODM751 powders. Fine grained regions contained greater concentrations of sub-μm particles, a large proportion of which were sub 100 nm oxides, than were observed in recrystallised areas. Broad sweep EDS line-scans recorded across these un-recrystallised swathes and the findings of EDS point analyses suggest that many of the particles located in the recovered grains were Y and Al oxides.

Due to limited supply of raw materials, measurements of recrystallised volume fractions in the MA956 were limited to 2 annealing temperatures. However, preliminary results suggest that the recrystallisation kinetics of MA956 powders were similar to those displayed by PM2000 (figure 4a). In contrast, annealing for 1 hour at 850 °C did not induce measurable recrystallisation in the ODM 751 powder. After annealing at 950 °C and 1050 °C recrystallisation was observed in ODM751 but the volume fraction measured was low compared to both MA956 and PM2000 alloy powders (figure 4a).

In a separate experiment, PM2000 powders were subjected to recovery anneals at 650 °C for 2 or 5 hours before the application of a second, 1 hour, heat treatment performed at higher temperatures. Although the kinetics of recrystallisation were reduced as a result of a 5 hour anneal at 650 °C prior to a 1 hour anneal at 850 °C, the effects became less noticeable when higher temperature recrystallisation treatments were performed. The duration of the recovery anneal also influenced recrystallisation kinetics e.g. a 2 hour anneal at 650 °C prior to a 1 hour anneal at 850 °C had little effect on the recrystallised volume fraction of powder (figure 4b).

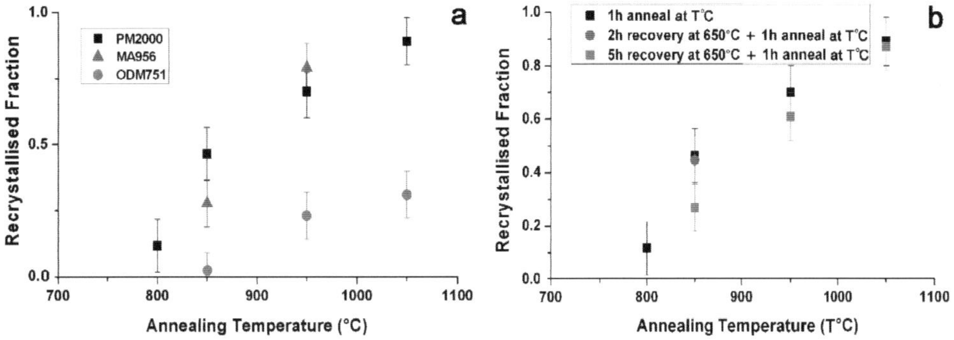

Figure 4. (a) Plot showing the recrystallised volume fraction as a function of isothermal annealing temperature and (b) a plot showing the effect of 650 °C recovery anneals of different durations on recrystallised volume fraction in PM2000 powders subsequently annealed for 1 hour at 850 °C.

TEM/STEM analysis of 850 °C and 950 °C annealed MA956 powders enabled examination of the mixed recrystallised and recovered microstructures in greater detail. Fine-grained material contained high number densities of dislocations and grains were often elongated; a legacy of the 'micro-forging' action of the MA process. Recrystallised grains, in contrast, contained negligible numbers of dislocations, but groups/stringers of particles which were aligned parallel to neighbouring elongated fine grains were observed (figure 5a); again, this was assumed to be carried over from the MA process.

In addition to the aforementioned aligned and relatively coarse particles, high number densities ($>10^{23}$ m^{-3}) of cube shaped nano-particles precipitated homogeneously in both fine grained and recrystallised regions of annealed MA956 (figure 5b). The nano-particles, which were enriched in yttrium, aluminium and oxygen, displayed identical growth habits, i.e. facets parallel to the ferrite {200} planes, to nano-particles which have been identified in PM2000 powders with similar annealing histories [5]. The same orientation relationship displayed by nano-particles and the ferrite matrices in annealed PM2000 powders were observed in both recrystallised and recovered grains of MA956 alloy powder.

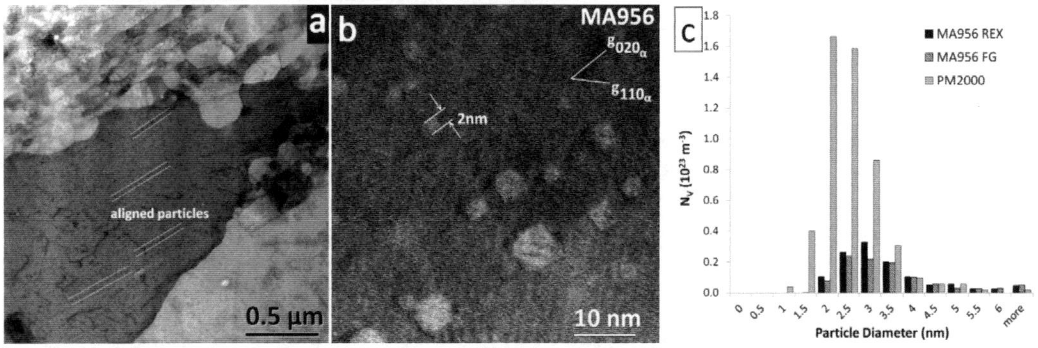

Figure 5. (a) BF-STEM image of recrystallising interfaces in annealed MA956 powder, (b) nano-particulates precpitated in MA956 powder (1 hour at 950 °C) and (c) a plot comparing the size distributions and number densities of nano-particles measured in MA956 with those observed in PM2000 powders after the same heat treatment.

The number densities and size distributions of precipitates measured in both recovered and recrystallised grains in annealed MA956 powder could not be differentiated; hence volume fractions of nano-particles in these two regions could be considered essentially equal. However, slight differences were recorded in precipitate populations measured between MA956 and PM2000. The mean diameter of nano-particles measured in powders annealed for 1h at 950 °C were 2.3 nm (±0.8 nm) and 3.3 nm (±1.6 nm) in PM2000 and MA956, respectively; and the particle number density was greater in the PM2000 alloy powders (figure 5c).

4. Discussion

The microstructures of and chemical distributions measured in as-MA ODM751, MA956 and PM2000 powders differed significantly. This suggests that the mechanical alloying processes and procedures adopted by the three manufacturers produced results which could be ranked in the following order: PM2000 powders appeared homogeneous, hence best mixed. MA956 powders were similar to PM2000 but Ti(C,N) and unmixed swathes of material revealed that slightly less homogeneous alloying had been achieved in the MA956 powder. In contrast, ODM751 powders appeared to be less thoroughly blended, with significant variations observed from one particle to another. It is likely that large particle-free, recrystallized grains that have been observed in consolidated ODM751 [9] may be the consequence of incorporation of such inhomogeneously milled powders. The coarse particles and poor distributions in ODM751 reported by Foxman et al. [10], were likely to have been caused by the incomplete dissolution of yttria, during the MA process, which has been observed in the present study. However, it is interesting to note that consolidated ODM751 offers superior creep strength to MA956 and matches the performance of PM2000 [8, 11, 12]. A possible explanation has been attributed to the addition of molybdenum made to ODM751[3] but it would be of significant interest to determine what level of inhomogeneities can be tolerated after the MA process whilst retaining acceptable levels of alloy performance.

SEM analysis of the wrought as-MA microstructures in MA956 and PM2000 powders revealed subtle differences in their deformation structures. A greater grain size distribution observed in the MA956 powder indicated a less complete/homogeneous mixing had been achieved during MA; as such, one might assume greater heterogeneities in stored energy would exist and play a role in the recrystallisation behaviour of these powders. In addition, more obvious differences included the numerous Ti(C,N) particles and occasional swathes of Cr-Al and Fe-enriched material observed in MA956; these too would be expected to affect the local kinetics of recrystallisation.

After annealing as-MA powders for 1 hour at 950 °C, elemental distributions within the FeCrAl and FeCrAlMo matrices were homogenised. However, Ti(C,N) particles were identified in MA956 and swathes of alloy showing enhanced aluminium-yttrium concentrations, probably in the form of oxides, were observed in ODM751. This indicates that intermediate temperature annealing treatments are capable of homogenising some elemental distributions but fail to dissolve more stable compounds.

The recrystallisation kinetics of ODM751 were slower than those of either PM2000 or MA956 at temperatures between 850-1050 °C. The difference in recrystallisation behaviour probably involved a combination of factors, including: a lower overall level of stored energy, as evidenced from analysis of the as-MA powders; and relatively coarse oxide particles left in more highly deformed areas of ODM751 as a result of failure to drive yttria into solution during MA.

All powders that were annealed at 950 °C showed some evidence of recrystallisation; yet their microstructures differed greatly. PM2000 powders displayed a bimodal grain size distribution comprising fine recovered and large recrystallised grains. The grain size distribution in recovered regions was narrow, with recovered grains easily distinguished from the much larger recrystallised grains. MA956 powders, in the same condition, displayed a similar recovered grain size distribution but the size range of recrystallised grains was broader than in PM2000 powder. ODM751 powders annealed for 1 hour at 950 °C were also partially recrystallised but the microstructure was significantly different to the PM2000 and MA956 powders. Recrystallised grains were smaller than those observed in the Plansee and Special Metals powders and it was sometimes difficult to easily differentiate between recrystallised and recovered regions. However, concentrations of larger oxide particles, present due to incomplete MA, were a constant and clear signature in recovered regions.

As first demonstrated in PM2000 alloy powder [5], this study shows that recrystallisation in MA956 powder progresses by passage of recrystallizing interfaces through material in which high number densities of nano-particles have already precipitated. In addition, the existence of an identical orientation relationship between nano-particles and matrix in both regions suggests that these particles dissolve during recrystallisation and re-precipitate behind the recrystallising interface.

Recrystallisation in such heavily precipitated material probably occurs due to a combination of the high residual stored energy from MA, despite recovery, together with the dissolution of particles at recrystallising interfaces. Although clear evidence of precipitate free zones adjacent to recrystallising interfaces was not observed in MA956 powders, it has been observed in PM2000 powders [5] and in other ODS alloy systems [13]. As such, dissolving precipitates might not exercise full Zener pinning on grain boundaries but exert a lesser force associated with solute drag. In contrast, larger precipitates pin recrystallising interfaces, with clear evidence observed of bowing boundaries pinned by 10 nm oxides. This explanation is consistent with evidence presented here, where recrystallisation commences in the presence of 2 to 3 nm particles, which can be dissolved by migrating boundaries, yet recrystallisation was suppressed in ODM751 in regions of relatively coarse oxide particles.

5. Conclusions

The microstructure of as-MA ODM751 powder differs significantly from those of as-MA MA956 and PM2000. ODM751 powders were incompletely homogenised by mechanical alloying; this could be due to factors which might include the adoption of a different processing route, process time, or use of pre-alloyed or elemental powders as the initial feedstock.

Annealing at 950 °C induced partial recrystallisation in all as-MA powders. High number densities of Y-Al-O nano-particles were identified in both PM2000 and MA956 powders.

Under the same annealing conditions, the V_f recrystallised differs between the three alloy powders studied; but in all cases, once recrystallisation initiated, V_f increased with annealing temperature.

The recrystallisation kinetics in MA PM2000 powder were reduced by the application of an initial 650 °C recovery annealing treatment. The change in kinetics was sensitive to the duration of recovery anneal and the temperature of the subsequent recrystallisation treatment.

Acknowledgements

We are pleased to acknowledge the financial support of this work through the Advanced Research Materials (ARM) Programme, U.S. Department of Energy, Office of Fossil Energy, managed by U.T.-

Batelle, LLC, and the Engineering and Physical Sciences Research Council – Materials for Fusion and Fission Power – grant EP/H018921/1.

We would also like to express our gratitude to both Special Metals and Bohumil Kazimierzak (Dour Metal s.r.o.) for providing the mechanically alloyed MA956 and ODM751 powders respectively.

References

[1] Benjamin J S 1970 *Metall. Trans.* **1** pp 2943-2951
[2] Laurent-Brocq M, Legendre F, Mathon M H, Mascaro A, Poissonnet S, Radiguet B, Pareige P, Loyer M and Leseigneur O 2012 *Acta Mater.* **60** pp 7150-7159
[3] Bhadeshia H K D H 1997 *Mater. Sci. Eng. A* **223** pp 64-77
[4] Suryanarayana C 2001 *Prog. Mater. Sci.* **46** pp 1-184
[5] Dawson K, Haigh S J, Tatlock G J and Jones A R 2015 *J. Nucl. Mater.* **464** pp 200-209
[6] Dawson K, Tatlock G J 2015 *Micron* **74** pp 54-58
[7] Giannuzzi L A, Kempshall B W, Schwarz S M, Lomness J K, Prenitzer B I and Stevie F A 2005 FIB Lift-Out Specimen Preparation Techniques: *Introduction to Focused Ion Beams*, Eds. Giannuzzi L A and Stevie FA (Springer US) pp 201-228
[8] Kazimierzak B, Prignon M, Lecomte-Mertens C and Coutsouradis D 1990 Fe Base ODS Alloys with Improved Mechanical Strength: *Proc. Int. Symp. High Temperature Materials for Power Engineering (Liege)* pp 131-142
[9] Jaeger D M and Jones A R 1994 Dispersoid Distributions in Fe-based ODS alloys made by Mechanical Alloying: *Proc. Int. Symp. Materials for Advanced Power Engineering (Liege)* p 1507
[10] Foxman Z, Sobol O, Pinkas M, Landau A, Hähner P, Krsjak V and Meshi L 2012 *Metallogr. Microstruct. Anal.* **1** pp 158-164
[11] Kazimierzak B, Prignon M, Lecomte-Mertens C and Coutsouradis D 1990 *Anti-Corros Method M* **37** pp 4-9
[12] Hurst R and Hähner P 2010 Materials qualification testing for next generation nuclear reactors: *Proc. Int. Symp. Baltica VIII: Life Management and Maintenance for Power Plants, (Helsinki)* pp 324-350
[13] Naka S, Octor H, Bouchaud E and Khan T 1989 *Scripta Metall.* **23** pp 501-505

IOP Conf. Series: Materials Science and Engineering **89** (2015) 012021 doi:10.1088/1757-899X/89/1/012021

Effects of interface roughness on the annealing behaviour of laminated Ti-Al composite deformed by hot rolling

Y Du[1,2], G H Fan[1], T Yu[2], N Hansen[2], L Geng[1] and X Huang[2]
[1]School of Materials Science and Engineering, Harbin Institute of Technology, Harbin, 150001 Harbin, China
[2]Danish-Chinese Center for Nanometals, Section for Materials Science and Advanced Characterization, Department of Wind Energy, Risø Campus, Technical University of Denmark, DK-4000 Roskilde, Denmark

E-mail: yandua@dtu.dk

Abstract. A laminated Ti-Al composite has been fabricated by hot compaction and hot rolling of alternate layers of commercial purity Ti and Al sheets with a thickness of 200 μm. The hot compaction temperature was 500°C and in a following step the composite has been reduced 50% in thickness by hot rolling. The fully consolidated composite has been annealed at 300°C and 500°C for different length of time. As a result of the differences in crystal structure and mechanical properties between Ti and Al protrusions and retrusions formed at the interface. A heterogeneous interface has thereby been created. The heterogeneity affected the recovery kinetics of the aluminium phase which at 300°C was faster near the interface than in the middle of the Al layer. This effect of a heterogeneous interface is of relevance when optimizing the thermomechanical processing of the composite to obtain high strength and formability for application.

1. Introduction

Efforts have been and would continue to be made in order to develop strong and light metals in response to the energy shortage. Both lightweight aluminium and titanium alloys have been developed and applied in the automobile industry. Significant progress has also been made in the development of laminated metal composites (LMCs), which have excellent mechanical properties including strength, toughness, high impact and fracture resistance [1-2]. The interface plays an important role in LMCs and has been studied in recent years for example in roll bonded composites. In such studies experimental observations have been analyzed also including numerical simulation of evolving interfaces [3-5]. It has also been reported that recrystallization and abnormal grain growth can occur in an Al layer after hot deformation and during subsequent annealing [6]. It has also been observed that roughness can form during deformation of polycrystalline metallic materials with FCC structure or HCP structure due to heterogeneity of plastic deformation [7-9]. This phenomenon has also been observed at the interfaces of LMCs produced by rolling [4] and this kind of heterogeneity can cause structural differences during annealing of the LMC. Thus, an investigation of interface roughness effects on microstructure of LMCs during deformation and annealing is important, especially for controlling the microstructure by interface design and optimizing deformation and annealing conditions.

In this study of a LMC, commercial purity Al (1060) sheets and commercial purity Ti (TA1) sheets are selected as materials to produce laminated Ti-Al composites by hot rolling. The interface roughness generated during hot rolling is analyzed and its effect on microstructural evolution of Al in the Ti-Al laminate both after hot rolling and during subsequent annealing is examined.

2. Experimental methods

2.1. Materials

Commercial purity titanium TA1 sheets with a purity of 99.4 wt% (0.25 wt% Fe, 0.2 wt% O, 0.1 wt% C) purchased from Tianrui Non-ferrous Metal Co., Ltd. (Baoji, China) and commercial purity aluminium 1060 O sheets with a purity of 99.2 wt% (0.35 wt% Fe, 0.25 wt% Si) purchased from Northeast Light Alloy Co., Ltd. (Harbin, China) were selected as starting materials. The as-received Ti sheets had a recrystallized structure (preferred orientation <0001> parallel to the ND) with an average grain size of 50 μm while the as-received Al sheets had a slightly deformed structure (random orientation) with an average grain size of 100 μm. Both the TA1 and 1060 O Al sheets had a thickness of 200 μm. Small pieces with a size of 100 mm×100 mm×200 μm were cut from the as-received Ti and Al sheets to fabricate laminated Ti-Al composites.

2.2. Processing and annealing treatment

The laminated Ti-Al composites were fabricated by three steps as shown in figure 1(a). i) Sheet preparation: the Ti and Al sheets were firstly treated by 10 vol.% HF liquor and 10 vol.% NaOH liquor, respectively, to remove contaminants and the oxide layer from the surface. Then, the sheets were roughened with a steel brush, cleaned in an ultrasonic bath of acetone for 5 min, dried and stored in air for later use. ii) Diffusion bonding: the sheets were stacked alternately like a sandwich with Ti sheets at both surface layers. The sandwich was put into a graphite die for confinement and pressed under a pressure of 40 MPa at 500°C for 1 h in a vacuum furnace to obtain LMCs. iii) Hot rolling: the compressed LMCs were held at 500°C for 10 minutes before each rolling pass and rolled through 5 passes with a total thickness reduction of 50% from 6 mm to 3 mm. The rolling geometry is shown in figure 1(b). Thickness reduction of each pass were controlled to ensure a macro homogeneous rolling condition, i.e. the ratio of contact length between the roll and the sheet over the average thickness *L/h* was between 0.5 and 5. This ratio can be expressed by the relationship:

$$\frac{L}{h} = \frac{\sqrt{R(h_0 - h_1)}}{(h_0 + h_1)/2} \tag{1}$$

where h_0 is the thickness of sheet before each rolling pass, and h_1 is the thickness of sheet after each rolling pass. The average thickness of sheet h is the average of h_0 and h_1 and R is the radius of the roll.

The hot rolled LMCs were annealed at 300 °C for 10 min, 30 min and 1 hour, and at 500°C for 1 hour to examine the thermal behaviour of the LMCs.

2.3. Characterization

The microstructure and the texture of hot-rolled Ti-Al LMCs were characterized in a scanning electron microscope (SEM, Zeiss Supra 35) equipped with an automated HKL electron backscatter diffraction (EBSD) facility. The samples were prepared by electropolishing and ion etching using a cross section polisher (JEOL IB-09020CP) to reveal various aspects of microstructures of the LMCs. Two step sizes of 2 μm and 0.1 μm were used for coarse and fine EBSD scans, respectively. Hardness of the Al layers in different conditions were measured to analyze the recovery and recrystallization behaviour of the aluminium phase in the LMCs. All tests were done in the ND-TD section where ND and TD stand for normal and transverse directions, respectively. The hardness measurement points were centered to the region 20 μm near the interface in the Al and in the middle of the Al layers. For each test, 5 selected points were measured.

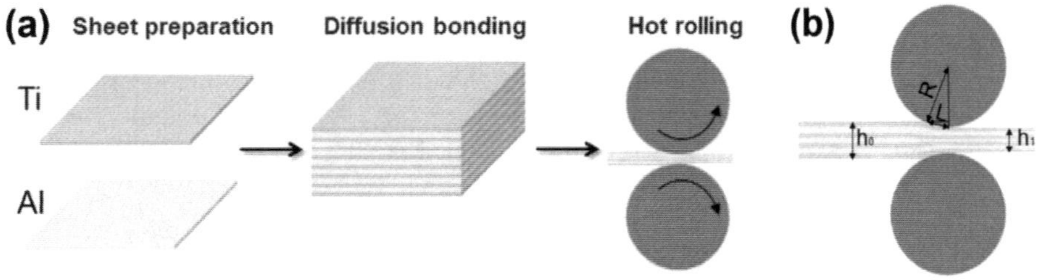

Figure 1. (a) Schematic illustration of three step processing of laminated Ti-Al composite; (b) rolling geometry.

3. Results and Discussion

3.1. Microstructure of deformed laminated Ti-Al composite

The microstructure of the LMCs before and after hot rolling were examined by SEM as shown in figure 2(a) and figure 2(b). It is clear that the interface of the LMC is flat before hot rolling and it turns rough after hot rolling. The microstructure of hot-rolled laminated Ti-Al was also examined by EBSD as shown in figure 3(a). It shows interface roughening and the formation of protrusions and retrusions at interfaces. The Al layers showed a hot deformed structure, which was confirmed by TEM observations. The Ti layer contains deformed grains with an average diameter of about 10 μm. Distributions of grain/subgrain sizes in Al and grain sizes in Ti layers are presented in figure 3(b) and 3(c), respectively. Since laminated Ti-Al composites were held at 500 °C for 10 min before each pass of rolling, the microstructure in the Al layers before rolling may be recrystallized. After the final pass, the hot-rolled Ti-Al was still kept at high temperatures, which provided energy for static recrystallization in Al. Several recrystallized Al grains with an average diameter of 20 μm were found near the interfaces as marked by arrows in figure 3(a). Most of these recrystallized grains are near the interface characterized by large protrusions. Similar microstructures were observed by EBSD in other Al layers in the ND-TD section of the hot-rolled laminated Ti-Al composite. For the Ti layers, the temperature of 500°C did not reach the recrystallization temperature. Thus, as shown in figure 3(c), the Ti layers contain deformed grains with a diameter ranging from 5 μm to 22 μm.

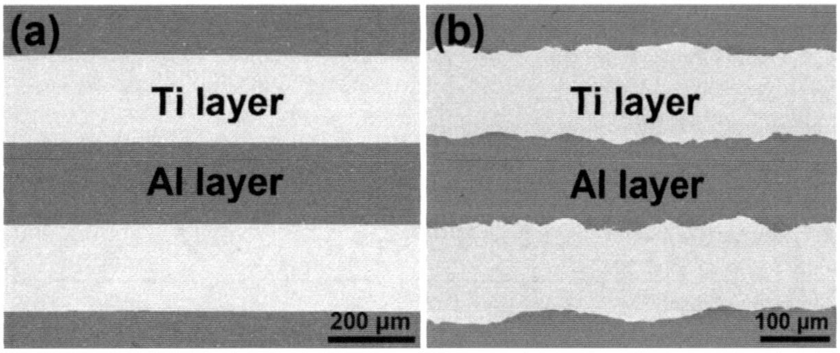

Figure 2. Microstructure in the ND-TD section of a LMC sample before and after hot rolling. (a) flat interface before hot rolling; (b) rough interface after hot rolling.

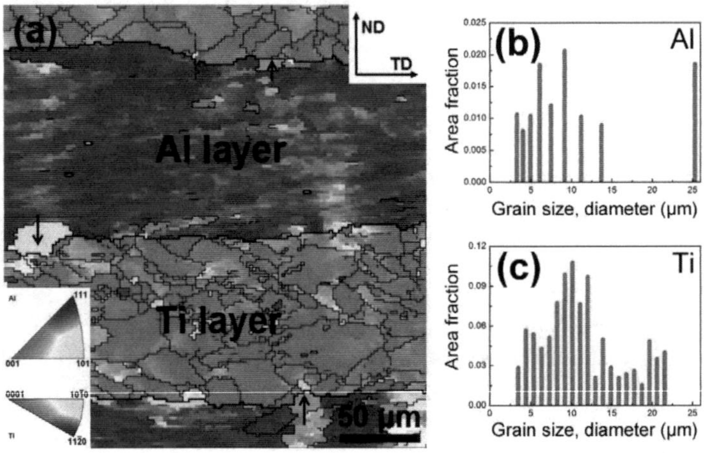

Figure 3. Microstructure in the ND-TD section of hot-rolled laminated Ti-Al. (a) EBSD map (inverse pole figure of the ND) of the LMC, showing recrystallized Al grains marked by arrows formed at the interface; (b) and (c) grain size distribution of Al and Ti, respectively.

During hot rolling both Al and Ti deformed by dislocation glide but their plastic deformation differ. Al with an FCC crystal structure can deform on many slip systems and thereby accommodate the rolling strain homogenously. In contrast, Ti with an HCP crystal structure has fewer available slip systems and the plastic deformation may be fairly heterogeneous. As a result protrusions and retrusions can form at the Ti-Al interfaces. The protrusions/retrusions create variations in plastic strain and stored energy along the interface and thereby in the potential for formation of recrystallization nuclei where a few are marked by arrows in figure 3(a).

3.2. Microstructure of annealed laminated Ti-Al composite

The microstructure of as-annealed laminated Ti-Al composites in the ND-TD section was examined by EBSD as shown in figure 4, where the colours are based on the crystallographic orientation of the ND as indicated in the inset. After annealing at 500 °C for 60 min as shown in figure 4(a), Ti still shows very strong preferred orientations parallel to the ND which is similar to the hot-rolled structure. The roughness of the interface between Ti and Al layers still exists. The Al layers contain big recrystallized grains with an average diameter of about 100 μm. Distribution of recrystallized Al grains compared to the hot-rolled state is shown in table 1. Both grain size and area fraction of recrystallized Al grains show very large increase after annealing at 500 °C for 60 min. The Ti layers still contain recovered grains with an average diameter of about 20 μm.

Table 1. Distribution of recrystallized grains in Al layers.

	Position	Area fraction (%)	Average grain size (μm)
Hot rolled	interface	0.74	28.5
500 °C, 60 min	whole layer	100	109

Figures 4(b)-(d) show examples of EBSD maps viewed from the ND-TD section of Al layers in LMCs annealed at 300 °C for 10 min, 30min and 60 min, respectively. As shown in figure 4(b)-(d), protrusions and retrusions are seen very clearly at the interfaces and more subgrains can be found at the protrusions and retrusions than at the flat interface. Also, there are more subgrains forming near the interface than in the middle of the Al layers. Similar deformed structures as in the hot-rolled state can still be found in the middle of the Al layers after annealing at 300 °C for 10 min and 30min as shown in figures 4(b) and 4(c), respectively. Subgrain coarsening is also very clear when comparing the microstructure between samples annealed at 300 °C for different times.

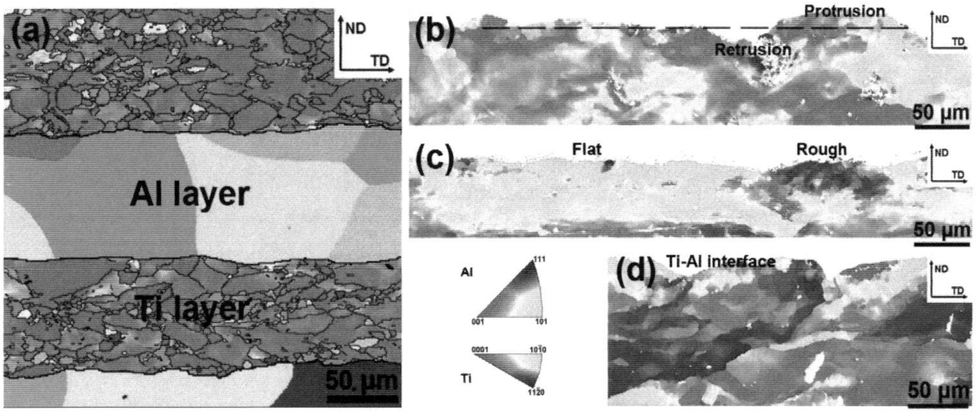

Figure 4. Microstructure in the ND-TD section of annealed laminated Ti-Al by EBSD (inverse pole figure of the ND). (a) EBSD map of the LMC annealed at 500°C for 60 min, showing large fully recrystallized Al grains; (b), (c) and (d) are EBSD maps of Al layers in LMCs annealed at 300°C for 10 min, 30 min and 60 min, respectively, showing that more subgrains had formed at the protrusions and retrusions than in the middle of the Al layer.

Statistical results of Al subgrains in the Al layer of laminated Ti-Al annealed at 300 °C for 10 min, 30min and 60 min are shown in figure 5. For each annealing treatment state, 4-5 EBSD maps including 8-10 Al layers were measured. Both the average grain size and number of equiaxed Al subgrains forming in every 100 µm length during annealing corresponding to the distance between interfaces in laminated Ti-Al were analyzed. Since there are protrusions and retrusions at the interfaces, the average interface line is defined as the average level of highest protrusion and lowest retrusion in each map scanned by EBSD. In general, the number of equiaxed Al subgrains in every 100 µm length of the interface increases with increasing annealing time. It is also observed that many subgrains expands from the interfaces to the middle of the Al layer with increasing time and for each annealing condition the density of subgrains decreases as the distance from the interface increases. For the 10 min and 30 min conditions, the average size of subgrains at the interfaces is bigger than in the middle of the Al layers. Nevertheless, the average size of subgrains from the interface to the middle of the Al layer is nearly the same after annealing at 300 °C for 60 min.

Isothermal annealing treatments at 300 °C lead to microstructural coarsening and softening. The softening is by change in hardness as illustrated in figure 6 and it also shows the hardness of Al annealed at 500 °C for 60 min and in the deformed state (annealing time of 0 min) for comparison. Randomly selected points were measured at each state and both areas near the interface and in the middle of Al layers were tested. Hardness of both hot deformed Al and Al annealed at 500 °C for 60 min does not change much from the interface to the middle of the Al layers due to uniform hot deformation and fully recrystallized microstructure as shown in figure 3(a) and figure 4(a), respectively. Hardness of deformed Al shows highest value of 29 HV, and it decrease to the lowest value of 22 after annealing at 500 °C for 60 min. For annealing treatments at 300 °C, hardness of Al near the interface is lower than that in the middle of Al layers generally. Hardness difference between interface and middle areas of Al annealed at 300 °C first increases and then decreases with increasing annealing time. This is because Al starts recovery from the interface in hot deformed laminated Ti-Al structure and the effect of recovery homogenizes gradually the structure with increasing annealing time.

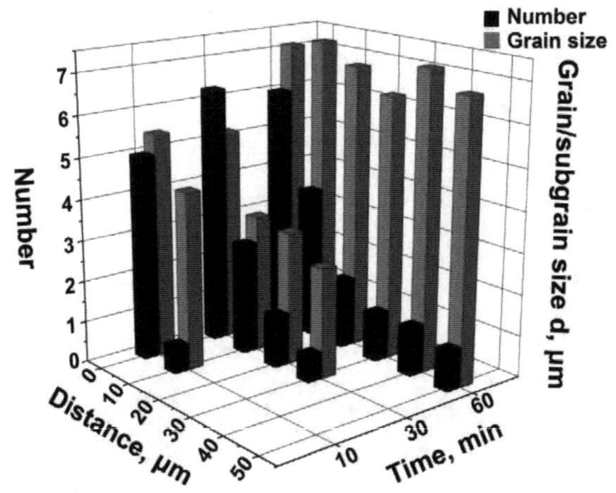

Figure 5. Statistical results of average grain/subgrain size and number in every 100 μm length of Al subgrains as a function of the distance from the interface in the LMC annealed at 10 min, 30min and 60 min at 300 °C according to the EBSD maps and measured in the ND-TD section of the samples.

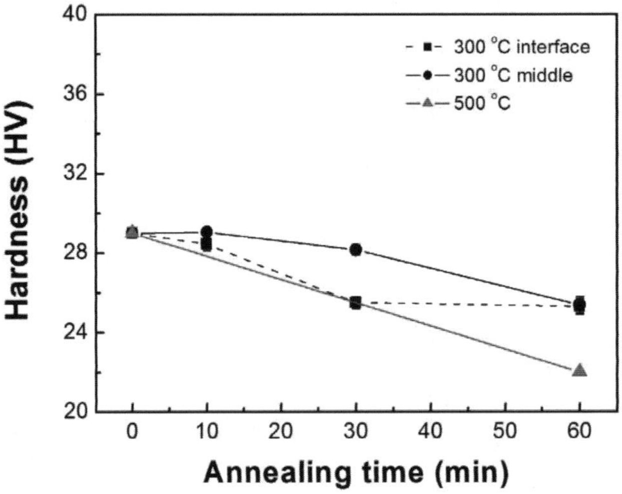

Figure 6. Change in hardness of Al with annealing condition in laminated Ti-Al composites (hardness of hot deformed Al is presented as annealing time of 0 min), showing that hardness in Al near the interface (black solid square) is lower than in the middle of the Al layer (black solid circle).

In an analysis of the thermal and mechanical behaviour of the laminated Ti-Al composite it must be taken into consideration the differences between Al and Ti when it comes to crystal structure and physical/mechanical/chemical properties. Some of these differences are illustrated in table 2. The difference in crystal structure between Al and Ti may give rise to the formation of protrusion/retrusion which is discussed in section 3.1. The difference in thermal expansion between Ti and Al may give rise to a thermal stress at the interface when the composite is cooled down from the hot compaction temperature. These may cause some elastic/plastic deformation near the interface. This effect may not be large due to the relatively low flow stress of aluminium. The difference in Young modulus between Al and Ti leads to differences in the elastic/plastic behaviour of the two phases and a formation of compressive stresses in Al. Also this effect is expected to be small. Finally there may be a chemical reaction between Ti and Al during processing and post process annealing as has been observed. The interaction between the different reactions at or near the interface is part of ongoing research.

Table 2. Materials parameters for the Ti and Al layers.

Layer	Elastic Modulus (GPa)	Shear Modulus (GPa)	Thickness (μm)	CTE (10^{-6}/°C)
Al	70	25	100	23
Ti	102	39	100	10.8

4. Conclusions

At present we conclude that the difference in plastic behaviour between Al and Ti due to their different crystal structures is the main cause of the observed heterogeneous deformation in Al near the interface. This difference results in a faster recovery in Al in the region near the interface than further away from the interface.

Our observations also show that the chosen Ti-Al system offers excellent possibilities for new research of scientific relevance. Such research will also guide a continued optimization of LMCs for future application for example in the car and aerospace industry.

Acknowledgments

The authors gratefully acknowledge the support from the Danish National Research Foundation (Grant No. DNRF86-5) and the National Natural Science Foundation of China (Grant No. 51261130091) to the Danish-Chinese Center for Nanometals, with which support this work has been performed. Author Y.D. also gratefully acknowledge the financial support from China Scholarship Council.

References

[1] Ortiz C, Boyce M C 2008 *Science* **319**: 1053-54
[2] Chaudhari G P, Acoff V 2009 *Compos. Sci. Technol.* **69** 1667-75
[3] Wang J, Beyerlein I J, Mara N A, Bhattacharyya D 2011 *Scr. Mater.* **64** 1083-86
[4] Lee S B, LeDonne J E, Lim S C V, Beyerlein I J, Rollett A D 2012 *Acta Mater.* **60** 1747-61
[5] Beyerlein I J, Mara N A, Carpenter J S 2013 *J. Mater. Res.* **28** 1799-1812
[6] Chang J, Takata K, Ichitani K, Taleff E M 2010 *Metal. Mater. Trans.* A **41** 1942-1953
[7] Becker R 1998 *Acta Mater.* **46** 1385-1401
[8] Wouters O, Vellinga W P, Tijum R V, Hosson J 2005 *Acta Mater.* **53** 4043-4050
[9] Stoudt M R, Levine L E, Creuziger A, Hubbard J B 2011 *Mater. Sci. Eng.* A **530** 107-116

Microstructure and hardness development in a copper-nickel diffusion gradient model system

L D L Duchstein[1,*], X Zhang[1] and N Hansen[1]

[1]Technical University of Denmark, Department of Wind Energy
Frederiksborgvej 399, Building 228, DK-4000 Roskilde, Denmark

*Email: ldld@dtu.dk

Abstract. Cu has been electrolytically coated with Ni and subsequently deformed by rotary swaging up to a strain of ε=2 to create a chemical gradient at the interface of the two elements. The extend of this chemical intermixing has been investigated through Energy Dispersive X-ray (EDX) spectroscopy in the Scanning and Transmission Electron Microscope (SEM and TEM). The depth, in which intermixing takes place, is about 1µm from the interface. Because of the uniform deformation, the structure does not get elongated but rather uniformly reduced in size. Microindentation hardness measurement shows a hardness increase from 120 to 135kp/mm^2 in the Cu phase with increasing strain. After annealing at 200°C for up to 4h the hardness first decreases, but raises above the value for the highly strained sample. The experimental findings are discussed with emphasis on surface mechanical alloying as a process of both scientific and technological interest.

1. Introduction

The modern world has a lot of requirements towards materials concerning strength, ductility, corrosion resistance, costs and often more like electrical conductivity or to be non-hazardous. Single unprocessed elements can often not fulfill these requirements, which is why new processing techniques, alloys and other material combinations are constantly developed to create hybrid materials [1-2]. This is an interesting way is to combine advanced processing techniques with the goal of at least partly alloying material together to make use of their traits but also creating a strong bonding between the two phases. Here we use the process of coating a metal onto another and deforming both of them to force the two metals to alloy thereby creating a chemical diffusion gradient. In one sample there will be different compositions of the two elements available while the whole sample will have received the same treatment. The coated and deformed samples can therefore be used as a modelsystem to investigate the microstructural effects depending on the different local alloying composition.

We have chosen to coat a Cu rod with Ni, since Cu and Ni form an intermixing crystal structure and can be alloyed together in every composition. This should make it easy to obtain a good bonding as well as induce a chemical interaction at the interface of Cu and Ni.

2. Experimental procedures

Ni was electrolytically coated onto Cu rods of 14mm diameter. The Cu rods were of commercial OFHC purity. The content of impurities is shown in table A of the appendix. Two levels of thickness of the coating were prepared. One with 50µm Ni coating around the rod and one with 1305µm Ni coating. In both cases, the rods have been etched to increase the adhesion of the Ni coating on the Cu rod. Table B in the appendix shows the contents of the solution that has been used to coat the Ni onto the Cu rod while Table C shows the contents of the solution used to etch the rods before the coating.

Both samples (with 50μm coating and 1305μm coating) have been deformed afterwards by rotary swaging. Rotary swaging (RTS) has been chosen to deform the samples, as it ensures a uniform strain along the axes of the rod. The same process has also been chosen for fabricating Al-Cu clad composites as well [3]. Furthermore it introduces compressive strain in the swaged samples, which may improve properties under fatigue load, to be investigated. It consists of rotating hammers that leave a circular gap in the middle to insert the rod and that deform the rods perpendicular to the axis of the rod inserted. The hammers are exchangeable to allow different diameters of the gap and therefore different strain. Both samples were deformed in several passes with decreasing hammer diameter and therefore increasing strain. Table 1 shows the hammer diameter used and the resulting true strain as well as the temperature reached after each pass.

Table 1: Hammer diameter and resulting sample diameter used for deforming the Ni coated Cu rods. The strain is calculated by using equation (1) and the temperature was measured directly after each pass of swaging deformation by using a pyrometer.

Cu rod (50μm coating thickness)				
Swaging pass	Diameter mm	swaging block gap used / mm	True strain ε	Temperature °C
0	14.1		0	
1	13.06*	12.7	0.15	
2	11.67	11.3	0.38	30
3	10.2*	9.9	0.65	55
4	8.16	8	1.09	106
5	7.12*	7	1.37	78
6	6.09	6	1.68	90
7	5.66*	5.5	1.83	58

*Samples taken

Cu rod (1305μm coating thickness)				
Swaging pass	Diameter mm	swaging block gap used / mm	True strain ε	Temperature °C
0	16.61		0	
1	14.04	14.3	0.34	
2	13.1*	12.7	0.47	30
3	11.68	11.3	0.70	28
4	10.23*	9.9	0.97	40
5	8.23	8	1.40	138
6	7.16*	7	1.68	116
7	6.14	6	1.99	115
8	5.7*	5.5	2.14	70

*Samples taken

The true strain ε has been calculated from the initial diameter of the rod and the resulting diameter after each pass using equation (1):

$$\varepsilon = \ln\left(\frac{S_0}{S_n}\right) \tag{1}$$

S_0 is the initial cross section and S_n is the measured cross section after each swaging pass. For the 50μm coated rod, samples have been cut after the 1st, 3rd, 5th and final 7th pass. For the 1305μm rod, samples have been taken after the 2nd, 4th, 6th and final 8th pass to maintain comparable diameters of the samples.

These samples have been subjected to investigation by Scanning Electron Microscopy (SEM), Transmission Electron Microscopy (TEM) and microhardness indentation.

3. Results and discussion

3.1. Chemical diffusion gradient

SEM scans have been taken on the samples in order to determine the thickness of the Ni phase after the swaging and also to investigate whether there have been cracks forming in one of the phases or the interfaces. Figure 1 shows an example for the 1305μm coated Ni samples after 2, 4, 6 and 8 passes of swaging.

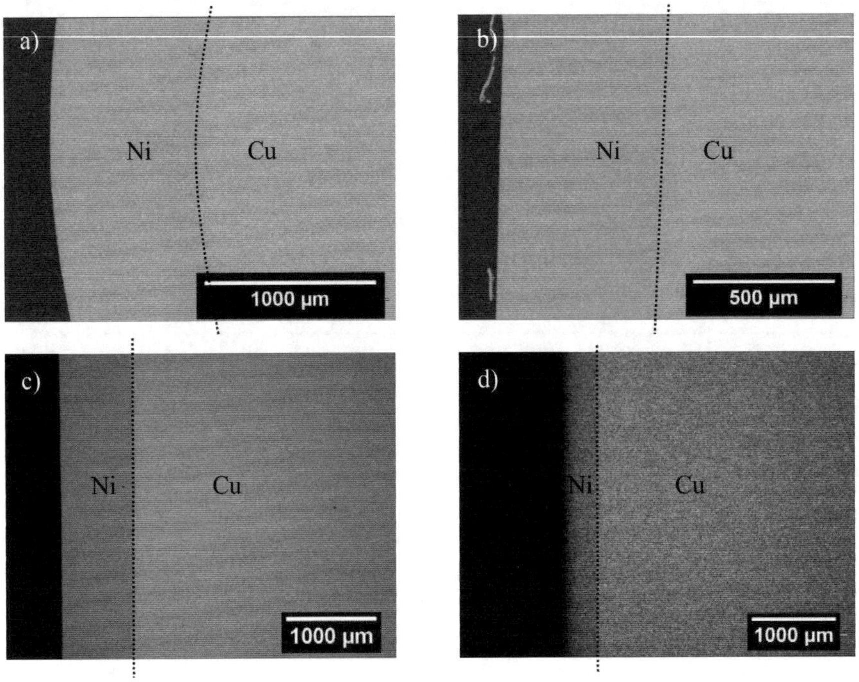

Figure 1. SEM images of the interface between the Cu core and the Ni coating after a) 2 passes (ε=0.47), b) 4 passes (ε=0.97), c) 6 passes (ε=1.68) and d) 8 passes of swaging (ε=2.14). The interface is marked with dotted line. There are no cracks visible at any of these stages both at the interface and within the core and the coating.

The SEM overview images show that the Ni coating decreases in thickness with increased number of swaging passes. This is expected as the total diameter of the coated rods decreases and instead elongates. No cracks were visible during the SEM investigation neither at the interface between the Cu core and the Ni coating, nor in the two phases itself.

Figures 2 and 3 show the results of an EDX scan over the Cu-Ni interface region of the 1305μm sample after 2 and 8 passes.

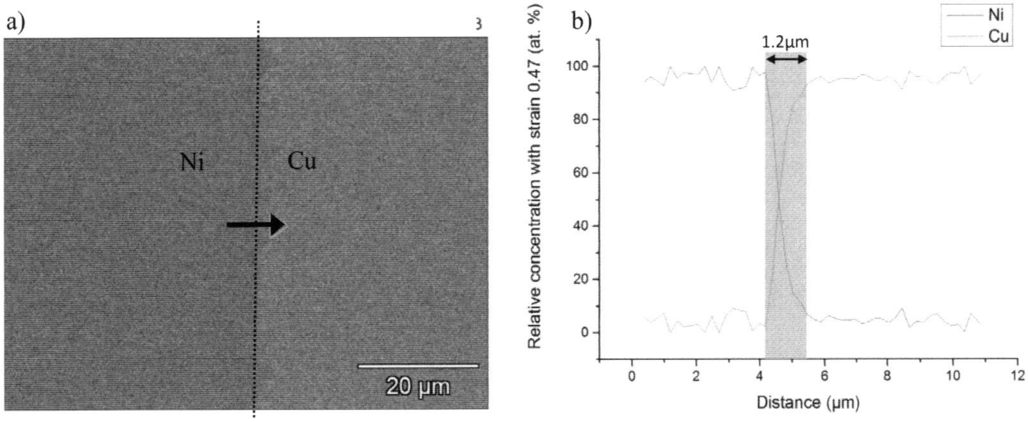

Figure 2. a) SEM image of the Ni-Cu interface region from the 1305μm coated sample after 2 passes of swaging (ε=0.47) taken at 15kV. b) EDX scan over the line, marked with an arrow in a), showing the Ni and Cu content for each point in at.%.

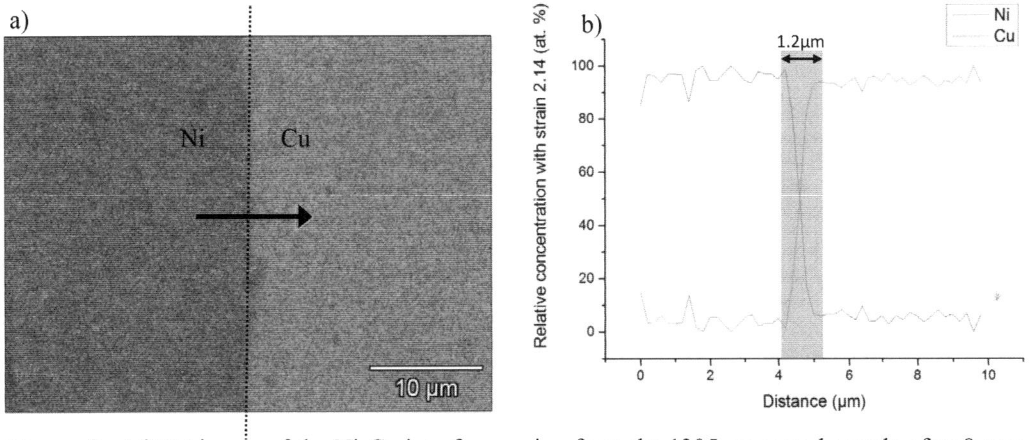

Figure 3. a) SEM image of the Ni-Cu interface region from the 1305μm coated sample after 8 passes of swaging (ε=2.14) taken at 15kV. b) EDX scan over the line selected in a) showing the Ni and Cu content for each point in at.%.

The chemical analysis shows the interaction between Cu and Ni existing in an interface region which reaches a width of about 1.2μm. Note that the extend of Cu into the Ni phase is of the same magnitude as the extend of Ni into the Cu phase and in the order of ~1μm. This correspondence has its cause in the similarity in atomic radii of both Cu and Ni and the full intermixability. Both scans show a similar sized intermixing region between the Ni and Cu. The area where both Ni and Cu is present is of the same width. There is no increase in the width from strain ε=0.47 to ε=2.14. We also have to be careful as the maximum spatial resolution of the EDX system is depending also on the thickness of the sample and the probesize [4] and in this case is around 1μm. From the data we can see that we are already close to this limit. The bonding itself, however, seems perfect and there are no crack visible even at the highest deformation.

3.2. Microstructure evolution

TEM investigations of the microstructure of the sample show the evolution of the deformation microstructure at increasing strain (see Fig.4). Cu in the as received condition is illustrated in Fig.4a, showing a very fine grained structure with single dislocations and dislocation boundaries within the

IOP Conf. Series: Materials Science and Engineering **89** (2015) 012022 doi:10.1088/1757-899X/89/1/012022

grain. The average grainarea is about 889µm², see Fig.4b, suggesting that the Cu in the as received condition is in a hot/cold deformed state. At a strain of ε=0.15, Fig.4c and Fig.4d show evolution of a fine deformation structure containing a network of dislocation boundaries forming a loose cell structure (Fig.4c). This structure evolution does refine during a strain increase to ε=0.65 and a well-defined cell structure has evolved. The average area of these cells is 201µm² (Fig.4e). In the structured characterization the grainarea can be surrounded by cell subgrain– and grain boundaries. In the text, grainarea is used as a descriptive phenomenon.

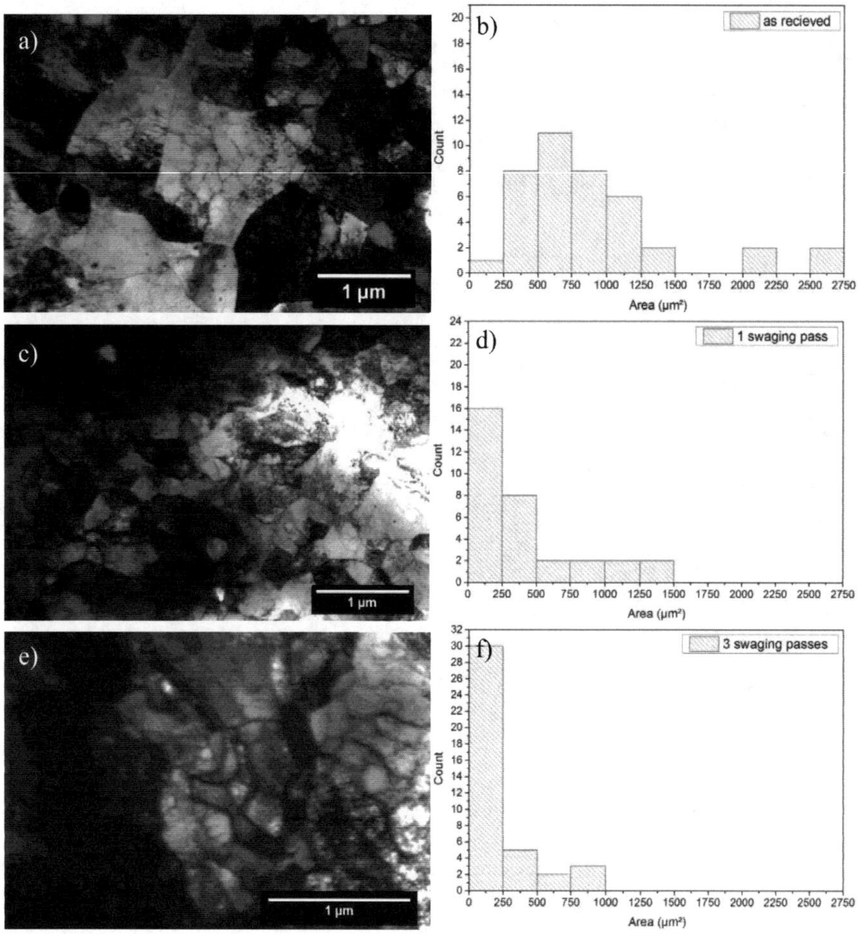

Figure 4. TEM micrographs and grain size distribution for a 50µm Ni coated Cu rod. a), b): as received sample; c), d): after 1 pass of swaging and a strain of ε=0.15; e), f): after 3 passes of swaging and a true strain of ε=0.65. The sample is taken from a position away from the Cu/Ni interface.

The TEM images in figure 4 a), c) and e) give a qualitative impression that with increasing strain, the structure refines. The quantitative analysis in figure 4 b), d) and f) also proves this. While for the unstrained sample the most common grainareas lay between 500µm² and 750µm², the dominant grainarea for that strained samples lays below 250µm², where after 3 passes of swaging and a true strain of ε=0.65 the percentage of small grains below 250µm² is larger than after 1 pass and a strain of ε=0.15.

A decreasing grainarea with increased strain is expected and also gives rise to an increase in hardness for the system.

3.3. Microhardness development

Microhardness indentation measurements according to the Vickers method have been conducted on the differently strained samples to correlate the microstructure evolution with the hardness development. The hardness is measured in the Cu phase in a position away from the Cu/Ni interface to allow comparison with the observed microstructure.

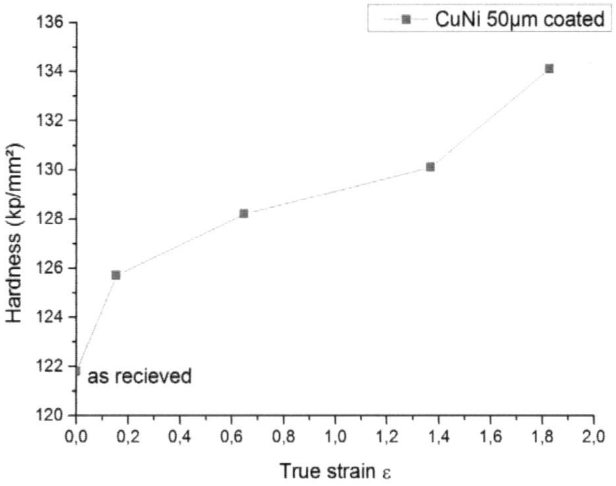

Figure 5. Microindentation hardness development of the 50µm Ni coated Cu rod when being strained up to a value of 1.83. Each point is the arithmetic mean value of 10 indentations taken from the same area in the Cu.

With increasing strain the hardness increases a limited extend from 120 to 134 kp/mm^2. The initial hardness indicates that the Cu rod as received is almost fully hardened in agreement with the fine microstructure shown in Fig.4a. As to the small hardening rise shown in Fig.5, it reflects typical stage IV hardening in Cu with an almost constant hardening rate.

Following the swaging deformation, the samples were annealed at 200°C for 0.5h, 1h, 2h and 4h (see Fig.6). A small hardness decrease is observed after annealing for 0.5h, which may be caused by annihilation of a small amount of dislocations. At larger annealing times the hardness increase with time suggesting that solutes in the Cu with an industrial purity (see Table A in the appendix) may segregate to the defects in the deformation structure and pin the dislocations. This segregation does apparently stabilize the deformed structure showing no sign of nucleation and recrystallization during the heat treatment.

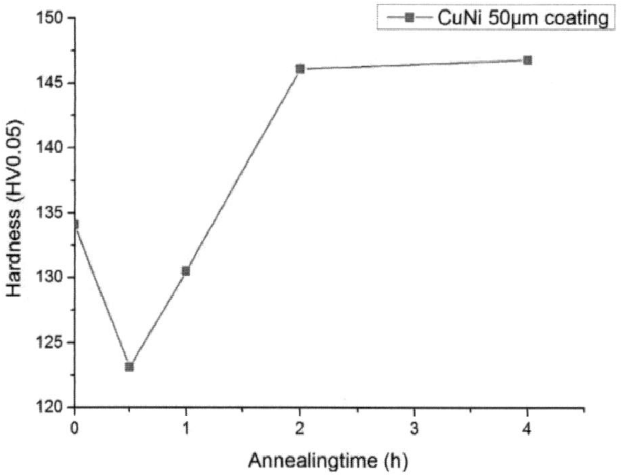

Figure 6. Microindentation hardness development for the 50μm Ni coated Cu rod after swaging and strain of 1.83 and subsequent annealing. Each point is the arithmetic mean value of 10 indentations taken from the same area in the Cu.

4. Concluding remarks

The application of rotary swaging has been successful in forming a defect free metallurgical bonding between a Ni coating and a solid Cu rod. The depth in which intermixing takes place is about 1μm creating an interface layer of about 1.2μm in thickness. By choosing the swaging process it is possible not only to produce a surface layer with improved corrosion resistance, but also to introduce a residual compressive stress in the sample, which may improve its fatigue resistance to be tested in a next step.

The intermixing interface layer is formed by mechanical alloying and it is of interest to study this layer in detail by Transmission Electron Microscopy (TEM) combined with 3D atom probe characterization. This is in order to shine light on the interaction between solute elements and the defects in the deformed microstructure in the form of dislocations, dislocation (cell) boundaries and deformations induced high angle boundaries. This interaction is important to understand fundamental processes such as dynamic and static recovery and recrystallization in metals containing a small amount of solutes, but also to optimize process and materials selection for advanced technological applications of a new approach.

Acknowledgements

The authors would like to thank Ms. Gitte Christensen for the preparation of the TEM samples, Mr. Preben Olsen for the preparation of the SEM samples and Mr. Lars Lorentzen for assisting with the swaging process. This work has been financed by The Danish Council for Independent Research under grant number: DFF-1335-00230.

References

[1] Ashby M F and Bréchet Y J M 2003 *Acta Mater.* **51** 5801–5821
[2] Sapanathan T, Khoddam S and Zahiri S H 2013 Journal of Alloys and Compounds **57** 85-92
[3] Kocich R, Machácková A, Kuncická L and Fojtík F 2015 Materials and Design **71** 36-47
[4] Bell D C and Garratt-Reed A J 2003 *Energy Dispersive X-ray Analysis in the Electron Microscope*; Garland Science
[5] Davis J R. 2001 *Copper and copper alloys*; (Novelty, OH: ASM international)

IOP Conf. Series: Materials Science and Engineering **89** (2015) 012022 doi:10.1088/1757-899X/89/1/012022

Appendix

Table A: Impurities present in the commercially pure copper rods used.

Percentage	Chemical element
99,995%	Cu
0.0003%	O
0.0012%	Ag
0.0012%	S
0.0002%	As
0.0005%	Pb
0.0005%	Sb
0.00003%	Bi
0.0001%	Se
0.00005%	Te
0.0001%	P
0.0005%	Ni
0.00005%	Zn

Table B: Content of the electrolytic solution used to coat Ni onto the Cu rod. The total volume used was 1 liter.

Weight	Percentage	Chemical formula	Scientific name
300g	78,5%	$Ni(II)SO_4\ (H_2O)_6$	Nickelsulfate-hexahydrate
45g	11,8%	$Ni(II)Cl_2\ (H_2O)_6$	Nickelcloride-hexahydrate
37g	9,7%	H_3BO_3	Boric acid
Water was added for 1 liter liquid			

Table C: Content of the solution used to etch the Cu rods before the swaging. The total volume used was 1 liter.

Volume	Percentage	Chemical formula	Scientific name
0,5l	50%	H_2SO_4	Sulfuric acid
0,2l	20%	HNO_3	Nitric acid
0,01l	1%	HCl	Hydrochloric acid
Water was added for 1 liter liquid			

Precession electron diffraction assisted orientation mapping of gradient nanostructure in a Ni-based superalloy

Z Q Feng[1,2], Y X Chen[2], G L Wu[1] and Y Q Yang[2]
[1] College of Materials Science and Engineering, Chongqing University, Chongqing 400044, China
[2] Shaanxi Materials Analysis and Research Centre, Northwestern Polytechnical University, Xi'an 710072, China

E-mail: zqfeng@cqu.edu.cn, yqyang@nwpu.edu.cn

Abstract. Surface mechanical grinding of a Ni-based superalloy can introduce a gradient microstructure in the surface layer with a grain size from nanoscale to microscale. In-depth investigation of the crystal orientation distribution of the surface nanostructured layer is more often, however, not an easy work by using the scanning electron microscope (SEM) based electron backscatter diffraction (EBSD) method due to its sensitivity to lattice distortions and spatial resolution limitation. Here we use a newly developed precession electron diffraction (PED) technique coupled with transmission electron microscopy (TEM) to investigate the microstructural and crystallographic characteristics of the surface gradient nanostructure, with particular emphasis on the topmost nanocrystalline layer. A strong shear texture and a minor Copper texture were identified according to orientation analyses of the 1.6 μm thick near-surface nanocrystalline layer. The PED technique is proved to be practical for two dimensional orientation mapping of severely deformed microstructures at the nanoscale.

1. Introduction

Surface integrity becomes increasingly vital for mechanical processing of structural components since the chemical, microstructural and crystallographic characteristics of the surface layers can crucially determine various kinds of mechanical properties in these regions, and greatly affect the final performance of these components [1-5]. Many efforts have been made in recent decades to optimize surface microstructure and properties, of which surface nanocrystallization is an effective and prevalent approach [4-10]. Through a surface mechanical attrition treatment, surface mechanical grinding treatment (SMGT), shot peening, laser shock processing as well as other machining treatments [1,2,4-10], the former coarse grains in surface-adjacent regions transform into a gradient microstructure with a grain size ranging from a few nanometers at the topmost surface to coarse grains in the bulk interior. The extent of the grain refinement depends on processing conditions and original microstructural features. The advent of such a surface gradient microstructure significantly and simultaneously improves tensile strength, fracture toughness, fatigue lifetime as well as other mechanical properties of the sample [3-6]. In-depth investigation of the surface nanocrystalline layer in some face centered cubic (FCC) metals to explore the origin of property optimization from a crystallographic perspective shows that a typical shear texture {001}<110> generally forms in the SMGT samples, based on the scanning electron microscope (SEM) based electron backscattered diffraction (EBSD) analyses of large amounts of nanocrystallines, ultrafine grains and coarse grains within a nearly half millimeter thick layer adjacent to the topmost surface [4]. Specifically, the texture determination of the nanocrystalline layer actually results from grains larger than 30-50 nm in diameter or spacing, and located 3-10 μm or more depth from the topmost surface, mainly because of the spatial resolution limitation of the SEM electron beam and sensitivity of EBSD to lattice distortion arising from severe plastic deformation (SPD) [11,12]. Such a situation indicates that grains smaller than 30-50 nm and located in the top 3-10 μm nanocrystalline layer actually do not contribute to the global texture analysis. Therefore, the individual texture information is missing in the texture analysis, leaving a question as to whether the top nanocrystalline layer of a few micrometers thick possesses a similar crystallographic feature as shown in large scale EBSD texture analyses. An effective texture analysis method with a higher spatial resolution and less sensitivity to lattice distortion, therefore, becomes an expanding demand for crystallographic analyses of SPD induced nanocrystallines, especially those in the topmost surface layer, with grain size smaller than 30-50 nm.

The electron beam in transmission electron microscope (TEM) generally has a high spatial resolution and its interaction with grains of polycrystalline materials can generate different orientated diffraction patterns at the back focal plane of the objective lens. Thus it is a good choice for nanoscale characterization of crystallographic orientations. However, typical dynamical effects, such as double diffraction, bending and thickness effects, may substantially reduce the sensitivity of a diffraction pattern to orientation variations [11-14]. While fast precessing focused incident electron beam of a small angle around the optical axis, reflections from zero and high order Laue zones and tilted reciprocal lattice may overlap, complement and finally present quasi-kinematical diffraction patterns [13-15]. By precisely and ultrafast template matching of these patterns with precalculated kinematical diffraction patterns corresponding to a series of orientations, the crystal orientation then can be determined. The precession electron diffraction (PED) currently used shows good prospect for phase and orientation mapping of polycrystalline materials [13-17].

In this study, we use PED coupled TEM to investigate microstructural and crystallographic characteristics of the surface nanostructured layer in the SMGT Ni-based superalloy, with particular emphasis on the topmost nanocrystalline layer. The texture feature is identified and the problems related with orientation indexing are further discussed.

2. Experimental

A commercial Ni-based superalloy (with nominal composition of Ni-18.9Cr-Fe-3.1Mo-5.3Nb-1.1Ti-0.5Al (wt.%)) plate (100 × 80 ×50 mm^3 in size) is used in the present investigation. Before SMGT, the plate surfaces were polished with silicon carbide papers, then aged at 720℃ for 8 h and furnace-cooled at the rate of 50℃/h to 620℃, further held at 620℃ for 8 h and air cooled. The microstructure of the annealed Ni-based superalloy is characterized by a coarse grain γ matrix with a high density of strengthening phase γ" and γ′ as well as randomly distributed δ particles. The average grain size of the γ matrix is 4-10 μm.

The precise SMGT of the pretreated plate was accomplished at ambient temperature on MM 7120A horizontal surface grinder under wet condition. The processing parameters were employed as follows: wheel speed 25 m/s, table speed 720 m/s, grinding depth 0.005 mm and feed rate 1.0 mm/rev. The feed direction was defined as SD, while the direction perpendicular to the grinding surface is defined as ND.

After SMGT, the cross-sectional TEM samples were prepared via an improved transverse TEM sample preparation method, to show microstructural evolution along the depth direction. Much attention has been paid to the sample preparation processes such as cutting, grinding, dimpling and ion milling to ensure the topmost surface of the SMGT sample was well protected. TEM microstructural characterization was performed on a 300kV field emission transmission electron microscope (TEM) Tecnai F30 G^2, equipped with a commercial PED system NanoMEGAS ASTAR (Brussels, Belgium). The crystal orientation analysis of the surface nanostructured layer was obtained through continuous scanning of the fast precessed electron beam on an area of interest with a precession angle of 0.6°, beam size of 5 nm and step size of 4 nm, simultaneously fast acquisition of PED patterns (Fig. 1a) and off-line ultrafast matching of these obtained PED patterns with the precalculated kinematical diffraction patterns (Fig. 1b and 1c). Further orientation mapping analysis and visualization were carried out using

ASTAR MapViewer, the information about orientation, index, reliability, misorientation and their combinations can be revealed in detail. To obtain more crystallographic information, the orientation data were also processed and visualized using a Channel 5 software package.

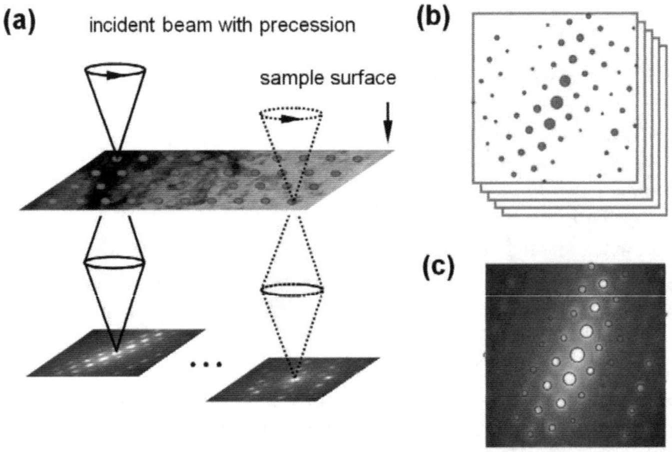

Fig. 1. (a) Schematic illustration of the precessed and focused electron beam and its continuous scanning on an area of interest near the sample surface, resulting in PED patterns corresponding to the grains with different orientations; (b) The simulated kinematical diffraction patterns in the crystal databank; (c) Template matching between the obtained PED patterns and their precalculated counterparts.

3. Results

Figure 2 shows the cross-sectional TEM image of the surface layer of the SMGT Ni-based superalloy. From low magnification TEM image of the top 6 μm thick surface deformation layer (Fig. 2a), the microstructural evolution along the depth direction from the sample surface can be clearly observed, i.e. the top tiny grain layer with thickness ranging from 150 to 600 nm (see red dotted line), a thin lamellar layer with irregularly elongated grains, and equiaxed coarse grain layer with increasing grain size with the depth from the sample surface. The coarse grain layer is characterized by intersecting slip traces on two typical slip planes, indicating the activation of multiple slip systems. Close observation of the top tiny grain layer distinctly demonstrates two sublayers (Fig. 2b, see yellow dotted line), one adjoining to the topmost surface is composed of equiaxed nanocrystallines of 4-10 nm in diameter, the other mainly consists of laminated grains with a boundary spacing of about 15-60 nm, as well as some scattered nanocrystallines with various sizes and morphologies. On the whole, such complicated but universal hierarchical nanostructures are a typical surface microstructure of the SMGT Ni-based superalloy as well as other materials after various kinds of surface mechanical treatments [4-10].

To further unravel the microstructural feature of the gradient nanostructure and its formation mechanism from a crystallographic perspective, the crystal orientation was analyzed on a PED platform integrated into a 300kV TEM. The operation procedures and parameters were detailed in Section 2. Through fast scanning of the focused precessing electron beam on a two dimensional area (1600 × 800 nm^2, as shown in the virtual bright field (VBF) image Fig. 3a) within ~1.6 μm thick near-surface layer, a series of PED patterns were acquired (Figs. 3b-d) and further indexed via ultrafast template matching. The obtained inverse pole figure (IPF) orientation map of the surface SPD layer is shown in Figs. 3e and 3f, from which both the morphological and crystallographic features of nanocrystallines can be distinguished, better than from the corresponding VBF image (Fig. 3a). The reliability of orientation indexing was roughly evaluated through point-to-point examination of template matching between the obtained PED patterns and probable precalculated patterns. Most of the PED patterns, such as diffraction pattern obtained from site A (Fig. 3b), are proved to be single crystal diffraction patterns and match well with corresponding orientation indexing results shown in the IPF orientation map on Fig. 3e. While the

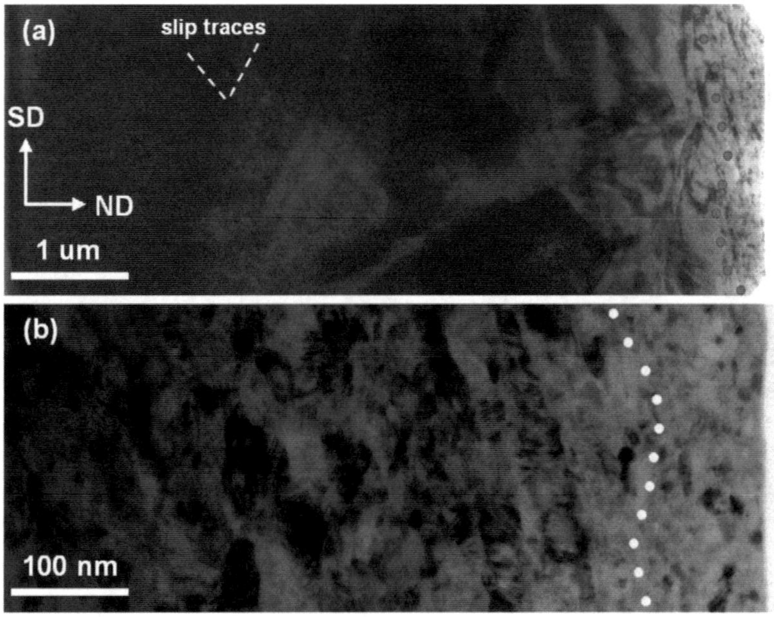

Fig. 2. (a,b) Cross-sectional TEM images showing surface gradient nanostructures in SMGT Ni-based superalloy sample. Note that the topmost surface of the samples are on the right.

rest are superimposed patterns, which gradually increase in number with decreasing grain size and thus reducing reliability of orientation indexing (Fig. 3g). Take the diffraction pattern (Fig. 3d) obtained from site C as an example, the superimposed PED pattern was actually constituted by four inclined diffraction patterns <011>, <214>, <215> and <114>, but only <215> was recognized as the most reliable orientation of the nanograin. This unique choice may probably cause misinterpretation of the true orientation of nanocrystallines.

Grain boundary misorientation analysis further demonstrates that some adjoining grains with planar boundaries (see arrows) show a misorientation of about 60° <111> (Fig. 3h), indicating the formation of nanoscale deformation twins in the surface SPD layer. In specific, nanotwin lamella (see dotted line in Fig. 3f) with a boundary spacing as small as 8 nm (Fig. 3h) can be clearly distinguished, implying a higher spatial resolution in PED assisted orientation mapping than that in commercial EBSD [11,12]. Based on orientation analyses, the texture components of the 1.6 μm thick near-surface SPD layer are identified as a major shear texture {001}<110> and a minor Copper texture {112}<111> (Fig. 3i).

4. Discussion

The IPF orientation maps from SD and ND directions (Figs. 3e and 3f) collectively revealed the microstructural and crystallographic features of the surface SPD layer, but some details of the same structure can also be detected different from part to part. As shown in the left part of Fig. 3e, the adjacent laminated grains show similar orientations when revealing by IPF in SD. However, these orientations show differences when revealing by IPF in ND (Fig. 3f). This implies that these laminated grains shared a common SD, forming a fiber texture.

The present PED assisted crystal orientation analysis of the surface nanostructures clearly identified a strong shear texture, {001}<110>, and a minor Copper texture. The shear texture was previously reported in the nanolaminated structure of SMGT pure Ni by large scale EBSD texture analyses [4]. It is reasonable that the shear texture was produced by the strong shear deformation during the SMGT process. While the present results show that there is a minor Copper texture within topmost layer. It was reported that Copper component can be produced by lattice rotation of {001}<110> orientation

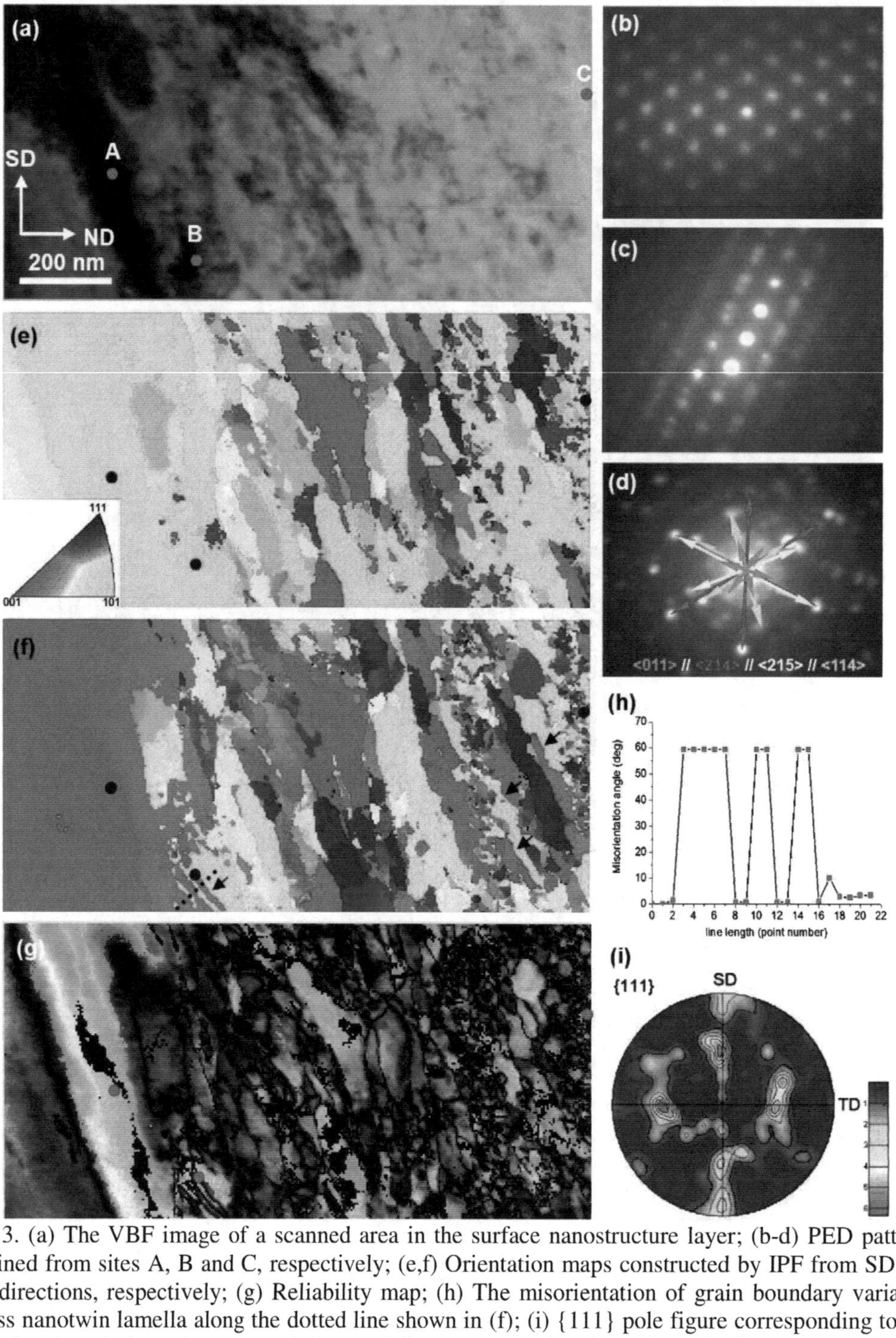

Fig. 3. (a) The VBF image of a scanned area in the surface nanostructure layer; (b-d) PED patterns obtained from sites A, B and C, respectively; (e,f) Orientation maps constructed by IPF from SD and ND directions, respectively; (g) Reliability map; (h) The misorientation of grain boundary variation across nanotwin lamella along the dotted line shown in (f); (i) {111} pole figure corresponding to (e). Note that (a,e-g) share the same scale bar, and the topmost surface is on the right side.

during plastic deformation [18]. In the IPF orientation map, nanotwin lamella with boundary spacing of 8 nm can be well distinguished in the surface SPD layer. PED coupled TEM orientation mapping is thus proved to be insensitive to severe lattice distortions and capable of suppressing various dynamical effects.

Despite the success, additional attention should also be paid to orientation indexing of nanocrystallines near the topmost surface. As shown in Fig. 3d, a superimposed PED pattern constituted by four possible diffraction patterns was obtained from site C (Fig. 3a). Three most possible diffraction patterns, including <214>, <215> and <114>, show similar reflection intensity and correlation index value, but only <215> was recognized as the most reliable orientation of a nanograin. This pattern indexing method, despite insufficient evaluation of indexing reliability, can still not avoid probable misinterpretations of superimposed PED patterns arising from grain boundaries and junctions, or overlapped grains. The typical example given by Wu and Zaefferer [19] vividly showed that two overlapped grains with equal thickness but different orientation can coincidentally generate superimposed diffraction pattern as the same as another single crystal diffraction pattern. For electron beam transparent polycrystalline sample, the smaller the grain size, the higher the overlapping probability of nanocrystallines along the incident beam direction, and the easier the mis-indexing of superimposed PED patterns. Similarly, as the focused precessing electron beam is often highly distorted, the equivalent spot size becomes several times larger than its counterpart without precession. Thus the spatial resolution would be dramatically reduced, along with an increasing probability of superimposed PED patterns resulting from adjoining grains. Specifically, as the average size of nanocrystallines is approximated to the spot size of the focused precessing electron beam, the reliability of pattern indexing of superimposed PED pattern becomes even worse. Considering the experimental parameters used in this study, i.e. an electron beam with a precession angle of 0.6°, equivalent beam size of 5 nm and scanning step size of 4 nm, conservatively we can estimate that orientation indexing of nanocrystallines in IPF orientation map with diameters or spacings larger than 8 nm can be trusted, since at least 2×2 pixel or 4 independent pattern indexing has the same orientation. This estimation was fully proved to be reasonable and reliable by orientation indexing of nanotwin lamella with boundary spacing of 8 nm in Figs. 3e and 3g. Last but not least, we should point out that this critical grain size for orientation indexing reliability is not constant but more often changes with local sample thickness, overlapped grain number and percentage thickness, diffraction condition and so on. Pattern indexing results, especially those for nanocrystallines with a grain size in the same order of magnitude of the beam size, need careful reevaluation to maximally avoid probable misinterpretations. The recently developed TEM based three dimensional orientation mapping technique [20], however, may be a powerful way to solve this problem and precisely reveal the crystallographic feature of the overlapped tiny nanocrystallines near the topmost surface. Anyway, the high spatial resolution, together with high efficiency of pattern acquisition and indexing, fully indicates that the PED assisted TEM orientation mapping is practical and promising for two dimensional crystal orientation mapping of severely deformed structures at the nanoscale.

5. Conclusions

With the help of a TEM based PED technique, the orientation mapping of a surface nanostructure in SMGT Ni-based superalloy was carried out in addition to conventional TEM characterization. Dislocation activity and deformation twinning play distinct but cooperative roles in grain refinement during SMGT, while dislocation based plasticity is still in effect for nanocrystalline as small as 5-10 nm. A strong shear texture was identified in the surface nanocrystalline layer, which is in agreement with large scale SEM-EBSD analysis results. However, a minor Copper texture was also detected in the layer. Despite the deficiency to resolve the superimposed diffraction patterns, PED assisted orientation mapping is proved to be capable of mapping nanocrystalline metals with grain size down to a few nanometers in diameter, far smaller than what can be resolved in the commercial EBSD technique.

Acknowledgements
Financial support from the Fundamental Research Funds for the Central Universities (grant No. CDJZR12130073, CDJZR14130002 and 106112015CDJXY130001) is gratefully acknowledged. GLW thanks the support from NSFC (grant No. 51471039).

References

[1] Bushlya V, Zhou J M, Lenrick F, Avdovic P, Stahl J E 2011 *Procedia Engineering* **19** 60
[2] Herbert C R J, Axinte D A, Hardy M C and Brown P D 2011 *Procedia Engineering* **19** 138
[3] Suresh S 2001 *Science* **292** 2447
[4] Liu X C, Zhang H W and Lu K 2013 *Science* **342** 337
[5] Lu K 2014 *Science* **345** 1455
[6] Fang T H, Li W L, Tao N R and Lu K 2011 *Science* **331** 1587
[7] Lu K and Lu J 1999 *J. Mater. Sci & Tech.* **15** 193
[8] Tao N R, Wang Z B, Tong W P, Sui M L, Lu J and Lu K 2002 *Acta Mater.* **50** 4603
[9] Tao N R, Wu X L, Sui M L, Lu J and Lu K 2004 *J. Mater. Res.* **19** 1623
[10] Li W L, Tao N R and Lu K 2008 *Scr. Mater.* **59** 546
[11] Zaefferer S 2000 *J. Appl. Crystal.* **33** 10
[12] Dingley J 2006 *Microchimica Acta* **155** 19
[13] Brons J G and Thompson G B 2014 *JOM* **66** 165
[14] Darbal A D, Gemmi M, Portillo J, Rauch E and Nicolopoulos S 2012 *Microscopy Today* **20** 38
[15] Moeck P, Rouvimov S, Hausler I, Neumann W and Nicolopoulos S 2011 Precession electron diffraction & automated crystallite orientation/phase mapping in a transmission electron microscope *11th IEEE International Conference on Nanotechnology (Oregon, Portland, August 15-18 2011)* pp 754-759
[16] Viladot D, Veron M, Gemmi M, Peiro F, Portillo J, Estrade S, Mendoza J, Llorca-Isern N and Nicolapoulos S 2013 *J. Microsc.* **252** 23
[17] D'Antuono D S, Gaies J, Golumbfskie W and Taheria M L 2014 *Scr. Mater.* **76** 81
[18] Wu G L, Godfrey A, Liu W and Liu Q 2001 *Scri. Mater.* **45** 1117
[19] Wu G L and Zaefferer S 2009 *Ultramicroscopy* **109** 1317
[20] Liu H, Schmidt S, Poulsen H F, Godfrey A, Liu Z, Sharon A and Huang X 2011 *Science* **332** 833

Simulating recrystallization in titanium using the phase field method

S P Gentry and K Thornton

Materials Science and Engineering Department, University of Michigan, Ann Arbor, Michigan, 48109, USA

E-mail: kthorn@umich.edu

Abstract. Integrated computational materials engineering (ICME) links physics-based models to predict performance of materials based on their processing history. The recrystallization phase field model is developed and parameterized for commercially pure titanium. Stored energy and nucleation of dislocation-free grains are added into a phase field grain-growth model. A two-dimensional simulation of recrystallization in titanium at 800°C was performed; the recrystallized volume fraction was measured from the simulated microstructures. Fitting the recrystallized volume fraction to the Avrami equation gives the time exponent n as 1.8 and the annealing time to reach 50% recrystallization ($t_{0.5}$) as 71 s. As expected, the microstructure evolves faster when driven by stored energy than when driven by grain boundary energy.

1. Introduction

Integrated computational materials engineering (ICME) is crucial in advancing material design and optimization. Linking models together will enable simulated parametric studies across length scales, incorporating microstructure into larger-scale component analysis. For example, simulations of static recrystallization are needed for the processing history of metals to be used in microstructural mechanical behavior simulations. As such, successful ICME requires accurate physics-based simulations that predict realistic microstructures.

When metals undergo plastic deformation, a network of dislocations is formed within the microstructure. Increasing the applied stress provides an additional driving force for dislocation motion and creates more dislocations. The increased dislocation density results in a stored energy (f_{stored}) within the microstructure, which is approximated as [1]

$$f_{stored} = \frac{1}{2}Gb^2\rho \tag{1}$$

where G is the shear modulus, b is the Burgers vector, and ρ is the dislocation density. Upon heat treatment, metals may undergo recrystallization, in which dislocation-free grains nucleate and grow in the microstructure. Reduction of the stored energy provides the driving force for growth of the new grains. The recrystallization process is both time- and temperature-dependent. Recrystallization is commonly described by an Avrami relation between the fraction recrystallized (X) and the annealing time (t) [1]. Specifically,

$$X = 1 - \exp\left(-\beta\left(\frac{t}{t_{0.5}}\right)^n\right) \tag{2}$$

where $\beta = -\ln(0.5)$, $t_{0.5}$ is the time to reach a fraction of 50% recrystallized grains, and n is the time exponent.

In this study, we focus on the microstructural evolution of titanium. Titanium is a lightweight metal that has high strength; predicting its recrystallization kinetics will facilitate alloy development. We seek to demonstrate that a recrystallization phase field model may be employed to predict the microstructural evolution of titanium during thermomechanical processing.

2. Phase Field Model

The phase field method was utilized to model recrystallization and grain growth in a deformed microstructure. The phase field grain-growth model employs a functional to describe the free energy of the system in terms of nonconserved order parameters [2]. Each grain is assigned an identifying number i and a corresponding order parameter η_i. A system with N grains has N independent order parameters $\eta_1, \eta_2, ...\eta_N$ that are tracked at every point and evolved according to Allen-Cahn dynamics. The total free energy of the system, $\mathcal{F}$, is the volume integral of the free energy density. The recrystallization model includes three contributions to the free energy density with the total free energy given by:

$$\mathcal{F} = \int_V \{f_{bulk} + f_{gradient} + f_{stored}\}\, dV \tag{3}$$

where f_{bulk} is the bulk free energy density, $f_{gradient}$ is the gradient energy density, and f_{stored} is the stored energy density. The bulk term is expressed as [3]

$$f_{bulk} = m_0 \left[\sum_{i=1}^{N} \left(-\frac{1}{2}\eta_i^2 + \frac{1}{4}\eta_i^4 \right) + \alpha \sum_{i=1}^{N} \sum_{j>i}^{N} \eta_i^2 \eta_j^2 + \frac{1}{4} \right] \tag{4}$$

where $\alpha = 1.5$ and m_0 is a constant related to the grain boundary energy and the diffuse interface width. The gradient energy density penalizes large gradients in the order parameter and is given by

$$f_{gradient} = \frac{\kappa}{2} \sum_{i=1}^{N} |\nabla \eta_i|^2 \tag{5}$$

where κ is the gradient energy penalty. The stored energy provides an additional driving force for microstructural evolution. Each grain i has a corresponding dislocation density ρ_i; the effective dislocation density interpolates the dislocation densities based on the values of the order parameters at that point. We interpolate the effective dislocation density as

$$\rho_{eff}(\eta_1, \eta_2, ..., \eta_N) = \frac{\sum_{i=1}^{N} \eta_i^2 \rho_i}{\sum_{j=1}^{N} \eta_j^2}. \tag{6}$$

This is similar to the work of Moelans, Godfrey, Zhang, and Juul Jensen [4, 5], but is generalized to an arbitrary number of grains. The stored energy, Eq. 1, is calculated with the effective dislocation density at a given point, ρ_{eff} above, resulting in

$$f_{stored} = \frac{1}{2} G b^2 \rho_{eff}(\eta_1, \eta_2, ..., \eta_N). \tag{7}$$

The evolution of the microstructure is given by Allen-Cahn dynamics [6]:

$$\frac{\partial \eta_i}{\partial t} = -L_i \left(\frac{\delta \mathcal{F}}{\delta \eta_i} \right) = -L_i \left(m_0 \left(-\eta_i + \eta_i^3 + 2\alpha\eta_i \sum_{j \neq i}^{N} \eta_j^2 \right) - \kappa(\theta)\nabla^2 \eta_i - G b^2 \frac{\eta_i (\rho_i - \rho_{eff})}{\sum_{j=1}^{N} \eta_j^2} \right). \tag{8}$$

The partial differential equations are discretized with the finite difference method and evolved using forward Euler time stepping.

Active parameter tracking [7, 8] is employed to reduce the computational expense of the phase field grain-growth model. Each order parameter is only stored where it is above 1×10^{-6}; otherwise the order parameter is assumed to be zero. The simulation in this work used an algorithm based on the data structure in McKenna and Voorhees's Anisotropic Phase Field Evolver [9, 10], with this work adding MPI parallelization, stored energy contributions, and recrystallization.

To simulate static recrystallization, new grains are added to the phase field model at the beginning of the simulation. A new order parameter is created for each recrystallized grain and assigned a dislocation density of zero. The order parameter is seeded with a circle at a randomly selected position on a grain boundary or on a triple point. The number of newly seeded grains is based on a predetermined number density of nuclei. For the first N_{hold} time steps, the order parameters of the new grains are held while the other order parameters adjust. When the hold period is complete, microstructural evolution is allowed to continue as normal.

3. Simulation Parameters

The model has been parameterized for commercially pure titanium that was uniaxially compressed by 20% at room temperature and then annealed at 800°C. The initial recrystallized nuclei density was estimated to be 5.5×10^{12} m^{-3} from experiments [11]. The grain boundary energy of titanium was taken as 0.7 J/m^2 for high-angle grain boundaries [12]. The grain boundary mobility at 800°C was extracted from recrystallization experiments and found to be 1.9×10^{-13} m^4/Js [11]. The shear modulus was taken to be 25 GPa at 800°C as approximated using the Voigt-Reuss-Hill model [13] from elastic constant measurements of single crystal titanium [14]. The Burgers vector was set to 0.295 nm for an $\vec{a}$ dislocation [15].

The dislocation density in the material can be approximated from [16]

$$\sigma = \sigma_y + M_T C_{stored} G b \sqrt{\rho} \tag{9}$$

where σ is the applied stress, σ_y is the yield stress, M_T is the Taylor factor, and C_{stored} is a constant that was approximated to be 0.15 from Ref. [17]. For titanium, M_T was taken to be 2.5 according to Ref. [18]. Compression of titanium reduced one dimension of the sample by 20%, corresponding to an applied stress of 582 MPa [11]. The yield stress from experiments was 325 MPa [11]. Using Eq. 9, the average dislocation density was calculated to be 8.6×10^{15} m^{-2}.

For the simulation, the initial microstructural domain was set to a random grain structure with an average grain size of 53 μm arranged on a 1500×1500 uniform grid with $\Delta x = \Delta y = 0.5$ μm. The time step, Δt, was set to 3.8×10^{-3} s. Each grain in the initial microstructure was randomly assigned a dislocation density within $\pm 25\%$ of the average dislocation density. The radius of each nucleus was $15\Delta x$ and the hold time (N_{hold}) was 200 time steps, which is sufficient for existing order parameters to adjust for the seeded order parameter. The nuclei number density was converted from three dimensions to two dimensions by assuming that the initial phase fractions of round recrystallized grains would be equal. The number density of nuclei in two dimensions was taken to be 5.5×10^7 m^{-2}.

4. Results and Discussion

The evolved microstructure is shown in Fig. 1 at four annealing times. For visualization, the dislocation density is scaled by the average dislocation density of the initially deformed microstructure ($\rho/\rho_{ave,t=0}$). Recrystallized grains are white (corresponding to a dislocation density of zero) while the existing grain structure is colored based on the scaled dislocation density. The recrystallized nuclei (16 μm-diameter circles) are visible in the deformed matrix in the first snapshot taken at 1.9 s (Fig. 1(a)). 70 s of annealing (Fig. 1(b)) results in significant growth in the recrystallized grains, while much less evolution is seen where there are no recrystallized grains. The microstructure is fully recrystallized after 220 s of annealing (Fig. 1(c)). The fraction recrystallized is plotted as a function of time, as shown in Fig. 2. The simulated data was fit to the Avrami equation (Eq. 2) using only the data points between $X = 0.1$ and $X = 0.9$. The best fit was obtained for n of 1.8 and $t_{0.5}$ of 71 s for the simulated two-dimensional microstructure.

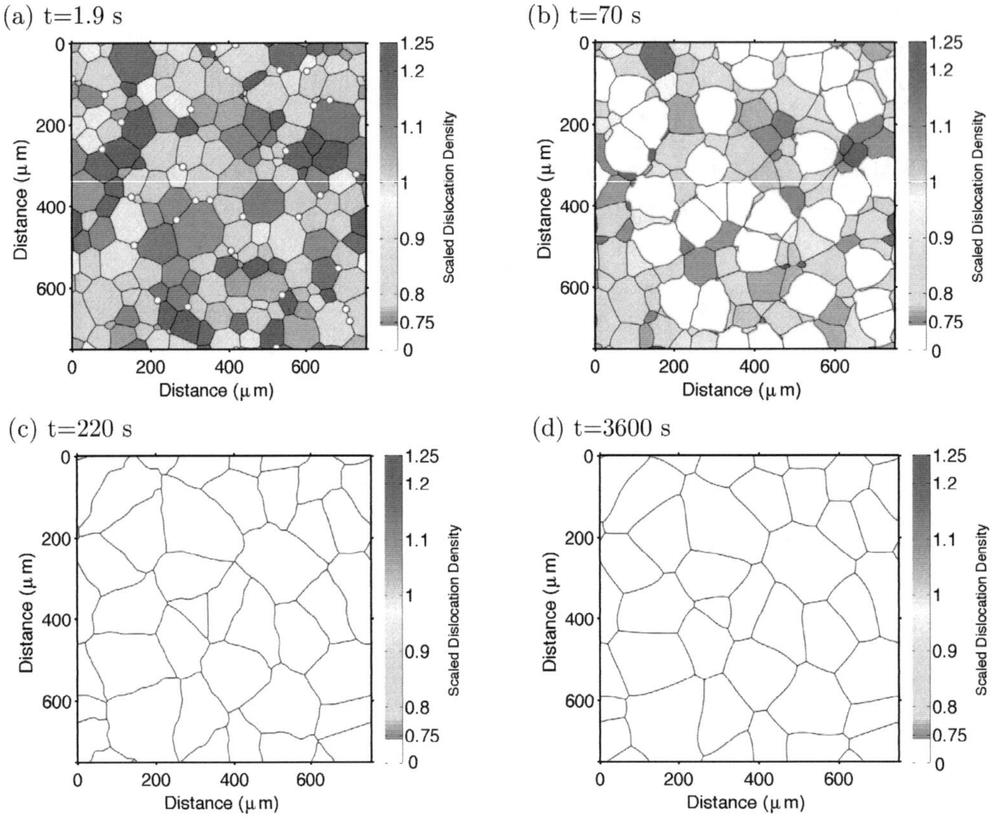

Figure 1. Simulated microstructural evolution after annealing at 800°C for 1(a) 1.9 s, 1(b) 70 s, 1(c) 220 s, and 1(d) 3600 s. The colors indicate the scaled dislocation density ($\rho/\rho_{ave,t=0}$); recrystallized grains are white.

During annealing there are two competing driving forces for microstructural evolution: stored energy and grain boundary energy. Stored energy differences between adjacent grains cause recrystallized grains to grow at the expense of the deformed grains. The grain boundary energy drives the large grains to grow at the expense of small grains, which reduces the total grain boundary length. During recrystallization, the simulated evolution is dominated by the stored energy contribution as evidenced by the growth of both the small recrystallized grains and larger grains with low stored energy (Figs. 1(a), 1(b), 1(c)). If the microstructural evolution of recrystallization was dominated by the grain boundary energy, then small dislocation-free grains would disappear to reduce the grain boundary length. When recrystallization is complete, only the grain boundary energy drives microstructural evolution because the stored energy is zero. Annealing a fully recrystallized microstructure is then governed by the reduction of the grain boundary energy. From 220 s (Fig. 1(c)) to 3600 s (Fig. 1(d)), the grains grow and the grain

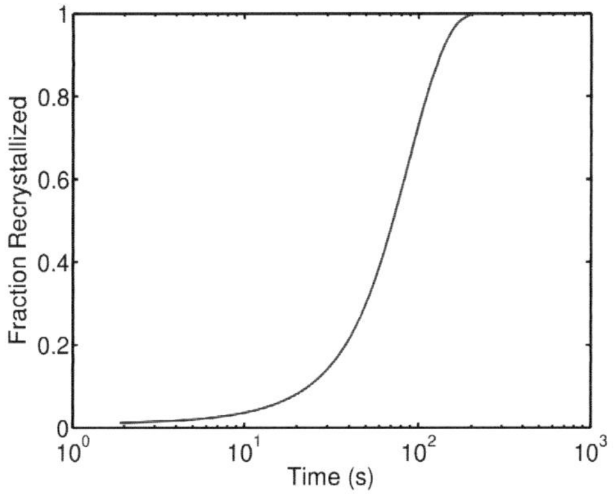

Figure 2. Area fraction of recrystallized grains as a function of time.

boundaries become straighter. The simulation of the microstructural evolution from 0 s to 3600 s shows the effects of both stored energy and grain boundary energy.

The competing driving forces result in different rates for microstructural evolution. The stored energy causes the recrystallized grains to grow quickly, with full recrystallization occurring in 220 s. However, grain growth that occurs after recrystallization is much slower. From 220 s (Fig. 1(c)) to 3600 s (Fig. 1(d)), there were only small changes in the grain sizes. These differences in growth rate occur as a result of the combined model; no changes are made to the evolution equation.

5. Conclusions

This paper presented a model and simulation results for recrystallization and grain growth in metals. The model is based on the phase field method, which takes into account the thermodynamics and kinetics of materials. A polycrystalline microstructure is modeled with a separate order parameter for each grain; these order parameters are evolved according to Allen-Cahn dynamics. The recrystallization phase field model incorporates stored energy contributions and the nucleation of new undeformed grains. The model also retains the driving force due to the reduction of the total grain boundary energy, which leads to grain growth.

The recrystallization phase field model was used to simulate microstructural evolution in two dimensions for deformed titanium. The model was parameterized for commercially pure titanium that was compressed to 20% uniaxial strain and then annealed at 800°C. The recrystallized grains grew at the expense of the deformed grains as expected, and the recrystallized fraction was calculated as a function of time. In the two-dimensional simulation presented, the time exponent of 1.8 and the time to reach 50% recrystallization of 71 s provided the best fit to the Avrami equation (Eq. 2). Recrystallization occured on a much faster time scale than growth of the dislocation-free grains because the driving force from stored energy is larger than the driving force from grain boundary energy. Recrystallization took 220 s, and annealing for an additional 3400 s only had a small effect on the microstructure.

This work involved simulations in two dimensions only; three-dimensional simulations are ongoing and will allow for direct comparison of experimental and computational microstructures. Simulated three-dimensional microstructures can also be used as inputs to other models such as those that predict microstructure-dependent mechanical behavior, and thus are essential for successful integration of the model into an ICME framework. These models can then be directly integrated so that processing conditions are linked to the stress analysis of a designed part.

Acknowledgments
This work was funded by a grant from the United States' Office of Naval Research (ONR Grant N00014-12-1-0013). Additional support for computational resources and services were provided by Advanced Research Computing at the University of Michigan, Ann Arbor, MI, USA.

References
[1] Humphreys F and Hatherly M 2004 *Recrystallization and related annealing phenomena* (Elsevier)
[2] Fan D and Chen L Q 1997 *Acta Mater.* **45** 611--622
[3] Moelans N, Blanpain B and Wollants P 2008 *Phys. Rev.* B **78** 024113
[4] Moelans N 2011 *Acta Mater.* **59** 1077--1086
[5] Moelans N, Godfrey A, Zhang Y and Juul Jensen D 2013 *Phys. Rev.* B **88** 054103
[6] Allen S M and Cahn J W 1979 *Acta Metall.* **27** 1085--1095
[7] Gruber J, Ma N, Wang Y, Rollett A D and Rohrer G S 2006 *Modelling Simul. Mater. Sci. Eng.* **14** 1189--1195
[8] Vedantam S and Patnaik B 2006 *Phys. Rev.* E **73** 016703
[9] McKenna I M 2010 *Three-dimensional anisotropic and isotropic grain growth simulations with comparisons to experiment* Ph.D. thesis Northwestern University Evanston, IL
[10] McKenna I M and Voorhees P W 2010 Anisotropic Phase Field Evolver URL `https://cosmicweb.mse.iastate.edu/wiki/display/software/Anisotropic+Phase+Field+Simulation`
[11] Trump A 2015 Private Communication
[12] Wu M S, Nazarov A A and Zhou K 2004 *Philos. Mag.* **84**
[13] Hill R 1952 *Proc. Phys. Soc.* A **65** 349
[14] Fisher E S and Renken C J 1964 *Phys. Rev.* **135** A482--A494
[15] Partridge P G 1967 *Metall. Rev.* **12** 169--194
[16] Saada G 1960 *Acta Metall.* **8** 841--847
[17] Zurob H S, Hutchinson C R, Brechet Y and Purdy G 2002 *Acta Mater.* **50** 3077--3094
[18] Morita M and Umezawa O 2011 *Mater. Trans.* **52** 1595--1602

Microstructural evolution of Ti-added interstitial free steel in high strain deformation by hot torsion

R Gholizadeh[1], A Shibata[1,2], D Terada[2,3] and N Tsuji[1,2]

[1] Department of Materials Science and Engineering, Kyoto University, Sakyo-ku, Kyoto 606-8501, Japan
[2] Elements Strategy Initiative for Structural Materials (ESISM), Kyoto University, Sakyo-ku, Kyoto 606-8501, Japan
[3] Department of Mechanical Science, Chiba Institute of Technology, Tsudanuma, Narashino, Chiba 275-0016, Japan

E-mail: gholizadeh.reza.77v@st.kyoto-u.ac.jp

Abstract. The dynamically evolved microstructure under high strain deformation condition does still have many debatable aspects, particularly in the case of easy-recovery metals like bcc-iron. In this research, microstructural evolution in high strain deformation by hot torsion of Ti-added interstitial free (IF) steel was systematically investigated. Torsion specimens were deformed up to an equivalent strain of ~ 7 at different temperatures (650 °C - 850 °C) and strain rates (0.01 s^{-1} - 1.0 s^{-1}), i.e., under various values of the Zener-Hollomon (Z) parameter. Immediately after the deformation, samples were water-quenched and microstructures were investigated by electron backscattering diffraction (EBSD) measurements and electron channelling contrast imaging (ECCI). Flow stress-strain curves of the IF steel under various deformation conditions showed typical flow curves of high stacking fault energy metals at low Z values, i.e., a peak stress followed by slight softening. On the other hand, under the high-Z deformation conditions, the specimens showed a larger stress drop after a certain amount of deformation. EBSD-based quantitative analysis was used to study the microstructural transition between high and low Z values. At low Z values, the occurrence of strain induced boundary migration (SIBM) as an initiation of dynamic recrystallization (DRX) was clearly observed. On the other hand, at high Z values, grain subdivision phenomena led to very fine and elongated structures.

1. Introduction

Dynamic recrystallization (DRX) is a type of microstructural evolution which leads to grain refinement of metals during hot deformation, especially those under low Zener-Hollomon (Z) parameter conditions. The Z parameter is expressed as $Z = \dot{\varepsilon} \exp(Q/RT)$, where $\dot{\varepsilon}$ is the strain rate, Q is the activation energy of deformation, R is the gas constant and T is the absolute temperature. Unlike metals with a low/medium stacking fault energy in which DRX easily happens, easy recovery metals with a high stacking fault energy including ferritic iron with a bcc structure do not show typical necklace-shape DRX morphologies along grain boundaries [1]. Although it is generally accepted that dynamic recovery (DRV) plays the main role in the dynamic restoration process of ferrite in this case, evidence of DRX phenomena happening in IF steel at low Z values has been also confirmed by some researchers [2,3].

On the other hand, surprisingly, it has been observed that grain refinement can also be achieved in medium and low temperature deformation or deformation under high Z condition by employing sufficient strains. The refined microstructure has been well explained by grain subdivision phenomena, which is mostly the case in severe plastic deformation (SPD) processes. During straining, initial grains are subdivided by deformation-induced boundaries into smaller domains with different operative slip systems as well as other fine micro-volumes so-called cell blocks and cells. All deformation-induced boundaries between these fragmented areas can be categorized into two groups; i.e., incidental dislocation boundaries (IDBs) and geometrically necessary boundaries (GNBs). With further straining, the misorientation angle across dislocation boundaries (especially across GNBs) tends to increase but the boundary spacing decreases [4-6].

Since different deformation mechanisms within a wide range of Z values seem to lead to grain refinement, the aim of the current research is to systematically study the microstructural evolution of highly strained IF steel at different deformation temperatures and strain rates. To do so, hot torsion deformation was employed to realize continuous deformation under precisely controlled deformation conditions.

2. Experimental

A Ti-added IF steel with a chemical composition of 0.002 C, 0.01 Si, 0.12 Mn, 0.003 P, 0.003 S, 0.034 Ti, 0.028 Al, and 0.002 N (wt%) was used in this study. The material was received as square bars (15 mm × 15 mm) with 300 mm-length cut from a hot-rolled/air-cooled plate with a fully recrystallized structure. Torsion specimens with a 4 mm gauge length and 8 mm gauge diameter (figure 1) were cut off from the as-received bars in a way that the specimen axis was parallel to the rolling direction (RD).

Hot torsion tests were carried out at temperatures of 650 ºC, 750 ºC and 850 ºC and strain rates of 0.01 s^{-1}, 0.1 s^{-1}, and 1 s^{-1} using a torsion machine equipped with a high-frequency induction heating system (Thermecmastor-TS). Under an Argon (Ar) atmosphere, the torsion specimens were heated up to a designed temperature below the bcc-to-fcc phase transformation at a heating rate of 2 ºCs^{-1} using an induction heating system. Then the samples were held at those temperatures for 5 minutes for homogenizing and subsequently deformed up to an equivalent strain of ~ 7. Additionally, torsion-deformations at lower strain levels were carried out to study the microstructural evolution of IF steel during the hot deformation. It should be noted that torsion deformation provides different strains and strain rates even within one specimen, both increasing from the center to the surface of the torsion specimen. For both the equivalent strain and strain rate mentioned in this research, the specimen surface was taken as the reference position for calculation. Immediately after the torsion deformation, samples were water-quenched within less than 1 second to stop any microstructural change related to post-dynamic restoration.

The microstructure was investigated using electron backscattering diffraction (EBSD) measurements and electron channelling contrast imaging (ECCI) carried out in field emission scanning electron microscopes (FE/SEM- JEOL JSM 7100F and 7800F, respectively). The microstructure of the deformed gauge part was studied on a tangential section beneath the specimen surface (~ 0.9R; R is the radius of gauge part). For EBSD observations, electro-polishing in a 10%$HClO_4$-90%CH_3COOH solution was employed after mechanical grinding the surfaces. A different surface preparation method was used for

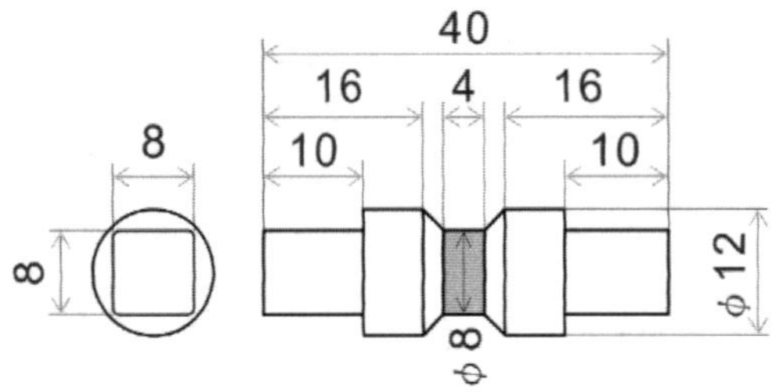

Figure 1. Shape and dimension of the torsion specimen (mm) with highlighted gauge part.

the channelling contrast imaging; i.e., mechanical grinding followed by mechanical polishing using 3 μm and 1 μm diamond paste and final polishing by a 0.04 μm colloidal silica suspension.

3. Results and discussion

3.1. Flow stress curves

Torque-rotation data obtained by the torsion test was converted into true shear stress-strain data by employing Fields and Backofen analysis [7] and subsequently von Mises criteria was used to attain equivalent stress-strain data [8]. Equations (1) and (2) show the relationships of equivalent strain (ε)-rotation number (N) and equivalent stress (σ)-torque (Γ), respectively, where R is the radius of the torsion specimen, L is the gauge length, n is the strain hardening coefficient (~ 0; near the peak and steady state condition) and m is the strain rate sensitivity (ranged from 0.1 at 650 °C to 0.2 at 850 °C in this study).

$$\varepsilon = \frac{2\pi R}{\sqrt{3}L} N \tag{1}$$

$$\sigma = \frac{\sqrt{3}\Gamma}{2\pi R^3}(3 + n + m) \tag{2}$$

Hot-torsion flow curves of the IF steel under different deformation conditions are depicted in figure 2(a). As is seen, they show typical flow curves of easy-recovery metals; i.e., a peak stress (σ_p) followed by slight softening before reaching a steady state stress. However large stress drops after the peak stress are observed for the deformation at 650 °C. In addition, it seems that the highest flow stress curve associated with the highest Z value (650 °C/1.0 s^{-1}) has not yet achieved a steady state condition even at a strain around 7. It is also noteworthy in hot torsion deformation of the IF steel in this study, that no matter under what condition a specimen was deformed, low stress levels (< 100 MPa) were recorded for all cases.

Based on the data derived from figure 2(a), peak stress and peak strain (ε_p) as characteristic parameters of the flow curves were measured. The dependence of these parameters on Z values is illustrated in logarithmic scales in figure 2(b). For calculating Z parameter, the activation energy of self-

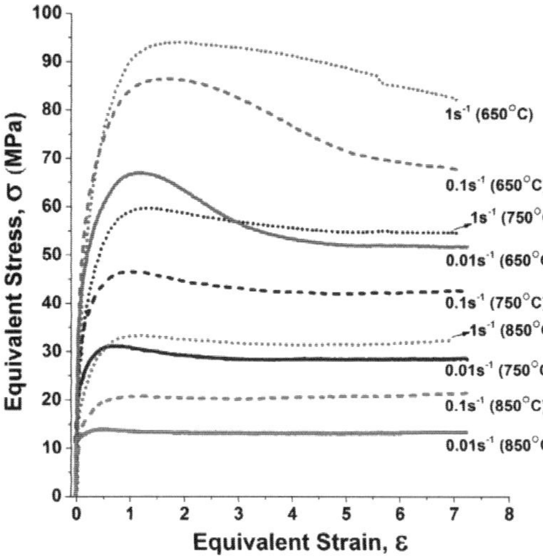

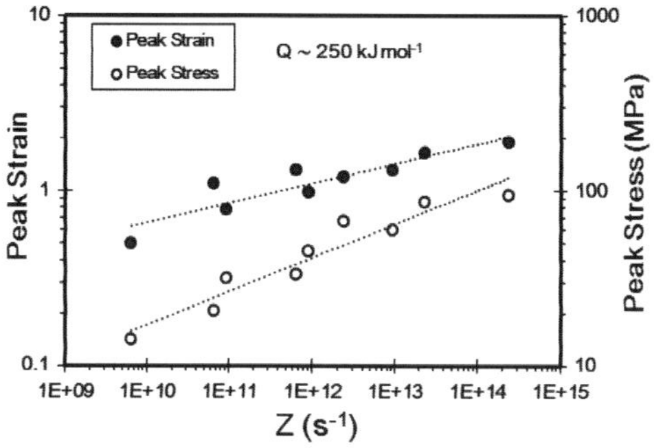

Figure 2. (a) Hot-torsion flow curves (von Mises equivalent stress-strain curves) of the Ti-added IF steel highly strained at different temperatures (650 °C - 850 °C) and strain rates (0.01 s^{-1}- 1.0 s^{-1}), (b) The dependence of peak strain (ε_p) and peak stress (σ_p) on the Zener-Hollomon (Z) parameter.

diffusion in bcc-iron (~ 250 $kJmol^{-1}$) was used as the activation energy (Q) of hot deformation. As is seen in figure 2(b), for both peak stress and peak strain, the dependence is well satisfied with power-law relationships on data. A power-law type relationship is generally the case of low flow stress levels in hot deformation [9].

3.2. Microstructures in high strain

Figure 3(a) and 3(b) show EBSD grain boundary maps of the deformed IF steel under two extreme conditions in the current study; i.e. the highest temperature/lowest strain rate (850 °C/0.01 s^{-1}) and the lowest temperature/highest strain rate (650 °C/1.0 s^{-1}), which correspond to the lowest and highest Z values, respectively. Between these two extremes, the refinement process was enhanced by decreasing the temperature and increasing the strain rate in a way that the range of mean linear intercept length for high angle boundaries ($\theta > 15°$) changed from ~ 35 μm (at 850 °C/0.01 s^{-1}) to ~ 1.5 μm (at 650 °C/ 1.0 s^{-1}). In general, it was observed that the structural refinement of the IF steel did not occur homogeneously which was previously reported by Tsuji et al. [3]. In other words, the deformed microstructure typically contained some large elongated grains with many low angle boundary segments ($2° < \theta < 15°$) inside, as well as relatively smaller elongated grains and newly developed-equiaxed grains.

Moreover, the fraction of high angle boundaries (F_{HABs}) and misorientation angle (θ) measured for all the deformation conditions are depicted versus the Z parameter in figure 4. According to this figure, at least two evolution regimes can be found for the deformation conditions in this study. Firstly, the specimens highly deformed at 850 °C show no Z-dependence of F_{HABs} and θ by varying strain rates (0.01 s^{-1}, 0.1 s^{-1}, and 1.0 s^{-1}), although microstructural refinement still occurs with increasing strain rates. The values of F_{HABs} and θ under these high temperature-deformation conditions are scattered around 0.4 and 19°, respectively. Furthermore, another trend is observed for the specimens highly deformed at 750 °C and 650 °C. Within this region, F_{HABs} and θ clearly increase with increasing Z parameter or decreasing temperature/increasing strain rate; i.e., F_{HABs}/θ increase from ~ 0.3/15° (at 750 °C/0.01 s^{-1}) to ~ 0.7/30° (at 650 °C/1.0 s^{-1}). This tendency was also found to be consistent with the microstructural observation. The microstructure of the specimen deformed under the lowest Z value condition in the region 2 of figure 4 (750 °C/0.01 s^{-1}) typically showed elongated grains with many subgrains inside. However, the specimen deformed under the highest Z value condition in this region (650 °C/1.0 s^{-1}) contained many fine grains with a smaller amount of detectable subgrains ($2° < \theta < 15°$) inside.

It is believed that the change in microstructural evolution in the region 2 suggests the occurrence of more effective grain subdivision phenomenon while in the region 1 with lower Z values, different deformation mechanism must be dominant. To prove this point, nevertheless, not only the highly

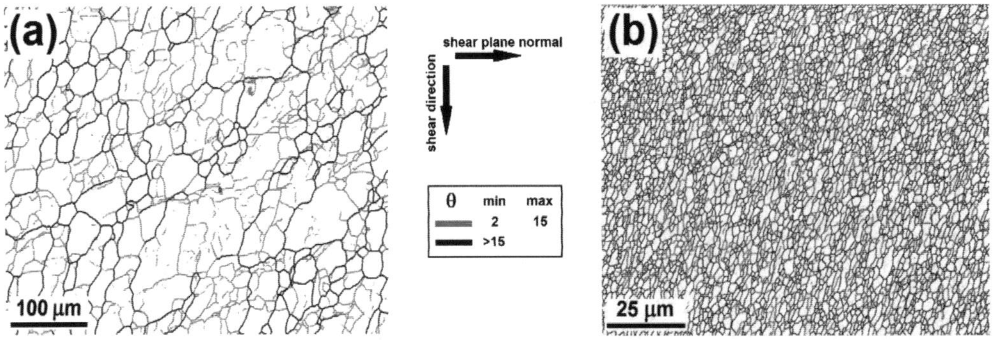

Figure 3. EBSD grain boundary maps of the hot torsion-deformed IF steel (a) at the lowest Z value (850 °C/0.01 s^{-1}) and (b) the highest Z value (650 °C/1.0 s^{-1}) in the current study. Red lines and black lines represent low angle boundaries ($2° < \theta < 15°$) and high angle boundaries ($\theta > 15°$), respectively. Data with misorientation angles below 2° were eliminated to prevent possible errors caused by the limitation on angular resolution of EBSD measurements.

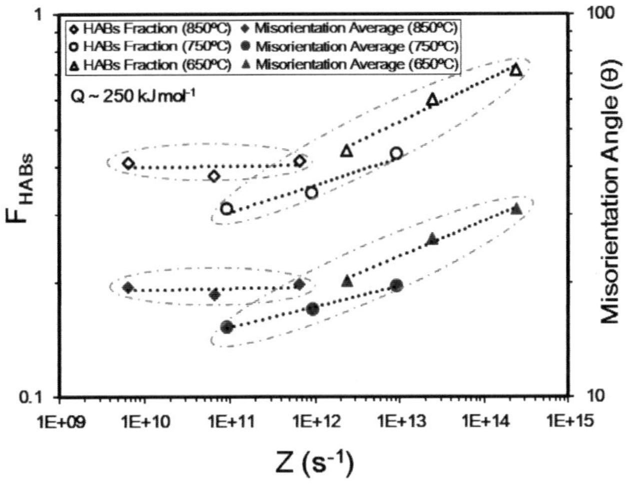

Figure 4. The dependence of high angle boundary fraction (F_{HABs}) and misorientation angle (θ) on Zener-Hollomon parameter.

deformed microstructures under steady state conditions are necessary but also microstructural evolution at different strain levels (before reaching steady state condition) must be investigated. This topic is discussed in the section 3.3.

3.3. Effect of strain on microstructural evolution

To study how the microstructure evolves from the initial microstructure during torsion-deformation, the specimens were gradually deformed at different strain levels especially within the region until the flow stress reached a steady state condition. For this purpose, three deformation conditions with a constant strain rate (0.01 s^{-1}) were chosen; i.e., 850 °C/0.01 s^{-1}, 750 °C/0.01 s^{-1}, and 650 °C/0.01 s^{-1} which hereafter are referred to as "low-Z", "medium-Z" and "high-Z" conditions, respectively. The results are discussed in the following sections 3.3.1 and 3.3.2.

3.3.1. Low-Z condition (850 °C/0.01 s^{-1}). Figure 5(a) and 5(b) show an EBSD inverse pole figure map and channelling contrast image of the water-quenched specimen deformed up to strain ~ 0.8. Since the peak strain measured in this deformation condition is about 0.4, the strain ~ 0.8 belongs to the post-peak region and is within the softening stage of flow stress curve until steady-state. It is seen that the microstructure shows elongated grains with relatively well-developed equiaxed subgrains inside. Additionally there exist very few small equiaxed grains which can be considered as newly-formed ones. It is also considered that many of those elongated high angle boundaries belong to the pre-existing grain boundaries at this strain level.

Furthermore, strong evidence for the phenomenon of so-called strain induced boundary migration (SIBM) [10] was found during straining before and after the peak stress. SIBM or bulging of pre-existing boundaries into adjacent substructures, shown by black arrows in figure 5(a), is a sign of the initiation of dynamic recrystallization and it is believed that many of those pre-existing grain boundaries in figure 5(a) play an important role in the restoration process at this stage. It should be noted that the boundary migration was clearer at the earlier stages of deformation while at the higher strain levels ($\varepsilon > 1$), a larger number of new equiaxed grains were observed in the microstructures.

3.3.2. Medium-Z (750 °C/0.01 s^{-1}) and high-Z (650 °C/0.01 s^{-1}) conditions. Microstructural evolution under the medium-Z and high-Z deformation conditions was also examined by gradual straining in torsion. It was observed that the microstructures, especially under the high-Z condition, adapted a different evolution regime from that which occurred under the low-Z condition (850 °C/0.01 s^{-1}). Figures

5(c) and 5(e) display EBSD inverse pole figure maps of the samples deformed up to a strain ~ 1.1 under a medium-Z condition and that deformed to ~ 1.8 under a high-Z condition. Similar to the low-Z case, these strain levels also belong to a softening stage of the flow curve after the peak stress.

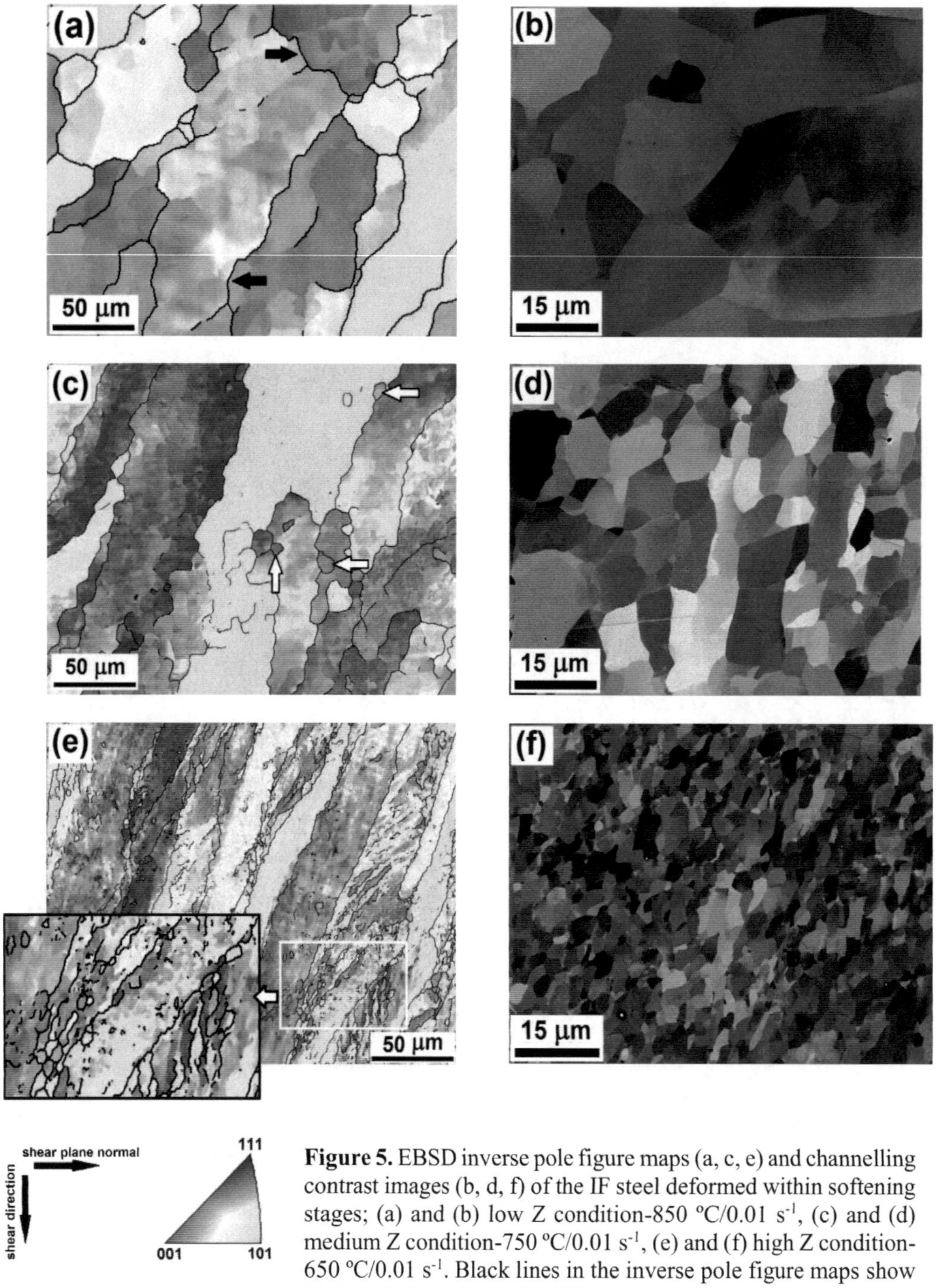

Figure 5. EBSD inverse pole figure maps (a, c, e) and channelling contrast images (b, d, f) of the IF steel deformed within softening stages; (a) and (b) low Z condition-850 °C/0.01 s^{-1}, (c) and (d) medium Z condition-750 °C/0.01 s^{-1}, (e) and (f) high Z condition-650 °C/0.01 s^{-1}. Black lines in the inverse pole figure maps show high angle boundaries (θ>15°).

Similar to the low-Z deformation condition, the microstructure after the peak of the medium-Z specimen in figure 5(c) shows elongated grains as well as a few small equiaxed grains (shown by white arrows). However, the substructures of the IF steel deformed under these two deformation conditions are completely different. The channelling contrast image of the medium-Z specimen in figure 5(d) shows the presence of dense subgrain structures resulting in a high fraction of low angle boundaries. This can be the reason why the specimen highly deformed at 750 °C/0.01 s^{-1} showed the lowest F_{HABs} (figure 4) among all the deformation conditions. Further microstructural observation at the early stages of deformation (before the peak stress) revealed that these fine subgrains originated from a type of substructures characterized by in-grain band patterns.

Moreover, figure 5(e) exemplifies a water-quenched microstructure in the middle of the softening region of the specimen deformed under a high-Z condition. The microstructure shows a significant difference from that of the medium-Z condition although dense substructures also appear under the high-Z condition (figure 5(f)). Unlike the medium-Z case, it is seen (figure 5(e)) that many new high angle boundary segments start to appear within the softening stage of the high-Z specimen. This event is seemingly associated with the noticeable stress drop in the flow curves of the specimens deformed at 650 °C (figure 2(a)). By decreasing the deformation temperature to 650 °C, numerous amounts of new high angle boundary segments were observed within the matrix where strong banding patterns or so-called deformation bands existed. By further straining under this condition, it was found that deformation bands with in-grain lamellar structures were gradually replaced by newly-developed grains and completely disappeared nearly at the end of softening stage ($\varepsilon \sim 5$).

As a rule of thumb, boundaries with a higher misorientation angle can provide higher migration mobility [11]. Based on this fact, it is reasonable to suggest that many high angle boundary segments previously formed within deformation bands can migrate and bulge out into new grains. This is similar to the case for static recrystallization of cold rolled and annealed IF steel where nucleation at an early stage of recrystallization was reported to occur on two intersecting sets of shear bands [12]. The band intersections which can provide new boundary segments with higher misorientation angles were also observed in this study at the early deformation stages ($\varepsilon \sim 0.6$) of high-Z specimen and became more frequent by further straining. It should be noted that in the medium-Z specimen, straining did not lead to significant production of high angle boundary segments within the band patterns.

4. Conclusions

Hot torsion deformation was used to study the microstructural evolution in a Ti-added IF steel in high strain deformation ($\varepsilon \sim 7$) at different temperatures (650 °C, 750 °C and 850 °C) and strain rates (0.01 s^{-1}, 0.1 s^{-1}, and 1.0 s^{-1}). The results are summarized as follows;

Hot-torsion flow curves of the IF steel showed a peak stress followed by slight softening although a larger stress drop after the peak stress was observed under deformation conditions at 650 °C. The dependence of both peak stress and peak strain on the Z parameter followed power-law type relationships.

Decreasing the temperature and increasing the strain rate enhanced the microstructural refinement process. The mean linear intercept length for high angle boundaries ($\theta > 15°$) ranged from ~ 35 μm at the lowest Z value (850 °C/0.01 s^{-1}) to ~ 1.5 μm at the highest Z value (650 °C/1.0 s^{-1}). In addition, the dependence of high angle boundary fraction (F_{HABs}) and misorientation angle (θ) on the Z parameter suggested at least two types of deformation regimes.

In the first deformation regime which happened in highly deformed specimens at 850 °C, F_{HABs} and θ did not show any strain rate dependence although microstructural refinement occurred with increasing strain rates. For the deformation condition at 850 °C/0.01 s^{-1}, the microstructure after the peak stress and within the softening region showed clear evidence for strain induced boundary migration (SIBM) accompanied with the formation of a few equiaxed grains newly formed, especially at the later stages of deformation.

In the second regime happening in specimens highly deformed at 750 °C and 650 °C, F_{HABs} and θ increased with decreasing temperature and increasing strain rate. It was suggested that this trend is caused by more effective grain subdivision phenomenon. Microstructures of the specimens deformed within the softening region at 750 °C and 650 °C (constant strain rate: 0.01 s^{-1}) more or less revealed the presence of dense substructures within the initial grains. Additionally, at a lower temperature (650 °C), the more frequent occurrence of deformation bands led to the formation of numerous numbers of high angle boundary segments.

Acknowledgements

This study was financially supported by the Grant-in-Aid for Scientific Research on Innovative Area, "Bulk Nanostructured Metals" (area No.2201), and the Elements Strategy Initiative for Structural Materials (ESISM) all through the Ministry of Education, Culture, Sports, Science and Technology (MEXT), Japan. The supports are gratefully appreciated by the authors. R. Gh. gratefully thanks the Japan Ministry of Education, Culture, Sports, Science and Technology (MEXT) for providing a Monbukagakusho scholarship.

References

[1] Sakai T, Belyakov A, Kaibyshev R, Miura H and Jonas J J 2014 *Prog. Mater. Sci.* **60** 130
[2] Najafi-zadeh A, Jonas J J and Yue S 1992 *Metall. Trans.* A **14** 2607
[3] Tsuji N, Matsubara Y and Saito Y 1997 *Scr. Mater.* **37** 477
[4] Kuhlmann-Wilsdorf D and Hansen N 1991 *Scr. Metall. Mater.* **25** 1557
[5] Hughes D A and Hansen N 1997 *Acta Mater.* **45** 3871
[6] Hansen N 2001 *Metall. Mater. Trans.* A **32** 2917
[7] Dieter G E 1988 ed D Bacon (London: McGraw-Hill Inc.) 344
[8] Shrivastava S C, Jonas J J and Canova G 1982 *J. Mech. Phys. Solids* **30** 75
[9] Sellars C M and McTegart W J 1966 *Acta Metall.* **14** 1136
[10] Humphreys F J 1997 *Acta Mater.* **45** 4231
[11] Rios P R, Siciliano Jr F, Sandim H R Z, Plaut R L and Padilha A F 2005 *Mater. Res.* **8** 225
[12] Quadir M Z and Munroe P R 2013 *Steel Res. Int.* **84** 1320

Gradient nanostructured surface of a Cu plate processed by incremental frictional sliding

C Hong*, X Huang and N Hansen

Danish–Chinese Center for Nanometals, Section for Materials Science and Advanced Characterization, Department of Wind Energy, Technical University of Denmark, Risø Campus, 4000 Roskilde, Denmark

*E-mail: chho@dtu.dk

Abstract. The flat surface of a Cu plate was processed by incremental frictional sliding at liquid nitrogen temperature. The surface treatment results in a hardened gradient surface layer as thick as 1 mm in the Cu plate, which contains a nanostructured layer on the top with a boundary spacing of the order of 100 nm. The boundary spacing increases with increasing distance from the surface, and is accompanied with a decrease in hardness from 2179±31MPa in the topmost surface layer to 568±10 MPa in the undeformed matrix.

1. Introduction

Surface modification of metallic materials by introducing a gradient nanostructured surface layer is an effective approach to enhance the strength and service lifetime of materials, as demonstrated in many investigations over the past decades [1]. Various techniques have been developed to produce such surfaces [2–6], but there is room for technique development in terms of flexibility in sample geometry, thickness of the hardened layer, efficiency of the processing, etc. This development must be followed up and underpinned by quantitative characterization and analysis of microstructures and properties, as is also part of the present study. In the present work, a technique called incremental frictional sliding is developed and applied to process a Cu plate, which results in thick surface layers with a gradient nanostructure.

2. Experimental

The experimental setup of the incremental frictional sliding has a pin-on-disk configuration as schematically illustrated in Figure 1a. A tool with a round hard tip is pressed against a Cu plate. While the Cu plate rotates, the tool repeats a back-and-forth movement in the radial direction of the plate. As a result, progressive deformation of the entire surface of the plate is achieved. The tool is constructed of a bearing ball of 3 mm in diameter, a steel tube which contains the ball, and a supporting steel rod inside the tube which presses the ball against the end of the tube where the inner diameter of the tube is made smaller than the diameter of the ball (Figure 1c). During the treatment only the ball on the tip of the tool contacts the plate. In the present configuration, the ball is fixed between the steel tube and the supporting rod without any relative rotation, and therefore it rubs the surface of the plate during the processing. For comparison, the tools in conventional deep rolling processes use a similar configuration, but in that case the ball is supported by high pressure fluid in the tube which allows the ball to rotate and roll on the surface of the plate during processing.

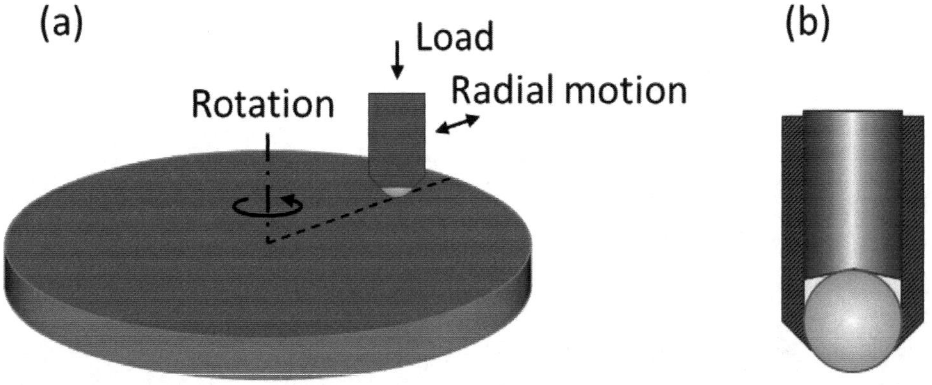

Figure 1. (a) A schematic illustration of the experimental setup for incremental frictional sliding of a Cu plate. (b) A schematic illustration the construction of the tool in (a). The containing tube is cut in cross-section to reveal the bearing ball and the support rod inside the tube.

The plate, 100 mm in diameter and 5 mm in thickness, is made of oxygen free high conductivity Cu with a purity of 99.95%. The plate was annealed at 600 °C for 2 hours prior to the surface processing, resulting in a fully recrystallized structure of tens of micrometers in grain size. During the processing the rotation speed of the plate is 200 rounds per second. To maintain low temperature, the processing is done in 25 interrupted sessions. For each session, the plate was first submerged in liquid nitrogen until the equilibrium temperature was reached, and then with its top surface exposed to the air, the plate is processed for a relatively short duration of 2 minutes during which the tool moved back and forth in the radial direction for about 100 times while the plate was rotating.

Vickers hardness tests and microstructural observations of the processed plate were carried out on the cross section perpendicular to the radius of the plate, at locations about 25 mm from the center of the plate (Figure 2a). Prior to preparation of metallurgical samples, the processed surface was protected by electroplating of Cu, which was done at 32–35 °C for 40 hours and using a solution composed of $CuSO_4.5H_2O$ (72.7 g/L), H_2SO_4 (225 g/L) and NaCl (98.7 mg/L). Metallurgical samples were prepared by mechanical polishing, with a colloidal silica suspension of 0.04 μm in particle size used in the final polishing step. Hardness tests were carried out on a Struers DuraScan-70, using a load of 0.245 N (25 gf). To represent the hardness at a given depth, 2–3 indents were measured. The smallest indent obtained in the present work has a diagonal length of 14.3 μm in the topmost layer, and the largest indent has one of 29.1 μm in the undeformed interior. Due to the limitation of minimum indent size, the smallest depth where hardness was possible to be measured is 8 μm from the surface. The microstructure was investigated using both electron channeling contrast (ECC) and electron backscattered diffraction (EBSD) on a Zeiss Supra 35 scanning electron microscope. A step size of 20 nm was used for EBSD at depths less than 59 μm, where the modal boundary spacing are less than 200 nm, and a step size of 50 nm was used at larger depths corresponding to larger modal boundary spacing.

3. Results

Processing of the Cu plate using the present technique results in substantial microstructural refinement associated with significant hardening below the surface of the plate. Figure 2b shows a typical ECC observation of the microstructure at a low magnification. Within about 100 μm below the surface, the microstructure is too fine to be resolved by ECC at this magnification so that it appears as a dark gray layer without detailed microstructural features. From about 100 μm to 1mm from the surface, the microstructure evolves from an obviously deformed grain structure to a barely deformed coarse-grained structure. Figure 2c shows the hardness in the plate as a function of depth from the surface.

The plate shows a steep hardness gradient down to a depth of 1 mm from the surface. The hardness has the highest value of 2179±31 MPa in the layer just below the surface, and decreases with increasing depth until it reaches the initial hardness of the undeformed material, 568±10 MPa, at a depth of about 1 mm.

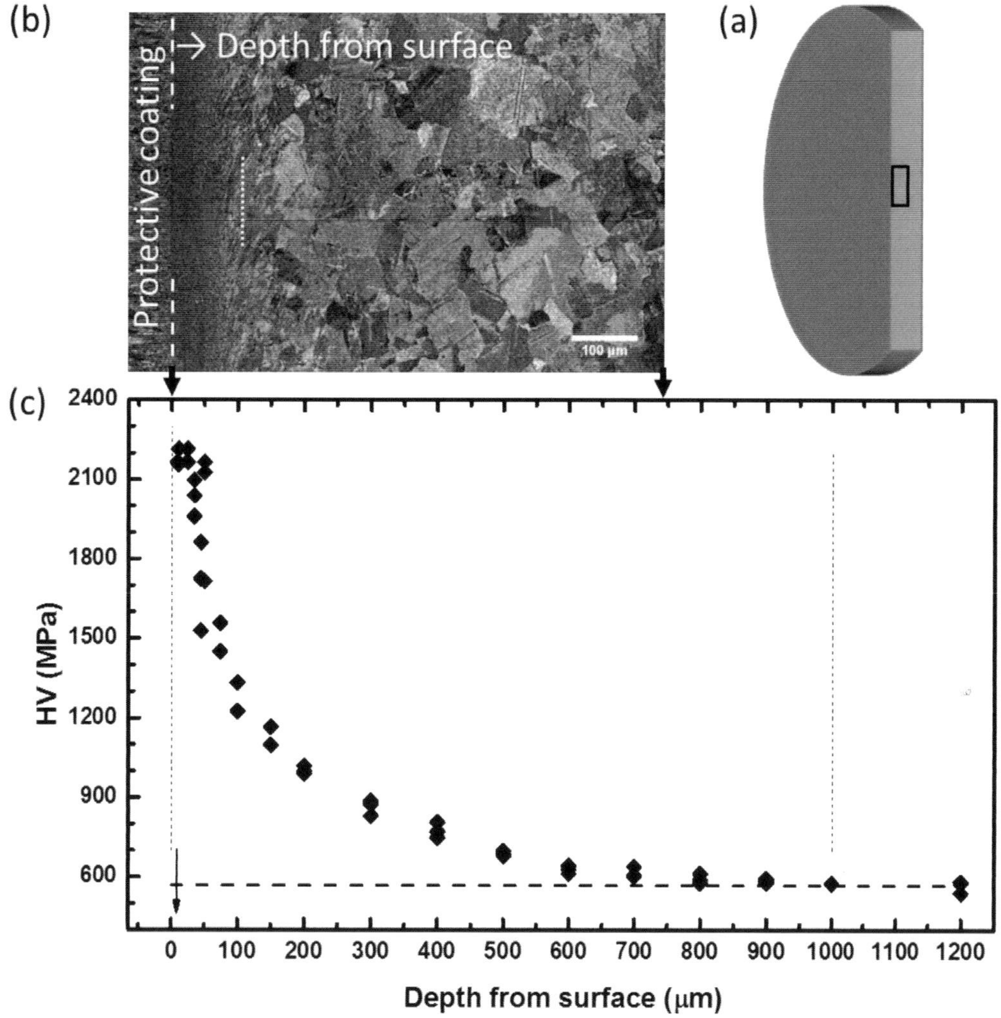

Figure 2. (a) A schematic illustration of the cross-section for the microstructural observations and hardness tests. The rectangle indicates the location of observations in the present work. (b) An ECC image showing the microstructure in the processed Cu plate. The vertical dashed lines mark the surface of the plate. The dotted line roughly marks the thickness range of the dark gray layer. (c) Vickers hardness as a function of depth from the surface. The horizontal dashed line indicates the initial hardness of 568±10 MPa. A pair of vertical dashed lines indicates the thickness range of the hardened layer. The arrow marks the smallest depth (8 µm) where hardness was measured. Note the correspondence between the image in (b) and (c) as marked by a pair of arrows.

The detailed microstructure in the dark gray layer in Figure 2b was further investigated by EBSD with high spatial resolution. In the topmost layer a nanograined structure is developed, as shown in the

EBSD map in Figure 3a. Most of the grains are slightly elongated along the direction of the plastic flow, which is parallel to the surface, and have an aspect ratio of 2.0–2.6. For measurement of boundary spacings, grains in EBSD maps were detected using a critical misorientation of 2° while grain boundary completion using a minimum misorientation of 1° was allowed. Figures 3b and c present statistics of the boundary spacing of the grains measured from Figure 3a. The statistics in Figure 3b use a probability represented by the number of grains, and the distribution shows a modal boundary spacing of 27 nm. The statistics in Figure 3c use a probability represented by the area fraction of grains (which is consistent with the volume fraction of the grains according to the Delesse principle), and the modal boundary spacing is 94 nm. Both statistical methods have been used for measurement of grain sizes of nanostructured materials. For a given microstructure, the modal boundary spacing determined using a probability based on the number of grains is necessarily smaller than the spacing determined based on the area fraction of grains, since larger grains consume a larger area than smaller grains given the same number of grains. However, for a material composed of grains with variation in sizes (and thus with variation in mechanical properties), since the overall mechanical properties are related to the properties of the composing grains via their volume fractions, by e.g. using the rule of mixtures, it is more reasonable to use the latter way of statistics when the mechanical property is of concern. Note that the smallest grains that can be detected by EBSD is limited by the step size (20 nm), and therefore the probability of grains with a boundary spacing smaller than 20 nm are expected to be smaller than the actual value. Nevertheless, both statistical methods show a modal boundary spacing less than 100 nm and indicate the existence of a well-defined nanostructure.

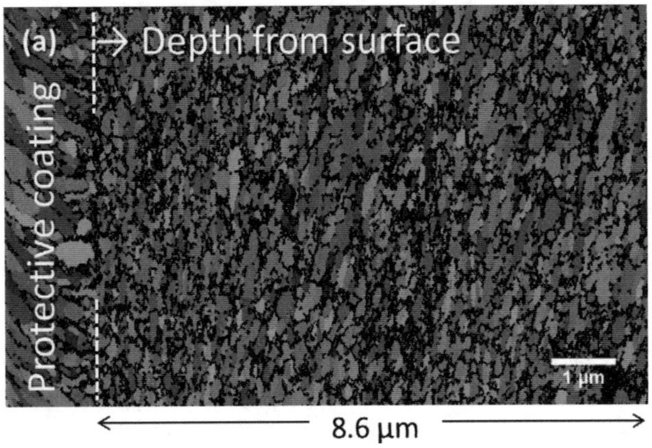

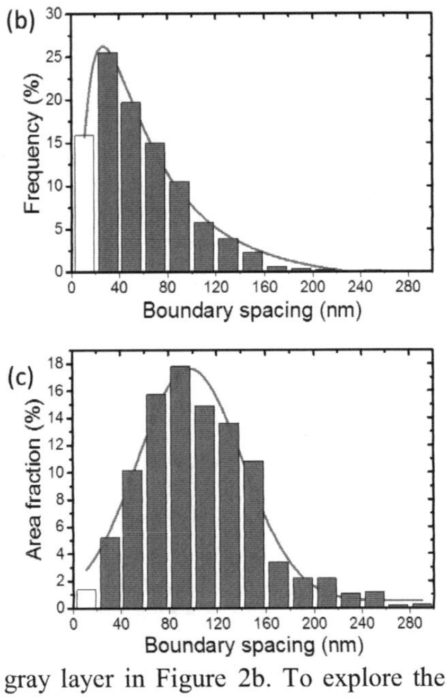

Figure 3. (a) An EBSD map showing the nanostructure in the topmost surface layer in the processed plate. The dashed lines mark the surface of the plate. (b and c) Statistics of the boundary spacing in (a), using a probability based on the frequency of grains and the area fraction of grains, respectively. The white columns in the histograms indicate uncertainty for grains smaller than the EBSD step size.

Figure 4a shows a closer ECC investigation of the dark gray layer in Figure 2b. To explore the variation of boundary spacings as a function of depth, EBSD maps at various depths were obtained. EBSD maps of three typical regions marked in Figure 4a, are shown in Figures 4c, d and e, which covers depth ranges of 0–8.6 μm, 27.6–56.3 μm and 59.0–74.5 μm, respectively. Based on the EBSD maps, the boundary spacing of grains as a function of depth were measured from statistics using a probability based on area fraction of grains. The result (Figure 4b) shows that, a nanograined layer with rather consistent modal boundary spacings ranging within 96±11 nm has developed in the depth range of 0–30 μm. From a depth of 30 μm to 70 μm, the boundary spacing gradually increases to about 300 nm.

For a depth beyond 70 μm, the boundary spacing increases more steeply as the depth increases, with a boundary spacing of 1.9 μm observed at the depth of 90 μm. Note that Figure 4 only covers depths down to 110 μm from the surface. The microstructure at deeper depths is also deformed up to a depth of about 1 mm but is in not covered in Figure 4.

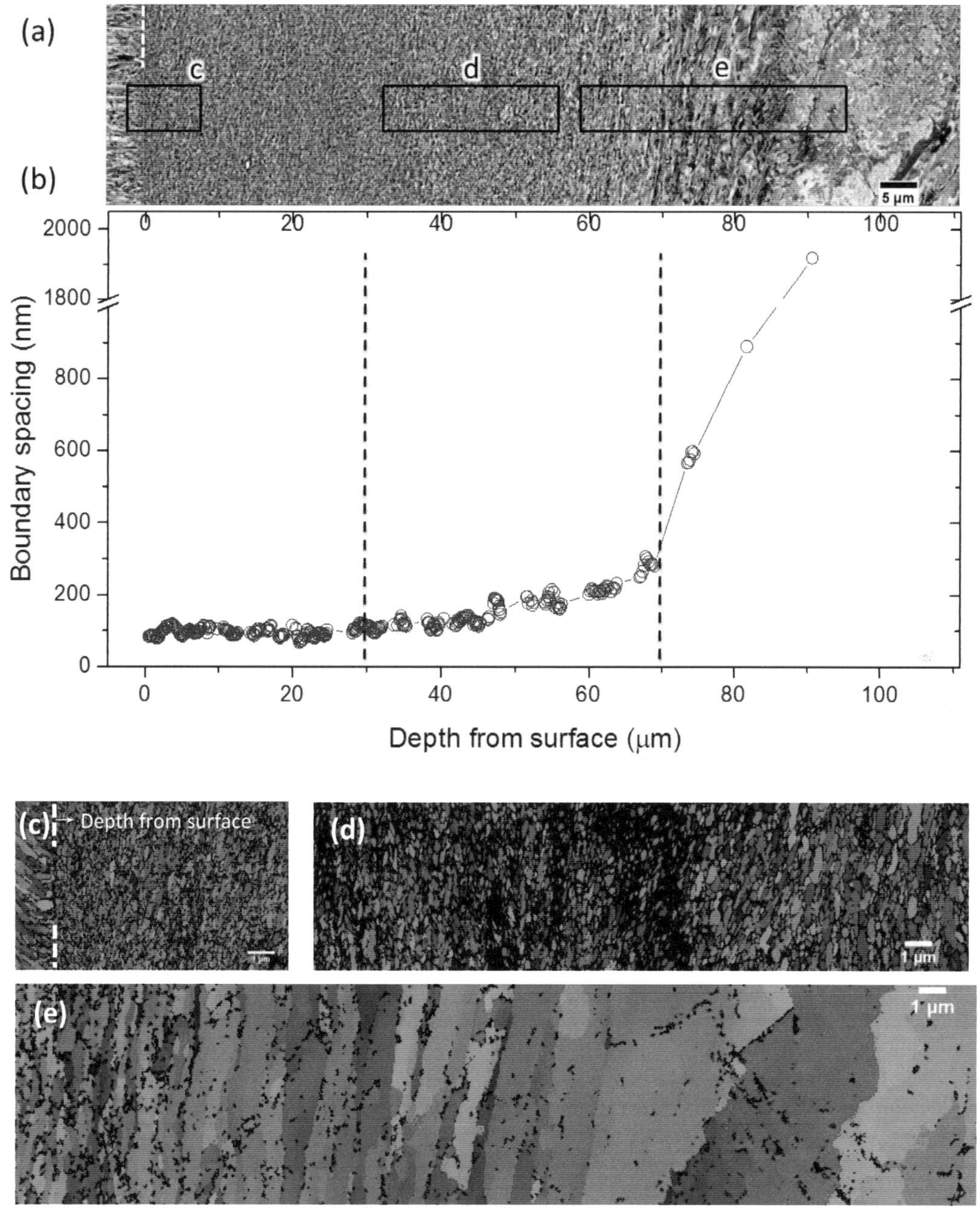

Figure 4. (a) An ECC image of the dark gray surface layer in the depth range of 0–110 μm from surface. The dashed line marks the surface of the plate. (b) Variation of the boundary spacings as a function of the depth from surface, as measured from EBSD maps. The two dashed lines separate regions of different gradients of the plot. Note that the image in (a) and the x-axis in (b) are on the

same scale. (c–e) Typical EBSD maps at various depths from regions marked by rectangles in (a). Note that (c–e) are on the same scale, and (c) is identical to Figure 3a.

The gradient nanostructured sample obtained in the present work may give opportunities to study the effects of microstructural parameters on the thermal behaviour. For example over a depth range of 1 mm the modal boundary spacing decreases from tens of micrometers to below 100 nm and the hardness increases to about 2.2 GPa which is 3.8 times the hardness of the fully recrystallized copper. As to the fraction of high angle boundaries (which are not shown in the present work) they also show a significant increase as the surface is approached. These microstructural changes reflect a significant increase in stored energy reducing the thermal stability in the surface and subsurface layers. This reduction may not be directly important for applications but it gives the opportunity to study kinetics of recovery and recrystallization in the finest scale and also as a function of the distance from the surface. To perform such a study microstructural information obtained by ECC/EBSD must be supplemented by transmission electron microscopy analysis, which can give information for example about triple junction motion [7] as the driving pressure will vary with the distance from the surface. Another study will be formation and growth of recrystallization nuclei into a microstructure with a gradient in driving pressure and differences in boundary characteristics. From an application point of view it may be possible by light annealing to soften the surface and increase its tribological properties while maintaining the advantages of a structure with a gradient from the subsurface to the center.

4. Conclusions

A technique called incremental frictional sliding has been developed and applied to process the flat surface of a Cu plate. The processed plate shows a hardened surface layer as thick as 1 mm with gradients both in hardness and in microstructure. In the topmost surface layer the hardness is increased to 2179±31MPa as compared with the initial hardness of 568±10 MPa, and a nanostructure is developed. This gradient microstructure is difficult to quantify when it comes to the relation between microstructure, strength, and thermal behavior. However new and important information can be gained about the general behavior of metals on a structural scale from the micrometer to the nanometer dimension.

Acknowledgments

The authors gratefully acknowledge the support from the Danish National Research Foundation (Grant No. DNRF86-5) and the National Natural Science Foundation of China (Grant No. 51261130091) to the Danish-Chinese Center for Nanometals, within which this work has been performed. C.H. also thanks support from the Danish Council for Independent Research (Grant number: DFF-1335-00230). C.H. thanks Lars Lorentzen and Johannes Steen Bang for help with the experimental setup for incremental frictional sliding treatment, and thanks Dr. Fengxiang Lin for help with polishing Cu samples.

References

[1] Lu K and Lu J 2004 *Mater. Sci. Eng. A* **375-377** 38–45
[2] Fang T H, Li W L, Tao N R and Lu K 2011 *Science* **331** 1587–90
[3] Zhang Y S, Li W L, Wang G, Zhang L C, Yao B and Han Z 2012 *Mater. Lett.* **68** 432–4
[4] Hughes D and Hansen N 2001 *Phys. Rev. Lett.* **87** 135503
[5] Li W L, Tao N R and Lu K 2008 *Scr. Mater.* **59** 546–9
[6] Zhang X, Hansen N, Gao Y and Huang X 2012 *Acta Mater.* **60** 5933–43
[7] Yu T, Hansen N and Huang X 2011 *Proc. R. Soc. A Math. Phys. Eng. Sci.* **467** 3039–65

Correlation between the deformation microstructure after rolling and the recrystallization nucleation of a non-oriented electrical steel

WC Hsu, PL Sun*, PW Kao and L Chang
Department of Materials and Optoelectronic Science, National Sun Yat-Sen University, Kaohsiung 80424, Taiwan

E-mail: plsun@mail.nsysu.edu.tw

Abstract. The characteristics of the cold-rolled (CR) microstructure of a non-oriented electrical steel after 50-90% cold rolling were studied using electron backscatter diffraction. The local heterogeneity of the deformed microstructure was found to consist of high misorientation regions (HMRs) and low misorientation regions (LMRs) according to local misorientation and size. The overall orientation distribution of HMRs was obtained accordingly and was compared to that of the fully recrystallized sample. Moreover, the CR and partially recrystallized states of the same area were analyzed and compared. The position and orientation of each HMR can be identified individually, and be correlated with the recrystallization nuclei. The HMRs were distributed near the original high angle boundaries at 50% rolling reduction, but extended into the grain interiors with increasing rolling reductions to 70% and 90%. As a result, the recrystallized nuclei for the 50% cold rolled sample were mainly distributed at the original high angle boundaries and the triple junctions, whereas extended nucleation at the deformation bands was observed for the high rolling reduction samples. Good correlation of the orientation between the HMRs and the recrystallized nuclei were observed.

1. Introduction

Texture evolution of non-oriented electrical steels from a deformed state to a fully recrystallized (RX) state during heat treatment has been an important issue in the academia and industry [1-3]. The cold rolling (CR) texture contains mainly a α-fiber, {hkl}<110>, which shows orientations from {001}<110> through {112}<110> to {111}<110>. Furthermore, a weak γ-fiber, {111}<uvw> has also been identified in the CR texture. The strong α-fiber develops during rolling has been noted to disappear after RX. On the contrary, the RX texture, comprising varying intensities of {001}<uvw>, {111}<uvw> and {011}<100> components, develops during annealing. It has been recognized experimentally that the RX nuclei evolve from fast growing subgrains surrounded by high angle boundaries (HABs) at high rolling reductions [4-7]. Additionally, Humphreys [8] proposed a unified theory determining whether RX is discontinuous during heat treatment according to the size and misorientation of the subgrains. Both experimental observations and theoretical calculations imply that the texture of the recrystallized steel evolve from the subgrains having high misorientation angles relative to the surrounding matrix in the deformed state.

The deformed microstructure is usually not homogeneous. For instance, extended planar dislocation boundaries forming cell blocks were observed in CR IF steel [3]. The presence of bands of localized micro-shearing was noticed to be stronger than in fcc aluminum and nickel [3]. RX nucleation sites are expected to closely depend on the heterogeneities in the deformed microstructure. It is therefore important to differentiate the structural heterogeneities and identify the subgrains, which have the potential to develop into RX nuclei, in the CR structure. Orientation image mapping (OIM) based on electron backscatter diffraction (EBSD) can give the spatial distribution of crystal orientations over large areas. Consequently, EBSD analyses are suitable for characterizing the heterogeneities in the deformed structures. HABs and the differences of stored energy across boundaries are related to the local misorientation in the deformed structure, which can be determined using EBSD with proper scan step sizes. Brewer *et al.* [9] showed that local misorientation distribution in CR structures can be illustrated using kernel average misorientation (KAM) method. Moreover, Mishin *et al.* [10-12] proved the use of EBSD to analyze local heterogeneity in deformed metals. The deformed microstructure can be considered as a two-phase mixture containing both high misorientation regions (HMRs) and low misorientation regions (LMRs). Subsequently, the LMR/HMR map can be constructed even when the specific characteristics of the low angle boundaries in the microstructure cannot be completely identified [10-12].

In the present work, the same areas in both the CR and RX states of an electrical steel were examined using EBSD. HMRs and LMRs can be distinguished in the CR state. It was revealed that the potential RX nucleation sites in the CR structure are the places located inside the HMRs and having high KAM values. Places satisfying the above criteria were the original grain boundaries in the 50% CR sample and extended to deformation bands in the 70% and 90% CR samples.

2. Experimental

A non-oriented electrical steel (ES) with the composition (wt%) of 0.0015 C, 0.86 Si, 0.31 Mn, 0.22 Al, 0.0021 N, 0.005 P and 0.005 S was used. The steel was hot rolled from 100 mm to 12 mm in thickness with a finishing temperature of 860°C and followed by annealing at 740°C for one hour. The hot rolled plate was subsequently cold rolled to 50%, 70% and 90% rolling reductions. RX temperatures of the three samples were determined from the microhardness value change as a function of the annealing temperature.

Texture analyses were carried out on the normal direction (ND) planes of the samples by X-ray diffraction (XRD) with Co-Kα radiation. The orientation distribution functions (ODFs) were calculated from three pole figures, (200), (220), and (211), using the series expansion method. Microtexture measurements were conducted on the transverse direction (TD) planes for the same areas of the CR states and partially RX states using EBSD. A step size of 0.25 μm was selected to scan four areas in each sample. The local misorientation distribution in the CR state was characterized using KAM approach with a selection of KAM > 10° as high-KAM. The heterogeneity of the CR structure was examined using LMRs and HMRs. It is essential to identify LMRs and the rest of the microstructures are HMRs. θ_0 is the maximum misorientation between adjacent pixels in a LMR. A_L defines the minimum size of LMR. In this work, θ_0 of 5° and A_L of 10 μm were chosen according to our previous analyses [13].

3. Results and Discussions

The ODF section at φ_2 = 45° for the CR and fully RX states are given in figure 1. It can be noted that the intensity of the α-fiber increases gradually, whereas that of the {111}<112> component decreases slightly, with increasing rolling reduction. The strongest orientation is close to {112}<110> along the α-fiber. These observations are consistent with the reports in the literature [1-4]. The fully RX states of the CR samples exhibit a considerably different texture. It can be detected that the intensity of the α-fiber diminishes drastically while the γ-fiber becomes the dominant texture. The intensity of the γ-fiber is noticed to become stronger with rolling reduction. The {111}<112> component along γ-fiber is the most intense orientation, while Goss orientation ({011}<100>) is also observed. However, the intensity of Goss is found to decrease with rolling reduction. The ODFs of the high-KAM sites inside the HMRs (denoted as K-H sites thereafter), which are extracted from the CR samples, are also

IOP Conf. Series: Materials Science and Engineering **89** (2015) 012027 doi:10.1088/1757-899X/89/1/012027

provided in figure 1. It is clear that the texture of the deformed substructures which fulfill both high-KAM and HMRs requirements correlates well with the texture of the RX grains. The intensity of the {111}<112> component increases with increasing the rolling reduction, whereas that of the Goss component decreases as the rolling reduction increases from 70% to 90%. The HMRs are regions containing high stored energy in which RX can proceed preferentially from the thermodynamic point of view. Moreover, the K-H sites exhibit high boundary mobility which allows a fast growth kinetically. The results therefore indicate that the potential nucleation sites in RX can be screened from the CR microstructure effectively.

The annealing temperatures for partial RX (~20 % RX) were determined from the microhardness tests. 650°C, 630°C and 610°C were chosen for the 50%, 70% and 90% CR samples, respectively. It is reasonable that the RX temperature decreases with increasing deformation strain since a higher stored energy is retained in the more severely deformed structure [14]. The ND inverse pole figure (IPF) maps of the 50% CR sample and the partial RX sample are shown in figure 2a and 2b. It can be noted that the deformed grains become slightly elongated and align with the rolling direction (RD). The average long axes and short axes of the grains were measured to be over 100 μm and about 20 μm, respectively. The CR grains are observed to show some extent of heterogeneity. The grains with orientations along the α-fiber are more elongated while the grains with γ-fiber orientations have smaller aspect ratio. Additionally, there are some deformation bands within the grain interiors. It can also be seen that some grains have more deformation bands subdividing them, which show multiple orientations while others (mostly the grains with orientations along α-fiber) are only slightly deformed. This is an obvious evidence of deformation heterogeneity. The partially RX IPF map of the same area clearly indicates the locations of RX nuclei (figure 2b) and can be directly compared to the CR structure. These RX grains are seen to locate at the original grain boundaries. More apparent confirmation will be given later.

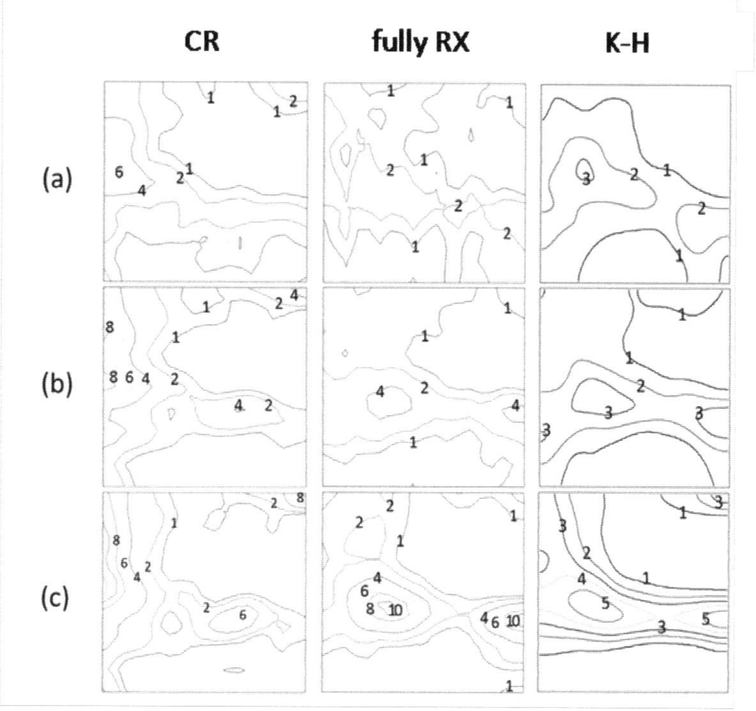

Figure 1 ODF φ_2 = 45° section for the (a) 50%, (b) 70% and (c) 90% CR samples. **CR**: XRD ODF for the as-rolled samples; **fully RX**: XRD ODF for the samples annealed at 710 °C for 60 s; **K-H**: EBSD ODF for the K-H sites screened from the CR samples.

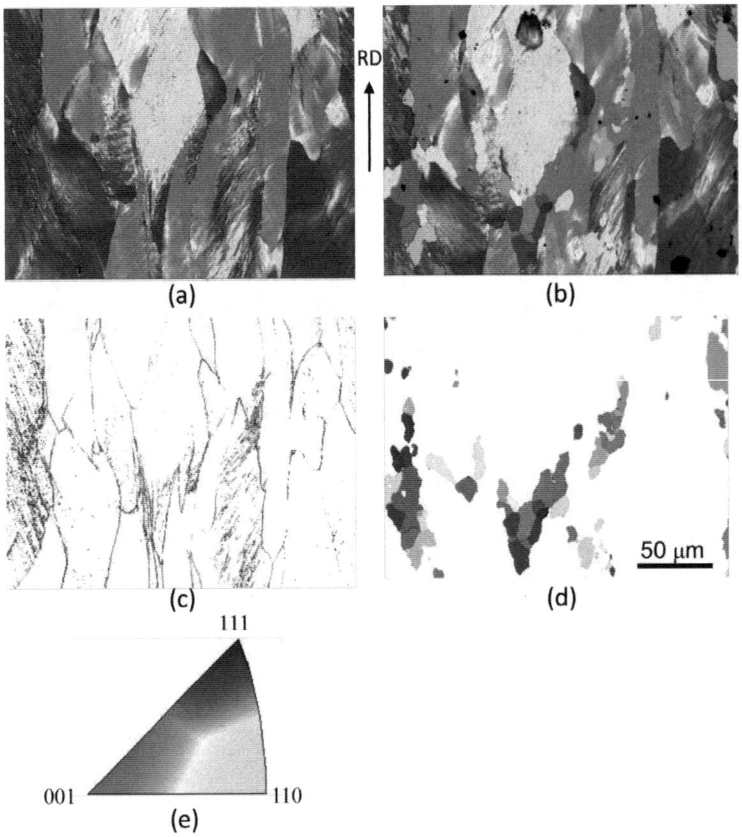

Figure 2 (a) ND IPF map of the CR state, (b) ND IPF map of the partial RX state, (c) K-H sites, and (d) the extraction of RX grains for the 50% CR sample. (e) color key as a reference for the IPF maps.

As the ES sample is cold rolled to a 70% reduction, the grains become more elongated (figure 3a). The long axes of the grains are measured to be over 200 μm, while the short axes are about 15μm. Similar to the 50% CR sample, many deformation bands with different orientations can be noticed within the original grains. Some grains are subdivided to a higher extent while others are not. An IPF map of the partial RX for the same area is presented in figure 3b, which can be used to make a direct comparison. The IPF maps for the K-H sites are revealed in figure 2c and 3c for the 50% and 70% CR samples, respectively. These figures indicate that the K-H areas are mainly distributed near and at the original grain boundaries and triple junctions with a small portion of them in the grain interiors in the 50% CR sample. Nevertheless, the K-H sites extend extensively to the interior of the γ-fiber grains in the 70% CR sample. The RX grains are extracted from figure 2b and 3b, and given in figure 2d and 3d. It is observed that the distribution of the RX grains coincides well with that of the K-H sites in both conditions. Moreover, a good correlation of texture between the K-H sites and the RX grains can also be observed. The use of HMR/LMR to identify the heterogeneity of the deformed microstructure was reported by Mishin *et al.* [10] in equal channel angular extrusion (ECAE) processed copper. These samples were kept at room temperature for eight years and RX occurred. It was found that RX grains appear typically within HMRs, which are located at shear bands, extended boundaries and the intersections of shear band and extended boundary. These observations are consistent with the present work.

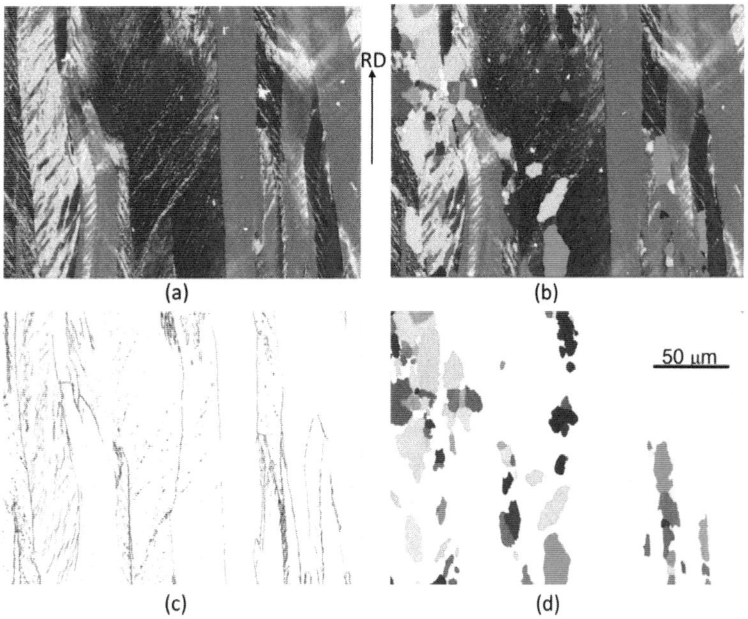

Figure 3 (a) ND IPF map of the CR state, (b) ND IPF map of the partial RX state, (c) K-H sites, and (d) the extraction of RX grains for the 70% CR sample.

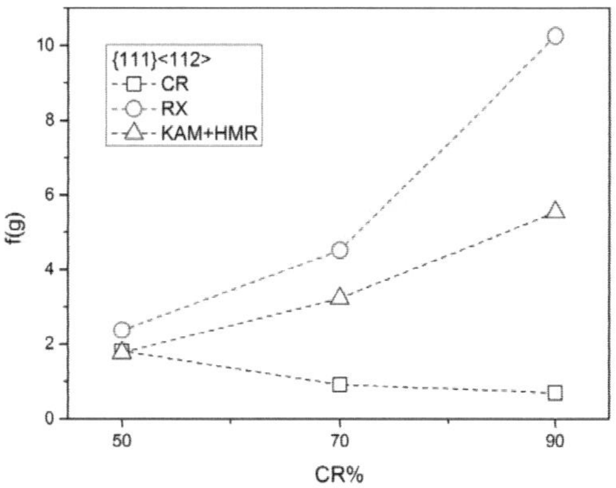

Figure 4 The intensity of {111}<112> orientation in the CR and RX states as well as K-H sites changes as a function of CR reduction.

As mentioned in the early section, {111}<112> is an important texture component observed in the RX state for the ES. However, the intensity of {111}<112> is relatively weak in the CR states. Furthermore, the intensity of {111}<112> is noticed to decrease with rolling reduction, as shown in figure 4. The decline is due to the formation of a strong α-fiber with rolling reduction. It is remarkable that the weak {111}<112> orientation becomes the dominant texture component in the RX state. The intensity of {111}<112> in the RX state is found to increase substantially with rolling reduction (figure 4). It is interesting to observe the large difference of intensity for the CR and RX states, especially in the 90% CR sample. It means that the small populations of {111}<112> orientation develop into the main component in the RX state. The intensities of {111}<112> of the K-H sites

IOP Conf. Series: Materials Science and Engineering **89** (2015) 012027 doi:10.1088/1757-899X/89/1/012027

obtained from the CR structures are also given in figure 4. It shows the same trend as the RX one with increasing rolling reduction, except that its intensity in the RX state is stronger than that in the K-H sites. The difference is mainly originated from competitive growth of nuclei in RX. Consequently, the combination of high-KAM with HMR criteria is an efficient method to predict the locations of RX nuclei and their orientations in the deformed structure. This is due to the deformation heterogeneity and then inhomogeneous distribution of stored energy. The approach can be employed to materials design, in which RX is important.

4. Conclusions

A non-oriented electric steel was cold rolled to different rolling reductions of 50-90%. EBSD was employed to analyze the deformed and partially RX microstructure and texture. The texture of the K-H sites correlated well to that of the RX samples of three reductions. Furthermore, both the texture and the spatial distribution of the K-H sites showed a good one-on-one correspondence to the nuclei formed in the early stage of RX. Therefore, the combination of high-KAM and HMR criteria is an efficient approach to predict the locations of RX nuclei and their orientations in the deformed structure.

Acknowledgements

This work is supported by the Ministry of Science and Technology, R. O. C (grant number MOST 103-2622-E-006-037). The authors also acknowledge the support from China Steel Corporation.

References

[1] Kestens L, Jonas JJ, Van Houtte P and Aernoudt E 1996 *Metall. Mater. Trans.* A **27** 2347
[2] Hutchinson B 1999 *Phil. Trans. R. Soc. London* A **357** 1471
[3] Li BL, Godfrey A, Meng QC, Liu Q and Hansen N 2004 *Acta Mater.* **52** 1069
[4] Humphreys FJ and Hatherly M 2004 Recrystallization of single-phase alloys *Recrystallization and Related Annealing Phenomena* (Oxford, UK: Elsevier) chapter 7 pp 215-267
[5] Doherty RD et al 1997 *Mater. Sci. Eng* .A **238** 219
[6] Ridha AA and Hutchinson WB 1982 *Acta Metall.* **30** 1929
[7] Rosen GI, Juul Jensen D, Hughes DA and Hansen N 1995 *Acta Metall. Mater.* **43** 2563
[8] Humphreys FJ 1997 *Acta Mater.* **45** 4231
[9] Brewer LN, Field DP and Merriman CC 2009 Mapping and assessing plastic deformation using EBSD *Electron Backscatter Diffraction in Materials Science* ed AJ Schwartz, M Kumar, BL Adams and DP Field (New York, US: Springer) chapter 18 pp 251-262
[10] Mishin OV and Godfrey A 2008 *Metall. Mater. Trans. A* **39** 2923
[11] Mishin OV and Bowen JR 2009 *Metall. Mater. Trans. A* **40** 1684
[12] Mishin OV, Godfrey A and Juul Jensen D 2009 Analysis of deformation structures in fcc materials using EBSD and TEM techniques *Electron Backscatter Diffraction in Materials Science* ed AJ Schwartz, M Kumar, BL Adams and DP Field (New York, US: Springer) chapter 19 pp 263-275
[13] Hsu WC, Chang L, Sun PL, Ho N, Kao PW and Hsiao YC 2015 submitted to ISIJ International
[14] Abbaschian R, Abbaschian L and Read-Hill RE 2010 *Physical Metallurgy Principles* (Stamford, CT: Cengage Learning) chapter 8 pp 216-260

Characterization of Si particles and their effects on and recrystallization in a nanostructured cold rolled Al-1%Si alloy

T L Huang[1], G L Wu[1], X Huang[2], N Hansen[2]

[1] College of Materials Science and Engineering, Chongqing University, Chongqing 400044, China

[2] Danish-Chinese Center for Nanometals, Section for Materials Science and Advanced Characterization, Department of Wind Energy, Technical University of Denmark, Risø Campus, DK-4000 Roskilde, Denmark

E-mail: huangtl@cqu.edu.cn

Abstract. An Al-1.08 vol.%Si alloy was cold rolled to a reduction of 98% (ε_{vM} = 4.5) and then annealed at different temperatures up to 210°C (0.52 Tm) for different times. The deformed structure is characterized by a nanoscale lamellar structure with the presence of Si particles of coarse (> 1μm), medium (100 nm – 1 μm) and fine (< 100 nm) sizes in the microstructure. Deformation zones are formed around the coarse Si particles and the boundary spacing is finer in the deformation zone than in the matrix. The medium Si particles have little effect on the morphology and boundary spacing. The fine Si particles are aligned along the lamellar boundaries indicating a stabilizing effect on the structural refinement during cold rolling. After annealing, enhanced recovery occurs in the deformation zones around the coarse Si particles. However the reduction in stored energy during recovery and the pinning effect of fine Si particles on the boundary migration prevent the advantage of particle stimulated nucleation (PSN) of coarse Si particles in the nanoscale lamellar structure. This study also demonstrates an important effect of the fine particles in delaying both recovery and recrystallization processes. This effect diminishes with increasing annealing temperature and coarsening the fine particles especially at triple junctions.

1. Introduction

The effect of second phase particles on the deformation microstructure and recrystallization of Al alloys has been extensively studied in the literature [1-7]. It is generally observed that the recrystallization kinetics in alloys containing large particles (typically with sizes > 1 μm) is accelerated while that in alloys containing fine particles is retarded. In the case of containing both coarse and fine particles, the overall kinetics of recrystallization could be retarded or accelerated depending on the relative strength of the effects from the two types of particles [3].

In a recent study [8], a nanostructured Al-1%Si (or Al-1.08% vol.%Si) alloy with an ultrahigh purity Al matrix (99.9996%) was produced by heavy cold rolling to 98% thickness reduction (ε_{vM} = 4.5). The nanostructured Al-1%Si alloy is characterized by a lamellar structure with an average lamellar boundary spacing of 230 nm, a fine dispersion of nanosized Si particles and a certain amount of large Si particles [8]. The Si particles of different sizes are expected to affect the structural and textural evolution during deformation and annealing of the nanostructured alloy. In this study, this effect has been analyzed separately for coarse and fine particles including their combined effect.

2. Material and experimental procedures

An Al-1%Si alloy ingot was cast of ultrahigh purity (99.9996%) Al and high purity Si. The ingot was hot deformed into a thick plate before being cold-rolled to produce a nanostructured sheet. A slab of dimensions of 300×300×50mm^3 was cut from the hot deformed plate and cold-rolled to a thickness reduction of 98% (ε_{vM} = 4.5). After each rolling pass, the sample was quenched into liquid-nitrogen-cooled alcohol. The rolled sample was annealed at temperatures of 100 – 210°C for different times to investigate the annealing effect on recovery and initiation of recrystallization.

The microstructure was characterized by means of scanning electron microscopy (SEM), electron backscatter diffraction (EBSD) and transmission electron microscopy (TEM) from the longitudinal plane (containing the normal direction and the rolling direction, RD-ND). The SEM/EBSD characterization was done on an FEI Nova 400 field emission gun scanning electron microscope (FEG-SEM). The SEM samples were polished in a 1:9 (vol. fraction) $HClO_4$: C_2H_5COOH solution at −20°C and 20V for 60 seconds. The electron channelling contrast (ECC) technique was used to observe the microstructure. The samples for EBSD were prepared using a Leica EM TIC 3C system. The TEM foils were prepared by a twin jet electropolishing method and examined with a JEOL 2000FX electron microscope operating at 120kV. An online Kikuchi-line analysis system installed in the microscope was used for crystallographic orientation measurements.

3. Results and discussion

3.1 Microstructure and Si particles in the deformed state

The microstructure of the Al matrix is characterized by a fine scale lamellar structure where the lamellar boundaries are low angle dislocation bounadries and high angle grain boundaries [8]. The fraction of high angle bounadries was measured to be 58.4% [9]. The addition of 1%Si is present as particles of various sizes as the soluability of Si in Al is almost zero at room temperature. The Si particles are separated into three goups depending on their sizes: coarse (>1 μm), medium (100 nm – 1 μm) and fine (<100 nm). Figs. 1-3 show examples of observations for coarse-, medium- and fine-sized Si particles. Quantitative measurements of size and volume fraction are listed in Table 1. Some of the large Si particles observed as clusters of medium sized particles are believed to be the result of breaking during cold rolling of large Si particles formed during casting (solidfication) [10]. The majority of the medium sized Si particles are 300 nm to 1 μm. The fine Si particles show a narrow distribution from 5 nm to 30 nm wih an average of 22 nm.

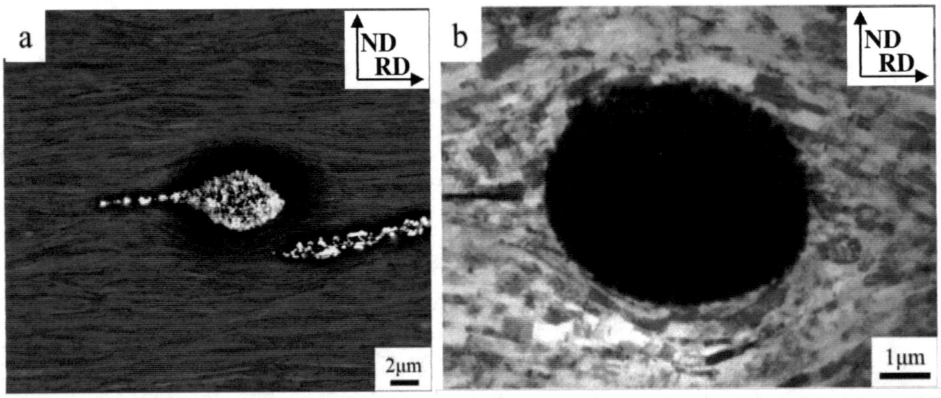

Figure 1. (a) SEM ECC image and (b) TEM image of large a Si particle in a 98% cold rolled Al-1%Si alloy showing the turbulent flow of a aluminum around the particles.

From figure. 1b it is seen that deformation zones of lentical shape are formed around large Si particles where the lamellar boundary spacings becomes finer when approching the particle. However, figure 2 shows that Si particles of medium sizes have little effect on the lamelalr morpholgy of the matrix structure and there are no clear deformation zones around the particles. For the fine particles, figure 3, they align along the lamellar boundaries and are also present in the volumes between boundaries. The alignment of fine Si at the lamellar boundaries indicates that they are fomeed during the cold rolling process. The crucial role played by these fine Si particles in producing a nanoscale structure by cold rolling has been discussed in [9, 10].

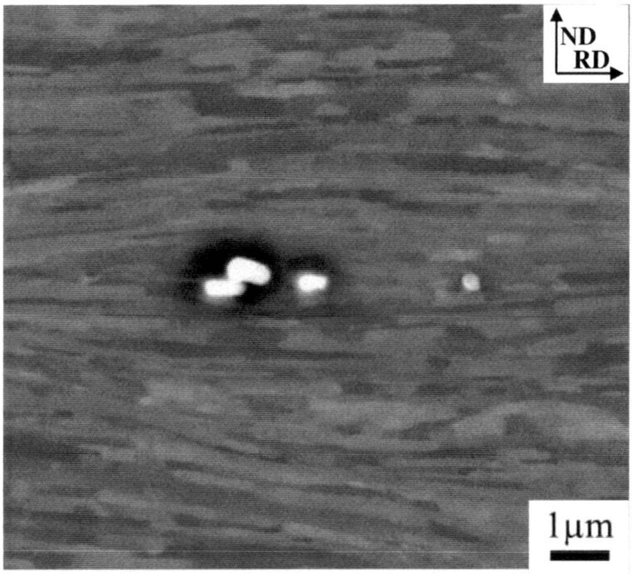

Figure 2. SEM ECC image showing the medium-sized Si particles in a 98% cold rolled Al-1%Si alloy.

To examine the effect of the large Si particles on the deformed structure, the lamellar boundary spacing was measured separatedly for the deformed zones around the large Si particles and for the matrix regions away from the large Si particles. The spacing was measured in the ND i.e. perpendicular to the lamellae and at the particles the spacings between five layers of lamellar structure was analyzed. The measured results are shown in figure 4. The average lamellar boundary spacing in the deformation zones is 180 nm (figure 4a), which is evidently finer than the average value 230 nm measured for the matrix regions (figure 4b).

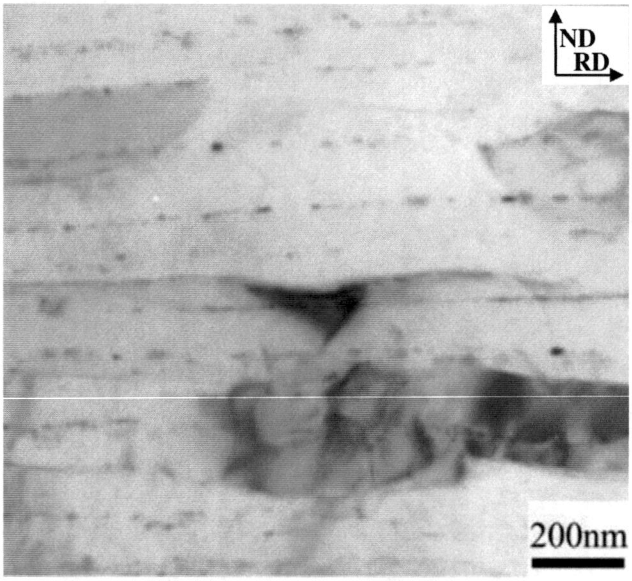

Figure 3. TEM image showing fine Si particles aligned along the lamellar boundaries in a 98% cold rolled Al-1%Si alloy.

Table 1. Size and volume fraction of Si particles in 98% cold rolled Al-1.08 vol.%Si alloy

Particle type	Volume fraction (%)	Average diameter
Coarse (> 1 μm)	0.27	6.80 μm
Medium (100 nm – 1 μm)	0.26	0.65 μm
Fine (100 nm)	0.55	21.5 nm

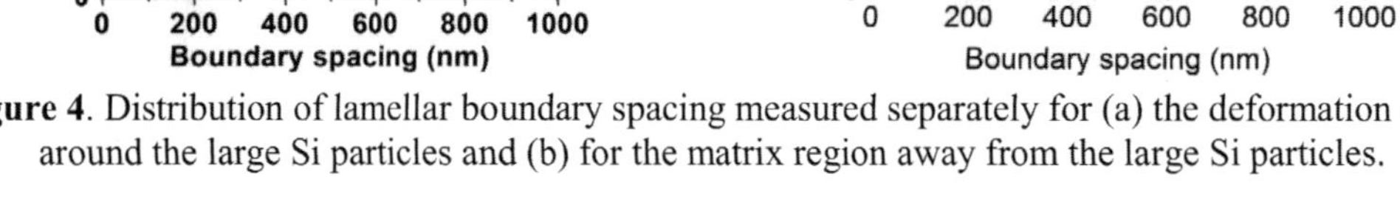

Figure 4. Distribution of lamellar boundary spacing measured separately for (a) the deformation zone around the large Si particles and (b) for the matrix region away from the large Si particles.

3.2 Recovery of the deformation zone around large Si particles

After recovery annealing, particular attention was paid to obseving microstructural changes in the deformation zones around large Si particles. Figure 5 shows two examples of such observations. Figure 5a shows no obvious change in the lemallar morphology in both the deformation zone and the matrix region after annealing for 1 hour at 150℃ as compared with the deformed state (figure 1b). However, quantification of lamellar bounadry sapcings showed that structural coarsenng has taken place by triple junction motion as observed in aluminum of commecial purity [11, 12]. Enhanced recovery in the deformatin zone around the large Si particle is seen after annealing for 1 hour at 200℃ , figure 5b-c. The original fine lamellar structures in the deformed state are now changed to more equiaxed suggrains that have sizes larger than the spacing of lamellar boundaries in the surrouding matrix. Coarsening of fine Si particles may also been seen from figure 5c, but the majority of the fine Si particles are still of sizes less than 50 nm, giving rise to an average of 35 nm. Note that the subgrain sizes around the large Si particles are still in the submicrometre regime, which could be due to the pinning effect of the fine particles present in the matrix as shown in figure 5c. The enhaced recovery of the deformation zones around large Si particles reduced the local stored energy and the pinning effect of fine Si in the same regions (figure 5c) may retard the nuclation proecss at the large Si particles.

3.3 Nucleation and the orientation of nuclei

SEM/ECC observations over large areas for the sample annealed for 1 hour at 200℃ revealed that nuclei or new grains of a few micrometre sizes have been formed, as shown in figure 6a. Some of the nuclei were indeed found to relate to the large Si particles. However, many other nuclei observed seem to have nothing to do with the large Si particles. To evalute statistically whether there is a relation between the formation of recrystallized grains and the large Si particles, several samples were annealed for different periods of time at a slightly higher temperature 210℃and examined by SEM/ECC. An example of such examinations is shown in figure 7b where there more and larger but still isolated recrystallized grains are seen. Again it is seen that the recrystallized grains are not necessarily related to the large Si particles. It is concluded that there is no advantage of particle stimulated nucleation (PSN) [13] at the large Si particles during annealing of the nanostructured Al-1% Si alloy.

The orientations of nuclei formed at the large Si particles and in the matrix were measured by EBSD in a sample that was annealed for 20 minutes at 210℃. More than 70 nuclei were analyzed for each case. Figure 7 shows that {111} pole figure of the orientations for all measured nuclei. The majoiry of the nuclei observed have ramdom orientaions irrespective of whether they formed at the large Si particles or in the matrix. Recently the textural evolution was investigated for the same nanostructured Al-1% alloy until the recrystallization was complete. It was found that the deformed state has a typical rolling texture which is replaced by a rather random texture after the recrystallization is complete [14]. The rather radom orientations of nuclei (figure 7) may explain the formation of the random recrystallization texture. However, clearly this is not particularly related to the effect of large Si paricles in stimulating the formation of nuclei of random orientations, rather the characteristic fine scale lamellar structure and the combined effect of fine and coarse particles on the random nucleation texture may play an important role.

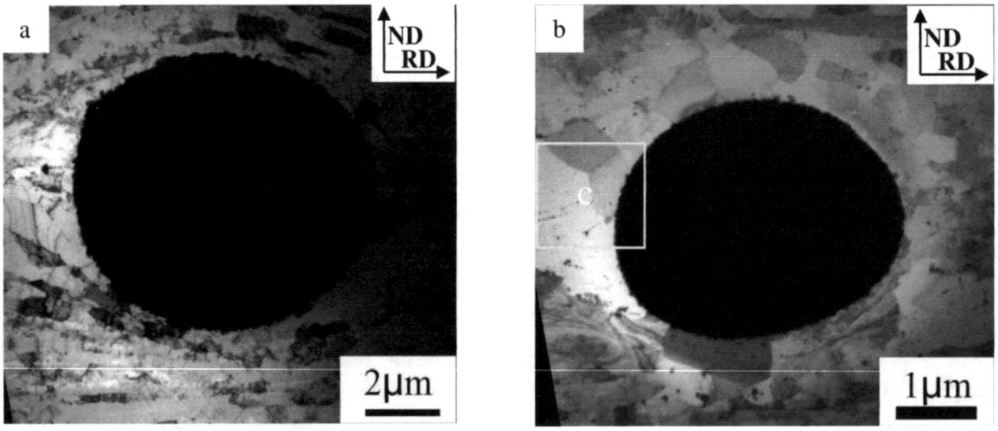

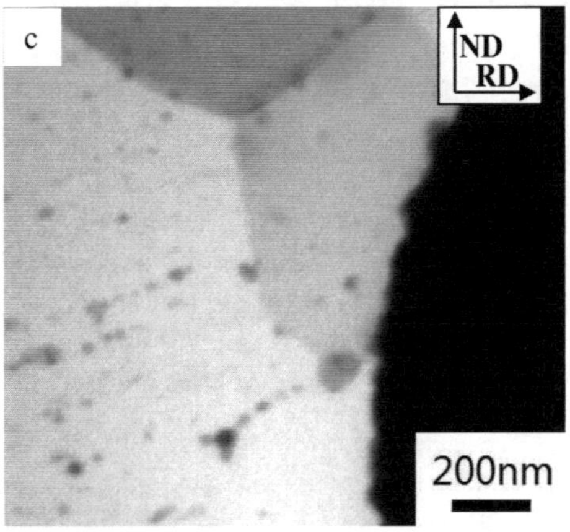

Figure 5. Recovery of the microstructure around large particles after annealing for 1h at different temperatures. (a) 150℃, (b) 200℃, c) Magnified image of the region framed in (b)

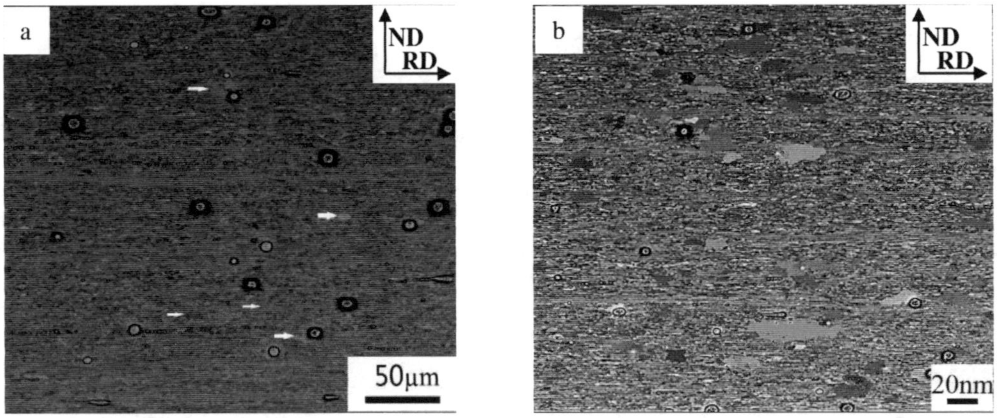

Figure 6. Nucleation at large Si particles and in the matrix observed in samples annealed (a) at 200℃ for 1 hour and (b) at 210℃ for 30 minutes

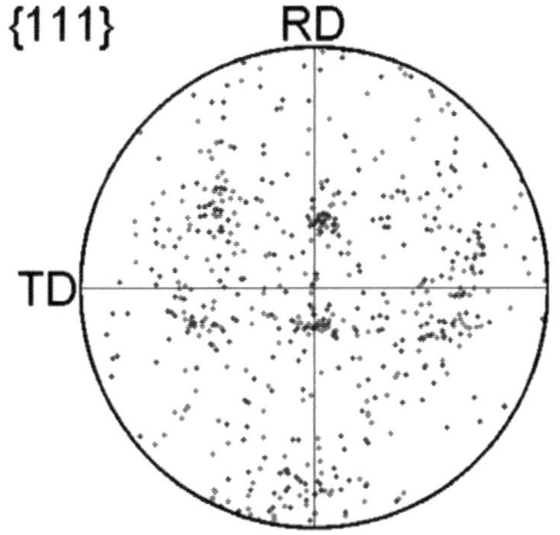

Figure 7. {111} pole figure of the recrystallized grains in Al-Si alloy annealed at 210℃ for 20min. Red points represent nuclei related to the large Si particles and blue points represent the nuclei that are not related to the large Si particles

4. Concluding remarks

The deformation, recovery and recrystallization behaviors have been studied in super-pure Al containing Si particles in the size range from tens of nanometres to several micrometres. The alloy has been cold rolled to 98% reduction (ε_{vM} = 4.5) and afterwards annealed at 150C° (0.45 Tm), 200°C (0.51 Tm) and 210°C (0.52 Tm) to study the recovery and initiation of recrystallization. The cold deformed structure is on the nanometre scale and contains a significant fraction of deformation induced high angle boundaries (>15°). The structure is homogeneous with the exception of regions at and near the large Si particles. The stress and strain in these regions are larger than in the Al matrix, they recover rapidly and PSN is observed. However PSN is not dominating as it competes with abundant nucleation in the Al matrix. It follows that the presence of the fine Si particles is of equal or large importance. The fine Si particles homogenize the deformed microstructure and they retard significant recovery and initiation of recrystallization due to the pinning of dislocation boundaries and high angle boundaries. The presence of fine particles both in the deformed and annealed states may therefore not only increase the thermal stability of the deformed microstructure but also be the cause of the uniform recrystallized grain structure at 250°C and above and the formation of a weak recrystallization texture.

Acknowledgements

The authors wish to thank the financial support from the Natural Science Foundation of China (NSFC, Grants No. 51471039, 51327805 and 51421001) and Fundamental Research Fund of Central Universities (Grant No. 10611205CDJZXZ138802). We also gratefully acknowledge the support from the Danish National Research Foundation (Grant No.DNRF86-5) and the NSFC (Grant No. 51261130091) to the Danish-Chinese Center for Nanometals within which part of the present work was carried out.

References

[1] Hansen N and Bay B 1974 *Scripta Metall.* **8** 1291
[2] Humphreys F 1977 *Acta Metall.* **25** 1323

[3] Hansen N and Bay B 1981 *Acta Metall.* **29** 65
[3] Jensen D J, Hansen N and Humphreys F J 1985 *Acta Metall.* **33** 2155
[4] Hansen N and Juul Jensen D 1984 The Metallurgical Society119
[5] Chan H and Humphreys F J 1984 *Acta Metall.* **32** 235
[6] Ferry M and Humphreys F J 1996 *Acta Mater.* **44** 3089
[7] Ferry M, Hamilton N and Humphreys F J 2005 *Acta Mater.* **53** 1097
[8] Huang T L, Dong Q, Gong X, Hansen N, Liu Q and Huang X 2012 *J. Mater. Sci.* **47** 7914
[9] Huang T L, Li C, Wu G L, Yu T, Liu Q, Hansen N and Huang X 2014 *J. Mater. Sci.* **49** 6667
[10] Huang T L 2014 Ph.D. thesis, Chongqing University
[11] Yu T, Hansen N, Huang X 2011 *Proc. R. Soc. A* **467** 3039
[12] Yu T, Hughes D A, Hansen N, Huang X 2015 *Acta Mater.* **86** 269
[13] Mishin, O. V., Godfrey, A., Jensen, D. J., & Hansen, N 2013 Acta Mater. 61 5354
[14] Chen Y L, Huang T L, Gong X, Wu G L, Liu Q and Huang X 2013 *Mater. Sci. Forum* **753** 2436

IOP Conf. Series: Materials Science and Engineering **89** (2015) 012029 doi:10.1088/1757-899X/89/1/012029

Investigation of SIBM driven recrystallization in alpha Zirconium based on EBSD data and Monte Carlo modeling

M Jedrychowski[1,2], B Bacroix[2], O U Salman[2], J Tarasiuk[1], S Wronski[1]

[1] AGH University of Science and Technology, Faculty of Physics and Applied Computer Science, al. Mickiewicza 30, 30-059 Krakow
[2] CNRS, Laboratoire des Sciences des Procédés et des Matériaux, UPR 3407, Université Paris 13, Sorbonne Paris Cité, 99 avenue J. B. Clément, 93430 Villetaneuse

E-mail: Mariusz.Jedrychowski@fis.agh.edu.pl

Abstract. The work focuses on the influence of moderate plastic deformation on subsequent partial recrystallization of hexagonal zirconium (Zr702). In the considered case, strain induced boundary migration (SIBM) is assumed to be the dominating recrystallization mechanism. This hypothesis is analyzed and tested in detail using experimental EBSD-OIM data and Monte Carlo computer simulations. An EBSD investigation is performed on zirconium samples, which were channel-die compressed in two perpendicular directions: normal direction (ND) and transverse direction (TD) of the initial material sheet. The maximal applied strain was below 17%. Then, samples were briefly annealed in order to achieve a partly recrystallized state. Obtained EBSD data were analyzed in terms of texture evolution associated with a microstructural characterization, including: kernel average misorientation (KAM), grain orientation spread (GOS), twinning, grain size distributions, description of grain boundary regions. In parallel, Monte Carlo Potts model combined with experimental microstructures was employed in order to verify two main recrystallization scenarios: SIBM driven growth from deformed sub-grains and classical growth of recrystallization nuclei. It is concluded that simulation results provided by the SIBM model are in a good agreement with experimental data in terms of texture as well as microstructural evolution.

1. Introduction

Hexagonal zirconium and its alloys bring scientific attention, mainly due to their importance in the nuclear industry. In consequence, they have been investigated in detail in order to understand and control microstructural evolution induced by recrystallization and grain growth phenomena.

In particular, it has been shown by Zhu et al. [1] that the microstructure of Zr-2Hf alloy strained up to 55% by plane strain compression exhibits high heterogeneity, which is characterized by the coexistence of highly deformed and less deformed zones. During annealing the classical nucleation occurs in the highly deformed areas, especially at grain boundaries and/or nearby twin intersections, then further growth of new recrystallization grains is observed. On the other hand, recovery and related continuous recrystallization occur in the less deformed areas. In this case, in-situ TEM observations revealed that the sub-grain structure can be formed due to annihilation and rearrangements of dislocations and further transformed through sub-grain coalescence. This, in turn, can facilitate formation and growth of recrystallization grains.

The presented description of possible recrystallization mechanisms has been refined further based on examination of rolled Zr-2Hf alloy (50% and 90% of strain) [2]. Taking into account an evaluation of stored energy and detailed analysis of texture evolution during annealing, it has been noted that strain induced boundary migration (SIBM) can play an important role, as the process responsible for growth of sub-grains formed inside grains during recovery. In the cited paper SIBM has been also used to explain competition between two main texture components: R -{*hkil*}<11-20> and D - {*hkil*}<10-10> observed over the grain growth stage.

A similar observation has been reported for Zr702α rolled by 80% of thickness reduction and further annealed, where SIBM facilitates consumption of large and slowly recrystallizing grains [3].

The main aim of this work is to contribute to the current understanding of recrystallization in zirconium. It is noted that all the above mentioned studies dealt with relatively high deformations (50% or higher) and its further influence on annealing behavior, which relies on a complex interplay between different mechanisms, classical nucleation in highly deformed zones and SIBM driven growth of sub-grains, for instance. Hence, it is difficult to separate each of them and analyze in detail.

However, in the opposite case of moderately deformed material (less than 20% of strain), SIBM driven recrystallization is expected to dominate [4], which gives an opportunity to focus on texture and microstructural evolution caused only by this mechanism.

Therefore, the hypothesis of SIBM driven recrystallization is analyzed and tested in the case of appropriately heat-treated zirconium. The investigation is based on EBSD analysis of experiments similar to the one conducted by Francillette et al. [5], which is then supported by computer simulations of a Monte Carlo Potts model.

2. Experimental procedure

Low alloyed zirconium (Zr702α) was formed by rolling and annealing into a recrystallized sheet, which was afterwards used to prepare parallelepipedic samples of dimensions 8.12 mm, 6.30 mm, 6.30 mm for channel-die compression, with the longest dimension parallel to the initial rolling direction (RD) of the sheet. Two compression directions were chosen, the transverse direction (TD) and the normal direction (ND). In both cases, elongation was restricted only to RD and the surface subjected to the load had the same dimensions. Using an Instron 1195 testing machine, the final achieved strain was 17%. It was observed that compression along ND required a slightly higher force than compression along TD. Because of that, the samples deformed in ND are referred to as hard (H), whereas samples deformed in TD are called soft (S). Then, all the samples were briefly annealed (615 °C, 15 minutes) in an air atmosphere to reach a partly-recrystallized state.

It has to be mentioned that this experiment is quite close to the one performed by Francillette et al. [5], which revealed a strong influence of the initial texture on the activated slip/twin systems, resulting hardening curve and level of macroscopic stress. They noticed that the soft samples correspond mainly to the activation of easy prismatic glide systems, which is related with a linear character of the hardening curve. Conversely, the hard samples are associated with the activation of prismatic glide and secondary hard systems, thus the hardening curve has a parabolic shape in this case.

Both, deformed and annealed states, were examined by the EBSD technique using a Cambridge S360 (WGUN) scanning electron microscope and Carl Zeiss Supra VP - Ultra High Resolution FEG-SEM, equipped with TSL/EDAX Data Acquisition software. The surface preparation procedure, required by EBSD, consisted of grinding with silicon carbide paper down to grade 4000, then grinding with an OPS solution from Struers, and finally electropolishing under 40 V with an A3 solution from Struers. EBSD topological maps were measured on the RD-TD plane for H samples and the RD-ND plane for S samples. The total area covered by EBSD maps was at least 1 mm^2 for each investigated sample. All the obtained EBSD data were analyzed with the OIM Analysis v5.3 software from TSL.

3. EBSD investigation

EBSD grains (deformed and recrystallized) were determined using standard algorithm provided by the software that we used as structures comprising of at least 10 EBSD points (taking into account an EBSD lattice step equal to 1.2 μm), which satisfy the condition that a misorientation angle between the nearest-neighboring points is lower than 15°. Then grain size was calculated using diameter and area methods. Textures were analyzed using Euler angles (φ_1, Φ, φ_2) and an Orientation Distribution Function (ODF) cross-section for $\varphi_1 = 0°$. The crystal coordinate system was defined as: X = [2-1-10], Y = [01-10], Z = [0001]. Calculations were conducted using harmonic method.

3.1. Deformed state.

The initial microstructure consists of completely recrystallized and equiaxed (in 3D) grains with an average area grain size of 240 μm^2 (or equivalent diameter of 16 μm).

Most of the grains have a c-axis located in the vicinity of ND. Further texture analysis reveals domination of a tilted {0001}<2-1-10> component, which appears in the cross-section of the ODF as strong symmetrical maxima (0°, 30°, 0°) and (0°, 30°, 60°).

Figure 1 presents example EBSD maps of the deformed state of the H and S samples strained by 17% and 9%, respectively. First one is used to explain the role played by orientation spread, whereas the second shows activity of twins.

Orientation spread appears within the grains (visualized by the color gradient), because of the local misorientations, which are associated with the formation of intragranular dislocations during deformation. In consequence, opposite parts of the deformed grain can vary in terms of orientation.

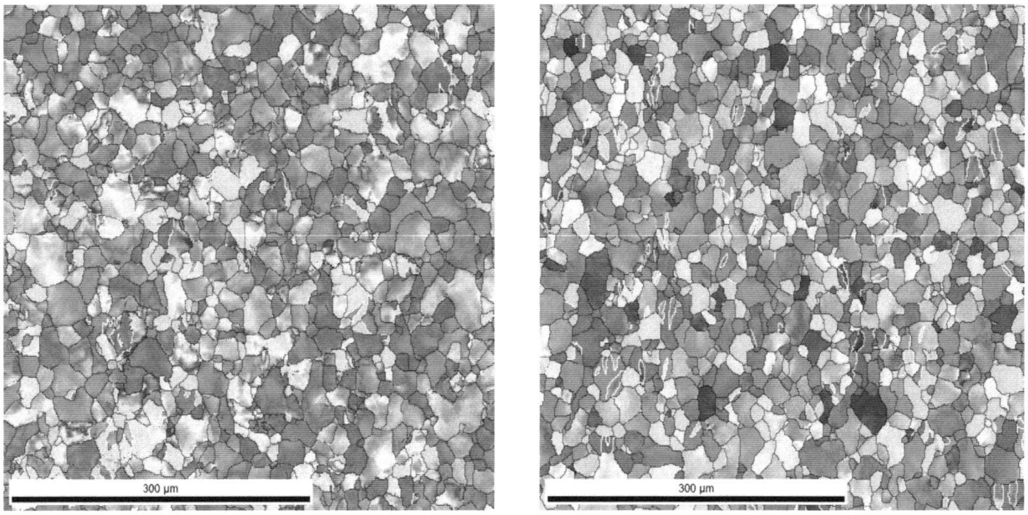

a) [001] IPF, RD-TD plane b) [010] IPF, RD-ND plane

Figure 1. EBSD maps (cropped area) of the deformed state of the H sample strained by 17% (a) and S sample strained by 9% (b). High angle grain boundaries (HAGBs) are represented by black lines, while white lines fulfill an axis/angle condition for tensile twins boundaries.

At the same time, the overall shape of the deformed grains, being discussed, is not changed significantly in comparison to the initial equiaxed microstructure. This can be explained by the fact that grains undergo accumulation of local misorientations instead of heavy fragmentation. Concerning the latter, the presented maps show that twinning, which is usually responsible for structural refinement in hexagonal metals, is not very active in the investigated samples. However, some subtle difference between them can be noticed at the beginning of the plastic regime, namely tensile twinning is much more frequent at the early stages (9% of strain) of TD compression – figure 1b, figure 2.

Regarding orientation gradients and local misorientations, both are investigated using grain orientation spread (GOS) and kernel average misorientation (KAM) statistics. The averaged GOS and KAM values reveal the same trend, which is a faster increase in H samples. This is most probably a direct consequence of different operating deformation modes and resulting hardening curves in H and S samples.

The described differences in misorientation accumulation process and slight influence of twinning observed in S samples have together a visible reflection in the final average size of deformed grains - Table 1.

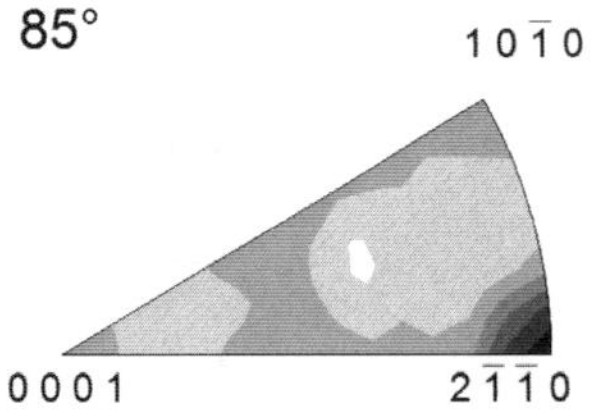

Figure 2. MDF plot for tensile twin misorientation angle (85°) in the case of S sample after 9% strain.

Table 1. Average grain size for Zr samples.

Sample	Strain [%]	Area [μm^2]	Diameter [μm] (Average/Std.Deviation)	
Initial	0	240	16.0	± 8.3
H	7	240	15.2	± 8.5
	17	312	17.0	± 10.6
S	9	200	14.0	± 7.8
	17	245	15.5	± 8.6

The expected decrease of grain size is only seen at the beginning of plastic deformation (9% of strain) in the S sample, which can be reasonably explained by the above mentioned presence of tensile twins. In the corresponding H sample (7% of strain) the average grain size is not changed at all. Then, further deformation results in an increased average grain size, especially in the H sample, which means that large grains have to be formed during compression. This surprising effect is a consequence of a grain coalescence process driven by a high density of local misorientations nearby grain boundaries and orientation spread within the grains.

Figure 3 presents example of an extremely big grain (purple color) determined by a grain reconstruction algorithm, which in fact is an area comprising of "partial-grains" delimited by low angle grain boundaries (LAGBs). The misorientation profile calculated for the trace marked on figure 3, shows that such large grain have a significant orientation spread, since misorientation from one side to another can reach 35° and despite that it is still treated as a single grain. Such examples were not observed in the initial state.

Therefore, it can be concluded that coalescence of deformed grains is caused by accumulation of local misorientations; they are responsible for transition from HAGB to LAGB, which in turn leads to sticking of the partial-grains. The described overall tendency of merging grains into larger areas is still valid, even if the threshold misorientation used for grain determination is changed to a lower value – 10° for instance. This mechanism is more pronounced in the H sample, thus it has more LAGBs and less HAGBs in comparison to the S sample. Moreover, it the latter case more HAGBs could be generated due to twinning at the beginning of the compression process.

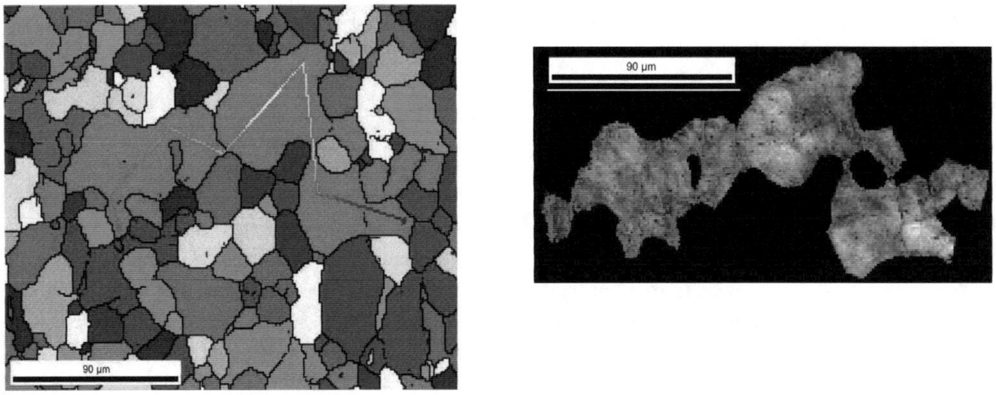

Figure 3. Fragment of deformed microstructure (H sample, 17% of strain) containing an extremely large grain - purple color (left) - and corresponding IQ map presenting LAGB structure – red lines (right).

3.2. Annealed state

Recrystallization is not complete in both of the investigated samples, since deformation substructures still appear in some of the grains. Therefore, the recrystallized fraction has been extracted based on the GOS analysis.

It is noted that recrystallization kinetics depends on the direction of the applied compression – the recrystallized area fraction is higher in the S sample: 0.7 in comparison to 0.58 in the H sample. However, in both samples recrystallized grains have a similar size – 160 μm^2 (12.7 μm). It means that in the S sample a higher number of recrystallized grains emerged from the deformed microstructure.

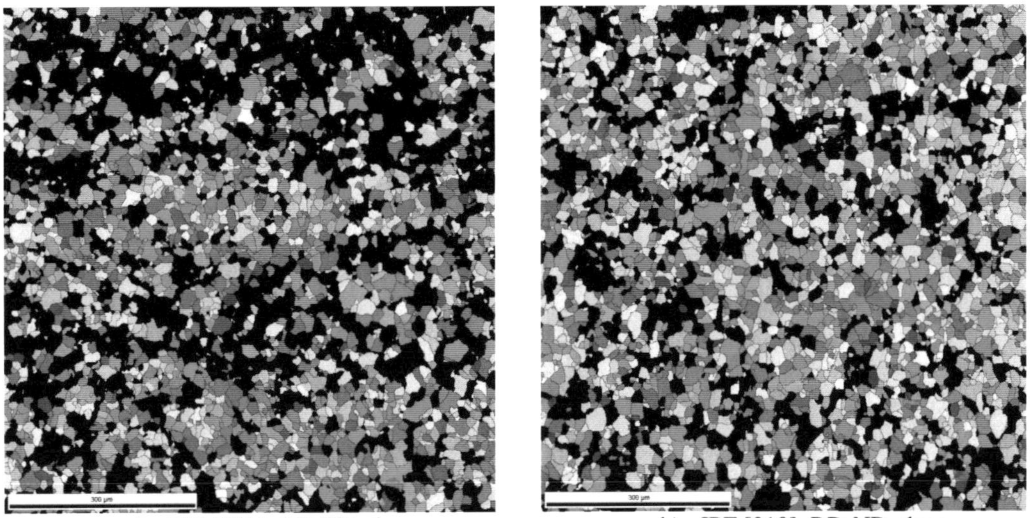

a) IPF [001], RD-TD plane b) IPF [010], RD-ND plane

Figure 4. EBSD maps (800 μm x 800 μm) of the recrystallized grains in annealed: H sample (a) and S sample (b).

An example EBSD map recrystallized partition extracted from the annealed state of H sample strained by 17% is shown in figure 4. It can be underlined that strain-free grains tend to gather in cluster-like structures instead of being randomly distributed in terms of spatial location.

3.3. Texture evolution

The performed texture analysis is shown only for H samples (figure 5), as the main motivation in this context is to compare results obtained with those reported in the literature for rolled zirconium, since both cases have similar strain geometry (compression direction parallel to ND).

Channel die compression results in the formation and strengthening of deformation {*hkil*}<10-10> component, which is mainly located along φ_2 = 30°. Hence, the ODF cross-section contains a very broad maximum, which can be described as superposition of four main orientations: (0°, 30°, 30°), (0°, 45°, 30°) and two symmetrically located (0°, 30°,15°) - (0°, 30°, 45°).

After annealing textures have been separately calculated for recrystallized and still deformed partitions. Based on this comparison, it can be noted that most of the grains with an orientation close to the main deformation components, described above, have been consumed and only those with an orientation (0°, 30°, 45°) retained in the annealed microstructure.

Texture evolution in recrystallized partition is very interesting, as the recrystallization component (0°, 30°, 0°), i.e. tilted{0001}<2-1-10> is formed, despite the fact that recrystallization is not complete, while it is often underlined that such a component should appear as a result of an additional grain growth process [2, 3]. Therefore, other mechanisms responsible for such a development of recrystallization texture should be expected in the considered case.

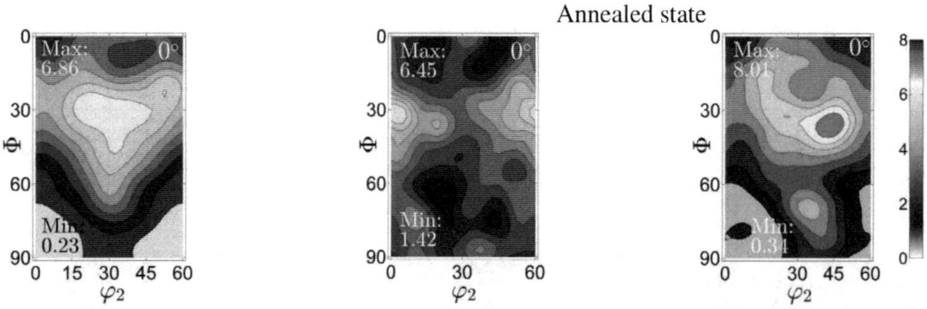

a) texture after deformation b) recrystallized partition c) still deformed partition

Figure 5. Texture evolution in H sample during thermo-mechanical treatment.

Additional insight into the presented texture evolution can be provided by analysis of grain boundary regions extracted from deformed microstructures. For that reason, EBSD points are specially selected based on misorientation criteria, as depicted in [6, 7]. Obtained data contain information about the orientation distribution of grain boundary regions – figure 6. Textures of Intragranular Boundary (IB) and LAGB regions, are of particular interest in this context, since these boundaries can be related with additional energy stored inside grains. These grains, in turn, should be replaced by a recrystallization front due to increased gradient of the stored energy.

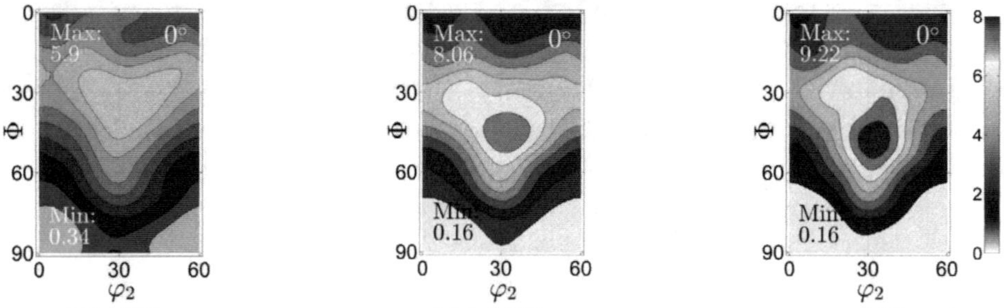

a) HAGB b) LAGB c) IB

Figure 6. Textures of grain boundary regions in deformed H sample.

The texture of the HAGB area follows that of the whole sample. More interesting observations concern regions adjacent to LAGBs and IBs, since they have a main maximum located at (0°, 45°, 30°) and two slightly less pronounced centered at (0°, 30°, 30°) and (0°, 30°, 15°). Please note that the deformation component (0°, 30°, 45°) in not present in this case.

After annealing it is clearly demonstrated that grains having a high fraction of IBs and LAGBs are disappearing first. Based on that, it can be understood why the orientation (0°, 30°, 45°) has remained in the texture of still deformed grains – figure 5c.

4. SIBM model and Monte Carlo simulations

Taking EBSD observations into account, a following scenario is proposed in order to explain the observed texture and microstructural evolution. First, SIBM is considered as the main mechanism of growth of an almost strain-free sub-grain separated from a deformed grain by HAGB. Then, it is assumed that grains having a substantial fraction of local misorientations (IBs and LAGBs) and the related dislocation density exhibit a higher stored energy and thus are consumed first as a consequence of the SIBM process activated on the HAGB interface adjacent to the above mentioned sub-grain. From this point of view, the higher fraction of HAGBs surrounded by local misorientations in the deformed sample, the higher driving force for sub-grain growth and therefore the higher number of recrystallized grains after annealing. This corresponds very well with the difference in recrystallized area fraction between annealed H and S samples. Moreover, this model can be used to explain texture evolution, if the main orientation of strain-free sub-grains is related with tilted {0001}<2-1-10> component.

In the next step, Monte Carlo Potts model [8, 9] has been employed in order to verify the suggested hypothesis. Simulations have been performed using EBSD microstructures of the deformed state (666x665 sites in the lattice) of H and S samples. However, results obtained are presented only for the first case. The stored energy (SE) distribution, which is a key parameter in simulations of recrystallization has been evaluated based on rescaled KAM parameter – SE $\in$ [0, 4.5]. Regarding boundary properties, well-known Read-Shockley and Huang-Humphreys formulas [10] have been applied in order to calculate the energy and mobility of IBs and LAGBs.

Several scenarios have been simulated, including the SIBM model and classical nucleation. In the latterit was assumed that spatial position of a nucleus is inside the HAGB structure, while its orientation was taken from the deformed sites located nearby or chosen randomly from the deformed texture. Simulations were stopped when a desired fraction of recrystallized area - 0.6 - was reached.

The recrystallized partition obtained from simulation with the SIBM model (figure 7a) consists of 2300 grains of an average size equal to 12.5 μm. These grains are not homogeneously distributed in terms of spatial location.

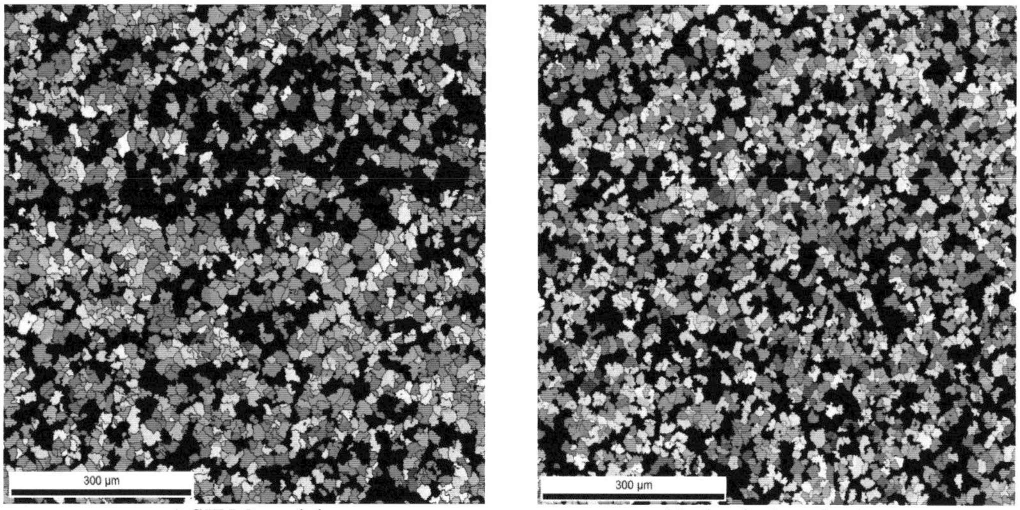

a) SIBM model   b) classical nucleation

Figure 7. Recrystallized grains obtained from simulation performed in the case of the H sample using: the SIBM model (a) or classical nucleation (b).

Also, a strong recrystallization component is developed and only orientation (0°, 30°, 45°) remains in the texture of the deformed partition – figure 8.

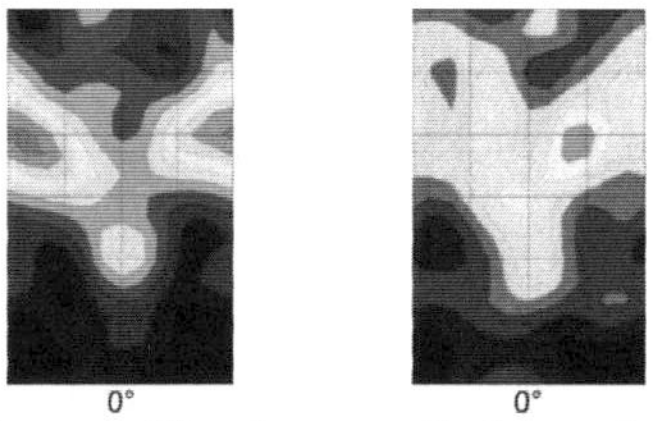

Recrystallized Partition   Deformed Partition

Figure 8. Texture (ODf cross-section for φ_1 = 0°) obtained from simulation of recrystallization in H sample.
Maximum value of ODF function (red color) is equal to 8.5.

All these observation are in very good agreement with experimental results presented for the annealed state of the H sample.

In contrary, simulation of a classical scenario does not match the experimental data: recrystallized grains are more homogeneously distributed in the microstructure (figure 7b) and texture evolution is wrongly predicted.

Finally, it should be emphasized that SIBM driven simulations also gave correct predictions in the case of the annealed S sample.

5. Conclusions

Hexagonal zirconium was channel-die compressed to reach 17% of final strain and briefly annealed to achieve a partially recrystallized state. Compression in ND (hard samples, H) requires a higher force than in TD (soft samples, S).

Deformation in the hard samples is strongly connected with the accumulation of dislocation cells inside the grains, which has a reflection in local misorientations (higher values of KAM, GOS value) as well as faster development of IBs and LAGBs. This, in turn, leads to grain coalescence – large areas are created from closely oriented grains due to transition from HAGB to LAGB.

In the S samples, in turn, grain fragmentation is easier to achieve at the beginning of the deformation due to activity of tensile twins, thus the fraction of HAGBs increases much more, while the development of the LAGB population is less significant. This difference in the HAGB population between samples was linked with recrystallization kinetics, which is faster in the S sample.

Transition from the tilted {0001}<10-10> to the tilted {0001}<2-1-10> component is seen in the texture of the H sample after partial recrystallization, which indicates that in this case the mechanism of formation of recrystallization texture is different from the one already reported in the literature for annealing of much higher deformations. The evolution described can be associated with SIBM driven recrystallization. Moreover, analysis of textures of grain boundary regions revealed the important role played by local misorientations located nearby HAGBs. Clearly, orientations that are present simultaneously in the texture of HAGBs, IBs, and LAGBs, disappear first during annealing. Such behavior also suggests a domination of the SIBM mechanism.

A physical model was formulated. It assumes sub-grain growth, which is caused by a SIBM process of HAGB in regions containing a high density of local misorientations. The proposed hypothesis was positively tested using a Monte Carlo modeling approach. The simulation results show very good agreement with the experimental data in terms of texture as well as microstructural evolution.

Acknowledgements

M. Jedrychowski has been partly supported by French Government scholarship. Authors gratefully acknowledge NCN for the grant number 2011/01/D/ST8/07399.

References

[1] Zhu K Y, Chaubet D, Bacroix B and Brisset F 2005 *Acta Mater.* **53** 5131–40
[2] Zhu K Y, Bacroix B, Chauveau T, Chaubet D and Castelnau O 2009 *Metall. Mater. Trans.* A, **40** 2423–34
[3] Dewobroto N, Bozzolo N, Barberis P and Wagner F 2006 *Int. J. Mat. Res.* **97** 826–33
[4] Humphreys F and Hatherly M 2004 *Recrystallization and Related Annealing Phenomena* (New York, Pergamon)
[5] Francillette H, Bacroix B, Gasperini M and Bechade J L 1998 *Acta Mater.* **46** 4131–42
[6] Jedrychowski M, Tarasiuk J, Bacroix B and Wronski S 2013 *J. Appl. Cryst.* **46** 483–92
[7] Jedrychowski M, Tarasiuk J, Bacroix B and Wronski S 2013 *Mater. Sci. Forum* **753** 93–6
[8] Rollett A and Raabe D 2001 *Comput. Mater. Sci.* **21** 69–78
[9] Miodownik M 2002 *J. Light Met.* **2** 125–35
[10] Humphreys F J 1997 *Acta Mater.* **45** 4231–4232

Non-destructive and three-dimensional measurement of local strain development during tensile deformation in an aluminium alloy

M Kobayashi[1], H Miura[1] and H Toda[2]

[1]Department of Mechanical Engineering, Toyohashi University of Technology, 1-1 Hibarigaoka, Tempaku-cho, Toyohashi, Aichi, JAPAN 441-8580
[2]Department of Mechanical Engineering, Kyusyu University, 744 Motooka, Nishiku, Fukuoka, JAPAN 819-0395

E-mail: m-kobayashi@me.tut.ac.jp

Abstract. Anisotropy of mechanical responses depending on crystallographic orientation causes inhomogeneous deformation on the mesoscopic scale (grain size scale). Investigation of the local plastic strain development is important for discussing recrystallization mechanisms, because the sites with higher local plastic strain may act as potential nucleation sites for recrystallization. Recently, high-resolution X-ray tomography, which is non-destructive inspection method, has been utilized for observation of the materials structure. In synchrotron radiation X-ray tomography, more than 10,000 microstructural features, like precipitates, dispersions, compounds and hydrogen pores, can be observed in aluminium alloys. We have proposed employing these microstructural features as marker gauges to measure local strains, and then have developed a method to calculate the three-dimensional strain distribution by tracking the microstructural features. In this study, we report the development of local plastic strain as a function of the grain microstructure in an aluminium alloy by means of this three-dimensional strain measurement technique. Strongly heterogeneous strain development was observed during tensile loading to 30%. In other words, some parts of the sample deform little whereas another deforms a lot. However, strain in the whole specimen was keeping harmony. Comparing the microstructure with the strain concentration that is obtained by this method has a potential to reveal potential nucleation sites of recrystallization.

1. Introduction

Metallic materials consist of a large number of individual grains which are separated by grain boundaries. The grains have crystallographic orientations. Depending on the crystallographic orientation, deformation behaviors can vary between grains. This deformation anisotropy causes concentration of deformation, i.e. local deformation variations on a mesoscopic scale (the grain size scale). The development of local plastic strain during a deformation process is one of the key issues to describe mechanical properties, such as formability and plastic anisotropy [1]. The local concentration of strain may also govern fracture. Furthermore, with regard to thermo-mechanical processing, investigations of the local plastic strain development are important for discussing recrystallization mechanisms, because the sites with higher local plastic strain seem to stimulate nucleation of recrystallization [2].

Recently, high-resolution X-ray tomography, which is a non-destructive inspection method, has been used for observations of material structures. The spatial resolution of current tomography techniques has been reduced to 1 μm, which is close to the limit of the resolution in a visible-light-conversion type detectors. By X-ray tomography using a monochromatic X-ray beam from synchrotron radiation facilities, more than 10,000 microstructural features, like precipitates, dispersions, compounds and hydrogen pores, can e.g. be observed in aluminium alloys. We have proposed employing these microstructural features as markers to measure local strains, and have developed a method to calculate the three-dimensional strain by tracking the microstructural features [3]. In this study, we report research on the development of local plastic strains affected by the grain microstructure in an aluminium alloy by this method.

2. Experimental method

An in-situ tensile experiment during tomography scans were performed at BL20XU at the Japanese synchrotron radiation facility, SPring-8. A small special test rig was installed at BL20XU. First, the specimen was scanned without loading. The tensile load was given to the specimen step by step. The tomography scans were done at each step. 6 time scans were achieved up to a maximum load in this study. Table 1 explains the loading steps and macroscopic strain. The macroscopic tensile strain was approximately 30% at the final step (Step 6). Note that the tensile displacement was controlled and a strain rate of 0.001 mm/s was used. After finishing the tensile test, the deformed specimen was removed from the test rig, and then the grain boundaries were visualized by means of a gallium application. Details of the sample and of the experimental method are given in the following.

Table 1 Loading steps and macroscopic strain.

Loading step	Load (N)	Macroscopic strain (%)
0 (undeformed)	0	0.0
1	67.9	3.2
2	82.3	7.1
3	91.5	11.4
4	96.2	16.9
5	97.9	23.3
6	97.2	29.6

2.1 Sample

The sample used in this study is an Al-4mass%Cu alloy that contains Al_2Cu precipitate particles employed as tracking markers to measure local plastic strains. To obtain both a relatively large grain size and a lot of Al_2Cu particles, a thermo-mechanical treatment was applied as shown in Fig. 1. A cast ingot of Al-4mass%Cu was manufactured by melting and casting base-materials of 3N pure aluminium and Al-33mass%Cu alloy. The ingot was cut into small billets and homogenized at 823 K for 36 ks with water cooling. The billets were processed with 80% cold rolling and heat-treated at 503 K for 7.2 ks as an intermediate annealing. Further 10% additional rolling was given to the intermediate annealed sample. The samples were then recrystallized at 803 K for 45 s in a salt-bath. After recrystallization, the samples were solution-heat-treated at 768 K for 18 ks and were then cooled for 900 s in a muffle furnace to obtain a dense population of precipitate particles on the grain boundaries. With this thermo-mechanical treatment, the average grain size and particle size were approximately 100 μm and 2.9 μm, respectively. Small specimens for tensile tests with cross section of 0.6 mm by 0.6 mm in the gauge (see Fig. 2) were cut from the sample by electro-discharge machining.

2.2 Synchrotron X-ray tomography

X-ray tomography was carried out at the experimental hutch at BL20XU. The set-up used in this study was the projection type microtomography as shown in Fig. 3 that possess a spatial resolution of approximately 1μm. Monochromatic X-ray with 20 keV energy was obtained by a (111)-(111) Si double crystal monochromater. The test rig was mounted on a high-precision rotation stage. A visible-light conversion type detector was used consisting of a scintillator ($Lu_2SiO_5:Ce^+$), ×10 optics lens and with a complementary metal oxide semiconductor (CMOS) camera (Hamamatsu photonics ORCA-Flash 2.8 1920×1440 pixels). A voxel size of $(0.5\mu m)^3$ was achieved by this set-up. The sample to detector distance was adjusted to appropriately 20 mm to make the object edge sharp by X-ray phase contrast enhancement. One thousand eight hundred radiographs with height and width of approximately 720μm and 960μm were captured during a 180 degree rotation (i.e. 0.1 degree step). The convolution back projection method was utilized for reconstruction. The linear absorption coefficient (LAC) was stored in the reconstructed slice image as a 16-bit grey depth. To reduce the image data size, a LAC value from 0 cm^{-1} to 31 cm^{-1} was converted into 8-bit grey depth (i.e. from 0 to 255). Note that LAC of aluminium, Al_2Cu particles were 8.56 cm^{-1} and 84.86 cm^{-1} at 20 keV, respectively, i.e. widely different.

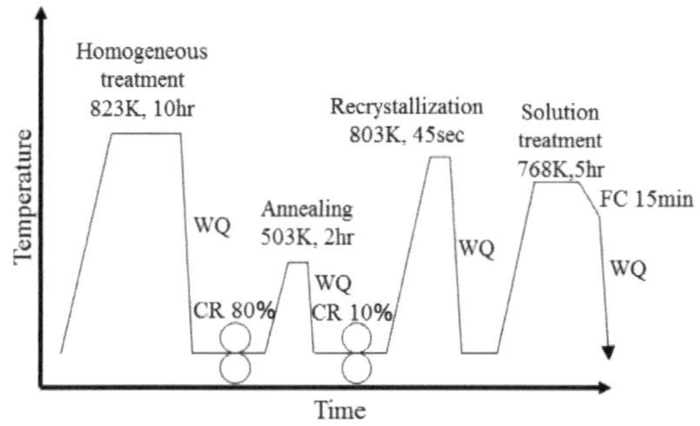

Figure 1. Diagram of thermo-mechanical processing for the sample preparation in this study

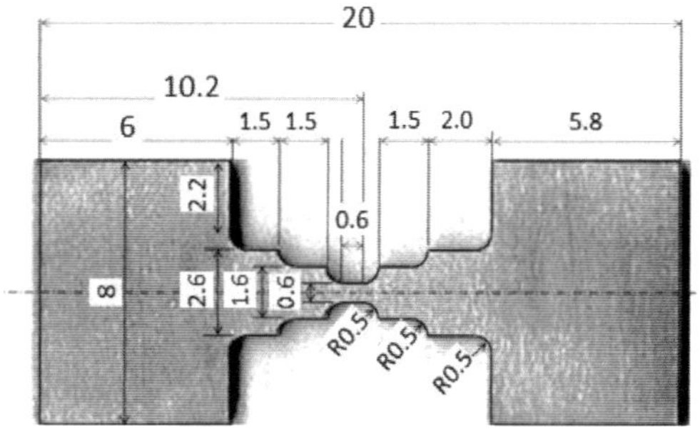

Figure 2. Dimensions of the tensile specimen. (mm)

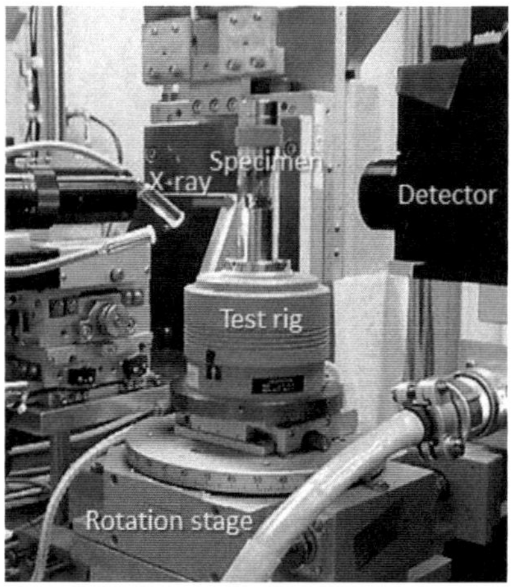

Figure 3. X-ray microtomography set-up at BL20XU

2.3 Visualization of grain boundaries

After finishing the X-ray tomography scans, the deformed specimen was removed from the test rig. The grain boundary network was visualized by applying gallium onto the specimen. It is well known that gallium with the low melting point, can penetrate aluminium grain boundaries easily and cause intergranular fracture because of liquid metal embrittlement. The LAC of gallium (239.27 cm^{-1}) is significantly higher than that of aluminium (8.56 cm^{-1}) at 20 keV. Therefore, gallium penetrated grain boundaries are seen as bright contrast within the aluminium matrix. Gallium was melted at 323 K on a hot-plate. The specimen was dipped into the liquid gallium and then the sample surface was scratched lightly by the tip of a pair of tweezers until gallium penetrates into the sample. The specimen was mounted in a specimen holder on a rotation stage of the X-ray instrument and was scanned again. The distance between the sample and detector was set to 20 mm to allow appropriate phase contrast. The LAC values of the images obtained by the gallium application were converted into 8-bit grey depth in the range from 0 cm^{-1} to 63 cm^{-1}. Further, images were binarized applying a threshold of 80 to determine the grain boundary positions correctly. Binarized objects with a volume under 3^3 voxel were deleted in binarized image, and then the distance transformation [4] and H-minima transformation [5] was applied to the binarized and modified images. $H = 8$ was used for the H-minima transformation in this study. A Watershed transformation [5] which is one of the segmentation methods was finally used. It was found that some grains were hyper fractionated by the Watershed transformation. Therefore, all results of the grain segmentation were checked, and were corrected by merging these grains into one. The three-dimensional volume rendered image after the Watershed transformation is shown in Fig. 4. 503 grains exist in the field of view. The average grain diameter of these grains is 64.5 μm. The numbers of grain boundaries, edges that form triple junction lines and vertices that form quadrupole points were 2871, 4416 and 1784, respectively. Details of this processing have been reported in ref [6].

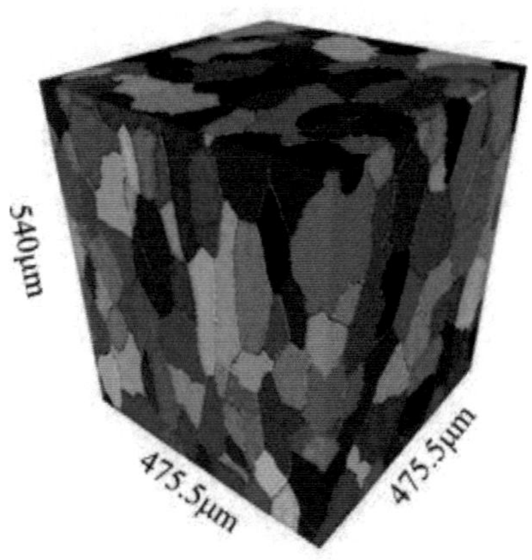

Figure 4. Three-dimensional image showing the grains in the sample

2.4. Strain calculation

A series of X-ray tomography images measured during tensile loading were binarized with a threshold value of 147. The positions (gravity centres), volumes and surface areas of all particles were measured in the binarized images by a homemade software. To remove uncertain markers, only particles with a volume larger than 3^3 voxel were accepted as tracking markers for the local plastic strain measurement. Since the specimen moved slightly during tensile loading, the marker positions in all volume images of different loading steps were converted into the load step 0 (before deformation) using an Affine transformation to make tracking easy. After conversion, markers were tracked by using a matching parameter method [3]. The position, surface area and volume were considered as matching parameters, M_p. The weighting of these were 0.65, 0.3 and 0.05, respectively. When The matching parameter of a marker is larger than the first threshold value, M_p^{th1} and the difference between the largest M_p and 2nd largest M_p is larger than the second threshold value, M_p^{th2}, those markers were matched. Approximately twenty thousand markers were tracked through all the loading steps. A Delaunay tessellation [7] was performed to make tetrahedra within the 3D space of the sample by utilizing the positions of the tracked markers. Here, we assume a linear displacement field inside the tetrahedra and displacements $u_i, v_i, w_i, u_j, v_j, w_j, u_k, v_k, w_k, u_l, v_l, w_l$ at tetrahedra vertices i, j, k, l in a rectangular coordinate system, then we can calculate normal strains, ε_x, ε_y, ε_z and shear strains, γ_{xy}, γ_{yz}, γ_{zx}, by the following equations;

$$\{\varepsilon_x \quad \varepsilon_y \quad \varepsilon_z \quad \gamma_{xy} \quad \gamma_{yz} \quad \gamma_{zx}\}^t = [\mathbf{B}_i \quad \mathbf{B}_j \quad \mathbf{B}_k \quad \mathbf{B}_l]\{u_i \quad v_i \quad w_i \quad u_j \quad v_j \quad w_j \quad u_k \quad v_k \quad w_k \quad u_l \quad v_l \quad w_l\}^t, \tag{1}$$

where $[\mathbf{B}_i]$ is given as

$$[\mathbf{B}_i] = \frac{1}{6V}\begin{bmatrix} o_i & 0 & 0 \\ 0 & p_i & 0 \\ 0 & 0 & q_i \\ p_i & o_i & 0 \\ 0 & q_i & p_i \\ q_i & 0 & o_i \end{bmatrix}, \tag{2}$$

$$\begin{cases} o_i = y_k z_j + y_l z_k + y_j z_l - y_k z_l - y_j z_k - y_l z_j \\ p_i = x_l z_j + x_j z_k + x_k z_l - x_j z_l - x_l z_k - x_k z_j \\ q_i = x_l y_k + x_j y_l + x_k y_j - x_j y_k - x_k y_j - x_l y_k \end{cases}$$

where x_j, y_j, z_j, x_k, y_k, z_k, x_l, y_l, z_l indicate the vertex positions of j, k, l. Shifting indices such as $i \to j, j \to k$, $k \to l$, and $l \to i$ brings $[\mathbf{B}_j]$, $[\mathbf{B}_k]$ and $[\mathbf{B}_l]$. The variable V means the tetrahedron volume. The calculated strains were arranged at the gravity centered positions of the tetrahedra. Three dimensional discrete strain data were converted into volume image data with a 5μm voxel size by linear interpolation to make comparison with the grain boundary images easy.

Table 2. Marker tracking result.

Loading step	Number of tracked markers	Average of local strain (%)
0-1	41278	3.2
0-2	34877	7.5
0-3	32246	12.0
0-4	24762	17.9
0-5	23271	25.9
0-6	19512	33.2

3. Results and discussions

Results of strain mapping at a macroscopic tensile strains of 3.2% (loading step 1) and 29.6% (loading step 6) are shown in Fig. 5. The strain along the tensile direction, ε_z, the other normal strain, ε_y and the shear strain, γ_{yz} are indicated in Fig. 5(a), (b) and (c), respectively. Some limited strain concentrations are seen at 3.2% strain. A very heterogeneous strain distribution is observed in the entire strain map at 29.6%. Regarding with the evolution of ε_z, the positions at which the strain concentration occurred at an early stage of deformation are found to develop a large deformation in the later loading step. Spots in which a large strain occurs thus seem to be defined in the early stage of deformation. The undeformed domains at the early strain step did not deform preferentially at later steps. Deformation spots did not propagate widely in the present case when the final macroscopic strain was approximately 30%. The grains that show large deformation may have a lower work hardening rate compared with other grains. In the map of the ε_y strain, compression strains can be confirmed as shown in Fig. 5 (b'). The compression strain, ε_y in Fig. 5(b') have a tendency to be large at the positions where large tensile strains are observed in Fig. 5 (a'). The strain distribution of ε_y seems to roughly correspond to that of ε_z.

The shear strain, γ_{yz}, in Fig. 5 (c') are seen not to be of the some magnitude as the ε_z strain. The direction of shear deformation is different in individual grains. Although we have not measured the crystallographic orientation of each grain, the dependence of deformation behaviour, like the deformation magnitude and shear direction is likely to depend in the grain orientation which may be determined by the non-destructive orientation measurement method, three dimensional X-ray diffraction [8]. The area having large shear strains are seen to propagate into low shear strain areas with increasing macroscopic tensile strain. This development of shear strains is different from that of the normal strain.

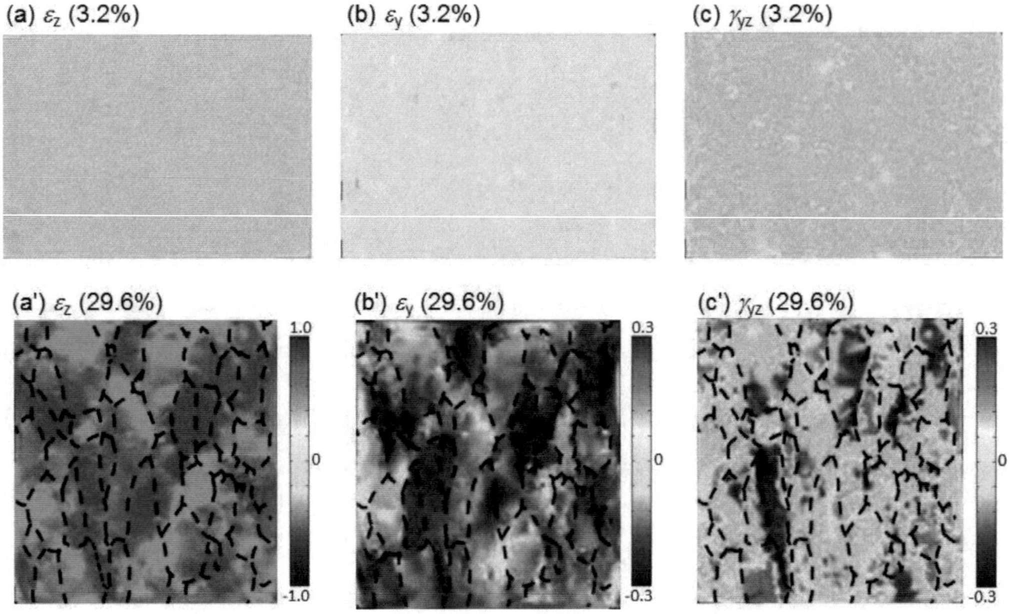

Figure 5. Local strain maps obtained by marker tracking in X-ray tomography. (a) and (a') show the ε_z strain which is along with tensile direction at 3.2% and 29.6% tensile deformation, respectively. (b) and (b') are the ε_y strain distribution at 3.2% and 29.6% tensile deformation, respectively. (c) and (c') show the shear strains γ_{yz} at 3.2% and 29.6% tensile deformation, respectively.

The distributions of local strains are shown in Fig. 6. Each strain distribution figure in Fig. 6 corresponds to the strain maps in Fig. 5. In the initial stage of tensile deformation, the distribution of local strains is narrow. Note that the distributions contain a measurement error due to the limitation of tomography resolution: because the average marker distance was 19.4 μm; if the spatial resolution of synchrotron radiation tomography is 1 μm, then the minimum measurable strain will be 5% (=1/19.4×100).

With increasing macroscopic tensile deformation, the distribution of local strains along the tensile direction shift towards positive (tensile) values and becomes wider gradually. In the final loading step 6 of 29.6%, the right tail of the distribution becomes long as shown in Fig. 6 (a'). In contrast, the bell-shaped distribution is kept in the ε_y strain (Fig. 6 (b')), a long left tail is observed in ε_x strain distribution. For the shear strain, the three components, γ_{zy}, γ_{yz} and γ_{zx} show a similar tendency. The maximum in the local shear strain did not change with tensile deformation. However, the widths increase with increasing macroscopic tensile deformation. The width of distribution in shear strains became wider than that of ε_z at the last step.

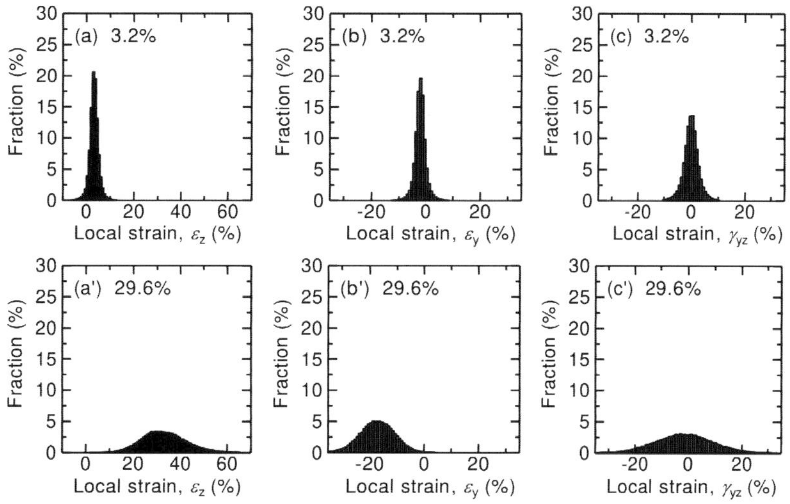

Figure 6. Distributions of local strains. Each figure correspond to the strain maps in Fig. 5.

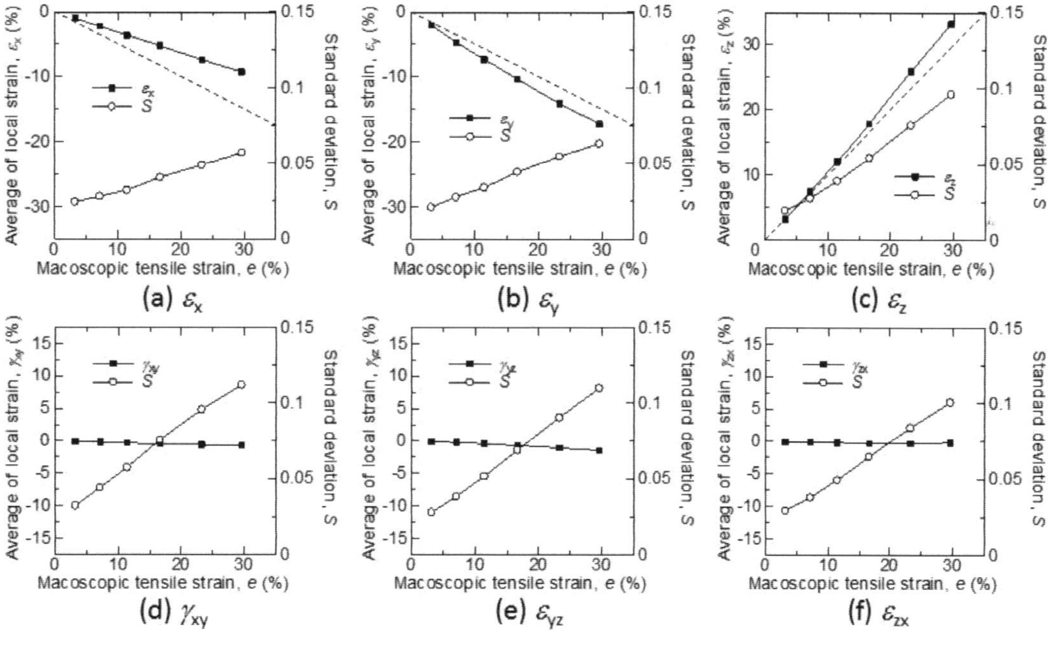

Figure 7. Changes in the average of local strain and standard deviations in the measured strain distributions. The six strain components (three normal strains and three shear strains) are plotted.

The averages of the local strains and standard deviations are plotted in Fig. 7. The dashed line shows when the local strain equals the macroscopic strain in the plot of ε_z (Fig. 7 (c)). A dashed line is also inserted in the plot of ε_x and ε_y, here it is assumed that $\varepsilon_x=\varepsilon_y=0.5e$, this is a valid assumption if the deformation causes no volume change and is isotropic in x and y. As seen in Fig. 7(c), the average of local strain, ε_z corresponds well with the macroscopic strain, e. This supports that the result of local strain measurement is reasonable. However, there are clear differences between ε_x (Fig. 7(a)) and ε_y (Fig.

7 (b)) and the macroscopic expectations. This could be due to the crystallographic texture of the sample. Although, some deviations from the dashed line are confirmed in all normal strain components, ε_x, ε_y and ε_z, the volume should not change. The volume change, $\Delta V=(1+\varepsilon_x)(1+\varepsilon_y)(1+\varepsilon_z)-1$ was checked for each loading step. The measured volume changes were less than 0.15% at all loading steps. Thus, the volume change was quite small. The standard deviation is largest for the ε_z strain. The standard deviation in ε_x and ε_y are similar and approximately half of that of ε_z at all strains. The averages of the local shear strains are close to zero for all shear strain components, γ_{zy}, γ_{yz} and γ_{zx} at all macroscopic strains. Therefore, although the local spatial distribution of the shear strains are very heterogeneous (see Fig. 5(c)), the average remains close to zero.

Local strains in the sample is very heterogeneously distributed spatially, the stored energy will therefore also vary significantly. It is suggested that the variations of local strains are considered when the aim is to pin point nucleation sites and understand nucleation mechanisms. Using the present method (without the gallium step) followed by annealing to study recrystallization is planned for the future.

4. Conclusions

The development of local strain distributions has been investigated in an aluminium alloy during tensile deformation by means of a three-dimensional strain measurement technique based on X-ray tomography. Strongly heterogeneous strain development was observed after 30% tensile loading. The normal strain components continued to develop at localized sites in the sample in which large strains occur already in the early stage of deformation. The development of shear strains was different to that of the normal strains. The area having a large shear strain propagated into low shear strain areas with increasing macroscopic tensile strain. Comparing deformation microstructures and local strain concentrations, obtained by this method, may be a powerful method to reveal nucleation positions of recrystallization.

Acknowledgements

The synchrotron radiation experiment was performed with the approval of JASRI through Proposal Nos. 2013B1537, 2013B1027 and 2014B1192. This work was undertaken partly with the support of a Grant-in-Aid for Scientific Research (S) and (C) from JSPS, through Subject No. 24226015 and 26420723. This research was also supported by Japan Science and Technology Agency (JST) under Industry-Academia Collaborative R&D Program "Heterogeneous Structure Control: Towards Innovative Development of Metallic Structural Materials". The authors also appreciate the financial assistance of the Light Metal Educational Foundation.

References

[1] Banabic D, Bunge H -J, Pöhlandt K, Tekkaya A E 2000 *Formability of metallic materials : plastic anisotropy, formability testing, forming limits* Springer

[2] Humphreys F J and Hatherly M 2004 *Recrystallization and related annealing phenomena, 2nd ed.* Elsevier

[3] Kobayashi M, Toda H, Kawai K, Ohgaki T, Uesugi K, Wilkinson D S, Kobayashi T, Aoki Y, Nakazawa M 2008 *Acta Materialia* **56** 2167

[4] Dougherty E R, Lotufo R A 2003 *Hands-on Morphological Image Processing* SPIE PRESS

[5] Ohser J, Schladitz K 2009 *3D Images of Materials Structures: Processing and Analysis* Wiley-VCH Verlag.

[6] Toda H, Ohkawa Y, Kamiko T, Naganuma T, Uesugi K, Takeuchi A, Suzuki Y, Kobayashi M 2013 *Acta Materialia* **61** 5535

[7] Barber C B, Dobkin D P, Huhdanpaa H T 1996 *ACM Transactions on Mathematical Software* **22** 469

Ultrafast analysis of individual grain behavior during grain growth by parallel computing

M Kühbach, L A Barrales-Mora, C Mießen, G Gottstein

Institute of Physical Metallurgy and Metal Physics, RWTH Aachen University, Kopernikusstrae 14, D-52056 Aachen, GERMANY

E-mail: kuehbach@imm.rwth-aachen.de

Abstract. The possibility to characterize in an automatized way the spatial-temporal evolution of individual grains and their properties is essential to the understanding of annealing phenomena. The development of advanced experimental techniques, computational models and tools helps the acquisition of real time and real space-resolved datasets. Whereas the reconstruction of 3D grain representatives from serial-sectioning or tomography datasets becomes more common and microstructure simulations on parallel computers become ever larger and longer lasting, few efforts have materialized in the development of tools that allow the continuous tracking of properties at the grain scale. In fact, such analyses are often left neglected in practice due to the large size of the datasets that exceed the available physical memory of a computer or the shared-memory cluster. We identified the key tasks that have to be solved in order to define suitable and lean data structures and computational methods to evaluate spatio-temporal grain property datasets by working with parallel computer architectures. This is exemplified with data from grain growth simulations.

1. Introduction

Understanding quantitatively annealing microstructures as they evolve during primary recrystallization (PRX), normal and abnormal grain growth (NGG, AGG) requires a characterization of the mechanisms and the impact of structural heterogeneities by in-situ experiments and/or numerical simulations [1, 2]. In that sense, much progress has been made in the past decade. In particular, this has been made possible by experimental techniques such as electron [3–8] and X-ray diffraction microscopy [9–13]. Additionally, progress in computational algorithms and methods for 2D and 3D microstructural evolution models in terms of representative volume elements (RVE) have allowed to study larger systems. Recent examples include sequential and parallel Monte Carlo [14–16], cellular automata [17], phase field [18], vertex models [19–22], Poisson-Voronoi networks [23], and level set approaches [24–26]. Owing to the success of parallel microstructural modeling, it is expected that even larger simulation studies will become feasible.

The main task in transforming numerical results into a quantitative physical understanding requires extensive data processing, filtering, and ultimatively a reduction of the information. Successful examples of this procedure [27] are, for instance, the DREAM.3D initiative [5, 28], diffraction analysis tools [29, 30], parallelized 3D microstructure reconstruction from synchrotron experiments [31], the MTex texture toolbox [32], and defect identification from molecular dynamics data [33, 34]. Clearly, such tools have become vital to evaluate and to analyze data from experiments and simulations.

The key problem of this process during the monitoring of grain boundary migration phenomena is the difficulty to determine a priori the grains or microstructural features, whose evolution becomes dominant for the system, for instance for the initiation of abnormal grain growth or the identification of critical stages during nucleation prior to recrystallization [35, 36]. A possible solution is the tracking of the evolution of the properties for all grains in an ensemble. This poses a challenge due to the high density of data required but also a great potential as hinted by results in 3D grain growth [37–39]. This motivated the present contribution, in which we suggest and provide first examples of a procedure that can be utilized to track in parallel single grain evolution.

For this task parallel computing architectures offer an excellent opportunity to analyze efficiently large datasets. However, the key limitation of these systems is the fact that each data transfer is directed to the central processing unit via a memory hierarchy which is much slower than the capacity of the central processing unit. Consequently, it is necessary to reduce these transfers to a minimum in order to keep the pieces of information as long and as close as possible to the physical computing processors [40–42]. Hence, making parallel codes scalable requires two steps. Namely, to utilize optimal sequential algorithms that are cache locality-aware [41] and to partition each analysis into as many as possible spatial and temporal independent tasks as possible. The term cache locality summarizes both the spatial and temporal proximity with which the processing unit can access data in its fastest memory (L1) during the execution flow. This concept was the key idea behind the algorithm that will be introduced in the following.

2. Analysis setup

The typical outcome of a sampled in-situ experiment or simulation is a collection $\mathbb{M}$ (dataset) that was sampled at time steps t_i with $i = 1, 2, \ldots, N_m$ into sets $\mathbb{M}_i$. Each of them comprises individual data, the most intuitive of which are the grains g_j^i with $j = 1, 2, \ldots, N_0^i$ and their characteristics. For these grains, we can distinguish scalar properties, such as a unique nametag I_j^i, the size V_j^i, and the orientation $\mathbb{O}_j^i$, the latter of which is commonly parameterized by the three Euler angles. Depending on the dimensionality of the analysis we distinguish further topological properties. Namely, the number of faces, referring to the linear segments in 2D as well as surfaces in 3D which enclose the grain g_j^i. Each face requires further a nametag that identifies the neighboring grain, and, in 2D, the length of the segment. In three dimensions it is necessary to track the area of each face to identify the bounding triple lines, their length, as well as nametags which identify the grains that meet at the triple lines, respectively. Moreover, quadruple junctions can be monitored in terms of their position and three nametags in order to specify the adjoining triple lines [23, 37].

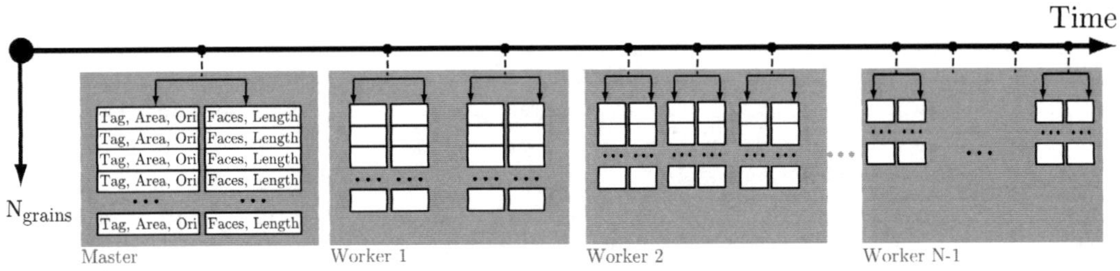

Figure 1. Association of data to the computing nodes.

Although several data analyses are of interest, in this contribution we focus on utilizing the data collection for answering the following questions:

(i) How can physical mechanisms be identified from property changes over time?
(ii) Which evolutionary differences allow certain grains to grow larger than others?
(iii) How is the size evolution of particular grains correlated with their environment?

Realizing that in the case of grain growth phenomena the number of grains decreases from an initial N_0^0 to N_{fin}^m (see figure 2), it is possible to approximate the size of $\mathbb{M}$ comprising N_m time steps. By making use of Euler's polyhedron formula [43], an average information size S_j^i that is necessary to store pieces of information e.g. from a 6-sided grain (in 2D), and a 3D grain with 14 faces on average can be taken as a reasonable average polyhedron to project the dataset size provided the underlying distribution is known. Storing into standard datatypes in 64-bit double precision and integer values would require (in 2D) to store for each grain its nametag, size, orientation, the number of adjoining faces, their neighbors, and line lengths. Hence, a total size of 112 Bytes per grain in 2D and 1106 Bytes in 3D is required.

Consequently, tracking of a dataset $\mathbb{M}$ from NGG simulations published in [24] would require $\approx$ 14 GB in 2D and only $\approx$ 5 GB in 3D as the latter was reported executed in fewer time steps than the 2D simulation. Larger population studies ($N_0^0 = 133110$, $N_{fin} = 14150$, $N_m = 300$, in 3D) [26] would range at $\approx$ 23 GB. Contrary to common perception this reveals that spatio-temporal analyses are not computationally prohibitive on contemporary workstations. However, for even larger simulations, longer lasting or finer resolved in time simulations, such as a hypothetical setup starting from $N_0^0 = 10^6$ grains that is simulated to coarsen to 1% of the population in $N_m = 10000$ steps would require $\approx$ 0.5 TB in 2D, and $\approx$ 5 TB in 3D. Clearly, this substantiates the upcoming necessity to rely on parallel analyses. Moreover, keeping all pieces of information that describe a grain and its neighbors would result in rather large data blocks per grain compared to contemporary processor data cache sizes (L1: 32 KB). In essence, a restructuring of information into e.g. topology-related and other pieces of information is unavoidable. Based on these facts, we propose the following analysis strategy, as depicted in figure 1. In the post-processing the sets $\mathbb{M}_i$ are per se mutually independent. However, the analysis requires pieces of information to be first filtered within the sets $\mathbb{M}_i$, and then re-organized into a temporal sequence and output into summary files. This renders the classical MapReduce algorithm inadequate [44]. In the following we detail further our analysis for a distributed memory computer system comprising N_{nodes} computing nodes. Each node is identified as $\tilde{N}_k$ with $0 \leq k < N_{nodes}$ defining the set $\mathbb{N}_{nodes}$ and for simplicity of similar performance characteristics, i.e. same local memory S_{crit} capacity and a connection to a mass storage device whose capacity S_{HDD} is significantly larger than S_{crit}. Within this framework, it is reasonable to assume that the nodes read and write this device at a rate of χ_r^k, and χ_w^k, respectively. The analysis of the data, as detailed schematically in figure 1, requires the following steps:

(i) Determination of the individual set sizes $S_{\mathbb{M}_i}$.
(ii) Grouping sets in time $t_i, t_{i+1}, ..., t_{i+n}, n \in \mathbb{N}$ under the constraint $\sum S_{\mathbb{M}_i} \leq S_{crit}$.
(iii) Loading and filtering in parallel by maximizing and balancing the total data transmission rates $\sum \chi_r^k$ and $\sum \chi_w^k$, respectively.
(iv) Collecting data along the time axis by explicit communication among the computing nodes.

For the partitioning of the set $\mathbb{M}_i$ on the computing nodes a solution to a knapsack-type optimization problem needs to be solved. Our strategy is to assign each computing node a collection of sets $\mathbb{M}_i$ as a consecutive interval along the time axis, resulting in a set $\mathbb{M}_{local}^k$ that is local to the k-th computing node. As the number of grains either reduces during NGG and AGG or increases during PRX it is further beneficial to keep the number of grains per node balanced [40]. For the loading and filtering of gigabyte-sized datasets in parallel, efficient strategies have been developed that can make use of parallel access to storage devices via plain parallel MPI I/O or similar schemes such as HDF5 or NetCDF. We utilized MPI I/O as it allows reading and writing in parallel.

In what follows, the tasks of filtering and collecting selected grains and properties are essentially trivial-parallel until the local results need to be aggregated into their final form. Specificially, we detail how to efficiently keep track of the evolving neighborhood for selected grains. Here, the key problem is that for writing collectively the characteristics of all neighbors into a file of fixed size, their total number and nametags have to be known before the output can be written. We solve this problem by identifying first, and in parallel, all disjoint nametags. Subsequently, these are sent one after another to a single computing node which eliminates all duplicates before finally redistributing a list of unique nametags to all. This strategy can further be improved by defining a communication hierarchy that permits the systematic elimination of duplicates.

3. Simulation setup

In order to utilize the proposed workflow, a first version of the algorithm was implemented as an MPI-parallel Visual C/C++ version that reads plain text files carrying the properties of all grains and their nearest neighbors by sampling typical 2D grain growth simulation studies. The analyzed dataset was extracted by sampling each integration step along a 2D grain growth simulation, which was conducted with the vertex model [19] as detailed and parameterized in [22]. Specifically, the simulation was initiated by a modified Voronoi tessellation mapping the orientation and grain size distribution from simulation results [22]. The grain boundary energies were assumed to depend on misorientation according to the Read-Shockley model [45]. The energy of high-angle grain boundaries ($\theta \geq 15°$) was considered constant $\gamma_{HAGB} = 0.9 J/m^2$. Their mobility was defined as $m = 10^{-13} m^4/Js$ whereas the mobility of low-angle grain boundaries was set to 10% of this value.

4. Results and Discussion

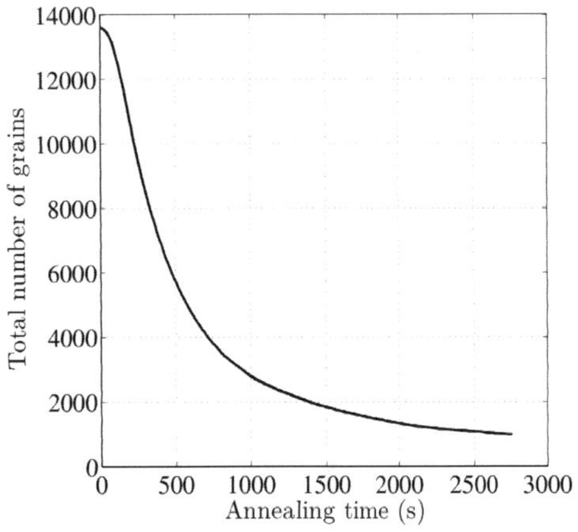

Figure 2. Population evolution.

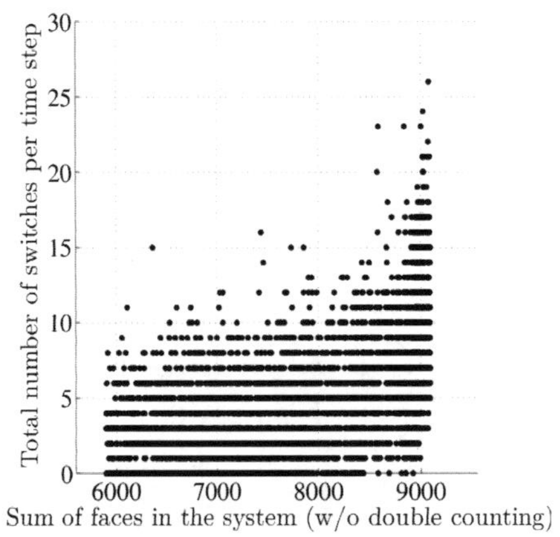

Figure 3. Total topological activity.

Figures 2 and 3 summarize the temporal evolution of the population size, and the topological activity in the network, respectively. As expected, the reduction of the population size was observed associated with a decreasing number of topological transformations per time step. The fact that the absolute number of individual topological changes was observed to be two orders of magnitude lower than the total number of faces, i.e. linear grain boundary segments, suggests a further potential in reducing the amount of topological data that have to be tracked during data acquisition.

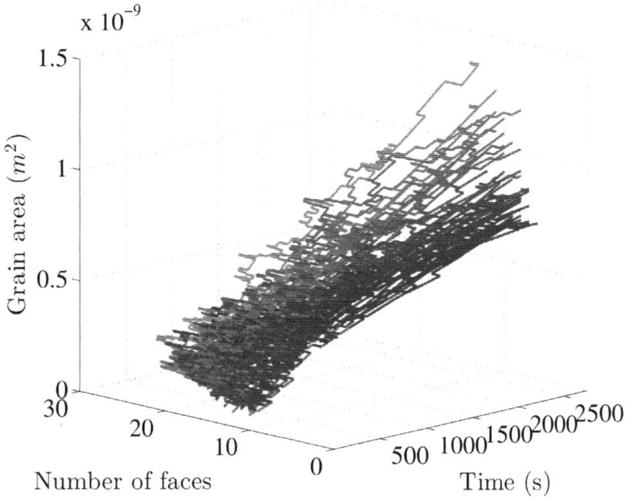

Figure 4. Evolution history of surviving grains that grew to at least $3\bar{V}$.

Application of the analysis scheme to the data allowed extracting the growth-topology history for selected grains which survived the coarsening process and grew to at least three times the average final grain size. This result, as shown in figure 4, suggests that the network evolves through a two stage transient. In the first stage, in particular some grains gain selectively faces whereas in the second stage this process becomes interrupted and followed by a transient phase along which the largest grains lose faces. This suggest an aborted onset of potential AGG.

A closer inspection of particular grains, namely one that became the largest of the population (figures 5, 6) and a significantly smaller grain experiencing stronger transient characteristics (figures 7, 8) were compared. For these grains, the development of their neighbors was tracked and correlated with their own evolution. The fact that with increasing annealing time more and more neighbors became not only topologically similar but as well comparably larger, resulted in the growth competition of grains that initially had a size advantage. As a consequence, the network evolution followed a transient back to normal grain growth. These quantitative results agree well with inspecting qualitatively selected snapshots of the network, such as those which are depicted in figures 9, 10, and 11.

To sum up, the approach of following individual grain behavior was shown to be a task that is computationally intense but not prohibitive when parallel computer architectures are utilized efficiently. This motivates future applications of the method, such as screening in an automatized manner simulation data in order to characterize the chain of events that have caused microstructure change. Particularly for characterizing rare events, such as the initiation of abnormal grain growth or inspecting the birth of recrystallization nuclei, this approach is beneficial.

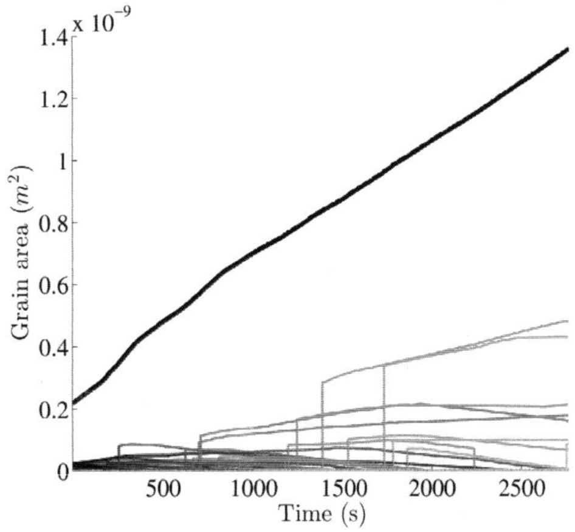

Figure 5. Largest grain in its competition with nearest neighbors.

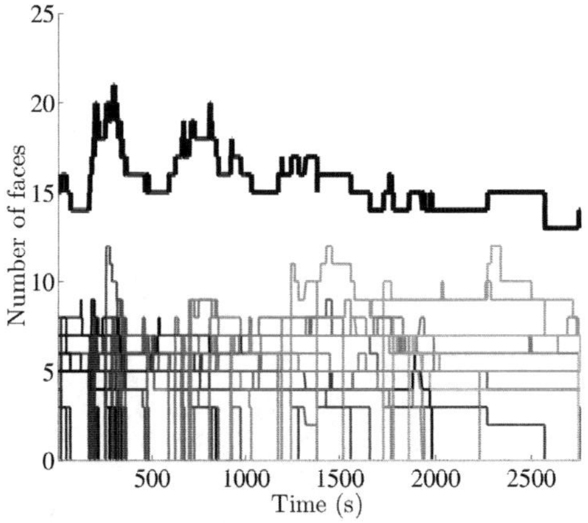

Figure 6. Topology of largest grain compared to nearest neighbors.

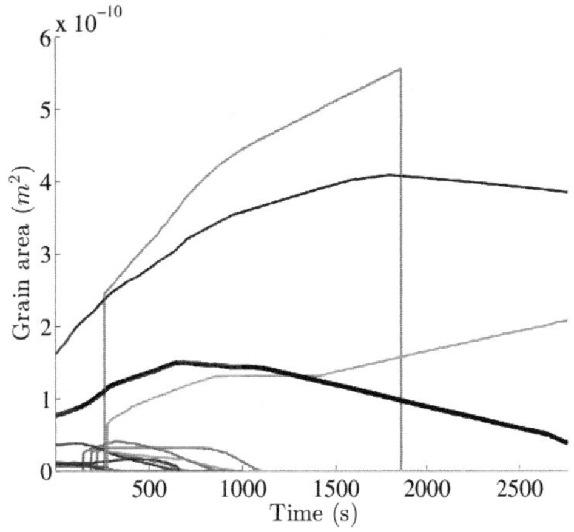

Figure 7. Size evolution.

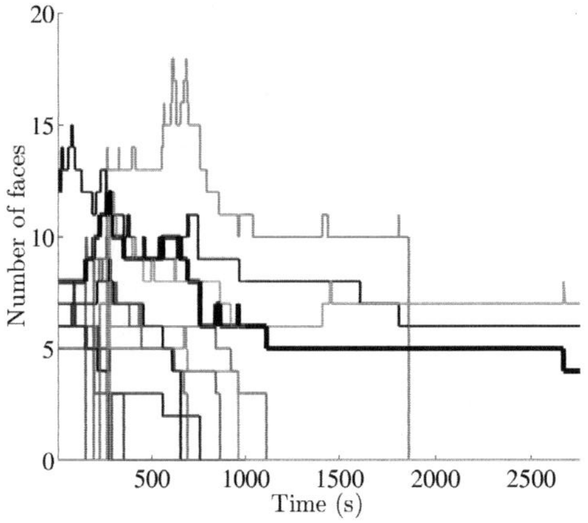

Figure 8. Topological evolution.

Figure 9. $t = 0s$

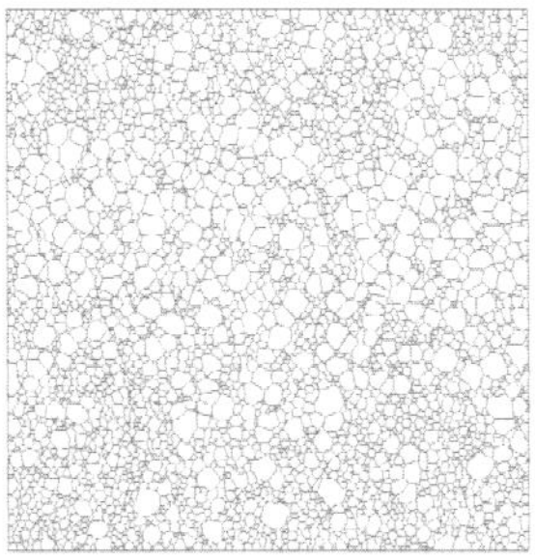

Figure 10. $t = 676s$

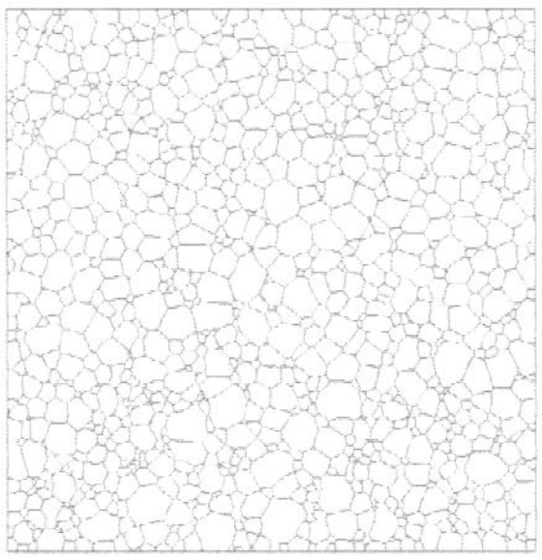

Figure 11. $t = 2754s$.

5. Conclusions

A scalable tool for the post-processing and the analysis of large datasets of microstructural evolution on a parallel computer was presented. The tool allows the tracking of local features of a microstructure in time and space. In particular, it was applied to a detailed investigation of a 2D grain growth vertex simulation in Ni-5at.%W. Contrasting in details the growth curves of selected grains against their immediate neighbors revealed that the progressive competition of similarly sized grains restricted the potential for triggering abnormal grain growth.

Acknowledgements

The authors gratefully acknowledge the financial support from the Deutsche Forschungsgemeinschaft (DFG) within the *Reinhart Koselleck-Projekt* (GO 335/44-1) as well as the FZ Jülich and the RWTH Aachen University for granting computing time within the frame of the JARAHPC project no 6687.

References

[1] Bay B, Hansen H, Hughes D A and Kuhlmann-Wilsdorf D 1992 *Acta Met. et Mat.* **40** 205–219
[2] Gottstein G and Shvindlerman L S 2010 *Grain Boundary Migration in Metals* 2nd ed (CRC Press, Taylor & Francis, Boca Raton)
[3] Hansen N, Leffers T and Kjems J K 1981 *Acta Met.* **29** 1523–1533
[4] Zhang C, A S, Ishimaru T and Enomoto M 2004 *Metall. Mat Trans. A* **35** 1927–1933
[5] Groeber M A, Ghosh S, Uchic M D and Dimiduk D M 2008 *Acta Mat.* **56** 1257–1273
[6] Rowenhorst D J, Lewis D C and Spanos G 2010 *Acta Mat.* **58** 5511–5519
[7] Tucker J C, Chan L H, Rohrer G S, Groeber M A and Rollett A D 2012 *Scripta Mat.* **66** 554–557
[8] Yu T B, Hansen N and Huang X X 2012 *Phil. Mag.* **92** 4056–4074
[9] Lauridsen E M, Poulsen H F, Nielsen S F and Jensen D J 2003 *Acta Mat.* **51** 4423–4435
[10] Gundlach C, Pantleon W, Lauridsen E M, Margulies L, Doherty R D and Poulsen H F 2004 *Scripta Mat.* **50** 477–481
[11] Toda H, Ohkawa Y, Kamiko T, Naganuma T, Uesugi K, Takeuchi A, Suzuki Y and Kobayashi M 2013 *Acta Mat.* **61** 5535–5548
[12] McKenna I M, Poulsen S O, Lauridsen E M, Ludwig W and Voorhees P W 2014 *Acta Mat.* **78** 125–134
[13] Pokharel R, Lind J, Li S F, Kenesei P, Lebensohn R A, Suter R M and Rollett A D 2015 *Int. J. Plast.* **67** 217–234
[14] Holm E A, Miodownik M and Rollett A D 2003 *Acta Mat.* **51** 2701–2716
[15] Madison J D, Tikare V and Holm E A 2012 *J. Nuc. Mat.* **425** 173–180
[16] Park C S, Na T W, Kang J K, Lee B J, Han C H and Hwang N M 2013 *Phil. Mag.* **93** 4198–4212
[17] Kühbach M, Barrales-Mora L A and Gottstein G 2014 *Modell. Sim. Mat. Sc. Eng.* **22**
[18] Suwa Y, Saito Y and Onodera H 2008 *Mat. Trans.* **49** 704–709
[19] Barrales-Mora L A, Gottstein G and Shvindlerman L 2008 *Acta Mat.* **56** 5915–5926
[20] Sprague A P, Patterson B R and Gandhi S 2010 *Met. Trans. A* **41** 592–602
[21] Barrales-Mora L A, Gottstein G and Shvindlerman L S 2012 *Acta Mat.* **60** 546–555
[22] Witte M, Barrales-Mora L A, Boer N D, Gilges S, Klöwer J and Gottstein G 2012 *Phil. Mag.* **92** 4398–4407
[23] Lazar E A, Mason J K, MacPherson R D and Srolovitz D J 2013 *Phys. Rev. E* **88** 063309
[24] Elsey M, Esedoğlu S and Smereka P 2009 *J. Comp. Phys.* **228** 8015–8033
[25] Elsey M, Esedoğlu S and Smereka P 2011 *Phil. Mag.* **91** 1607–1642
[26] Elsey M, Esedoğlu S and Smereka P 2011 *Proc. Roy. Soc. A* **467** 381–401
[27] Simmons J, Bouman C, Fahrroo F, Kalidindi S R, Knopp J and Voorhees P 2014 *Session at the 2014 TMS*
[28] Groeber M A and Jackson M A 2014 *Integrating Mat. and Manu. Inno.* **3**
[29] Bergbäck K E, Sørensen H O, Wright J P, Goret G and Kieffer J 2013 *J. Appl. Cryst.* **46** 537–539
[30] Schmidt S 2014 *J. Appl. Cryst.* **47** 276–284
[31] Wozniak J M, Sharma H, Armstrong T G, Wilde M, Almer J D and Foster I 2014 Big Data Staging with MPI-IO for Interactive X-ray Science Tech. rep. University of Chicago, Argonne National Laboratory
[32] Bachmann F, Hielscher R and Schaeben H 2010 *Solid Stat. Pheno.* **160** 63–68
[33] Panzarino J F and Rupert T J 2014 *JOM* **66** 417–428
[34] Stukowski A 2014 *JOM* **66** 399–406
[35] Humphreys F J 1992 *Scripta Met. et Mat.* **27** 1557–1562
[36] Weygand D, Bréchet Y and Lépinoux J 2000 *Phil. Mag. B* **80** 1987–1996
[37] Patterson B R, Rule D J, DeHoff R T and Tikare V 2013 *Acta Mat.* **61** 3986–4000
[38] Patterson B R, Rowenhorst D J, Tikare V, DeHoff R T and Kaub T M 2014 *Acta Mat.* **79** 411–420
[39] Keller T, Cutler B, Lazar E A, Yauney G and Lewis D J 2014 *Acta Mat.* **66** 414–423
[40] Rabenseifner R, Hager G and Jost G 2009 *17th Euromicro Int. Conf. on Parallel, Distributed and Network-Based Processing, Weimar, Germany, Febr 18th - 20th, 2009* p 427–436
[41] Williams S, Waterman A and Patterson D 2009 *Communications of the ACM* **52** 65–76

[42] Hennessy J L and Patterson D A 2012 *Computer Architectures* 5th ed (Morgan Kaufmann, Amsterdam)
[43] Euler L 1752 Elementa Doc. Soli., Novi Comm. Acad. Sc. Imp. Petro. Rep. in Opera, Vol. 26, pp. 71-92
[44] Dean J and Ghemawat S 2004 *OSDI'04: 6th Symp. on Op. System Design and Impl., San Francisco, CA*
[45] Read W T and Shockley W 1950 *Phys. Rev.* **78** 275–289

Recrystallization textures and microstructures of Al-0.3%Cu alloy after deformation to high strains

X R Li[1], A Wakeel[1], T L Huang[1], G L Wu[1,*] and X Huang[2]
[1] College of Materials Science and Engineering, Chongqing University, Chongqing 400045, China
[2] Danish-Chinese Center for Nanometals, Section for Materials Science and Advanced Characterization, Department of Wind Energy, Technical University of Denmark, Risø Campus, 4000 Roskilde, Denmark

E-mail: wugl@cqu.edu.cn

Abstract. An Al-0.3%Cu alloy was deformed to high strains by cold rolling. The as-deformed samples were annealed at different temperatures until complete recrystallization. The cold rolling textures were determined by X-ray diffraction while the recrystallization textures and microstructures were characterized by electron backscatter diffraction. It was found that the rolling texture was characterized by a strong Brass component. After complete recrystallization Goss and Cube textures were developed. The effects of deformation strain and annealing temperature on the recrystallization textures are discussed.

1. Introduction

The recrystallization texture of aluminium alloys has been studied for many decades due to its importance for the mechanical properties and formability of Al alloys. It is known that the development of a recrystallization texture is strongly dependent on the deformation texture. As a typical face centred cubic (FCC) alloy with a high stack fault energy, Al alloys reveal a well-known rolling texture called the β -fibre, which runs from Brass {110}<112> through S {123}<634> to Copper {112}<111> texture components during plastic deformation [1] The Brass component is usually weaker than the S and Copper components.

During subsequent annealing, recrystallization textures develop usually having components of Cube {001}<100>, retained rolling texture and/or random orientation [2]. In some cases, P {110}<111> and/or $Cube_{ND}$ {001}<310> components are formed in Al alloys containing particles [3].

However, in our recent experiment a strong Goss {011}<100> texture was found in the centre layer of the Al-0.3%Cu sheets after being cold rolled to 98% and annealing at 300℃ [4]. In this paper, the Al-0.3%Cu alloy was cold rolled to different strains and annealed at different temperatures until complete recrystallization. The textural evolutions during deformation and annealing were investigated.

2. Experimental

The material used in the present investigation was an Al-0.3%Cu alloy, which was casted using 5N Al and an OFHC Cu. The ingot was forged at 200℃ to obtain a plate 50mm thick. The material has a very coarse grain size of about 1.5-2mm. The as-forged plate was then cold rolled to 80%, 90%, 95%

and 98% reductions in thickness. To ensure homogeneous deformation for each rolling pass, the contact length to mean thickness ratio (l/h) was controlled to be ~2.5 [5]. All the cold rolled samples were annealed at 200, 300 and 400 ℃ in a tube furnace for different times until complete recrystallization. The samples were inserted into the furnace when the furnace temperature reached the desired temperature. Then it took about 5 minutes for the temperatures to be stabilized at the setting points.

The textures of the as-forged and cold-rolled samples were measured at the centre layer of the samples by X-ray diffraction on the rolling plane. The samples were 10 mm long along the rolling direction (RD) and 8 mm wide along the transverse direction (TD). (1 1 1), (2 0 0), and (2 2 0) pole figures were measured up to a maximum tilt angle of 70° by the Schulz back-reflection method using Cu-K α radiation. The orientation distribution functions (ODF) were calculated. Volume fractions of different texture components were calculated from the ODFs. The microstructures and textures of the annealed samples were measured on the longitudinal section, using an Oxford Aztec electron backscatter diffraction (EBSD) detector attached to a TESCAN MIRA3 scanning electron microscope. Texture components were defined as within 15° from their ideal orientations.

3. Results

To ease the description of the results, Figure 1 illustrates the texture components that are relevant for the present study. As an example, figure 2 shows the ODF for the as-forged state. The orientation of the highest intensity does not correspond to a typical texture component although a relatively weak Goss component is seen. The high intensity is caused by the presence of a limited number of grains in the measured area due to the large grain sizes (1.5-2mm). Several similar X-ray measurements showed the occurrence of high intensity at different orientations. These observations indicate that the as-forged sample has a weak texture.

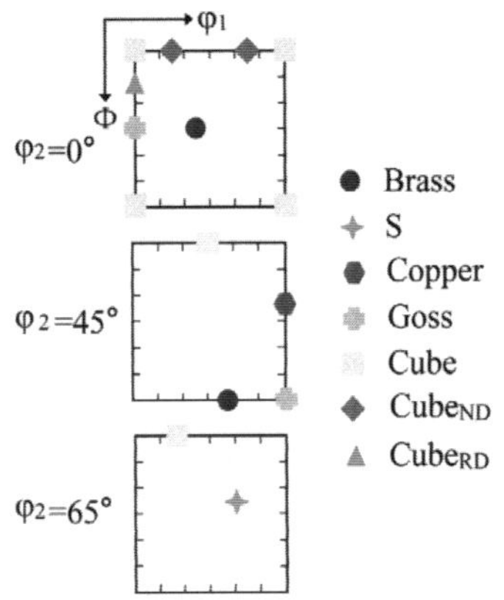

Figure 1. ODFs position of the texture components relevant for the present study.

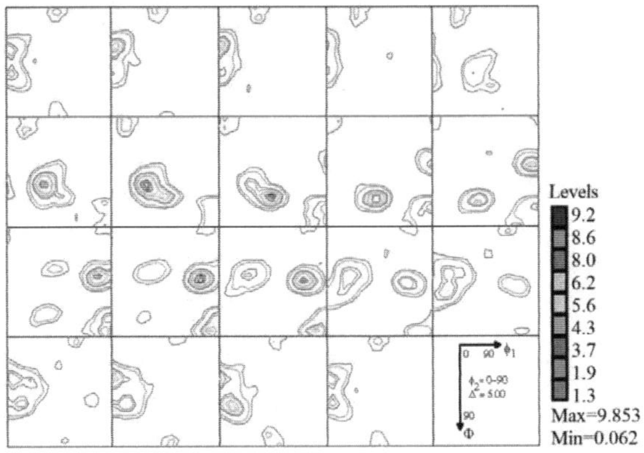

Figure 2. ODFs of the as-forged sample

Figure 3 shows the texture evolution of the Al-0.3%Cu alloy cold rolled to 80%, 90%, 95% and 98% reductions in thickness, respectively. It is seen that a typical rolling texture is developed after cold rolling to 80%. Along the β-fiber of the rolling texture, the Brass component shows a higher intensity than the S and Copper components.. Beside these typical rolling texture components, there exists a weak Goss component at all strains.

Figure 4 shows the evolution of the volume fractions of individual texture components during rolling. It is seen that the volume fraction of Brass component increases while those of S and Copper compnents decrease with increasing rolling reduction over the range of 80 -98%. At all strains, the

Brass component has much larger volume fractions than S and Copper components. Besides, there are about 5-10% Goss grains at all strains.

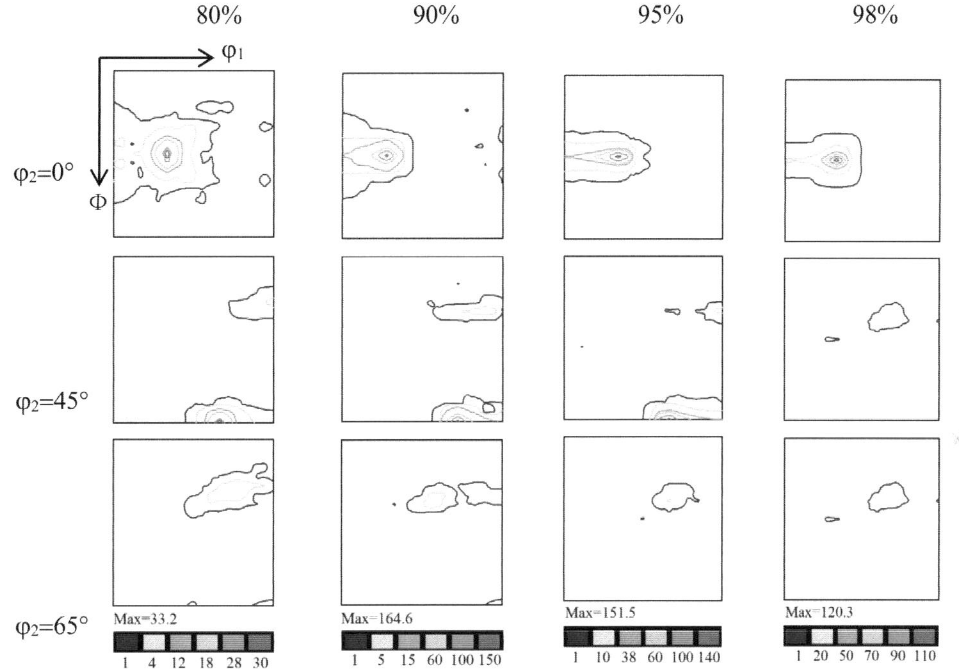

Figure 3. Sections $\phi 2 = 0°$, $45°$, $65°$ in the ODF of the samples cold rolled to thickness reductions of 80, 90, 95 and 98.

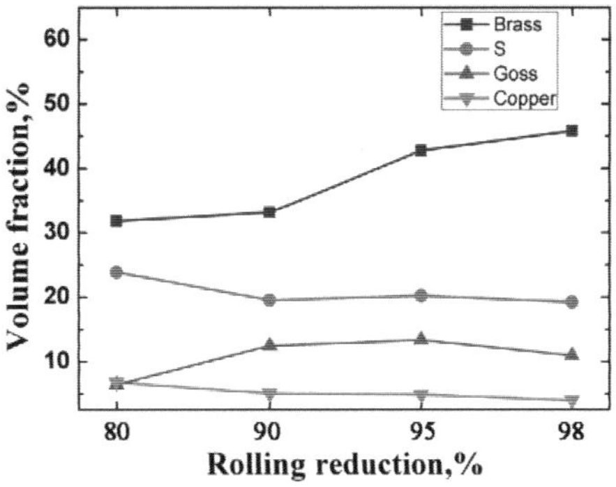

Figure 4. Evolution of the volume fractions of individual texture components during rolling.

The microstructures and textures after complete recrystallization are shown in Figure 5 and Figure 6, respectively. After recrystallization, the textures are mainly $Cube_{RD}$ (Cube rotated about RD), Goss and $Cube_{ND}$ (Cube rotated about ND), all seen in the section of $\phi 2=0°$ of the ODF. Therefore only this section is shown in Figure 6. After completely recrystallized at 200℃, the recrystallized grains are elongated along RD (see Figure 5(a)-(d)). Cube and $Cube_{RD}$ components are developed in the 80% deformed and annealed sample. $Cube_{RD}$ still dominates the recrystallization texture in the 90%

IOP Conf. Series: Materials Science and Engineering **89** (2015) 012032 doi:10.1088/1757-899X/89/1/012032

deformed and annealed sample. As the rolling reduction increases, Cube and $Cube_{RD}$ components almost disappear while a Goss component is increasing (see Figure 6(a)-(d)). It indicates that the recrystallization texture transforms from Cube and $Cube_{RD}$ components to a Goss component when annealing at 200 ℃ . Figure 5(e)-(h) show equiaxed recrystallized grains after complete recrystallization at 300℃. $Cube_{ND}$, Goss and $Cube_{RD}$ components are observed except for the 98% deformed sample, which only exhibits a strong Goss texture with a weak $Cube_{ND}$ component (Figure 6(e)-(h)). The intensity of the Goss texture increases with increasing rolling reduction. After complete recrystallization at 400℃, very coarse grains are formed in the samples (see Figure 5(i)-(l)). The recrystallization texture is relatively random with weak textures around Cube and Goss in the 80% and 90% deformed samples, while for the samples deformed to reductions of 95% and 98% a strong $Cube_{ND}$ are observed after recrystallization (see Figure 6(i)-(l)).

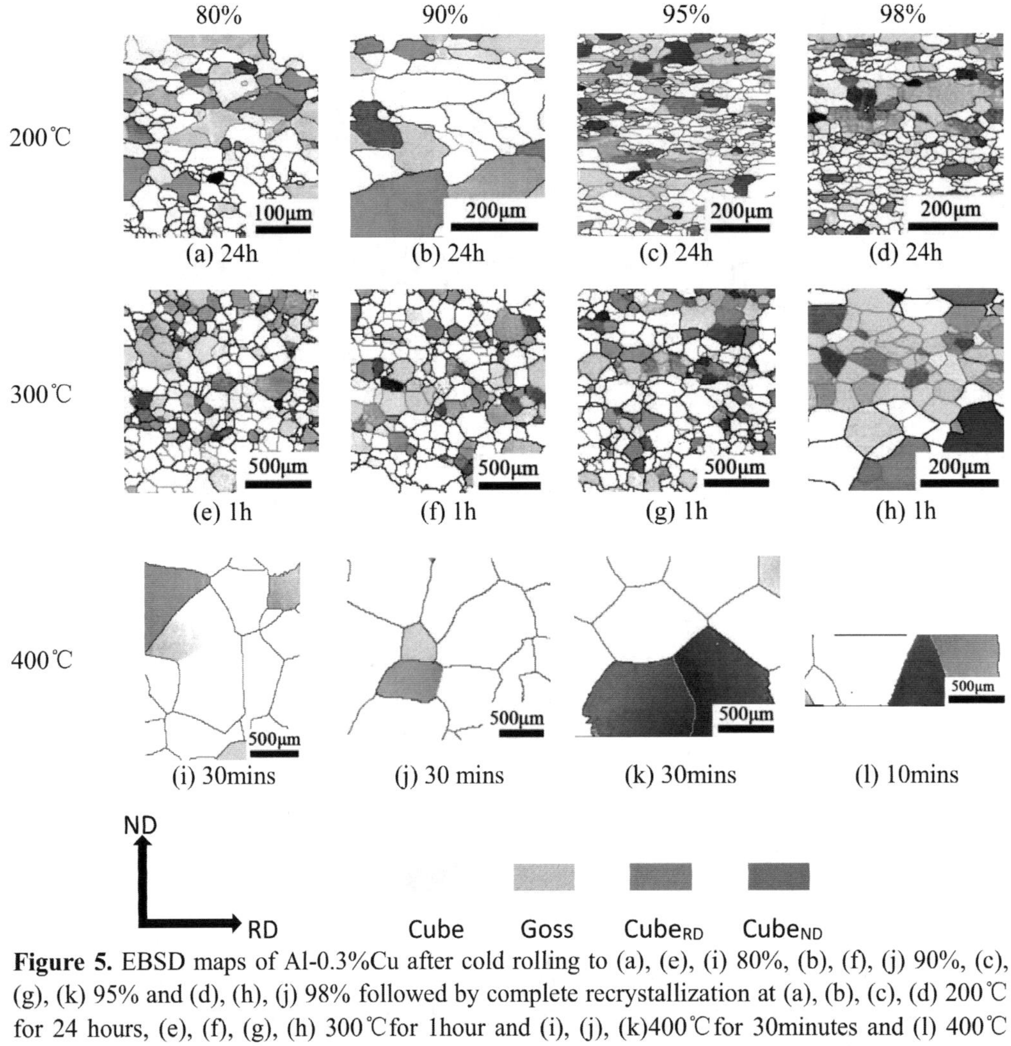

Figure 5. EBSD maps of Al-0.3%Cu after cold rolling to (a), (e), (i) 80%, (b), (f), (j) 90%, (c), (g), (k) 95% and (d), (h), (j) 98% followed by complete recrystallization at (a), (b), (c), (d) 200℃ for 24 hours, (e), (f), (g), (h) 300℃for 1hour and (i), (j), (k)400℃for 30minutes and (l) 400℃ for10minutes, respectively. In the maps, light grey lines and black lines represent grain boundaries ⩾2° and ⩾15°, respectively.

IOP Conf. Series: Materials Science and Engineering **89** (2015) 012032 doi:10.1088/1757-899X/89/1/012032

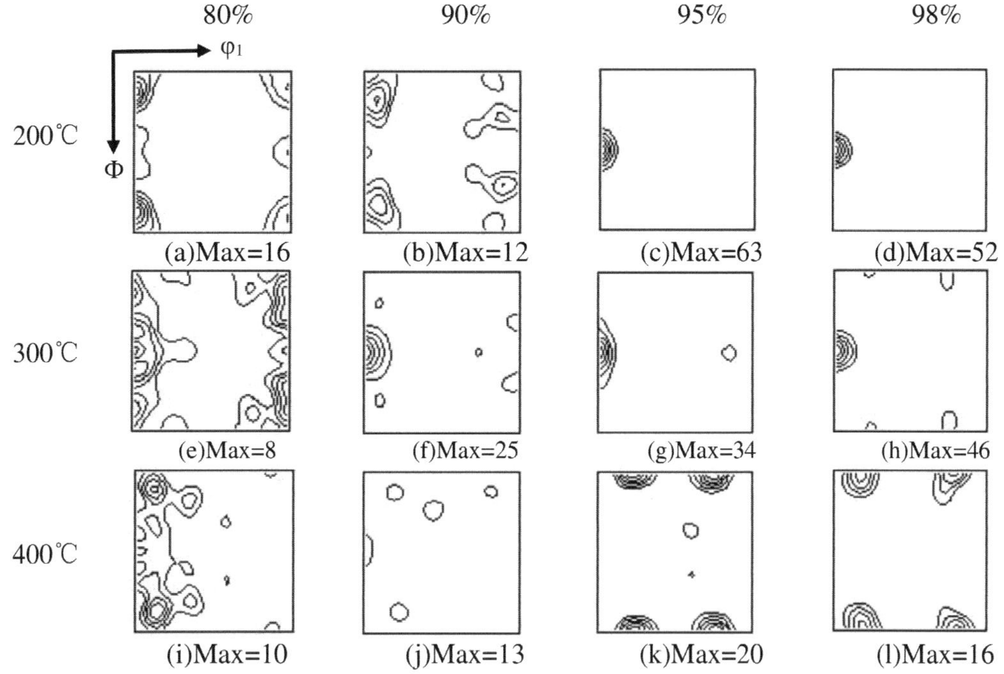

Figure 6. φ2 = 0° sections of ODFs of Al-0.3%Cu cold rolled to (a), (e), (i) 80%, (b), (f), (j) 90%, (c), (g), (k) 95% and (d), (h), (j) 98% followed by complete recrystallization at (a), (b), (c), (d) 200℃, (e), (f), (g), (h) 300℃ and (i), (j), (k), (l) 400℃.

Figure 7 shows the effects of annealing temperature and rolling reduction on the recrystallization texture for the Al-0.3%Cu alloy. It is found that the volume fraction of the Goss component increases with increasing rolling reduction when the samples were annealed at 200℃ and 300℃, whereas the Goss orientation almost vanished after recrystallization at 400℃(Figure 7(a)). $Cube_{ND}$ orientation is obtained only in highly deformed and high temperature annealed samples (Figure 7(b)). $Cube_{RD}$ decreases in volume with increasing rolling reduction and annealing temperature (Figure 7(c)).

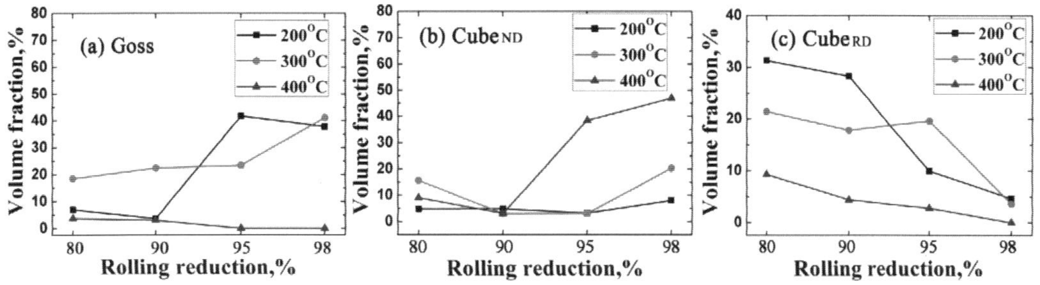

Figure 7. Evolution of the texture components (a) Goss, (b) CubeND, (c) CubeRD

4. Discussion

The present results revealed that the deformation texture of the Al-0.3%Cu alloy exhibits a strong Brass orientation (Figure 3). Usually S orientation has been reported as the strongest texture component and Copper the second strongest in Al alloys cold rolled to high strains. However, the Brass orientation was also found to be the dominant component in a hot rolled Al alloy containing fine particles, especially at high strains. It was suggested that the strong Brass texture was associated with differential dynamic grain growth, which was principally driven by different substructural energy densities in different grain orientations. The presence of Zener pinning effect made this process more selective. Deformation geometric and dynamic recovery enhanced the growth process [6]. However, nano-spaced dislocation boundaries were formed after 98% rolling in the present material. It was proposed that the dislocation boundaries were stabilized by the pinning effect of Cu solute element [7]. Therefore it is unlikely that extensive recovery occurred during rolling in the present experiment. There is no particle in the present alloy, either. One possible explanation for the strong Brass texture in the present alloy is that the Brass texture is rotated from orientations near Goss and Brass as seen in the as-forged sample (Figure 2). These orientations would rotate to Goss and Brass upon rolling, leading to an increase in the volume fraction of the Brass component. It was also proposed that Cu element in Al alloys promotes the formation of microshear bands resulting in a strong Brass texture [1]. Microshear bands were characterized by EBSD in the present alloy after cold rolled to 98% reduction [7].

Normally the main component of recrystallization texture in Al alloys is Cube. It has been suggested that deformation cube bands act as nucleation sites for the recrystallized cube grains so that the Cube orientation nucleates more frequently than other orientations [12]. The near 40°<111> relationship between the cube orientation and S deformation matrix may give the cube orientation a growth advantage compared to other recrystallization textures [8]. In this study, the Cube and $Cube_{RD}$ textures form in the 80% deformed sample annealed at 200℃. The formation of these textures may be caused by the reasons mentioned above. The $Cube_{ND}$ texture could also form in an Al alloys [9-11]. The formation of $Cube_{ND}$ was found to be associated with particles, and the 40°<111>orientation relationship between $Cube_{ND}$ and Copper enhances the development of the $Cube_{ND}$ texture. However, the $Cube_{ND}$ texture is formed only in highly deformed samples annealed at a high temperature. It is thus clear that particle stimulated nucleation is unlikely to be the dominating factor in the present results. The grain sizes are very coarse after short time annealing so that extensive grain growth might occur; and there is almost no Copper component in the material after deformation. So the formation of the present $Cube_{ND}$ texture might be caused by grain growth after recrystallization.

Instead of the Cube texture, the material exhibits a strong Goss texture in the 95% and 98% rolled samples after annealing at 200 and 300℃. It has been proposed [13] that microshear bands cut through the Cube bands and thus decrease the probability of Cube nucleation, and that the shear bands can act as nucleation sites for the Goss orientation, which strengthen the Goss texture [13]. In a recent study [4] on the same material, it was found that upon annealing both lamellar structures and shear bands nucleate Goss grains. There are ~10% volume fractions of Goss texture after rolling. Thus, these Goss orientations present in the deformed state may act as nucleation sites for Goss grains during annealing. Goss grains appear to have a growth advantage when they grow into the matrix of Brass orientations, leading to a strong Goss texture after complete recrystallization [4]. With increasing deformation strain, the stored energy increases and the critical nucleation size decreases. Therefore the nucleation of the Goss orientation during annealing increases with an increase in the rolling strain. As a result, the intensity of the Goss texture is high in annealed samples deformed to high strain.

5. Summary

An Al-0.3% Cu alloy was cold rolled to 80, 90, 95 and 98% reductions in thickness, and subsequently annealed at different temperatures until completely recrystallized. The following conclusions are reached:

- The cold-rolled material is characterized by a rolling texture with a strong Brass component.

- A $Cube_{RD}$ texture was formed in the 80% rolled sample annealed at 200℃. The volume fraction of the $Cube_{RD}$ decreases with increasing deformation strain and annealing temperature.
- A strong Goss texture was developed especially in 95 and 98% rolled samples after complete recrystallization at 200℃ and 300℃. The intensity of Goss texture increases as the rolling strain is increased.

Acknowledgements
The authors wish to thank the financial support of the Natural Science Foundation of China (NSFC, Grants No. 51471039, 51327805 and 51421001) and Fundamental Research Fund of Central Universities (Grant No. 10611205CDJZXZ138802). We also gratefully acknowledge the support from the Danish National Research Foundation (Grant No.DNRF86-5) and the NSFC (Grant No. 51261130091) to the Danish-Chinese Center for Nanometals.

References
[1] Engler O, Hirsch J and Lücke K 1989 *Acta Metall. Mater.* **37** 2743
[2] Humphreys F J and Hatherly 2003 *Recrystallization and related annealing phenomena*, Oxford, UK, ELSEVIER
[3] Liu W C and Morris J G 2007 *Scripta mater.* **56** 217
[4] Wakeel A, Huang T L, Wu G L, Mishin O V and Huang X 2015 *IOP Conf. Series: Mater Sci. & Eng.* **82** 012050
[5] Mishin O V, Bay B, Winther G and Juul Jensen D 2004 *Acta Mater.* **52** 5761
[6] Bate P S, Huang Y and Humphreys F J 2004 *Acta Mater.* **52** 4281
[7] Wakeel A 2015 Mechanical and Thermal Behavior of Nanostructured Al-0.3%Cu Alloy. Ph.D. Thesis, Chongqing University
[8] Vatne H E, Shahani R and Nes E 1996 *Acta Metall.***44** 4447
[9] Daaland O and Nes E 1996 *Acta Mater.* **44** 1413
[10] Wang X F, Guo X X, Chapuis A, Luo J, Zhang J S and Zhuang L Z 2015 *Sci. Eng.* A **633** 46
[11] Engler O, Yang P and Kong X W 1995 *Acta Mater.* **44** 3349
[12] Duggan B J, Lücke K, Köhlhoff G D and Lee C S 1993 *Acta Metall. Mater.* **41** 1921
[13] Engler O 2000 *Scripta Mater.* **44** 229

IOP Conf. Series: Materials Science and Engineering **89** (2015) 012033 doi:10.1088/1757-899X/89/1/012033

Effects of structural heterogeneity of nanostructured copper on the evolution of the sizes of recrystallized grains during annealing

F X Lin[1], Y B Zhang[1], W Pantleon[2] and D Juul Jensen[1]

[1]Danish-Chinese Center for Nanometals, Section for Materials Science and Advanced Characterization, Department of Wind Energy, Technical University of Denmark, Risø Campus, 4000 Roskilde, Denmark

[2]Section for Materials and Surface Engineering, Department of Mechanical Engineering, Technical University of Denmark, 2800 Kongens Lyngby, Denmark

E-mail: lnfe@dtu.dk

Abstract. Recrystallization in copper deformed by dynamic plastic deformation was investigated using electron backscatter diffraction. The recrystallized grains show a broad size distribution. The kinetics of grains of different sizes is observed to be different: In the beginning of recrystallization, the area fraction of small recrystallized grains increases rapidly. At later stages of recrystallization, the area fraction of small recrystallized grains is stable, while the area fractions of medium and large recrystallized grains increase. Correlation between the broad grain size distribution (and its evolution) and the heterogeneous deformed microstructure is discussed.

1. Introduction

After deformation at cryogenic temperatures or with high strain rates, copper often has different mechanical properties, textures and microstructures, compared with room temperature low-strain-rate deformation [1-3]. During low temperature or high-strain-rate deformation, mechanical twinning occurs, depending on the crystallographic orientations of the grains [4]. Nano-sized mechanical twins have been reported to favour the formation of shear bands [5]. Both mechanical twins and shear bands increase the structural heterogeneity of the deformed sample, which affects subsequent annealing behaviour. In earlier work, we studied copper deformed in liquid nitrogen by dynamic plastic deformation (DPD) with strain rates in the range of 10^2-10^3 s^{-1}, and found that the structural heterogeneity significantly affected the recrystallization kinetics [6]. In the present work, we investigate the correlation between the structural heterogeneity of the deformed sample and the recrystallized microstructure.

2. Experimental

A cylindrical copper sample (99.995% purity) with an initial grain size of 200 μm was immersed in liquid nitrogen and deformed by DPD to a strain of 2.1 ($\varepsilon = \ln(L_0/L_f)$, where L_0 and L_f are the sample height before and after deformation, respectively). The samples were deformed at Institute of Metal Research, Shenyang, China, transported to Denmark, and kept at -18 °C for a few months. A volume fraction of around 18% was already recrystallized before we started the present investigation and conducted furnace annealing [6]. This initial partial recrystallization has to be considered in the analysis of the present results, but is not expected to affect the conclusions reached.

The samples were isothermally annealed at 120 °C for time intervals from 2 minutes to 4 h to achieve various recrystallized volume fractions. Cross sections of the samples were ground and electro-polished using a D2 electrolyte. Microstructures of the cross section were characterized using electron backscatter diffraction (EBSD) with a step size of 0.5 μm, and for each annealing time, two to seven maps with a typical size of 250 × 300 μm^2 each were obtained. For the sample annealed for 5 minutes, three extra scans over relatively large regions (~1000 × 300 μm^2 each) using a step size of 1 μm were conducted, in order to reveal the sample heterogeneity on a larger scale. Also for the sample annealed for 5 minutes, an extra specimen was prepared by mechanical polishing of the cross section using 40 nm oxide polishing suspension to obtain a better surface quality for a better characterization of the deformed matrix.

The in-house code DRG was applied to identify recrystallized grains [7]. The recrystallized grains were defined as grains with internal misorientations less than 1°, fully or partly surrounded by high angle boundaries, with misorientation angles larger than 15°, and with a grain size larger than a critical value. The proper critical grain size is affected by the scanning step size. In this work, an equivalent circular diameter (ECD) of 3 μm and 1.5 μm was chosen as the critical size for maps scanned with a step size of 1 μm and 0.5 μm, respectively. During detection of recrystallized grains, twin boundaries were ignored; thus annealing twins and the parent grain are treated as one grain.

3. Results and discussion

3.1. Partially recrystallized microstructure

Figure 1 shows the recrystallized part of the sample after 5 minutes of annealing in a relatively large sample area (1380 × 300 μm^2) scanned with a step size of 1 μm. The area shown in figure 1 is 39% recrystallized. When 3 maps of sizes comparable to figure 1 are averaged, the sample was found 34% recrystallized. Recrystallized grains tend to form bands almost perpendicular to the compression axis, with band widths from a few μm to more than 200 μm. In general, the spatial distribution of the recrystallized grains is very heterogeneous.

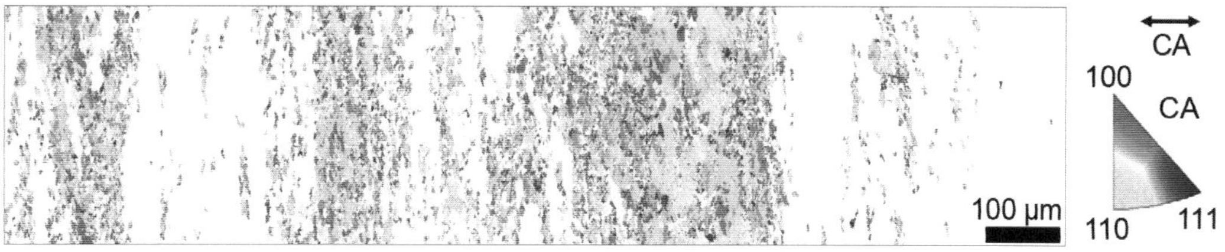

Figure 1. Orientation map showing the recrystallized grains in the sample annealed for 5 minutes. The compression axis (CA) is horizontal. The white regions are unrecrystallized, and other colours represent the crystallographic direction of the recrystallized grains along CA.

In the sample annealed for 5 minutes, the recrystallized grains show a broad grain size distribution. The number weighted average ECD of recrystallized grains is 6.6 μm, whereas the maximum ECD is 60 μm, about 9 times the average grain size. The size distribution of recrystallized grains is plotted in figure 2 using both number and area weighted density functions, f_N and f_A:

$$f_N(D_i) = \frac{F_N(D_i)}{\Delta D_i} \tag{1}$$

$$f_A(D_i) = \frac{F_A(D_i)}{\Delta D_i} \tag{2}$$

where $F_N(D_i)$ and $F_A(D_i)$ are the number and area fractions, respectively, and ΔD_i is the bin size, which is 1 μm for figure 2. In the number density plot, only one peak appears, since the number of small grains is much larger than that of large grains. In the area density plot, a long tail is clearly seen for grains larger than 20 μm.

A lognormal distribution is usually observed for the grain sizes after recrystallization [8]. For this sample, lognormal functions were used to fit f_N and f_A (figure 2). For the number density distribution, significant deviations from the lognormal function appear at around 10 μm and above. Clearly, the sizes of the recrystallized grains in this sample do not follow the lognormal distribution. On the other hand, a reasonable description of the area density distribution can be achieved by a lognormal distribution.

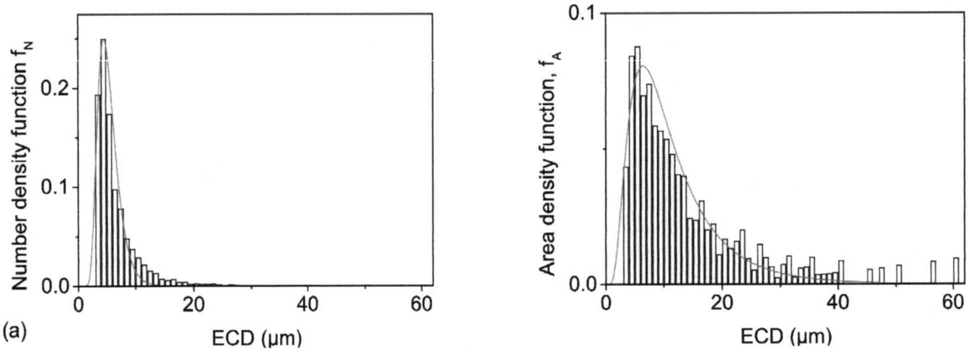

Figure 2. Number (a) and area (b) weighted grain size density functions for recrystallized grains in the sample annealed for 5 minutes. Lognormal functions were used to fit the density functions, which are shown by the red lines.

To analyse possible differences between grains of different sizes, the recrystallized grains are divided into three groups: small grains (ECD < 10 μm), medium grains (10 μm ≤ ECD < 20 μm), and large grains (ECD ≥ 20 μm). The three groups of grains in figure 1 are shown separately in figure 3. The large grains are severely clustered, and most of them are located along three bands. It is noted that the locations of these large grains correspond to the three thickest bands (>100 μm wide) in figure 1. The small grains tend to form thin bands, which are a few to tens of μm wide. The spatial distribution of small grains is more dispersed than that of the medium and large ones, but clearly not homogeneous: there are several locations, where thin bands cluster to form thicker ones, and several other locations, where recrystallized grains are rarely observed in bands of 50 μm to 200 μm wide.

As shown by the inverse pole figures corresponding to the compression axis (figure 3), the recrystallized grains show a weak fibre texture for all the three groups of grains with the maximum intensity shifted from <110> towards <100>. The orientations of the large grains are not significantly different from those of the small grains. This result can be compared to a heavily cold-rolled and annealed copper sample with an initial grain size of 22 μm, in which a broad size distribution is also observed [9]. In that sample, it has been shown that the large grains are all cube oriented grains, whereas small grains have various orientations [9]. Different to the rolled sample, the large sizes of a few grains in the present sample are not associated with a special orientation different from that of the small grains. It is rather suggested that the formation of large grains in the present sample strongly depends on the local deformed microstructure, as evidenced by the severe clustering.

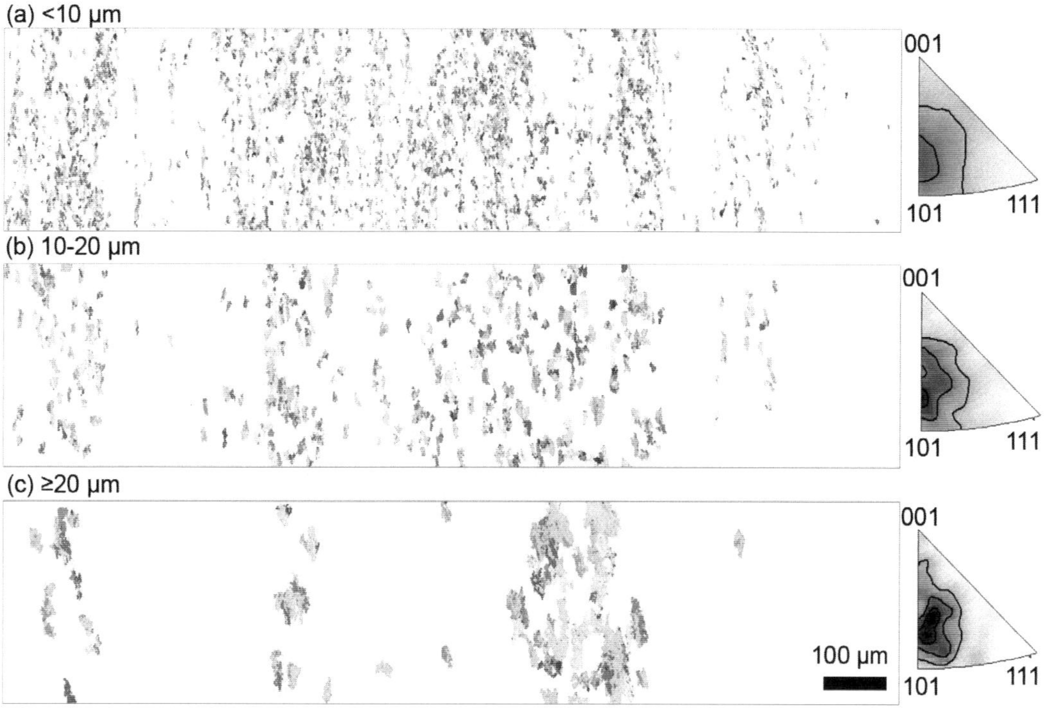

Figure 3. Orientation maps and inverse pole figures (IPFs, corresponding to CA) of the recrystallized grains belonging to different size groups in the sample annealed for 5 minutes (same as figure 1): (a) ECD < 10 μm, (b) 10 μm ≤ ECD < 20 μm, (c) ECD ≥ 20 μm. It should be noted that twin boundaries were ignored when defining grains. Thus a single grain in the map often has several parts with different colours. Contour levels in the IPFs are multiples (1, 2, 3) of random intensity.

3.2. Kinetics

The area fractions of recrystallized grains belonging to the three size groups are plotted in figure 4 as a function of annealing time. The sample without furnace annealing was already 18% recrystallized, and the recrystallized parts contain almost equal area fractions of small, medium and large grains. During furnace annealing, the area fraction of small grains increases with annealing time until approximately 30 minutes. After 30 minutes, the area fraction of small grains tends to be stable, and accounts for around 40% of the characterized area. The area fraction of medium grains increases slowly until 1 h, but from 1 h to 4 h the area fraction of medium grains doubles. Similarly, the area fraction of large grains is almost stable up to 1 h, but increases from 1 h to 4 h.

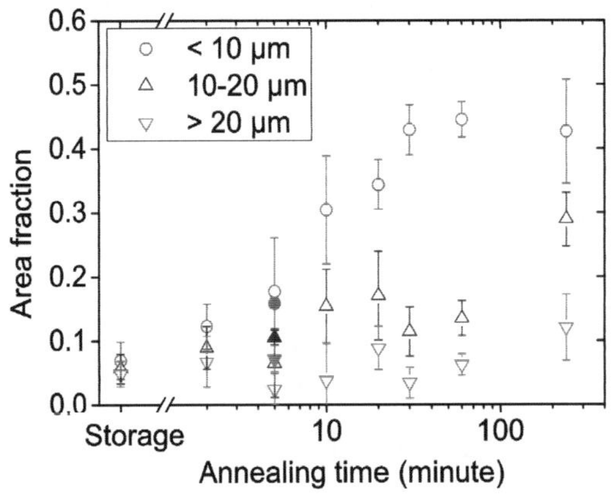

Figure 4. Evolution of area fraction of different sized recrystallized grains as a function of annealing time, determined based on EBSD maps scanned with a step size of 0.5 μm. For the sample annealed for 5 minutes, an extra set of data, shown with solid marks, was determined from larger EBSD maps scanned with a step size of 1 μm (the same EBSD data as used in Section 3.1). The error bars show the standard errors.

3.3. Correlation with the heterogeneity of the deformed microstructure

Based on our previous EBSD characterization (step size 50 nm) of the same DPD sample prepared by electropolishing, the deformed microstructure of the DPD sample was found to consist of three types of regions [6]. Type A regions belong to a <110> fibre texture, and contain mainly low angle boundaries. Type B regions belong to a <111> fibre texture, and consist of nanoscale twins. Type C regions could not be indexed. In the present work, the surface quality was improved using mechanical polishing, after which Type C regions can also be indexed when a small step size was used. The map shown in Figure 5a illustrates this and contains all three types of microstructures. The green region in the left part of figure 5a is a typical Type A structure, while the blue region in the right part is a typical Type B region with twins. Type C regions are marked with blue borders in figure 5b, and they are bands with a width less than 3 μm, containing very fine structures. According to the morphology, most of the Type C regions in figure 5b are shear bands, like those observed by transmission electron microscopy (TEM) [10]. These shear bands cluster in the centre of figure 5b. It should be noted that figure 5 only covers a small region, and Type C regions in fact include not only shear bands, but also other regions of very fine structures, such as fragmented twins.

When a larger scanning step size is used, the indexed ratio decreases significantly even with a good surface quality, as shown in figure 5c. Type A and B regions can still be relatively well indexed, although the internal microstructures are not resolved. Most of the non-indexed points correspond to Type C regions, but some parts of Type B regions are not indexed either, most likely because of an overlap of EBSD patterns. As an approximation, it is possible to determine area fractions of the three types of deformed microstructures from EBSD maps scanned with a relatively large step size based on the different orientations, although this method tends to underestimate the amount of Type B regions.

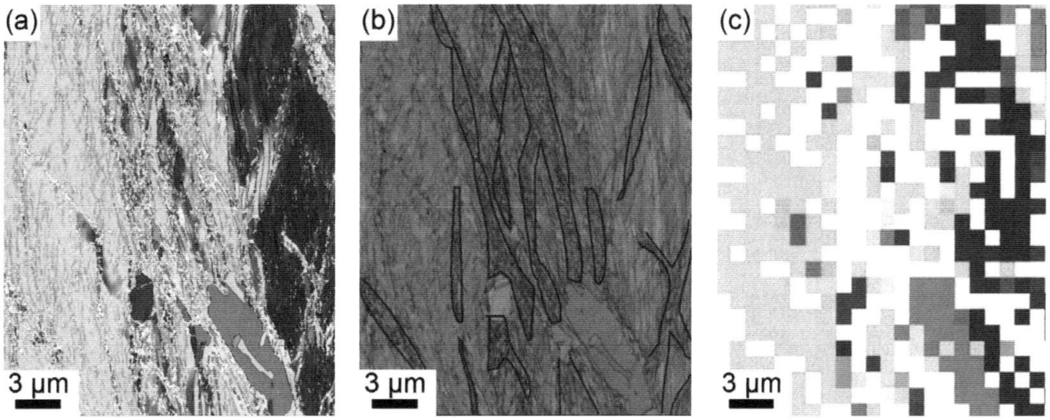

Figure 5. Orientation maps of a small region in the sample annealed for 5 minutes, with a surface prepared by mechanical polishing. CA is along the horizontal direction. Maps (a) and (b) are from the same EBSD data scanned with a step size of 30 nm, and Map (c) is the same area but scanned with a step size of 1 μm. The colours in (a) and (c) represent the crystallographic direction along CA, and the white regions are non-indexed points. Black, grey and red lines in (a) represent boundaries with misorientations higher than 15° and 2°, and twin boundaries, respectively. Map (b) shows the band contrast (BC). Regions in (b) bounded by blue lines indicate regions with a very fine structure. Note that these fine-structure regions correspond approximately to the non-indexed regions in Map (c).

Figure 6 shows the deformed microstructure on a larger scale. The map was obtained also with a coarse step size of 0.5 μm, and thus contains a relatively large area fraction of non-indexed points. The indexed parts mainly belong to a <110> fibre texture (green colour), and some parts belong to a <111> fibre texture (blue colour). Very few isolated points that belong to <100> fibre texture (red colour) are also observed. As discussed above, regions with a <110> and <111> texture are considered as Type A and B, respectively. All the non-indexed points are considered as Type C. The few isolated <100> texture points are also grouped into Type C regions.

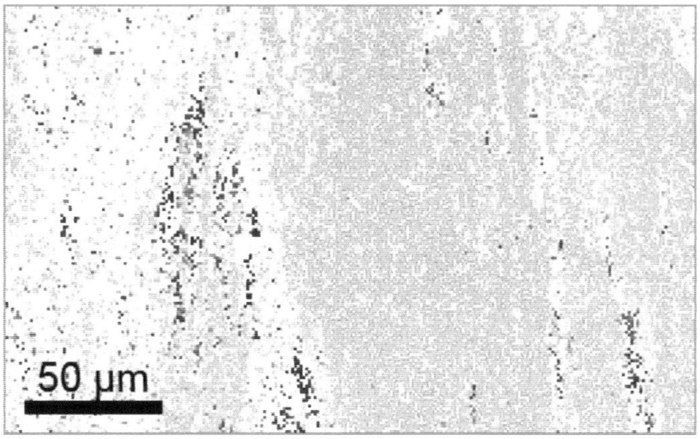

Figure 6. Orientation map obtained with a step size of 0.5 μm, showing the deformed microstructure in the sample without furnace annealing. CA is horizontal. White points are non-indexed, and other colours represent the crystallographic direction along CA. The microstructure is divided into three types according to the local orientations. Type A: <110> texture, Type B: <111> texture, Type C: non-indexed points and the few isolated <100> oriented points.

Figure 7 shows the area fractions of the three types of deformed microstructures determined this way in samples annealed for different time periods. In the sample without furnace annealing, the area fractions of Type A and C regions are similar, and are around 40% each. The area fraction of Type C regions decreases quickly during furnace annealing, whereas the area fraction of Type A regions decreases much more slowly. After 1 h annealing, when the sample is 64% recrystallized, only few Type C regions are left, but there is still an area fraction of around 30% of Type A regions. The area fraction of Type B regions is about 5% in the sample without furnace annealing. Considering the small amount of Type B regions, they do not significantly affect the recrystallization process, and will not be discussed further.

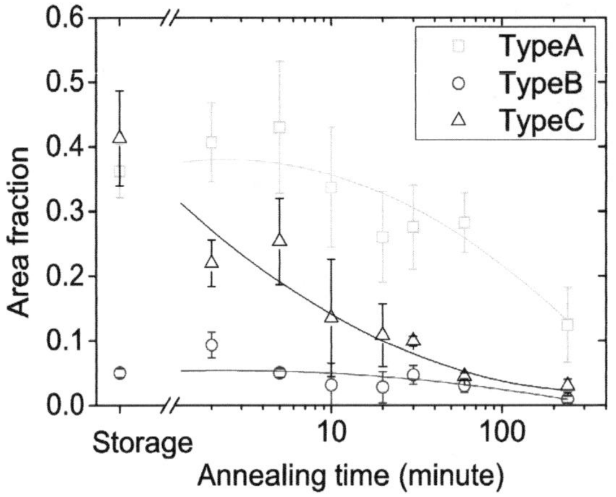

Figure 7. Evolution of area fraction of the three types of deformed microstructures as a function of annealing time, determined based on EBSD data obtained with a step size of 0.5 μm. Type A/B/C regions were determined according to the local orientations, as demonstrated in figure 6. The error bars show the standard errors. Second order polynomial functions were fitted to the data points, shown by the curves, to clearly visualize the trends for the different types of deformed microstructures.

The decrease in area fractions of Type C and Type A regions is due to recrystallized grains nucleating and growing in the these regions. One may speculate whether previous non-indexed regions become indexed after annealing, namely Type C regions being erroneously treated as other types. But it would require significant recovery, which has not been observed by TEM in a similar material [11]. By comparing figures 4 and 7, it is seen that before 30 minutes, the major area decrease in the deformed matrix occurs in Type C regions, and at the same time, the area fraction of small recrystallized grains increases. After 1 h, few Type C regions are left, and the area fraction of small recrystallized grains stabilizes. These two observations suggest that most of the small recrystallized grains form in Type C regions, and are limited to this type of regions and do not grow into the neighbouring Type A/B regions. As Type C regions have a very fine structure, and thus also high stored energy [12], the nucleation density is high in this type of regions. Recrystallized grains will thus quickly impinge, resulting in a small recrystallized grain size. These interpretations are supported by the fact that the area fraction of small recrystallized grains is close to that of Type C regions after storage.

From 1 h to 4 h, the increase in the recrystallized area is mostly by growth of medium and large recrystallized grains. At the same time, the area fraction of Type A regions decreases in the deformed matrix. The Type A regions are thus likely to be the places where medium and large recrystallized grains develop. The internal microstructure of Type A regions contains only low angle boundaries, and thus Type A regions may have only a few nucleation sites. Therefore, Type A regions are more likely to develop grains of larger sizes. The stored energy of Type A regions is also lower compared with Type C regions, and the recrystallization kinetics is thus slower.

As shown in figure 4, the area fraction of large grains is more or less constant before 1 h. This means that possibly most of the large grains observed in the beginning of recrystallization, such as those shown in figure 3c, form during the storage. Whether a fresh deformed sample will also behave like this or the phenomenon only occurs during long-term low temperature storage needs further investigations.

4. Conclusions

We investigated recrystallization in copper deformed by DPD using EBSD. The deformed sample contains three types of microstructure: Type A with a {110} fibre texture, Type B with twins and Type C with very fine structures. In the sample without furnace annealing, similar area fractions of Type A and C regions are observed, while the area fraction of Type B regions is very small. The recrystallized grains were found to have a broad size distribution. The sizes of grains are not related to the orientations of the recrystallized grains, but to the deformed matrix where the recrystallized grains grow. The area fraction of small recrystallized grains (< 10 μm) increases rapidly from the beginning of recrystallization, and saturates at later stages of recrystallization. The small recrystallized grains are considered to develop mainly in Type C regions. At the later stages of recrystallization, medium (10–20 μm) and large (> 20 μm) recrystallized grains still increase, and they are considered to develop at Type A regions.

Acknowledgements

The authors are grateful to Drs N. Hansen and N.R. Tao for helpful discussions, Mr. X. Si for DPD sample preparation at IMR, and Dr. G.H. Fan for help with mechanical polishing. The authors gratefully acknowledge the support from the Danish National Research Foundation (Grant No. DNRF86-5) and the National Natural Science Foundation of China (Grant No. 51261130091) for the Danish-Chinese Center for Nanometals, within which this work was performed.

References

[1] Leffers T and Ray R K 2009 *Prog. Mater. Sci.* **54** 351
[2] Bhattacharyya A, Rittel D and Ravichandra G 2005 *Scr. Mater.* **52** 657
[3] Luo Z P, Mishin O V, Zhang Y B, Zhang H W and Lu K 2012 *Scr. Mater.* **66** 355
[4] Hong C S, Tao N R, Lu K and Huang X 2009 *Scr. Mater.* **61** 289
[5] Morri K, Meching H and Nakayama Y 1985 *Acta Mater.* **33** 379
[6] Lin F X, Zhang Y B, Tao N R, Pantleon W and Juul Jensen D 2014 *Acta Mater.* **72** 252
[7] Wu G L and Juul Jensen D 2008 *Mater. Charact.* **59** 794
[8] Pande C S 2000 *Proc. 21st Risø Inter. Symposium on Mater. Sci.* pp 491
[9] Lin F X, Zhang Y B, Pantleon W and Juul Jensen D submitted
[10] Li Y S, Tao N R and Lu K 2008 *Acta Mater* **56** 230
[11] Li Y S, Zhang Y, Tao N R and Lu K 2008 *Scr. Mater.* **59** 475
[12] Yan F and Zhang H W 2012 *J Mater. Sci. Technol.* **28** 289

IOP Conf. Series: Materials Science and Engineering **89** (2015) 012034 doi:10.1088/1757-899X/89/1/012034

Dynamic recrystallization of Zr–1Sn–0.3Nb alloy during hot compression

B F Luan, Q H Zeng, W Zeng and Q Liu
College of Materials Science and Engineering, Chongqing University, Chongqing 400044, China

Email: bfluan@cqu.edu.cn, zengqinghui@cqu.edu.cn

Abstract. In this paper, using a hot-rolled and recrystallized Zr-1Sn-0.3Nb alloy sheet, five strain levels (10, 20, 30, 45 and 60%) of uniaxial compression experiments were performed at 700°C with a strain rate of 0.001s^{-1} using a Gleeble 1500D thermal simulator. The morphologies, distributions and misorientations of the grain boundaries at different strains were characterized by an electron backscatter diffraction (EBSD) technique. The EBSD maps indicated that dynamic recrystallization occurred during the high-temperature deformation. The processes, namely, grain boundary bulging, strain induced subboundaries around the bulged grain boundaries and their transformation from low misorientation to high misorientation, were considered as the formation mechanisms of recrystallized grains. At the early stage of deformation, low angle boundaries (LABs) formed in grains, the original grain boundaries become serrated, and LABs formed around the serrated boundaries. With increasing strain, the LABs transformed into high angle boundaries (HABs), leading to nucleation of new grains.

1. Introduction

Dynamic recrystallization (DRX) as one of the important mechanisms to obtain fine grains occurring under hot deformation is widely studied in cubic metals (e.g. aluminum and iron) [1-5]. Belyakov et al. [6] indicated that the DRX of the 304 type austenitic stainless steel occurred mainly through dynamic bulging of original grain boundaries under the conditions of high temperature, which can be accelerated by the progress of serrated grain boundaries and strain induced dislocation subboundaries. Furthermore, by studying the multiple compression at high strains of above 4 of a polycrystalline copper, Belyakov et al. [7] reported that the dynamic new grains generated by increasing misorientations of deformation-induced dislocation subboundaries with increasing strain, this process accompanied with dynamic recovery at 473K under a strain rate of 10^{-3} s^{-1}. However, few studies focusing on DRX in hexagonal close-packed (HCP) metals have been done. Yang et al. [8] reported that new grains generated at corrugated initial grain boundaries or the kink bands in grain interiors by increasing misorientation of kink bands boundaries to a saturation of around 43°, and the final grain size is almost the same as the average size of the regions fragmented by kink bands in deformed magnesium alloy AZ31 at 673K. By means of EBSD technique, Gerspach et al. [9] found that oriented nucleation occurred during recrystallization in the case of low strain (40%) cold cross-rolled Zr702 samples, whereas in the case of highly cold rolled Zr702 samples, non-oriented nucleation occurred during recrystallization. Additionally, by using in situ high voltage electron microscopy (HVEM), Zhu and his colleagues [10] observed that subgrains result from rearrangement of dislocation cells, and then grow to form nuclei, finally nuclei grow by high angle boundary migration. Perez-Prado et al. [11] pointed out that geometric dynamic recrystallization (GDRX) occurred during creep at temperatures from 400 to 800°C. This process can be described as follows: original grain boundaries become serrated resulting from subgrain formation, and with increasing deformation, "pinching-off" of the serrated grains occur, leading to fine grains.

All of the above works focused on the influence of a grain boundary on DRX in cubic metals and magnesium. There are few studies in zirconium alloys, and the dynamic softening mechanisms in zirconium alloys during hot deformation have not been well understood. In this work, cylindrical specimens of a Zr alloy were isothermally compressed to five strain levels (10, 20, 30, 45 and 60%), and the morphologies, distributions and misorientations of the grain boundaries at different strains were obtained by EBSD, in order to study the dynamic softening mechanisms.

2. Experimental

The as-received material was a hot-rolled and recrystallized Zr-1Sn-0.3Nb alloy plate of 9.5 mm in thickness. The chemical composition of the material is given in Table 1. Cylindrical compression specimens with 6 mm in diameter and 6 mm in height were cut from the plate, with the compression direction parallel to the normal direction (ND) of the original plate. In order to evaluate the evolution of grain boundary during deformation, uniaxial compression experiments were performed to five strain levels (10, 20, 30, 45 and 60%) at 700 °C using a Gleeble 1500D thermal simulator at a strain rate of $0.001s^{-1}$. The experimental process is shown in Figure 1. After hot compression, the deformed specimens were immediately quenched into water to retain the deformation microstructure.

Table 1. Nominal composition of the Zr-1Sn-0.3Nb alloy (wt.%).

Sn	Nb	Fe	Cr	O	Zr
1.06	0.36	0.30	0.10	0.13	Bal.

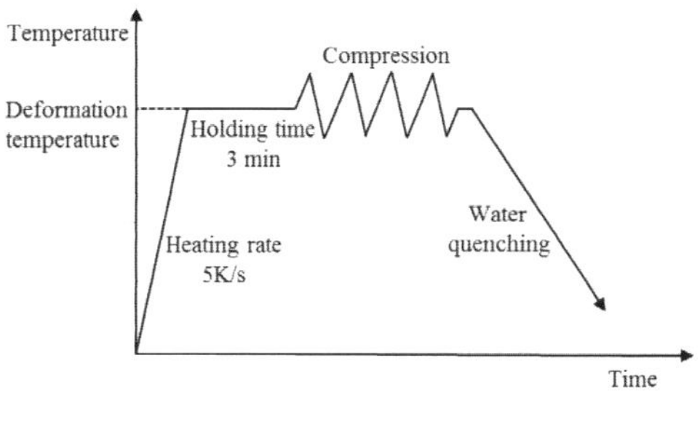

Figure 1. Hot deformation process

Figure 2. EBSD grain map of the initial specimen

For microstructural analyses, the specimens were cut along the direction of the compression axis, and prepared by mechanically polished and electropolishing. The microstructure was characterized using a field emission gun scanning electron microscope (TESCAN MIRA 3 XMU) equipped with a backscatter electron detector and an EBSD analysis system (AZtec, Oxford Instruments). The step size of EBSD scan was 0.2μm. The initial microstructures of the hot-rolled and recrystallized Zr-1Sn-

0.3Nb alloy plate is shown in Figure 2, where the black lines and the gray lines represent HABs (> 15°) and LABs (< 15°), respectively.

3. Results and Discussion

3.1. High-temperature deformation behaviour

The true stress-strain curve is shown in Figure 3. It can be seen from Figure 3 that the flow stress decreases after reaching the maximum, which is caused by high temperature dynamic softening, and then increases slowly after a minimum. The stress increase is caused by grain refinement resulting from DRX, which will be discussed in the next section in detail.

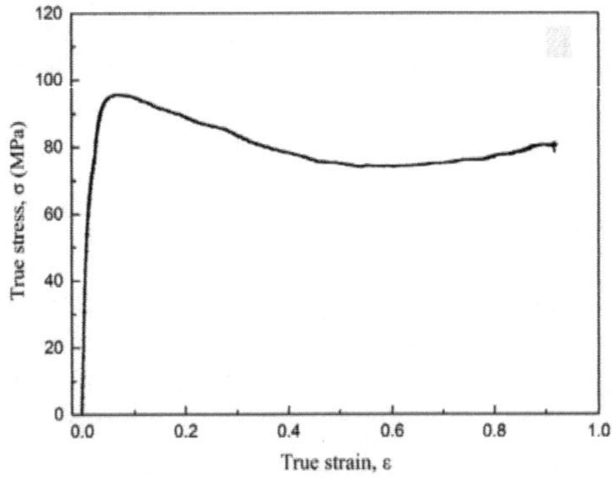

Figure 3. Compression true stress-true strain curve of zirconium alloy deformed to various strains at 700°C and at $0.001s^{-1}$.

3.2. Effects of compression reduction on microstructural evolution

The EBSD maps of specimens at different reductions in thickness are shown in Figure 4, in which different colours represent different crystal orientations indicated in the inverse pole figure. It can be seen from Figure 4 that the reduction in thickness significantly influences the morphology of grains and grain boundaries. Compared with the original microstructure in Figure 2, the grains are elongated at the reduction of 10%. Meanwhile, the grain boundary becomes serrated, and a number of LABs appear inside the grain with a short morphology. With increasing strain, more and more fine grains are produced along the elongated grain boundaries. Such a phenomenon may be ascribed to the formation of long LABs, separating the grain into several parts. Subsequently, the long LABs caused by the reaction of short LABs transform to HABs. As the reduction increases to 60%, the grains become fine and uniform, and the serrated grain boundaries are replaced by a smooth morphology. Based on the above analysis, it can be concluded that the grain refinement can be attributed to DRX, including the formation of LABs and HABs, as well as their migration.

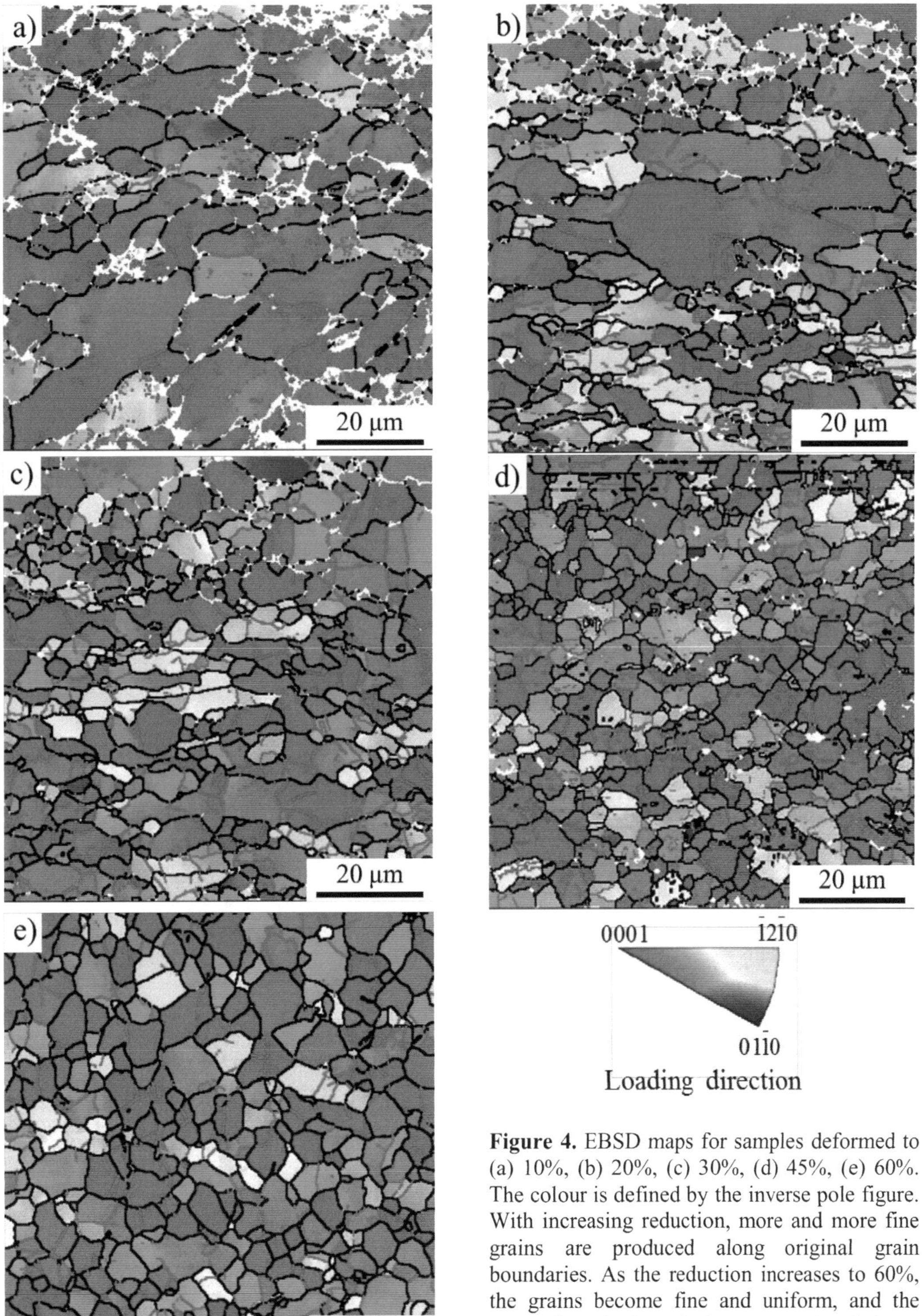

Figure 4. EBSD maps for samples deformed to (a) 10%, (b) 20%, (c) 30%, (d) 45%, (e) 60%. The colour is defined by the inverse pole figure. With increasing reduction, more and more fine grains are produced along original grain boundaries. As the reduction increases to 60%, the grains become fine and uniform, and the serrated grain boundaries are replaced by a smooth morphology.

3.3. Dynamic nucleation and progress of new grains

Figure 5a shows the EBSD map of the specimen after 20% compression. The point to point misorientation (θ) and the accumulative misorientation (Σθ) were measured along the lines L1 and L2 indicated in Figure 5a and are shown in Figure 5b and 5c. The θ represents the misorientation between two adjacent scanned points (0.2μm distance), and the Σθ represents the misorientation between the first point and a point-n. On the line L1, θ ranges from 0 to 1°, while the Σθ continuously increases from 0° to 11° along the direction indicated by the arrow, demonstrating the existence of an orientation gradient. Line L2 crosses several LABs, and θ ranges from 0 to 12°. Σθ changes discontinuously at the same positions which correspond to low misorientation peaks. It can be concluded that the generation of subgrains was caused by the orientation gradient.

Figure 6a shows the EBSD map of the sample after 30% reduction. The grains became flattened and the grain boundaries became serrated. Figure 6b shows the distribution of θ and Σθ along the line L1 in Figure 6a, Σθ continues increasing from 0° to 4.5° along the arrow direction, demonstrating the existence of an orientation gradient. In comparison with Σθ in Figure 5b and 5c, Σθ in Figure 6b is lower, signifies that the orientation gradient presence in this region is lower than that in the region crossed by line L1 in Figure 5a. Strain induced subboundaries enclosing the bulged grain boundary in this area did not evolve into HABs, which is partly due to the low orientation gradient. Another typical case of nucleation of new grains near grain boundary is illustrated in Figure 6a indicated by line L2. Miura et al. [12] pointed out the serrated grain boundaries and a non-uniform strain gradient along the boundaries were induced by grain boundary shearing. High temperature deformation is easier to make a grain boundary serrated than cold deformation. LABs formed around the bulged grain boundaries separate the bulge boundaries from the initial grain, leading to nucleation of new grains. In addition to the serration of grain boundaries, the strain induced LABs is also important.

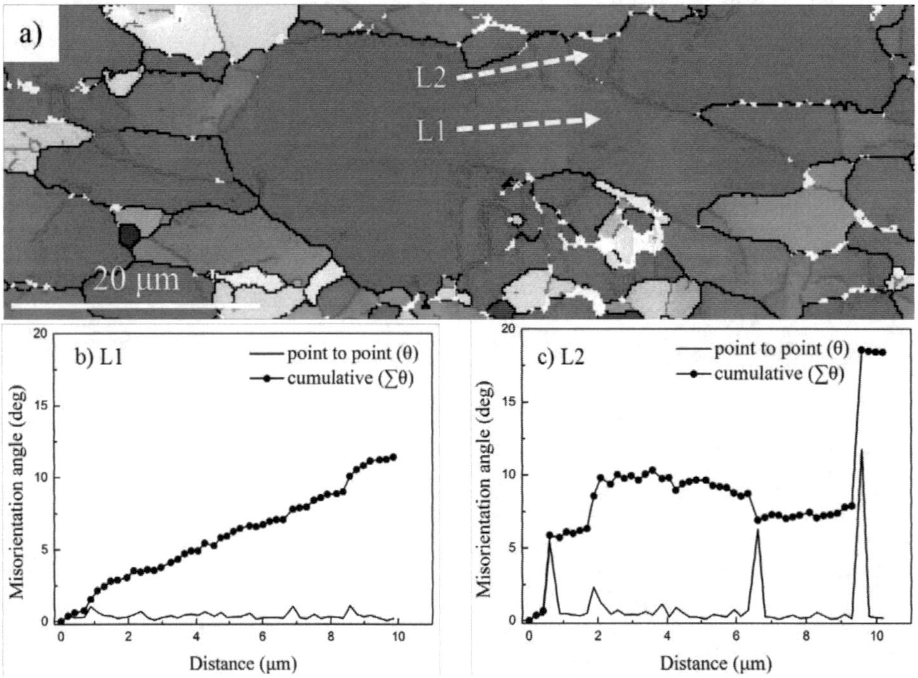

Figure 5. (a) EBSD maps for the sample deformed to 20%, the colour is defined by the inverse pole figure.in figure 4. Point to point misorientations θ and cumulative misorientations ∑θ along the lines (b) L1 and (c) L2 marked in (a). Increasing ∑θ along line L1 demonstrates the existence of an orientation gradient in this region.

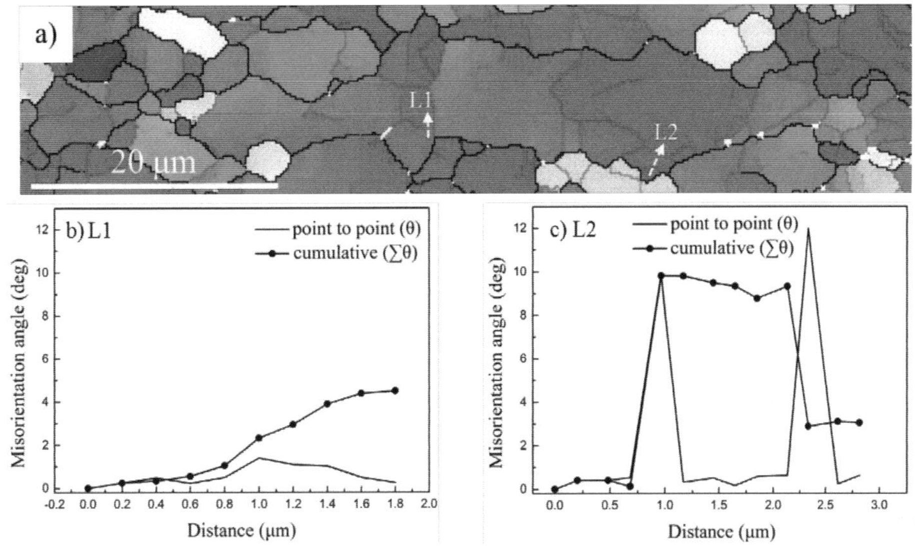

Figure 6. (a) EBSD maps for the sample deformed to 30%, the colour is defined by the inverse pole figure in Figure 4. Point to point misorientations θ and cumulative misorientations ∑θ along the lines (b) L1 and (c) L2 marked in (a).

The phenomenon that "pinching-off" of serrated grains to form fine grains in zirconium alloys proposed by Perez-Prado et al. [11] was not found in this work. Perez-Prado et al. [11] claimed that, with increasing strain, original grain boundaries become serrated, original grains elongate and became thinner in the direction perpendicular to the tensile direction. Finally, when the thickness decreases to a threshold value, grain boundary serrations come in contact with each other, grain pinching-off can occur. As a result, fine grains are formed. In this study, strain induced subboundaries in the regions which have an orientation gradient near original bulged boundaries evolve into HABs with increasing strain, finally leading to nucleation of new grains in these regions. This phenomenon is similar to the rotational recrystallization (RRX) in which the subgrains around original grain boundaries transform into HABs, partly due to the presence of high local stresses [13]. Loge et al. [14] claimed that, at temperatures higher than 600°C, DRX occurs at a high strain rate, while dynamic recovery predominates at low strain rates in Zircaloy-4.

As a conclusion in this study, the structural evolution is characterized by various types of strain induced subboundaries evolved around serrated original grain boundaries. This process can be described as follows: firstly, LABs are formed in grains by rearrangements of dislocations in the initial stage of deformation. Secondly, with increasing strain, the original grain boundaries become serrated, a strain gradient forms around grain boundaries because of grain boundary shearing. Thirdly, strain induced LABs form around the serrated boundaries resulting from the strain gradient. Finally, as the misorientation increases by the accumulation of the dislocations, the LABs transform into HABs. Hence, the recrystallized grains form in the deformed grains. This process is illustrated in Figure 7.

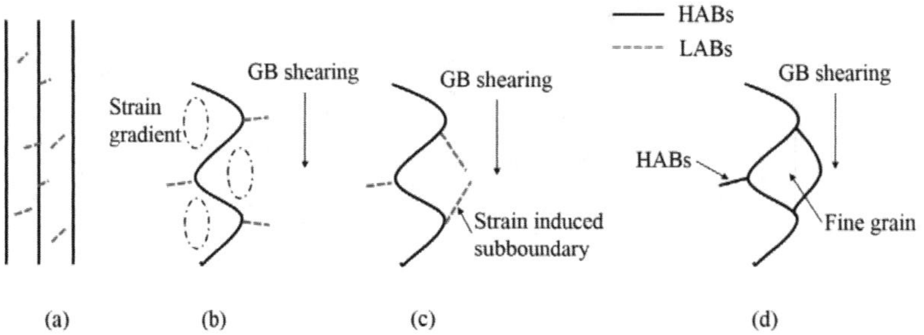

Figure 7. Proposed microstructural evolution during DRX: (a) Generation of LABs due to the rearrangement of dislocations, (b) Boundary corrugation, partial grain boundary shearing, leading to the development of a strain gradient, (c) Formation of strain induced subboundaries, (d) Generation of DRX grains by LABs transforming into HABs.

4. Conclusions

Based on the EBSD observation and analysis, the deformation behavior and structural evolution of a zirconium alloy compressed at 700°C at a strain rate of $0.001s^{-1}$ were studied. The main results can be summarized as follows:

(1) With increasing strain, more and more fine grains are produced along the boundaries of elongated grains.

(2) The structural evolution is characterized by various types of strain induced subboundaries evolved around serrated original grain boundaries. At the early stage of deformation, LABs form within grains, and the original grain boundaries become serrated. The LABs formed around the serrated boundaries, gradually transform into high angle grain boundaries, leading to nucleation of new grains.

Acknowledgements

This study was financially supported by the National Science Foundation of China (51171213).

References

[1] McQueen H J and Jonas J J 1975 *Treatise on Materials Science and Technology* vol 6, ed Arsenault R J (New York: Academic Press) p 393
[2] Sakai T and Jonas J J 1984 *Acta Metall.* **32** 189
[3] Ghosh A K and Gandhi C 1985 *Strength of Metals and Alloys (ICSMA7)* ed McQueen H J, Bailon J P, Dickson J I, Jonas J J and Akben M G (Oxford: Pergamon Press) p 2065
[4] Hales S J, McNelley T R and McQueen H J 1991 *Metall. Trans. A* **22A** 1037
[5] Humphreys F J and Hatherly M 1996 *Recrystallization and Related Annealing Phenomena* (Oxford: Pergamon Press) p 363
[6] Belyakov A, Miura H and Sakai T 1998 *Mater. Sci. Eng., A* **255** 139
[7] Belyakov A, Gao W, Miura H and sakai T 1998 *Metall. Mater. Trans. A* **29A** 2957
[8] Yang X Y, Miura H and Sakai T 2003 *Mater. Trans., JIM* **44** 197
[9] Gerspach F, Bozzolo N and Wagner F 2009 *Scripta Mater.* **60** 203
[10] Zhu K Y, Chaubet D, Bacroix B and Brisset F 2005 *Acta Mater.* **53** 5131
[11] Perez-Prado M T, Barrabes S R, Kassner M E and Evangelista E 2005 *Acta Mater.* **53** 581
[12] Miura H, Aoyama H and Sakai T 1994 *J. Jpn. Inst. Met.* **58** 269
[13] Del Valle J A, Perez-Prado M T and Ruano O A 2003 *Mater. Sci. Eng., A* **355** 68
[14] Loge R E, Signorelli J W, Chastel Y B, Perrin M Y and Levensohn R A 2000 *Acta Mater.* **48** 3917

IOP Conf. Series: Materials Science and Engineering **89** (2015) 012035 doi:10.1088/1757-899X/89/1/012035

Structural coarsening during annealing of an aluminum plate heavily deformed using ECAE

O V Mishin[1], Y B Zhang[1] and A Godfrey[2]
[1] Danish-Chinese Center for Nanometals, Section for Materials Science and Advanced Characterization, Department of Wind Energy, Technical University of Denmark, Risø Campus, 4000 Roskilde, Denmark
[2] Key Laboratory of Advanced Materials (MOE), School of Materials Science and Engineering, Tsinghua University, Beijing 100084, China

E-mail: olmi@dtu.dk

Abstract. The microstructure and softening behaviour have been investigated in an aluminum plate heavily deformed by equal channel angular extrusion and subsequently annealed at 170 °C. It is found that at this temperature the microstructure evolves by coarsening with no apparent signs of recrystallization even after 2 h of annealing. Both coarsening and softening are rapid within first 10 minutes of annealing followed by a slower evolution with increasing annealing duration. Evidence of triple junction (TJ) motion during coarsening is obtained by inspecting the microstructure in one region using the electron backscatter diffraction technique both before and after annealing for 10 minutes. The fraction of fast-migrating TJs is found to strongly depend of the type of boundaries composing a junction. The greatest fraction of fast-migrating TJs is in the group, where all boundaries forming a junction are high angle boundaries. The fraction of fast-migrating TJs is significantly lower for junctions containing at least one low angle boundary.

1. Introduction

It has been reported that annealing of materials cold rolled to high plastic strains can result in significant structural coarsening during recovery [1–5], which can occur by migration of both deformation-induced boundaries and triple junctions (TJs). It is considered that the process of coarsening during recovery of heavily deformed materials is facilitated by the presence of a large fraction of high angle boundaries (HABs) produced by deformation. As the fraction of HABs in a deformed microstructure increases with increasing plastic strain [6], it is expected that in a material deformed to a very high strain, the fraction of TJs containing HABs should also be high.

Equal channel angular extrusion (ECAE) is a well-known deformation technique which enables very high strains by multiple passes through a die with two intersecting channels, and which has been widely used for producing heavily deformed samples of rod, bar and plate shapes [6–11]. In our previous work [12], ECAE was applied for severe plastic deformation of 15 mm thick plates of commercial purity aluminum. By comparing the microstructure in the center of two plates processed either with or without sequential 90° rotations about the plate normal between passes, it was found that deformation with 90° rotations resulted in a fairly homogeneous microstructure with a high fraction of HABs, 70% as measured using the electron backscatter diffraction technique (EBSD) [12]. The evolution of this ECAE-processed microstructure during recovery annealing is investigated in the present work, and the results obtained are compared with those reported previously for the heavily-rolled material [1–4].

2. Experimental

An AA1050 aluminum plate with dimensions of 15 × 75 × 75 mm^3 was extruded at room temperature using an ECAE die with a sliding floor, a sharp outer corner and a 90° angle between channels (see [12] for more details). For this die configuration the strain per pass was 1.15, and the plate was deformed by 8 ECAE passes with 90° sequential rotations between passes. The deformed material was then annealed at 170 °C and samples were taken after 10 minutes, 30 minutes, 60 minutes and 2 h of annealing.

Vickers hardness measurements and EBSD analysis of these samples were conducted near the middle plane in the polished longitudinal section which contained the extrusion direction and the plate normal. The hardness was measured applying a load of 200 g and a dwell time of 15 s. A step size of 70 nm was used for statistical EBSD analysis of the average subgrain size and the HAB fraction in each deformed and annealed condition. In addition, one region with an area of 180 μm^2 was investigated with a step size of 25 nm both in the deformed condition and after 10 minutes of annealing at 170 °C without re-polishing. Because of the limited angular resolution of the EBSD technique [13,14], misorientations less than 2° were ignored in the EBSD data. Therefore, low angle boundaries (LABs) were defined as boundaries with a misorientation angle $\theta = 2–15°$. Boundaries with $\theta > 15°$ were defined as HABs.

3. Results

The deformed microstructure obtained after ECAE is shown in figure 1a. The average subgrain size d in this microstructure is ~0.5 μm and the fraction of HABs is 70%. The subgrains are mostly elongated although some fairly equiaxed subgrains are also present in the microstructure. The average aspect ratio in this as-deformed condition is almost 2, and the Vickers hardness is 55 HV0.2. During annealing at 170 °C comparatively rapid coarsening to $d = 0.63$ μm takes place within the first 10 minutes followed by a slower increase in the average subgrain size (figure 1 and figure 2a). However, the average subgrain size remains in the submicrocrystalline range, ~0.8 μm, even after 2 h of annealing.

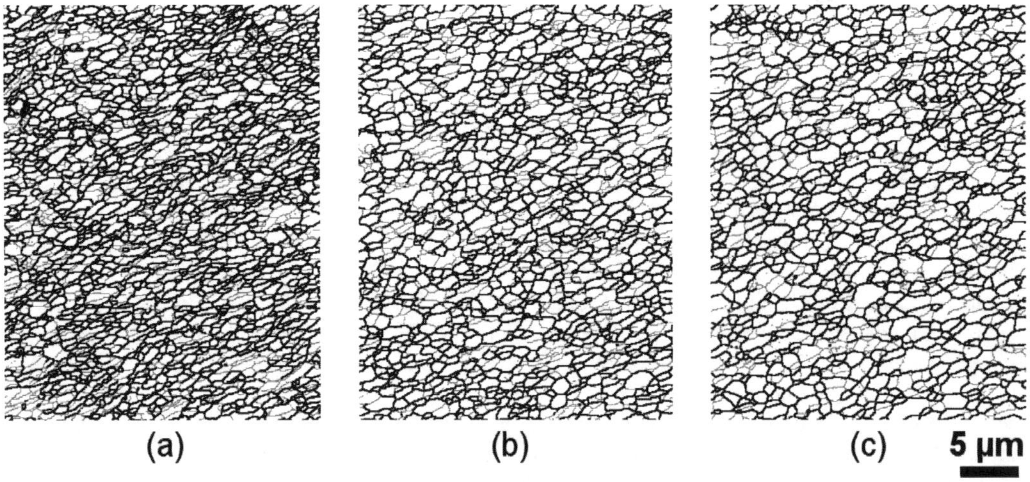

Figure 1. EBSD maps representing the microstructure of aluminum after ECAE (a) and subsequent annealing at 170 °C for 10 minutes (b) and 2 h (c). Gray and black lines show LABs and HABs, respectively. The extrusion direction is parallel to the scale bar.

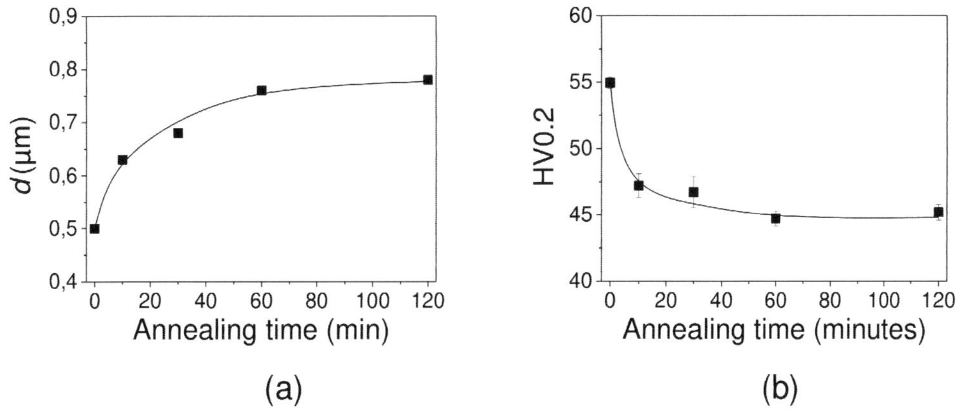

Figure 2. The effect of annealing at 170 °C on the average subgrain size (a) and Vickers hardness (b) in the ECAE-processed aluminum plate.

It is apparent that in each annealed condition the microstructure is morphologically similar to the deformed microstructure. Compared to the as-deformed material, the average aspect ratio of the subgrains decreases only by 10% (to ~1.8) after 2 h of annealing. The variation in the fraction of HABs during annealing is also very small, being in the range, 71–74%. The structural coarsening is accompanied by softening. Similar to the evolution in the average subgrain size, the largest change in hardness is observed after the first 10 minutes at 170 °C followed by only minor additional softening during further annealing (figure 2b).

To monitor changes in the deformed microstructure during the period of rapid coarsening in more detail, figure 3 shows fragments of the EBSD maps obtained in one region with a step size of 25 nm both before annealing and after 10 minutes of annealing. It is evident from this figure that boundary migration and TJ motion took place in many locations of this region. To quantitatively analyze which junctions are more prone to migration than the others, 653 TJs identified in the inspected region were first grouped into four types: HHH, HHL, HLL and LLL, where the number of "H" and "L" corresponds to the number of HABs and LABs in each junction. Thus, HHH junctions contain three HABs, HHL junctions contain two HABs and one LAB, etc. The TJs were then sorted out into two categories: (i) fast-migrating, i.e. junctions migrating over at least 100 nm; and (ii) all other TJs, which were considered slow or stationary.

The results of this categorization approach are presented in figure 4. It is seen that in the deformed microstructure the largest group of TJs was of the HHH type. HHL junctions were slightly less frequent (see figure 4a). 112 TJs were LLL and only 25 of all TJs were of the HLL type. The frequencies of rapidly migrating TJs (green in figure 4a) in the inspected region can be ranged in the same order: HHH, HHL, LLL and HLL. However, when the frequency of rapidly migrating junctions is normalized to the number of TJs within each group, it is clear that the fraction of migrating HLL-type junctions is considerably higher than that of the LLL-types (see figure 4b). Thus, the percentage of fast-migrating TJs is highest within the HHH group, i.e. when all boundaries are HABs, and this percentage consistently decreases with increasing number of LABs present in TJs, being lowest for the LLL group.

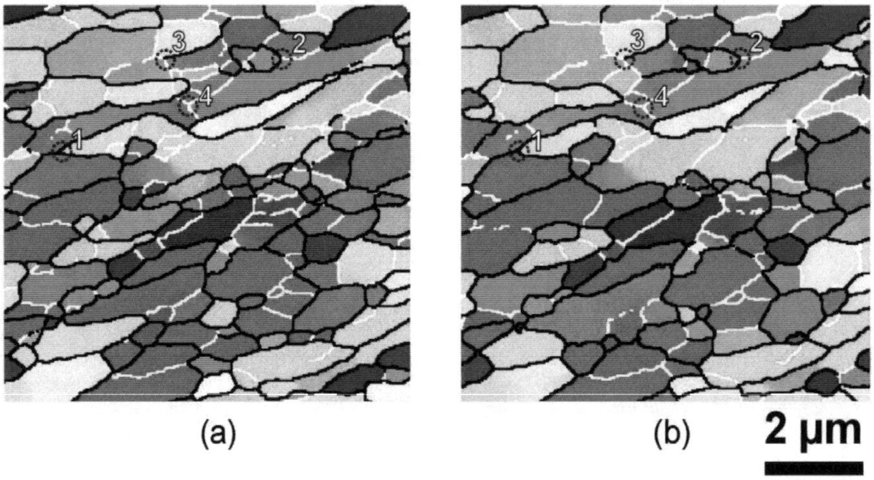

Figure 3. Changes in one region mapped using the EBSD technique in the deformed condition (a) and after annealing at 170 °C for 10 minutes (b). White and black lines show LABs and HABs, respectively. Examples of four different TJ types considered in this work are marked by blue circles and numbers: 1 – HHH, 2 – HHL, 3 – HLL and 4 – LLL. The extrusion direction is parallel to the scale bar.

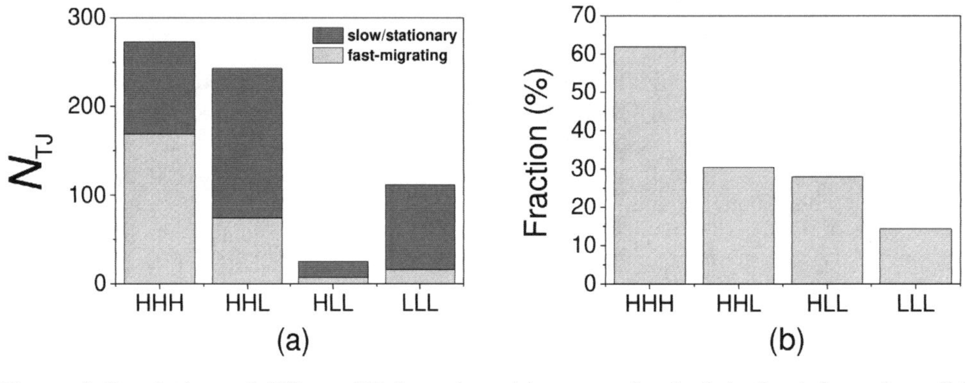

Figure 4. Populations of different TJs investigated in one region both in the deformed condition and after 10 min at 170 °C: (a) frequencies of TJs classified as either fast-migrating or slow/stationary; (b) fractions of fast-migrating junctions in each TJ group.

4. Discussion

Structural coarsening taking place in the heavily ECAE-processed AA1050 plate during annealing at 170 °C increases the average subgrain size from 0.5 μm in the as-deformed condition to ~0.8 μm after 2 h of annealing, and softens the material (see figure 2). The distinct feature of this coarsening process is that the increase in the average subgrain size does not appreciably affect the morphology of the microstructure. In this regard, the coarsening process in the ECAE-processed AA1050 plate is phenomenologically similar to that previously reported for heavily rolled aluminum [1–4], where a lamellar morphology obtained by rolling was preserved during coarsening, which proceeded via lateral motion of TJs. For ECAE-processed aluminum investigated in the present work it is also found that TJ motion plays a significant role during coarsening of the deformed microstructure at least within the first 10 minutes at 170 °C.

Categorization based on types of boundaries composing TJs, conducted in our study, clearly indicates that the susceptibility to rapid TJ motion during this initial period of coarsening depends on the number of LABs and HABs in a TJ. HHH-type TJs are found to be much more prone to rapid TJ motion than any other TJ type (figure 4b). Taking into account that the HHH-type TJs are also most frequent in the deformed microstructure of the given ECAE-processed aluminum plate (figure 4a), it can be concluded that HHH junctions play a dominant role in the process of TJ motion of the microstructure investigated in the present work. It should however be mentioned that since the TJs of the present sample have been characterized only using surface inspection, it is possible that the observed migration may also be influenced by the motion of boundary segments and TJs below the surface.

Whereas TJs of the HHL-type are also rather frequent in the deformed microstructure and thus can appreciably contribute to coarsening via TJ motion (see figure 4a), the total number and the frequency of fast-migrating boundaries of the HLL- and LLL-type TJs are too small to have a significant contribution to the coarsening process. The fact that TJs containing LABs migrate less than HHH-type TJs is not surprising as the mobility of LABs is known to be generally lower than that of HABs (except for coherent twin 60° ⟨111⟩ boundaries, which are however very rare in Al). Considering this, it appears more surprising that some LLL-type TJs were able to migrate quickly. Analysis of the 16 fast-migrating LLL junctions indicated that 11 of them contained at least one boundary with a misorientation of above 10°. Therefore, their properties can be considered similar to those of HHL and HLL-type TJs. Nevertheless, the remaining 5 LLL junctions formed by boundaries with misorientations less than 10° also migrated over fairly large distances.

A small subset from figure 3a is shown in figure 5a. This subset contains three migrating LLL-type junctions with $\theta \approx 6°$ across boundary 1–4; $\approx 7°$ across boundaries 1–2 and 2–3; $\approx 9°$ across boundaries 2–4, 3–4, 5–6; and approximately 2° and 11° across boundaries 1–5 and 1–6, respectively. After 10°minutes at 170 °C subgrain 2 becomes rather large, thus extending the length of LABs 1–2 and 2–4, whereas subgrains 3 and 8 disappear (see figure 5b). The disappearance of these subgrains results in the removal and the formation of several HABs and LABs in the microstructure. Since subgrains 2 and 8 were almost equaixed in the as-deformed condition, their evolution during annealing may be described as a result of conventional curvature-driven subgrain growth.

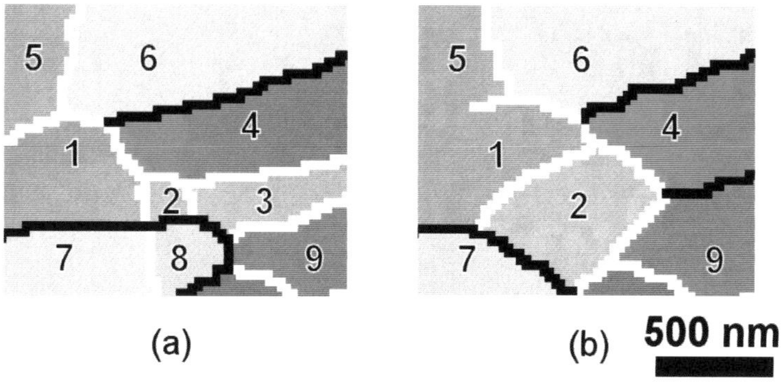

Figure 5. An example showing migrating LLL-type junctions (a) before and (b) after annealing at 170 °C for 10 minutes. White and black lines show LABs and HABs, respectively. Numbers mark subgrains present in this area.

It is apparent that coarsening in the ECAE-processed microstructure affects all boundary types, thus maintaining the proportion of HABs and LABs in the coarsened microstructure similar to that in the as-deformed sample. The latter contrasts our previous observations made in heavily rolled materials, where coarsening was found to result in considerable reductions in the fraction of HABs [3–5]. This difference in the evolution of the HAB fraction in either rolled or ECAE-processed samples may result from the difference in crystallographic textures produced by the two deformation processes: i.e. between a very strong rolling texture in the heavily rolled material [4,5] and a rather weak texture in the ECAE-processed plate [15].

Another difference between heavily rolled aluminum analyzed in [1–4] and the ECAE-processed sample studied in the present work is the extent of annealing regime where TJ motion dominates. For Al cold-rolled to a strain of 6, coarsening is such that the average boundary spacing measured along the normal direction increases 3 times before a significant amount of curvature-driven coarsening is observed [3]. In contrast, for the ECAE material of the present study some evidence of curvature-driven subgrain coarsening is seen when the average subgrain size increases only approximately 1.3 times (after 10 minutes of annealing at 170 °C, see figure 5). This suggests that for the ECAE sample there is a greater competition between regular coarsening and TJ motion. This difference, as compared to the rolled samples, may arise both from the pronounced lamellar morphology of the rolled microstructure characterized by a higher aspect ratio [1] and from differences in the spatial arrangement of HABs and LABs in each sample.

5. Summary

EBSD investigation of an ECAE-processed aluminum plate reveals the importance of TJ motion in microstructural coarsening during annealing at 170 °C. Coarsening by TJ motion in this material is phenomenologically similar to that seen during annealing of aluminum rolled to similarly high plastic strains, although the extent of this coarsening is smaller for the ECAE sample. Direct evidence for coarsening via TJ motion is provided by examination of the same region before and after annealing for 10 minutes at 170 °C. In this experiment it is found that junctions formed by three HABs play a dominant role in the process of TJ motion in the given material, and that the presence of even one LAB has a significant effect on the likelihood of rapid TJ motion.

Acknowledgements
The authors gratefully acknowledge the support from the Danish National Research Foundation (Grant No. DNRF86-5) and the National Natural Science Foundation of China (Grant No. 51261130091) to the Danish–Chinese Center for Nanometals.

References
[1] Yu T, Hansen N and Huang X 2011 *Proc. Roy. Soc. A* **467** 3039
[2] Yu T, Hansen N and Huang X 2013 *Acta Mater.* **61** 6577
[3] Mishin O V, Godfrey A, Yu T, Hansen N and Juul Jensen D 2015 *IOP Conf. Ser: Mater. Sci. Eng.* **82** 012083
[4] Mishin O V, Godfrey A, Juul Jensen D and Hansen N 2013 *Acta Mater.* **61** 5354
[5] Tian H, Suo H L, Mishin O V, Zhang Y B, Juul Jensen D and Grivel J-C 2013 *J. Mater. Sci.* **48** 4183
[6] Mishin O V, Juul Jensen D and Hansen N 2003 *Mater. Sci. Eng. A* **342** 320
[7] Segal V M 1995 *Mater. Sci. Eng. A* **197** 157
[8] Kamachi M, Furukawa M, Horita Z and Langdon T G 2003 *Mater. Sci. Eng. A* **361** 258
[9] Segal V M 2004 *Mater. Sci. Eng. A* **386** 269
[10] Segal V M 2008 *Mater. Sci. Eng. A* **476** 178
[11] Mishin O V and Bowen J R 2009 *Metall. Mater. Trans. A* **40** 1684
[12] Mishin O V, Segal V M and Ferrasse S 2012 *Metall. Mater. Trans. A* **43** 4767
[13] Humphreys F J 2001 *J. Mater. Sci.* **36** 3833
[14] Mishin O V, Östensson L and Godfrey A 2006 *Metall. Mater. Trans. A* **37** 489
[15] Li S and Mishin O V 2014 *Metall. Mater. Trans. A* **45** 1689

IOP Conf. Series: Materials Science and Engineering **89** (2015) 012036 doi:10.1088/1757-899X/89/1/012036

Cube texture formation during the early stages of recrystallization of Al-1%wt.Mn and AA1050 aluminium alloys

M M Miszczyk and H Paul
Polish Academy of Sciences, Institute of Metallurgy and Materials Science, Krakow, Poland

Email: m.miszczyk@imim.pl

Abstract. The cube texture formation during primary recrystallization was analysed in plane strain deformed samples of a commercial AA1050 alloy and an Al-1%wt.Mn model alloy single crystal of the Goss{110}<001> orientation. The textures were measured with the use of X-ray diffraction and scanning electron microscopy equipped with an electron backscattered diffraction facility. After recrystallization of the Al-1%wt.Mn single crystal, the texture of the recrystallized grains was dominated by four variants of the S{123}<634> orientation. The cube grains were only sporadically detected by the SEM/EBSD system. Nevertheless, an increased density of <111> poles corresponding to the cube orientation was observed. The latter was connected with the superposition of four variants of the S{123}<634> orientation. This indicates that the cube texture after the recrystallization was a 'compromise texture'. In the case of the recrystallized AA1050 alloy, the strong cube texture results from both the increased density of the particular <111> poles of the four variants of the S orientation and the ~40°(~<111>)-type rotation. The first mechanism transforms the S^{def}-oriented areas into S^{rex} ones, whereas the second the near S-oriented, as-deformed areas into near cube-oriented grains.

1. Introduction

The crystallographic aspects of the new orientation generation during recrystallization have been the subject of discussion for several years. This concerns both the description of the mechanisms 'controlling' the texture transformation as well as the relations between the 'texture images' of the deformed and recrystallized states. It is widely accepted that the starting point of new orientations are the components found in the deformed state, e.g. [1-6]. It means that the orientation of each new grain originates from the 'family' of orientations in the deformed state. However, a clearly observable change of the nuclei orientation with respect to the as-deformed orientations, e.g. [7,8] suggests a connection with the dislocation removal from a given region, e.g. [7-13]. The literature on recrystallization pays a lot of attention to the role of the high-angle boundaries (HABs) and the conditions of their migration. Nevertheless, literature is unable to provide a comprehensive answer to the question about the mechanisms controlling the formation of nuclei inside the deformed structure and especially to explain the different nuclei orientation with respect to the deformed matrix. It is clear that, in deformed polycrystalline materials, the fraction of the HABs is generally very large, and these boundaries play a key role in nucleation. However, nucleation by way of migration of the deformation-induced HABs has not been reported as unique or universal. For instance, in the case of face centred cubic (fcc) crystallites with stable orientations in plane strain compression ({110}<112>

IOP Conf. Series: Materials Science and Engineering **89** (2015) 012036 doi:10.1088/1757-899X/89/1/012036

or {110}<001>), HABs are not present or are only very rare, yet nucleation still occurs, e.g. [8,14,16,17]. Therefore, in deformed materials, where the fraction of low-angle boundaries (LABs) is still large, the mechanisms based on the migration of the LABs are also needed for the early stages of recrystallization. It is assumed that the thermally activated migration of the LABs combined with the movement of the dislocations stored inside the (sub)grains and/or the diffusion of vacancies, e.g. [18, 19] may play the key role in the orientation change of a 'chosen area' with respect to the deformed matrix, and in the formation of HABs (in fact, it is the 'transformation' of the LABs into high-angle ones). Still, the nature of this process is not clear.

The understanding of the nucleation mechanisms is critically important for the description of the origin of the cube texture in fcc metals. In particular, the recrystallization texture of aluminium alloys is often represented by a strong cube{100}<001> orientation with some scattering in the rolling direction towards the Goss{110}<001> orientation. Some authors stated that the cube-oriented grains in highly deformed aluminium alloys are nucleated from deformation, transition or shear bands preferentially located in the Cu or ND-rotated Cu-texture components, e.g. [20]. Yet the details of the transformation that leads to the extreme sharpness of the cube texture are not understood.

The aim of this study is to examine crystallographic aspects of the strong cube texture formation during the early stages of recrystallization of two aluminium alloys. Particular attention was paid to the description of the misorientations across the recrystallization front and on possible dislocation mechanisms during the primary recrystallization. The use of aluminium alloys with a high stacking fault energy (SFE) practically excludes mechanical and recrystallization twinning, and it facilitates identifying the mechanisms responsible for the texture transformation during the initial annealing stages. The strong cube texture formation were analysed by means of a scanning electron microscopy equipped with an electron backscattered diffraction (SEM/EBSD) facility as well as (for more global scale analyses) by X-ray diffraction.

2. Experimental

The materials were a commercial polycrystalline AA1050 alloy and a (single phase) Al-1%wt.Mn single crystals of the Goss(110)[00-1] initial orientation. Samples of 10 x 10 x 10 mm^3 were plane strain compressed (PSC) in a channel-die. The multi-stage deformation was performed at ambient temperature at the initial strain rate of 0.02 s^{-1} to the logarithmic strain of 0.51 (40%). In order to limit the friction between the sample, the punch and the walls of the channel-die, each sample was wrapped in 0.2 mm thick Teflon™ tape. The deformed samples were annealed for 1h at temperatures ranging between 100°C and 450°C with the purpose to obtain different stages of recrystallization. The deformed and recrystallized microstructures were analyzed in the ND/ED section (where: ND and ED denote, the normal and the extension directions, respectively) with the use of SEM - QUANTA 3D FEG, equipped with a field emission gun and an EBSD facility. The SEM was working in the automatic beam scanning mode and the maps were created with step sizes ranging between 100 nm and 700 nm. The numbers of correctly indexed points was always above 97%. In each case, the orientation maps covered the area of at least 100 µm x 100 µm. In the {111} pole figures corresponding to the SEM orientation maps, the indices {*hkl*}<*uvw*> represent the texture component which has the {*hkl*} plane parallel to the ED/TD plane (where: TD denote the transverse direction) and the <*uvw*> direction parallel to ED. At the sample scale, the textures were measured on the ED/TD plane with the use of the Philips X Pert PW-1830 X-ray diffractometer.

3. Results and discussion

SEM/EBSD and X-ray diffraction were applied to reveal the inherent characteristics of the microstructure and texture development during the PSC and subsequent annealing. The use of single crystals of a stable orientation enables one to precisely define the orientation relationship developed when the nuclei are formed in the as-deformed structures free from HABs. Those results are related to the mechanism responsible for the new orientation formation observed in the commercial AA1050 alloy. In the two materials, the orientations of the nuclei were similar, despite the fact that the deformation textures were quite different.

3.1. Deformation and recrystallization behaviour of an Al-1%wt.Mn single crystal of Goss{110}<001> orientation.

3.1.1. Microstructure and texture of the as-deformed state. At a logarithmic strain of 0.51, the microstructure of the Goss-oriented crystal consist of two complementary sets of elongated microbands delineated by dislocation walls (Fig. 1a). Both sets of dislocation walls were inclined very close to the expected traces of the active {111} slip planes (Fig. 1b). The adjacent elongated cells displayed opposite rotation senses, as reported earlier [16, 17, 21]. Most of the rotations occurred around the axes lying close to TD, both away from and almost back to, the initial crystal orientation. The misorientation angles were significantly below 10°. Consequently, the textures of the areas corresponding to the above microstructures confirm the stability of the initial orientations at a logarithmic strain of 0.51, as shown by the {111} pole figures of the mapped areas.

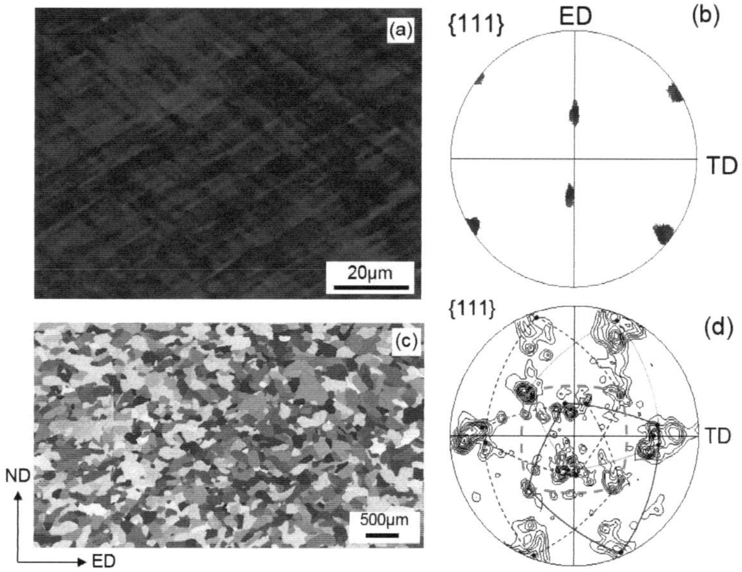

Figure 1. Microstructures and corresponding {111} pole figures observed in a Goss(110)[00-1] oriented single crystal of Al-1%wt.Mn alloy, (a, b) as-deformed state and (c, d) after recrystallization at 688°C for 480 s. In (d) the black lines mark the four positions of S{123}<634> orientations, whereas bold blue (dashed) line - cube orientation. For better visualized the microstructures details in the deformed and recrystallized states the 'cubic' and IPF (||Y) color codes were applied in (a) and (c), respectively.

3.1.2. Recrystallization behaviour. Recrystallized grains were analysed inspects large sample areas (Fig. 1c). Only a small fraction of the grains was observed to be twinned. For the statistical analysis of the misorientation relation between the deformed material and the recrystallized grains, the few twinned grains were ignored. The misorientation relation was calculated between the 'average' (stable) orientation of the deformed state and the orientations of the recrystallized grains.

Figures 1b and d show the orientations of the as-deformed crystal and the recrystallized grains. The grain orientations form groups, which are symmetric with respect to the external directions. Those groups of orientations can be described, in an idealized form, by the four symmetric S{123}<634> orientations. This symmetry results from the positive and negative rotations of the new grain orientations around the axes grouped near the normals of the {111} planes, highly active during deformation. A comparison of the pole figures of the deformed and the recrystallized states shows that the <111> poles of the recrystallized grains do not coincide with the <111> poles of the deformed areas or, more precisely, overlapping of the <111> rotation axes was only rarely observed (Fig. 1d). It is important to note that no new grains of the near Goss orientation were observed.

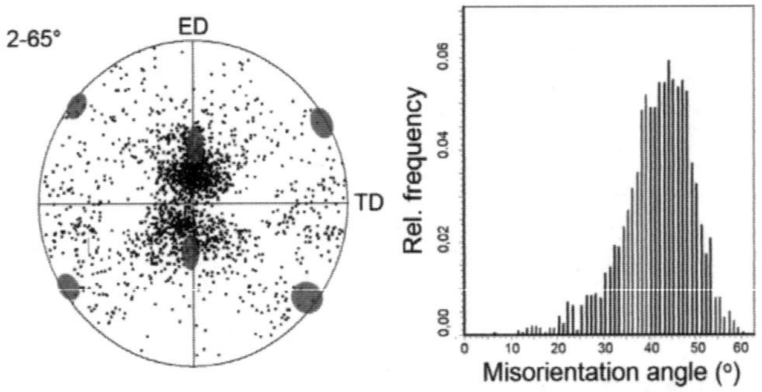

Figure 2. The distribution of misorientation axes and angles between the average deformed orientation and recrystallized grains (taken from the orientation map presented in Figure 1c in the Goss(110)[00-1] oriented single crystal of Al-1%wt.Mn alloy). Red circles mark the orientation in the deformed state.

The frequency distribution of the misorientation angles and axes between the new grains and the 'average' as-deformed orientations is presented in Fig. 2. The plot shows that most of the angles across the migrating recrystallization front are within the range of 25-55°. The scattering of the misorientation axes is larger than that of the orientations in the deformed state. The spread of the {111} poles of the as-deformed state is schematically marked by the red areas. Basically, the majority of the misorientation axes across the recrystallization front are concentrated around the <111> poles of the two highly stressed slip planes during the deformation. A 'grouping' of rotation axes near the two remaining the <111> poles is weaker. Only occasionally, other misorientation axes were observed.

3.1.3. Cube component. It is clear that the Goss{110}<001>-oriented areas in the as-deformed state are transformed during the recrystallization into four variants of $S^{(I-IV)}$-oriented grains. The transformation, which occurs during the formation of the new grain could be described as a result of a (+/-) ~30-40° rotation of the crystal lattice of the as-deformed areas around the selected <112> axes near the normal of the most active {111} slip planes, and can be schematically presented as:

$$\text{Goss}\{110\}\langle 001\rangle^{deformed} \rightarrow (\sim 30\text{-}40^{\circ}\langle 112\rangle \text{ rotation}) \rightarrow \sim S\{123\}\langle 634\rangle^{(I\text{-}IV)\ recrystallized}$$

The description of the mechanisms responsible for the formation of recrystallization grains with new orientation, given by Miszczyk et al. [8], combines the rapid growth along specific planes with the rotation of the crystal lattice of new grains around the specific axes. The thermally activated movement of the dislocations can lead to the creation of a twist or tilt grain boundary, due to the screw or edge character of the dislocations, respectively. In an idealized form, if the dislocations stored along the plane of growth are pure screw (and their Burgers vectors are orthogonal), their movement leads to a pure twist boundary with a rotation axis perpendicular to the boundary plane. If the dislocations are pure edge, their movements leads to a pure tilt boundary with a rotation axis parallel to the boundary plane, as discussed in [8]. It is assumed here that the rotation around the <112> direction may result from the thermally activated (pure edge) dislocations movement along the {111}<011> 'systems'. This mechanism leads to tilt boundary formation and can be responsible for rapid growth. A similar explanation of the phenomenon was given in [8-11], based on the analysis of the grains showing growth anisotropy. The cube grains (within 20°) are observed very rarely after the recrystallization. However, an increased density of <111> poles in the positions typically occupied by the cube orientation was observed. The explanation of this phenomenon can be based on the 30-40°<112> crystal lattice rotation during the recrystallization from the Goss to the S orientation and superposition of four variants of the S orientation. It can be clearly seen that one of the <111> poles of each of the four variants of the S orientation marked the position of the cube orientation. This indicates that the cube texture after the recrystallization was a 'compromise texture'. This 'artefact', was even more clearly observed in the case of the polycrystalline AA1050 alloy (to be discussed in the next section).

3.2. *Recrystallization behaviour of commercial AA1050 alloy*

3.2.1. Microstructure and texture of the as-deformed state. Figure 3a shows the microstructure of the sample deformed to 40%. The microstructure has a large fraction of HABs (misorientations greater than 15°) and LABs (misorientations ranged between 2° and 15°). The microstructure is composed of elongated (in fact, flat) grains subdivided by a network of LABs. The width of the grains (measured in the ND/ED section) range between 15 μm and 30 μm, whereas the length is between 50 μm and 100 μm. The longer axis of the grains are nearly parallel to ED. The development a typical copper-type rolling texture composed of four variants of the S{123}<634> orientation with scattering towards the brass{110}<112> and C{112}<111> orientations was observed, as presented in Fig. 3b. Based on the X-ray diffraction and the SEM/EBSD measurements, it is clear that the cube component is not present in the deformed state (Fig. 4a). The orientations of the dominant fraction of the deformed grains have the <111> direction, nearly parallel to TD. Very often, neighbouring layers were observed to be strongly misoriented and a near-twin orientation relationship sometimes occurred between those areas.

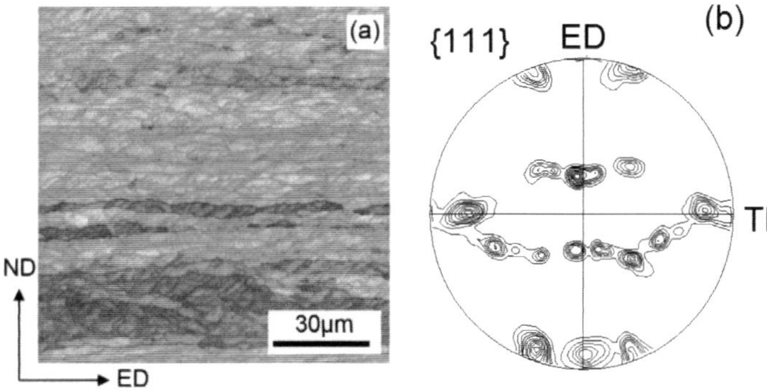

Figure 3. Microstructure and texture of the AA1050 alloy deformed to a logarithmic stain of 0.51 (40%). (a) Orientation map and (b) corresponding {111} pole figure. A step size of 100 nm was used for the SEM/EBSD measurements.

3.2.2. Microstructure and texture after annealing. In our recrystallization experiments, clear separation was observed between the deformed (or recovered) and the recrystallized areas. The nuclei were observed as single isolated ones or grouped into parallel chains reflecting the 'directionality' of the deformed microstructure. It is not clear, at the moment, why some layers of a given variant of the S orientation undergo recrystallization, whereas others are still only recovered.

The textures of the as-deformed and annealed states (in sample scale) were analysed by X-ray diffraction. For annealing at temperatures up to 250°C the {111} pole figures show that, the intensity and the exact position of the main texture components after annealing were similar to those observed for the as-deformed state (see Fig. 4a). At temperatures of 300°C and above, the development of a strong cube texture component is observed, as presented in Fig. 4b. Figure 5a shows an orientation map of the sample partly recrystallized at 300°C for 1h. The four variants of the S orientation are marked in different colours: S1 - (123)[6-43] – maroon, S2 - (231)[3-46] – navy blue, S3 - (213)[-3-6-4] – lime green and S4 - (231)[-34-6] – fuchsia. Additionally, the cube-oriented areas are marked in yellow. In all the cases, a scattering of 20° from the ideal position was applied. Basically, the nuclei were largely confined to particular layer(s). However, for longer annealing times, the grains grew also in the ND direction to widths larger than those of the layers (Fig. 5a), respect from just before. A majority of the new grain orientations was classified as cube, but some of the grains were identified as S-oriented. Also a small number of grains were classified as 'other orientations' (Fig. 5b). Possible mechanisms leading to nucleation of the near S-, cube- and Goss- oriented grains are discussed below.

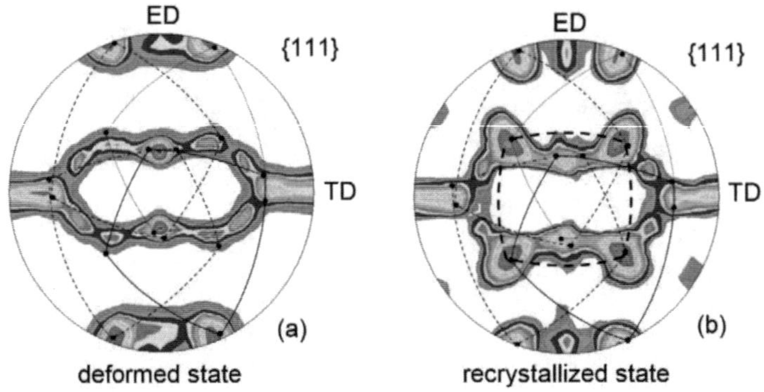

Figure 4. The {111} pole figures showing global textures measured by X-ray diffraction of: (a) the deformed sample of AA1050 and (b) after annealing at 250°C for 480 s. Thin black lines mark the four positions of S orientations, whereas bold black (dashed) line – the cube orientation.

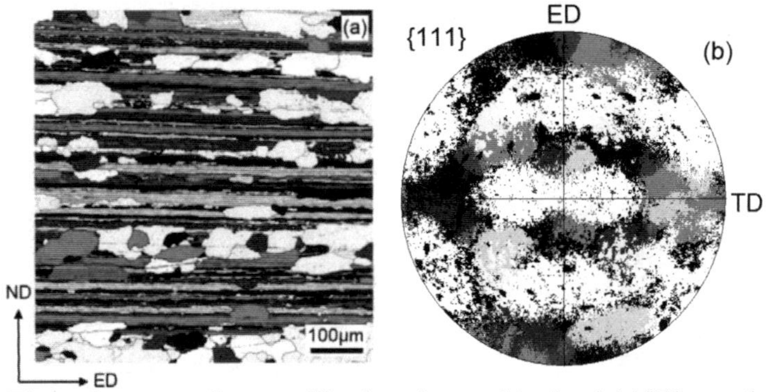

Figure 5. The early stages of recrystallization observed in the AA1050 sample deformed to logarithmic stain of 0.51 and subsequently annealed at 300°C for 1h. (a) SEM/EBSD orientation map and corresponding {111} pole figure. The four variants of the S orientation are marked in different colors: S1 - (123)[6-43] – maroon, S2 - (231)[3-46] – navy blue, S3 - (213)[-3-6-4] – lime green and S4 - (231)[-34-6] – fuchsia, whereas the cube(100)[001] oriented areas are marked in yellow. The step size was 700 nm.

S-oriented grains. Figure 6a shows a typical case of the S-oriented grain growing along the structure of the deformed/recovered grains. The new grain orientation is similar to one of the deformed/recovered grain (Figs. 6b and c). Therefore, the misorientation calculated between the deformed/recovered areas and the new grain is low (misorientations <10°). It is clear that the growth along ND is hindered. The misorientation is relatively large with respect to the other two S-oriented grains (two other variants of the S orientation) where, a HABs of the ~30-40° <111>-type is formed across the recrystallization front. This enables a rapid growth along ED. One may assume that the new, recrystallized grain is growing from the as-deformed areas of nearly the same orientation which is often discussed in literature, e.g. [20]. However, if the results of the recrystallization experiments on single crystals of a stable orientation (§ 3.1.2. and, e.g. [7,8,23]) where a new grain orientation similar to one in the deformed state has never been observed also applies for the AA1050 polycrystalline sample, other explanation have to be analysed. One may therefore speculate that the new grains nucleate from one of the two other 'S-oriented' areas of the deformed state and grew quickly into those S-oriented bands. In this way, the formation of the new grain of the S orientation (rotation around the normal of the most active {111} slip plane - HA) can be schematically presented as:

$$\sim S\{123\}\langle 634\rangle^{(I\text{-}IV)\,deformed} \rightarrow (\sim 30\text{-}40^{\circ}\ \langle 111\rangle^{HA}\ \text{rotation}) \rightarrow \sim S\{123\}\langle 634\rangle^{(I\text{-}IV)\,recrystallized}$$

This means that the S-oriented grains of the given variant nucleate inside the as-deformed areas of the other S variant. In this case, again, a twist boundary is formed at the head of the recrystallization front.

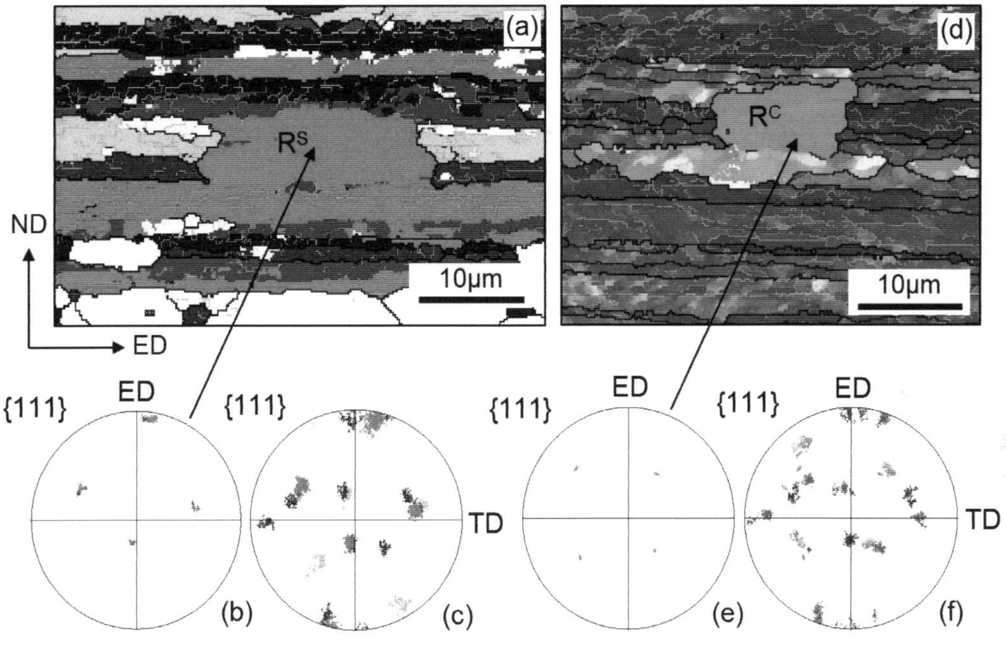

Figure 6. Crystallography of grain growth at the early stages of recrystallization observed in a sample of AA1050 alloy deformed to a logarithmic stain of 0.5 and then recrystallized at 300°C for 1h. Details from map presented in Fig. 5. (a) and (d) Orientation maps and corresponding (b, c) and (e, f) {111} pole figure. R^S – S oriented new grain, R^C – cube-oriented new grain. ('Arbitrary' color codes was applied).

Cube-oriented grains. Figure 6d shows a cube-oriented grain growing into S-oriented areas of the as-deformed state. A new grain orientation is presented in Fig. 6e. The bottom part of the grain is in contact with a layer in which the size of the cells forming the internal structure is significantly larger than the one observed in the other as-deformed areas. This suggests a more advanced stage of recovery here. Moreover, those cell orientations differ from the orientations typically observed in the deformed state (Fig. 6f). The spread of the orientations of strongly recovered cells mark the transition from the near S to the cube orientations and suggests the possible rotation mechanism (rotation around the normal of the less active {111} slip plane - LA) that can be described schematically as:

$$\sim S\{123\}\langle 634\rangle^{deformed} \rightarrow (\sim 40\text{-}50^{\circ}\ \langle 111\rangle^{LA}\ \text{rotation}) \rightarrow \sim \text{cube}\{100\}\langle 001\rangle^{recrystallized}$$

The new grain is elongated along the boundaries of the as-deformed flat grains. Therefore, it is likely that the preferred growth is connected with rapid migration of the HAB along ED, i.e. the new grain grows preferentially along the boundaries of the flat grains.

Other recrystallized grain orientations. There is also a small quantity of grains characterized by other orientations. Those grain orientations are not random and a majority of them represents the ED- or ND- rotated cube{100}<001> orientation. However, the grouping some of the new grain orientations near the Goss position can suggest a mechanism as:

$$\sim S\{123\}\langle 634\rangle^{(I\text{-}IV)\ deformed} \rightarrow (40\text{-}45^{\circ}\langle 112\rangle^{LA}\ \text{rotation}) \rightarrow \sim \text{Goss}\{110\}\langle 001\rangle^{recrystallized}$$

Which is in a way the reverse transformation type to that seen in the single crystal. To sum up, the present analysis suggest that, despite the fact that the new grains have a 40°<111> misorientation relation with respect to the deformed areas, no direct evidence was found supporting the idea that possible 40°<111> relations occurs at a higher rate than relations with other misorientation relations.

4. Conclusions

Based on local orientation measurements and X-ray diffraction, the present results provide detailed information about the development of the recrystallization microstructures and textures in a commercial AA1050 alloy and a single crystal of a model Al-1%wt.Mn alloy it was found that:

(i) Although the deformation texture of the two materials were different, in both materials nuclei of new grain orientations developed during recrystallization are similar, but different from those in the deformed matrix.

(ii) After recrystallization of the Al-1%wt.Mn single crystal of the Goss orientation, the texture of the recrystallized grains was dominated by four variants of the S{123}<634> orientation. Cube grains were only sporadically detected by the SEM/EBSD system. Nevertheless, an increased density of the <111> poles corresponding to the cube orientation was observed. This effect was connected with the formation of nuclei with orientations near the four variants of the S{123}<634> orientation.

(iii) In the case of the recrystallized AA1050 alloy, the strong cube texture after recrystallization results from both the increased density of the particular <111> poles of the four variants of the S orientation and the ~40°(~<111>)-type rotation of the as-deformed orientations. The first mechanism transforms the S^{def}-oriented areas into (another variant) S^{rex} ones, whereas the second, the near S-oriented, as-deformed areas into near cube-oriented grains.

Acknowledgments:

This work was partially supported by the National Centre of Science and the Polish Ministry of Science and Higher Education, projects: 2014/13/B/ST8/04291 and 'Iuventus Plus' IP2011 036471, respectively. M.M. acknowledges the Region Rhone Alpes for a graduate student scholarship.

References

[1] Doherty R D, Hughes D A, Humphreys F J, Jonas J J, Juul Jensen D, Kassner M E, King W E, McNelly T R, McQueen H J and Rollett A D 1997 *Mater Sci Eng A* **238** 219
[2] Humphreys F J and Hatherly M 2004 *Recrystallization and related annealing phenomena (Pergamon)* 2nd Edition
[3] Engler O, Kong X W and Lücke K 2001 *Acta Mater* **49** 1701
[4] Godfrey A, Juul Jensen D and Hansen N 2001 *Acta Mater* **49** 2429
[5] West S S, Winther G and Juul Jensen D 2011 *Metall Mater Trans A* **42** 1400
[6] Fan G H, Zhang Y B, Driver J H and Juul Jensen D 2014 *Scripta Mater* **72-73** 9
[7] Paul H, Driver J H and Lens A 2005 *Mater Sci Forum* **495-497** 1249
[8] Miszczyk M, Paul H, Driver J H and Maurice C 2015 *Acta Mater* **83** 120

[9] Gottstein G, Murmann H, Renner G, Simpson C and Lucke K 1978 *Proc. Fifth Inter. Conf.* on 'Textures of Materials', ed. G. Gottstein and K. Lücke **1** pp 521
[10] Kohara S, Parthasarathi M N and Beck P A 1958 *J Appl Phys* **29** 1125
[11] Parthasarathi M N and Beck PA 1961 *Trans Metall Soc AIME* **221** 831
[12] Ridha A A and Hutchinson W B 1984 *Acta Metall* **30** 1929
[13] Inoko F, Kashihara K, Tagami M and Okada T 2004 *Mat Sci Forum* **467-470** 57
[14] Godfrey A, Juul Jensen D and Hansen N 1998 *Acta Mater* **46** 823
[15] Albou A, Driver J H and Maurice C 2010 *Acta Mater* **58** 3022
[16] Borbély A, Maurice C, Piot D and Driver J H 2007 *Acta Mater* **55** 487
[17] Paul H, Maurice C and Driver J H 2010 *Acta Mater* **58** 2799
[18] Rios P R and Sandim H R Z 2005 *Mat Res* **8/3** 225
[19] Paul H, Morawiec A and Baudin T 2012 *Metall Mater Trans A* **43/12** 4777
[20] Hjelen J, Orsund R and Nes E 1991 *Acta Metal Mater* **39** 1377
[21] Ferry M and Humphreys F J 1996 *Acta Mater* **44** 3089
[22] Paul H. 2003 *Mat Chem Phys* **81/2-3** 531
[23] Miszczyk M, Paul H, Driver J H and Maurice C 2011 *Arch Metall Mater* **56** 933

A phase-field simulation study of irregular grain boundary migration during recrystallization

N Moelans[1], Y B Zhang[2], A Godfrey[3], D Juul Jensen[2]

[1]Department of Materials Engineering, KU Leuven, Kasteelpark Arenberg 44, Box 2450, B-3001 Leuven, Belgium

[2]Danish Chinese Center for Nanometals, Section for Materials Science and Advanced Characterization, Department of Wind Energy, Risø Campus, Technical University of Denmark, DK-4000 Roskilde, Denmark

[3]Key Laboratory of Advanced Materials (MOE), School of Materials Science and Engineering, Tsinghua University, Beijing 100084, Peoples Republic of China

E-mail: nele.moelans@mtm.kuleuven.be

Abstract. We present simulation results based on a phase-field model that describes the migration of recrystallization boundaries into spatially varying deformation energy fields. Energy fields with 2-dimensional variations representing 2 sets of dislocation boundaries lying at equal, but opposite, angles to the moving boundary are considered. The simulations show that the shape and overall migration rate of the recrystallization front is considerably affected by spatial variations in the deformation microstructure. It is seen that, depending on characteristics of the variations in the deformation microstructure, highly asymmetrical protrusions and retrusions can develop on the migrating recrystallization front resulting in a migration velocity considerably larger than that expected from standard recrystallization models. It is also seen that, when the wavelength of the variations in a deformation microstructure along the grain boundary is larger than the wavelength of the variations in the direction of migration, parts of the boundary show a stop-and-go type of migration, resulting in a lower overall migration rate. These simulations thus reproduce and explain many of the typical features observed in recrystallization experiments. They give new insights in the way deformation microstructures can affect the migration behavior of recrystallization boundaries and can lead to a stop-and-go type of migration of the recrystallization boundary even in pure materials.

1. Introduction

During recrystallization new almost defect free grains grow in a deformed matrix containing dislocations and other defects. In recrystallization studies, it is generally assumed that the driving force for recrystallization is homogeneous along the recrystallization boundary and that the boundary moves smoothly with a velocity v_{gb} proportional to the (homogeneous) driving force for recrystallization P_{rex}:

$$v_{gb} = MP_{rex}, \tag{1}$$

where M is called grain boundary mobility. Extensive studies [1] have however shown that local variations exist in the deformed microstructures and that grain boundary movement during recrystallization is often very irregular, even in pure materials [2–6].

For instance, the arrangement of the dislocations within a deformed grain typically shows significant local variations. Often dislocations organize themselves into boundaries in the form of a cell block microstructure, where dislocation boundaries of larger misorientation angles delineate volumes containing interconnecting dislocation boundaries of smaller misorientation [7–9]. Local variations in the spacing between and misorientation angle across the dislocation boundaries are generally observed. The driving force for recrystallization can consequently vary significantly even along a short length of a recrystallization boundary.

In-situ experiments of the movement of a recrystallization boundary have indeed shown that individual grain boundary segments move with different rates [2, 3, 5]. For example, in-situ observations and ex-situ microstructural studies of recrystallization in materials with medium to high stacking fault energy, such as Ni and Al, clearly show the presence of protrusions (parts of the boundary that have advanced more than other parts) and retrusions (parts of the boundary that are behind) of a few up to tens of μm along recrystallization boundaries. Moreover the boundary profile is dynamic: protrusions can disappear or transform into retrusions and vice-versa as a migrating recrystallization boundary moves into a deformed matrix. Often parts of the boundary also show a stop-and-go type of movement.

Zhang et al. [10, 11] have measured the curvatures at protrusions and retrusions formed along recrystallization boundaries and related these values to the local stored deformation energy in front of the moving boundary. It appeared that the magnitude of the local curvature based driving force for grain boundary migration at protrusions and retrusions is comparable to that of the stored energy within the deformed microstructure and thus cannot be neglected when characterizing the local migration rate of a recrystallization boundary. Although, large variations in stored deformation energy may result in protrusions and retrusions with high curvatures, the size and shape of the protrusions and retrusions seem to be correlated in a complex way to the spatial variations in the stored deformation energy [10].

Recently, we presented an approach based on a phase-field model to simulate the formation of protrusions and retrusions during migration of a recrystallization boundary into a deformation microstructure containing variations in stored energy [12]. The deformation energy field for the model is assumed to be constant in time and constructed to represent characteristics of typical deformation microstuctures obtained for medium-to-large strains.

As the simplest model to capture the typical variations in stored energy in a deformation microstructure, we considered energy fields with 1-dimensional variations representing a single set of dislocation boundaries. One dimensional functions with maxima at the positions of the cell block boundaries and minimum energy in the cell block interior were considered.

The model was able to reproduce the experimentally observed asymmetry between the protrusion and retrusion curvatures. Furthermore, we found that the characteristics of the spatial variations in stored deformation energy density have strong effects on the shape of the boundary and, more importantly, the overall boundary migration rate. For variations with low amplitude or small wavelength, the protrusions and retrusions are almost symmetric and the overall boundary velocity scales with the average stored deformation energy density, in agreement with generally accepted theories of recrystallization (assuming a homogeneous driving force along the recrystallization boundary). For large amplitudes and/or large wavelengths of the variations, an asymmetrical grain boundary shape is obtained with much larger curvatures at the retrusions than at the protrusions. In this case, the velocity scales with the maximum of the deformation energy density along the variation, resulting in a considerably larger velocity than that obtained assuming that the same energy would be distributed homogeneously in the material. The net driving forces due to curvature obtained in the simulations are similar in magnitude to those calculated from experimental micrographs.

The model can however easily treat more complex variations in stored deformation energy. Our long term aim is a systematic study of the effect of different characteristics of the spatial variations in deformation energy on the shape and velocity of a migrating recrystallization boundary, considering gradually more and more complex energy distributions.

In this paper, a more complex, but still idealized, case is considered, namely the migration into deformation energy fields with 2-dimensional variations representing dislocation boundaries in 2 directions. The simulation results are discussed and the findings are compared with those from the simulations with 1-dimensional variations.

2. Simulation details

We refer to our previous paper [12] for a detailed description of the phase-field model used for the simulations. The spatial variations in stored deformation energy are assumed to be constant in time (i.e. no recovery is included in the simulations) and are included in the model description using a free energy density term $f_{def}(x, y)$, which is a function of the spatial coordinates and can be of any continuous or discontinuous form that describes the deformation field.

In the present study, the movement of an initially planar boundary into energy fields with 2-dimensional variations representing 2 sets of dislocation boundaries lying at equal, but opposite angles to the boundary are considered. The considered deformation energy fields are shown in figures 1a and 2a using a color scaling, with black representing the highest energy density and white an energy density equal to 0. They are constructed as follows

$$f_{def}(x, y) = \max(f_{def,1}(x, y), f_{def,2}(x, y)) \tag{2}$$

with

$$f_{def,1}(x, y) = A_{def}\left[1 + \sin\left(\frac{\lambda_{def}}{|\sin(\alpha)|}\left(x - \frac{y}{\tan(\alpha)}\right)\right)\right] \tag{3}$$

$$f_{def,2}(x, y) = A_{def}\left[1 + \sin\left(\frac{\lambda_{def}}{|\sin(\alpha)|}\left(x - \frac{y}{\tan(-\alpha)}\right)\right)\right] \tag{4}$$

with α and $-\alpha$ the angles between the x-axis and the orientations of the 2 sets of dislocation boundaries, λ_{def} the spacing between the dislocation boundaries (i.e. the wavelengths of the variations measured perpendicular to the dislocation boundaries) and A_{def} the amplitude of the variations. The movement of initially flat grain boundaries in either the x- or y-direction into these fields is simulated.

In our previous study considering 1-dimensional variations, values for A_{def}, λ_{def} and the grain boundary energy σ_{gb} were chosen based on experimental information [9–11, 13] for aluminium and nickel: the amplitude of the variations A_{def} was varied between 0.1 and 20 MJ/m^3, the wavelength λ_{def} was taken 1.6, 3.2 and 6.4 μm for the symmetrical variations and the grain boundary energies of σ_{gb} = 0.32 J/m^2 (aluminium) and σ_{gb} = 0.87 J/m^2 (nickel) were used. The remaining model and simulation parameters were taken based on numerical considerations, such that grain boundary motion is resolved with a numerical error smaller than 5 % [14–16].

Similar parameter values are used in this work. Two energy fields are considered with A_{def} =1.3416 MJ/m^3 and $\alpha = \pi/12$ and for one field λ_{def} = 1.66 μm and for the other field λ_{def} = 6.18 μm. The average deformation energy present in the material is then $f_{def,av}$ = 1.8850 MJ/m^3 for both fields. One simulation was also performed assuming a grain boundary migrating into a homogeneous deformation energy field with energy density $f_{def} = f_{def,av}$ = 1.8850 MJ/m^3. A grid spacing of Δx = 0.05 μm and time step of Δt = 0.1 s were used. The grain boundary mobility was arbitrarily taken as $M = 1.7 \times 10^{-15}$ m^4/(Js), since experimental values for the grain boundary mobility are hard to find and strongly dependent on solute content. The value of M however only affects the time scaling of the simulations.

3. Simulation results

The images in figure 1 show the boundary positions obtained every 1000 time steps in the simulations (Starting from time step 40000 to be in the regime where the grain boundary

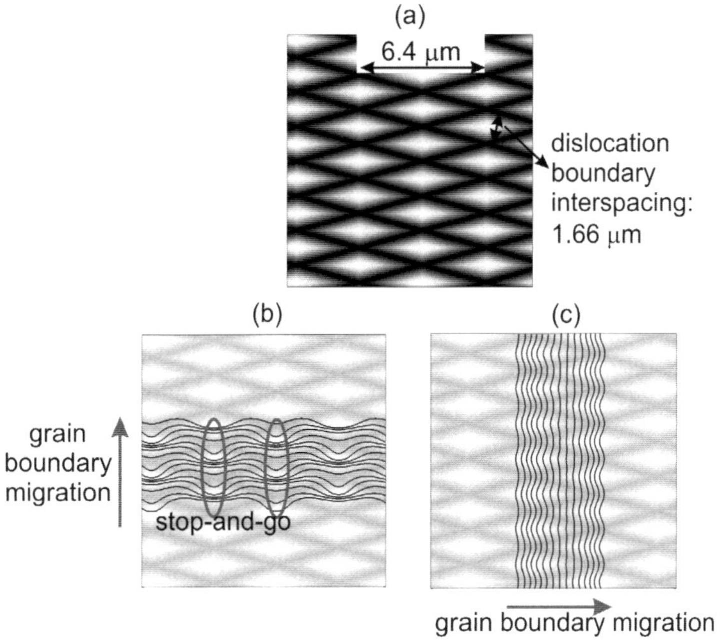

Figure 1. Simulation results obtained for a recrystallization boundary migrating in a deformation energy field with variations given by equations (2)-(4) with $A_{def} = 1.3416$ MJ/m^3, $\alpha = \pi/12$ and $\lambda_{def} = 1.66\mu$m. (a) Representation of the considered energy field using a color scaling with black representing the highest energy density and white an energy density equal to 0. (b), (c) Grain boundary positions as a function of time obtained in the simulations, plotted every 1000 time steps starting from time step 40000. (b) Boundary moving from bottom to top. Parts of the boundary, indicated with the red ovals, show a stop-and-go type movement: boundary parts near minima in the energy density field remain for a longer period at the same position and then advance suddenly to the next minimum in the energy density. (c) Boundary moving from left to right. Almost symmetrical protrusions and retrusions develop along the migrating recrystallization front. Protrusions transform into retrusions and vice-versa in a continuous way while the grain boundary is migrating, due to the spatial variations in deformation energy density. (a), (b) and (c): Only part of the simulation domain is shown.

shape varies periodically in time. The initial migration behavior starting from a flat boundary will be analysed in later work) for the 2-dimensional field described by equations (2)-(4) with $A_{def} = 1.3416$ MJ/m^3, $\alpha = \pi/12$ and $\lambda_{def} = 1.66\mu$m, considering the migration of a recrystallization boundary in the y- (figure 1a) and x-direction (figure 1b) of the simulation domain. Protrusions and retrusions clearly form along the migrating recrystallization boundary. The migration behavior in the y-direction (from bottom to top) is very different from that in the x-direction (from left to right). For the boundary migrating in the y-direction, parts of the grain boundary located near a minimum in the energy field (where the energy density is almost 0) almost stop moving for some time and then advance quickly to the next position with minimum energy density, very similar to the stop-and-go behavior seen in experiments. The movement of the boundary migrating in the x-direction, on the other hand, is smooth. Protrusions and retrusions develop symmetrically with a shape that is very similar to the protrusion/retrusion shapes seen in the simulations with 1-dimensional variations with small

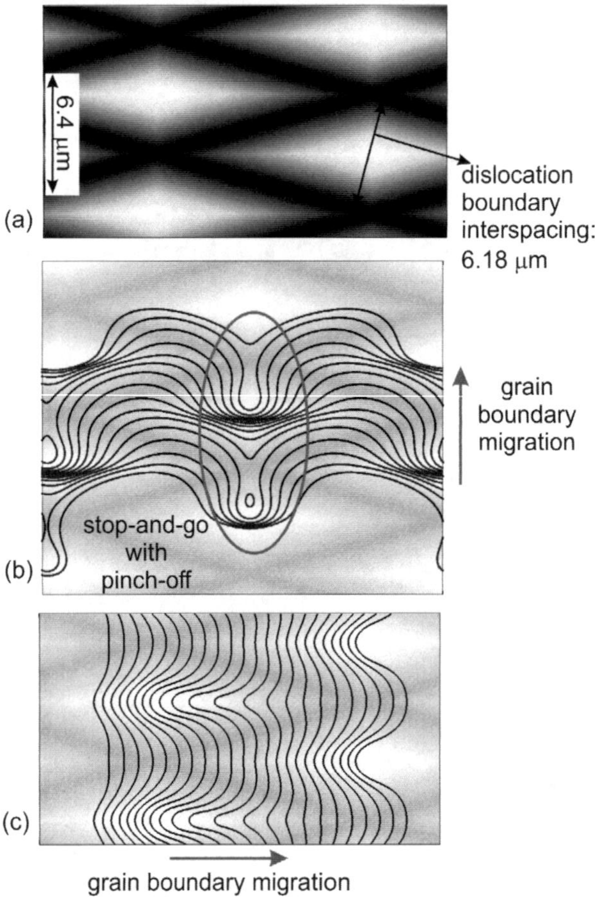

Figure 2. Simulation results obtained for a recrystallization boundary migrating in a deformation energy field with variations given by equations (2)-(4) with $A_{def} = 1.3416$ MJ/m^3, $\alpha = \pi/12$ and $\lambda_{def} = 6.18\mu$m. (a) Representation of the considered energy field using a color scaling with black representing the highest energy density and white an energy density equal to 0. (b), (c) Grain boundary positions as a function of time obtained in the simulations, plotted every 2000 time steps starting from time step 60000. (b) Boundary moving from bottom to top. Parts of the boundary, indicated with the red oval, show a stop-and-go type of movement with pinch-off: boundary parts near minima in the energy density field remain for a longer period at the same position, while the rest of the boundary continues to advance. At a certain point, the stagnated parts are pinched off, leaving behind a small part of deformed material. (c) Boundary moving from left to right. Asymmetrical protrusions and retrusions develop along the migrating recrystallization front with a considerably higher curvature at the retrusions than at the protrusions. Protrusions transform into retrusions and vice-versa in a continuous way while the grain boundary is migrating. (a), (b) and (c): Only part of the simulation domain is shown.

wavelength. Different, however, from the simulations with 1-dimensional variations, no steady-state is reached. Protrusions transform into retrusions and vice-versa during grain boundary migration due to the variations in stored deformation energy experienced by the boundary.

The boundary positions obtained every 2000 time steps in the simulations (starting from time

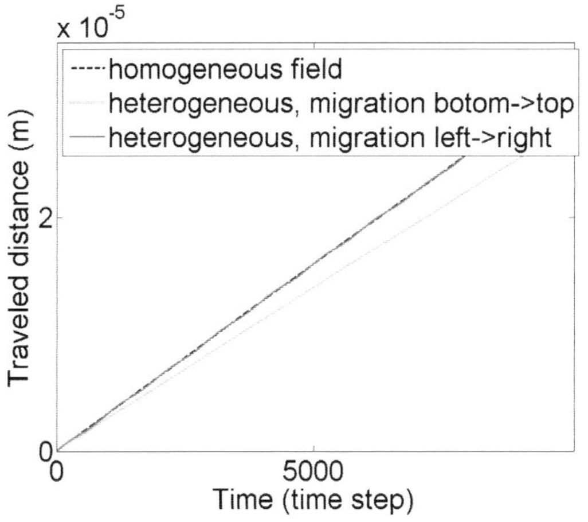

Figure 3. Average travel distance as a function of time obtained for recrystallization boundaries migrating into a varying energy field given by equations (2)-(4) with $A_{def} = 1.3416$ MJ/m^3, $\alpha = \pi/12$ and $\lambda_{def} = 1.66$ μm, in either the y- (bottom to top) or x-direction (left to right). The travel distance of a recrystallization boundary migrating in a homogeneous energy field with the same average energy density is also shown for comparison.

step 60000) with a 2-dimensional energy field described by equations (2)-(4) with $A_{def} = 1.3416$ MJ/m^3, $\alpha = \pi/12$ and $\lambda_{def} = 6.18\mu$m, are shown in figure 2. Again protrusions and retrusions clearly develop on the migrating recrystallization boundaries. For the boundary migrating in the y-direction, parts of the boundary located near a minimum in the stored deformation energy field stagnate and are pinched off from the boundary after some time. Because of the larger spacings between the dislocation boundaries, the moving parts of the boundary can bow out so widely that the parts of the moving grain boundary on both sides of the stagnated part can join and pinch off a part of the boundary, leaving behind an island of unrecrystallized material. The boundary migrating in the x-direction migrates smoothly. For this field, asymmetrical protrusions and retrusions with a much higher curvature at the retrusions than at the protrusions form along the boundary. Periodically, protrusions transform into retrusions and vice-versa.

Comparison of figures 1 and 2 shows that for one given value of the amplitude in the variations A_{def}, boundary movement becomes more irregular and the protrusions and retrusions become more asymmetric for increasing distance between the dislocation boundaries λ_{def}. The increasing asymmetry between the protrusion and retrusion curvature with increasing wavelength of the variations is consistent with our findings from the simulations with 1-dimensional variations. The stagnation of parts of the grain boundary at positions where the stored deformation energy is close to 0, was not observed in the simulations with 1-dimensional variations. From the considered cases, we conclude that stagnation of parts of the grain boundary occurs when the wavelength of the spatial variations along the grain boundary is larger than that of the variations in the direction of the movement. In contrast, when the wavelength of the variations in the direction of movement is larger than that of the variations along the grain boundary, the energy variations experienced by the grain boundary look similar to a 1-dimensional variation over a considerable distance and therefore protrusions and retrusions, with a shape very similar to those seen in the simulations with 1-dimensional variations, can develop.

Figure 3 gives the average distance travelled as a function of time of the migrating recrystallization boundaries considered in figure 1 ($\lambda_{def} = 1.66\mu$m). The graphs show that for this small distance between the dislocation boundaries and medium stored deformation energy, the stop-and-go behavior observed for parts of the boundary when migrating in the y-direction of the simulation domain, results in a smaller travelled distance at each time than expected (green curve) than expected from equation (1) with $P_{rex} = f_{def,av}$ used in classical recrystallization studies. For the boundary migrating in the x-direction, the curvatures at the protrusions and retrusions remain almost symmetric. Consistent with observations in the simulations with 1-

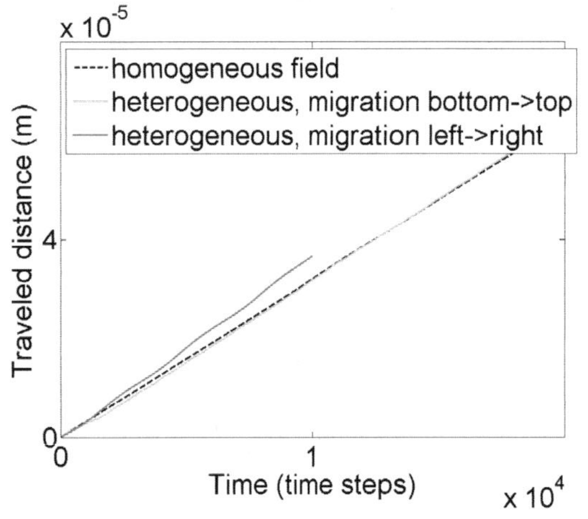

Figure 4. Average travel distance as a function of time obtained for recrystallization boundaries migrating into a varying energy field given by equations (2)-(4) with $A_{def} = 1.3416$ MJ/m^3, $\alpha = \pi/12$ and $\lambda_{def} = 6.18$ μm, in either the y- (bottom to top) or x-direction (left to right). The travel distance of a recrystallization boundary migrating in a homogeneous energy field with the same average energy density is also shown for comparison.

dimensional variations, the average displacement (magenta curve) then equals that of a boundary migrating in a homogeneous deformation energy field with the same average energy density, except for a small periodic variation in velocity, resulting from the continuous transformation of protrusions into retrusions and vice-versa.

Figure 4 gives the average travel distance as a function of time for the recrystallization boundaries considered in figure 2 ($\lambda_{def} = 6.18\mu$m). For this stored deformation energy field, the average displacement (green curve) of the grain boundary migrating in the y-direction as a function of time is close to that obtained for a recrystallization boundary moving in a homogeneous field with the same average energy density, although parts of this boundary show a stop-and-go type of migration. Because of the wider spacing in between the dislocation boundaries, the moving parts of the boundary can still advance considerably when other parts of the boundary are pinned. Moreover, the moving parts of the grain boundary can bow out sufficiently to pinch off the stagnated boundary parts, reducing the effect of the stagnation of these parts on the average grain boundary velocity. For the considered conditions, the resulting grain boundary velocity appears to be close to that observed for a homogeneous field with the same average energy density. It would be interesting, however, to study in detail as a function of the amplitude of the variations and the distance between dislocation boundaries how the stop-and-go type of movement of parts of a recrystallization boundary affect its average velocity. This will be done in a future investigation. On the boundary migrating in the x-direction, protrusions and retrusions with a considerably larger curvature at the retrusions develop. Consistent with the findings from our previous simulations with 1-dimensional variations, a larger average velocity (magenta curve) is obtained in this case compared to the case of a boundary migrating into a homogeneous deformation field with the same average energy density. The grain boundary velocity is thus considerably larger than that expected from standard recrystallization models. Since the curvatures at the retrusions result in an extra driving force in the direction of grain boundary migration, this higher velocity can be explained by the higher curvature at the retrusion tips than at the protrusion tips, resulting in a net curvature driving force in the direction of grain boundary migration. As the size of the protrusions and retrusions varies periodically in time, there is a small perodic variation in the velocity with time.

4. Conclusions

A previously developed phase-field model has been applied to study the migration behavior of recrystallization boundaries into deformation energy fields with 2-dimensional variations, representing 2 sets of dislocation boundaries, lying at equal, but opposite, angles to the moving recrystallization front. Compared to the results from previous simulations with 1-dimensional variations, more complex migration behavior, very similar to that observed in experiments, is obtained in these simulations. Nevertheless the conclusions from the simulations with 1-dimensional variations are still relevant.

The simulations show that the deformation microstructure and the orientation of the recrystallization boundary have a large effect on the shape and overall velocity of the migrating boundary. Although the grain boundary properties were chosen to be isotropic, very different migration behavior was obtained for migration in the x- and y-direction of the simulation system for the same deformation energy field. When the wavelength of the variations perpendicular to the grain boundary is larger than that of the variations along the boundary, smooth migration of the recrystallization boundary is observed with the development of protrusions and retrusions along the grain boundary. The protrusion/restrusion shapes are in this case very similar to those observed in the simulations with 1-dimensional variations. For a large enough distance between the dislocation boundaries, the curvatures obtained at the retrusions are much larger than those obtained at the protrusions, resulting in an average grain boundary displacement considerably larger than that obtained for a grain boundary moving in a homogeneous field with equal average stored deformation energy density. When the wavelength of the variations perpendicular to the grain boundary is smaller than that of the variations along the boundary, parts of the boundary undergo a stop-and-go type of movement. Moreover, if the distance between the dislocation boundaries is large enough, the stagnated parts of the recrystallization boundary can pinch off, leaving behind a small domain of unrecrystallized material. The stop-and-go type of movement of parts of the boundary generally results in a lower average grain boundary velocity.

These insights highlight the need for reliable experimental data on stored deformation energy on the scale of the local heterogeneities in deformation microstructures. The simulation results also motivate a detailed quantitative study of the effect of characteristics of 2-dimensional variations in stored deformation energy density on the shape and migration velocity of recrystallization boundaries.

Acknowledgments
NM and AG acknowledge support for Bilateral Scientific Cooperation between Tsinghua University & KU Leuven (Grant BIL10/19T). Simulations were performed using the VSC Flemish Supercomputer Center, funded by the Hercules foundation and the Flemish Government department EWI.

References
[1] Hansen N and Juul Jensen D 2011 *Materials Science and Technology* **27** 1229–1240
[2] Beck P and Sperry P 1950 *J. Appl. Phys.* **21** 150
[3] Schmidt S, Nielsen S F, Gundlach C, Margulies L, Huang X and Juul Jensen D 2004 *Science* **305** 229–232
[4] Xu W, Quadir M Z and Ferry M 2009 *Metall. Mater. Trans. A* **40A** 1547–1556
[5] Zhang Y B, Godfrey A, Liu Q, Liu W and Juul Jensen D 2009 *Acta Mater.* **57** 2631–2639
[6] Martorano M A, Sandim H R Z, Fortes M A and Padilha A F 2007 *Scripta Mater.* **56** 903–906
[7] Godfrey A, Juul Jensen D and Hansen N 1998 *Acta Mater.* **46** 823–833
[8] Godfrey A, Juul Jensen D and Hansen N 1998 *Acta Mater.* **46** 835–848
[9] Godfrey A, Hansen N and Juul Jensen D 2007 *Metall. Mater. Trans. A* **38A** 2329–2339
[10] Zhang Y B, Godfrey A and Juul Jensen D 2009 *CMC-Computers Materials & Continua* **14** 197–207
[11] Zhang Y B, Godfrey A and Juul Jensen D 2011 *Scripta Mater.* **64** 331–334
[12] Moelans N, Godfrey A, Zhang Y B and Juul Jensen D 2013 *Phys Rev B* **88** 054103
[13] Knudsen T, Cao W Q, Godfrey A, Liu Q and Hansen N 2008 *Metall. Mater. Trans. A* **39A** 430–440
[14] Moelans N, Blanpain B and Wollants P 2008 *Phys. Rev. B* **78** 024113
[15] Moelans N, Wendler F and Nestler B 2009 *Comp. Mater. Sci.* **46** 479–490
[16] Moelans N 2011 *Acta Mater.* **59** 1077–1086

About quantitative EBSD analysis of deformation and recovery substructures in pure Tantalum

C Moussa[1], M Bernacki[1], R Besnard[2] and N Bozzolo[1]

[1] MINES ParisTech, PSL - Research University, CEMEF - Centre de mise en forme des matériaux, CNRS UMR 7635, CS 10207 rue Claude Daunesse 06904 Sophia Antipolis Cedex, France.
[2] CEA DAM Valduc, F-21120 Is-sur-Tille

E-mail: charbel.moussa@mines-paristech.fr

Abstract. The aim of this work is to present a quantitative analysis of features involved in recovery during annealing of deformed Tantalum. In pure metals where crystalline defects usually have high mobility, dislocation annihilation and rearrangement occur to a great extent prior to recrystallization. Therefore a complete understanding of recrystallization cannot be accomplished without an advanced knowledge of the recovery phenomenon. Depending on whether dislocations induce a measurable curvature in the crystal lattice or not, they are called Geometrically Necessary Dislocations (GNDs) or Statistically Stored Dislocations (SSDs) respectively. In the present work only GNDs are considered. For this purpose electron backscatter diffraction (EBSD) is an advantageous technique to obtain statistically representative results when compared to Transmission Electron Microscopy (TEM). However, a quantitative analysis of GNDs from EBSD data is not straightforward. Since local misorientations are induced by the curvature of the crystal lattice caused by GNDs, GNDs analysis can be done using local misorientations. However the values obtained from this analysis are step size dependent and influenced by the measurement noise. Reasoning on the basis that when the step size tends to zero, local misorientation should also tend to zero, measurement noise can be estimated [1]. The measurement noise appears to notably be very much dependent on the amplitude of local misorientations, which must be considered in the perspective of GND density calculation.

1. Introduction

When a crystalline material undergoes a plastic deformation, the crystalline lattice rearranges itself to compensate the geometric changes in the material, which in most cases introduces line defects called dislocations. Plastic deformation of a polycrystal may appear homogenous at a macroscopic scale but can be very inhomogeneous when comparing two different grains [2] and even inside one grain [3].

The free energy of a crystalline material increases by the presence of dislocations [4] and therefore, dislocations do not only play a major role in plastic deformation but also in recovery and recrystallization phenomena. Knowledge of the density and distribution of dislocations is then necessary for the understanding of many physical mechanisms. Several methods allow for observing dislocations. A good review of these methods is presented in the book entitled "introduction to dislocations" [5], Transmission Electron Microscopy (TEM) and X-Ray diffraction are the most used. The major drawback of TEM methods is their local nature; they may lead to relatively high accuracy results but with only poor statistical relevance because they work at the nanoscale and the microscale. This can be a major issue since, at the polycrystal scale, deformation and dislocations distribution are inhomogeneous. With X-ray diffraction methods, dislocation densities are indirectly estimated from the shape of the diffraction peaks based on models, and their topological distribution can not be assessed.

Dislocations can be separated into two families, Geometrically Necessary Dislocations (GNDs) and Statistically Stored Dislocations (SSDs). When plastic deformation occurs, some dislocations allow for accommodating the lattice curvature [6,7] which develops as a result of non-uniform strain at the crystal scale, those are called GNDs by Ashby [8,9]. The SSDs are on the contrary dislocations with no cumulative effect on the crystal lattice curvature. From a practical point of view, the lattice curvature corresponds to the measurable local intragranular misorientation. It is worth pointing out that the accuracy of the orientation measurement and the spatial resolution have thus a direct influence on the discrimination between GNDs and SSDs.

The present study is a preliminary step before investigating static recovery during the annealing of a cold deformed Tantalum. As a BCC material, Tantalum exhibits a total of 48 slip systems: 12 {110} planes, 12 {112} planes and 24 {123} planes with <111> slip directions and a Burger vector magnitude $b=2.86\times10^{-10}m$. In pure metals where crystalline defects usually have high mobility, dislocation annihilation and rearrangement occur to a great extent prior to recrystallization. Tantalum has a relatively high stacking fault energy, $220\ mJ.m^{-2}$ [10], and recovery phenomena occur before recrystallization. Therefore a complete understanding of recrystallization cannot be accomplished without an advanced knowledge of the recovery phenomenon. For this purpose it is necessary to describe the dislocation substructures in the plastically deformed material and their evolution during subsequent annealing. In the present study, Electron BackScatter Diffraction (EBSD) technique is used to estimate the GNDs density and distribution.

2. Estimating GND density from EBSD data

The EBSD technique has been automated and intensively developed during the two last decades in terms of indexing speed and accuracy [11]. Local misorientations are induced both by the curvature of the crystal lattice associated with GNDs and by the effect of the elastic strain field. Since the effect of the elastic field is much smaller than the lattice rotation [12], one can assume that it is negligible and therefore the lattice curvature can be related to the presence of GNDs only. EBSD is an advantageous technique to estimate GND density based on the analysis of intragranular misorientations. A quantitative analysis of GNDs from EBSD data can be done with two main approaches.

The first approach is based on the calculation of the dislocation density tensor proposed by Nye [6]. From EBSD data obtained from a 2D section of the microstructure, it is possible to calculate directly three components of the Nye's tensor and to estimate two other components [13]. The four remaining components that are not experimentally accessible can then be estimated by minimizing the overall dislocation density [13–17]. The main drawback of this method is that multiple solutions can be found and the minimization of the total dislocation density to estimate the full Nye's tensor may be time consuming.

The second approach is simpler, based on the estimation of the GND density from the misorientation angle between measuring points located at a given distance x from each other. Based on the strain gradient model proposed by Gao et *al.* [18], Kubin and Mortensen [19] considered the case of a simple torsion and related the misorientation angle θ to GND density with the following equation:

$$\rho_{GND} = \frac{\alpha\theta}{bx} \tag{1}$$

where x is unit length and α is constant equal to 2 in the considered case. This formulation was also proposed by Liu et *al.* [20] with α equals 2 for a pure tilt boundary or 4 for a pure twist boundary (since a tilt misorientation can be accommodated by one set of edge dislocations while two sets of screw dislocations are necessary for a twist configuration). It is worth mentioning here that since the α parameter is used for estimating a 3D value (GND density) from a 2D data set (EBSD map), it implicitly encompasses also topological assumptions. The Kernel Average Misorientation (KAM), which is easily obtained from EBSD data, is the average misorientation angle of a given point with all its neighbors. KAM is generally used for the misorientation angle value θ in equation (1), it must then be calculated with a fixed neighbor distance, and not including all points of the kernel.

Comparisons between results obtained with the two approaches can be found in the literature, and results obtained with the second approach tend to slightly underestimate the results obtained with the first approach [21,22]. However, the second approach gives a more even distribution of GNDs [21]. This may be explained by the better statistics associated with the second approach. In fact, the maximum number of orientation gradients used to determine Nye's tensor are 3 as for KAM it is calculated as the average misorientation of 4 to 16 values depending on the choice of the neighboring regions [21].

The major drawback of estimating GND density from EBSD data is that the 3D information is calculated from a 2D measurements. However, comparing between EBSD 2D and 3D, Calcagnotto *et al.* [21] reported that GND estimation from EBSD 2D data is a reasonably reliable method. A second drawback is that the results are step size dependent and can sometimes be on the order of magnitude than the measurement noise. The measurement noise is closely linked to the quality of the Kikuchi patterns from which the orientation is determined. Among the most obvious factors controlling that quality, the material itself, the crystal orientation and misorientations, the quality of the sample surface preparation can be listed. EBSD acquisition settings also condition the pattern quality and the accuracy of the Kikuchi band detection: e.g. integration time, number of frames per point, resolution of the Kikuchi pattern image and resolution of the Hough space. The step size and the measurement noise influence are treated in the present study, for fixed acquisition settings.

3. Material, EBSD acquisition stings and data processing details

The studied material is a high purity Tantalum. Two samples were used. The first one, "Tantalum a", was submitted to a compression test at room temperature up to a strain of 36% and then annealed at 900°C for 2 hours. The second one, "Tantalum b", was cold-rolled to 92% reduction. The samples surfaces were prepared by standard mechanical polishing with 800, 1200, 2400 and 4000 grit SiC paper. Mechanical – chemical polishing was done using a colloidal silica solution with an average diameter of silica particles of 20 nm. Finally, vibration polishing for 2 hours was done using the same solution. EBSD maps were acquired using a QUANTAX EBSD system from the Bruker company (with a e-Flash^{HR} EBSD detector controlled by the ESPRIT software package), mounted on a Zeiss Supra40 FEG SEM operated at 20 keV.

The EBSD map of figure 1 (a) was obtained on "Tantalum a" with a step size of 1.44µm and an indexing rate of 98%. It reveals a recovered and partially recrystallized microstructure. KAM can be calculated using the first, the second …order neighbors. For the calculation of KAM only neighboring pixels with a misorientation angle lower than a threshold are considered in the calculation in order not to include misorientations due to grain boundaries. KAM with a 5° threshold angle is presented in figure 1(b). KAM is a good indicator of the local misorientation and, hence, of GND density. Even though the θ value of equation (1) is divided by the unit length, the choice of the kernel radius may have a big influence on the calculated GND density, depending on the characteristic length of the intragranular substructures. For quantitative GND analysis, a kernel size should be chosen even though that choice can hardly be *a priori* justified.

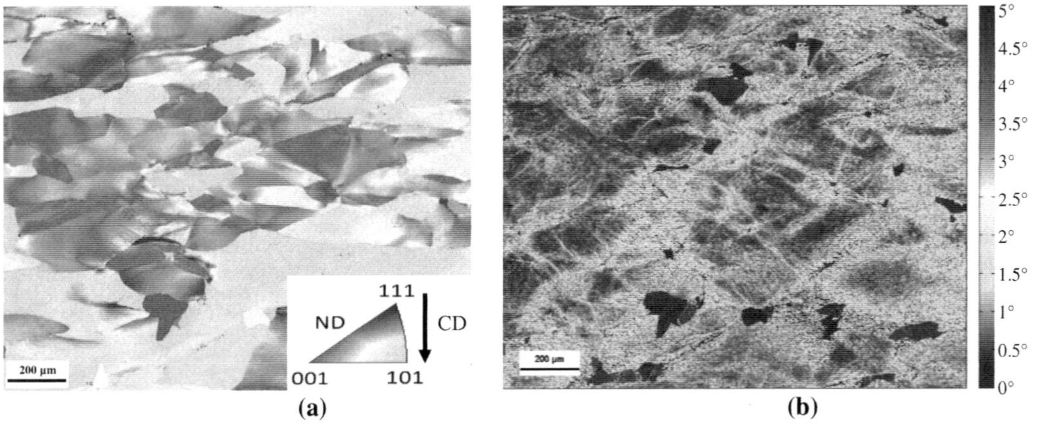

Figure 1. (a) Orientation color-coded EBSD map of a partially recrystallized Tantalum (normal direction to the analyzed section projected onto the standard triangle) and **(b)** kernel average misorientation to the 1st neighbor with 5° threshold angle.

Kamaya [1] proposed an interesting method for estimating the measurement noise. If the measurement noise is zero, local misorientation is only induced by the orientation gradient. The local misorientation should then tend to zero when the step size tends to zero. Using the EBSD data presented in figure 1(a), the average KAM values of all measurement points are calculated for the first five neighbors and presented as a function of the absolute kernel radius value in figure 2.

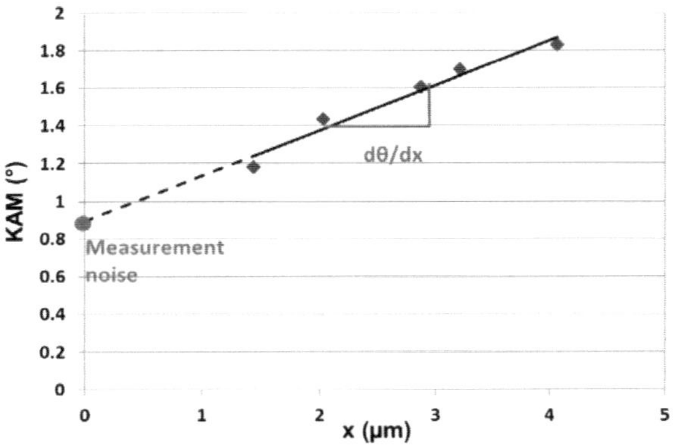

Figure 2. Average KAM as a function of the kernel radius (first to fifth neighbors).

The variation of the average KAM values with the kernel radius is linear and intersects the axis at 0.89° which is different from zero. This value can be considered as an estimate of the average measurement noise. For a fully recrystallized pure Tantalum sample, this value was determined to be about 0.5° under the same EBSD acquisition settings, which is fully consistent with the classical value for the EBSD angular resolution. It is worth mentioning that the variation of θ with x is not always linear. It is linear if the orientation gradient is regular at the kernel length scale. If dislocations form substructures that are smaller than the maximal considered kernel radius, then the KAM value may vary non-evenly with x. The slope of the linear evolution is the misorientation per unit length $d\theta/dx$ and can be used in equation (1) to replace θ/x. Thus no assumption is required for choosing the considered neighborhood range. Even more, Kamaya [1] demonstrated that $d\theta/dx$ is not affected by the step size of the EBSD map. The observation made in figure 2 is also valid on a local basis for each measurement point. Using equation (1) with $\alpha=2$, the GND density can be estimated, as an average value for the whole map, or as values associated to each individual pixel to build a GND density map.

4. Estimation of the GND density with removal of the measurement noise contribution

The measurement noise and GND density maps are presented in figure 3 (a) and (b) respectively. The minimal values of the measurement noise and GND density correspond to the few recrystallized grains (some arrowed on figure 3), as expected since the measurement noise is drastically affected by the crystalline lattice distortions, minimal in recrystallized grains. As it can be seen in Figure 3 (a) the measurement noise can go up to several degrees where the crystalline lattice is highly distorted.

The average GND densities calculated using eq. (1) with $\alpha=2$ and KAM (considering the first, second or third neighbors) or using Kamaya's method are presented in table 1. From these values, the influence of the choice of the kernel radius without applying Kamaya's method is obvious. On the other hand, the result obtained with Kamaya's method is smaller than the one calculated with KAM. This difference is mainly due to the measurement noise that leads to overestimating GND densities if not considered, and this is especially true when the crystal lattice is highly distorted due to a high GND density. The threshold angle has also an influence on the GND density as presented in table 1. This parameter should be chosen carefully, in our case with 15° exclusion angle some grain boundaries was accounted for intragranular misorientation which tends to increase GND density.

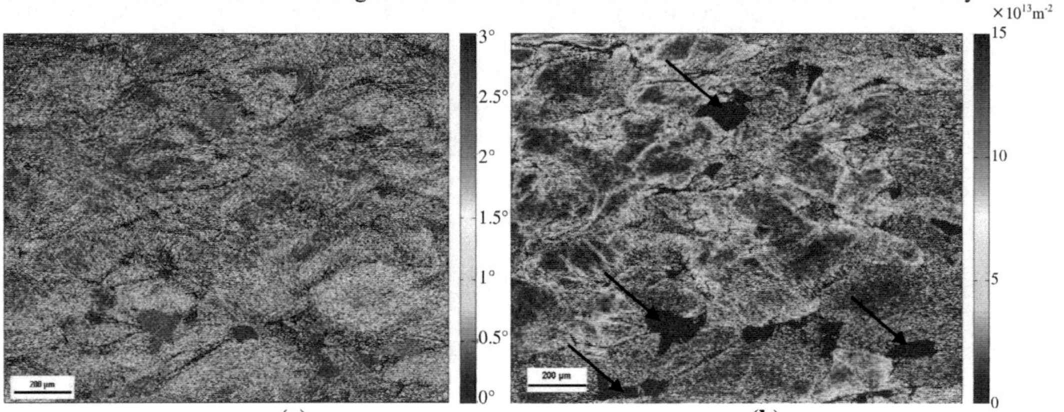

Figure 3. **(a)** measurement noise and **(b)** GND density maps.

Table 1. GND densities calculated using either KAM or Kamaya's method.

Threshold angle	KAM 1st neighbor	KAM 2nd neighbor	KAM 3rd neighbor	$\frac{d\theta}{dx}$
5°	$1\times10^{14}m^{-2}$	$1.22\times10^{14}m^{-2}$	$1.36\times10^{14}m^{-2}$	$3.12\times10^{13}m^{-2}$
15°	$1.1\times10^{14}m^{-2}$	$1.36\times10^{14}m^{-2}$	$1.56\times10^{14}m^{-2}$	$4.1\times10^{13}m^{-2}$

Hall-Petch analysis on an annealed Tantalum was done by Stüwe *et al.* [23]. The authors reported a critical dislocation density for spontaneous nucleation of 3×10^{14} m^{-2}. In figure 3 (b) we can see that the maximal GND densities are around 1.5×10^{14} m^{-2} which is in the same order of magnitude of the critical value reported by Stüwe *et al.*.

In order to verify the ability of Kamaya's method to characterize the GND cell structures, it is applied on "Tatalum b". An EBSD map was acquired with a step size of 38 nm and an indexing rate of 99%, see figure 4 (a). The corresponding measurement noise and estimated GND density are presented in figure 4 (b) and (c) respectively. The threshold angle was set to 15° in this case because of the highly disoriented cell structures that should be considered as intragranular. The cell structures can be clearly observed in figure 4 (b) and (c). The cell walls are regions with a high disorientation and high GND density, see figure 4 (c). Consistently, the cell walls are the regions with the highest measurement noise, see figure 4 (b). From figure 4 (c) one can see that the shape and size of cell structures is non-homogenous and depends on the grain orientation, as a result of crystal plasticity anisotropy. Once again we can see that the measurement noise can locally be much higher than the classical value of 0.5° for EBSD angular resolution and should be considered for a better accuracy of GND density estimation. A great advantage of Kamaya's method is that the measurement noise is estimated locally for each measurement point and not assumed to be uniform over the whole EBSD map.

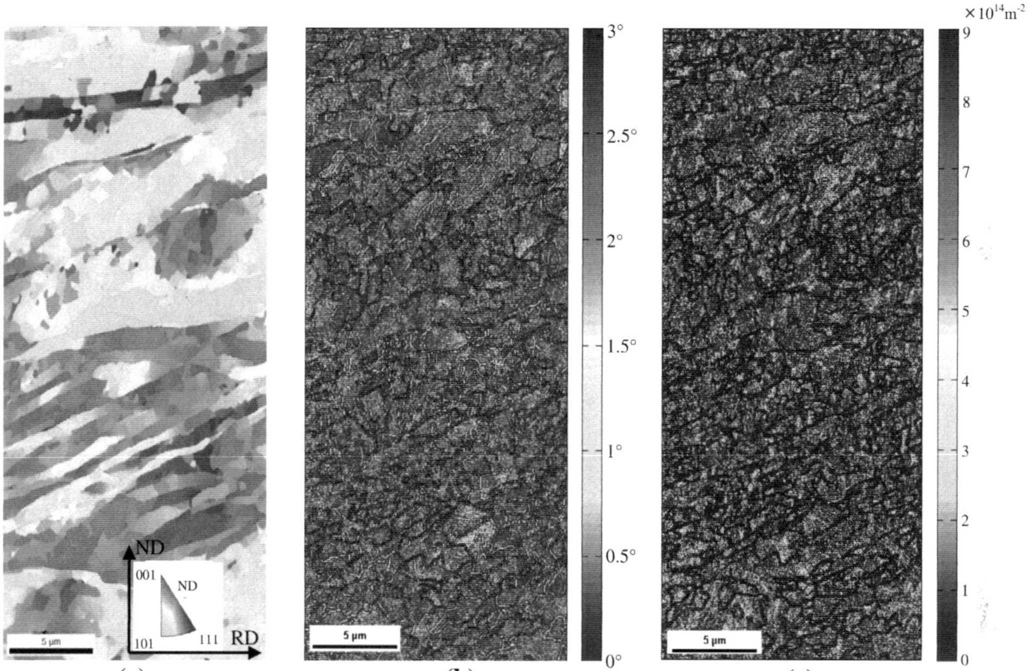

(a) **(b)** **(c)**

Figure 4. **(a)** ND inverse pole figure EBSD map of a 92% cold-rolled Tantalum sample, **(b)** measurement noise and **(c)** GND density maps

5. Conclusions

The present work evaluates the advantage of applying Kamaya's method for estimating GND densities from EBSD maps, in the frame of a further research work dedicated to recovery in pure Tantalum. This method is independent on step size and avoids the contribution of the measurement noise to be taken into account in the GND density calculation. The quantification of the GND density remains dependent on the choice of the factor α in equation (1), which corresponds to an assumption on the dislocation types. Nevertheless, the estimated values of critical dislocation density for nucleation of recrystallized grains in Tantalum are in good agreement with those reported in the literature.

References

[1] Kamaya M 2011 *Ultramicroscopy* **111** 1189
[2] Sun S, Adams B L and King W E 2000 *Phil. Mag. A* **80** 9
[3] Schwarts A J *et al* 1999 Journal of Engineering Materials and Technology **121** 178
[4] Humphreys F J and Hatherly M 2004 *Recrystallization and Related Annealing Phenomena* Elsevier Ltd (second edition) chapter 6 p 169
[5] Hull D and Bacon D J 2001 *Introduction to dislocations* ed BH (fourth edition) chapter 2 p 22

[6] Nye J F 1953 *Acta Metall.* **1** 153
[7] Kröner E 1958 *Kontinuumstheorie der Versetzungen und Eigenspannungen* (Berlin: Springer)
[8] Ashby M F 1970 *Phil. Mag.* **21** 399
[9] Gao H Huang Y 2003 *Scripta Mater.* **48** 113
[10] Pegel B 1968 *Physica Status Solidi (b)* **28** 603
[11] Dingley D 2004 *Journal of Microscopy* **213** 214
[12] Wilkinson A J and Randman D 2010 *Phil. Mag.* **90** 1159
[13] Pantleon W 2008 *Scripta Mater.* **58** 994
[14] Beausir B and Fressengeas C 2013 *Int. J. Solids Struct.* W 2008 *Scripta Mater.* **50** 137
[15] Demir E, Raabe D, Zaafarani N and Zaefferer S 2013 *Acta Mater.* **57** 559
[16] Field D P, Merriman C C, Allain-Bonasso N and Wagner F 2012 *Modelling Simul. Mater. Sci. Eng.* **20** 024007
[17] Ruggles T J and Fullwood D T 2013 *Ultramicroscopy* **133** 8
[18] Gao H, Huang Y, Nix W D, and Hutchinson J W 1999 *J. Mech. Phys. Solids* **47** 1239
[19] Kubin L and Mortensen A 2003 *Scripta Mater.* **48** 119
[20] Liu Q, Juul Jensen D, Hansen N 1998 *Acta Mater.* **46** 5819
[21] Calcagnotto M, Ponge D, Demir E and Raabe D 2010 *Mater. Sci. Eng. A* **527** 2738
[22] Jiang J, Britton T B and Wilkinson A J 2013 *Ultramicroscopy* **125** 1
[23] Stüwe H, Padilha A and Siciliano Jr F 2013 *Mater. Sci. Eng. A* **333** 361

IOP Conf. Series: Materials Science and Engineering **89** (2015) 012039 doi:10.1088/1757-899X/89/1/012039

Characterization of microstructural changes in near pearlitic steels using orientation imaging microscopy - influence of pre-deformation on local sensitivity to thermal degradation

D Nikas and J Ahlström

Department of Materials and Manufacturing Technology, Chalmers University of Technology, SE-412 96 Gothenburg, Sweden

Email: nikas@chalmers.se

Abstract. The focus of this study is the degradation of a near pearlitic microstructure under combined mechanical and thermal loadings leading to changes in mechanical properties. More specifically, it is examined how the orientation gradients inside the pearlite colonies, affect the spheroidisation. Samples were extracted from virgin near pearlitic railway wheels and pre-strained, thereafter heat treated at different temperatures. Microstructural characterization by scanning electron microscopy (SEM) and Electron Backscatter Diffraction Analysis (EBSD) was performed and evaluated. Results showed that spheroidised areas appear to have lost their initial orientation gradients after spheroidisation and obtain a more uniform orientation. More sub-grain boundaries are present after exposure to higher temperatures.

1. Introduction

During train operation, frictional heating occurs upon acceleration, braking, curving and occasional slippage. The temperature in the wheel tread can rise to 550 °C during freight operation with some extreme cases going up to 800 °C [1]. Apart from thermal loadings, plastic deformation in the wheel tread surface ("rim") changes the material microstructure and leads to an increased sensitivity to thermal damage [2]. Since the initial microstructure is predominantly pearlitic, the strength and wear properties of the material depend mainly on the interlamellar spacing of the pearlite [3, 4, 5]. Because of the loadings, spheroidisation of the pearlite is observed and is found to vary between different grains. One reason could be that plastic straining is inhomogeneous, which should give differences in dislocation density and thus orientation gradients in the pearlite colonies. It is known that pearlite colony size and other microstructural characteristics can affect pearlite spheroidisation [6, 7]. However, the influence of orientation gradients within the pearlite colonies on spheroidisation has not been studied extensively.

A previous work [8] has showed that orientation gradients exist within the pearlitic ferrite and also presumably within the cementite that suggests a density of geometrically necessary dislocations is present in the pearlite colonies. It was also shown that colonies within the same nodule have a similar interlamellar spacing [9].

The aim of the present study is to examine the microstructure of pearlitic wheel steels and evaluate how the orientation gradients of the pearlite colonies affect the spheroidisation of the lamellae in pre-strained material. The microstructural changes (and following effects on strength) after exposure to elevated temperature from 250 °C up to 650 °C for three different time durations (4 minutes – 4 hours) was investigated with scanning electron microscopy (SEM) in a previous study [10]. Although lamellae start to break up around 400 °C, spheroidisation is limited below 500 °C [10, 11]. After exposure at

650°C for sufficient time, most areas are spheroidised. In the range 500–650 °C there is a mixture of the above conditions. So these two extremes were considered suitable to study the spheroidisation behaviour of this material using orientation imaging analysis of the pearlite colonies in the material microstructure. This will also help in understanding the effect of microstructural changes on the mechanical properties of this type of steels.

2. Experimental

The material studied is the railway wheel steel grade ER8 according to the standard EN13262 [12]. Its nominal composition is shown in Table 1. Here the commonly used name R8T will be used. The "T" added to the designation marks that the wheels have been rim chilled during production; the wheel tread and flange are cooled with water jets after austenitisation. This creates a fine-pearlitic microstructure with some 7–10% pro-eutectoid ferrite close to the rim with a slight increase in free ferrite and interlamellar spacing with increasing depth [10].

Table 1. Chemical composition of R8T wheel material, maximum levels, in wt%.

	C	Si	Mn	Mo	Cr	Ni	S	P	V	Fe
R8T	0.56	0.40	0.80	0.08	0.30	0.30	0.015	0.020	0.06	Bal.

Samples were extracted from unused (''virgin") wheels from around 15 to 20 mm below the wheel tread surface; this material has a representative microstructure for volumes exposed to elevated temperatures and mechanical loading during operation. The flange and the side closer to the rim face were excluded as they have shown to have a variation in hardness due to different cooling rates experienced in the rim chilling process. Tensile bars with thickness 5 mm and "dog bone" shape (see figure 1, dimensions in mm) were taken from the same depth. These bars were pre-strained using an Instron electro-mechanic tensile machine to 6.5% longitudinal strain. Two extensometers were used to prove an even strain distribution. Tests were run in strain control at a strain rate of 10^{-4} s^{-1}. The extension and cross section in the waist were measured before and after some of the tests to verify pre-straining levels measured by the extensometers. Samples were taken from the waist and cut into pieces around 5x8x8 mm^3 large, to be used for heat treatment experiments and then microstructural analysis.

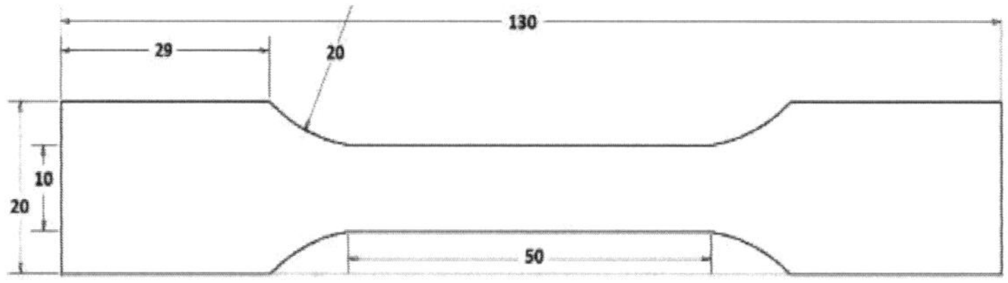

Figure 1. Pre-strain tensile specimen geometry (in mm)

Two samples heat treated for 4 hours were selected for EBSD analysis. The first was heat treated at 500 °C; here some regions are spheroidised while other pearlite colonies are still intact. The second sample was heat treated at 650 °C where most of the pearlite is almost fully spheroidised. Representative microstructures of the two temperatures are shown in figure 2.

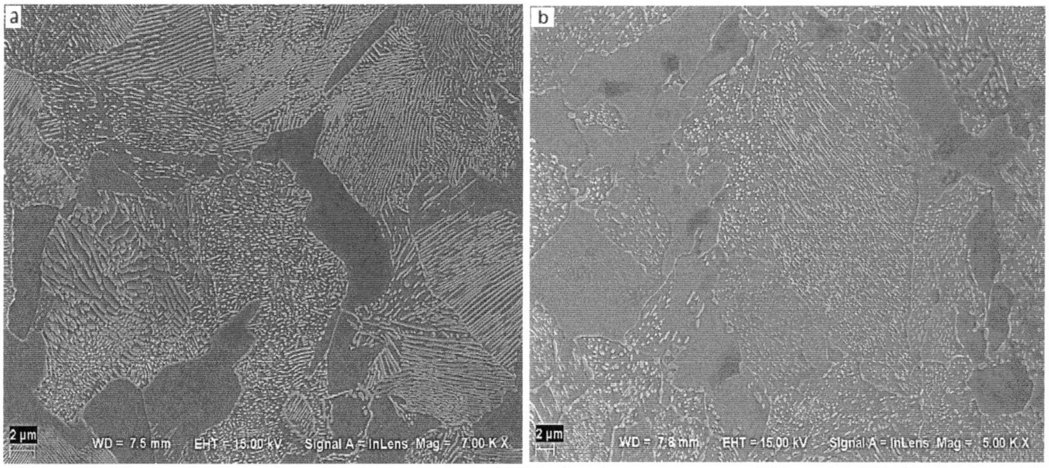

Figure 2. SEM micrographs of the pre-strained R8T after heat treatment (4h) at (a) 500 °C (b) 650 °C

The specimens were mechanically ground and polished to 0.04 µm using a colloidal silica suspension. Etching was done using Nital (3% HNO_3 in ethanol) to gain some topographical contrast and thus be able to map regions of interest. The EBSD measurements were carried out in a LEO 1550 high resolution field emission scanning electron microscope (FEG-SEM). The system was equipped with an EBSD detector (Nordlys, Oxford instruments), a high speed camera for EBSD pattern recording and software for crystal orientation mapping (AZtecHKL).The samples were tilted at 70° and the SEM was operating at an accelerating voltage of 20kV. Crystallographic orientation maps were taken on the pearlite colonies as well as on the pro-eutectoid ferrite with a step size of 130 nm which is around the interlamellar spacing of the pearlite in this material [11, 10]. The EBSD patterns of the pearlitic cementite were of insufficient quality and were thus omitted from the analysis.

One useful map type for this investigation is the kernel average misorientation map (KAM). This type of map reveals short range orientation gradients and allows to distinguish local orientation changes and sub-boundaries. For this calculation the average misorientation between a point on the measurement grid and its neighbours is measured. Misorientations between all neighbouring points within the kernel are then averaged and to exclude the grain boundary effect a 2°-5° sub-grain boundary angle was used as a filter to only focus on small rotations that could also indicate ä high dislocation density. For this study a square scanning grid was used as shown in figure 3 and from that the 2nd neighbour shell (5x5) was used to evaluate the material.

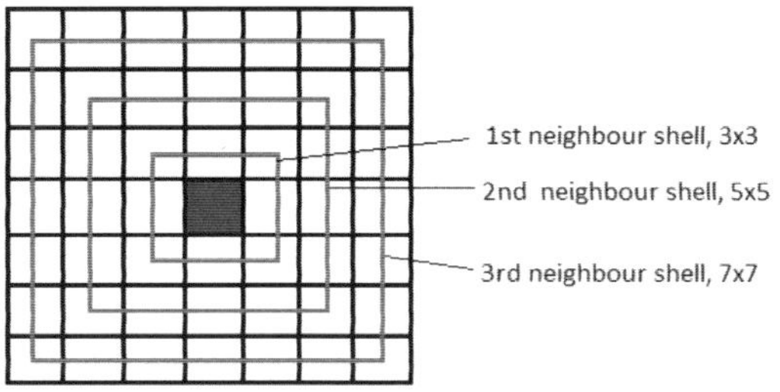

Figure 3. Kernel shells using a square scan grid

3. Results and Discussion

Figure 4a shows a SEM micrograph from the specimen heat treated at 500 °C. It is possible to find areas that are spheroidised as well as areas with apparently intact pearlitic colonies. The inverse pole figure map (figure 4b) reveals the crystallographic orientation of the grains with respect to the sample normal direction. From the colour coding of the IPF map it can be observed that the pro-eutectoid ferrite grains have random orientations with very few substructure boundaries and minor orientation gradients. Larger orientation gradients exist within the pearlite colonies. The bottom triangle shows the relative clustering density of the poles for this field of view, where red indicates a high cluster density and blue a low cluster density. The texture is mainly remaining from the production, as pre-straining to 6.5% does not give large rotations [13]. The grain boundaries can be observed better in figure 4c where only the boundaries are shown; thick lines indicate the large angle grain boundaries and thin/red lines the low angle grain boundaries. For this study the grain boundaries are separated into large angle boundaries with a misorientation higher than 10° and low angle boundaries with a misorientation between 2° and 10°. The pro-eutectoid ferrite grains are separated mostly by large angle boundaries. The intact and the spheroidised pearlitic areas on the other hand both contain low angle boundaries. The areas that appear grey in the IPF map are the unresolved pearlitic cementite regions that were omitted from the map.

The KAM map in figure 4d reveals the local misorientation of the sample. Since it was pre-strained there is a variation in kernel average misorientation. The intact pearlite colony in the middle of the image appears to have stronger local orientation gradients than the partially spheroidised area in the left. The ferrite grains seem to have smaller orientation gradients (marked with arrows).

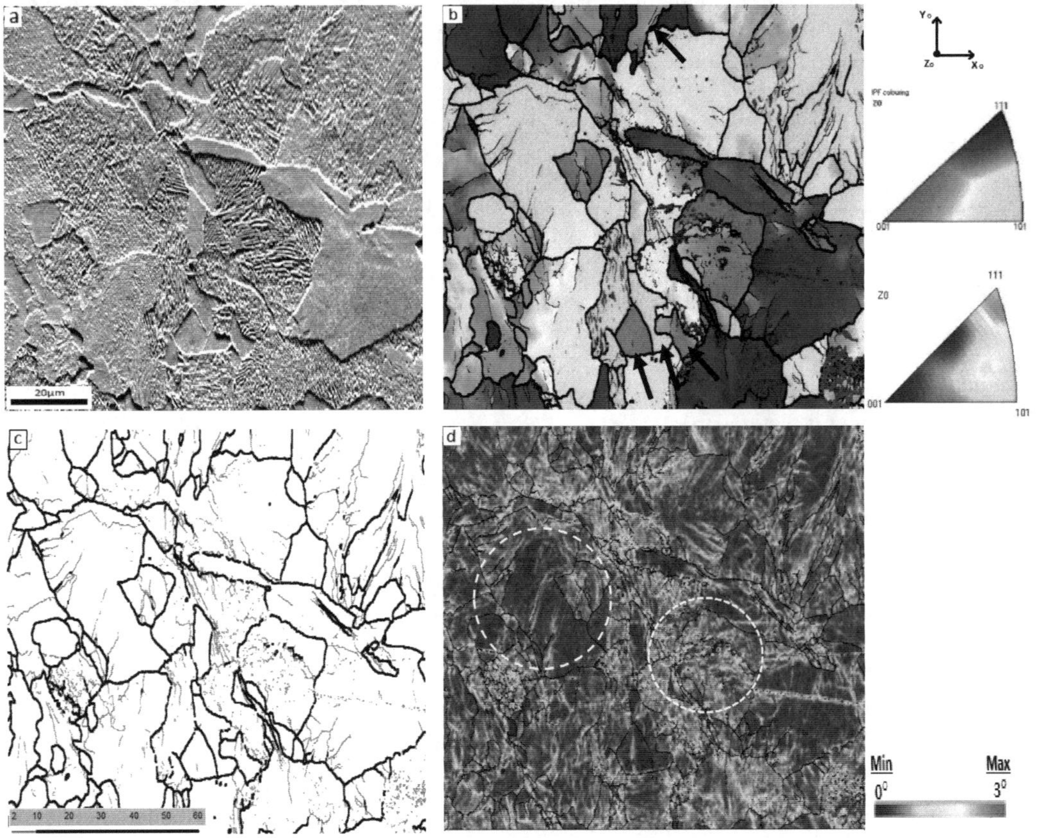

Figure 4. Microstructure of a sample heat treated at 500 °C (a) SEM micrograph (b) Inverse pole figure map (IPF) (c) Grain boundary map (d) Kernel average misorientation map using up to 2nd neighbour shell

In figure 5 the sample heat treated at 650 °C is shown. Strong spheroidisation is evident in most pearlitic areas of the microstructure (figure 5a). The inverse pole figure (figure 5b) reveals the crystallographic orientation of the grains with respect to the sample normal direction. Larger orientation gradients appear inside the pearlite colonies with less spheroidised microstructure (circled areas in IPF map). However, most of the spheroidised areas appear with even colours in the IPF map which indicates a more uniform orientation. It is also possible that during the annealing, carbon can diffuse from previous pearlite colonies to the pro-eutectoid ferrite grains and precipitate [14]. These grains would then appear as spheroidised pearlite colonies.

A slightly larger amount of sub-grain boundaries is also visible (red lines) but still in lesser extent inside the pro-eutectoid ferrite grains (figure 5c). The kernel average misorientation seen on the KAM map (figure 5d) is more equally distributed over the field of view. This correlates with a larger fraction of the microstructure being spheroidised. Some pro-eutectoid ferrite grains appear to have a lower local misorientation as in the 500 °C sample.

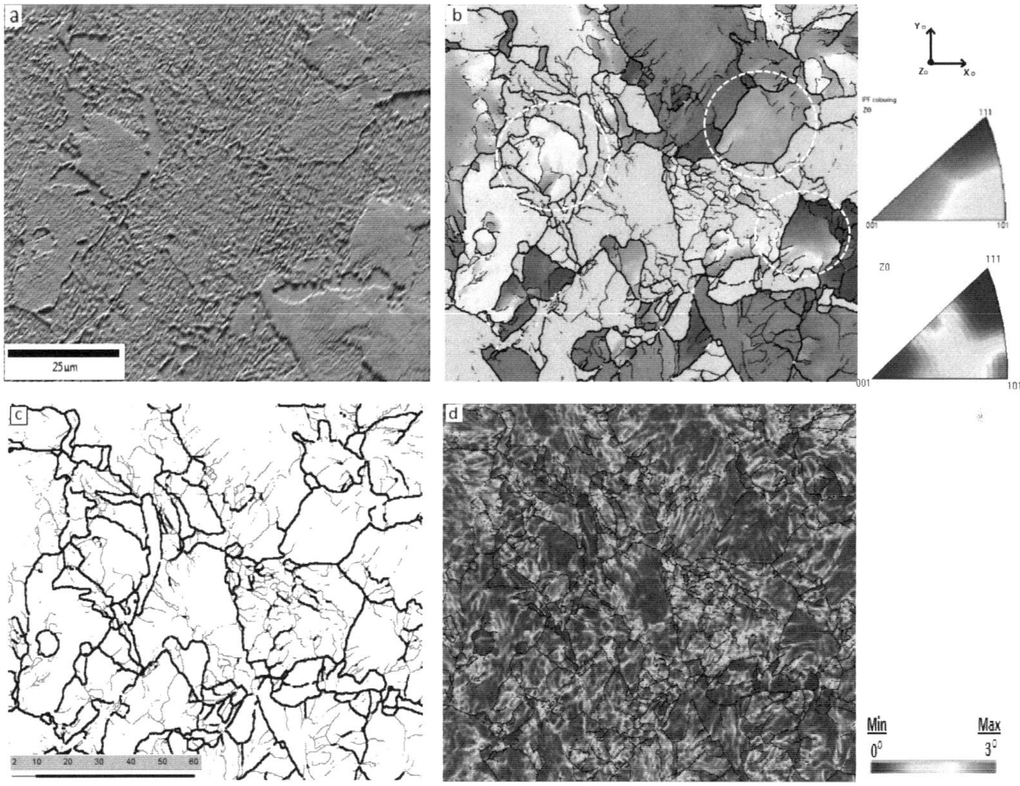

Figure 5. Microstructure of a sample heat treated at 650 °C (a) SEM micrograph (b) Inverse pole figure map (IPF) (c) Grain boundary map (d) Kernel average misorientation map using up to 2nd neighbour shell

4. Conclusions

The effect of pre-straining and high temperature annealing on the spheroidisation behaviour of a near pearlitic steel was evaluated. Samples were heat treated at various temperatures and durations. Two samples that were heat treated for 4 hours, at 500 °C and at 650 °C, were selected for EBSD examination.

1. Pearlite colonies appear to have orientation gradients presumably both from pre-deformation and initial formation of the pearlite while ferrite grains have a more uniform orientation.
2. Spheroidised pearlite colonies appear to have lost their initial orientation gradients and obtain a more uniform orientation after spheroidisation.
3. A higher annealing temperature introduces more sub-grain boundaries in the material, mostly in the pearlite colonies.

From this initial study it is concluded that further investigation is needed to fully understand the spheroidisation mechanism with respect to the orientation of the pearlite colonies. The next step would be to examine the undeformed specimens of the same material and compare between the two states. A more advanced technique to consider is in-situ EBSD with a heating element inside the SEM. The EBSD maps could then show the initial orientation state of the pearlite colonies and the orientation gradients after the spheroidisation along with intermediate stages.

References

[1] Walther F and Eifler D 2004 *Mater. Test.* **46** 158
[2] Cvetkovski K, Ahlström J and Karlsson B 2011 *Mater. Sci. Tech.* **27** 648
[3] Modi O P, Deshmukh N, Mondal D P, Jha A K,Yegneswaran A H, and Khaira H K *Mater. Charact.* **46** 347
[4] Hyzak J and Bernstein I 1976 *Metal. Trans. A* **7** 1217
[5] Clayton P and Danks D 1990 *Wear* **135** 369
[6] Tian Y L and Kraft R W 1987 *Metal. Trans. A* **18A** 1403
[7] Caruso M and Godet S 2010 **89-91** 79
[8] Takahashi T, Ponge D and Raabe D 2007 *Steel Res. Int.* **78** 38
[9] Guo N and Liu Q 2012 *J. Microscopy* **246** 221
[10] Nikas D, Ahlström J and Malakizadi A 2015 *Proc. 10th Int. Conf. on Contact Mechanics, (Colorado Springs)*
[11] Nikas D and Ahlström J 2014 *Proc. 35th Risoe Int. Symp. on Materials science (Roskilde)*
[12] European committee for standarization 2009 *Railway applications - Wheelsets and bogies - Wheels - Product requirements*
[13] Gurao N P and Suwas S 2014 *Sci. Rep.* **4** 5641
[14] Song D, Ponge D, Raabe D and Kaspar R 2004 *Acta Mater.* **53** 845

Distributions of energy storage rate and microstructural evolution in the area of plastic strain localization during uniaxial tension of austenitic steel

W Oliferuk[1] **and M Maj**[2]
[1]Bialystok University of Technology, Wiejska 45C, 15-351 Bialystok, Poland
[2]Institute of Fundamental Technological Research, PAS, Pawinskiego 5B,02-106 Warsaw, Poland

E-mail: wolif@ippt.gov.pl

Abstract. The presented work is devoted to an experimental determination of the energy storage rate in the area of strain localization. The experimental procedure involves two complementary techniques: i.e. infrared thermography (IRT) and visible light imaging. The results of experiments have shown that during the evolution of plastic strain localization the energy storage rate in some areas of the deformed specimen drops to zero. To interpret the decrease of the energy storage rate in terms of micro-mechanisms, microstructural observations using Transmission Electron Microscopy (TEM) and Electron Back Scattered Diffraction (EBSC) were performed. On the basis of microstructural studies it is believed that a 0 value of energy storage rate corresponds to the state in which only two dominant components of the texture appear, creating conditions for crystallographic shear banding.

1. Introduction

When a material deforms plastically, a part of the mechanical energy w_p expended on plastic deformation is converted into heat q_d while the remainder e_s is stored within the material as an energy related to defects of crystal lattice, mainly dislocation structure:

$$e_s = w_p - q_d.$$

The stored energy represents a change in internal energy of the deformed material and it is an essential measure of its cold-worked state. This energy was discovered in calorimetric tests performed by Taylor and Quinney [1]. Further experimental studies have shown that the ratio of the stored energy to the plastic work depends on the deformation level of the tested material [2–14]. Thus a measure of energy conversion at each instant of the plastic deformation is the rate of energy storage Z defined as the ratio of the stored energy increment Δe_s to the plastic work increment Δw_p:

$$Z = \frac{\Delta e_s}{\Delta w_p}.$$

The stored energy increment is equal to the difference between Δw_p and the increment of energy dissipated as heat, Δq_d, $\Delta e_s = \Delta w_p - \Delta q_d$. Therefore:

$$Z = \frac{\Delta e_s}{\Delta w_p} = 1 - \frac{\Delta q_d}{\Delta w_p}. \tag{1}$$

The deformation process, from the initial state to the fracture of the tested specimen, consists of two stages: homogeneous deformation and a heterogeneous one. There are two indicators of the appearance of plastic strain localization: a non-uniform temperature distribution and a non-uniform strain on the surface of the tested specimen.

In the previous works of Oliferuk and co-workers it has been shown that during heterogeneous deformation (localization of plastic strain) of polycrystalline material, the energy storage rate rapidly decreases reaching the 0 value and then becomes negative [13, 15]. But up to now, only the average value of the energy storage rate for the gauge part of deformed specimen was estimated. The questions appear: a) What is the energy storage rate distribution along the gauge length of the strained specimen during development of plastic strain localization? b) What is the relation between the distribution of energy storage rate and microstructural evolution? The purpose of the present work is to answer these questions. To reach the purpose a new method of energy storage rate determination has been designed. The method allows one to obtain the distribution of this quantity along the gauge length of the specimen. The results obtained are related to the microstructure evolution.

2. The method of determining distribution of energy storage rate

The energy storage rate can be calculated by a determination of the plastic work increment Δw_p and the corresponding increment of energy dissipated as heat Δq_d, see (1). An experimental technique for simultaneous measurements of temperature, and strain distributions on the surface of tested specimen during tensile deformation was used [16]. This procedure involves two complementary imaging techniques: a CCD camera and infrared thermography (IRT). In order to determine the strain distribution, markers in the form of graphite dots were plotted on one surface of the specimen (figure 1). In this way, the surface was divided into sections, whose sizes are specified by the distance between the dots.

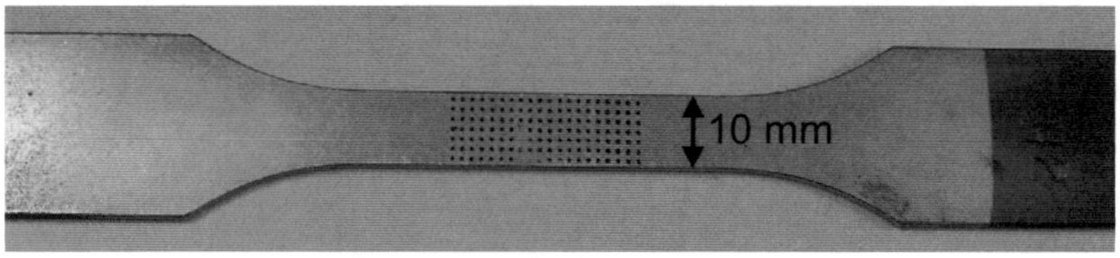

Figure 1. The graphite dots on the gauge part of the specimen.

Displacements of the dots were recorded by a CCD camera during the deformation process. Taking into account the reference distance l_0 between the centers of the dots and recording the distance between these centers as a function of deformation time and vertical coordinate y_n of the centre of the section n, $l(t, y_n)$, the true local strain $\varepsilon_y(t, y_n)$ in the direction of tension was calculated:

$$\varepsilon_y(t, y_n) = \ln\left(\frac{l_y(t, y_n)}{l_0}\right).$$

The true stress σ_y was determined by dividing the load by the current cross-sectional area of the specimen corresponding to the given section. Assuming that a dependence of Young's modulus E on strain is negligibly small, the local plastic strain $\varepsilon_y^p(t, y_n)$ was obtained from the following formula:

$$\varepsilon_y^p(t, y_n) = \varepsilon_y(t, y_n) - \frac{\sigma_y(t, y_n)}{E}.$$

The strain and stress distributions along the axis of the tested specimen were used to calculate the surface distribution of specific plastic work

$$w_p(t, y_n) = \frac{1}{\rho(t, y_n)} \int_0^{\varepsilon_y^p(t,y_n)} \sigma_y(t, y_n)\, \mathrm{d}\varepsilon_y^p, \tag{2}$$

where $\rho(t, y_n) = \frac{m_s}{V(t,y_n)}$ is the current mass density of a particular section, where $m = \mathrm{const} = \rho_0 \cdot V_0 = \rho(t, y_n) \cdot V(t, y_n)$ is the mass of a section and $V(t, y_n)$ is the current volume of a particular section.

The energy dissipated as a heat causes an increase in the temperature of the specimen. This increase depends on the strain rate. In the area of the strain localization, the distribution of the strain rate is not uniform. This heterogeneity is manifested by a non-uniform temperature field on the specimen surface. The temperature field during the deformation process was recorded by means of an IR thermography system and the average temperature $T(t, y_n)$ as a function of time for each n-th section along the axis of the specimen was determined. Such a selection of positions of sections allows avoiding the influence of the specimen edges on the measured temperature values. The initial size of each, considered section is identified by the distance between the centres of the dots and the thickness of the specimen. Thus, the initial volume V_0 of each section was equal to $(1.3 \times 1.3 \times 2)\,\mathrm{mm}^3$ and the initial cross-section S_0 was $(1.3 \times 2)\,\mathrm{mm}^2$. Dimensions of the sections were changing during the tensile process. These changes were determined by tracking dot centres in the visible range by means of the CCD camera. All these data have been used to calculate the heat $\Delta q_d(t, y_n)$ generated during a time interval Δt in the n-th section. The interval Δt is the reciprocal of the sampling frequency. In order to avoid the influence of the heat convection phenomenon, the displacement rate $2000\,\frac{\mathrm{mm}}{\mathrm{min}}$ was chosen. Nevertheless, despite the relatively high displacement rate the deformation process was non-adiabatic mainly due to the heat transport to the grips of the testing machine. Thus, taking into account the phenomena considered above and the heat expended for compensation of the drop in temperature caused by the thermo-elastic effect, the heat $\Delta q_d(t, y_n)$ generated during time Δt in the n-th section has been calculated according to the following formula:

$$\begin{aligned}\Delta q_d(t, y_n) = c_w \cdot \Delta T(t, y_n) + \frac{\alpha T(t, y_n)\Delta\sigma_y(t, y_n)}{\rho(t, y_n)} + \\ + \frac{S_{n/n+1}(t, y_n) \cdot \lambda \cdot \Delta t}{\rho_0 \cdot V_0} \cdot \frac{T(t, y_n) - T(t, y_{n+1})}{y_{n+1} - y_n} + \\ + \frac{S_{n/n-1}(t, y_n) \cdot \lambda \cdot \Delta t}{\rho_0 \cdot V_0} \cdot \frac{T(t, y_n) - T(t, y_{n-1})}{y_n - y_{n-1}},\end{aligned} \tag{3}$$

where α is the coefficient of linear thermal expansion, $T(t, y_n)$ is the mean absolute temperature of the section considered, $\Delta\sigma_y(t, y_n)$ is the increase in the stress during subsequent time intervals Δt, λ is the coefficient of thermal conductivity, c_w is the specific heat and V_0 is the volume of the considered section. $S_{n/n+1}(t, y_n)$ and $S_{n/n-1}(t, y_n)$ are cross-section areas between neighbouring sections $n/n+1$ and $n/n-1$. The first component of equation (3) expresses the heat needed to raise the temperature of a unit mass of the tested material by ΔT. The second describes the heat expended for compensation of the temperature drop due to the thermo-elastic effect. The third and fourth components, according to Fouriers law, take into account the heat transfer between the given section n and its neighbours $n/n+1$ and $n/n-1$, respectively. It is worth noticing that Fouriers law refers to a heat flux. Thus, in order to determine the heat related to a mass unit, the heat flux should be divided by the mass of the section and multiplied by Δt and by $S_{n/n+1}(t, y_n)$ or $S_{n/n-1}(t, y_n)$ respectively.

Adding up the heat increments $\Delta q_d(t, y_n)$ in the successive time intervals, the time dependence of the energy dissipated as heat $q_d(t, y_n)$ for each tested section has been obtained. The energy storage rate for a particular section is defined as:

$$Z = 1 - \frac{\Delta q_d(t, y_n)}{\Delta w_p(t, y_n)}. \tag{4}$$

During non-uniform deformation in the equal time intervals Δt, the increments of plastic work Δw_p are not equal. Thus, Δq_d was determined for an equal increments of Δw_p. In further analysis $\Delta w_p = 2\frac{\mathrm{J}}{\mathrm{g}}$ was taken. Such a value was selected in order to reduce the dispersion of ratio $\frac{\Delta q_d}{\Delta w_p}$. Having $w_p(t, y_n)$ and $q_d(t, y_n)$ for the considered sections and using the procedure mentioned above the distribution of the energy storage rate was determined.

3. Experiments and results

Experiments were performed on austenitic stainless steel (304L). From sheets of the annealed material the specimens for tensile testing were cut out using electro-spark machining. In order to remove the technological surface layer, the specimens were electro-polished. Optical metallographic and TEM observations of tested material indicated complete recrystallization (figure 2) and an average grain size of about 50 μm. The dislocation density in the specimens before deformation was low; dislocations were randomly distributed in the matrix and in the grain boundaries, and no regular dislocation arrangements were observed. Grains were randomly oriented (figure 3).

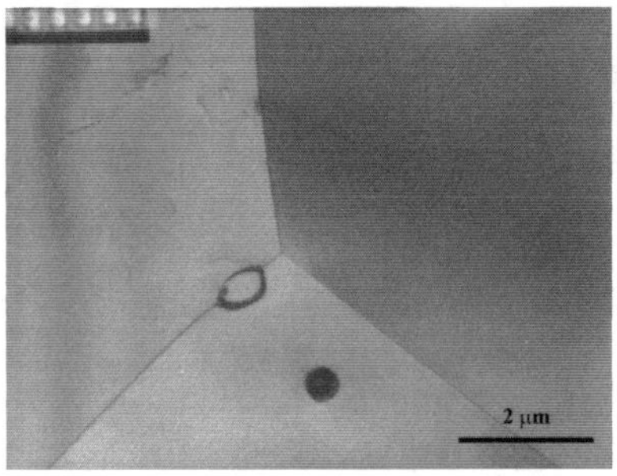

Figure 2. Typical microstructure of the initial state of the tested steel.

Figure 3. Distribution of grain orientation in a specimen before deformation.

All specimens were strained using an MTS 858 testing machine. During the tension the displacement rate was controlled and equal to 2000 $\frac{\mathrm{mm}}{\mathrm{min}}$. For such a displacement rate and given geometry of the specimen the corresponding mean value of the strain rate was equal to 0.6 s^{-1}.

In the course of the deformation process infrared and visible range image sequences were recorded simultaneously. The experimental setup is presented in figure 4.

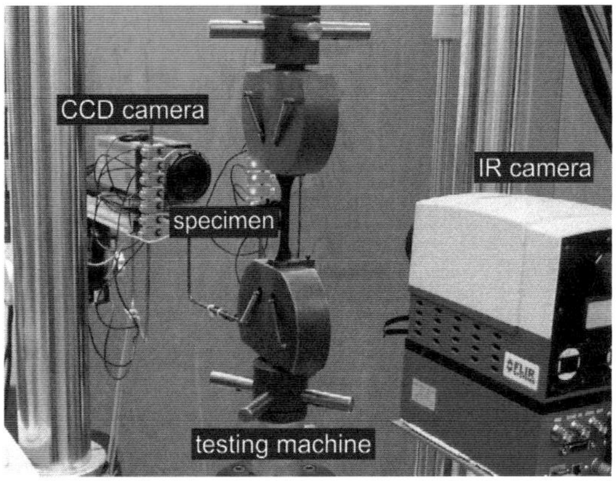

Figure 4. Experimental setup for simultaneous measurement of displacement and temperature fields.

The taking frequency of both cameras was equal to 538 Hz. Such a frequency was selected in order to take into account the heat transfer between neighbouring sections. Simultaneously, the straining force as a function of time t was measured. The analysis of the displacement fields (based on markers tracking) and the assignment of them to the appropriate temperature distributions were performed using MATLAB software. From the experimental data, tensile curves for selected local sections: A, B, C, D, E, lying on the axis of the specimen, were calculated (figure 5).

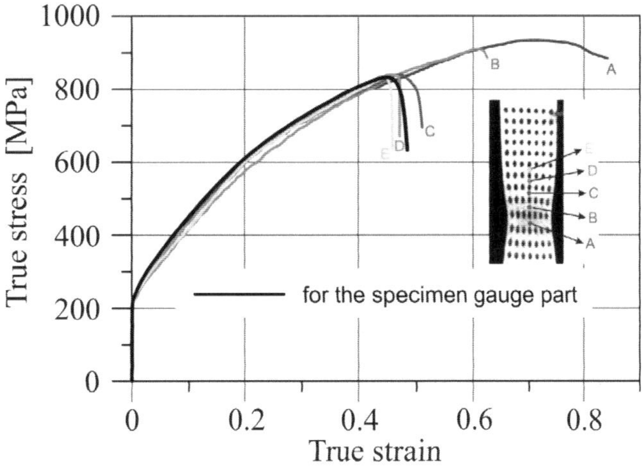

Figure 5. Stress-strain curves for selected sections of the specimen.

It is easy to see that with an increase in strain, individual sections of the specimen stop to deform, whereas others are still deforming. This is a result of the localization of plastic deformation. On the basis of the stress-strain dependences, the plastic work as a function of time for the selected sections lying on the specimens axis were calculated (equation 2). From the surface temperature field, the distribution of energy dissipated as heat along the specimen axis was determined using a local form of the heat equation (equation 4). Then, using formula (1) the energy storage rate Z for each of the considered sections was obtained. The results, as a function of true strain, are presented in figure 6. The results obtained for different sections are marked by different symbols with different grey intensities, corresponding to the stress-strain curves shown in figure 5. It is easy to see that the energy storage rate decreases with strain for all the considered sections. During the evolution of plastic strain localization some sections cease to deform, which causes a halt in the energy storage process in this region, whereas the energy storage rate in the other sections (where deformation still proceeds) drops to zero.

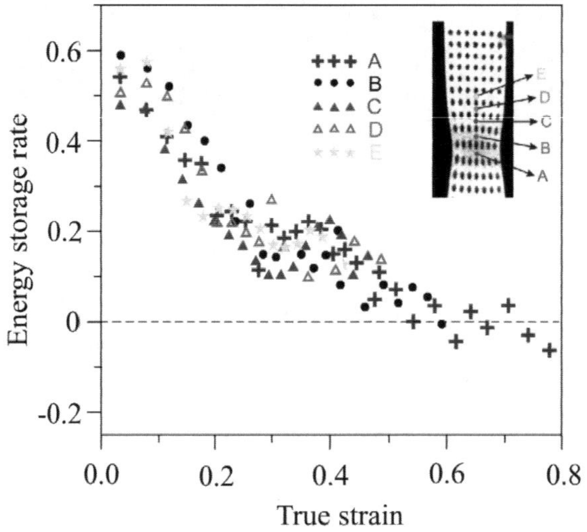

Figure 6. Energy storage rate as a function of true strain for selected sections of the specimen.

The zero value of the energy storage rate can be treated as a plastic instability criterion based on energy conversion. Negative values of this quantity could be explained by a release of the energy stored in previous deformation stages.

To identify micro-mechanisms corresponding to the appearance and evolution of plastic strain localization microstructural characterization was performed by transmission electron microscopy (TEM), and electron back scattered diffraction (EBSD) with orientation imaging microscopy. Microstructural observations were performed on specimens in different states of deformation namely: at the onset of strain localization and in the state just before fracture. The results are presented in figures 7–10.

The non-uniform strain distributions was used as an indicator of the onset of plastic strain localization. The characteristic elements of the microstructure corresponding to this state of deformation are deformation bands activated by deformation twins in the neighbouring grains (figure 7) and evolution of grain orientation in the direction of the two dominant components of the texture (figure 8).

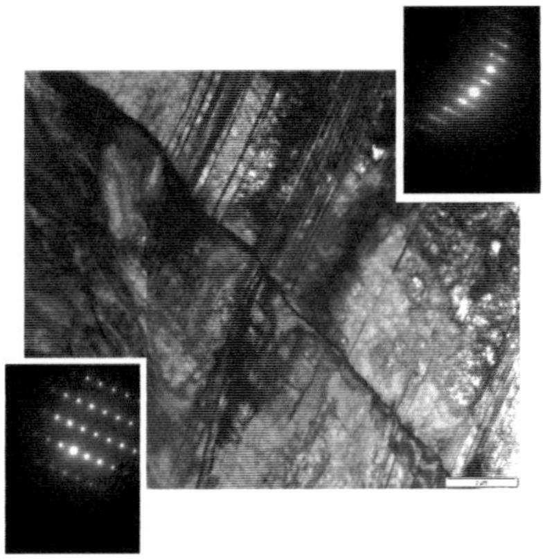

Figure 7. Typical microstructure of the onset of plastic strain localization zone.

Figure 8. Distribution of grain orientation at the onset of plastic strain localization zone.

The microstructure of the material, just before the fracture of the stretched specimen is shown in figure 9.

Figure 9. Typical microstructure of the zone of advanced plastic strain localization.

It is easy to see two sets of intersecting deformation twins in the background of the dislocation cell structure. It means that the contribution of twinning in the deformation process is growing. The development of strain localization is accompanied by further rotation of individual grains in the direction of the two texture components (figure 10).

Rotation of grains proceeds in such a way, that the {111} type planes become parallel to the planes of maximum shear stress. The angle between the trace of these planes and the tension direction is about 50° (figure 10). This type of texture seems to be a condition for the formation of shear bands, which corresponds to the loss of stability of plastic deformation. The above analysis shows that in the necking zone, just before the fracture (see figure 10), the macroscopic shear of the localized area is possible without a change of a crystallographic plane. In other

Figure 10. EBSD image for advanced localization of plastic strain (just before fracture).

words, the conditions for the crystallographic shear banding are created [17]. The macroscopic manifestation of such a state is the zero value of the energy storage rate. The preliminary results presented have shown, the that the zero value of the energy storage rate may be treated as the criterion of plastic instability.

4. Conclusions

The experimental determination of the energy storage rate distribution in the area of heterogeneous has been presented.

In the plastic strain localization area the material reaches the state, where the energy storage rate reaches the 0 value and then becomes negative. This means that the material loses an ability to store the energy. Although an energy supply to the specimen, its internal energy decreases, thus, the 0 value of the energy storage rate could be regarded as the plastic instability criterion based on the energy conversion.

On the basis of microstructural observations it is believed that the 0 value of the energy storage rate corresponds to the state in which only two dominant components of the texture appear, creating conditions for crystallographic shear-banding.

References

[1] Taylor G, Quinney H 1934 *Proc. R Soc.* **143** 307
[2] Bever M B, Holt D L, Titchener A L 1973 *Progress. Mater. Sci.* **17** 23
[3] Oliferuk W, Gadaj S P, Grabski M W 1984 *Mater. Sci. Eng.* **70** 131
[4] Chrysochoos A, Maisonneuve O 1985 *C R Acad. Sci.* **300** 985
[5] Oliferuk W, witnicki W, Grabski M W 1993 *Mater. Sci. Eng.* **A161** 55
[6] Chrysochoos A, Maisonneuve O, Martin G, Caumon H, Chezeaux J C 1989 *Nuclear Eng. Des.* **114** 323
[7] Oliferuk W, witnicki W, Grabski M W 1995 *Mater. Sci. Eng.* **A197** 49
[8] Oliferuk W, Korbel A, Grabski MW 1997 *Mater. Sci. Eng.* **A224-226** 1122
[9] Rittel D 1999 *Mech. Mat.* **31** 131
[10] Chrysochoos A, Louche H 2000 *Int. J. Eng. Sci.* **38** 1759
[11] Hodowany J, Ravichandran G, Rosakis A J, Rosakis P 2000 *Exp. Mech.* **40** 113
[12] Oliferuk W, Korbel A, Bochniak W 2001 *Mater. Sci. Eng.* **A319-321** 250
[13] Oliferuk W, Maj M, Raniecki B 2004 *Mater. Sci. Eng.* **A374** 77
[14] Oliferuk W, Maj M 2004 *Mater. Sci. Eng.* **A387-389** 218
[15] Oliferuk W, Maj M, 2007 *Mater. Sci. Eng.* **A462** 363
[16] Chrysochoos A, Wattrise B, Muracciole J M, El Kaim Y 2009 *J. Mech. Mater. Str.* **4** 245
[17] Paul H, Driver J H 2007 *Mater. Sci. Forum* **550** 521

IOP Conf. Series: Materials Science and Engineering **89** (2015) 012041 doi:10.1088/1757-899X/89/1/012041

Grain refinement of 2Mn-0.1C steel by repetitive heat treatment and recrystallization

M H Park[1], A Shibata[1,2], N Tsuji[1,2]

[1] Department of Materials Science and Engineering, Kyoto University, Sakyo-ku, Kyoto, 606-8501, Japan
[2] Elements Strategy Initiative for Structural Materials (ESISM), Kyoto University, Sakyo-ku, Kyoto, 606-8501, Japan

Email: park.heom.25v@st.kyoto-u.ac.jp

Abstract. Grain refinement in metals is well-known as one of the most effective methods to enhance their strength without addition of other elements. In this study, repetitive heat treatment combined with subsequent cold-rolling and recrystallization were investigated to obtain ultrafine-grained ferrite. Ultrafine-grained (UFG) ferritic structure having a mean grain size smaller than 1 μm was fabricated by repetitive heat treatment at 810 °C for 180 s and cold rolling by 90 % plus a recrystallization heat treatment at 600 °C. Starting from this UFG ferrite, fine-grained dual phase (DP) steel composed of ferrite and martensite phases with grain sizes smaller than 3 μm could be also obtained by intercritical heat treatment at 740 °C for 30 minutes followed by water-quenching. The mechanical properties of the ultrafine-grained ferritic and DP structured specimens were evaluated by tensile test. Results of the tensile test showed that fine-grained ferritic and DP structures had higher yield and tensile strength than the coarse-grained ferritic and DP structures of the same steel.

1. Introduction

Dual phase (DP) steels composed of ferrite and martensite phases are widely used for sheet materials in the automotive industry due to their good mechanical properties, such as high strength, adequate ductility and good formability. It is generally known that several factors, such as grain size of each phase, volume fraction and distribution of martensite, change the mechanical properties of DP steels. Among them, it is expected that grain refinement is an effective way to further improve the mechanical properties of DP steels. Recently, several studies reported that fine-grained DP steels showed excellent mechanical properties compared with DP steels with conventionally coarse grain sizes [1-5]. In order to fabricate bulky metals with ultrafine-grained structures, severe plastic deformation (SPD) processes, such as Equal Channel Angular Pressing, Accumulative Roll-Bonding and High Pressure Torsion, have been widely used [6-9]. The ultrafine-grained DP steels have also been obtained successfully using SPD processes [10, 11]. However, it seems difficult to apply SPD processes to mass production of steels because of the limited dimensions of the SPD processed materials. Several alternative processes for refining grain size other than SPD have been also proposed, for example, repetitive heat treatment [12, 13] and recrystallization of cold-rolled DP steel [14]. In the present study, grain refinement of ferrite and DP structures through repetitive heat treatment and subsequent cold-rolling and recrystallization was investigated.

2. Experimental Methods

The material used in the present study is a 2Mn-0.1C (wt. %) steel. The detailed chemical composition of the steel studied is shown in Table 1. The heat treatment pattern conducted is shown in Fig. 1. Firstly, the specimen was austenitized at 830 °C for 6 h and then water-quenched (W.Q.) or furnace-cooled (F.C.) to obtain a coarse martensitic structure ((1) in Fig.1) or a coarse ferrite-pearlite structure ((2) in Fig.1), respectively. The austenite grain size of the specimen was refined through repetitive heat treatment of austenitizing at 810 °C for 180 s and water-quenching. In the last cycle of the repetitive heat treatment, the specimen was water-quenched to obtain a fine martensitic structure ((3) in Fig. 1) or air-cooled (A.C.) to obtain a fine ferritic structure ((4) in Fig. 1). After the repetitive heat treatment, the specimen was annealed at 740 °C (between A_{e1} and A_{e3}, ferrite + austenite two-phase region) for 30 minutes and water-quenched to obtain a DP structure composed of ferrite and martensite ((5) in Fig. 1). In order to further refine the grain size of DP structure, the specimen was cold rolled by 90 % and annealed at 600 °C (below A_{e1}) for different holding times ranging from 1 minute to 20 minutes to obtain a fine recrystallized ferrite structure ((6) in Fig. 1). Finally, the specimen was heat-treated at 740 °C for 30 minutes to obtain a fine DP structure ((7) in Fig. 1).

The microstructure of the specimen at each stage of the heat treatment shown in Fig. 1 was observed by scanning electron microscopy (SEM) and electron backscattering diffraction patterns (EBSD) analysis under an accelerated voltage of 15 kV. For the SEM and EBSD observations, the specimens were polished mechanically and then electrolytically in a solution of 10 % $HClO_4$ and 90 % CH_3COOH. The mechanical properties of the specimens were examined by a uniaxial tensile test at a strain rate of 8.3×10^{-4} s^{-1} at room temperature (RT).

Table 1 Chemical composition of the steel studied (wt. %).

C	Mn	Si	P	S	Fe
0.103	2.00	0.01	<0.002	0.0008	Balance

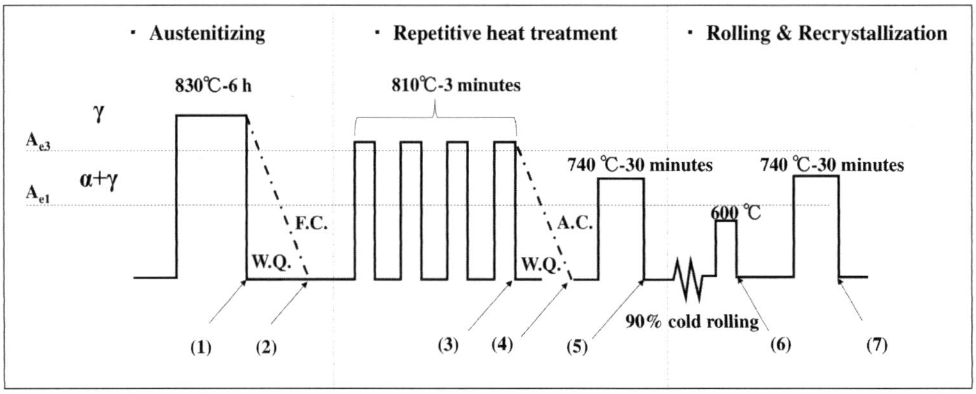

Figure 1 Schematic representation of the heat treatment route for the 2Mn-0.1C steel.

3. Results and Discussion

3.1. Grain refinement of austenite and ferrite by repetitive heat treatment

Figure 2 shows EBSD orientation color maps of (a, b) the specimens after austenitization ((a): W.Q., (b): F.C., corresponding to the stages (1) and (2) in Fig. 1, respectively), and (c, d) the specimens after the repetitive heat treatments (4-cycles) ((c): W.Q., (d): A.C, the stages (3) and (4) in Fig.1). In Fig. 2 (a) and (c), the prior austenite grain boundaries determined through orientation analysis were drawn in black dashed lines. The average prior austenite grain size was decreased from 28.6 μm to 6.1 μm by the repetitive heat treatment. The martensite structure contains a high density of lattice defects, i.e., dislocations and high angle boundaries. These lattice defects would act as nucleation sites for reverse transformation to austenite, resulting in the refinement of austenite grains by the repetitive heat treatment. However, further repetition of the heat treatment above 4-cycles did not change the austenite grain size. The grain size of ferrite in Fig. 2 (d) was 4.5 μm, which was much smaller than that in Fig. 2 (b) (10.2 μm). This indicates that the grain size of ferrite depends on the prior austenite grain size.

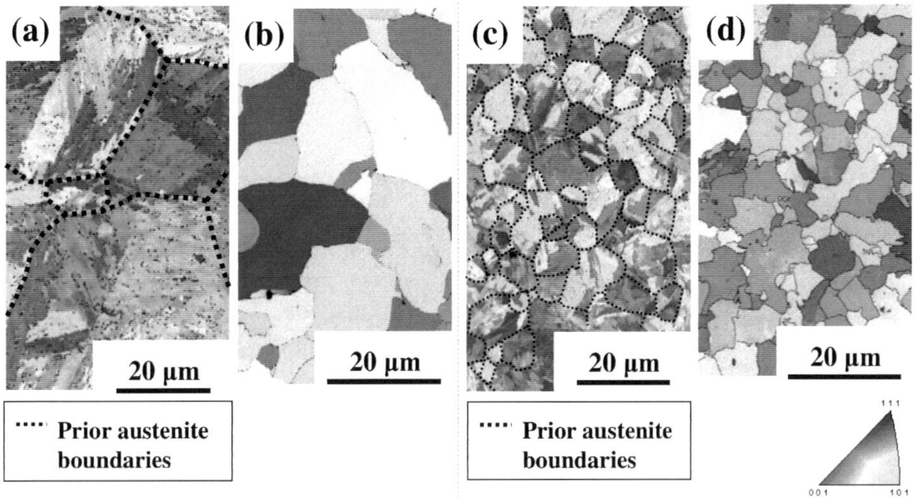

Figure 2 EBSD orientation color maps of the specimens heat-treated at 830 °C for 6 h and then (a) water-quenched (stage (1) in Fig.1), or (b) furnace-cooled (stage (2)), repetitive heat-treated at 810 °C for 180 s for 4-cycles and then (c) water-quenched (stage (3)) or (d) air-cooled (stage (4)).

3.2. Grain refinement of ferrite by combination of cold rolling and recrystallization

As reported by Okitsu et al. [14], deformation and recrystallization of a DP structure can effectively refine finally obtained ferrite. The specimen with a DP structure fabricated by the repetitive heat treatment followed by the intercritical annealing at 740 °C was cold-rolled by 90 % and then heat-treated at 600 °C (below A_{e1} temperature) for different periods of holding.

The EBSD orientation color map of the 90% cold-rolled specimen with DP structure is shown in Fig. 3 (a). The colors correspond to the crystallographic orientation parallel to the transverse direction (TD). The high angle boundaries with a misorientation above 15 degrees and low angle boundaries with misorientation ranging from 2 degrees to 15 degrees are drawn in black and white lines, respectively. The observation direction of Fig. 3 was parallel to the transverse direction (TD) of the sheet specimen. Almost all grains (including the martensite phase) in Fig. 3 (a) were elongated along the rolling direction (RD). After the heat treatment at 600 °C for 1 minute, 2 minutes, 5 minutes, 10 minutes and 20 minutes, corresponding to the stage (6) in Fig.1, (Fig. 3 (b), (c), (d), (e) and (f), respectively), on the other hand, ferritic structures with equiaxed grains were observed. The average ferrite grain sizes of these recrystallized specimens (b), (c), (d), (e) and (f) are 0.7 μm, 0.9 μm, 1.2 μm, 1.6 μm and 2.0 μm, respectively. The fraction of high angle grain boundaries (HAGBs) of those specimens (b-f) was 73.6 %, 62.6 %, 67.3 %, 88.7 % and 85.6 %, respectively. This indicates that the grain size and fraction of HAGBs in the ferrite increase with increasing holding time at 600 °C.

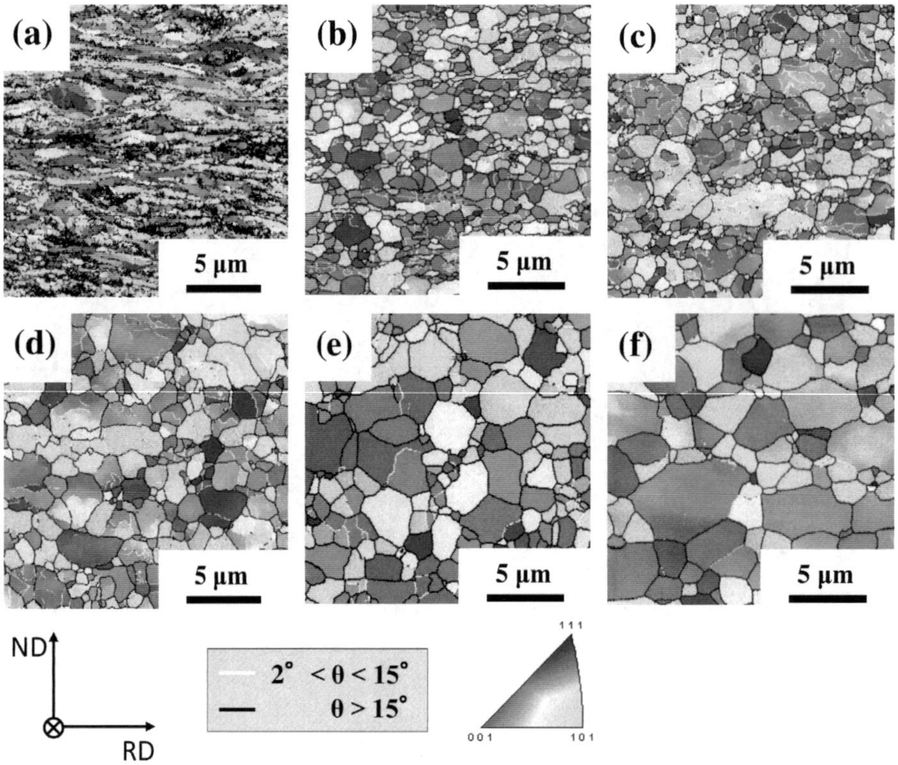

Figure 3 EBSD orientation color maps (overlapped with grain boundary maps) of (a) the 90% cold-rolled DP specimen, and the specimens annealed at 600 °C for (b) 1 minute, (c) 2 minutes, (d) 5 minutes, (e) 10 minutes and (f) 20 minutes after 90% cold-rolling.

3.3. Fabrication of dual-phase steel with different ferrite grain sizes

In order to fabricate fine-grained DP steels, the ferrite structures obtained in Fig. 3 (stage (6) in Fig. 1) were intercritically annealed in the ferrite + austenite two-phase region and then water-quenched. Figure 4 displays EBSD Image Quality (IQ) maps of the DP structures obtained by the intercritical annealing at 740 °C for 30 minutes followed by water-quenching of the specimens (a) at the stage (2) in Fig. 1, (b) at the stage (4) and (c) at the stage (6), respectively. The grain size of each ferritic structure was 10.2 μm, 4.5 μm and 1.2 μm, respectively. IQ map expresses the quality of Kikuchi-lines obtained at each scanning spot. Thus, the martensite phase including many lattice defects is expressed in darker contrast, while the ferrite phase in lighter contrast. It was found that almost all the martensite in Fig. 4 formed along grain boundaries of ferrite. The fraction of martensite in Fig. 4 (a), (b) and (c) was 44.4 %, 43.8 % and 42.2 %, respectively. The grain sizes of ferrite in the DP structured specimen of (a), (b) and (c), on the other hand, were 8.1 μm, 4.1 μm and 2.6 μm, respectively. This result indicates that the grain size of the ferrite phase in DP steels strongly depends on the grain size of the ferritic structure prior to the intercritical annealing.

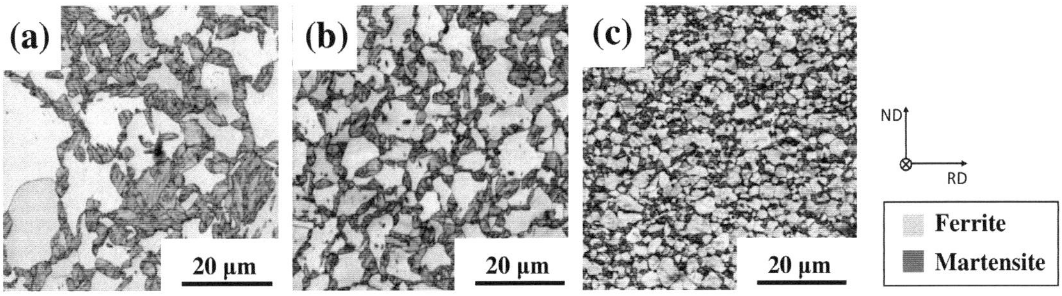

Figure 4 EBSD Image Quality maps of the DP structures obtained by the intercritical annealing at 740 °C for 30 minutes followed by water-quenching of the specimens (a) at the stage (2) in Fig. 1, (b) at the stage (4) and (c) at the stage (6).

3.4. Mechanical properties of ferritic and DP structures with different grain sizes

In order to evaluate the mechanical properties of the ferritic structures obtained and DP structures with different grain sizes, tensile tests were carried out at a strain rate of 8.3×10^{-4} s^{-1} and room temperature. Figure 5 shows obtained stress-strain curves of (a) the ferritic structures and (b) the DP structures with different ferrite grain sizes. The grain size of ferrite is indicated on each curve in Fig. 5.

In Fig. 5 (a), it was found that the yield strength as well as the tensile strength increased with decreasing the ferrite grain size, and the yield-drop and Lüders deformation were obviously recognized in the specimens. The uniform elongations of the specimens with ferrite grain sizes of 1.2 μm, 4.5 μm and 10.2 μm are 0.008, 0.195 and 0.220, respectively. The result indicates that the UFG (1.2μm) ferritic structure has limited uniform elongation compared with the specimens of 4.5 μm and 10.2 μm grain sizes, while the UFG specimen shows a high tensile strength. The UFG (1.2μm) ferritic structure showed early plastic instability, as has been reported in various UFG materials [15-17].

In Fig. 5 (b), on the other hand, the DP structure with a smaller grain size exhibits a higher elongation as well as higher strength. The uniform elongation of the specimens having the ferrite grain sizes of 8.1 μm, 4.1 μm, and 2.6 μm are 0.09, 0.12 and 0.12, respectively, which are not different to each other. On the other hand, the total elongation significantly increased from 0.15 to 0.23 with decreasing the grain size. From this result, it can be concluded that the fine-grained DP specimen with a ferrite grain size of 4.1 μm shows very good performance in both ductility (especially total elongation) and strength, compared with other specimens.

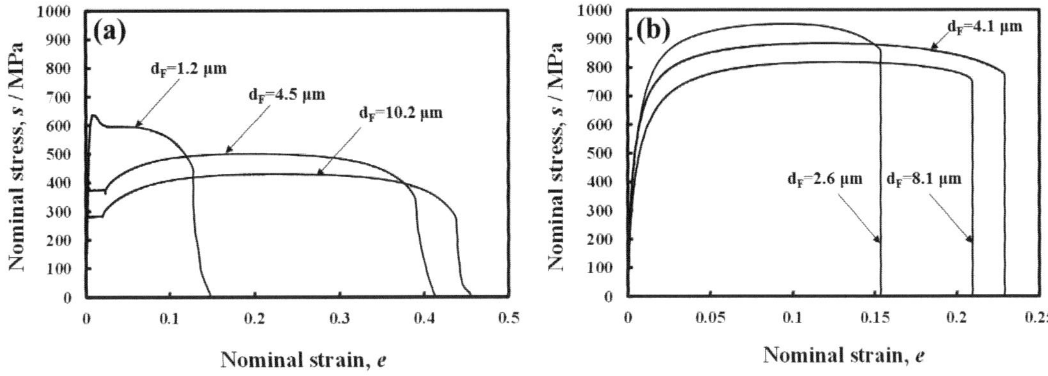

Figure 5 Nominal stress-strain curves of the specimens of (a) ferritic structures and (b) DP structures with different ferrite grain sizes.

4. Conclusions

Grain refinement of a 2Mn-0.1C steel was tried by a repetitive heat treatment combined with cold rolling and recrystallization. SEM / EBSD analysis and tensile tests were carried out for the specimens obtained with various structures. The main results obtained in this study are as follows.

The grain size of austenite decreased down to about 6 μm by 4-cycles of the repetitive heat treatment (810 °C for 180 s plus water-quenching) without deformation. By using this process, the ferrite grain size also decreased down to about 4 μm by air-cooling after the 4-cycle repetitive heat treatment.

90% cold rolling of the fine DP structure followed by annealing at 600 °C (below A_{e1}) was effective to obtain ultrafine-grained (about 1-2 μm) ferritic structures. Recrystallized ferrite became coarse with increasing the holding time at 600 °C.

Fine-grained DP structures were obtained by intercritical annealing at 740 °C of the UFG ferrite fabricated through the repetitive heat treatment and 90% cold-rolling + annealing. The grain size of the ferrite phase in the DP steel depended on the grain size of ferrite prior to the intercritical annealing.

In the specimens with ferritic structures, the yield and tensile strengths increased with decreasing the ferrite grain size. However, early plastic instability happened in the UFG ferrite with a grain size of 1.2 μm. On the other hand, the DP specimens having smaller grain sizes showed larger tensile elongation as well as higher strength. The DP specimen with 4.1 μm ferrite grain size showed very good performance in ductility (especially total elongation) as well as strength.

Acknowledgements

This study was financially supported by the Grant-in-Aid for Scientific Research on Innovative Ares, "Bulk Nanostructured Metals" (area No.2201), and the Elements Strategy Initiative for Structural Materials (ESISM) all through the Ministry of Education, Culture, Sports, Science and Technology (MEXT), Japan. M.P. was supported by the Japanese Government Scholarship. All the supports are gratefully appreciated.

References

[1] Saeidi N, Ashrafizadeh F and Niroumand B 2014 *Mater. Sci & Eng. A.* **599** 148
[2] Delince´ M, Jacques P J and Pardoen T 2006 *Acta Mater.* **54** 3395
[3] Calcagnotto M, Pndge D and Raabe D 2010 *Mater. Sci & Eng. A.* **527** 7832
[4] Calcagnotto M, Adachi Y. Ponge D and Raabe D 2011 *Acta Mater. 59* 662
[5] Okayasu M, Sato K, Mizuno M, Hwang D Y and Shin D H 2008 *Int'l J Fatigue.* **30** 1360
[6] Azushima A, Kopp R, Korhonen A, Yang D Y, Micari F, Lahoti G D, Groche P, Yanagimoto J, Tsuji N, Rosochowski A and Yanagida A 2008 *Manufacturing Tech.* **57** 716
[7] Stolyarov V, Zhu T, Lowe C and Valiev Z 2001 *Mater. Sci & Eng. A.* **303** 82
[8] Tsuji N, Ueji R and Minamino Y 2002 *Scripta Mater.* **47** 69
[9] Mohamed I F, Yonenaga Y, Lee S, Edalati K and Horita Z 2015 *Mater. Sci & Eng. A.* **627** 111
[10] Son Y I, Lee Y K, Park K T, Lee C S and Shin D H 2005 *Acta Mater.* **53** 3125
[11] Tsuji N, Kamikawa N, Ueji R, Takata N, Koyama H and Terada D 2008 *ISIJ Int'l.* **48** 1119
[12] Furuhara T, Kikumoto K, Saito H, Sekine T, Ogawa T, Morito S and Maki T 2008 *ISIJ Int'l.,* **48** 1040
[13] Shibata A, Daido S, Terada D and Tsuji N 2013 *Mater. Trans.* **54** 1571
[14] Okitsu Y, Takata N, Tsuji N 2009 *Scripta Mater.* **60** 78
[15] Ding Y, Jiang J and Shan A 2009 *J Alloy Compd.,* **487** 520
[16] Sun C, Ma J, Yang Y, Hartwig K T, Maloy S A, Wang H and Zhang X 2014 *Mater. Sci & Eng. A.* **597** 416
[17] Ma E, 2003 *Scripta Mater.* **49** 664

Microstructure evolution during isochronal annealing of a 42% cold rolled TRIP-TWIP steel

S Pramanik[1,2*], A A Saleh[2], D B Santos[3], E V Pereloma[1,2] and A A Gazder[1]
[1]Electron Microscopy Centre, University of Wollongong, New South Wales 2500, Australia
[2]School of Mechanical, Materials and Mechatronic Engineering, University of Wollongong, New South Wales 2522, Australia
[3]Metallurgical and Materials Engineering Department, Federal University of Minas Gerais, Belo-Horizonte, Brazil

*Email: sp345@uowmail.edu.au

Abstract. A high manganese TRIP-TWIP steel was cold rolled to 42% thickness reduction and isochronally annealed between 600 to 900 °C for 300 s. The microstructural evolution during annealing was studied by high resolution electron back-scattering diffraction. After cold rolling, the steel comprised predominant fraction of α′-martensite, a small fraction of blocky ε-martensite and a trace fraction of retained austenite (γ). During annealing, the reversion of ε and α′ martensite to γ was followed by the recrystallisation of γ. While the processes of reversion to and recrystallisation of γ were completed by 700 °C, further annealing between 750 to 900 °C led to γ grain growth. A novel method to delineate the γ-γ grain boundaries was developed in order to accurately quantify γ grain size and subsequently the activation energy for γ grain growth.

1. Introduction

Stringent environmental and safety regulations has led steelmakers and downstream automotive industries to develop new grades of advanced high-strength steels (AHSS) that improve on fuel economy through savings in vehicle weight and provide greater passenger safety. The high mechanical strengths (>1100 MPa) and large ductilities (≥ 55%) of this family of AHSS steels is due to their capacity to work harden considerably to large imparted strain levels [1, 2]. AHSS steels containing 15-25 wt.% Mn along with 2-4 wt.% of Si and Al comprise a metastable face centred cubic (fcc) austenite (γ) phase that accommodates deformation via a combination of twinning (TWIP) and transformation (TRIP) induced plasticity effects. γ phase transforms to either hexagonal close packed (hcp) ε-martensite (γ → ε) or body centred cubic (bcc) α′-martensite. The latter can occur via two possible routes: γ → α′ or γ → ε → α′ [3].

Upon annealing, the reversion of deformation-induced ε and α′ martensite to γ occurs [4, 5]. Thereafter the γ undergoes recrystallisation which is manifested by the growth and polygonisation of γ grains. The process of γ recrystallisation is typically accompanied by the formation of annealing twins. With increasing annealing temperatures, thermally-induced ε and α′ martensites also manifest within recrystallised γ grains as parallel or intersecting ε plates with α′ martensite grains located between the ε martensite plates.

The driving force behind γ grain growth is an overall reduction of the total grain boundary area in order to reduce the total energy associated with grain boundaries [6]. The rate of grain growth is dictated by boundary mobility and the activation energy for grain growth. In turn, boundary mobility depends on alloy composition, annealing temperature and the nature of the growing boundary. For example, the addition of alloying elements leads to an overall decrease in grain growth rate due to the solute drag effect [7], an increase in annealing temperature results in higher boundary mobility and coincident site lattice boundaries are less affected by the solute drag effect compared to other boundaries [8, 9]. The fundamental parameter controlling overall grain growth behaviour is the activation energy which is defined as the overall resistance of a material to grain growth. It follows that alloys with higher activation energy values compared to the pure metal returns a lower rate of grain growth.

The standard method to compute the activation energy for grain growth is by tracking the change in grain size with annealing time or temperature. In the present steel, γ grain sizes can be measured from optical and secondary electron micrographs or electron back-scattering diffraction (EBSD) maps. However, it is difficult to accurately distinguish the various phases using the former two methods. On the other hand, while EBSD successfully identifies all phases, the returned grain size information is based on a user-defined critical boundary misorientation angle that is uniformly applied to all phases. As a result, the γ grain size is markedly underestimated due to the inability to strictly define γ-γ boundaries. The latter is caused by the boundary contributions from thermally induced ε and α′ martensites (i.e. – the existence of γ-ε and γ-α′ martensite interphase boundaries). In turn, the underestimated γ grain sizes lead to the calculation of erroneous activation energy values. Thus a method is needed to accurately determine γ grain sizes by using γ-γ boundaries only.

With respect to the above outlook, the aim of this work is two-fold: (i) to characterise the evolution of the various phases during isochronal annealing and (ii) to quantify the activation energy for γ grain growth by accurately determining γ grain sizes from γ-γ boundaries only.

2. Experimental procedure

An Fe-17Mn-2Si-3Al-1Ni-0.06C wt.% steel was slab cast and homogenised at 1100 °C for 7200 s. The homogenised slab was hot-rolled at 1100 °C to 52.5% thickness reduction in 4 passes. Cold rolling of the hot-rolled strip was subsequently done to 42% thickness reduction in 11 passes. Samples were machined from the centre of the cold-rolled strip and isochronally annealed between 600 and 900 °C. Annealing involved a heating period of 240 s to the set temperature, a holding period of 300 s followed by immediate water quenching. Ø3 mm diameter discs were punched out from the normal direction (ND) -rolling direction (RD) mid-plane and twin-jet electropolished using a solution of 90% methanol and 10% perchloric acid in a Struers Tenupol-5 operating at 30 V (~150 mA) and -25 °C.

High resolution EBSD was undertaken on a JEOL JSM-7001F field emission gun - scanning electron microscope operating at 15 kV accelerating voltage, ~5.5 nA probe current, 12 mm working distance and equipped with a Nordlys-II(S) EBSD detector interfacing with the Oxford Instruments AZtec software suite. EBSD maps were acquired with step sizes of 30 nm for the cold-rolled and 600 °C annealed samples and 100 nm for the 700 to 900 °C annealed samples.

3. Analytical procedure

3.1 EBSD map post-processing

The raw EBSD maps of the 42% cold rolled and isochronally annealed samples returned an average indexing rates of 64.0% and 86.3±14.1%, respectively. Post-processing of the maps was undertaken using the Oxford Instruments Channel-5 software package using well-established procedures (see Ref. [10] and the references therein) by eliminating any potential wild orientation spikes and filling in zero solutions via cyclic extrapolation from 8 to 5 neighbours. As a consequence, the above procedure rules out the introduction of any artificial grains. In all the maps, (sub)grain structures are defined by a minimum of three pixels and are bound by misorientations (θ) ≥ 2°. Here $2° \leq \theta < 15°$ are classified as low angle grain boundaries (LAGBs) whereas $\theta \geq 15°$ denote high angle grain boundaries (HAGBs). First order twin boundaries (TB_γ) are defined as Σ3 = 60° <111> with a maximum deviation ($\Delta\theta$) of 6° following the Palumbo-Aust criterion ($\Delta\theta \leq 15° \Sigma^{-5/6}$) [11].

3.2 Boundary map image processing

The sample annealed at 900 °C is used to describe the procedure applied to identify the γ-γ grain boundaries and to calculate the γ grain sizes (Figure 1). It is pointed out that the methodology described in the following paragraphs relies on the image processing of grain boundary maps and does not utilise any orientation data.

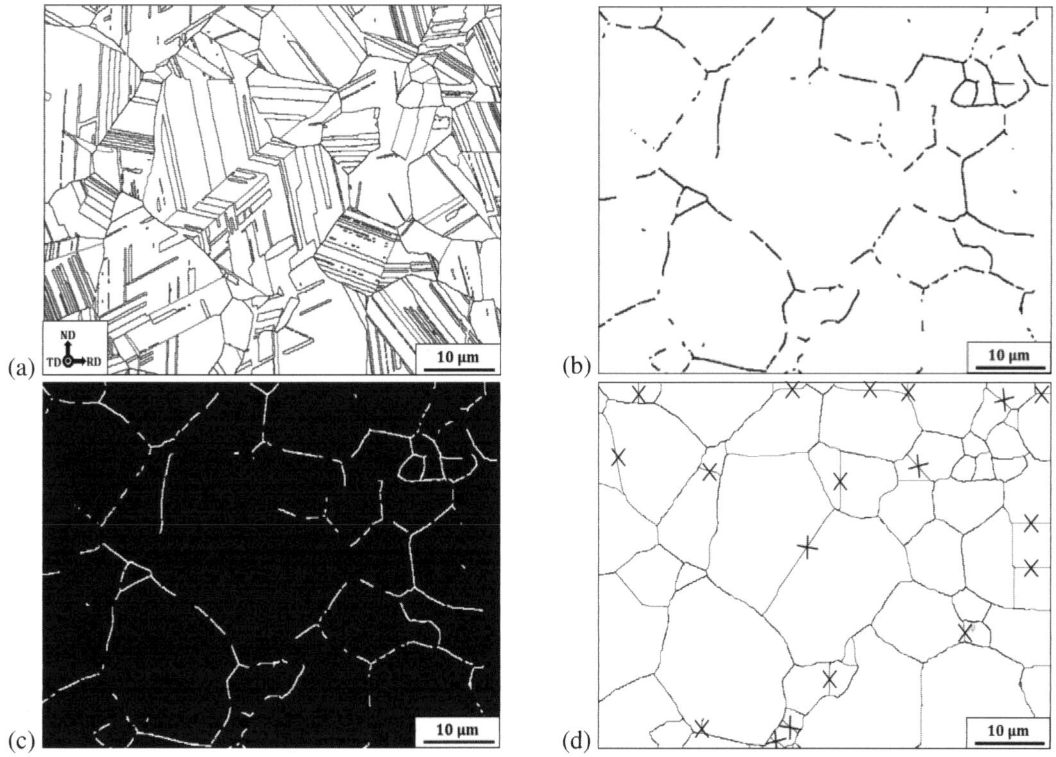

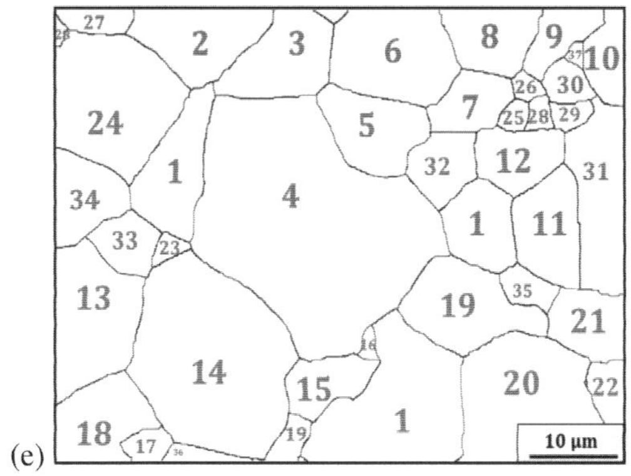

Figure 1 Representative boundary maps after annealing at 900 °C showing (a) γ-γ (black), TB_γ (red) and ε-γ and α′-γ interphase (blue) boundaries, (b) incomplete γ-γ boundaries, (c) 8 bit binary, inverted and thresholded incomplete γ-γ grain boundaries, (d) complete γ-γ boundaries after applying the watershed algorithm (artificially introduced boundaries are shown by an "X") and (e) identifying individual γ grains to compute the equivalent circle diameter. Rolling direction (RD) = horizontal, transverse direction (TD) = vertical.

As shown in Figure 1a, the procedure began by creating a map comprising γ-γ high-angle grain boundaries (black), 60° <111> twin boundaries of the γ phase (red) and ε-γ, α′-γ interphase boundaries (blue). Following this, the colour of the TB_γ and interphase boundaries was re-assigned to white in order to return the incomplete γ-γ grain boundary map (Figure 1b).

The incomplete γ-γ boundary map was saved as a TIFF image and imported into the ImageJ software package [12]. A pixel-based image scale was defined, then the image was cropped to remove legend information. The cropped RGB image was then converted to an 8 bit binary image and inverted such that the pixels comprising the incomplete γ-γ grain boundaries were white while the pixels denoting the γ grain interior were black (Figure 1c). The above procedure was necessary to reveal the incomplete γ-γ grain boundaries and to enable image thresholding into two distinct pixels sets of 0 (or black) and 255 (white) colour channels. Next, the watershed algorithm was applied to complete the γ-γ boundaries by connecting the end pixels of incomplete γ-γ boundaries to their nearest neighbouring pixels by the linear shortest distance (Figure 1d) [13].

Following this, Figures 1b and 1d were visually compared in order to manually remove any artificially introduced boundaries (see the boundaries marked with an "X" in Figure 1d). Once completed, the γ-γ grain boundary outlines were detected and the number of pixels within each outlined grain was quantified in order to determine the grain areas. The grain areas were then used to compute the equivalent circle diameter (ECD).

It follows that the ECD value corresponds to that diameter of a circle whose area is equivalent to the area of an γ grain [14]. Lastly, the average and standard deviation of the ECD distribution is computed and the procedure was repeated for all the annealing conditions (Figure 2).

3.3 Calculation of the activation energy

The activation energy for γ grain growth was calculated by using the equation given below [15]:

$$d_T^{1/N} - d_0^{1/N} = K_0 t^n e^{(-Q/RT)} \quad (1)$$

where d_T is the γ grain size (in microns) at a given annealing temperature T (in K), $d_0 = 1.85 \pm 1.4$ μm is the initial average γ grain size (determined at 700 °C), N is the grain growth exponent, K_0 is a kinetic constant that is equivalent to the y-intercept, t = 300 s is the holding time during annealing, n = 1 is a constant, Q is the computed activation energy for γ grain growth (in kJ/mol) and R = 8.314 J/molK is the universal gas constant.

The value for the grain growth exponent measured from a 1000 °C isothermal annealing experiment undertaken on an Fe-29Mn-5Al-0.06C TWIP steel was found to be N = 0.34 [15]. Due to the scarcity of data on high Mn steels, the change in N with homologous temperature ($T_h = T/T_m$ with T_m = 1550 °C) was estimated using data from austenitic stainless steel [16, 17]. Using N = 0.34 for $0.69T_h$, 1000 °C as a reference point [15], the extrapolated N-values for the $0.53T_h$ to $0.64T_h$ or 700 to 900 °C temperature range that corresponds to our isochronal annealing experiments varies between 0.29 and 0.33, respectively. Since abnormal grain growth is absent in the present steel, drastic changes to the N-value are not expected [17]. Consequently, N = 0.29 and 0.33 are used to estimate the range of activation energies for γ grain growth.

When using Eq. (1) it is also implied that: (i) grain morphology is equiaxed and the microstructure is of uniform size and (ii) no inclusions or chemical segregation are present at grain boundaries [18]. The activation energy (Q) is calculated from the slope of $\ln(d_T^{1/N} - d_0^{1/N})$ vs $1/T$ as shown in Figure 3b for N = 0.29.

4. Results and Discussion

4.1 Microstructural evolution during annealing

Figure 2a is a representative phase distribution map of the 42% cold rolled microstructure. It comprises large aspect ratio α′ martensite (green) grains along with blocky ε martensite (violet) and untransformed γ (grey) grains. It is self-evident that α′ martensite is the dominant phase followed by ε martensite and only a trace amount of untransformed γ. Gazder *et al.* [19] reported similar results for 66% cold rolled TRIP-TWIP steel. Annealing at 600 °C results in martensite reversion such that the map comprises a dominant γ phase, a remanent α′ martensite phase and an absent ε martensite phase. The last observation suggests that ε martensite reverted to γ at temperatures lower than 600 °C [20].

Annealing at 700 °C results in the development of equiaxed γ grains with a large fraction of annealing twin boundaries and small fractions of thermally-induced ε and α′ martensites (Figure 2c). Further annealing between 750 to 900 °C leads to γ grain growth such that a wide distribution in γ grain sizes is noted (Figures 2c-g). The microstructures are a mix of fine γ grains located at the boundary triple junctions of the coarse γ grains. This is expected, considering that: (i) the concurrent processes of α′ martensite reversion and subsequent γ nucleation occur over a range of annealing temperatures and (ii) the growth of previously recrystallised γ grains (as a result of higher boundary mobilities across some interfaces) will also result in coarser γ grains in the microstructure.

The formation of thermally-induced ε and α′ martensites in relatively coarse γ grains after 750 to 900 °C annealing (Figures 2c-g) agrees with previous reports by Refs. [21] and [22]. The nucleation and growth of ε martensite plates was explained by Jiang *et al.* [21] on the basis of their increased probability of nucleating at the intersections of stacking faults in coarse γ grains. In the case of fine γ grain sizes (classified here as 5-15 μm), ε martensite comprises morphologically similarly-oriented plates propagating across entire grains (dashed white circle in Fig 1(g)) whereas in the case of coarse γ grains (>15 μm) (solid white circle in Fig 1(g)), a network of intersecting ε martensite plates is promoted. The above observation agrees with Takaki *et al.* [22] who also reported that ε martensite plates tend to propagate across smaller γ grains (<30 μm) or consists of intersecting plates in larger γ grains (>130 μm).

4.2 Activation energy for γ grain growth

To-date, conventional approaches to γ grain size calculations in the presence of microstructural features or secondary phases have tended to underestimate the average γ grain size. In the present steel, the underestimation is caused by the inclusion of TB_γ, γ-ε and γ-α′ martensite interphase boundaries when computing the γ grain size (white circles in Figure 3a). However, when twin and interphase boundaries are excluded and the γ-γ boundaries are strictly defined via the procedure detailed in Section 3.2, a more accurate measure of the γ grain size can be obtained (black circles in Figure 3a).

When the above grain sizes are used in conjunction with Eq. (1), the activation energies for γ grain growth can be computed (Figure 3b). When twin and interphase boundaries are considered and N = 0.29, the activation energy was estimated as 232.0±23.6 kJ/mol (white circles in Figure 3b). The relatively poor linear fit (R^2 = 0.80) results in a higher value of standard deviation in this case (±23.6 kJ/mol). Alternatively, when γ-γ boundaries are strictly delineated using the method described in Section 3.2, a more robust linear fit (R^2 = 0.99) results in an activation energy of 265.8±9.9 kJ/mol. In order to compute the possible range of activation energy, N = 0.33 (0.64T_h, 900 °C) was used while considering only γ-γ boundaries and an activation energy of 243.5±10.5 kJ/mol was estimated.

(a) 

(b) 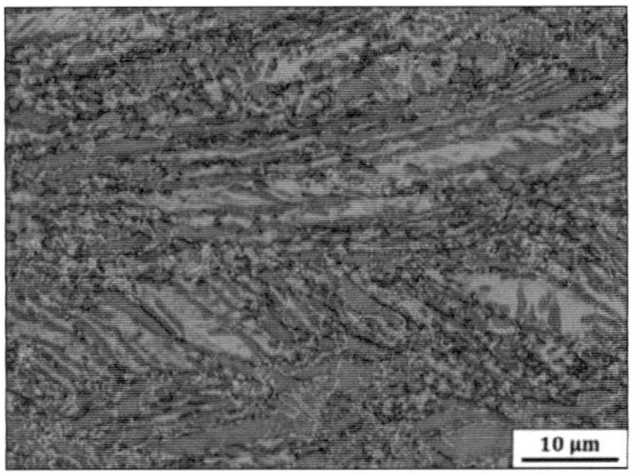

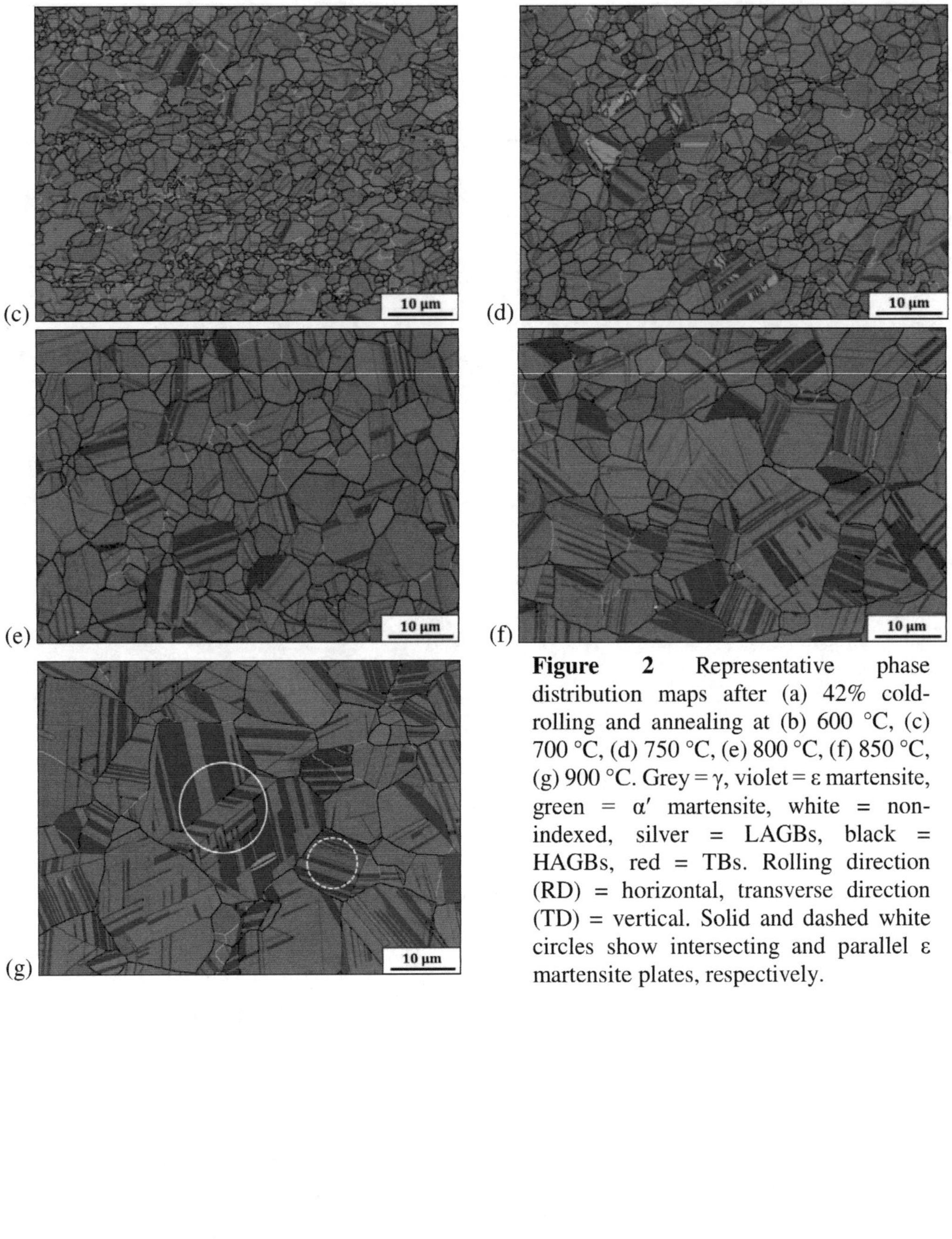

Figure 2 Representative phase distribution maps after (a) 42% cold-rolling and annealing at (b) 600 °C, (c) 700 °C, (d) 750 °C, (e) 800 °C, (f) 850 °C, (g) 900 °C. Grey = γ, violet = ε martensite, green = α' martensite, white = non-indexed, silver = LAGBs, black = HAGBs, red = TBs. Rolling direction (RD) = horizontal, transverse direction (TD) = vertical. Solid and dashed white circles show intersecting and parallel ε martensite plates, respectively.

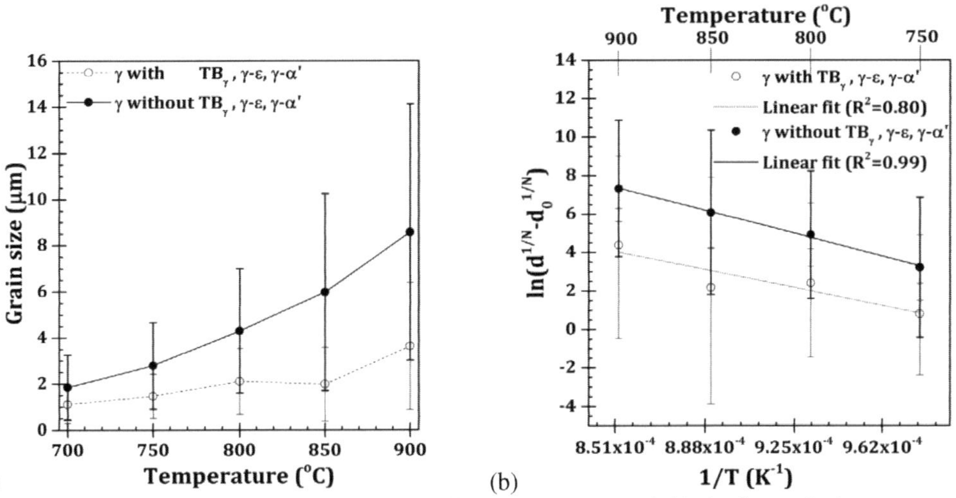

(a) (b)

Figure 3 (a) Variation of γ grain size with annealing temperature and (b) the linear fit that computes the activation energy for γ grain growth using N = 0.29 when $TB_γ$, γ-ε and γ-α′ boundaries are included and excluded (white circles with thin dashed lines and black circles with thick solid lines, respectively).

The above calculated values can be compared to the activation energies determined from the γ grain size data of Refs. [23, 24]. For the Fe-18Mn [24] and Fe-18Mn-1.5Si-0.6C [23] steels, N values of 0.29 ($0.53T_h$, 700 °C) and 0.30 ($0.56T_h$, 750 °C) resulted in activation energy estimates of 283.83 and 263.4±11.7 kJ/mol, respectively. It follows that the activation energy for γ grain growth for this class of TRIP-TWIP steels is significantly higher than the activation energy for grain boundary diffusion in γ iron (Q_b = 149 kJ/mol) [25] and tends to be close to the self-diffusion of γ (Q_s = 270 kJ/mol) [5]. In turn, this indicates that lattice diffusion may be a key mechanism for γ grain growth rather than grain boundary diffusion.

5. Conclusions

The microstructural evolution of a TRIP-TWIP steel subjected to 42% cold-rolling and isochronal annealing in the temperature range from 600 to 900 °C for 300s was studied by high resolution electron back-scattering diffraction.

1. After cold rolling, the steel contained predominantly α′ martensite, a small fraction of blocky ε martensite and a trace amount of untransformed γ. Annealing resulted in the reverse transformation of ε and α′ martensites to γ, the subsequent recrystallisation and growth of γ and the development of annealing twins. Formation of thermally-induced ε and α′ martensites within recrystallised γ grains occurred on cooling.

2. A method to accurately determine the ECD-based grain size of recrystallised γ was developed. It involved the removal of twin and interphase boundaries and strictly defining γ-γ boundaries only. In turn, the γ grain sizes were used to compute the activation energy for γ grain growth. The computed activation energy of 265.8±9.9 kJ/mol indicates that lattice diffusion may be a key mechanism for γ grain growth.

Acknowledgements

The authors acknowledge the use of the JEOL–JSM7001F FEG-SEM purchased with the financial support from the Australian Research Council (LE0882613). This work was supported by the Engineering Materials Research Strength at UOW.

References

[1] Grässel O, Krüger L, Frommeyer G and Meyer L W 2000 *Int. J. Plast.* **16** 1391-409
[2] Frommeyer G, Brux U and Neumann P 2003 *ISIJ Int.* **43** 438-46
[3] Tomota Y, Strum M and Morris J W 1986 *Metall. Trans. A* **17** 537-47
[4] Lü Y, Hutchinson B, Molodov D A and Gottstein G 2010 *Acta Mater.* **58** 3079-90
[5] Lü Y, Molodov D A and Gottstein G 2011 *Acta Mater.* **59** 3229-43
[6] Huang Y and Humphreys F J 2012 *Mater. Chem. Phys.* **132** 166-74
[7] Suehiro M, Liu Z K and Ågren J 1996 *Acta Mater.* **44** 4241-51
[8] Simpson C J, Aust K T and Winegard W C 1971 *Metall. Trans.* **2** 987-91
[9] Aust K T and W R J 1959 *Trans. AIME* **125** 119-27
[10] Gazder A A, Al-Harbi F, Spanke H T, Mitchell D R and Pereloma E V 2014 *Ultramicrosc.* **147** 114-32
[11] Palumbo G and Aust K T 1990 *Acta Metall. Mater.* **38** 2343-52
[12] Schneider C A, Rasband W S and Eliceiri K W 2012 *Nat. Methods* **9** 671-5
[13] Pérez J M M and Pascau J 2013 *Image Processing with ImageJ* (Birmingham UK: Packt Publishing)
[14] Jennings B R and Parslow K 1988 *Proc. Roy. Lond.: Math. Phys. Eng. Sci.* **419** 137-49
[15] Razmpoosh M H, Zarei-Hanzaki A, Haghdadi N, Cho J H, Kim W J and Heshmati-Manesh S 2015 *Mater. Sci. Eng. A* **638** 5-14
[16] Hu H and Rath B B 1970 *Metall. Trans.* **1** 3181-4
[17] Mizera J, Wyrzykowski J W and Kurzydłowski K J 1988 *Mater. Sci. Eng. A* **104** 157-62
[18] Rajasekhara S, Karjalainen L P, Kyröläinen A and Ferreira P J 2008 Kinetics of grain growth in ultra-fine grained AISI 301LN stainless steel. In: *Proc. of 6th European Stainless Steel Conference,* ed Karjalainen P and Hertzman S (Helsinki, Finland) pp 511-7
[19] Gazder A A, Saleh A A, Nancarrow M J B, Mitchell D R G and Pereloma E V 2015 *Steel Res. Int.* **86** doi: 10.1002/srin.201500089
[20] Escobar D P, Dafé S S F and Santos D B 2014 *J. Mater. Res. Tech.* **4** 162-70
[21] Jiang B H, Sun L, Li R and Hsu T Y 1995 *Scr. Metall.* **33** 63-8
[22] Takaki S, Nakatsu H and Tokunaga Y 1993 *Mater. Trans. JIM* **34** 489-95
[23] Jo S Y, Han J, Kang J H, Kang S, Lee S and Lee Y K 2015 *J. Alloy Compd.* **627** 374-82
[24] Jun J H and Choi C S 1998 *Mater. Sci. Eng. A* **257** 353-6
[25] Medina S F and Quispe A 2001 *ISIJ Int.* **41** 774-81

Hardening by annealing: insights from different alloys

O Renk[1*], A Hohenwarter[2], B Schuh[2], J H Li[3] and R Pippan[1]

[1]Erich-Schmid-Institute of Materials Science, Jahnstraße 12, 8700 Leoben, Austria
[2]Departement of Materials Physics, Montanuniversität Leoben, Jahnstraße 12, 8700 Leoben, Austria
[3]Institute of Casting Research, Montanuniversität Leoben, Franz-Josef-Strasse 18, 8700 Leoben, Austria

Email: oliver.renk@oeaw.ac.at

Abstract. In contrast to the general notion about the annealing behavior of coarse grained materials, hardening phenomena in nanocrystalline materials can occur. Although the phenomena have already been recognized several years ago, the mechanisms behind are still controversially discussed. For example, the influence of solutes segregated to grain boundaries on the strengthening mechanism is unclear. We present a combination of atom probe tomography and mechanical data to reveal the role of segregations to the strengthening. The results show that despite large modifications of the boundary chemistry the mechanical behavior remains widely unaffected. Additionally, it will be shown that hardening upon annealing can only occur below a material-specific grain size threshold value.

1. Introduction

It is well known that cold worked coarse grained materials (where precipitation hardening does not occur) undergo softening during annealing due to recovery and recrystallization. Interestingly, nanocrystalline (nc) materials can show a different behavior - hardening upon annealing instead of a softening. Such phenomena occur at temperatures where significant grain growth of the nc structure can be impeded [1-2]. Although this behavior has been frequently observed, different hypothesis about the mechanisms and the origin of the hardening have been reported, see for example [1-4]. While the first reports, focusing on very pure metals, attributed the observed strengthening to a reduction in the defect densities during annealing [1-2], the role of impurities or segregated solute in alloyed systems remained unclear or was expected to cause the hardening [2-4].

Since with decreasing grain size intergranular dislocation sources are thought to become dominant over intragranular sources [5], segregated solute might suppress dislocation emission from the boundaries [4]. Furthermore, as in nc materials dislocations are emitted from the boundary regions and may relax immediately at the opposite boundary, also the relaxation process could be affected by the presence of segregated solute, leading to a different mechanical response.

Nowadays nc and ultrafine-grained materials (UFG) are frequently processed by methods of Severe Plastic Deformation (SPD). After the heavy deformation processes, lattice dislocations will not be completely absorbed at the boundaries and are often stored in the form of dipoles [2]. These dislocations, present in the material before the heat treatment, require lower stress levels for plastic deformation when compared to the activation energy of grain boundary sources [1-2]. During annealing treatments such segments or dipoles from the interior can be fully absorbed in the boundary leaving a crystal with a low content of mobile defects behind. Additionally, the structure of high angle grain boundaries (HAGB) will change, making the emission of dislocations more complicated, which was also shown in MD simulations [6]. Also the structure of LAGB, mostly present as so called interconnecting boundaries in severely deformed metals will change, and their movement or extraction of dislocations from the boundary will become more complicated [7]. All these processes should lead to increased stress levels to activate dislocation sources after a heat treatment and to the observed hardening. Both, the annihilation of lattice defects and the segregation of impurities or solute will take place simultaneously thus, a separation of their contribution to the hardening seems challenging. Nevertheless, the role of segregations to the strengthening should be investigated further. If they play a key role, an identification of the most effective alloying elements and the interplay between them might lead to the development of ultrastrong materials.

In this study we present a combination of mechanical data of a nc austenitic steel with insights into its boundary chemistry gained by atom probe tomography (APT). These results should lead to a deeper understanding of the role of segregation to the nc-related hardening mechanism. Furthermore, as hardening upon annealing is not observed in all ufg/nc materials, additional suitable experiments were carried out to gain further insights as to why certain materials do not show the mentioned phenomena.

2. Hardening by annealing – is segregation necessary?

The idea to answer the raised question was to compare mechanical data of nc samples with similar grain size but different levels of grain boundary segregation. To do so, we used a standard 316L austenitic stainless steel. Nc structures were obtained by high pressure torsion (HPT) processing. Further experimental details can be found in Ref. [8]. The as HPT deformed structures consist of crystallites with a size of about 50 nm as calculated from several TEM micrographs by the line interception method. The corresponding microhardness was found to be 5 GPa. An isochronal annealing treatments for 30 minutes in a temperature range up to 973 K showed pronounced hardening up to 6.1 GPa at a temperature of 823 K. Measured room temperature hardness values as a function of the annealing time for the isochronally annealed samples (30 minutes) are shown in Fig. 1a). For comparison the hardness of heavily cold rolled samples is also shown in Fig. 1a. To check the thermal stability of the material and to see how fast the hardening process takes place, isothermal treatments were carried out at 823 K, the temperature at which the maximum hardness was obtained. The results are summarized in Fig. 1b). The strengthening occurs very rapidly, within 5 minutes the peak hardness values are almost reached. For longer annealing times the hardness only changed slightly before levelling off in a plateau even for long annealing times. Selected area diffraction patterns (SAED) do not reveal any additional Debye Scherrer rings of other phases formed during annealing. As can be expected from the hardness measurements in Fig. 1b significant grain growth did not occur over the wide range of annealing times. Selected bright field images of the as deformed and the annealed conditions are shown in Fig. 2. Since grain size and hardness (see Fig. 1) have not drastically changed but the amount or type of segregated solute might change over time, isothermally annealed samples could be suitable to clarify if segregation contributes to the strengthening. Therefore selected samples were analyzed by APT. A combination of APT data and the mechanical response of the samples should allow one to comment on the contribution of segregated solute to the strengthening. In the following, the major results of the study are summarized, further details of the experiments are outlined in Ref. [8].

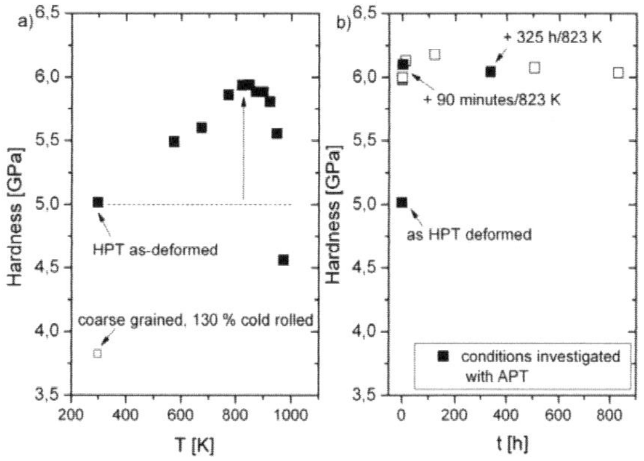

Figure 1. Room temperature microhardness of a) isochronally annealed samples; b) isothermally annealed nc 316L samples [8].

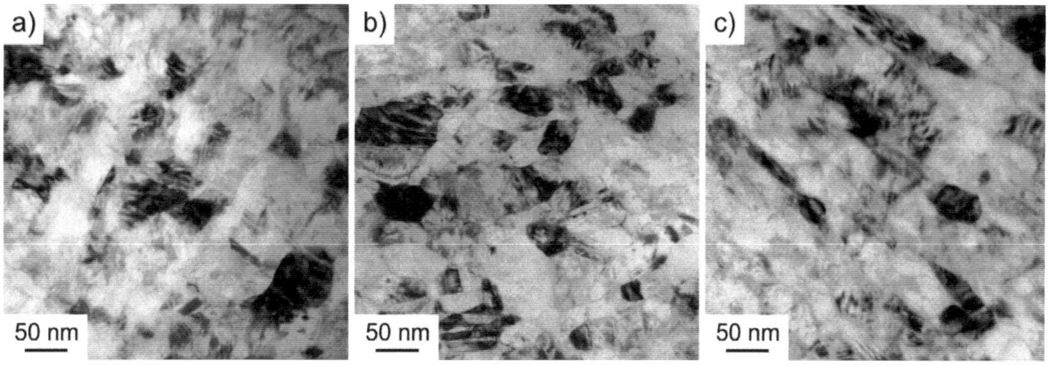

Figure 2. TEM micrographs of a) as HPT deformed b) 30 minutes/823 K annealed c) 325 h/823 K annealed samples. No significant grain growth was observed during annealing.

Three different sample conditions were investigated by APT, the as HPT deformed samples, samples annealed for 90 minutes and 325 h at 823 K (see also Fig. 1b). APT results of these three samples are summarized in Fig. 3. It can be seen that even in the case of the as HPT deformed sample a slight average Si excess at the grain boundaries of $\Gamma = 0.15$ atoms nm^{-2} can be found. Short time annealing of 90 minutes at 823 K leads to an increase of the average Si excess to $\Gamma = 0.40$ atoms nm^{-2}. Apart from the Si, only traces of other alloying elements were found for both samples, the as deformed as well as the 90 minutes annealed sample. For the long term annealed sample (325 h) boundary chemistry has drastically changed. Additionally to a similar Si excess of $\Gamma = 0.36$ atoms nm^{-2} than in the 90 minutes annealed samples, second phase particles could be identified at the grain boundaries. These particles contain mainly Si, Mo and Cr as can be seen in the proxigram in Fig 3. Although the temperature range where these particles have formed would suggest sigma phase formation, the average composition of the particles is far from those reported for 316L steels [9]. Further identification of the second phase was not possible from SAED patterns, which only showed reflections that would be referred to a fcc structure.

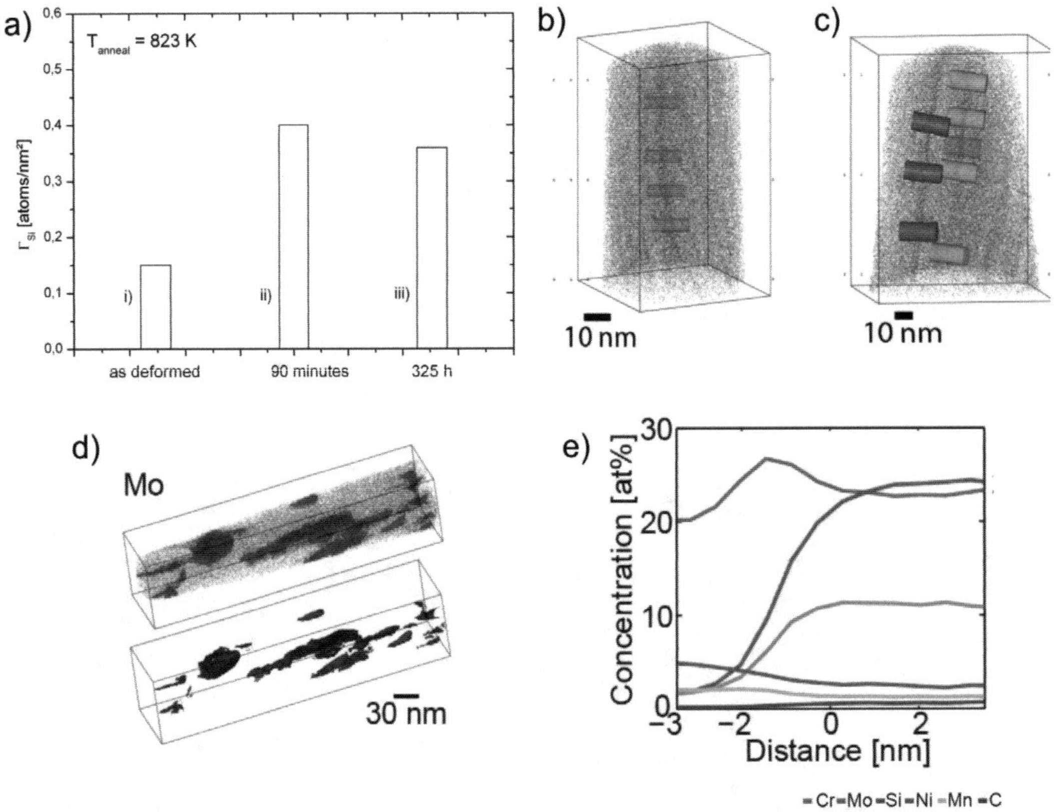

Figure 3. Results of the samples investigated by APT. a) The average Si excess of several boundaries is displayed for the analyzed samples. b) - d) Reconstruction of the APT tips with Si atoms colored in grey (b and c) and Mo atoms in red (d). d) For the 325 h annealed samples additional particles formed at the grain boundary. e) Their average composition can be seen in the plot [8].

Between annealing times of 90 minutes and 325 h in addition to the segregated Si, Mo and Cr segregate and form particles. Despite this large variation of the boundary chemistry, the mechanical properties remain widely unchanged (see Fig. 1b). From that it can be concluded that second phase particles at the grain boundary do not influence the strength of a nc material. Furthermore large variations of the type of the segregated element seems also not to have a strong effect on the strength of the material. However, possible effects of small excess values still remain unclear because during the first 90 minutes, both, the hardness but also the Si excess increased. For further clarification samples either annealed for a very short time or at lower temperatures should be suitable to capture the effects of small excess values. Experimentally, lower temperatures but longer times are easier to realize as one can expect a large scatter when samples are annealed just for very short time intervals.

In Fig. 4 we present the first results of the hardness evolution as a function of the annealing time for temperatures lower than 823 K. It can be seen that the hardness increase is smaller for lower annealing temperatures. Independent of the annealing temperature, a similar behavior to that at 823 K (see Fig. 1b) is observed. The hardness rises quickly before it seems that it levels off at a certain plateau value. Surprisingly even at the lowest annealing temperature of 343 K (~ 0.2 T_m) a significant hardness increase can be observed. Although it is very unlikely that segregation takes place at such a low annealing temperature within 5 minutes, a detailed study has to be carried out to clarify this result.

Apart from this study, several other results also suggest, that segregation is not essential for the hardening. For example, nc samples showed pronounced hardening when cyclically loaded at, for nc metals, relatively large plastic strain amplitudes [10]. This behavior was also observed for the nc austenitic steel used in the present study [11]. The reason for the cyclic hardening seems to be an exhaustion of easy mobile dislocations with increasing number of cycles. An in-situ straining TEM study on UFG Al points in the same direction. Stresses to operate a grain boundary source were found to be much larger than those necessary to overcome an obstacle, most likely an impurity [12].

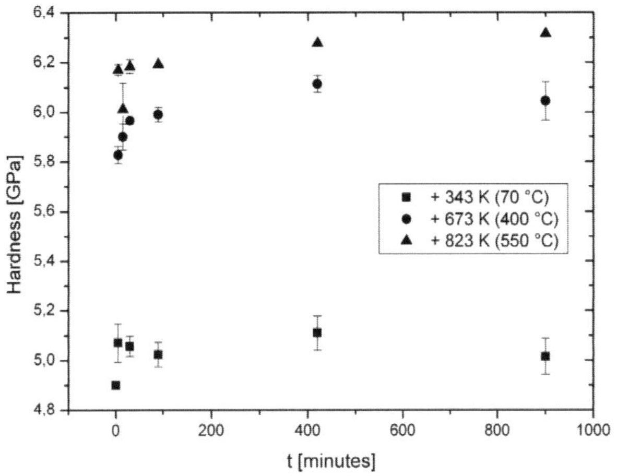

Figure 4. RT hardness of isothermally annealed nc 316L samples as a function of the annealing time.

3. Hardening by annealing – a grain size related phenomena?

As mentioned earlier, hardening by annealing was reported for a wide range of pure metals as well as alloys. Despite this fact, several severely deformed materials do not show a hardness increase upon annealing. In contrast, they show rather constant hardness values until grain growth takes place. One example is UFG high purity Ni (99,99 %) processed by HPT, where no hardening occurs during annealing, see Fig. 5. When Ni powder (3 -7 µm particle size) was consolidated and subsequently HPT deformed a slight hardening during annealing takes place. The higher initial hardness of the powder consolidated samples indicates a smaller grain size for this set of samples, as impurities and oxides on the powder surface slow down boundary mobility, thus shifting the minimum grain size achievable by HPT to smaller sizes. Annealing of the Ni powder in air (400 °C for 10 minutes) results in an increase of the oxide content of the powder and to a further reduction of the grain size after HPT. Interestingly, these samples showed an even larger hardness increase upon annealing. From these results one could expect hardening by annealing to be a grain size dependent phenomenon. However, the impurity level of the powder and the bulk material were different. Although the results in section 2) indicate that this might not explain the different extent of hardening, experiments on samples of the same impurity/alloying content would be more meaningful to interpret. As the minimum grain size achievable in HPT can also be modified by changing the deformation temperature, samples of identical nominal composition but a different grain size can easily be processed. To do so, a Pt5Ru (concentration in wt-%) alloy produced by arc melting was deformed by HPT at 77 K, RT and 573 K. The alloy contains only a single phase after deformation as confirmed by TEM. SAED patterns captured after annealing also did not show signs of the formation of additional phases. Grain sizes of the three samples were calculated by analyzing EBSD data. After deformation, the samples having different grain sizes but the same composition were subjected to isochronal heat treatments (30 minutes) in a temperature range up to 773 K. The results are summarized in Fig. 6.

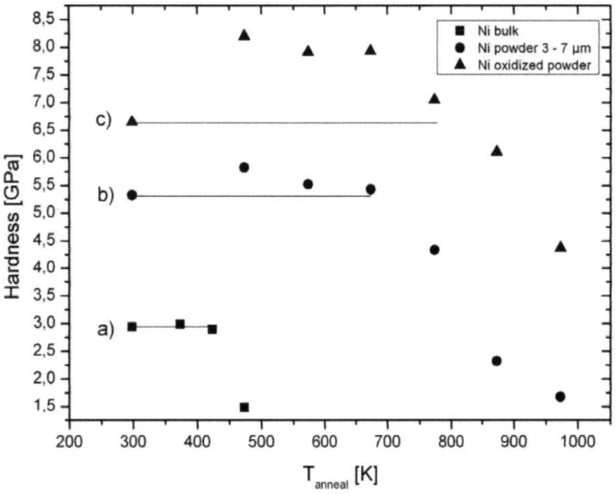

Figure 5. Room temperature hardness of HPT deformed a) bulk Ni b) Ni powder c) oxidized Ni Powder samples as a function of the annealing temperature. For all temperatures the annealing time was kept to 30 minutes.

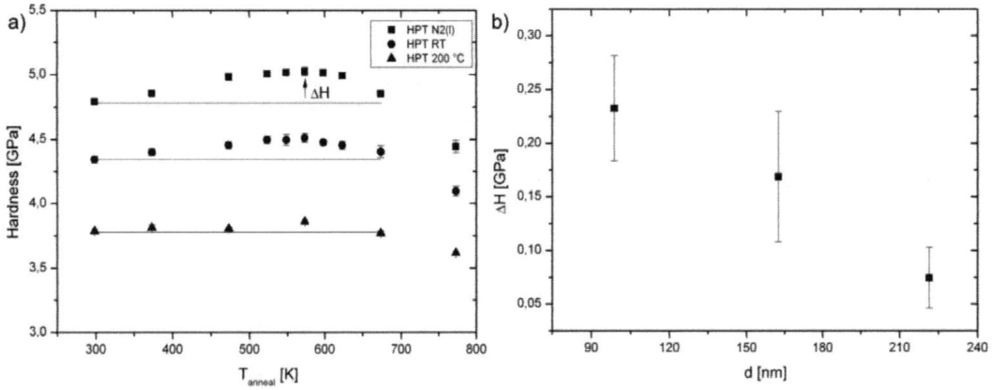

Figure 6. a) RT hardness values for the isochronally annealed Pt5Ru samples. b) Hardness increase as a function of the grain size.

It can be seen that with a decrease of the deformation temperature also the hardness of the alloys increases, which is in good agreement with grain size measurements. Upon annealing, for the samples deformed at 77 K and RT pronounced hardening can be seen, whereas for the samples deformed at 573 K the hardness increase is almost negligible. Similar to the results in Fig. 5, the relative hardness increase ΔH is increasing with decreasing grain size. By plotting the hardness increase ΔH as a function of the grain size, a clear dependency can be seen as shown in Fig. 6b). These results strongly indicate that hardening by annealing will only occur below a threshold value of the grain size, which is 220 nm for the Pt5Ru alloy under investigation. As can be expected from Fig. 5, also for Ni, hardening will occur only below a certain grain size.

The occurrence of the hardening mechanism only below a specific grain size might further rule out a strong contribution of segregation to the strengthening. Since for all Pt5Ru samples the global composition remains the same, only the excess values of segregated solute might be different. For a given annealing temperature one can expect the diffusion length, i.e. the distance from which solute can be transferred to the grain boundaries, to be similar for all three sets of samples. Only the surface area over which the solute can be distributed drastically increases with decreasing grain size (~ 1/d). Based on this estimation one would expect larger solute excess values for the samples having the largest grain size. The estimation of 'smaller is cleaner' was also experimentally validated [13]. In the case of a strong correlation between segregation and strengthening, one would expect a significantly enhanced strengthening of the samples having the largest excess of segregated solute – the opposite of what is observed, see Fig. 6.

The reason for this grain size dependent behavior is yet not clear. It seems that this transient behavior must be related to intrinsic features of crystal plasticity being dominant in the grain size regimes below and above the observed threshold value. One could expect, for grains coarser than a critical value, intragranular sources to be still active after the heat treatment, while with decreasing grain size a transition to intergranular dislocation sources takes place. As outlined above, this requires much higher stress levels.

4. Conclusion and Outlook

Hardening by annealing is frequently observed in nc materials. Although this phenomenon has been known for several years the role of segregated impurities or solute to the strengthening is yet not clear. Additionally, several UFG materials show hardening by annealing while others do not. The present study presents experimental findings to clarify the raised questions. The findings lead to the following conclusions:

1) Formation of second phase particles and large variations of the type of segregated solute do not have an influence on the mechanical properties of the investigated nc steel

2) Hardening by annealing can occur at relatively low homologous temperatures

3) Hardening by annealing occurs only in materials with grain sizes below a material specific threshold value

Although the combination of APT and mechanical datasets provided deeper insights to rationalize the origin of the strengthening, the effect of small excess values still remains unclear. However, as significant hardening within short time intervals occurs, further in depth studies are promising to resolve this question. Apart from these observations the impact of this hardening mechanism, specifically of different boundary chemistries, on other mechanical properties such as ductility or fracture toughness will bear further scrutiny.

Acknowledgements

Financial support by the FWF Austrian Science Fund within project number 24141-N19 and the European Research Council under ERC Grant Agreement No. 340185 USMS is gratefully acknowledged.

References

[1] Huang X, Hansen N, and Tsuji N 2006 *Science* **312** 249
[2] Ma E, Shen T and Wu X 2006 *Nat. Mater.* **5** 515
[3] Shen T D *et al* 2007 *Acta Mater.* **55** 5007
[4] Valiev R Z *et al* 2010 *Scripta Mater.* **63** 949
[5] Van Swygenhoven H, Derlet P M and Frøseth A G 2006 *Acta Mater.* **54** 1975
[6] Hasnaoui A, Van Swygenhoven H and Derlet P M 2002 *Acta Mater.* **50** 3927
[7] Li J C M 2006 *Phys. Rev. Lett.* **96** 215506
[8] Renk O *et al* 2015 *Scripta Mater.* **95** 27
[9] Chih-Chun Hsieh and Wu W 2012 *ISRN Metallurgy* **2012** 732471
[10] Moser B *et al* 2006 *Scripta Mater.* **54** 1151
[11] Renk O, Hohenwarter A and Pippan R 2012 *Adv. Eng. Mater.* **14** 948
[12] Mompiou F *et al* 2013 *Acta Mater.* **61** 205
[13] Terwilliger C D and Chiang Y 1995 *Acta Metall. Mater.* **43** 319

Characterization of boundary roughness of two cube grains in partly recrystallized copper

J Sun[1], Y B Zhang[1], A B Dahl[2], K Conradsen[2] and D Juul Jensen[1]
[1] Danish-Chinese Center for Nanometals, Section for Materials Science and Advanced Characterization, Department of Wind Energy, Technical University of Denmark, Risø Campus, 4000 Roskilde, Denmark
[2] Section for Image Analysis and Computer Graphics, Department of Applied Mathematics and Computer Science, Technical University of Denmark, Richard Petersens Plads, 2800, Kgs. Lyngby, Denmark

E-mail: jusu@dtu.dk

Abstract: Protrusions and retrusions typically form on recrystallizing boundaries and thus the boundaries often appear rough. Characterization of the boundary roughness is necessary in order to evaluate the effects of protrusions and retrusions on boundary migration. In the current work, a variable termed area integral invariant is employed to provide quantitative information of individual protrusions/retrusions on boundaries surrounding two selected recrystallizing grains in partly recrystallized copper as well as of the overall roughness of the boundaries.

1. Introduction

During the past few years, in-situ 3-dimensional X-ray diffraction (3DXRD) experiments have been used to observe the growth of recrystallizing grains in the bulk of deformed metals [1, 2]. It has been revealed that the nuclei do not grow in a homogeneous manner with constant moving speed as described in classic models, but occur with variations at the local scale: most boundary segment move in a jerky stop-go mode, with locally large protrusions and retrusions forming and disappearing as the boundary migrates [1, 2]. Ex-situ electron backscattered diffraction (EBSD) and electron channeling contrast (ECC) studies of local boundary migration have also revealed similar migration behavior [3-5]. Quantitative characterizations of local protrusions and retrusions have shown the importance of these local structural variations [4, 5]. Results from phase field simulations of boundary migration during recrystallization have shown that the protrusions and retrusions forming on the migrating boundaries may contribute an additional driving force resulting from the boundary curvature and change the migration kinetics at both the local and global scale [6]. Protrusions and retrusions have been observed in both high-purity metals and alloys, also under different thermal-mechanical processing conditions [7-9]. For analysis of the effects of protrusions/retrusions on the local boundary migration kinetic or for better knowledge of the effects of macro parameters such as purity of the metal, annealing temperature, deformation strain etc. on the formation of protrusions/retrusions, statistical quantification of protrusions/retrusions on many recrystallization boundaries in partly recrystallized microstructures is required. One or more variables should be developed to describe the characteristics of individual protrusions/retrusions such as wavelength and amplitude as well as the overall roughness of many boundaries in a given sample.

2. Method to characterize boundary roughness

Efforts have been made in previous studies to provide a quantitative description of protrusions/retrusions. For example, Martorano et al. [10] have used a sinusoidal function to simulate boundary shape in a numerical analysis of growth during recrystallization. Zhang et al. [11] have applied both sinusoidal and polynomial functions to fit experimental data for protrusions/retrusions to calculate the local curvature driving forces. These methods are capable of quantitatively describing individual protrusions/retrusions. However, they are based on manual selections and are not suitable for statistical quantification of many protrusions/retrusions on a number of boundaries in a sample. Fractal analysis has been used in characterizing irregular and rough morphological features in the microstructure of materials [12, 13], since Mandelbrot made the correlation between the calculated fractal dimension of fracture surfaces and the impact energy for metals [14]. Whereas it is clear that neither fracture surfaces nor recrystallizing grain boundaries follow self-similarity which is the basis for obtaining fractal dimensions, analysis based on the fractal methods may be useful for characterization of recrystallizing boundaries as described in our previous work [15]. In the present work, a fully automatic method is presented and used to quantify the roughness of the boundaries surrounding two selected grains in a partly recrystallized copper sample. The same two grains were previously characterized using fractal analysis [15], however the new method presented here provides a significantly clearer description and is suitable for statistical quantification.

The area integral invariant (*AII*) is one of the integral invariants used in the digital image computing discipline for applications such as shape matching, geometry processing etc. [16,17]. In the current study, *AII* is modified to be a variable to quantify the boundary morphology. The method is schematically shown in figure 1: *AII* works by drawing a circle with a specified radius, termed *sampling radius*, and with the center of the circle placed on the boundary. Then the area of the circle is separated into two parts, the part within the recrystallized grain (marked as red in figure 1) and the other part within the deformed structure. The *AII* at a position is calculated with a specified sampling radius as:

$$AII = \frac{Circle\ area\ within\ recrystallized\ grain}{Area\ of\ the\ entire\ circle} \tag{1}$$

A number with a value between 0 and 1 is obtained: *AII* smaller than 0.5 indicates a protrusion at that position, and the smaller the value, the sharper is the protrusion; conversely for a retrusion; *AII* equals 0.5 for planar boundaries.

By measuring *AII* with a specified sampling radius along the boundary, a distribution of *AII* values for the whole boundary can be obtained and further statistical analysis can be done to characterize the roughness of the boundary. Several roughness parameters are further developed based on the distribution of *AII* to characterize various aspects of the boundary roughness (see table 1). The roughness of the boundary can be defined as the deviation of *AII* values from 0.5. As shown in table 1, the boundary roughness R_a is therefore calculated very similarly to the descriptive statistical parameter "coefficient of variation" but replacing the mean value with 0.5. $L_{boundary}$ is the total length of the boundary calculated by counting the pixels forming the boundary. Using a similar calculation process for only the *AII* data for protrusions or retrusions, the protrusion and retrusion roughness can be obtained as R_p and R_r, respectively. The L_p and L_r in the equation represent the length of boundary segments that are composed of pixels with $AII < 0.5$ (AII_p) and $AII > 0.5$ (AII_r), respectively. The *Range* is calculated as the difference between the maximum and minimum *AII* values. Finally the fraction of protrusions/retrusions is calculated as the total number of *AII* values that are smaller than 0.4 or larger than 0.6 divided by the boundary length. This parameter provides an estimation of the fraction of rough features with respect to the entire boundary.

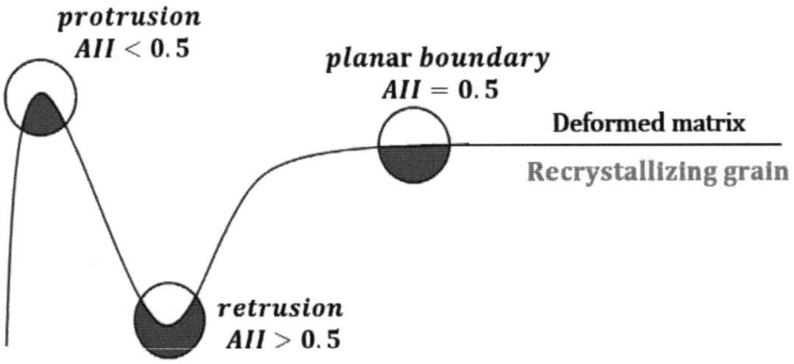

Figure 1. Sketch demonstrating the principle of how the *AII* is calculated. The deformed matrix is separated from the recrystallizing grain by the migrating boundary. The circle area is divided by the boundary into two regions and the *AII* is calculated as the area marked in red divided by the entire circle area.

Table 1. Roughness parameters

Roughness parameter	Calculation equation
Boundary roughness	$R_a = \left\{\frac{\sum_{i=1}^{N}(AII_i - 0.5)^2}{L_{boundary}}\right\}^{\frac{1}{2}} /0.5$
Fraction of protrusions/retrusions	$R_f = \frac{Number\ (AII < 0.4\ or\ AII > 0.6)}{L_{boundary}}$
Protrusion roughness	$R_p = \left\{\frac{\sum_{i=1}^{Np}(AIIp_i - 0.5)^2}{L_p}\right\}^{\frac{1}{2}} /0.5$
Retrusion roughness	$R_r = \left\{\frac{\sum_{i=1}^{Nr}(AIIr_i - 0.5)^2}{L_r}\right\}^{\frac{1}{2}} /0.5$
Range	$Range = AII_{max} - AII_{min}$

3. Experimental materials and boundary measurement

Two cube oriented grains in partly recrystallized microstructures are investigated. Both grains are from oxygen free high conductivity (OFHC) copper (99.9% purity) cold rolled to 90% reduction and annealed for 60 minutes at 150 °C. The EBSD images for both grains are shown in figure 2 as well as the boundaries of the grains after image processing with the interior twin boundaries removed. The two grains are designated as CuSC1 and CuSC2.

Visual inspection indicates that both CuSC1 and CuSC2 have rough boundaries, with protrusions and retrusions of various sizes formed on the boundaries. However, CuSC1 appears more irregular than CuSC2 as CuSC1 has some narrow branch-like features, whilst CuSC2 can be simply described as an elongated grain with some smaller protrusions/retrusions on the boundary.

The EBSD mapping was conducted using a step size of 0.1 μm, implying that 1 pixel in the EBSD maps corresponds to 0.1 μm. The sampling radius for calculating the *AII* determines the length scale of the rough features to be characterized, and the calculation is using pixel as unit in the digital image. So the correlation between the physical length in μm and the length in the digital image measured by pixels should be clarified. The rough features with a small length scale on the boundary require small sampling radius. But if the sampling radius is smaller than 20 pixels, the error from pixelization of drawing the circles would be significant. To reduce this error, the original images have been enlarged by a factor of two using the MATLAB image processing toolbox, i.e.1 pixel in the new images corresponds to the physical length of 0.05 μm. The features on the boundary are preserved during this image-resize processing. As mentioned above, sampling radius no smaller than 20 pixels is then used to calculate *AII* with the new images. The boundaries used for calculation of *AII* are 1 pixel wide and pixels on the boundaries are connected with 4-neighbor connectivity.

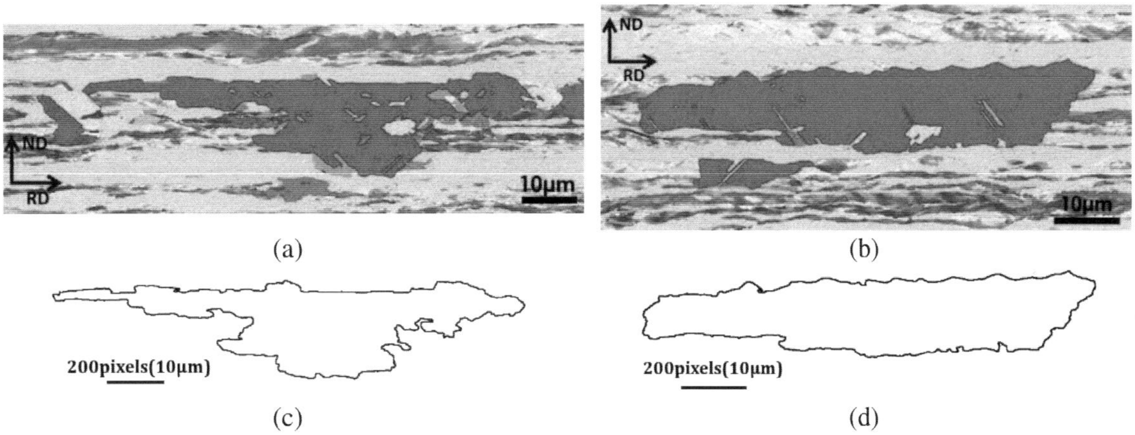

Figure 2. Electron backscattered diffraction (EBSD) images of CuSC1 (a) and CuSC2 (b). The boundaries surrounding the two grains are shown in (c) and (d). Interior twin boundaries are ignored during grain detection. Note that the boundaries showing in (c) and (d) are after pixel doubling of the original boundaries determined from EBSD maps.

The *AII* was calculated for both CuSC1 and CuSC2 at each pixel position along the boundaries using multiple sampling radii. In the following text, most results are obtained with a sampling radius of 40 pixels, corresponding to a physical length of 2 μm.

4. Results and discussion

Histograms of *AII* calculated for both CuSC1 and CuSC2 using sampling radius of 40 pixels (2 μm) are shown in figure 3. Both distributions show a clear peak at 0.5, which means that a major fraction of the two boundaries are relatively smooth at this length scale. The protrusions are revealed as the left tail of the distribution and the retrusions as the right. It is clear that the tails of the distribution are longer and bigger on both sides for CuSC1 than for CuSC2, which indicates that the rough features for CuSC1 are more prevalent than for CuSC2. Both distributions are not symmetric, and have longer tails on the retrusion side.

Larger retrusions than protrusions on a boundary have been observed from both experimental data and simulation [4, 6, 11]. The simulation has shown that in such a case the overall boundary migration rate can be affected [6]. Therefore the present asymmetry in the *AII* distribution shall be analyzed in more detail, for one protrusion and one retrusion with exactly the same morphology, $AII_r = 1 - AII_p$. Thus the distribution of $1 - AII_p$ could be directly compared with the distribution of AII_r.

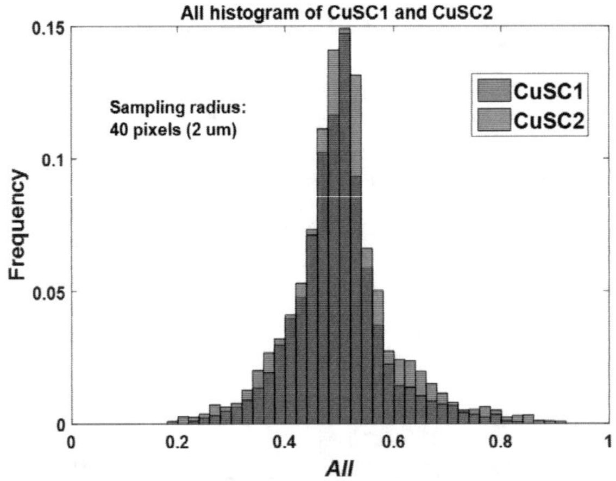

Figure 3. Histogram of *AII* for CuSC1 and CuSC2 calculated with a sampling radius of 40 pixels corresponding to 2 μm.

This comparison for CuSC1 and CuSC2 is shown in figure 4. For both boundaries, the tails of AII_r distribution are longer than those of $(1 - AII_p)$, implying that retrusions are sharper than protrusions and there are more sharp-retrusions than protrusions on both boundaries. More values of $(1 - AII_p)$ closer to 0.5 on both boundaries indicate that at the applied sampling scale the protrusions are "smoother" than retrusions. Figure 2(a) and (b) reveal sharp retrusions at the right bottom region of CuSC1 and at the bottom part of the boundary for CuSC2. By comparing figure 4(a) and (b), the tails for both $(1 - AII_p)$ and AII_r are longer for CuSC1 than for CuSC2, showing that both protrusions and retrusions are sharper on the boundary of CuSC1 than CuSC2 at the applied sampling scale, which corresponds with the visual impression from figure 2.

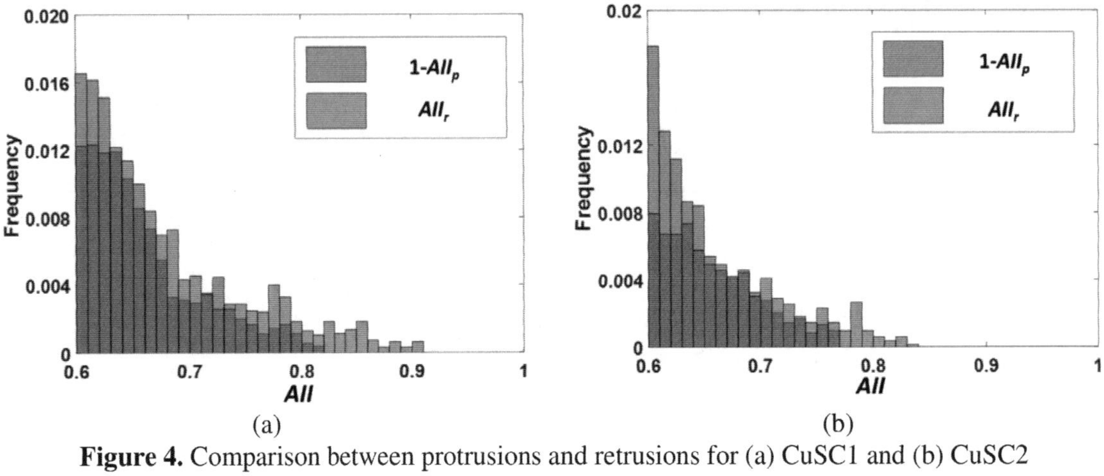

Figure 4. Comparison between protrusions and retrusions for (a) CuSC1 and (b) CuSC2

IOP Conf. Series: Materials Science and Engineering **89** (2015) 012044 doi:10.1088/1757-899X/89/1/012044

The roughness parameters calculated based on the *AII* values shown in figure 3 for CuSC1 and CuSC2 are listed in table 2. The roughness parameters allow a direct quantitative comparison between the two boundaries from various perspectives. For all the roughness parameters listed, CuSC1 has larger values than CuSC2, which implies that on all aspects considered here, on the sampling scale of 40 pixels (2 μm), boundary CuSC1 is rougher than CuSC2. Also for both CuSC1 and CuSC2, the R_p values are smaller than R_r, consistent with the analysis above.

Table 2. Roughness parameters of CuSC1 and CuSC2 (sampling radius: 40 pixels corresponding to 2 μm)

Parameter	R_a	R_f	R_p	R_r	*Range*
CuSC1	0.198	0.255	0.329	0.390	0.721
CuSC2	0.157	0.165	0.303	0.363	0.602

To investigate if the roughness of CuSC1 and CuSC2 also differs on other length scales, roughness parameters are calculated with a sampling radius from 20 pixels (1 μm) to 250 pixels (12.5 μm). The results are shown in figure 5. It can be seen from figure 5(a) that at each length scale investigated, the boundary roughness R_a of CuSC1 is larger than that of CuSC2 and the difference increases at a larger length scale. With the increasing sampling radius, the boundary roughness of CuSC1 increases continually. This is because protrusions/retrusions of larger sizes are revealed with a larger sampling radius. While for CuSC2, R_a reaches a local maximum at about 50 pixels (2.5 μm), which means the largest protrusions/retrusions are mostly around this length scale. Referring to the morphology of CuSC2 (figure 2(d)), the protrusions/retrusions observable on the top and bottom parts of the boundary are approximately around this scale. R_a of CuSC2 starts to increase again with sampling radius larger than 150 pixels (7.5 μm), as both ends of the elongated grains are then tracked as protrusions by the *AII* calculation.

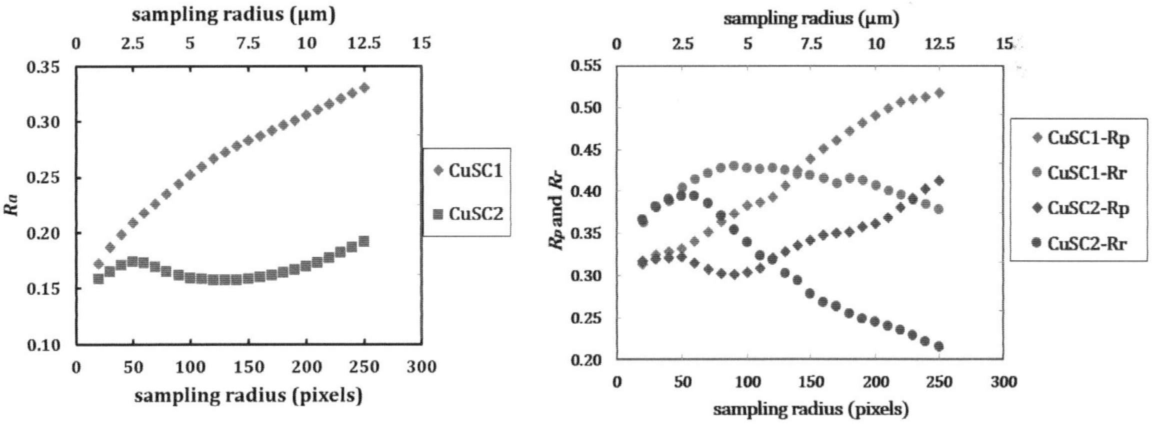

(a) (b)

Figure 5. Calculated roughness parameters for CuSC1 and CuSC2: (a) R_a and (b) R_p and R_r as a function of sampling radius.

The roughness variation considering only protrusions or retrusions is shown in figure 5(b). As discussed above at the sampling radius of 40 pixels (2 μm), the protrusions are "smoother" than retrusions. While with increasing sampling radius, the protrusions become the dominant rough features. As shown in figure 5(b), for CuSC1 R_p become larger than the R_r with a sampling radius above 140 pixels (7 μm), also for CuSC2,

R_p exceeds R_r with the sampling radius above 120 pixels (6 μm). The reason why R_p is higher for CuSC1 than CuSC2 at sampling radii above 50 pixels (2.5 μm) is that there are on CuSC1 several large branch-like features that have a scale larger than 2.5 μm on this boundary. It can also be observed that the retrusion roughness of CuSC2 (red circle) has a peak at a sampling radius of 60 pixels (3 μm), followed by a rapid decrease with increasing sampling radius, which leads to the observed local maximum in figure 5(a). While for CuSC1, the retrusion roughness is relatively steady, with a moderate peak at a sampling radius of 90 pixels (4.5 μm). The three calculated roughness parameters all demonstrate that CuSC1 is more irregular than CuSC2 at the length scales investigated. The visual difference of morphology and irregularity of the two grains as shown in figure 2 is clearly revealed from the roughness parameters. And this result corresponds well with the results obtained from previous work of boundary fractal analysis [15].

The EBSD images used for the current study were taken from the RD-ND plane for both specimens. Since in reality the recrystallizing boundaries are rough surfaces, one might wonder whether the characterization of protrusions/retrusions from a 2-dimensional image would represent the characteristics of real 3-dimensional structures. This of course cannot be verified for the present data, however, an 3-dimensional characterization of the recrystallizing boundaries in partly recrystallized aluminum has revealed the protrusions/retrusions appeared as ridges prolonged along TD [18], so the 2D microstructural characterization well indicates the rough features of the recrystallizing boundary.

5. Conclusions and outlook

A method is developed to quantitatively characterize the roughness of recrystallizing boundaries in partly recrystallized microstructures. Using the area integral invariant (*AII*) as a variable to obtain morphological information of the boundary, several parameters are calculated to quantify the boundary roughness. The potentials of these parameters in characterizing roughness are evaluated by comparing two copper grains with large visual differences in irregularity. It is found that:

(1) The visual impression of the grain CuSC1 being much more irregular than CuSC2 is quantified by the *AII* and the therefrom calculated roughness parameters;
(2) The roughness parameters are well suited to quantify possible differences in curvatures of the protrusions and retrusions. For the two grains investigated in the present work, it is statistically documented that larger curvatures exist at retrusions than at protrusions. This is important for the overall migration rate;
(3) By changing the sampling radii when measuring the *AII* it is possible to get information on the size distribution of both protrusions and retrusions.

Future efforts will be on further standardization of the method for boundary roughness quantification, and using this method to analyze the effects of protrusions/retrusions on boundary migration kinetics. Also the factors influencing the roughness of recrystallization boundary will be investigated.

Acknowledgement

The authors would like to thank the support from the Danish National Research Foundation (Grant NO.DNRF86-5) and the NSFC of China (Grant No. 51261130091) to the Danish-Chinese Center for Nanometals.

References

[1] Schmidt S, Nielsen S F, Gundlach C, Margulies L, Huang X X and Juul Jensen D 2004 *Science* **305** 229–32
[2] Van Boxel S, Schmidt S, Ludwig W, Zhang Y B, Juul Jensen D and Pantleon W 2014 *Mater. Trans.* **55** 128–36

[3] Zhang Y B, Juul Jensen D and Godfrey A 2012 *Mater. Sci. Forum* **715-716** 329–32
[4] Zhang Y B, Godfrey A and Juul Jensen D 2011 *Scripta Mater.* **64** 331–4
[5] Zhang Y B, Godfrey A and Juul Jensen D 2014 *Metall. Mater. Trans. A.* **45** 2899–905
[6] Moelans N, Godfrey A, Zhang Y B and Juul Jensen D 2013 *Phys. Rev. B.* **88** 1–10
[7] Juul Jensen D, Lin F X, Zhang Y B and Zhang Y H 2013 *Mater. Sci. Forum* **753** 37–41
[8] Martorano M, Sandim H, Fortes M and Padilha F 2007 *Scripta Mater.* **56** 903–6
[9] Drury, M R and Humphreys F J 1986 *Acta Metal.* **34** 2259–2271
[10] Martorano M, Fortes M and Padilha F 2006 *Acta Mater.* **54** 2769–76
[11] Zhang Y B, Godfrey A and Juul Jensen D 2009 *CMC* **3** 197-207
[12] Li J 2003 *Image-based Fractal Description of Microstructures*, Kluwer Academic Publisher
[13] Bérubé D and Jébrak M 1999 *Comput. Geosci.* **25** 1059–71
[14] Mandelbrot B, Passoja D and Paullay A 1984 *Nature* **308** 721–2
[15] Sun J, Zhang Y B, Dahl A B, Conradsen K and Juul Jensen D 2015 *IOP Conf. Ser.: Mater. Sci. Eng.* **82** 012006
[16] Manay S, Cremers D, Hong B W, Yezzi A J and Soatto S 2006 *IEEE.* **28** 1602–17
[17] Pottmann H, Wallner J, Huang Q X and Yang Y L 2009 *Comput. Aided Geom. Des.* **26** 37–60
[18] Zhang Y B and Juul Jensen D 2015 *Proc. 36th Risø Int. Symp. on Mater. Sci (Roskilde, DK, 2015)*, ed S Fæster et al (Roskilde: Risø).

IOP Conf. Series: Materials Science and Engineering **89** (2015) 012045 doi:10.1088/1757-899X/89/1/012045

Effect of initial grain size on inhomogeneous plastic deformation and twinning behavior in high manganese austenitic steel with a polycrystalline microstructure

R Ueji[1, *], N Tsuchida[2], K Harada[3], K Takaki[3] and H Fujii[1]
[1] Joining and Welding Research Institute, Osaka University, 11-1 Mihogaoka, Ibaraki, Osaka 567-0047, Japan
[2] Faculty of Engineering, University of Hyogo, 2167 Shosya, Himeji, Hyogo 671-2280, Japan
[3] Faculty of Engineering, Kagawa University, 2217-20 Hayashi-cho, Takamatsu, Kagawa 567-0047, Japan

Email: *ueji@jwri.osaka-u.ac.jp

Abstract. The grain size effect on the deformation twinning in a high manganese austenitic steel which is so-called TWIP (twining induced plastic deformation) steel was studied in order to understand how to control deformation twinning. The 31wt%Mn-3%Al-3%Si steel was cold rolled and annealed at various temperatures to obtain fully recrystallized structures with different mean grain sizes. These annealed sheets were examined by room temperature tensile tests at a strain rate of 10^{-4} /s. The coarse grained sample (grain size: 49.6μm) showed many deformation twins and the deformation twinning was preferentially found in the grains in which the tensile axis is parallel near to [111]. On the other hand, the sample with finer grains (1.8μm) had few grains with twinning even after the tensile deformation. The electron back scattering diffraction (EBSD) measurements clarified the relationship between the anisotropy of deformation twinning and that of inhomogeneous plastic deformation. Based on the EBSD analysis, the mechanism of the suppression of deformation twinning by grain refinement was discussed with the concept of the slip pattern competition between the slip system governed by a grain boundary and that activated by the macroscopic load.

1. Introduction

Austenite is a useful phase to manage both strength and elongation in steel [1, 2]. Among several kinds of typical austenitic steels, high manganese austenitic steel gathers much attention from materials scientists during the last decade [3-6]. Especially, recent research works focus on the high manganese steel with alloying of aluminum in order to prevent the dynamic phase transformation from austenite to ε phase within HCP structure. This alloy usually has a relatively low stacking fault energy of around 40 mJ/m^2 at ambient temperature and it is reported that deformation twinning occurs occasionally so that this steel was widely called a Twinning Induced Plasticity (TWIP) steel [3]. Previous papers reported that the deformation twinning preferentially occurs when the load direction is near parallel to [111] direction at the tensile deformation [6]. The deformation twinning provides new boundaries for the object against the dislocation slip, which provides a higher work hardening rate followed by a larger uniform elongation. It was also clarified that the adequate ductility of the TWIP steel keeps even at a

IOP Conf. Series: Materials Science and Engineering **89** (2015) 012045 doi:10.1088/1757-899X/89/1/012045

high strain rate deformation up to 10^2/s [7, 8]. The excellent properties of TWIP steel seems to be attractive also for industrial applications [3].

By the way, grain refinement is one of the strengthening methods without any additional alloying. It is well-known that when the mean grain size becomes as small as a few micrometers in the BCC steels or single phase aluminum, the strength increases however, the elongation decreases [9]. On the other hand, for the 31%Mn-3%Al-3%Si TWIP steel, the grain refinement provides strengthening while keeping adequate elongation, although the deformation twinning is suppressed by the grain refinement. Ueji et al. [6] clarified the crystallographic tendency that [111] parallel to the tensile direction is preferable for the twinning even in a fine grained sample. However, the mechanism of the suppression of the twinning has been not clarified yet.

The relationship between initial grain size and deformation twinning in polycrystalline metal has been studied for example by Barnett et al.[10] They studied the grain size effect in a Mg alloy and clarified the mechanism of the suppression by the difference between the Hall-Petch slope for the dislocation slip and that for the twinning. It is reasonable because deformation twinning in the magnesium alloy occurs at plastic yielding. However, the twinning in the TWIP steel generally appears after the leading plastic deformation by dislocation slip [11] as is found in other FCC metals [12, 13]. This fact implies the importance of the investigation on the change of dislocation structure by the grain refinement. Consequently, this study was conducted to examine the deformation induced substructure in the fully recrystallized TWIP steels with different grain sizes and to discuss the mechanism of the twinning suppression by the grain refinement.

2. Experimental Procedures

TWIP steel with a chemical composition of 31%Mn-3%Si-3%Al-Bal.Fe was used. The ingot was hot rolled at 1200 °C to obtain a plate with a thickness of 12mm. The hot rolled sheet was cold rolled to a reduction in thickness of 12 mm followed by annealing at 900 °C for 9.2 ks for homogenization. The heat treated sample was cold rolled to a reduction of 83% and subsequently annealed at 700 °C, 800 °C or 1000 °C for 1.8ks.These annealed samples have a fully recrystallized microstructures whose mean grain sizes were 1.8 μm, 7.2 μm or 49.6 μm, respectively. This thermomechanical treatment is just the same as performed in the previous work [6]. The tensile specimens with a gage part of 25mm length x 5mm width x 1.2mm thickness were cut from these annealed samples and tensile test were conducted at an ambient temperature at a strain rate of 10^{-4}/s. The samples were strained to various strains and these were measured by the Electron Back Scattering Diffraction (EBSD) technique in a Scanning Electron Microscope (SEM). The EBSD data were analyzed mainly focused on local misorientation angles. The EBSD scan step size was 0.05 μm at all scans except for the measurements shown in Fig.4(c) and Fig.5(b) where the coarse grained sample were measured with a step size of 0.25 μm.

3. Results and discussion

Figure 1 shows the stress - strain curves of the annealed TWIP steels. The end points of the curves equal to the point where the load reached the maximum which corresponded to the starting point of the necking. All of the curves show a somehow liner morphology, which is typically found in some of the FCC metals with a low stacking fault energy. With decreasing mean grain size, the yield stress and the flow stress increase because of the grain refinement strengthening.

Figure 2 shows the gray scale maps for the distribution of Kernel Average

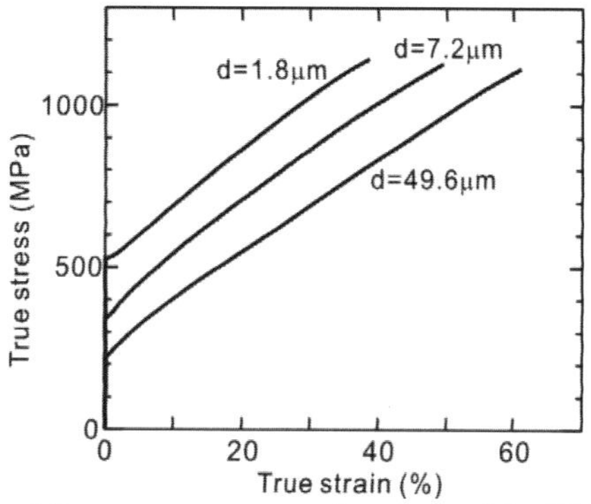

Figure 1 True stress - true strain curves of the 31%Mn-3%Al-3%Si TWIP steels with different mean grain sizes (d).

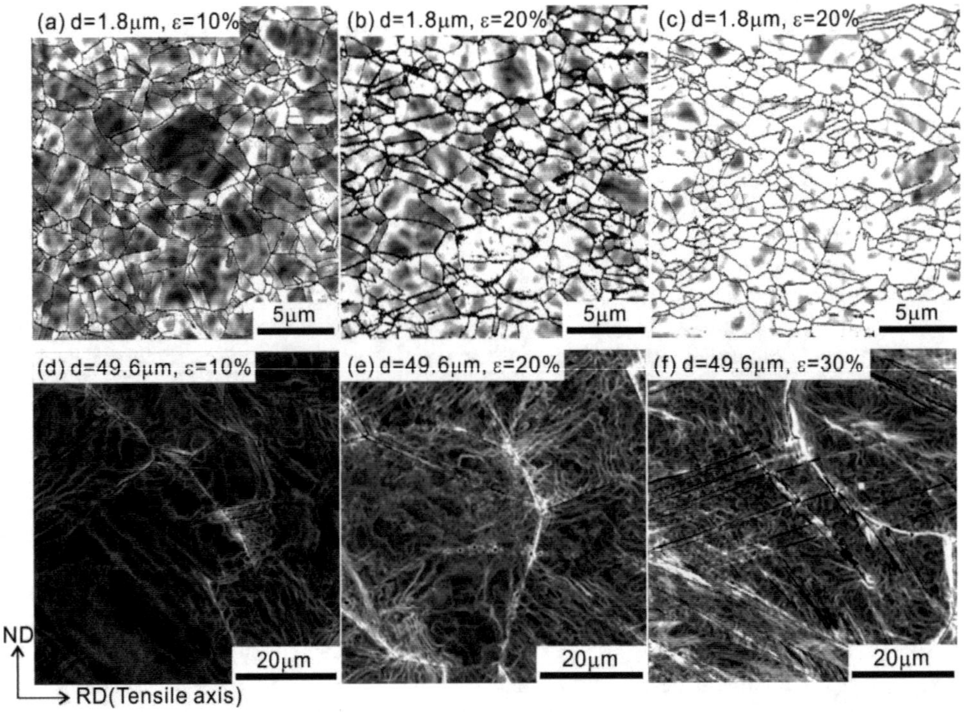

Figure 2 Grey scale maps for the distribution of KAM angles in the 31%Mn-3%Al-3%Si TWIP steels with initial grain size, d of 1.8 μm (a, b, c) or 49.6 μm (d, e, f) tensile tested to a true strain, ε of 10% (a, d), 20% (b, e) or 30% (c, f). The EBSD measurements were conducted on the cross-sectional plane parallel to the normal direction (ND) and the rolling direction (RD).

Misorientation (KAM) angles in the samples with initial grain sizes of 1.8μm (a, b, c) or 49.6μm (d, e, f) tensile tested to a strain of 10%, 20% or 30%. The KAM angle equals to the mean value of all the misorientation angles between the measurement position and the EBSD scan points which have a constant distance (KAM step) from the measurement position [14]. The main purpose for the KAM evaluation is the examination of substructure, so that the vicinity point belonging to the different grain from that of the measurement position is eliminated to estimate the KAM angle. In addition, the EBSD data from the position adjacent of the initial grain boundaries are not evaluated. In the gray scale map, the larger KAM angle are indicated by the whiter color.

In the case of the 1.8 μm sample, 10% tensile deformation provides a local increase of the KAM angle especially at the locations near the grain boundaries or the triple junctions of grain boundaries. It should be noted that no deformation twinning was found and that these substructures were due to the slip deformation of the dislocations. With increasing strain, the area with higher KAM angle extends more widely. On the other hand, a different morphology was observed in the coarse grained sample: as shown in the KAM maps of the 49.6 μm sample, many bands having a larger KAM angle are additionally found also in the interior of the grains. These bands extend along different orientations and the intersection of these band produces a kind of cell structure. These cell structures show a different morphology in the different grains. When the coarse grained sample was tensile deformed to a strain of 30%, a lot of deformation twinning occurred in some of the grains although the twinning cannot be found in the fine grained samples. This result confirmed the suppression effect of the grain refinement on the deformation twinning which was reported already [6].

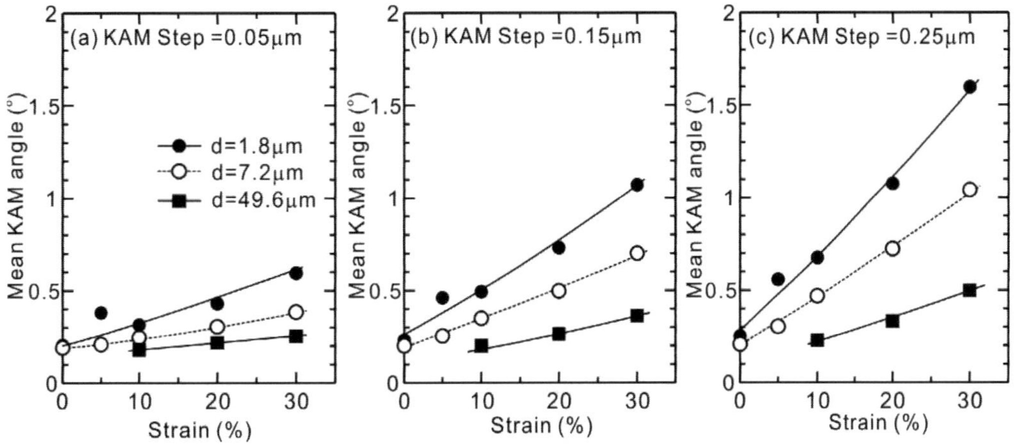

Figure 3 Mean KAM angles in the 31%Mn-3%Al-3%Si TWIP steels with various mean grain sizes (d) tensile strained. The mean KAM angle was calculated with different KAM steps.

The mean value of all the KAM angles in each scan of the samples is shown in Fig.3. Although various kinds of the KAM steps, 0.05 μm (a), 0.15 μm (b) and 0.25 μm (c) were adopted, the basic tendency is not changed by the change of KAM step. When compared with the same sample with an identical initial grain size, the mean KAM value increases with increasing of the tensile strain. This result indicates that the magnitude of the KAM angle is correlated with the magnitude of the accumulation of the dislocations. In addition, with decreasing mean grain size, the increasing rate of the mean KAM angle becomes larger, indicating a rapid increase of dislocations which contributes the local misorientation. Concerning with the effect of the KAM step, when comparing the data with the same initial grain size and the same strain, the mean KAM angle becomes larger with increasing KAM step.

As mentioned in the introduction of this paper, deformation twining show a strong orientation dependence. That is, the [111] is a more preferable tensile direction than [001] and this tendency is not changed even in the fine grained TWIP steel. Consequently, the orientation dependence of the KAM angle change is required in the study to link the dislocation structure to twinning if it exists. For the first trial, the EBSD scan data were separated between grains and the mean KAM angle of each grain were calculated to evaluate the substructure evolution related to the tensile orientations. Figure 4 shows inverse pole figures showing the crystallographic orientations parallel to the tensile direction of each grain. The grayscale of a data point indicates the mean KAM value of the corresponding grains. Concerning the distribution of the tensile directions, the sample with smaller grains has randomly distributed orientations; whereas in the coarse grained sample the tensile direction preferentially distributes along the oblique line indicating the orientation from [001] to [111]. This change of the texture evolution by grain refinement can be confirmed by the texture plots as inverse pole figures shown in Fig.5. As is found in a previous paper [6], the initial textures showed no strong orientation concentration and the maximum intensity decreased with the progress of grain growth. This tendency is opposite to what was found in this work. This result clearly indicates the difference of the lattice rotation between the coarse and fine grained samples.

Concerning the mean KAM angles, it is difficult to find the orientation dependence although these mean KAM angles were evaluated with the larger KAM step (0.25 μm) which seems to be better to show the change of substructure evolution as shown in Fig.3. More careful observation clarified that in the 7.2 μm sample (b), there are some concentrations of the points with relatively larger mean KAM angles at the orientations from [111] to [112]. This tendency seems to be related with the previous study reported especially by the research group in Denmark [15-18]. They studied the substructure evolution

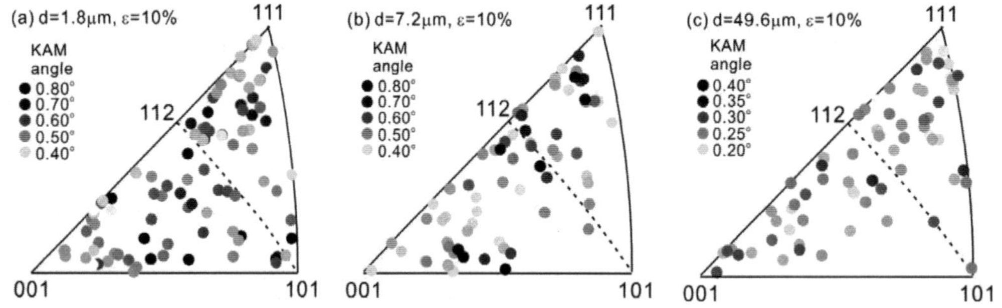

Figure 4 Inverse pole figures showing the mean crystallographic orientations parallel to the tensile axis of the grains in the 31%Mn-3%Al-3%Si TWIP steels with various mean grain sizes, d tensile strained to a strain, ε of 10%. The mean KAM angle was calculated with KAM step of 0.25 μm and illustrated by the grey scale.

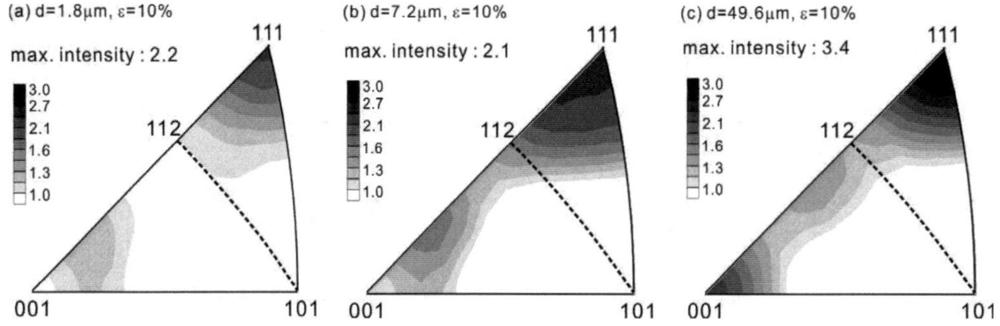

Figure 5 Inverse pole figures showing the tensile axis orientations in the 31%Mn-3%Al-3%Si TWIP steels with various mean grain sizes, d tensile strained to a strain, ε of 10%.

during the tensile deformation of FCC metals with medium to high stacking fault energies and they found that the dislocation density related to local misorientation preferentially increases when the tensile direction is closely parallel to [111]. However, a more clear orientation dependence of the mean KAM angle cannot be found in the other samples. This result indicates that conditions for the evolution of the mean KAM angle are needed to clarify the orientation dependence.

As shown in Fig.3, with increasing of the tensile strain and/or the KAM step, the mean KAM angle tends to increase. The larger value of the KAM angle should be preferable to detect the change of the orientation dependence. Figure 6 shows the mean KAM angles evaluated with a relatively large KAM step which does not exceed both the grain size and the cell size in the samples tensile strained to various strains. The samples with a mean grain size of 1.8μm (a) or 49.6μm (b) were examined with a KAM step of 0.5μm or 2.5μm, respectively. In order to clarify the orientation dependence, the EBSD scan data were separated in tensile orientation was close to the [111] direction or not. The threshold angle is set as 30° and the orientation whose deviation angle less than 30 ° is almost the same as the orientation with whom the grains have twinning preferentially when the initial grain size is coarse. When the tensile strain is limited to 10%, it is difficult to find the orientation dependence of the mean KAM value in both the samples as is similar to be found in Fig. 4. However, when the samples were deformed more than 20%, the coarse grained sample shows the orientation dependence, that is, the mean KAM value becomes smaller when the orientation is near to [111]. On the other hand, the fine grained sample still has no clear orientation dependence. Although this orientation dependence is not in agreement with the previous finding, the result at least supports the disappearance of the orientation dependence of the substructure evolution by the grain refinement to a grain size of about 1 μm.

By means of the experimental works in this study, three significant characteristics of the grain size effect on the substructure were found:

(1) When the sample has a coarse grained structure, the KAM angle becomes preferentially large both near the grain boundary and at the interior of the grain; whereas, in the fine grained sample, only the grain boundary preferentially accelerates the increase of the KAM angle. (Fig.2)

(2) The rate of increase of the mean KAM angle becomes larger with decreasing of the initial grain size. (Fig.3)

(3) The orientation dependence of the mean KAM angle can be found only in the coarse grained sample. (Fig.6)

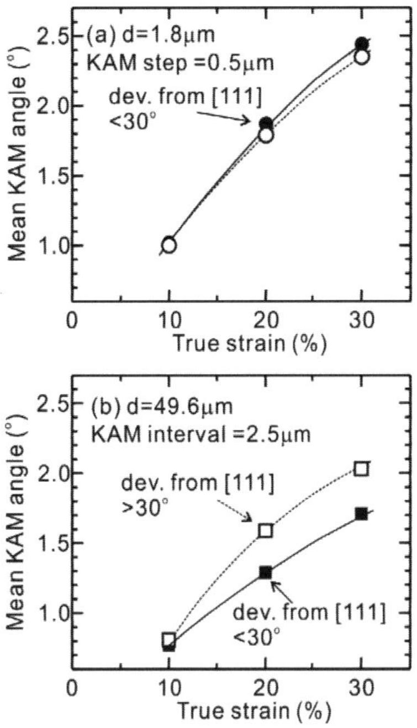

Figure 6 Mean KAM value of the 31%Mn-3%Al-3%Si TWIP steels with various mean grain sizes, d tensile strained to various strains, ε. The mean KAM angle was calculated from the data separated by the classification whether the deviation angle between tensile axis and [111] is less than 30 ° or not.

These behaviors can be explained by the difference between the slip pattern at the area near the grain boundaries and that at the grain interior. The term, slip pattern means the combination of the active slip systems and this is related to the substructure development [15-18]. When the grain size is large enough, the slip pattern is determined by the macroscopic load axis and as a result the development of the mean KAM angle shows an orientation dependence. At the area near the grain boundaries, the constraint effect on the plastic deformation [19] is supposed to be accelerated and provide the different slip pattern which includes activation of additional slip systems. In this case, the KAM angle can become larger. When the initial grain size becomes smaller, the slip pattern governed by the grain boundaries becomes dominant and weakens the orientation dependence of the KAM angle. In other words, a slip pattern competition changes with the grain refinement.

The concept of the slip pattern competition between the slip system governed by grain boundary and that activated by the macroscopic load can explain the suppression of twinning by grain refinement. As mentioned in the introduction of this paper, deformation twinning in FCC metals should need some additional force from the dislocation structure which evolves by the slip deformation before the twinning. Actually, Narita et al. [12, 13] proposed a quantitative model to estimate the additional stress and Karaman et al. [11] modified their theory to explain the deformation twinning behavior in single crystal high manganese steel. In these studies, the active slip systems can be predicted with the orientation of deformation axis. That is to say, the slip pattern activated by the macroscopic load is needed for the deformation twinning. The grain refinement extends the area where the slip system are governed by grain boundaries and suppresses the slip systems activated by the macroscopic load. Consequently, deformation twinning should be prohibited by the grain refinement.

For a further study in order to clarify the effect of slip pattern competition quantitatively, the numerical information of the development of the mean KAM value as shown in Fig.6 should be useful. However, the reason why the mean KAM angle increases rapidly in the grains whose [111] orientation is near to the tensile axis has not been clarified yet, although the previous work [15-18] supported the opposite tendency. This seems to be due to the difference of the materials with different stacking fault energy or the difference of the mythology of microstructural characterization, which is EBSD/SEM or transmission electron microscopy (TEM). However, the FCC metals with a low stacking fault energy hardly have the conventional dislocation cell structure, so that it is difficult to consider the critical method to clarify the substructure development in TWIP steel and further study should be needed.

4. Summary

The mechanism of the suppression effect of the grain refinement on the deformation twinning was proposed based on the EBSD analysis of the grain size dependence of the substructure formed by dislocation slip in the 31%Mn-35Al-3%Si TWIP steel. The summary can be drawn as following:

(1) When the grain size is large enough, the evolution of an inhomogeneous substructure which can be detected by KAM angle changes whether the observation location is the grain interior or the vicinity of the grain boundaries. Near the grain boundary, the substructure is formed mainly by the slip pattern which can easily produce higher KAM angles. On the other hand, at the interior of the grain, the slip pattern should be governed by the macroscopic load direction, so that the evolution of the KAM value shows an orientation dependence when the grain size is large enough.

(2) The grain refinement provides the wider area where the slip pattern controlled by grain boundaries operates and spontaneously the slip pattern governed by the macroscopic load direction is weakened.

(3) The mechanism of the twinning suppression by the grain refinement can be explained by the competition between the slip pattern governed by the macroscopic load direction and that controlled by the grain boundaries, because the deformation twinning needs an additional stress which is due to substructure produced by the slip pattern related to the macroscopic load direction.

Acknowledgements

The authors would like to thank Prof. N. Tsuji in Kyoto University, Japan for useful discussion. This work was supported by a Grant-in-Aid for Scientific Research on Innovative Area "Bulk Nanostructured Metals" (ID: 22102002) from the Japan Society for the Promotion of Science (JSPS) and the Japan Science and Technology Agency (JST) under Collaborative Research Based on the Industrial Demand, "Heterogeneous Structure Control: Towards Innovative Development of Metallic Structural Materials".

References

[1] Angle J 1954 *J. Iron Steel Inst.*, **177** pp 165–174
[2] Adler P H, Olson G B and Owen W S 1986 *Metall. Mater. Trans. A* **17A** pp 1725–1737
[3] Bouaziz O, Allain S, Scott C P, Cugy P and Barbier D 2011 *Curr. Opin. Solid State Mater. Sci.* **15** pp 141–168
[4] Park K -T, Jin K G, Han S H, Hwang S W, Choi K and Lee C S 2010 *Mater. Sci. Eng. A* **527** pp 3651–3661
[5] Rahman K M, Vorontsov V A and Dye D 2015 *Acta Mater.*, **89** pp 247-257
[6] Ueji R, Tsuchida N, Terada D, Tsuji N, Tanaka Y, Takemura A and Kunishige K 2008 *Scr. Mater.*, **59** pp 963–966
[7] Ueji R, Harada K, Tsuchida N and Kunishige K 2007 *Mater. Sci. Forum* **561-565** pp 107-110
[8] Ueji R, Harada K, Takemura A and Kunishige K 2008 *Mater. Sci. Forum* **584-586** pp 673-678
[9] Tsuji N, Ito Y, Saito Y and Minamino Y 2002 *Scripta Mater.*, **47** pp 893-899
[10] Barnett M R, Keshavarz Z, Beer A G and Atwell D 2004 *Acta Mater.* **52** pp 5093-5174
[11] Karaman I, Sehitoglu H, Gall K, Chumlyakov Y I and Maier J 2000 *Acta Mater.*, **48** pp 1345-1359
[12] Narita N and Takamura J 1992 *Dislocations in Solids* **9** pp 135-189

[13] Miura S, Takamura J and Narita N 1968 *Trans. JIM* **9** pp 555-561
[14] Choi J Y, Ji J H, Hwang S W and Park K -T 2012 *Mater. Sci. Eng. A* **535** pp 32-39
[15] Hansen N and Huang X 1998 *Acta Mater.* **46** pp 1827-1836
[16] Hansen N, Huang X and Hughes D A 2001 *Mater. Sci. Eng. A*, **317** pp 3-11
[17] Huang X and Winther G 2007 *Phil. Mag.* **87** pp 5189-5214
[18] Winther G and Huang X, 2007 *Phil. Mag.* **87** pp 5215-5235.
[19] Higashida K, Takamura J and Narita N 1986 *Mater. Sci. Eng.* **81** pp 239-258

IOP Conf. Series: Materials Science and Engineering **89** (2015) 012046 doi:10.1088/1757-899X/89/1/012046

Combined effect of rapid nitriding and plastic deformation on the surface strength, toughness and wear resistance of steel 38CrMoAlA

B Wang[1], Z Q Lv[2,3], Z A Zhou[1], S H Sun[1,4], X Huang[3], W T Fu[1*]

[1] State Key Laboratory of Metastable Materials Science and Technology, College of Materials Science and Engineering, Yanshan University, Qinhuangdao 066004, China
[2] College of Mechanical Engineering, Yanshan University, Qinhuangdao 066004, China
[3] Danish-Chinese Center for Nanometals, Section for Materials Science and Advanced Characterization, Department of Wind Energy, Technical University of Denmark, Risø Campus, DK-4000 Roskilde, Denmark
[4] College of Sciences, Yanshan University, Qinhuangdao 066004, China

[*] E-mail: wtfu@ysu.edu.cn

Abstract. The combined treatment of pressurized gas nitriding and cold rolling is proposed as a new approach to rapid preparation of a strong and tough nitrided layer for steel 38CrMoAlA. The microstructural characteristics and properties of the modified surface layer in comparison with those of the conventionally gas nitrided sample have systematically been evaluated. The results show that the hardness and toughness of the nitrided surface layer can be significantly improved by the combined treatment. Especially, the wear resistance of nitrided surface layer under heavy loads was greatly enhanced. It can provide a new approach to rapidly preparing a nitrided layer with high strength and toughness.

1. Introduction

Gas nitriding (GN), one of the well-established commercial surface modification techniques, has been widely used for improving the mechanical properties and wear/corrosion resistance of alloys and steels [1-3]. During GN, a gradient composite structure with a hard surface compound layer and a tough diffusion layer, is formed via thermochemical treatment in an NH_3 containing atmosphere. However, conventional GN processes are usually performed at 500~590°C for a long duration (20~80 h), which results in high energy consumption and low production efficiency in industrial applications [1-3]. Moreover, compared with the diffusion layer, the compound layer consisting of a heterogeneous mixture of ε-$Fe_{2-3}N$ and γ'-Fe_4N phase, is very thin and contains high internal stresses, so that the nitrided surface layer is friable and brittle [4-9]. This will affect the performance of the nitrided components subjected to severe service environments involving high shear, compressive and impact loading conditions [4,5,7].

Up to now, many rapid GN technologies have been developed to improve GN efficiency such as a surface nanocrystallization pretreatment [10,11], an activated GN by adding ammonia chloride [12,13], as well as a cathode sputtering pretreatment [14]. Meanwhile, some methods of nitriding such as two-stage GN and controlled GN with the nitriding potential below the critical value were developed to obtain a single γ′-Fe_4N phase or to form a nitrided surface without a compound layer [4,5,15]. However, it would take a longer duration when nitriding under the lower nitriding potential using an extremely low ammonia flow rate. Although the brittleness of the surface nitrided layer could be eliminated by means of the two-stage nitriding process, consisting of conventional nitriding and decomposition of the compound layer by ion sputtering, serious deterioration of the surface hardness and wear resistance of the nitrided steel will be induced [4]. Therefore, it is necessary to develop a technical method that can not only get a compound-free nitrided layer with excellent toughness on the steel surface, but also have a high production efficiency.

The previous works show that pressurized gas nitriding (PGN) can significantly accelerate the nitriding efficency due to the faster reaction rates, the sufficient and stable nitrogen surface concentration under higher pressure [16,17]. It is noted that the nitriding potential of PGN can be adjusted by the nitriding pressure and ammonia flow rate. This implies that it is possible to rapidly prepare a compound-free nitrided layer by means of a PGN treatment using continuous NH_3. In fact, it has been reported that the hardness of this tough nitrided layer can be further improved by means of cold rolling (CR) or a surface mechanical attrition treatment (SMAT) without leading to spalling on the steel surface [4]. But what effect of the combined treatment (PGN + CR) on the surface performance for the nitrided steel has not been clarified so far.

The present work aims to examine the effects of a combined treatment (PGN + CR) on the microstructure and surface properties of the nitrided steels, and compare with those in the CGN and PGN sample, and to provide theoretical and experimental warrants for developing a rapid preparation of a highly strengthened and toughened nitrided layer for steels.

2. Experimental procedure

The tested steel 38CrMoAlA was cut from a hot-rolled bar with the chemical composition (mass%) of 0.39 C, 1.62 Cr, 0.22 Mo, 1.04 Al, 0.31 Si, 0.42 Mn, 0.021 P, 0.018 S and bal. Fe. Before nitriding, the samples were oil quenched from 940°C, and tempered at 650°C for 1.5 h to produce the desired mechanical properties (with a hardness of 310 HV). All of the treated samples were cut to 60 mm × 50 mm × 3 mm, surface ground and washed carefully with acetone and absolute alcohol, then immediately hung in a pressurized GN furnace for nitriding [16]. For the nitriding treatment, high-purity NH_3 (99.99 vol.%) gas was used as the nitriding medium. The CGN was performed at 530°C for 30 h using a nitriding potential of r_N=1.96 atm$^{-1/2}$ (NH_3 flux: 0.5 l/minute) under 1 atm pressure, whereas the PGN was performed at 530°C for 5 h using a nitriding potential of r_N=0.26 atm$^{-1/2}$ (NH_3 flux: 0.2 l/minute) under 5 atm pressure. Nitriding potential is defined as $r_N = p_{NH_3} / p_{H_2}^{3/2}$, where p denotes partial pressure [1-3]. Note that, on applying this nitriding potential for PGN, hardly any iron nitrides can develop in the case of the nitrided sample, according to the Lehrer diagram [18]. After the PGN treatment, the PGN samples were cold rolled by 30%.

Cross-sectional observations of the nitrided samples were performed by optical microscopy (OM). The phases in the surface layer were identified by means of a D/MAX-PC 2500 X-ray diffractometer (XRD) using CuKα radiation, operated at 40 mA and 40 kV. A JEM-2010 TEM was employed to characterize the microstructural features in the surface layer of the PGN + CR (PGNC) sample. The depth-dependent hardness of the nitrided sample was measured using an FM-ARS 9000 micro-Vickers hardness tester with a load of 200 g and a holding time of 10 s. Before the toughness test, a mirror finish was performed to ensure minimal loss of surface layer, the surfaces of the treated samples were slightly polished using 1 μm diamond paste. The polished surfaces were then subjected to a Vickers hardness test using a load of 10 kg.

The wear tests were carried out in a pin-on-disc machine at room temperature under 1 atm pressure. The GCr15 steel pin (with a hardness of 63~65 HRC) was turned in an anti-clockwise motion at 50 rpm in contact with the nitrided sample (disc) under two loads of 100 and 400 N, respectively. After a sliding distance of about 170 m, the experiment was stopped, and then the worn surfaces were examined using an S-4800 scanning electron microscope (SEM).

3. Results and discussion

3.1. Microstructural characterization of the surface layer

Fig. 1a shows the cross-sectional microstructure of the steel 38CrMoAlA sample after CGN at 530°C for 30 h. It was clearly observed that the nitrided layer was subdivided into a bright compound layer (CL) and a gray diffusion zone. The XRD patterns revealed that the CL with about 15 μm in thickness was composed of ε-$Fe_{2\text{-}3}N$ and γ′-Fe_4N, as well as a few oxide phases (Fig. 1b). The presence of the oxide phases suggested that slight oxidation had occurred during the long GN duration. After the PGN treatment at 530°C under 5 atm for 5 h, the nitrided surface layer almost consisted of α_N phases except for a very small amount of γ′-Fe_4N phase (Fig. 1d), which was consistent with the OM observation of the PGN sample (Fig. 1c). Moreover, it is worth noting that the thickness of the nitrided layer in the PGN sample is about 282 μm, which is almost equal to that (~ 316 μm) after CGN for 30 h. The above results demonstrated that a compound-free nitrided layer can be rapidly prepared by PGN with a controlled nitriding potential, which was obtained through adjusting the nitriding pressure and ammonia flow rate.

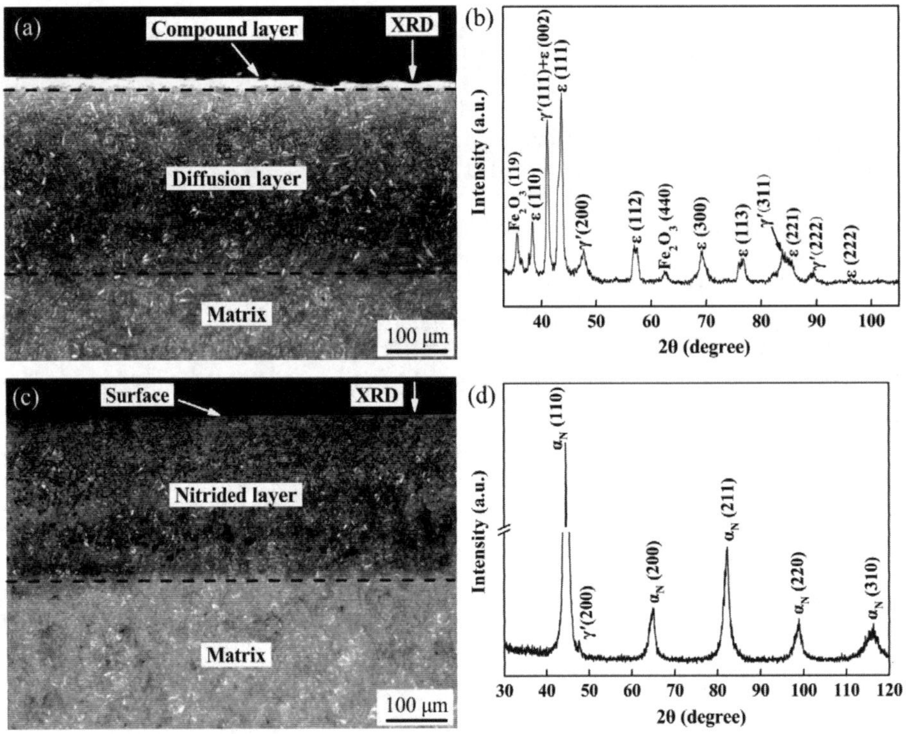

Figure 1. Cross-sectional optical micrographic observations of (a) CGN sample and (c) PGN sample, and the XRD patterns of (b) CGN sample and (d) PGN sample.

After the CR treatment to the PGN sample, a continuous flat surface without spalling was formed, and elongation of the microstructure along the rolling direction could be observed in the cross-sectional microstructure of the PGNC sample, as shown in Fig. 2a. The thickness of the plastic deformation nitrided layer is approximately 170 μm. A TEM image and SAED pattern of the nitrides in the top surface layer of the PGNC sample are shown in Fig. 2b. Besides a high-density of dislocations, a few particles of about 100-200 nm sizes were also found in the matrix of the top surface layer. The corresponding electron diffraction pattern verified that these particles are γ′-Fe_4N nitrides.

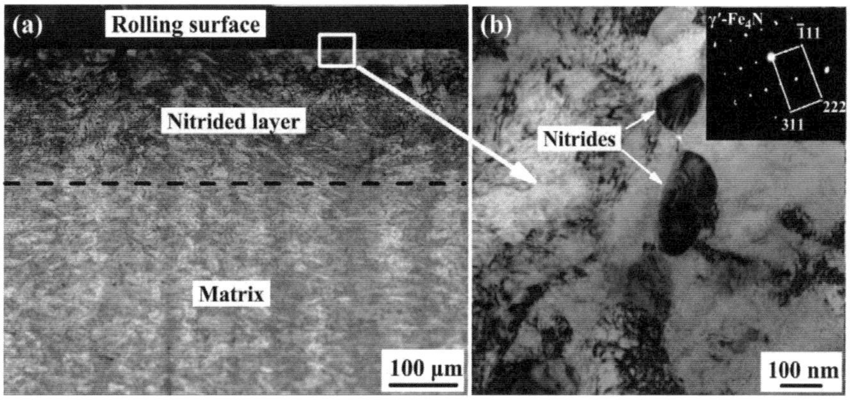

Figure 2. (a) Cross-sectional microstructure of PGNC sample. (b) A TEM image and electron diffraction pattern (inset) of the nitrides in the top surface layer of the PGNC sample.

3.2. Hardness and toughness of the nitrided layer

Previous investigations have shown that the CL consisting of ε-$Fe_{2\text{-}3}N$ and γ'-Fe_4N phases presents a relatively high hardness with a high brittleness, whereas a thin mono-phase layer of γ'-Fe_4N or a compound-free diffusion layer has a relatively low hardness with a good toughness and supports a high dynamical force [6,8,19]. Fig. 3 shows the hardness variation with depth below the surface of the steel 38CrMoAlA sample treated by different process. It was found that the surface microhardness of PGNC sample could reach about 1165 HV, which was notably higher than that of the PGN sample (1080 HV), and almost equal to that of the CGN sample (1160 HV). It was also evident that in the first 50 μm of the nitrided layer, the hardness of the PGNC sample was higher than that of the CGN sample. This phenomenon can be explained by the dislocations and other defects induced by CR in the α phase [11]. Based on the previous work, the effective case depth of the nitrided sample can be calculated as the case hardness value reaches the substrate hardness plus 50 HV [4]. The effective case depths are approximately 349, 306 and 198 μm for the CGN sample, PGN sample and PGNC sample, respectively. The results demonstrated that the nitriding kinetic process could still be significantly accelerated by PGN, although under a lower nitriding potential.

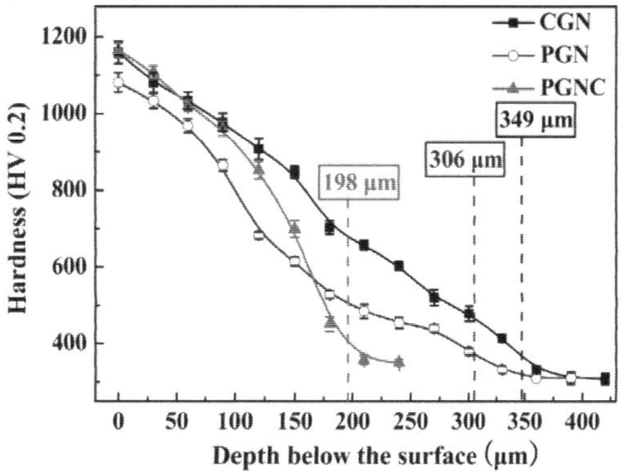

Figure 3. Variations in the microhardness along the depth from the top surface layer for the CGN sample, PGN sample and PGNC sample.

It has been shown that the morphological feature of the indentation produced by the hardness test reflects the bearing capacity and the toughness of the nitrided layers [4,20]. In order to qualitatively evaluate the toughness of the surface layer in each treated sample, Fig. 4 shows the indentation morphologies of the surfaces under a load of 10 kg. There are apparently different morphological features immediately noticed in these optical micrographs. It can be seen that, for the CGN sample (Fig. 4a), obvious linear cracks extending outward along each corner of the indentation could be noted, and serious damage near an edge was also observed. For comparison, there existed a regular indentation with no cracks in the PGN sample and PGNC sample after the test, as shown in Fig. 4b and c. The results indicated that the CGN sample presented the lowest bearing capacity and toughness. Hence, it can be confirmed that a compound-free nitrided layer with a relatively high hardness and satisfactory toughness can be rapidly prepared by the combined treatment (PGN+CR).

Figure 4. Optical micrographs of Vickers indentation for (a) CGN sample, (b) PGN sample and (c) PGNC sample under a load of 10 kg.

3.3. Wear properties

Fig. 5 shows the weight losses of wear scars after a wear test with a sliding distance of 170 m and at a load of 100 N and 400 N for the three treated samples, respectively. It can be seen that the wear weight loss measured at 100 N was significantly smaller than that obtained under the load of 400 N, but the corresponding wear behaviors were apparently different. It should be noted that the CGN sample, which exhibited the best wear resistance under the load of 100 N, had the largest wear loss amongst all the treated samples when the applied load was up to 400 N. Additionally, the wear loss of the PGNC sample was similar to one half of that of the CGN sample under the load of 400 N, indicating that excellent wear resistance could be obtained in the PGNC sample. The prior investigations have shown that the wear characteristics of the nitrided sample depend on many factors such as the compound layer composition, compound layer toughness, mode of mechanical loading, etc [5-8]. The compound layer will be flaked and formed into hard abrasive particles in the initial stage of sliding when the contact pressure exceeds a critical value [9]. These hard abrasive particles that stay in the contact zone between the pin and the disc will accelerate the abrasive wear and promote an increasing of the weight loss [5-7]. This would explain the wear behavior found for the nitrided samples in the present study. In order to further explain the wear characteristics occurring in each surface of the nitrided sample, the surface morphologies of the nitrided samples were presented in Fig. 6.

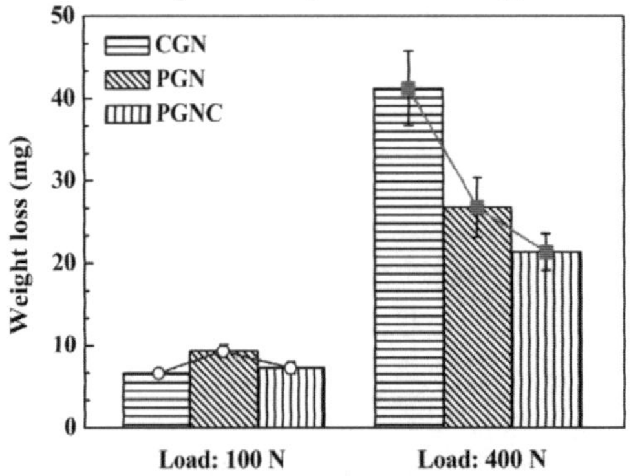

Figure 5. Wear weight loss for different samples at the applied loads of 100 N and 400 N.

Fig. 6a-c show the morphologies of the worn surfaces after a wear test with a load of 100 N and sliding distance of 170 m. Apparently, when the wear load was 100 N, no obvious traces of abrasive wear had been found in all of the sample surfaces. The worn surface of the CGN sample was very smooth with some thin grooves (Fig. 6a), whereas the worn surface of the PGN sample was relatively rough with more grooves and spalling (Fig. 6b). The PGNC sample shared a similar wear mechanism with the CGN sample, and a small amount of surface layer removal was observed in this case (Fig. 6c). Increasing the wear load from 100 N to 400 N caused an obvious increase in the surface wear rate, as shown in Fig. 6d-f. Three different wear morphologies, indicative of different wear mechanisms, could be distinguished. The worn surface of the CGN sample is shown in Fig. 6d. It is clear that the compound layer was being removed from the wear track during testing, indicating that abrasive wear had occurred in this case. Traces of adhesive wear could be clearly observed on the worn surface of the PGN sample (Fig. 6e). The original surface had been worn away, and wear debris was agglomerated in the machined grooves and then formed the discontinuous compacted surface. The worn surface of the PGNC sample consisted of a large number of thin abrasion grooves and pits along the sliding direction, and no obvious surface layer removal was observed (Fig. 6f), demonstrating a minor weight loss during sliding. This emphasizes the fact that a good combination of hardness and toughness of the coating is crucial for good wear resistance [21]. The hardness distribution and compressive residual stress in the compound-free nitrided layer can be further improved by means of plastic deformation, which is attributed to the existence of a lot of defects [4,11]. Especially, the compressive residual stress is recognized to hinder the crack initiation and propagation [4,21]. Therefore, it is reasonable to believe that the wear resistance under heavy loads can be improved on the PGNC sample in comparison to the PGN sample.

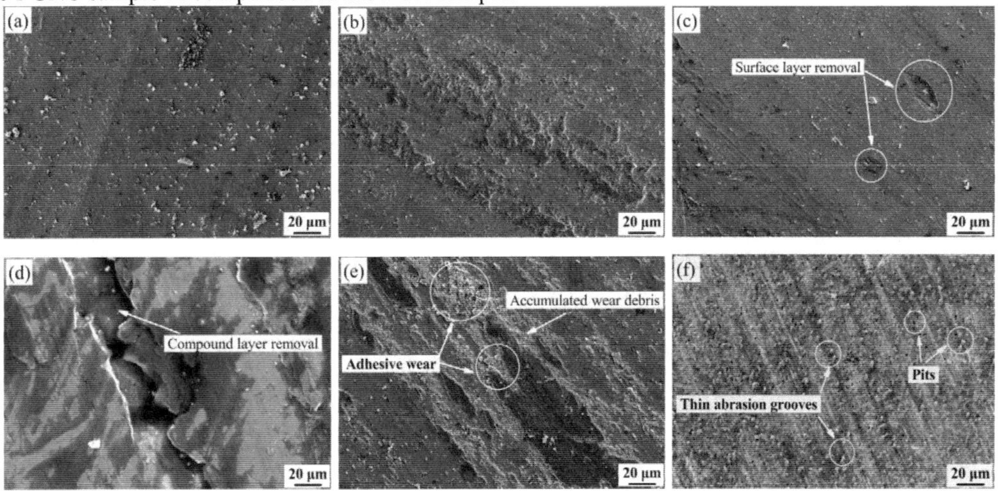

Figure 6. SEM images of the wear scars formed at an applied load of 100 N in (a) CGN, (b) GN and (c) PGNC sample and at 400 N in (d) CGN, (e) PGN and (f) PGNC sample.

4. Conclusions

A compound-free nitrided layer on steel 38CrMoAlA steel with a relatively low hardness but good toughness can be rapidly prepared by PGN with a controlled nitriding potential. After a CR treatment to the PGN sample, the surface hardness can be further improved without loss of toughness. The CGN sample treated for 30 h showed excellent wear resistance under light loads due to its higher surface hardness, but showed a serious deteriorated performance under heavy loads, which can be attributed to the fracture of the compound layer leading to the formation of the hard abrasive particles. The greatly enhanced hardness as well as the surface toughness contributed to the best bearing capacity and wear resistance under heavy loads in the combined treated sample (PGN for 5 h + 30% CR). The formation of the compound-free nitrided layer by using the PGN and CR technique not only provides a new approach to improve the surface hardness, toughness and wear resistance of the nitrided layer, but also increases the manufacturing efficiency.

Acknowledgements
This work was supported by the Natural Science Foundation of Hebei Province for Distinguished Young Scholar (E2011203131) and the Natural Science Foundation of China (51171161).

References
[1] Bell T 1977 *Source book on nitriding* (Ohio: ASM international)
[2] Mongis J, Peyre J P, Tournier C 1984 *Heat Treat. Met.* **3** 71-75
[3] Mittemeijer E J 2013 *Fundamentals of nitriding and nitrocarburizing* ed Dossett J and Totten G E (Ohio: ASM International) pp 619-646
[4] Sun J, Tong W P, Zhang H, Zuo L, Wang Z B 2012 *Surf. Coat. Technol.* **213** 247-252
[5] Podgornik B, Vizintin J, Leskovsek V 1998 *Surf. Coat. Technol.* **108** 454-460
[6] Podgornik B, Vizintin J, Leskovsek V 1999 *Wear* **232** 231-242
[7] Karams M B 1991 *Wear* **147** 385-391
[8] Karams M B 1993 *Mater. Sci. Eng. A* **168** 49-53
[9] Castro G, Fernandez V A, Cid J 2007 *Wear* **263** 1375-1385
[10] Tong W P, Tao N R, Wang Z B, Lu J, Lu K 2003 *Science* **299** 686-688.
[11] Tong W P, Han Z, Wang L M, Lu J, Lu K 2008 *Surf. Coat. Technol.* **202** 4957-4963
[12] Kochmański P, Nowacki J 2006 *Surf. Coat. Technol.* **200** 6558-6562
[13] Kochmański P, Nowacki J 2008 *Surf. Coat. Technol.* **202** 4834-4838
[14] Baranowska J, Szczeciński K, Wysiecki M 2002 *Surf. Coat. Technol.* **151** 534-539
[15] Schwarz B, Meka S R, Schacherl R E, Bischoff E, Mittemeijer E J 2014 *Acta Mater.* **76** 394-403
[16] Fu W T, Wang B, Zhao P, Fu R D, Zheng Y Z, Liu T Y 2012 *China Patent* 201210530358.6
[17] Hoffmann F et al 1991 *Härterei-Tech. Mitt.* **46** 354-360
[18] Lehrer E 1930 *Z. Electrochem.* **36** 383-392
[19] Demchenko L, Sidorenko S I, Tinyaev V 2003 *Def. Diff. Forum* **216** 87-92
[20] Chen W L, Wu C L, Liu Z R, Ni S, Hong Y, Zhang Y, Chen J H 2013 *Acta Mater.* **61** 3963-3972
[21] Tang B, Zhu X, Hu N, He J 2000 *Surf. Coat. Technol.* **131** 391-394

IOP Conf. Series: Materials Science and Engineering **89** (2015) 012047 doi:10.1088/1757-899X/89/1/012047

Phase dependent restoration mechanisms of TC8 titanium alloy during hot compression in the two phase region

K Wang[1,2], B F Luan[1], M Q Li[2] and Q Liu[1]
[1] College of Materials Science and Engineering, Chongqing University, Chongqing 400045, China
[2] School of Materials Science and Engineering, Northwestern Polytechnical University, Xi'an 710072, China

E-mail: study_ke@cqu.edu.cn

Abstract. The TC8 titanium alloy was isothermally compressed at 1133 K and 1213 K in the (α+β) two phase region. The microstructural evolution and restoration mechanism in the α and β phases were characterized by optical microscopy and transmission electron microscopy. The results show a significant effect of phase content on the microstructural evolution and restoration mechanism. The grain refinement occurs in the α phase at both temperatures, but in the β phase only at the higher temperature of 1213 K. This difference in microstructural evolution is attributed to the different temperature dependence of restoration mechanisms in the two phases. A significant increase in the volume fraction of β phase makes the restoration mechanism in the β phase change from dynamic recovery (DRV) to dynamic recrystallization (DRX), which subsequently induces the β grain refinement.

1. Introduction

The α/β titanium alloys have been widely used in the aeronautical industries due to their high strength, low density, good toughness and good high-temperature properties [1, 2]. Optimizing mechanical properties of titanium alloys can be achieved by controlling the microstructure obtained via complex thermomechanical processing routes [3, 4]. However, the deformation process of α/β titanium alloys is difficult to control due to the sensitive microstructural evolution to small fluctuations in both temperature and strain rate [5, 6]. Therefore, a better understanding of the effect of processing parameters on the microstructural evolution and restoration mechanism is essential for process optimization and microstructure control of α/β titanium alloy.

Significant efforts have been made to investigate the microstructural evolution and restoration mechanism of titanium alloys. Some investigators [7, 8] pointed out that dynamic recovery (DRV) dominated if the apparent activation energy for deformation was close to the activation energy for self-diffusion. Hua et al. [9] pointed out that dynamic recrystallization (DRX) was more noticeable at a temperature of 1173 K. Some investigators [10, 11] also found that the restoration mechanism of α and β phase depended on the temperature. Chen et al. [10] investigated hot deformation microstructures of an α/β titanium alloy with an equiaxed structure, and found that the β phase bore the main deformation and presented DRX at a high temperature near the β transus, while the α phase bore the main deformation and presented DRX, and the β phase deformed to match the deformation of α phase at a low temperature. He et al. [11] studied the deformation mechanism and microstructural evolution during hot deformation of Ti-6Al-2Zr-1Mo-1V alloy, and found that the high angle grain boundaries were mainly formed via dislocation accumulation, subboundary sliding and subgrain rotation at the temperatures ranging from 973 K to 1073 K, while DRX was the dominant mechanism at a high temperature of 1123 K. It is therefore can be concluded that the restoration mechanism depends on the phase content of α and β phase, which needs further investigations.

In this work, an α/β titanium alloy (TC8) is isothermally compressed in the two phase region where the phase content varies with the temperature. Based on the microstructure examination using an optical microscope (OM) and a transmission electron microscope (TEM), the dynamic restoration mechanisms of the α and β phases are studied.

2. Experimental

The TC8 titanium alloy used in this study was received in the form of rod with a diameter of 25 mm and had the chemical composition (wt.%) of 6.5 Al, 3.3 Mo, 0.3 Si, 0.06 Fe, 0.01 C, 0.002 H, 0.075 O, 0.005 N, and a balance of Ti. The microstructure of the as-received TC8 titanium alloy is shown in figure 1. The β transus temperature was measured to be 1273 K [12]. Cylindrical specimens that were 8.0 mm in diameter and 12.0 mm in height were machined from the TC8 titanium alloy rod. The specimens were isothermally compressed on a Gleeble 3500 thermal simulator at temperatures of 1133 K and 1213 K, a strain rate of 30 s^{-1}, and a height reduction of 60%. Before compression, the specimens were heated to the pre-defined temperature and held for 3 min to establish a uniform temperature. After compression, the specimens were cooled in air to room temperature.

To observe the microstructural evolution, the compressed specimens were sectioned along the compression axis and prepared for microstructure examination by standard metallographic techniques. For the OM examination, the sectioned specimen was prepared following standard grinding/polishing procedures and etched in a solution of 5% HF, 15% HNO_3, and 80% H_2O. For the TEM examination, the sectioned specimen was mechanically ground to 60~80 μm followed by twin-jet electropolishing. An OLYMPUS GX71 OM and a Tecnai F30 G^2 TEM were used to examine the microstructure.

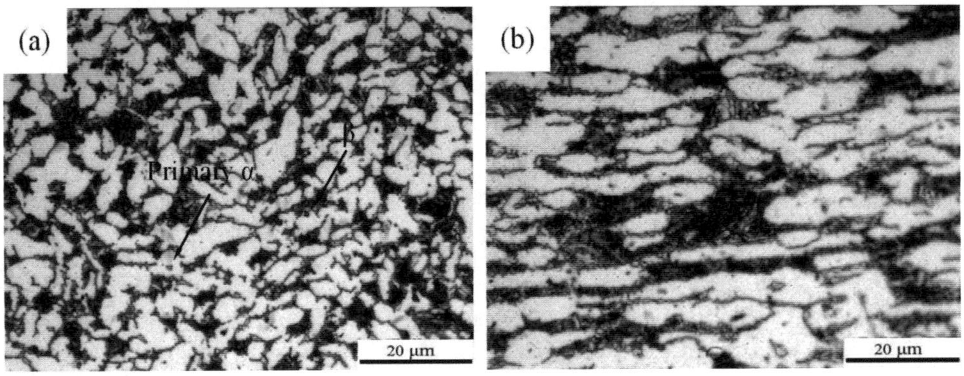

Figure 1. Microstructures of the as-received TC8 titanium alloy: (a) transverse section, (b) longitudinal section.

3. Results and Discussion

3.1 Microstructural evolution

To compare the difference in the microstructure between the compressed and uncompressed TC8 titanium alloy, two specimens were only heated to 1133 K and 1213 K and held for 3 min followed by air cooling. Figure 2 shows that the volume fraction of α phase decreases significantly from 58% to 30% as temperature increases from 1133 K to 1213 K, and the volume fraction of β phase attains 70% and becomes a dominant phase at 1213 K. Figure 3 shows the microstructure of the isothermally compressed TC8 titanium alloy at temperatures of 1133 K and 1213 K, a strain rate of 30 s^{-1} and a height reduction of 60%. It can be seen from figures 2 and 3 that, after the isothermal compression, only the α phase appears with grain refinement at 1133 K, which agrees well with some previous investigations [13, 14], but both α and β phase show grain refinement at 1213 K. This contrast in microstructural evolution between the two temperatures is closely related to the different restoration mechanisms, which will be discussed in the next section.

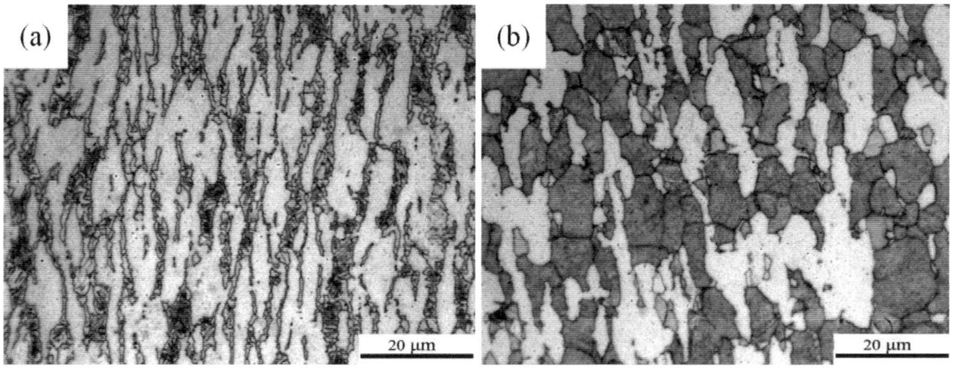

Figure 2. OM micrographs of the uncompressed TC8 titanium alloy heated for 3 minutes at (a) 1133 K, (b) 1213 K.

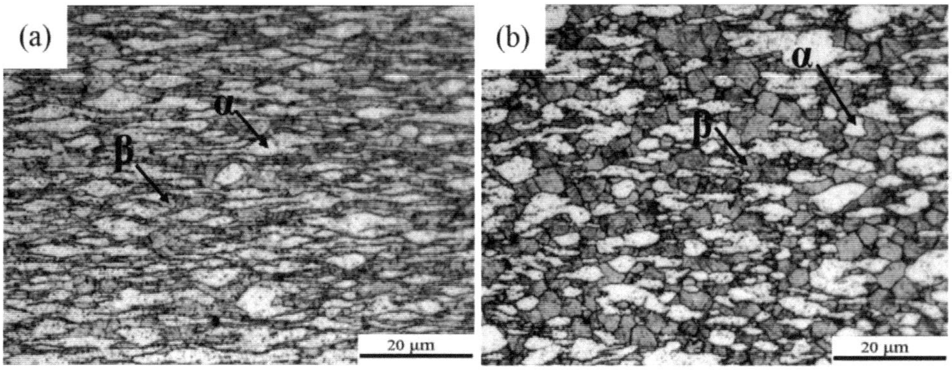

Figure 3. OM micrographs of the TC8 titanium alloy isothermally compressed at (a) 1133 K, (b) 1213 K.

3.2 Restoration mechanism

The restoration mechanism of the α phase could be analysed based on TEM observations in the isothermally compressed TC8 titanium alloy. As shown in figure 4, the temperature has a significant effect on the microstructural evolution. At a low temperature of 1133 K (figure 4(a)), the α phase forms a DRX grain, which is bounded by distinct grain boundary with a high angle and contains few intragranular dislocations, and subgrains, which are bounded by the fuzzy subboundaries with a low angle and contain many intragranular dislocations. While at a high temperature of 1213 K, many subboundaries with long and fuzzy morphology (figure 4(b)) and subgrains (figure 4(c)) are formed in the α phase. This phenomenon indicates that both the DRX and DRV occur in the α phase and induce the occurrence of DRX grains and subgrains at a low temperature of 1133 K, while the DRV acts as a dominant restoration mechanism and induces the occurrence of subboundaries and subgrains at a high temperature of 1213 K. Therefore, the grain refinement of the α phase at the two temperatures can be interpreted by a β phase wedging model [15, 16], i.e. the wedging of the β phase into the α phase along

IOP Conf. Series: Materials Science and Engineering **89** (2015) 012047 doi:10.1088/1757-899X/89/1/012047

the subboundary or DRX grain boundary in the α phase would break the elongated α grain to get refinement eventually.

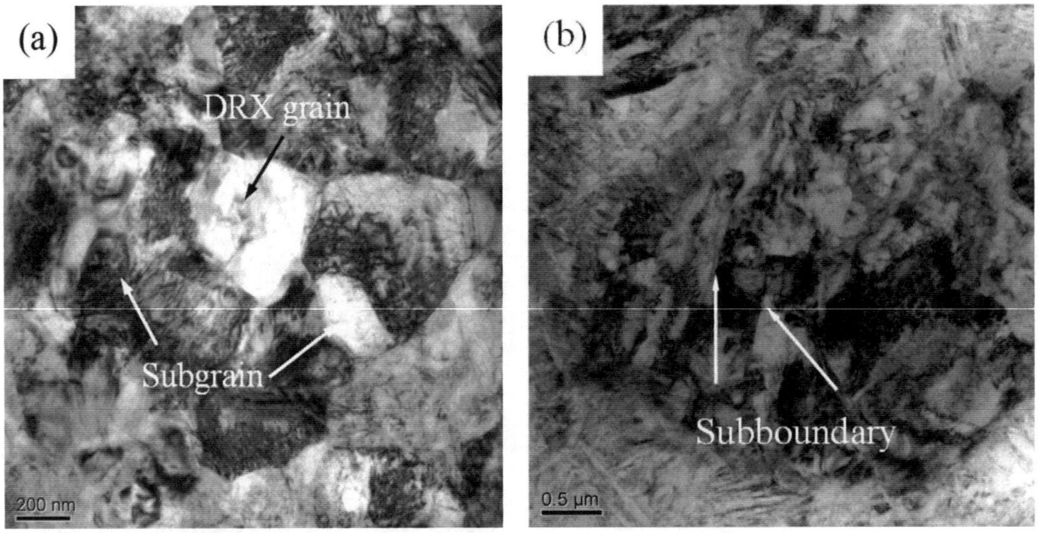

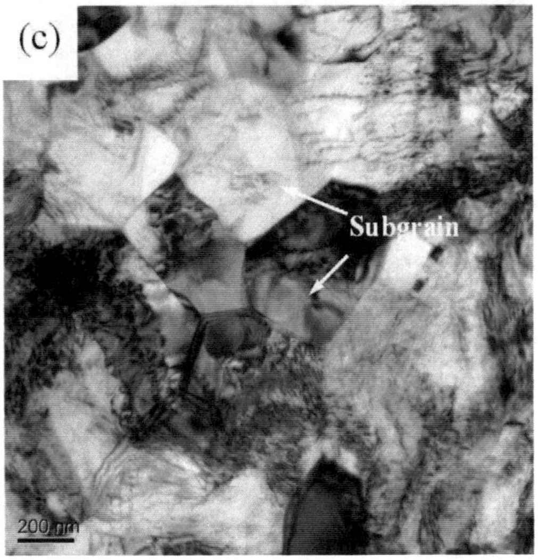

Figure 4. Typical TEM micrographs of the α phase in the TC8 titanium alloy isothermally compressed at (a) 1133 K, (b) and (c) 1213 K.

It is worth noting that the information on crystal defects, such as dislocations and subboundaries, cannot be acquired well in the β phase, due to the quick precipitation of secondary α phase in the β matrix phase, inducing a difficulty to reveal the restoration mechanism of the β phase. However, as shown in figure 5, this work finds that a lot of dislocations are generated in the β grain and arranged along the α-β phase interface, and the α grain with surrounded dislocations appears cleanly with few intragranular dislocations at a high temperature of 1213 K. The generation of dislocations means that the β phase bears large plastic deformation in the isothermal compression, and then provides sufficient stored energy for the DRX of β phase, which could explain the grain refinement of the β phase well, as shown in figure 3(b). However, the little morphology change of the β phase indicates a dominant restoration mechanism of DRV in the β phase at a low temperature of 1133 K.

Therefore, it can be concluded that the volume fraction of the β phase plays a key role in the dynamic restoration mechanisms of the α and β phases in the isothermal compression of TC8 titanium alloy. Li et al. [16] has pointed out that, the deformation would mainly concentrate on the β phase due to the following two reasons: (i) the β phase with a bcc structure has more slip systems than the α phase with a hcp structure; (ii) the β presents a dominant phase in the titanium alloy. In this work, the high temperature (1213 K) makes the β phase become the dominant phase to bear the main plastic deformation, which subsequently induces sufficient stored energy for the DRX of the β phase, and while less stored energy makes DRV dominant in α phase. However, at a low temperature of 1133 K, a low volume fraction of β phase makes the β phase mainly play an accommodation role in the deformation of the α phase [10], and then the DRV dominates in the β phase due to the less stored energy, while both DRX and DRV occur in α phase due to the sufficient stored energy. This is the main reason for the contrast in grain morphology and restoration mechanism between the two temperatures in this work.

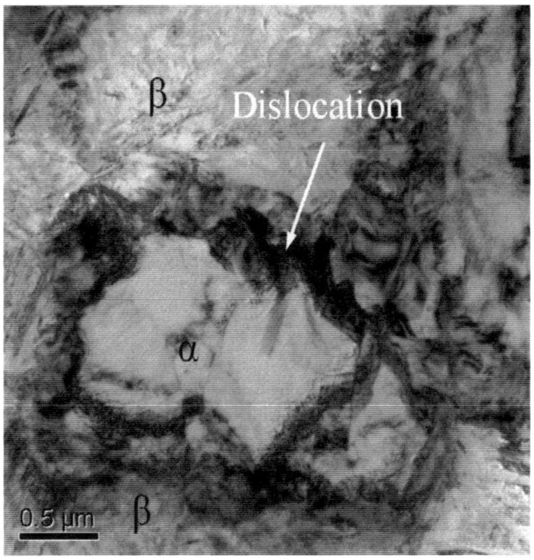

Figure 5. Typical TEM micrograph of the β phase in the TC8 titanium alloy isothermally compressed at 1213 K.

4. Conclusions

Based on the OM and TEM examinations, this work has investigated the restoration mechanisms of the α and β phases of the TC8 titanium alloy isothermally compressed in the two phase region, and the following conclusions can be reached:

(1) Only the α phase shows a grain refinement at 1133 K, but both α and β phases show grain refinement at 1213 K.

(2) Both the DRX and DRV occur in the α phase and induces the occurrence of DRX grains and subgrains at 1133 K, while the DRV acts as a dominant restoration mechanism and induces the occurrence of subboundaries and subgrains at 1213 K.

(3) A high volume fraction of β phase makes the β phase bear the main deformation, inducing sufficient stored energy for the DRX of the β phase to occur.

Acknowledgement

This work is supported by Fundamental Research Funds for the Central Universities (Grant No. 106112015CDJXY130014).

References

[1] Liang H Q and Guo H Z 2015 *Mater. Lett.* **151** 57-60
[2] Okulov I V, Wendrock H, Volegov A S, Attar H, Kühn U, Skrotzki W and Eckert J 2015 *Mater. Sci. Eng. A* **628** 297-302
[3] Jackson M, Jones N G, Dye D and Dashwood R J 2009 *Mater. Sci. Eng. A* **501** 248-254

[4] Jones N G, Dashwood R J, Dye D and Jackson M 2008 *Mater. Sci. Eng. A* **490** 369-377
[5] Xu J W, Zeng W D, Sun X and Jia Z Q 2015 *J. Alloys Comp.* **637** 449-455
[6] Gao P F, Yang H, Fan X G and Yan S L 2012 *J. Mater. Process. Tech.* **212** 2520-2528
[7] Wanjara P, Jahazi M, Monajati H, Yue S and Immarigeona J P 2005 *Mater. Sci. Eng. A* **396** 50-60
[8] Jia W J, Zeng W D, Zhou Y G, Liu J R and Wang Q J 2011 *Mater. Sci. Eng. A* **528** 4068-4074
[9] Hua K, Xue X Y, Kou H C, Fan J K, Tang B and Li J S 2014 *J. Alloys Comp.* **615** 531-537
[10] Chen H Q and Cao C X 2012 *Trans. Nonferrous Met. Soc. China* **22** 503-509
[11] He D, Zhu J C, Lai Z H, Liu Y and Yang X W 2013 *Mater. Des.* **46** 38-48
[12] Wang K and Li M Q 2013 *Scripta Mater.* **68** 964-967
[13] Sun S D, Zong Y Y, Shan D B and Guo B 2010 *Trans. Nonferrous Met. Soc. China* **20** 2181-2184
[14] Li H, Li M Q, Han T and Liu H B 2012 *Mater. Sci. Eng. A* **546** 40-45
[15] Chen H and Cao C 2011 *Rare Met. Mater. Eng.* **40** 946-950
[16] Li L, Luo J, Yan J J and Li M Q 2015 *J. Alloys Comp.* **622** 174-183

Atomistic simulations of the effect of helium clusters on grain boundary mobility in iron

A T Wicaksono, M Militzer and C W Sinclair

Centre for Metallurgical Process Engineering, The University of British Columbia, Vancouver, Canada, V6T 1Z4

E-mail: matthias.militzer@ubc.ca

Abstract. A series of molecular dynamics simulations was performed in this work to investigate the kinetic interaction between helium clusters and grain boundaries in iron. Helium cluster formation and size distributions were found to be markedly different in the bulk compared to the region of a stationary boundary. Upon reaching a steady-state cluster distribution, the spatial fluctuation of cluster-enriched boundaries was analyzed to determine the grain boundary mobility using the random walk method. Segregated clusters reduced the boundary mobility, the drag effect of clusters increasing as the bulk solute concentration increases. The drag effect was further rationalized by employing Cahn's solute drag model using the effective binding energy of He clusters and the grain boundary diffusivity of a single He atom, their magnitudes having been determined from the segregation level and from monitoring the trajectory of a solute atom in the investigated grain boundaries, respectively. The model is found to provide a satisfactory explanation of the simulation results in the zero velocity limit.

1. Introduction

The kinetic interaction between impurities and grain boundaries is conventionally treated within the framework of either solute drag models [1] for mobile solute atoms or particle pinning models [2] for immobile particles. Impurities forming clusters constitute an intermediate case because clusters of a few interacting solutes migrate at a much lower rate than individual atoms. Further, the effect of clusters on grain boundary mobilities has not been extensively explored in spite of its technological importance, e.g. embrittlement in irradiated steel vessels due to He clusters [3] or grain refinement of high-strength low-alloy steels due to Nb clusters [4]. The lack of studies on cluster-boundary interaction may be associated with both the length and time scales of the problem. The atomistic length scale requires experimental and/or simulation techniques with atomic resolution whereas the time scale extends from the order of atomic frequency to that of diffusion.

Atomistic computational tools can provide a valuable insight to tackling such a multi-scale problem. In this work, the interaction between grain boundaries in iron (Fe) and clusters of helium (He) in the limit of a dilute solution will be explored using molecular dynamics simulation. The formation and distribution of He clusters in the bulk matrix and grain boundaries of Fe will first be analyzed. Segregation behaviour of He will be characterized using a classical model to determine the effective binding energy. The mobilities of cluster-enriched grain boundaries in the limit of zero-velocity will then be extracted. This will permit an evaluation of the overall drag effect due to segregated clusters and its dependence on properties such as solute diffusivity and binding energy. Finally, the simulation results will be assessed against a classical model on solute-grain boundary interaction.

2. Simulation methodology

Simulations were run using LAMMPS [5] with a time step of 1 fs. Visuals were produced using AtomEye [6]. The interaction among Fe atoms, among He atoms and between Fe and He atoms were modelled using the Ackland-04 [7], the Aziz-95 [8] and the Gao-11 [9] EAM potentials, respectively.

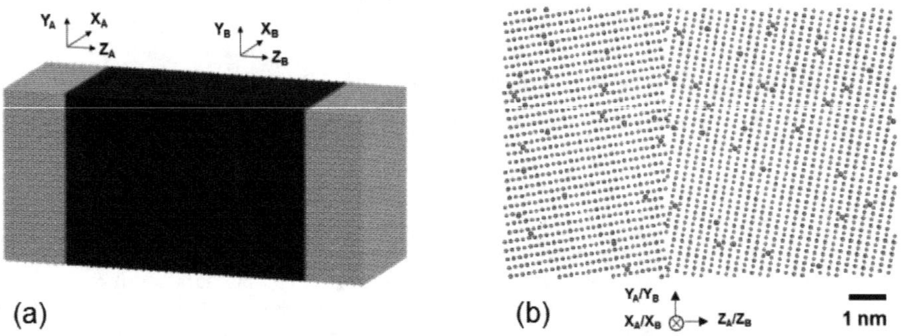

Figure 1. (a) A bicrystal cell with planar boundaries separating two grains, their orthogonal axes indicated in Table 1, (b) A magnified two-dimensional view of the bicrystal cell, populated with multiple He atoms (coloured pink) occupying tetrahedral interstitial sites.

The simulation box is a bicrystal cell, see Fig. 1(a), containing two crystals, A and B, of different orientation being separated by planar grain boundaries and having a total number of Fe atoms N_{Fe} in excess of 600,000 (see Table 1). Four types of boundaries were simulated, their geometrical features being given in Table 1. Each boundary is identified by their inclination angle measured relative to a symmetric coherent twin boundary, i.e. a grain boundary with $\mathbf{Y_A}/\mathbf{Y_B}$ axes of $\langle 111\rangle/\langle 11\bar{1}\rangle$ and $\mathbf{Z_A}/\mathbf{Z_B}$ axes of $\langle \bar{1}\bar{1}2\rangle/\langle 112\rangle$.

The interaction between grain boundaries and He atoms was studied by introducing He atoms into each bicrystal using two types of configurations. In the first configuration, a He atom was placed in a tetrahedral site at the boundary. The binding energy profile of a single He atom was then obtained from molecular statics simulations, see [10] for details. The grain boundary width δ was also extracted by determining the distance within which the binding energy is less than -10^{-3} eV. Molecular dynamics simulations were then run at 1000 K for 3 ns to track the trajectory of the He atom along the boundary. By computing its mean-squared displacement, $\langle r^2\rangle$, as a function of time-interval, τ, the single-atom boundary diffusivity, D_{gb}, can be determined via $\langle r^2\rangle = 6D\tau$, see [11] for example.

Table 1. Geometrical setup of bicrystals and their solute contents. Each bicrystal shares a common $\mathbf{X_A}/\mathbf{X_B}$ axis, $\langle 1\bar{1}0\rangle$, with a dimension L_X of 20.2 nm.

Inclination	Orthogonal axes, see Fig. 1(b)				Dimensions [nm]		N_{Fe}	N_{He} [atoms]		
φ [°]	$\mathbf{Y_A}$	$\mathbf{Z_A}$	$\mathbf{Y_B}$	$\mathbf{Z_B}$	L_Y	L_Z	[atoms]	0.1at%	0.5at%	1.0at%
19.5	$\langle 112\rangle$	$\langle \bar{1}\bar{1}1\rangle$	$\langle 55\bar{2}\rangle$	$\langle 115\rangle$	20.0	17.8	646000	647	3256	6543
44.7	$\langle 118\rangle$	$\langle \bar{4}\bar{4}1\rangle$	$\langle 17\,17\,4\rangle$	$\langle \bar{2}\bar{2}17\rangle$	20.9	19.7	712200	709	3560	7157
70.5	$\langle 115\rangle$	$\langle \bar{5}\bar{5}2\rangle$	$\langle 11\bar{1}\rangle$	$\langle 112\rangle$	22.2	16.8	642000	648	3230	6485
90.0	$\langle 112\rangle$	$\langle \bar{1}\bar{1}1\rangle$	$\langle 11\bar{2}\rangle$	$\langle 111\rangle$	21.0	17.8	642000	645	3230	6485

In the second configuration, multiple He atoms were placed randomly in tetrahedral interstitial sites, see Fig. 1(b). Several levels of bulk solute concentration , C_0, were initiated, i.e. 0.1 at%, 0.5 at% and 1.0 at%. Table 1 lists the number of He atoms, N_{He}, that correspond to each concentration. Periodic boundary conditions were applied in all directions. Simulations were run at 1000 K under zero-pressure condition for a duration of 20 to 40 ns. Atomic positions were recorded every 1 ns during which time the cluster distribution was analyzed.

To characterize and quantify the size distribution of clusters, this work employs the Ward's technique [12] implemented via the Lance-Williams algorithm [13]. It is chosen since the number of clusters are not known prior to the analyses. A detailed discussion of this technique is available elsewhere [14]. Clustering is deemed to reach steady state if the distributions from two snapshots taken within a 3 ns interval differ by less than 20%. Additionally, to investigate the effect of grain boundaries on clustering, clusters were classified into boundary and bulk clusters. Clusters are considered to belong to the boundary if 60% of their constitutents are within $\delta/2$ from the average boundary position.

After steady state clustering has been achieved in the bulk and boundary portion of the bicrystal, the average position of the boundary, $\bar{h}$, is monitored over time to determine the boundary mobility using the random-walk (RW) technique [15]. A detailed discussion of the technique adapted to compute $\bar{h}$ is available elsewhere [16]. This technique can be applied since the boundary is rough at 1000 K, i.e. it performs a random walk that exhibits diffusive characteristics. The evolution of mean-squared displacement, $\langle \bar{h}^2 \rangle$, of the average position of the boundary yields a coefficient that is directly related to the boundary mobility, M, via [15]

$$\langle \bar{h}^2 \rangle = 2Mk_BT\tau/A \tag{1}$$

where τ is the time-interval, A is the boundary area, and k_BT is the product of Boltzmann's constant and absolute temperature.

3. Results

3.1. Grain boundary interaction with a single He atom

The interaction between grain boundaries and a single He atom at 0 K is represented by the single-atom binding energy profile, a typical example shown in Fig. 2. The binding energy, represented by the maximum E_{b0} in Table 2, is in a first order approximation 1.35 eV independent of the boundary inclination. The boundary width, δ, determined from the binding energy profile increases with the boundary inclination from 7 to 9 Å, see Table 2.

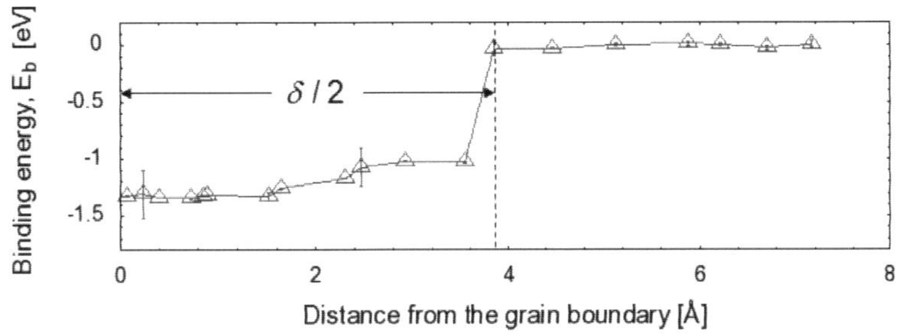

Figure 2. The profile of binding energy of a He atom to the 44.7°-boundaries at 0 K, the regime of non-zero binding energy being defined as the boundary width, δ.

The kinetic behaviour of a He atom is also significantly affected by the grain boundary. At 1000 K and for all investigated boundaries, the He atom at the boundary performed a random walk motion but remained within the boundary width at all times. Table 2 provides the single-atom boundary diffusivity determined from these simulations, D_{gb}. The bulk diffusivity D_{bulk} at 1000 K is also presented in this table, and was obtained from performing the same type of random-walk analysis on a simulation where the He atom occupied tetrahedral sites of a single b.c.c. crystal. D_{gb} is 2 to 3 orders of magnitude lower than D_{bulk}, i.e. the grain boundary is not a fast diffusion path for interstitial He atoms. This observation is consistent with simulations [17] and experiments [18] for interstitial diffusion of H, C and He in Fe grain boundaries.

Table 2. Relevant material properties computed in this work.

Property	Bulk	19.5° boundary	44.7° boundary	70.5° boundary	90.0° boundary
E_{b0} [eV]	0	1.31	1.34	1.34	1.38
δ [Å]	-	6.8	7.7	7.8	8.8
D [m^{2}s^{-1}]	1.24×10^{-8}	1.53×10^{-11}	3.10×10^{-11}	1.60×10^{-10}	1.37×10^{-10}
E^*_{b0} [eV]	0	0.48	0.50	0.53	0.52

3.2. Helium clustering and segregation

The behaviour of multiple He atoms in a bicrystal at 1000 K is now presented. He cluster formation occurs within a few ns, as is evident from Fig. 3(a), due to the high diffusivity of He atoms in the Fe matrix. There is, however, a markedly different clustering trend in the bulk and in the boundary, both in terms of their size distribution and their evolution. Figs. 3(b) and (c) illustrate the cluster distribution in the bulk and boundary, respectively, having been obtained from the Ward's technique applied at different times during the simulation run.

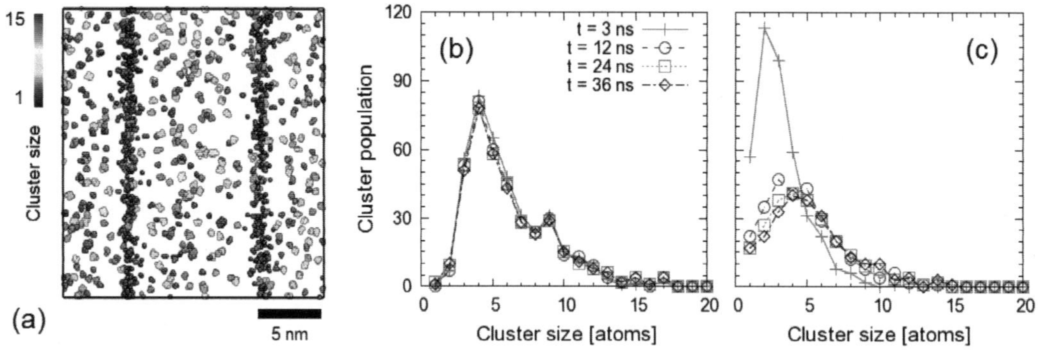

Figure 3. (a) A snapshot of 44.7°-bicrystal containing 0.5 at% He taken at 3 ns, showing only He clusters, coloured according to their size, i.e. the number of atoms, (b) and (c) evolution of cluster size distribution in the bulk and boundary portion of the bicrystal in (a), respectively.

The bulk distribution, see Fig. 3(b), shows a negligible variation over a 36-ns simulation time, suggesting a steady state cluster distribution. Additionally, these bulk clusters are essentially immobile. This implies that, within the same period of time, no bulk clusters migrated to the boundary. Clusters found at the boundary form during the initial transient time (< 3 ns), due to the segregation of single fast-diffusing He atoms to the boundary. Solute segregation takes place rapidly and reaches steady state at approximately the same instance as the bulk clustering does.

In contrast to the bulk cluster distribution, the distribution at the boundary exhibits a notable evolution, see Fig. 3(c). Segregated individual He atoms diffuse along the boundary to form larger clusters, thereby changing the cluster size distribution at the boundary. The slow formation of boundary clusters compared to that of bulk clusters is consistent with the single-atom boundary diffusivity being 2 to 3 orders of magnitude lower than the bulk diffusivity, see Table 2. The single-atom boundary diffusivity can thus be considered as the rate-controlling process of cluster formation at the boundary. While the cluster distribution at the boundary continues to change over time even after 36 ns, the variation becomes increasingly negligible at longer times. The boundary distribution is deemed to have reached steady state at 12 ns, for the 44.7° boundary containing 0.5 at% He, because the distribution captured at that time differs only by 15% with that captured at 15 ns. Similar observations can be made for the other three boundaries studied here.

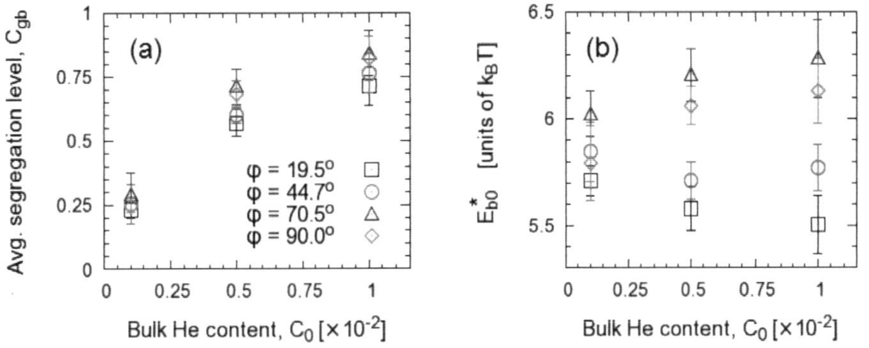

Figure 4. (a) Average segregation for different boundaries as a function of C_0, error bars being the standard deviation, (b) the effective binding energy according to the Langmuir-McLean model, Eq. (2).

The segregation level at the boundaries is quantified by normalizing the number of segregated He atoms, i.e. the area under the graphs in Fig. 3(c), with the total number of boundary sites, i.e. the number of tetrahedral sites available in a perfectly planar grain boundary. The segregation level was computed in intervals of 3 ns to obtain the average segregation level. Fig. 4(a) shows the segregation level C_{gb} as a function of bulk content C_0. The minor variation in C_{gb} with inclination angle φ indicates that the sink strength of these four boundaries is similar. Furthermore, the effective binding energy of these boundaries, E^*_{b0}, can be estimated from C_{gb} using the Langmuir-McLean model [19], i.e.

$$E^*_{b0}/k_B T = \ln\left[C_0(1 - C_{gb})/((1 - C_0)C_{gb})\right] \tag{2}$$

and is plotted as a function of C_0 in Fig. 4(b). The effective binding energy varies moderately with C_0, indicating that the model assumption of non-interacting solute atoms [19] is a simplification for the Fe-He system. The magnitude of this variation is about 5% (or $0.3k_BT$) for all boundaries. The model is thus considered a reasonable first approximation for E^*_{b0}, which is taken as the average of the values obtained from three different C_0, see Table 2.

These binding energies, E^*_{b0}, are about three times smaller than the single-atom binding energies, E_{b0}, see Table 2. The single-atom binding energy E_{b0} has been obtained from 0 K simulations and would lead to saturated segregation, i.e. $C_{gb} = 1$, at 1000 K, overestimating the actual segregation in Fig. 4(a). E^*_{b0} can thus be considered a more representative parameter than E_{b0} to describe the interaction between grain boundaries and He clusters.

3.3. Mobilities of cluster-enriched grain boundaries

Upon clusters reaching a steady-state distribution at the boundary, simulations were run for another 20 to 30 ns, during which the boundary position, $\bar{h}$, was recorded every 0.1 ns. Fig. 5(a) shows an example of the distribution of boundary displacement $\Delta\bar{h}(\tau = 0.2\,\text{ns})$ for the pure and cluster-enriched (C_0 = 0.5at%) 44.7°-boundaries. Both distributions are Gaussian, their width indicating the extent of boundary fluctuation. The average mean-squared displacement (MSD) at different τ for the pure and enriched boundaries are plotted in Fig. 5(b) and (c), respectively. The boundary mobility is obtained from the slope of linear regression of the MSD via Eq. (1).

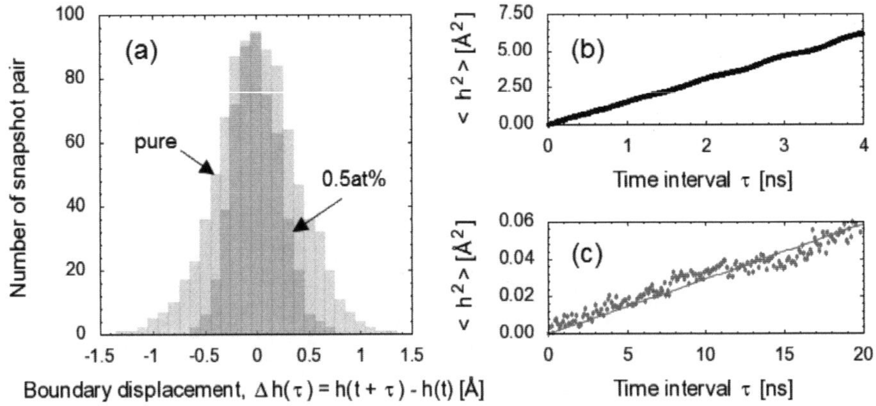

Figure 5. (a) Distribution of boundary displacements at an interval τ of 0.2 ns, for the pure and cluster-enriched (C_0 = 0.5at%) 44.7°-boundaries; the mean-square displacement as a function of τ of (b) the pure boundary, and (c) the cluster-enriched boundary.

The same procedure was repeated for different boundaries and solute contents to compute the mobilities of these boundaries, see Fig. 6(a). These mobilities are labelled as an effective quantity, M_{eff}, because they arise from the intrinsic mobility of the boundary, M_{pure}, and the contribution of cluster segregation, M_{clust}. The presence of clusters reduces the mobility by about 3 to 4 orders of magnitude as well as the effect of φ on the mobilities.

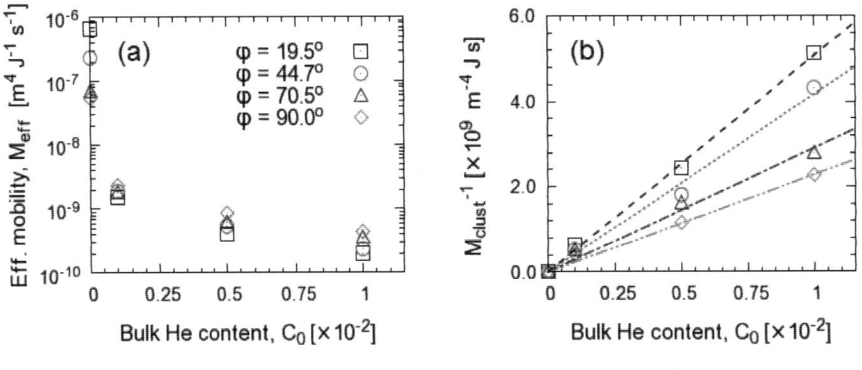

Figure 6. (a) Effective mobility of pure and enriched grain boundaries, (b) Cluster drag contribution, M_{clust}^{-1} in Eq. (3), as a function of bulk solute content.

The effective mobility can be further decoupled into its intrinsic and extrinsic component following, for example, Cahn [1], i.e.

$$M_{\text{eff}}^{-1} = M_{\text{pure}}^{-1} + M_{\text{clust}}^{-1} \tag{3}$$

The cluster drag, i.e. the inverse of cluster contribution to the boundary mobility, is plotted in Fig. 6(b) as a function of C_0. Within the range of solute content investigated here, the cluster drag appears to show a linear increase with increasing C_0, which is consistent with Cahn's solute drag model [1]. This linear relationship, i.e. $M_{\text{clust}}^{-1} = \alpha C_0$, is represented by the dashed lines in Fig. 6(b) where α denotes the drag coefficient.

4. Discussion

To further analyze the cluster drag coefficient α in Fig. 6(b), the solute drag model of Cahn [1] is adopted. The model postulates that, in the limit of low velocity, the drag coefficient, α, can be obtained from

$$\alpha = 4N_v k_B T \int_{-\delta/2}^{+\delta/2} \frac{\sinh^2\left[E_b(z)/2k_B T\right]}{D(z)} dz \tag{4}$$

where N_v is the bulk atomic concentration (8.4×10^{28} m^{-3} [20]), $E_b(z)$ and $D(z)$ are the profile of binding energy and solute diffusivity along the boundary normal z, respectively.

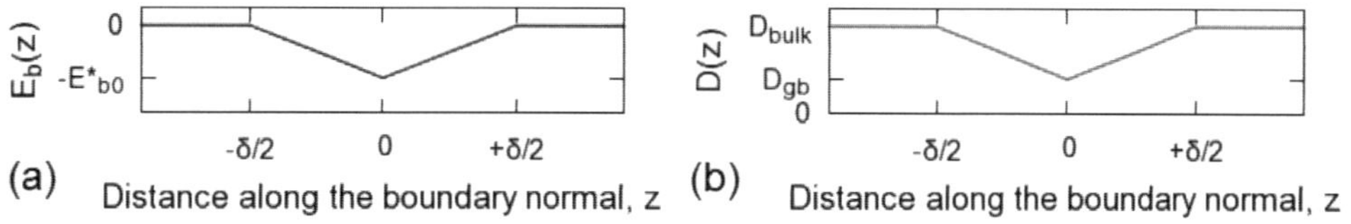

Figure 7. Assumed profile of (a) binding energy and (b) diffusivity to estimate α from Eq. (4).

The parameters $D(z)$ and $E_b(z)$ specified in Cahn's model correspond to properties that represent the interaction between the grain boundary and a single solute atom. In order to adapt this model for the drag effect of He clusters, the effective binding energy of clusters was employed, instead of the single-atom binding energy, to obtain an enrichment level consistent with the simulation results. The single-atom grain boundary diffusivity was also implemented because the kinetics of this process controls the size distribution of clusters at the boundary. Both parameters were integrated into an assumed triangular profile, see Figs. 7(a) and (b), to obtain a closed-form solution of the drag coefficient α via Eq. (4).

Table 3. Cluster drag coefficients α obtained from simulations and the Cahn's model.

α [$\times10^{11}$ m^{-4} J s]	19.5° boundary	44.7° boundary	70.5° boundary	90.0° boundary
Simulations [Fig. 6(b)]	5.1	4.2	2.9	2.3
Cahn's model [Eq. (4)]	4.1	2.4	1.8	1.1

Table 3 compares the drag coefficients estimated from Eq. (4) with those extracted from molecular dynamics simulations, shown in Fig. 6(b). Both drag coefficients are in the same order of magnitude. The drag coefficient obtained from Cahn's model is smaller by a factor of 0.5 to 0.8, but the trends with boundary inclination are similar in both cases. Given the simplifications employed in the continuum model analysis, this is a rather reasonable agreement.

5. Conclusions and Outlook

The effect of solute clusters on grain boundary mobility in the zero velocity limit for dilute solutions in the Fe-He system has been investigated via atomistic simulations. The cluster drag effect observed from simulations can, in a first order approximation, be rationalized using Cahn's solute drag model.

Properties obtained from these simulations have been shown to provide valuable insight in analyzing the cluster-boundary interaction. The single-atom boundary diffusivity has been identified as the rate-controlling process in cluster formation at the boundary. The cluster drag effect presented in this work is, however, only applicable in the zero velocity limit. From a practical perspective, it is of interest to investigate the case of a migrating boundary. This topic is the focus of a further study where He clusters interact with a capillarity driven migrating boundary. In such a situation, cluster formation at the boundary may never reach steady state. As a consequence, the cluster drag effect of a driven boundary may not be consistent with predictions from the classical solute drag model.

Acknowledgments

The authors gratefully acknowledge the financial support from Natural Sciences and Engineering Research Council of Canada and the computing resources provided by WestGrid, a division of Compute Canada. ATW thanks Dr. J. J. Hoyt and H. Song for useful discussions and their warm hospitalities during his visit to McMaster University.

References

[1] Cahn J W 1962 *Acta Metall.* **10** 789–798
[2] Smith C S 1949 *Trans. AIME* **175** 15
[3] Wang C A, Grossbeck M L and Chin B A 1995 *J. Nucl. Mater.* **225** 59–68
[4] DeArdo A J 2003 *Int. Mater. Rev.* **48** 371–402
[5] Plimpton S J 1995 *J. Comput. Phys.* **117** 1–19
[6] Li J 2003 *Modell. Simul. Mater. Sci. Eng.* **11** 173–177
[7] Ackland G J, Mendelev M I, Srolovitz D J, Han S and Barashev A V 2004 *J. Phys.: Condens. Matter* **16** S2629
[8] Aziz R A, Janzen A R and Moldover M R 1995 *Phys. Rev. Lett.* **74**(9) 1586–1589
[9] Gao F, Deng H, Heinisch H and Kurtz R 2011 *J. Nucl. Mater.* **418** 115 – 120
[10] Tschopp M A, Gao F, Yang L and Solanki K N 2014 *J. Appl. Phys.* **115** 033503
[11] LeSar R 2013 *Introduction to Computational Materials Science* (Cambridge University Press)
[12] Ward Jr J 1963 *J. Am. Stat. Assoc.* **58** 236–244
[13] Lance G and Williams W 1967 *The Computer Journal* **9** 373–380
[14] Gan G, Ma C and Wu J 2007 *Data Clustering: Theory, Algorithms and Applications* (SIAM)
[15] Trautt Z T, Upmanyu M and Karma A 2006 *Science* **314** 632
[16] Wicaksono A T, Sinclair C W and Militzer M 2015 *(to be submitted)*
[17] Deng H Q, Hu W Y, Gao F, Heinisch H L, Hu S Y, Li Y L and Kurtz R J 2013 *J. Nucl. Mater.* **442** S667–S673
[18] Teus S M, Mazanko V F, Olive J M and Gavrilijuk V G 2014 *Acta Mater* **69** 105–113
[19] McLean D 1957 *Grain Boundaries in Metals* (Clarendon Press)
[20] Mendelev M I and Mishin Y 2009 *Phys. Rev. B* **80** 144111

IOP Conf. Series: Materials Science and Engineering **89** (2015) 012049 doi:10.1088/1757-899X/89/1/012049

Post processing effects on GND calculations from EBSD-based orientation measurements

S I Wright[1], D P Field[2] and M M Nowell[1]
[1]EDAX, 392 E 12300 S, Draper, Utah, USA
[2]School of Mechanical and Materials Engineering Washington State University, Pullman, Washington, USA

Email: stuart.wright@ametek.com

Abstract. Electron Backscatter Diffraction (EBSD) has shown great utility in characterizing the aspects of microstructure related to crystallographic orientation. Such information is critical to understanding deformation in crystalline materials as well as the impact of deformation induced structural variations on recrystallization. Small angle rotations induced by the production of dislocations and their movement through the structure can be well captured by EBSD. Geometrically Necessary Dislocations (GND) can be derived from the measurement of these local variations in orientation. However, these local orientation variations are often right at the limit of angular precision that can be achieved by EBSD. Various post-processing tools have been developed to improve the angular precision. However, this is generally achieved through point-to-point smoothing of the orientation data within the measurement grid. The impact of such various filtering method are explored in terms of their impact on GND calculations. A new post processing approach which improves the EBSD indexing rate will also be presented along with results on its influence on local orientation variations. Fortunately, the general conclusion drawn from the reduction results is that these approaches generally improve the overall GND measurements.

1. Introduction

Electron Backscatter Diffraction (EBSD) in the Scanning Electron Microscope has become one of the preferred tools for the study of microstructural evolution particularly for deformation based metal forming processes. We have surveyed the EBSD literature using various online search tools and found that deformation is the most common application area for EBSD based analysis. Two EBSD analytical tools frequently used in the study of deformation by EBSD are first Kernel Average Misorientation (KAM) [1] and second Geometrically Necessary Dislocations (GND) [2-4]. The KAM measurement is the average misorientation between a point in the EBSD scan grid and its neighbors (usually just the nearest neighbors). The movement of dislocations through the material results in small angle rotations which can be captured by the KAM measurement. However, this measurement is very much sensitive to the step size used in the original scan data. A more quantitative measure of the dislocations is provided by the GND calculation. The GND calculation takes into account the slip systems in the material and provides some compensation for the step size. However, studies have shown that the GND measurements are still sensitive to the step size [5-6]. These measurements are not only sensitive to step size but are also sensitive to the precision of the orientations measured by EBSD. Various "filtering" approaches have been developed to mitigate the effects of orientation "noise" [7-10] in the EBSD measurements. The impact of these approaches on the GND measurements are characterized by adding artificial noise to high-fidelity EBSD measurements and then comparing the original results with those obtained after applying the noise reduction filters..

2. Step size

In order to confirm the step size dependence of EBSD measurements, we have done some studies on the step size dependence of GND calculations on deformed brass materials used in a study of the orientation precision of EBSD measurements [11]. In this study, orientations were measured by EBSD on grids with a fine step (20 nm) and a coarse step (250 nm). Four samples were investigated with four different levels of cold-rolling reduction – 4.5, 11, 20 and 30%. These scans were then coarsened by removing every other point in the scan grid. The average GND density was then calculated. The results, shown in Fig. 1, clearly demonstrate a dependence on the step size used with the exception of the one anomalous point for the 4.5% 20 nm original step size.

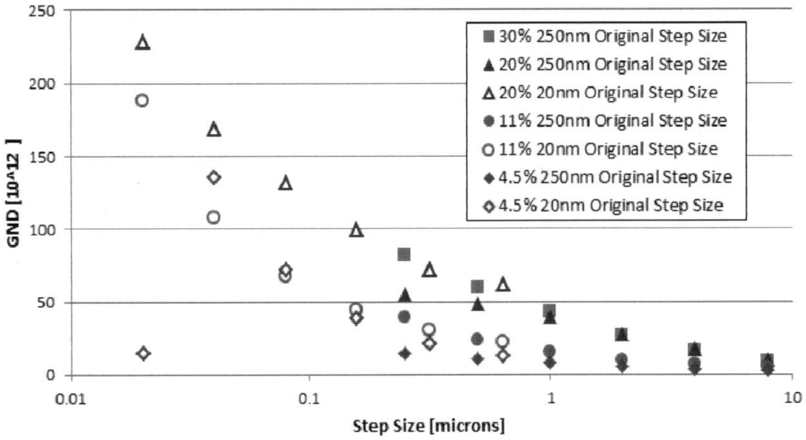

Figure 1. GND measurements of deformed brass showing general decrease in measured GND density with increasing step size.

One of the challenges of measuring local orientations using EBSD on deformed materials is that the deformation generally results in lower-quality diffraction patterns than would be observed in the recrystallized material. It is assumed that the effects of noise in the deformed would manifest itself by higher KAM and GND values but that this effect would decrease with increasing step size. In order to verify this, Gaussian noise was added to EBSD patterns collected on a rolled copper sample. 4.35 million measurements were collected on a hexagonal scan grid with 50 nm spacing. An example of an as-collected pattern and a pattern after adding noise is shown in Fig. 2.

The noisy patterns were then re-indexed, the indexing success rate [12] prior to the addition of the noise was found to be 99.7% and 94.7% after. As expected, the GNDs measured from the noisy data are higher for the smaller step sizes but then drop to match those of the original data at 200 nm. These results are shown in Fig. 3. It should be noted that adding more noise to the patterns lead to much lower indexing success rates and then data did not follow the expected trend likely due to so many missing points.

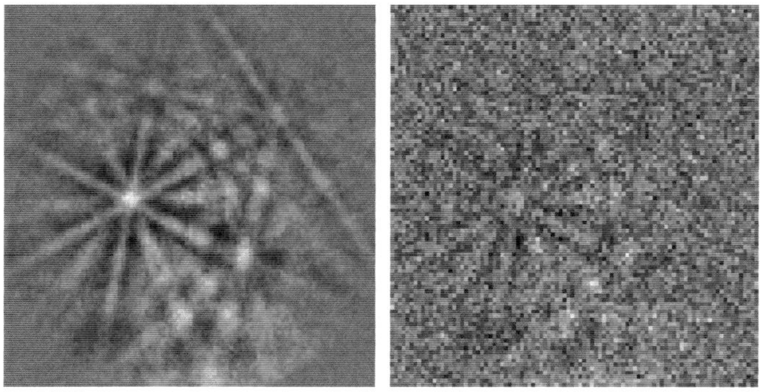

Figure 2. Pattern from a rolled copper sample before and after the addition of Gaussian noise.

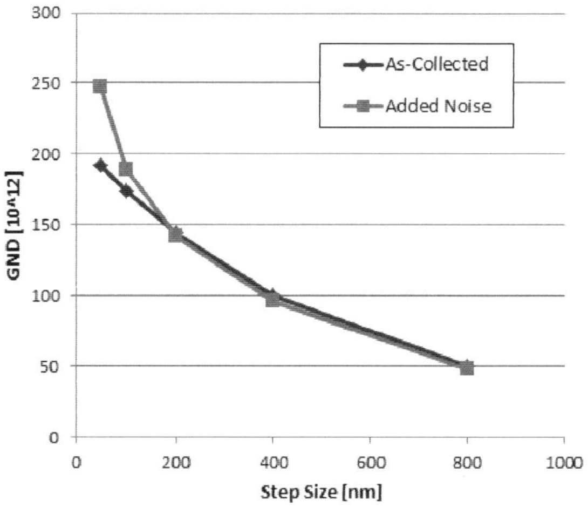

Figure 3. GND measurements from rolled copper before and after the addition of Gaussian noise to the patterns.

3. Noise reduction

Four different orientation noise reduction methods were investigated: (1) the Kuwahara filter [8], (2) a Bilateral smoothing filter [9], (3) a Gaussian smoothing filter and (4) a neighbor pattern averaging filter followed by re-indexing (NPAR) [10].

The Kuwahara is an edge retention filter where, in this application, the edges are essentially sub-grain boundaries. The Bilateral and Gaussian filters are smoothing filters. The pattern averaging filter averages the pattern at each point on the scan grid with the patterns of the neighboring points. These neighbor averaged patterns are then re-indexed in the standard manner. It requires that patterns be recorded during the original scanning process at the SEM. The four noise reduction schemes were all applied to the rolled copper data obtained after adding Gaussian noise to the as-collected EBSD patterns. Fig. 4 shows the results.

The Kuwahara filter produced an average GND value which most closely matches that of the original scan data while the NPAR approach most closely matches the original GND distribution. However, all four filtering approaches brought the GND distribution generally closer to that calculated on the original scan data. Corresponding KAM maps are shown in Fig. 5. The many points colored red in the Kuwahara and Bilateral filtered maps are primarily points which are either unindexed or points without any neighbors within 5° in orientation.

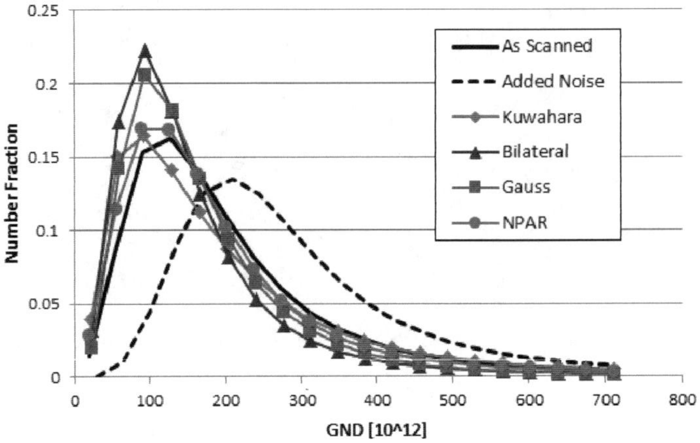

Figure 4. GND measurements from rolled copper with orientation noise reduction filters applied.

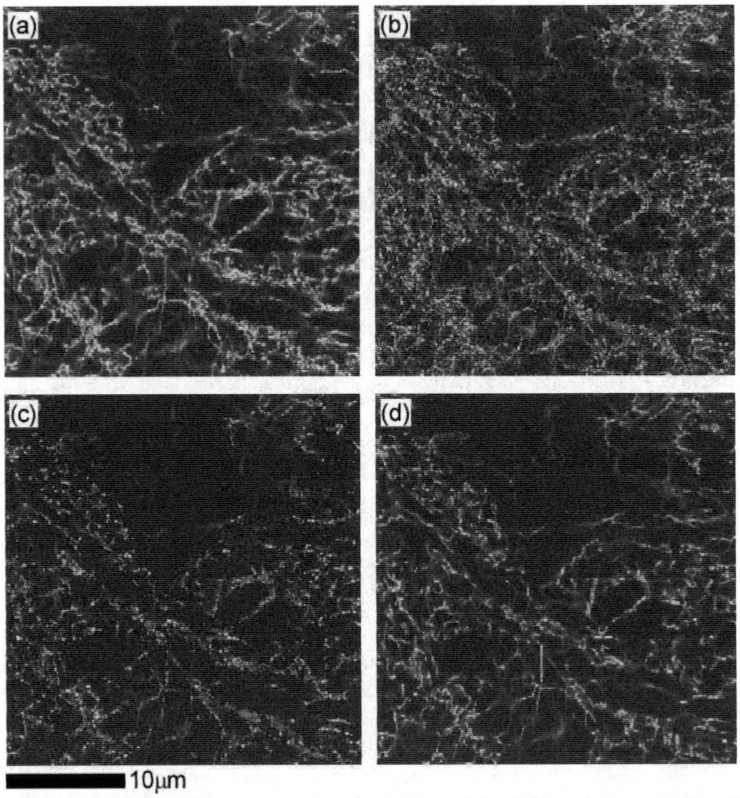

Figure 5. KAM maps for a 30□m square sub-region of the full copper scan data for the (a) original scan and the following filters applied to the noisy data - (b) the Kuwahara filter, (c) the Bilaterial smoothing filter and (d) NPAR.

It should be noted that cursory results on data from other materials, at different strain levels and step sizes relative to the visible structure suggest that the different noise reduction filters may not always be as successful as those shown here and warrant further investigation.

4. Conclusions
The results presented confirm the step size dependence of the GND calculations from EBSD scan data. The results also demonstrate that GND values tend to increase with decreasing orientation precision but that this effect is most apparent at the smaller step sizes as expected. Furthermore, the effects of reduced orientation precision can be mitigated through the application of noise reduction methodologies at least in terms of overall GND distributions.

References
[1] Wright SI, Nowell MM and Field DP 2011 *Microsc. Microanal.* **17** 316
[2] Pantleon W 2008 *Scripta Mater.* **58** 994.
[3] Wilkinson AJ and D. Randman D 2010 *Philos. Mag.* **90** 1159.
[4] Field DP, Merriman C, Allain-Bonasso N and Wagner F 2012 *Model. Simul. Mater. Sc.* **20** 024007
[5] D. Field, C. Merriman, J. Smith, 2007 *Microsc. Microanal.* **13** 920.
[6] J. Jiang, T. Britton, and A. Wilkinson 2013 *Ultramicroscopy* **125** 1.
[7] Godfrey A., Wu GL and Liu Q 2002 *Mater. Sci. Forum*, **408-412** 221.
[8] Humphreys FJ 2001 *J. Mater. Sci.* **36** 3833.
[9] Chen D, and Kuo JC 2010 *Ultramicroscopy* **110** 1297.
[10] Wright SI, Nowell MM, Lindeman SP, Camus PP, De Graef M and Jackson M 2015 *Ultramicroscopy* under review.
[11] Wright SI, Nowell MM, de Kloe R and Chan L 2014 *Microsc. Microanal.* **20** 852.

Microstructure and texture of asymmetrically rolled aluminium and titanium after deformation and recrystallization

M Wronski[1,2], K Wierzbanowski[1], S Wronski[1] and B Bacroix[2]

[1]AGH University of Science and Technology, Faculty of Physics and Applied Computer Science, al. Mickiewicza 30, 30-059 Kraków, Poland
[2]CNRS, Université Paris 13, 99, av. J.B. Clement, 93 430 Villetaneuse, France

E-mail: mwronskii@gmail.com, krzysztof.wierzbanowski@fis.agh.edu.pl, wronski.sebastian@gmail.com, brigitte.bacroix@univ-paris13.fr

Abstract. Asymmetric rolling is used to modify material properties and to reduce forces and torques applied during deformation. This geometry of deformation is relatively easy to implement on existing industrial rolling mills and it can provide large volumes of a material. The results of the study of microstructure and crystallographic texture in asymmetrically rolled aluminium 6061 and titanium (grade 2) are presented in this work. These characteristics were determined using the EBSD technique and X-ray diffraction. The rolling asymmetry was realized using two identical rolls, driven by independent motors, rotating with different angular velocities. It was found that asymmetric rolling leads to microstructural refinement and texture rotation (around the transverse direction). The impact of asymmetric rolling on microstructural refinement appears also in recrystallized samples of both materials. On the other hand, texture rotation, caused by asymmetric rolling, persists after annealing in titanium but not in aluminium samples.

1. Introduction

Asymmetric rolling (AR) is a promising deformation process, which can have many practical applications. It reduces the applied rolling forces and torques, changes the rolled plate shape and modifies the microstructure and crystallographic texture (e.g., [1-5]). The rolling asymmetry was realized in the present work using two identical rolls driven by independent motors, rotating with different angular velocities: ω_1 and ω_2. Therefore, the degree of process asymmetry is defined as: $A=\omega_1/\omega_2$ (asymmetry ratio). The main goal of this work was to study the influence of AR on crystallographic texture and microstructure of aluminum (6061 type) and titanium (grade 2). Two experimental techniques were used: electron backscatter diffraction (EBSD) and X-ray diffraction. The material characteristics were studied for the rolled samples and also for the recrystallized material in order to check, which modifications of material parameters persist after recrystallization.

2. Microstructure of asymmetrically rolled aluminium

The selected material was the 6061 type Al alloy (containing 2.4% Mg). The initial samples were prepared from as-delivered sheet of material (sample were not annealing before deformation). Rolling deformation was performed in one pass up to 36% reduction. The following degrees of rolling asymmetry were considered: A=1, A=1.1, A=1.3 and A=1.5. Two sets of samples were examined: the

rolled samples and the samples rolled and annealed in air at 450°C for 15 minutes. The microstructure was examined by the EBSD technique using a Cambridge S360 (W-GUN) electron microscope. The presented Inverse Pole Figures (IPF) maps for ND were determined on the RD-TD plane (where: RD - rolling direction, TD - transverse direction and ND - normal direction). They were determined in the central sample layers. For initial and deformed sample the maps of the size 700x400 μm^2 with the step of 1 μm were measured. On the other hand, for the deformed and annealed samples the maps of the size of 650x650 μm^2 with the step of 1.2 μm were determined.
The maps for the initial material, for the sample after symmetric rolling (SR) and for the sample rolled and then annealed - are presented in figure 1.

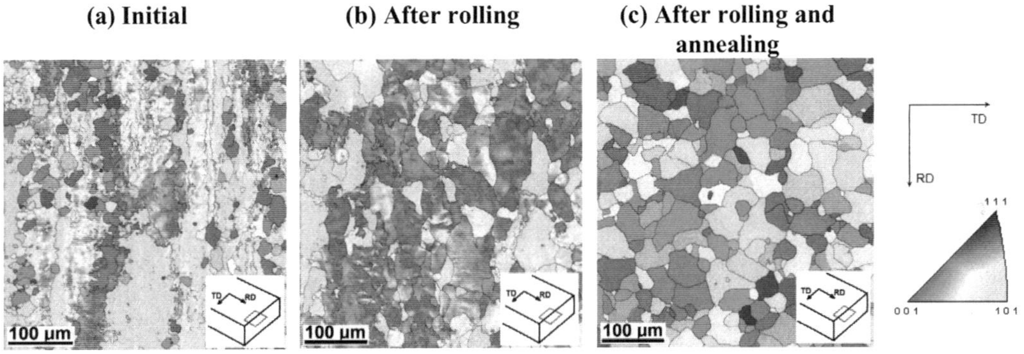

Figure 1. EBSD maps for the initial material (a), for the sample rolled symmetrically(A=1) to the reduction of 36% (b), and for the sample annealed after symmetric rolling at 450^{0}C for 15 minutes (c). IPF maps of the size 450x450 μm^2 for centre layers are shown.

Based on the measured maps the characteristic microstructural parameters like: average grain size and Kernel average misorientation (KAM) were calculated and are presented in function of asymmetry ratio (A) in figure 2. The following definition of grain was used in our analysis: a grain is defined as a set of points with misorientation angles between neighbouring measurement pixels smaller than some threshold value (15^0 in the present case) and it has to contain at least 5 measurement points.
After AR we observe fragmentation of grains, which is reflected in the increase of KAM vs. A (figure 2a). It should be noted that in the case of the annealed sample the grain structure was rebuilt and one observes equiaxed grains (figure 1c). The grain refinement in the annealed state, caused by AR, is reflected in the decrease of the average grain size with A - figure 2b.

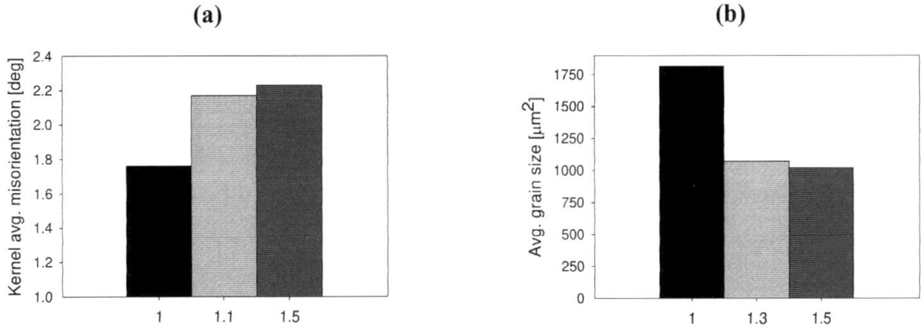

Figure 2. Influence of rolling asymmetry (A=1, A=1.1, A=1.3 and A=1.5) on: a) Kernel average misorientation for samples rolled to 36% reduction, b) average grain areas (μm^2) for samples rolled to 36% reduction and annealed at 450^0 C for 15 minutes. Results for central layers are shown.

IOP Conf. Series: Materials Science and Engineering **89** (2015) 012050 doi:10.1088/1757-899X/89/1/012050

3. Texture variation in asymmetrically rolled aluminium

Crystallographic textures were determined in the centre layers of the rolled samples using the X-ray diffraction technique (INEL goniometer with Co radiation). The texture of the initial material depicted in figure 3 is very sharp and contains Brass (*B*) and Cube (*W*) orientations. This material was next rolled symmetrically (A=1) and asymmetrically (A=1.5). The orientation distribution function (ODF) [6] sections at $\varphi_2=0^0$ presenting rolling textures, are shown in figure 4. A full analysis showed that the texture of the SR sample contained the following components: *W* (dominant one), *B*, rather weak *C* orientation and traces of *S* and *G* orientations [4]. A strengthening of W component is somewhat surprising. However, the shifts of B orientation are essential in the present study and we will focus our attention on this effect.

Characteristic effects appearing during SR and AR can be followed examining the *B* component which is visible in $\varphi_2=0^0$ section. In the case of the SR material four equivalent B component maxima lie on the horizontal blue line drawn at $\Phi=45^0$ (figure 4a). In the case of the AR material one observes the shifts of B component maxima. Namely, the first two maxima of the B component (in the [0^0, 180^0] range of φ_1) are shifted up and the next two maxima - are shifted down (figure 4b). Graphical explanation of this texture modification after AR is depicted in figure 4c. It was checked that the observed shifts of texture maxima result from a texture rotation around TD [5].

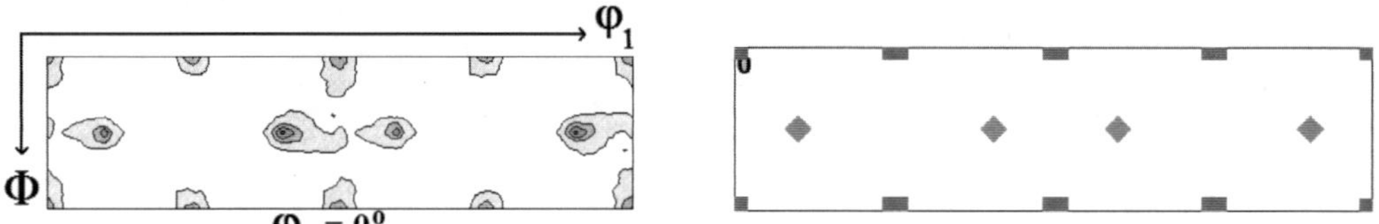

Figure 3. The crystallographic texture of the initial aluminium sample determined by X-ray diffraction. $\varphi_2=0^0$ sections are shown and orientations W (■) and B (◆) are marked.

(a) A=1 (b) A=1.5

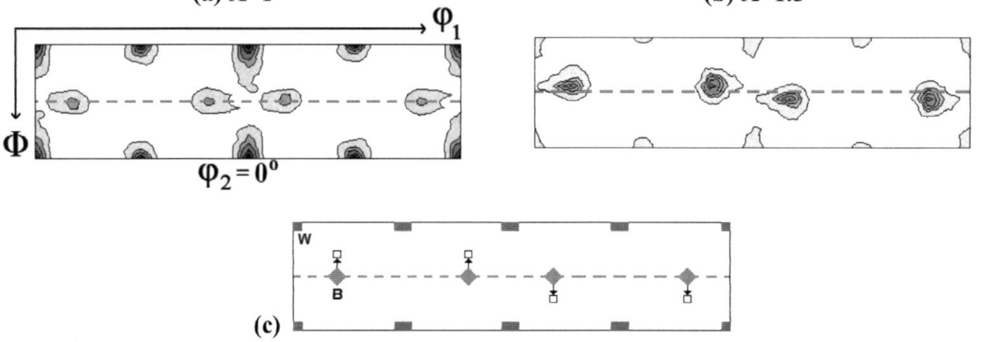

Figure 4. Textures of (a) symmetrically (A=1) and (b) asymmetrically (A=1.5) rolled samples determined in central material layers. $\varphi_2=0^0$ sections are shown and blue lines are located at $\Phi=45^0$. W (■) and B (◆) orientations are marked and (c) graphical explanation of texture modification after asymmetric rolling is shown. Textures were determined by X-ray diffraction and are presented using levels: 2.40, 8.00, 12.22, 17.61, 22.00, 35.00

The rolled aluminium samples were annealed next at 450 °C for 15 minutes. The recrystallization textures corresponding to SR (A=1) and AR (A=1.5) are shown in figure 5. They both contain only Cube (*W*) component. Therefore, we do not observe an influence of the rolling asymmetry ratio, *A*, on the annealing textures.

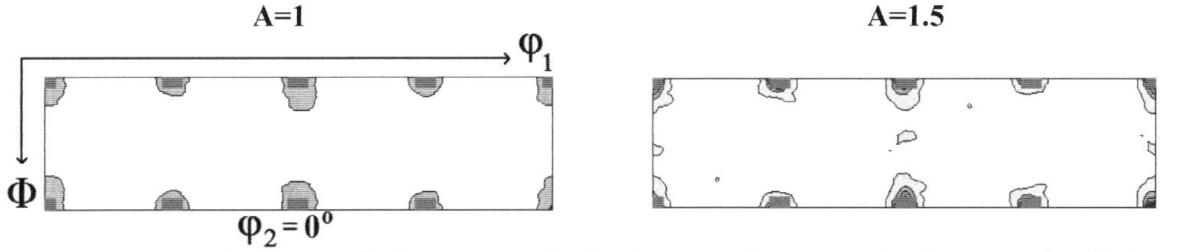

Figure 5. ODF sections ($\varphi_2=0^0$) for symmetrically (A=1) and asymmetrically (A=1.5) rolled and recrystallized aluminium 6061 samples (annealed at 450^0 for 15 minutes). Textures of centre layers are shown and the W(■) component is marked. Textures were determined by X-ray diffraction and are presented using levels: 2.40, 8.00, 12.22, 17.61, 22.00, 35.00

4. Microstructure of asymmetrically rolled titanium

The second examined material was commercially pure titanium (Ti grade 2). The samples were prepared from as-delivered sheet of material (without annealing). Two set of samples were examined: after low deformation (20% rolling reduction) and after higher deformation (60% rolling reduction) and annealing in air at 550 °C for 1 h. Rolling deformation was performed in many passes with 5% reduction per pass. For deformed sample (20%) the maps of the size 350x300 μm^2 with the step of 0.3 μm were measured. On the other hand, for the deformed (60%) and annealed samples the maps of the size of 400x200 μm^2 with the step of 0.5 μm were determined.

The EBSD maps for the initial material and for the samples rolled up to a 20% reduction are shown in figure 6. The maps for symmetric ($A=\omega_1/\omega_2=1$) and asymmetric rolling ($A=1.5$) are shown in this figure.

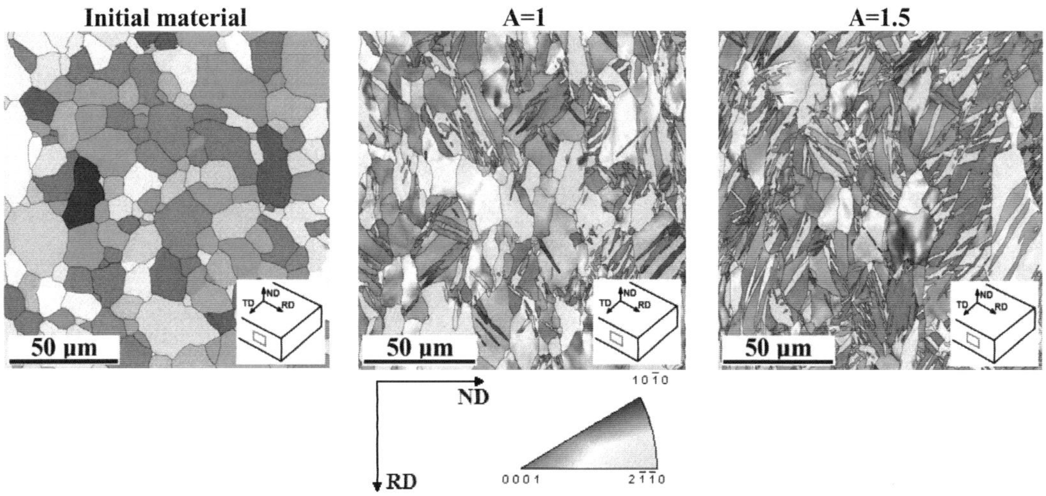

Figure 6. EBSD maps of the size 150x150 μm^2 for the initial material (titanium grade 2) and for the samples rolled symmetrically(A=1) and asymmetrically (A=1.5); the rolling reduction was 20%. Maps were measured from the central layer of the sample.

It is already visible that the microstructure is finer in the AR case. This is confirmed by the average grain size vs. A plot - figure 7a; the results for three degrees of asymmetry are shown: A=1, A=1.3 and A=1.5. It is visible that the grain size decreases with asymmetry ratio. The other factor, namely KAM, which characterizes the fragmentation of the microstructure, is also presented in this figure. Similarly, like in the case of aluminum, KAM increases vs. the degree of rolling asymmetry - figure 7b. In can be therefore concluded that AR produces smaller and more fragmented grains.

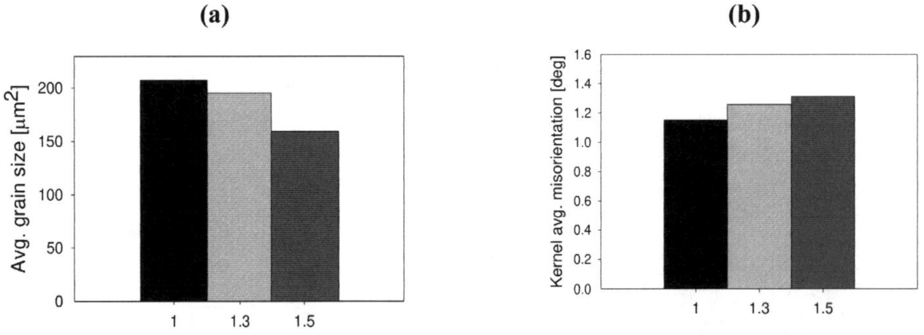

Figure 7. Influence of rolling asymmetry (A=1, A=1.3 and A=1.5) on: a) average grain size (μm^2) and b) Kernel average misorientation. Results for 20% rolling reduction and for central titanium layers are shown.

It is common industrial practice that an annealing is applied after rolling in order to recover some material properties, e.g., ductility. Therefore, the microstructural characteristics were determined for titanium samples rolled to 60% reduction and then annealed for 1 h at 550 ^{0}C. The EBSD maps measured in the central layers of SR and AR samples are presented in figure 8a,b. Based on the measured maps the average grain size for the cases A=1, A=1.3 and A=1.5 was calculated and is presented in figure 8c. The conclusion is that the average grain size of titanium samples decreases with asymmetry ratio (A).

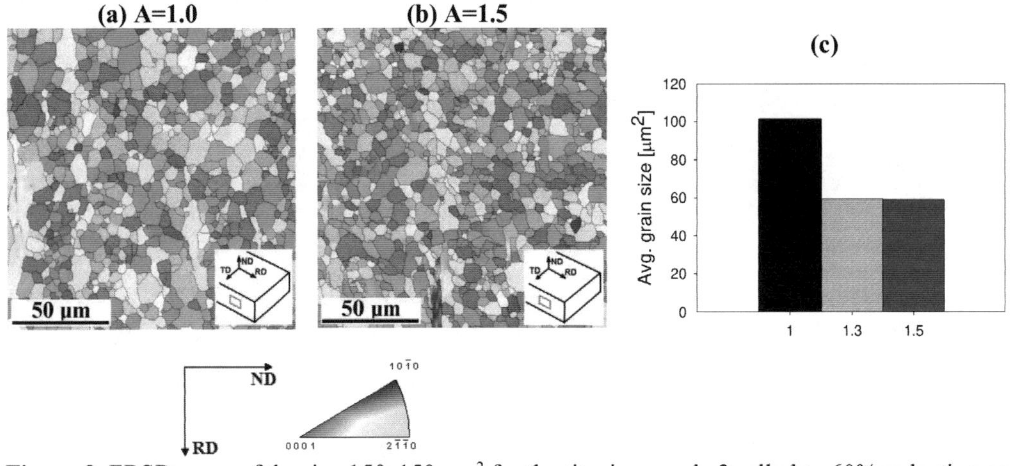

Figure 8. EBSD maps of the size 150x150 μm^2 for the titanium grade 2 rolled to 60% reduction and annealed for 1 h at 550°C: (a) after SR (A=1) , (b) after AR (A=1.5). Figure (c) presents a bar chart showing the average grain size (μm^2) vs. A relation for the rolled (A=1, A=1.3 and A=1.5) and annealed samples .

5. Texture variation in asymmetrically rolled titanium

Crystallographic textures of SR and AR titanium samples rolled to 40 % were examined. The measured pole figures (determined from EBSD maps) were not symmetrised and the orientation distribution functions (ODFs) were calculated. The $\varphi_1=0$ ODF sections for the initial material and for the symmetrically (A=1) and asymmetrically (A=1.3) rolled samples are presented in figures. 9a and b. We note that the main texture maximum of the initial material as well as of the SR material has a symmetric position with respect to the line plotted at $\varphi_2 =30^0$ (blue dashed line). In contrast, the AR process leads to a shift of the main texture maximum to the right - figure 9b. This effect is related again to a texture rotation around TD and is caused by an occurrence of a strong shear stress component, Σ_{13}, in the sample during the AR process (x_1=RD, x_2=TD and x_3=ND). This result was confirmed by texture modeling using the Finite Element Method [7] with an implemented crystalline model [8]. It should be mentioned that AR also leads to texture homogenization across the sample thickness [7]. Moreover, the residual stresses determined by diffraction method (see e.g., [9,10]) are generally lower in the AR samples [7,11]. The ODF section for the titanium sample rolled asymmetrically (*A*=1.3) and then annealed for 1h at 550 °C is shown in figure 8c. One finds that the characteristic shift of the main texture component, appearing after AR, persists after annealing.

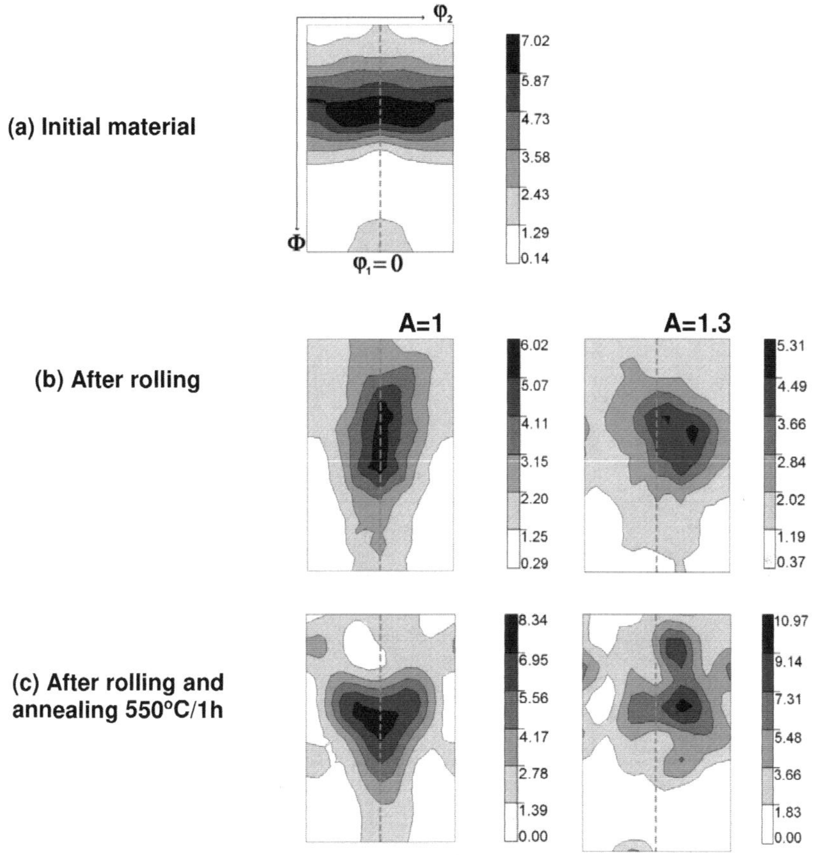

Figure 8. Texture of titanium grade 2 determined by the EBSD technique: a) initial material, b) SR (A=1) and AR (A=1.3) samples rolled 40% reduction, c) textures of SR and AR samples annealed at a temperature 550 °C for 1 h; φ_1=0 sections are shown.

6. Conclusions

The results presented, obtained for asymmetrically and symmetrically rolled aluminium 6061 and titanium grade 2 samples, lead to the following conclusions:

- average grain size is smaller and grains are more fragmented after asymmetric rolling (compared with the symmetric rolling mode),
- an impact of asymmetric rolling on the average grain size appears also after recrystallization of titanium and aluminium samples,

- textures after asymmetric rolling are rotated around the transverse direction; this rotation leads to characteristic shifts of maxima in aluminium and titanium textures,
- in the case of titanium samples - the shift of the texture maximum, caused by asymmetric rolling, persists also after recrystallization.

Acknowledgements
This study was financed by the Polish National Centre for Science (NCN) under decision numbers: DEC-2011/01/D/ST8/07399 and DEC-2013/11/B/ST3/03787.

References
[1] Gao H and Chen G 1998, *Iron and Steel* **33** 63
[2] Lee S H and Lee G N 2001, *Int. J. Mech. Sci.* **43** 1997
[3] Chhann S, Solas D, Etter A L, Penelle R and Baudin T 2007, *Mater. Sci. Forum* **550** 551
[4] Wronski S, Ghilianu B, Chauveau T and Bacroix B 2011, *Mater. Charact.* **62** 22
[5] Wronski S, Wierzbanowski K, Bacroix B, Wróbel M, Rauch E, Montheillet F and Wronski M 2009, *Arch. Metall. Mater.* **54** 89
[6] Tarasiuk J, Wierzbanowski K and Baczmanski A 1998, *Cryst. Res. Technol.* **33** 101
[7] Wronski M, Wierzbanowski K, Wronski S, Bacroix B and Lipinski P 2014, *Int. J. Mech. Sci.*, **87** 258
[8] Wierzbanowski K, Baczmanski A, Lipinski P and Lodini A 2007, *Arch. Metall. Mater.* **52** 77
[9] Baczmanski A, Wierzbanowski K, Tarasiuk J, Ceretti M and Lodini A 1997, *Rev. Metall-Paris,* **94** 1467
[10] Baczmanski A, Tidu A, Lipinski P, Humbert M, and Wierzbanowski K 2006, *Mater. Sci. Forum,* **524-525** 235
[11] M. Wroński, K. Wierzbanowski, M. Wrobel, S. Wronski, B. Bacroix 2015 *Met. Mater. Int.*, **21** (5) - to appear

In-situ observations of nucleation in Al-0.1Mg

G L Wu[1], H S Ubhi[2], M Petrenec[3] and D Juul Jensen[4]
[1] College of Materials Science and Engineering, Chongqing University, Chongqing, China
[2] Oxford Instruments, Nano Analysis, High Wycombe, HP12 3SE, UK
[3] Tescan Orsay Holdings a.s, Brno, Czech Republic
[4] Danish-Chinese Center for Nanometals, Section for Materials Science and Advanced Characterization, Department of Wind Energy, Technical University of Denmark, Risø Campus, DK-4000 Roskilde, Denmark

E-mail: wugl@cqu.edu.cn

Abstract. A tensile sample of an Al-0.1Mg alloy was in-situ tested in a SEM followed by in-situ annealing to develop recrystallizing nuclei/grains. The evolution of microstructure and crystallographic orientations were characterized using the EBSD technique. Changes in the same area within the sample during deformation and annealing steps were closely followed. During the deformation process the original grains remained recognizable, but became elongated and developed sub-structures. Over the annealing process nuclei with old and new orientations develop at high angle grain boundaries as well as at interior regions with large orientation spreads. Recrystallization twinning was also observed to occur in this Al alloy sample. Furthermore, some localized grain growth during the recrystallization was observed.

1. Introduction

Recrystallizing nuclei form in deformed metals and alloys during annealing. The orientations of nuclei to a large extent determine the final texture of recrystallized metals and alloys [1]. The relationships between the orientations of nuclei and the deformation matrix in which they form are not clear in particular when the large variations in orientations over small distances in the deformed matrix are considered. Classic nucleation mechanisms, predict that the nuclei will have orientations as those present at the matrix sites where they form [2]. It has however also been reported that recrystallizing nuclei with new orientations different from those in the deformed matrix are observed experimentally [3-9]. In most of the above studies, post-mortem analyses were used. One of the difficulties of these studies is that once the nucleation has occurred the pre-existing matrix has disappeared. This is also referred to as the lost evidence problem [10]. Direct characterizations of the microstructural evolution during the nucleation process are rare but considered essential for understanding of the nucleation mechanisms.

Recently, three-dimensional X-ray diffraction microscopy (3DXRD) was used for characterization of nucleation in 3D. Both Cu and Al samples were studied in this way [11, 12]. In both experiments nuclei with new orientation were observed. This type of experiment does not suffer from the lost evidence problem. However, the spatial resolution of those 3DXRD experiments was limited so the deformed microstructures could not be well resolved. Only the orientation distributions in the deformed materials were precisely determined. Another disadvantage of the 3DXRD is the restricted access to synchrotron sources so that only few experiments are possible. Therefore in the present work we have used in-situ heating in a scanning electron microscope (SEM) coupled with electron backscatter diffraction (EBSD) measurements for in-situ studies of nucleation with the aim of pin-pointing nucleation sites and orientation relationships between nuclei and parent matrix. Such in-situ measurements have been used before fore recrystallization studies, e.g. Ref. [13-18]; new in the present work is that we also follow the deformation in-situ before the annealing.

2. Materials and methods

The material used in the present study was a fully annealed 2mm thick Al-0.1%Mg rolling sheet. A dog bone tensile sample with a gauge size 2mm long and 2mm wide was cut from the sheet, with the tensile direction (TD) of the sample being parallel to the transverse direction of the rolling sheet. The tensile sample was ground to 4000 grit SiC paper followed by electropolishing in a perchloric acid based solution. Then the sample was mounted in a TESCAN MIRA XM FEGSEM using a GATAN Microtest EH2000EW in-situ tensile and heating stage. The SEM was equipped with an Oxford HKL NordlysMax EBSD detector for microstructure characterization. The clippers of the in-situ stage were 70° pre-tilted so that the normal surface of the sample (containing TD and short transverse direction, SD, perpendicular to the normal direction, ND) to be analyzed by EBSD was 20° aligned to the phosphor screen of the EBSD detector. Tensile elongation was driven and controlled by a step motor. Two heating elements were mounted on the clippers of the stage with one element on each clipper. The sample was heated directly by heat conduction from the heating elements by an electric current.

An area approximately at the center of the tensile sample was first characterized by EBSD. Then the sample was elongated to a true strain of about 0.58 in 8 steps at room temperature. During the in-situ tensile test, the scanned area was elongated along the tensile direction and was compressed perpendicular to the tensile direction. The corresponding area was characterized by EBSD after each tensile step so that the evolutions of microstructure and orientations of the individual grains within the area were followed in-situ. Then, the sample was heated to 290℃. It took about 30 minutes to reach 290℃. After collecting an EBSD scan of the selected area, the sample was held at 290℃ for 1 hour. EBSD scans were obtained every 10 minutes of the same area to follow the microstructural evolution during the in-situ annealing. Finally, the recrystallizing grains were detected from the EBSD maps by the DGR method [19]. The recrystallizing grains detected in the first two annealing steps (i.e. heated to 290℃ and hold for 10 minutes) were considered as recrystallization nuclei and are analysed in the present work.

3. Results and Discussion

3.1 Initial microstructure

Figure 1a shows the initial microstructure of the area that was followed during the in-situ tensile deformation. The area is 550μm by 1mm. The average grain size is about 90 μm. The initial texture as shown by the pole figure consists of cube, {001}<100>, and rotated Goss, {110}<011>, orientations.

3.2 Evolutions of microstructure and orientations during in-situ tensile test

Figure 2a shows the microstructure after step 4 (corresponding to a true strain of 0.33). Most of the initial grains can be recognized. Low angle dislocation boundaries are formed within the grains. Some grains develop relatively large local orientation variations. In a few areas, the microstructures cannot be resolved due to poor pattern quantity caused by large local plastic deformation and severe surface relief. Figure 2b shows that cube grains have large orientation spreads and relative small overall lattice rotations, whereas the Goss grains rotated about one of {111} poles so that they developed relative small orientation spreads but large lattice rotations.

Figure 3a shows the microstructure after in-situ tension to a true strain of 0.58. Although many areas cannot be resolved by EBSD due to local plastic deformation and surface roughing, some initial grains can still be recognized. Figure 3b shows the corresponding pole figure. As compared to the pole figure after deformation to 0.33 in Fig. 2b, it is seen that all the grains developed large orientation spreads and small overall lattice rotations during the tensile deformation from 0.33 to 0.58.

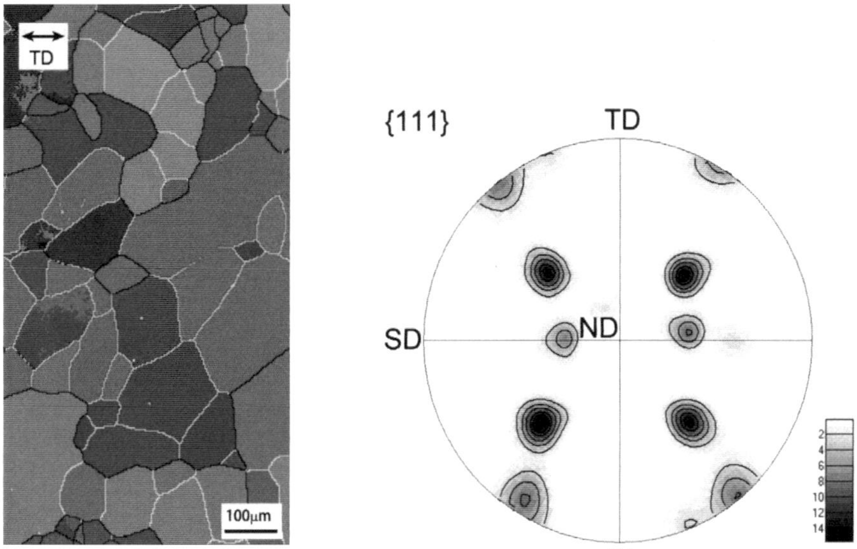

(a) **(b)**

Figure 1. (a) The initial microstructure and (b) the initial texture of the area to be followed during the in-situ tensile deformation. In the EBSD map, white and black lines represent grain boundaries ≥ 2° and ≥ 15°, respectively. The colours of the EBSD map are so-called Euler colours, in which the three Euler angles by Bunge notion for the orientation of each datapoint are coloured by red, green and blue, respectively.

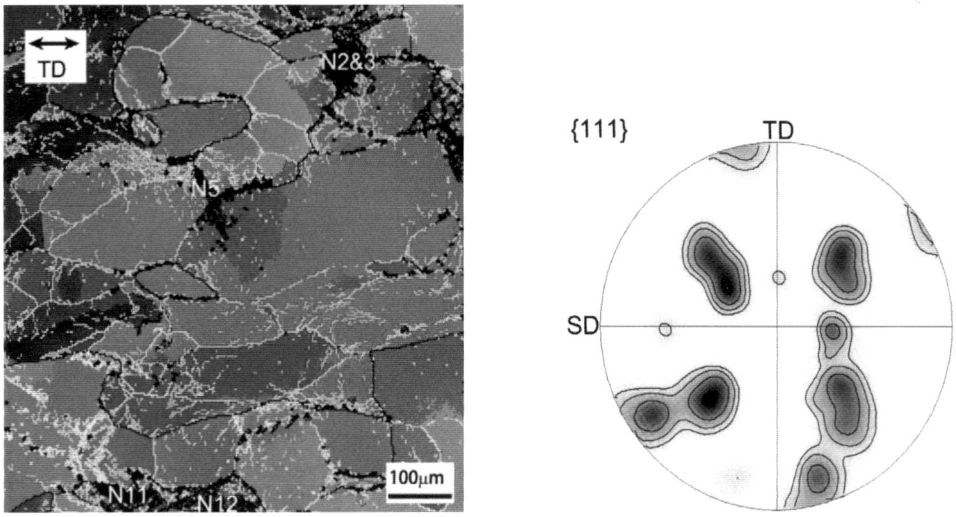

(a) **(b)**

Figure 2. (a) EBSD map and (b) {111} pole figure after in-situ deformation to 0.33. In the EBSD map, the black pixels are non-indexed data points. Some positions where nuclei are found to develop upon annealing are numbered by N_i. The intensity levels in the pole figure are as in Fig. 1b.

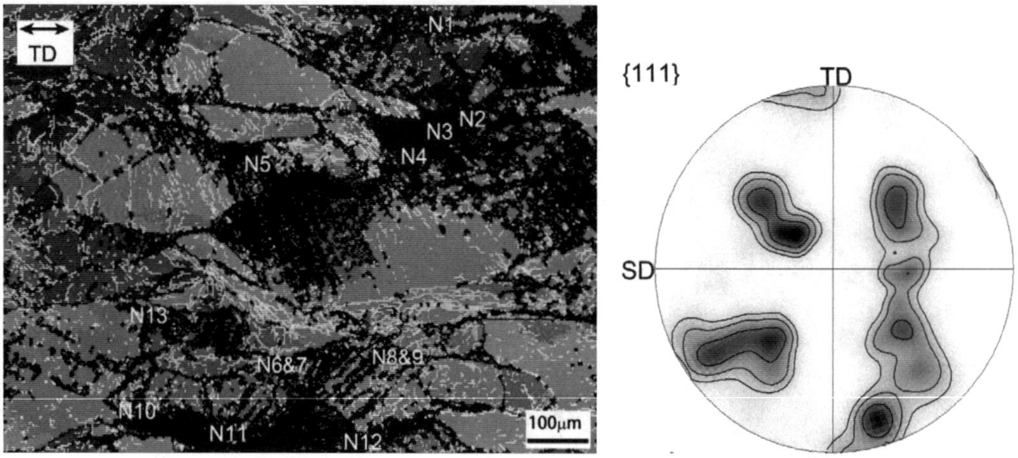

(a) **(b)**

Figure 3. (a) EBSD map and (b) {111} pole figure after in-situ deformation to 0.58. The positions where nuclei are found to develop upon annealing are numbered by N_i. The intensity levels in the pole figure are as in Fig. 1b.

3.3 Evolutions of microstructure and orientation during in-situ annealing.

Figure 4a and 4b show the microstructures after the sample has been heated to 290℃ and being held at the temperature for 10 minutes, respectively. In the area, 13 recrystallization nuclei can be clearly identified. All the nuclei sites are marked in Fig. 4. After carefully tracking the grain structure during the in-situ tensile test, the nucleation sites of all the nuclei were successfully determined in the original structure and subsequent tensile steps (see Figs.2 and 3). It is seen that all nuclei are formed at areas inside the original grains associated with large deformation or at original grain boundaries.

Grain boundaries are thus confirmed to be effective nucleation sites during annealing for nucleation in agreement with many previous observations [e.g. 20]. In interior areas with large localized plastic deformation which are also found to be potential nucleation sites in the present work, the lattice rotation and/or lattice curvatures are large and thus the local orientation spreads are also large. High deformation energy will therefore be stored in these areas, which will facilitate the formation of nuclei.

After comparing the orientations of the nuclei to the orientations in the deformed matrix where they developed (as shown in Fig. 3a), it is found that 8 nuclei have orientations fully within the orientation spread of the deformed matrix or their misorientations to the deformed matrix are smaller than 20°.These orientations are considered to have "old orientations", already presented at these sites in deformed structures. The reason why deviations as large as 20° are included for nuclei of old orientations is that in certain areas the EBSD pattern quality is low with many non-indexed pixels. So the 20° deviation is accepted and is considered a conservative criterion. Even applying this conservative criteria, there are 6 (N2, N3, N5, N6, N11 and N12) nuclei with misorientations to the deformation matrix greater than 20° (i.e. with "new orientations"). It is found that the nucleation sites of five of these six nuclei with new orientations are interior areas with large local deformation, where the EBSD pattern qualities are comparable low. The nucleation sites for these 5 nuclei are therefore also marked in Fig. 2a.

Figure 5a shows the pole figure of the nucleus N6. N6 is about 45°<110> misoriented to one of its parent deformed grains. However, it is found that there is a thin twin to the right of N6 (indicated by the arrow). The twin of N6 is 59°<433> misoriented to the deformed grain. Like N6, the tiny grain to the right of N5 (indicated by the arrow) is also twin related to the deformed grain. Therefore, it is plausible that N5 and N6 are formed by twinning during annealing. Figure 5b shows the pole figure of the nucleus N3 and its deformed parent grains. The misorientation between N3 and the pixels near the nucleation site in the deformed grains are approximately 42°<343>, 31°<340> and 25°<110>, respectively. It is seen that the nucleus has a ~40°<111> misorientation to the deformed grain lying to the right of the nucleus. Similarly, N2, N11 and N12 also have a ~40°<111> misorientation relationships to their parent deformed grains.

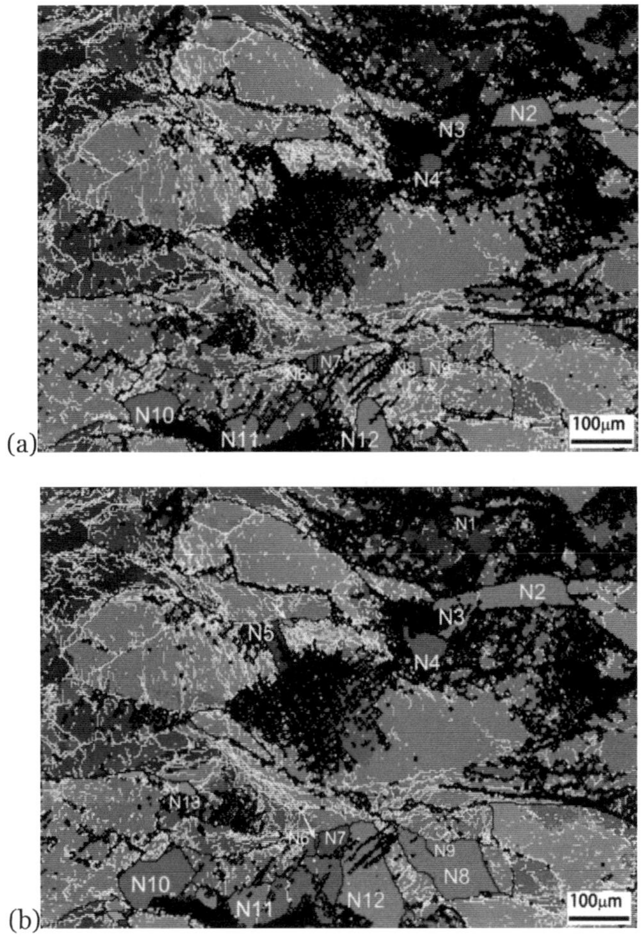

Figure 4. EBSD maps of the selected area after the sample was (a) heated to 290℃ and (b) hold at 290℃ for 10 minutes.

For the local areas in the deformed parent grains leading to the formation of nuclei, the average orientation spreads of the areas at nucleation sites (expressed by the average misorientation angle between each data point within the area and the mean orientation of the area) were calculated. It is found that areas/grains with small orientation spreads only form nuclei with old orientations. Grains having large orientation spreads may also form nuclei with old orientations. However, deformed areas/grains developing nuclei with new orientations all show large orientation spreads. It should be noted that especially in these areas many pixels on the EBSD map were not indexed and the local orientation spreads are therefore expected to be even larger than that measured.

Although most of nuclei were found to have old orientations, there are 4 nuclei which are neither twin-related to the deformed matrix nor close to the orientations within deformed parent grains. Nuclei with old orientations may just develop in the deformed parent grains with small orientation spreads by the processes such as subgrain growth, since the sizes of subgrains are normally large in these areas. The original grain boundaries will promote the growth of the subgrains due to high mobility of the high angle boundaries between the subgrains across the original grain boundaries. Concerning the formation of nuclei with new orientations, the 40°<111> misorientation relationship seems to dominate in the present work. However, the statistic is weak and also the lack of knowledge about what is going on below the investigated surface holds us from further analysis of such misorientation relations based on the present data.

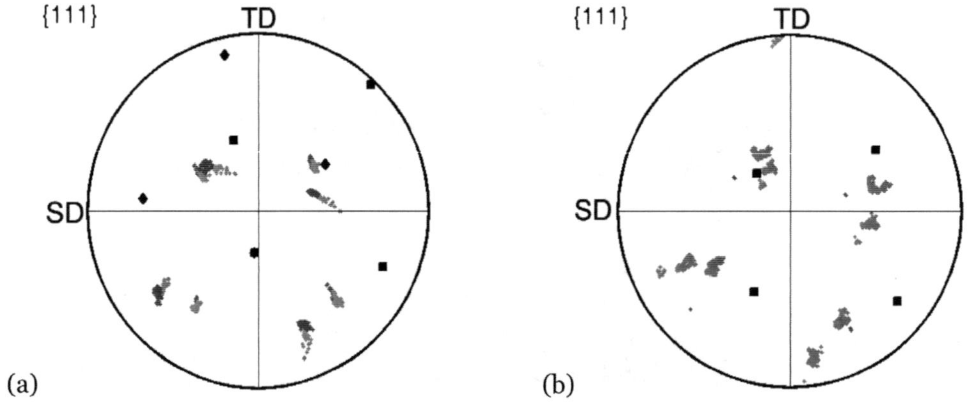

Figure 5. Pole figures of (a) N6 and (b) N3 and the pixels at the nucleation sites in their deformed parent grains. The squares represent orientations of the nuclei, diamonds twins, and colour points deformed grains. The colours in the pole figures are the same as in the EBSD maps.

Annealing twins are often observed in Ni and Cu due to the low stacking fault energies. However, annealing twinning has also been observed in Al [12] with high stacking fault energy although less frequently. In the present work twinning are observed for nuclei N5 and N6. It cannot be determined whether these twins were formed directly from the deformed matrix or from an already existing recrystallizing nucleus/grain. But one interesting point is that these twins (having twin orientation relationship to the deformed matrix) are very small and are observed to be consumed by their neighbouring grains during subsequent annealing; and after complete recrystallization these twins have disappeared. For example, Fig. 6a and 6b show the EBSD maps of grains near N5 after annealing at 290℃ for 30 minutes and 40 minutes, respectively. After annealing for 30 minutes, N5 and the grain twin-related to its deformed parent grain (indicated by the arrow in Figs 4b and 6a) are seen. However, the twin disappeared after annealing for 40 minutes. From the morphologies of the grain boundary between N5 and its neighbouring grain after annealing for 30 minutes and 40 minutes, it is likely that the twin was consumed by the neighbouring grain. These results emphasize that in-situ annealing is an important method to understand nucleation mechanisms, as even more evidence may be lost than just the deformed matrix.

By following the microstructural evolution during all in-situ annealing steps, it is also possible to identify the growth of individual nuclei. It is found that N9 was consumed by N8 after annealing at 290℃ for 40 minutes, at which state the material is only partly recrystallized and N8 and N9 are both partly surrounded by the deformed matrix (see Fig. 7). It is seen in Fig. 4b that the grain boundary between N8 and N9 was highly curved. The curvature might have contributed to the growth of N8 into N9, and it may be concluded that some localized grain growth is also possible during recrystallization in the present material (also see the example shown in Fig. 6).

It may be argued that recrystallization observed by the surface inspections as the present in-situ annealing study, might be different from the bulk behaviour. This of course cannot be ruled out. However, it is at least reassuring that the orientations of the recrystallizing nuclei/grains measured on the surface in the present work are similar to those determined in bulk.

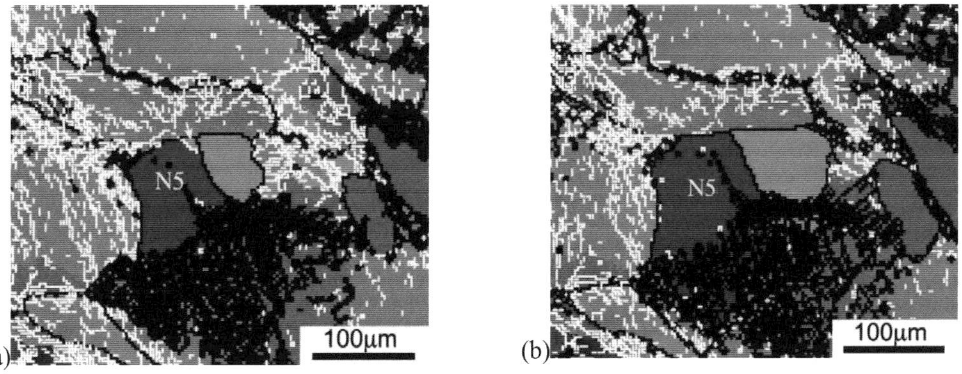

Figure 6. EBSD maps of N5 after annealing at 290℃ for (a) 30 minutes and (b) 40 minutes.

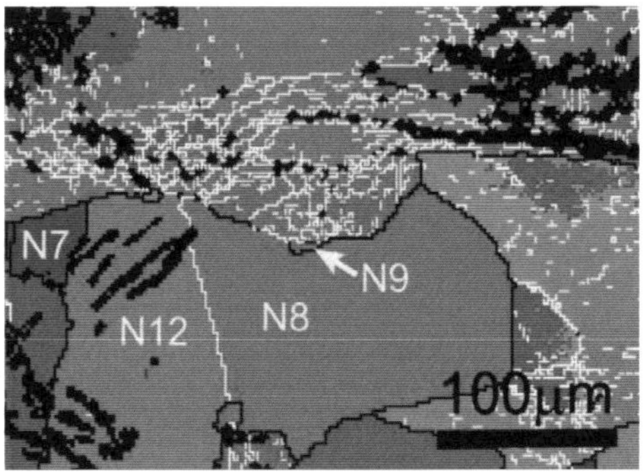

Figure 7. EBSD map of N8 and N9 after annealing at 290℃ for 40 minutes. The grain N9 is indicated by the arrow in the map.

For the studies of recrystallization nuclei, most published results are based on statistic investigations of deformed and annealed samples. The whole deformation history is thus generally not followed. The deformation history including the lattice rotation of grains and the strain accommodation across grain boundaries may influence the nucleation process. In this sense, the in-situ tensile and heating offers an excellent opportunity to follow the microstructural evolution leading to nucleation, which will be analysed in a future study of the present experiment.

4. Conclusions

An Al-0.1Mg alloy sample was in-situ tensile strained to a true strain of 0.58 and then in-situ annealed at 290℃ for 1 hour in a SEM. EBSD was employed to characterize the microstructure and local orientations within the same area on the surface of the sample during the in-situ deformation and annealing processes. After annealing for 10 minutes recrystallizing nuclei/grains developed at original grain boundaries and within interior grain areas with large orientation spreads. Most of the nuclei were found to have orientations close to the orientations within the parent deformed grains and preferentially form at original grain boundaries between grains with relative small local orientation spreads. Nuclei with orientations not observed in the deformation matrix were also detected. These nuclei formed at interior sites with large local orientation spreads. Recrystallization twins were also found in the Al-0.1Mg alloy. The twins, at least for the two cases in the present results, were observed to disappear during the subsequent recrystallization process. Furthermore some localized grain growth was observed to occur before the sample was completely recrystallized due to highly curved grain boundaries. Although the formation of nuclei with new orientations cannot be analysed in detail in the present results, the present in-situ deformation and annealing experiments provide a powerful approach to study nucleation mechanisms.

Acknowledgments

Financial supports from the National Natural Science Foundation of China (NSFC, Grants No.51471039, 51327805 and 51421001) and Fundamental Research Fund of Central Universities of China (Grant No. CDJZR14135502 and 10611205CDJZXZ138802) are gratefully acknowledged. The authors would also greatly thank the supports from Danish National Research Foundation (Grant No. DNRF86-5) and NSFC (Grant No. 51261130091) to the Danish-Chinese Center for Nanometals.

References

[1] Doherty R D, Hughes D A, Humphreys F J, Jonas J J, Juul Jensen D, Kassner M E, King W E, McNelly T R, McQueen H J and Rollet A D 1997 *Mater. Sci. Eng.* **A 238** 219-74
[2] Humphreys, F J and Hatherly M 2003 *Recrystallization and phenomena 2nd edition*, Oxford, UK, ELSEVIER
[3] Barrett C S 1940 *Trans. AIME.* **137** 128
[4] Beck P A, Sperry P R and Hu H 1950 J. *App. Phy.* **21** 420-25
[5] Inoko F and Miga G 1987 *Scripta Metall.* **21** 1039-1045
[6] Paul H, Driver J H, Maurice C and Jasieński Z 2002 *Acta Mater.* **50** 4339-55
[7] Sabin T J, Winther G and Juul Jensen D 2003 *Acta Mater.* **51** 3999-4011
[8] Okada T, Huang X, Kashihara K, Inoko F and Wert J A 2003 *Acta Mater.* **51** 1827-39
[9] Wu G L and Juul Jensen D 2007 *Acta Mater.* **55** 4955-64
[10] Chung C Y and Duggan B J 1996 *Proc. of 11th Inter. Conf. on Texture of Mater.* (Xi'an, China) 399-404
[11] Larsen A W, Poulsen H F, Margulies L, Gundlach C, Xing Q F, Huang X and Juul Jensen D 2005 *Scripta Mater.* **53** 553-7
[12] West S S, Winther G and Juul Jensen D 2011 *Metall. Mater. Trans.* **42A** 1400-8
[13] Humphreys F J and Ferry M 1996 *Mater. Sci. Forum.* **217** 529-34
[14] Springer F and Radomski M 1997 *Mater. Sci. Forum.* **273** 497-502
[15] Lens A, Maurice C and Driver J H 2005 *Mater. Sci. Eng.* **A 403** 144-53
[16] Nakamichi H, Humphreys F J and Bate P S 2007 *Mater. Sci. Forum.* **550** 441-6
[17] Bozzolo N, Jacomet S and Logé R E 2012 *Mater. Charact.* **70** 28-32
[18] Kerisit C, Logé R E, Jacomet S, Llorca V and Bozzolo N 2013 *J. Microscopy.* **250** 189-99
[19] Wu G L and Juul Jensen D 2008 *Mater. Charact.* **59** 794-800
[20] Vandermeer R A 1995 *Proc. of 16th Risø Inter. Symp. on Mater. Sci.* (Roskilde, Denmark) 193-245

Microstructure and mechanical properties of ARB processed Mg-3%Gd alloy

J Q Wu[1], S Huang[1], Y H Wang[2], G L Wu[1,*], N Hansen[3] and X Huang[3,*]
[1] College of Materials Science and Engineering, Chongqing University, Chongqing 400045, China
[2] National Engineering Research Center for Equipment and Technology of Cold Strip Rolling, Yanshan University, Qinhuangdao 066004, China
[3] Danish-Chinese Center for Nanometals, Section for Materials Science and Advanced Characterization, Department of Wind Energy, Technical University of Denmark, DK-4000 Roskilde, Denmark

E-mail: wugl@cqu.edu.cn

Abstract. Mg alloys have various advantages. However, the low formability due to the poor ductility of Mg alloys limits their engineering applications. In this study, an Mg-3%Gd alloys was chosen to explore processing approaches for improving its strength and ductility combination. The alloy was processed by accumulative roll-bonding (ARB) at 400℃ to 4 cycles followed by annealing at various temperatures. The microstructures after annealing were characterized by the electron backscatter diffraction technique and the mechanical properties were measured by a tensile test. It was found that the alloy has a good combination of strength and ductility after 2 cycle ARB processing followed by annealing at 290℃ for 1h. The strength is 2.3 times higher than that of the fully annealed coarse grained alloy, and the elongation is comparable with that of fully annealed coarse grained counterpart. The good mechanical properties were related to the fine-sized heterogeneous microstructures and weakened texture.

1. Introduction

There has been an increasing demand for magnesium (Mg) alloys as light-weight structural materials in automobile industries during the last decade, mainly due to their various advantages, such as high specific strength, light weight and good dump resistance. However, commercial magnesium alloys usually exhibit poor ductility at room temperatures because of a lack of sufficient number of slip systems. The most effective method to overcome this major drawback of Mg alloy is modifying their microstructures and textures. As is well known, severe plastic deformation (SPD) techniques, such as equal channel angular extrusion, differential speed rolling, large strain hot rolling and accumulative roll-bonding (ARB), have been proved to be effective to weaken the basal texture or to obtain a modified microstructure in Mg alloys. However, ARB is the most prospective SPD method, which could be adopted by industries as a continuous process to produce large sheets with ultrafine grains [1]. The microstructures and texture of the AZ-series Mg alloys processed by ARB method have been well studied in the past decade [2-4]. Recently, hot processing and post-deformation annealing has been proved be a significantly effective to weaken the textures of Mg alloys containing rare earth (RE) elements, such as Nd, Ce, Gd and Y [5, 6]. In this study, an Mg-3%Gd alloy was selected as an experimental material and processed by ARB. The mechanical behaviour and microstructural evolution upon annealing of this alloy are studied.

2. Experimental

The material used in this study was an Mg-3%Gd alloy containing 96.86wt.% Mg, 3.0wt.% Gd and 0.14wt.% Al. The cast ingot was extruded at 500℃ into ARB starting sheets with a thickness about 1mm, and then annealed at 500℃ for 3h. After degreasing and wire-brushing followed by preheating at 400℃ for 8 minutes, two stacks of starting sheets were rolled to a 50% thickness reduction. Then, the rolled sheet was cut, stacked and preheated at 400℃ for 8 minutes, and then roll-bonded again. Such a procedure was repeated for up to 4 cycles, ε_{vM} = 3.2. Rolling was performed in a rolling mill with two 230mm diameter rolls that rotate at 35r/min. The rolls were neither lubricated nor heated and samples were air-cooled after rolling. The subsequent annealing was performed for 1 hour at four different temperatures 190℃, 250℃, 290℃ and 400℃.

The microstructures were observed in a Zeiss Auriga dual beam station. The specimens were etched in a solution of 5g picric acid, 10ml H_2O, 70ml ethanol and 10ml acetic acid and local crystallographic orientations were characterized using an Oxford HKL Channel 5 electron backscatter diffraction (EBSD) detector attached to the microscope. For EBSD measurements, the specimen surface was mechanically polished followed by electrochemical polishing in the AC2 solution. Tensile testing was carried out at room temperature under a constant cross-head speed of 0.5mm/minutes. Tensile specimens with 10mm in gauge length and 5mm in width were cut from the ARBed sheets, with the gauge length direction being parallel to the rolling direction (RD). Vickers microhardness (Hv) was measured on the longitudinal cross-section of the samples by imposing a load of 3N dwelling for 10s.

3. Results

3.1. Starting material

The microstructure of the starting sheets before ARB process (after annealing) is shown in Figure 1a. It is seen that the grains are equiaxed with an average grain size of about 40μm. The (0002) pole figure (Figure 1b) of the starting sheets determined by EBSD shows a basal texture, which is typical for Mg alloys.

3.2. Mechanical properties of ARBed and annealed samples

Figure 2 shows the microhardness of the material after ARB processing. It is seen that the microhardness of the sample increases from 43 to 62 during the first pass of the ARB processing; and a very small change is observed during the following passes. Therefore, it is believed that grain refinement has taken place mainly during the first pass, as also shown in [2] that an ultrafine grained microstructure (d < 1μm) in AZ91 was obtained during the ARB process under similar rolling condition.

Figure 3a shows the tensile curves of the ARBed sheets. Similar to the tendency of the microhardness curve, the yield strength increases dramatically during the first pass, and then shows little change during the following passes. It is noted that the elongation after the 2nd ARB cycle is slightly better than other cycles.

Figures 3b-d show the tensile curves of the sheets after 1, 2, and 4 ARB cycles and annealed at different temperatures for 1 hour. For comparison, the curve of the starting material is also included in these figures. The low yield strength of the starting sample is attributed to the large grain sizes (Figure 1a). After ARB processing, the yield strength of the material increases greatly and the elongation decreases, which are commonly observed in metals and alloys deformed by ARB. However, the strength increases to even higher values after annealing at 190℃ and 250℃. The possible reason could be the segregation of Gd on twin boundaries [7] or grain boundaries [8]. After annealing at 290℃, the

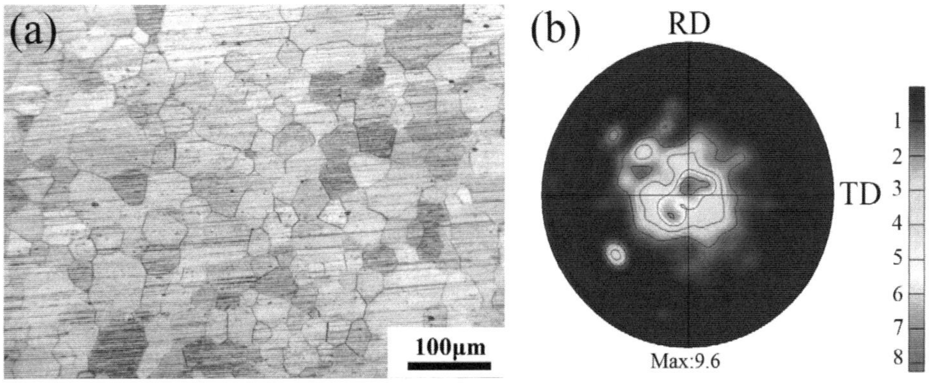

Figure 1. OM micrograph (a) and (0002) pole figure (b) of the starting material for ARB processing.

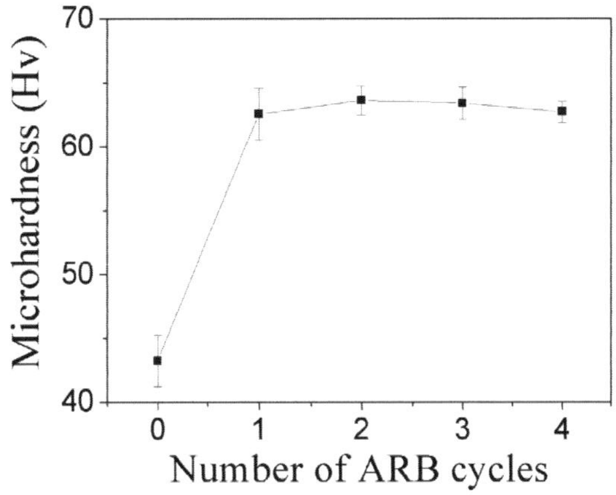

Figure 2. Curve of microhardness of the material after ARB processing.

yield strength of the sample decreases slightly compared with the deformed state. However, the elongation of the sheets increases significantly, close to that of the starting alloy. After annealing at 400℃ the yield strength decreases a lot, close to the as-annealed state, and the elongation is almost constant. The mechanical properties of the 2nd and 4th ARBed materials (shown in Figs. 3c and 3d, respectively) are very similar to those to 1st ARBed one. The only difference is that the 2nd ARBed material has a slightly larger elongation after ARB and annealing at 290℃ and 400℃.

3.3. Microstructure and texture of annealed samples

It is rather difficult to obtain EBSD patterns with good quality on Mg alloys after deformation to large strains. No EBSD map is available on the ARBed and low temperature annealed samples. Since the mechanical behaviours of the alloy are very similar after ARB processing and subsequent annealing, only the microstructures and texture of the 2nd ARBed sample were characterized.

Figure 4a shows the microstructure from one surface to the center of the sample after annealing at 290℃. A typical heterogeneous microstructures with a well recovered matrix mixed with fine-sized recrystallized grains can be observed in the sample. The size of the large grains is about 10μm whereas the small grains have grain sizes of about 1μm. Figure 4b shows that the sample is fully recrystallized

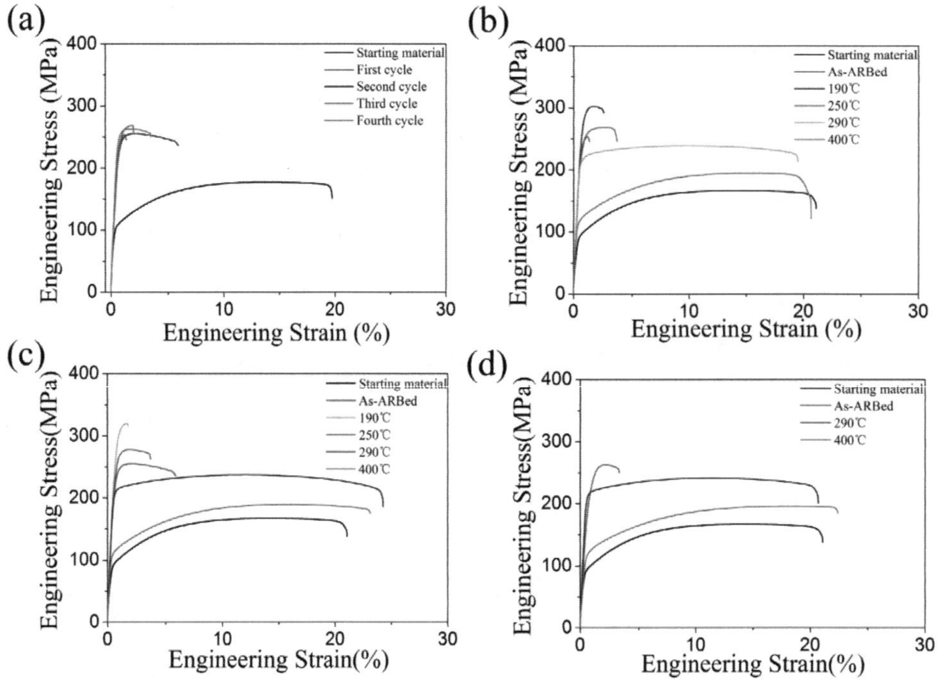

Figure 3. Tensile curves of (a) the ARBed sheets and 1st (b), 2nd (c) and 4th (d) ARBed sheets after annealing at different temperatures.

after annealing at 400℃, exhibiting equiaxed grains with an average size of ~15μm. As shown in Figure 4, the interface introduced by the first cycle can still be observed. Although the size of grains at the interface is smaller than those in other areas, there is no obvious gradient of grain size across the thickness of the samples annealed at 290℃ and 400℃. Figure 5a shows that the texture of the sample annealed at 290℃ is the basal texture. Compared with the texture of the starting sheet (Figure 1b), the c-axis of the maxima rotates 15° away from the ND towards the negative RD. As observed in the Figure 5b, a rather weak basal texture is formed in the sample after annealing at 400℃and the texture is much more spread.

4. Discussion

4.1. ARB processing effect on mechanical properties

The present results demonstrate that the strength and hardness were enhanced due to the large strain imposed on the sheet mainly during the first cycle (Figure 2 and Figure 3a). It is known that grain growth in Mg alloys is rapid when the ARB temperature is higher than 0.5Tm (the melting point) or 461K or 188℃, thus the repeated heating at 400℃ between rolling cycles is above the recrystallization temperature of magnesium can result in static recrystallization and partially eliminate the accumulated strain. It has found that the interval reheating of pure Mg can lead to full recrystallization and a large grain size [9, 11]. In this study, the intermediate annealing leads to saturation of hardening already after one cycle.

ARB introduces a large shear strain in the sheet thus the microstructure is greatly refined. At the same time the texture is also weakened during the subsequent annealing at 290℃ and 400℃ due to recovery and static recrystallization. After annealing at 290℃, the microstructure is greatly refined.

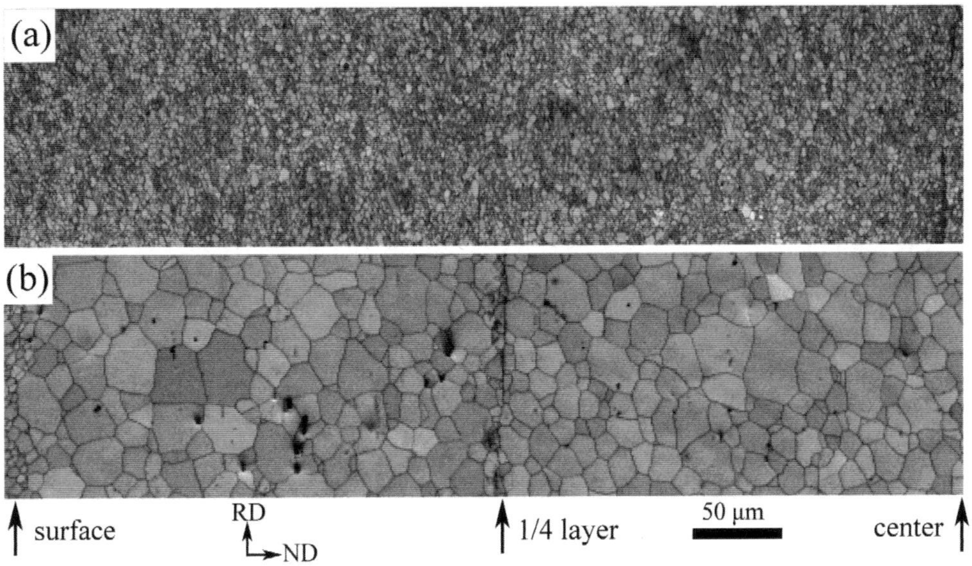

Figure 4. Microstructures of the materials annealed at 290℃ (a) and 400℃ (b).

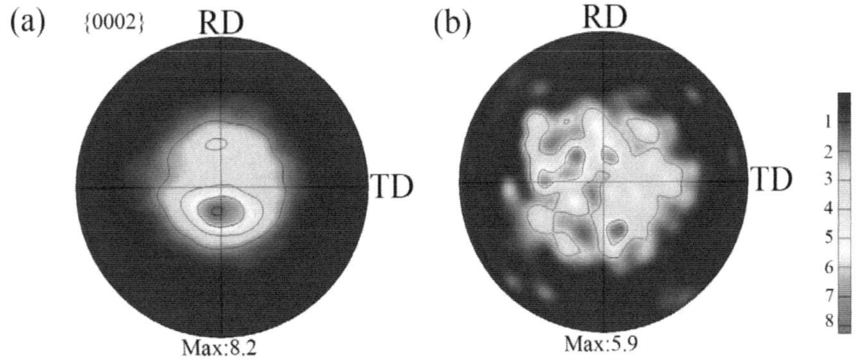

Figure 5. (0002) pole figure of the samples annealed at 290℃ (a) and 400℃ (b).

The best combination of elongation and strength is achieved, with the yield strength being 210 MPa and total elongation 24%. Compared to the mechanical properties of the starting Mg-3%Gd alloy, the yield strength is 2.3 times larger while the elongation is about the same (even slightly larger). In another study, the yield strength and elongation of an Mg-3(wt.) %Gd with an average grain size of 76μm after solute treatment at 535℃ for 1.5 hours are 71MPa and 13%, respectively [10]. For an Mg-Gd-Zn alloy with similar compositions (3%Gd, 1%Zn), the highest yield strength of 143MPa and elongation of 30% were obtained after warm rolling at 430℃ to 85% reduction and annealing at 350℃ [11]. Therefore, the combination of strength and elongation of the present Mg-3%Gd alloy is greatly improved by ARB processing and subsequent annealing. Compared with other ARB processed commercial Mg alloys, such as AZ31, the present Mg-Gd alloy has much better mechanical properties. For example, optimized yield strength and total elongation are 160 MPa and 20% in an AZ31 alloy after 2 cycle ARB processing at 450℃ and annealing at 350℃ [1].

4.2. Grain size and texture effect on mechanical properties

The elongation of the ARB processed Mg-3%Gd alloy does not increase through annealing at temperatures below 290℃. Greatly improved elongation is obtained after annealing at 290℃. This is attributed to the occurrence of static recrystallization and the presence of new fine recrystallized grains in the microstructure. Extensive static recrystallization takes place in the microstructure at 290℃. Thus, the increased volume of fine sized recrystallized grains from the previous ARB process will improve the elongation of the alloy, which is better than the elongation of the sheet annealed at 400℃ with a grain size of 15μm. It is interesting to note that the sheet annealed at 290℃ has a strong basal texture while the 400℃ annealed sheet has a more spread and weak basal texture. It was suggested that the modified texture would introduce an excellent room temperature ductility in Mg-RE alloys sheets and that the grain size has a negligible effect on room-temperature ductility compared with texture [12]. However the present study shows the beneficial effect of grain size is worth a further investigation.

Figure 6 shows the EBSD maps taken from the necking zones close to the fracture. Figure 7 shows the misorientation angle distribution of the sample before and after the tensile test. Both figures show that twinning is activated during the tensile test of the sample annealed at 400℃. It is well known that grain size has a significant influence on twinning activity [13]. The effect of grain size on ductility in AZ31 alloy processed by equal channel angular extrusion and rolling was studied, and it was found that a transition from slip-dominated to twinning-dominated flow occurs with increasing grain size [14]. They have shown that the twinning activity of the rolled AZ31 sheets obviously increases when the grain size increases from 4.2 μm to 17 μm. When the grain size is larger than 20 μm, the deformation mode of the AZ31 sheet even turns to be twinning-dominated flow. The 400℃ annealed sample has a coarse grain size. In this sense, the twin activity of the 400℃ annealed sample fits with the results in the literature. Although the 400℃ annealed sample with a more random texture favours slip during deformation, the deformation mode is twinning-dominated flow after deformation due to the large grain size.

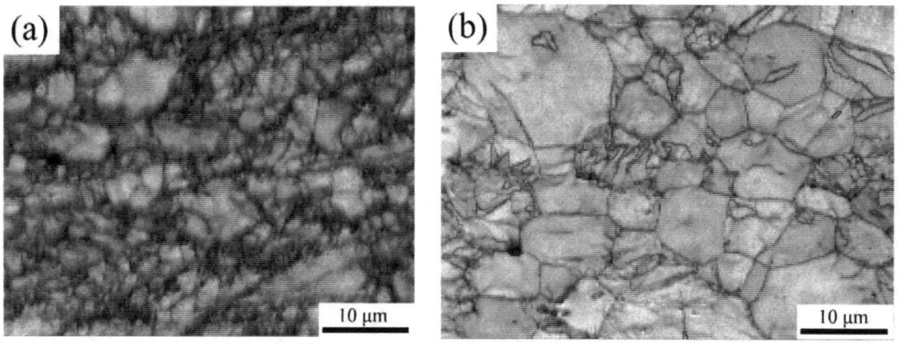

Figure 6. The EBSD maps of (a) the 290℃ annealed sample and (b) the 400℃ annealed sample taken from the necking zone close to the fracture. The red line, blue line and purple line indicate extension twinning, double twinning and contraction twinning, respectively.

It is seen that the frequency of low angle grain boundaries in the 290℃ annealed sample is higher than that of the fully annealed sample, except for the first data point, which is believed to be related to the rather low angular resolution limitation of the EBSD technique (Figure 7b). In comparison with the 400℃ annealed sample, the twinning activity of the sample annealed at 290℃ is negligible. This suggests that slip of dislocations is the deformation mechanism. Although the basal texture is rather strong in the 290℃ annealed sample, the fine grain size may facilitate slip activity.

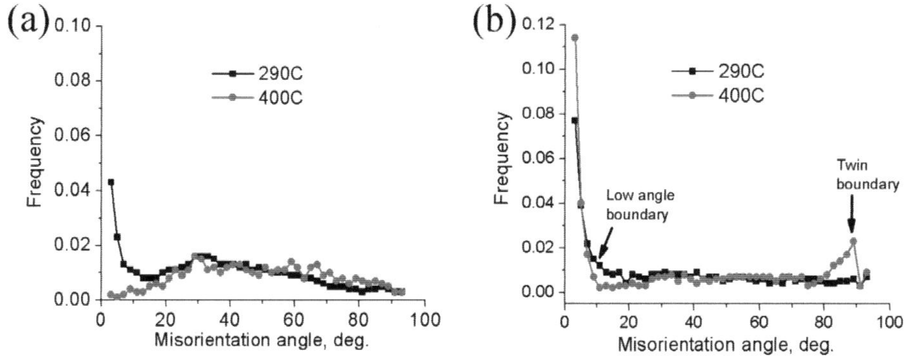

Figure 7. The misorientation angle distribution of the sample before (a) and after (b) the tensile test.

5. Summary

A magnesium alloy containing 3wt.% Gd and 0.14wt.% Al alloy was processed by ARB at 400℃ up to 4 cycles. The microstructure, texture and mechanical properties of the ARBed Mg-3%Gd sheets were studied and the main findings are the following:

- Mechanical properties (yield strength and tensile elongation) were improved dramatically through ARB and post-process annealing at 290℃.
- The sample annealed at 290℃ develops a heterogeneous microstructure, a mixture with fine (1μm) and coarse grains (10μm). It has a 2.3 times higher yield strength than that of the fully annealed coarse grained alloy; and the elongation is almost the same, about 24%.
- It is suggested that there is a transition from twinning-dominated deformation to slip-dominated deformation when the grain size changes from 15 μm to 2.6 μm in the present Mg-3%Gd alloy.

Acknowledgments

The authors gratefully acknowledge the support from the National Natural Science Foundation of China (NSFC, Grants No.51471039, 51327805 and 51421001), Fundamental Research Fund of Central Universities of China (Grant No. CDJZR14135502 and 10611205CDJZXZ138802) and Danish National Research Foundation (Grant No.DNRF86-5) to the Danish-Chinese Center for Nanometals.

References

[1] Roostaei A A, Zarei-Hanzaki A, Abedi H R and Rokni M R 2011 *Mater. Design* **32** 2963
[2] Perez-Prado M T, Valle J.A del and Ruano O A 2004 *Scr. Mater.* **51** 1093
[3] Chang H, Zheng M Y, Wu K, Gan W M, Tong L B and Brokmeier H G 2010 Mater. *Sci. Eng.* A **527**
[4] Zhan M Y, Li Y Y, Chen W P 2008 *Trans. Nonferrous Met. Soc. China* **18** 309
[5] Stanford N and Barnett M 2008 *Scr. Mater.* **58** 179
[6] Stanford N, Atwell D and Barnett M R 2010 *Acta. Mater.* **58** 6770
[7] Nie J F, Zhu Y M, Liu J Z and Fang X Y 2013 *Science* **340** 957
[8] Bugnet M, Kula A, Niewczas M and Botton G A 2014 *Acta Mater.* **79** 66
[9] Yi S B, Schestakow I and Zaefferer S 2009 *Mater. Sci. Eng.* A **516** 58

[10] Gao L, Chen R S and Han E H 2009 *J Alloy and Comp.* **481** 379
[11] Wu D, Chen R S, Tang W N and Han E H 2012 *Mater. Design* **41** 306
[12] Wu W X, Jin L, Wang F H, Wang F H, Sun J, Zhang Z Y, Ding W J and Dong J 2013 *Mater. Sci. Eng.* A **582.** 194
[13] Barnett M, Keshavarz A, Beer A and Atwell D 2004 *Acta Mater.* **52** 5093
[14] Valle J, Carreno F and Ruano O 2006 *Acta. Mater.* **54** 4247

A preliminary investigation of dislocation cell structure formation in metals using continuum dislocation dynamics

S X Xia and A El-Azab
Purdue University, West Lafayette, IN 47906, USA

E-mail: aelazab@purdue.edu

Abstract. A continuum dislocation dynamics model capable of capturing the cellular arrangements of dislocations in deformed crystals is presented. A small strain formulation of the model is given, followed by sample results of stress-strain behaviour, dislocation density evolution, dislocation cell pattern, lattice rotation, and geometrically necessary dislocation density and strain energy density distributions. An important finding of the current work is that dislocations form patterns under all circumstances due to their long range interactions. It is found, however, that the famous cell structure pattern forms when cross slip is activated. It is also found that cells are 3D sub-regions surrounded by dislocations walls in all directions, and they form, disappear, and reappear as a result of the motion of cell walls and formation of new walls by cross slip. It is further found that the average cell size is connected with the applied resolved shear stress according to the similitude principle observed in related experiments. The importance of these results is briefly discussed in the context of recrystallization.

1. Introduction

Cellular dislocations patterns in deformed metals are important in understanding recrystallization because these patterns define the mesoscale energy and dislocation density distributions required to understand nucleation and crystal growth behaviour. In spite of many attempts to model cell structure formation in metals from the first principles of dislocation mechanics, capturing this structure remained a challenge to the mechanics and materials science communities for a long time. Early research on dislocation based plasticity was carried out from a continuum mechanics point of view, where dislocations were regarded as a measure of incompatibility of plastic distortion [1]. In this view, a tensor field called dislocation density tensor is proposed to characterize the dislocations ensembles in crystals [1,2]. Discrete Dislocation Dynamics (DDD) simulation [3] was introduced later as a means to model plasticity but it proved to be limited computationally. The ideas proposed earlier by Kröner [1] and Nye [2] regarding the representation of dislocations by a tensor density were later elaborated by Acharya and co-workers [4,5] into a field theory of dislocation mechanics in which the evolution equation for the dislocation density tensor is closed with constitutive specification of the velocity field. This work was adopted by Chen and co-workers [6] who captured a fractal dislocation structure in crystals relaxed after introducing random plastic distortions. Another approach to model dislocations from the perspective of statistics was carried out for 2D [7,8] and 3D [9,10] situations. In the 3D cases, the dislocation line orientation was an essential ingredient of the formulation because it defines cross slip processes and short range reactions such as annihilation [9,11].

In this paper, we summarize a newly developed 3D model of Continuum Dislocation Dynamics (CDD) [12] and report some results that are relevant to recrystallization. The model is based on a vector representation of the dislocation densities corresponding to individual Burgers vectors. The dislocation density evolution is governed by a set of kinetic equations, one for each Burgers vector. These equations are coupled to crystal mechanics via Orowan's law for plastic shear strain, which is expressed in terms of the dislocation density and velocity, and the stress field driving the dislocation density evolution. The paper is organized as follows. The model is briefly summarized in Section 2 followed by a brief discussion of the numerical scheme in Section 3. Sample results are given in section 4, including dislocation density evolution, stress-strain behaviour, cell structure, and the lattice rotation and elastic energy distributions at the mesoscale. The paper ends with some remarks related to the importance of the current work in the context of recrystallization.

2. Continuum dislocation dynamics model

Consider a crystal undergoing small deformation where the kinematics of deformation is linearized. Consider also dislocations on a single slip system. The (partial) dislocation density tensor $\boldsymbol{\alpha}$ associated with dislocations on this slip system can be defined as the curl of the plastic distortion, $\boldsymbol{\beta}^{\mathrm{P}}$, associated with these dislocations:

$$\boldsymbol{\alpha} = -\nabla \times \boldsymbol{\beta}^{\mathrm{P}} . \tag{1}$$

While this equation was originally derived for the overall dislocation density and plastic distortion tensors [1], it applies to individual slip systems in the case of linearized kinematics. The time derivative of the above equation gives the change of the dislocation density tensor:

$$\dot{\boldsymbol{\alpha}} = -\nabla \times \dot{\boldsymbol{\beta}}^{\mathrm{P}} . \tag{2}$$

Assuming that at a continuum point there are dislocations of the same line orientation, an assumption that is relaxed latter, the rate of plastic distortion can be obtained by Orowan's law in tensor form:

$$\dot{\boldsymbol{\beta}}^{\mathrm{P}} = -\mathbf{v} \times \boldsymbol{\alpha} , \tag{3}$$

where $\mathbf{v}$ is the local dislocation velocity. Insertion of equation (3) into equation (2) generates the equation that governs the evolution of dislocation density tensor in terms of the velocity field:

$$\dot{\boldsymbol{\alpha}} = \nabla \times (\mathbf{v} \times \boldsymbol{\alpha}) . \tag{4}$$

Because the dislocation density tensor locally contains dislocations of the same line orientation, called here a dislocation bundle, it can be written following Nye [2]:

$$\boldsymbol{\alpha} = \boldsymbol{\rho} \otimes \mathbf{b} . \tag{5}$$

where $\mathbf{b}$ is Burgers vector and the vector $\boldsymbol{\rho}$ is defined as:

$$\boldsymbol{\rho} = \rho \boldsymbol{\xi} , \tag{6}$$

where ρ represents the number density of dislocation lines within the bundle and $\boldsymbol{\xi}$ is the local line direction. Equation (4) and (5) can be combined to yield the kinetic equation in terms of the dislocation density vector $\boldsymbol{\rho}$,

$$\dot{\boldsymbol{\rho}} = \nabla \times (\mathbf{v} \times \boldsymbol{\rho}) . \tag{7}$$

Equation (7) preserves the form of equation (4) and serves as the main kinetic equation in the model. It is pointed again here that equation (7) only applies to one slip system under the assumption that at every point there exists only one line direction. The kinetic equation derived here tracks dislocation bundles gliding on their slip planes. Cross-slip can be easily incorporated into this description by simply changing the direction of velocity field.

The plastic distortion $\boldsymbol{\beta}^{\mathrm{P}}$ is the quantity that connects the dislocation kinetics with the crystal deformation. For a given slip system, it can be written as:

$$\boldsymbol{\beta}^{\mathrm{P}} = \gamma \mathbf{S}, \tag{8}$$

where γ is the plastic shear strain and $\mathbf{S}$ is Schmidt tensor expressed as: $\mathbf{S} = \mathbf{n} \otimes \mathbf{b} / b$, with $\mathbf{b}$ and $\mathbf{n}$ being Burgers vector (with magnitude b) and slip plane normal. The evolution law for plastic slip is given by Orowan's relation:

$$\frac{\dot{\gamma}}{b}\mathbf{n} = -\mathbf{v} \times \boldsymbol{\rho}. \tag{9}$$

Equation (7) and equation (9) are used together in the model to evolve the dislocation density field and the corresponding plastic distortion tensor. The latter gives the eigenstrain [13] required to solve the elastic boundary value problem for stress.

In order to include cross-slip, $\boldsymbol{\rho}$ is separated into two parts: $\boldsymbol{\rho}^{\mathrm{CS}}$ and $\boldsymbol{\rho}^{\mathrm{G}}$ with the former corresponding to cross-slip density and the latter corresponding to the gliding density. For screw dislocations whose density vector is $\boldsymbol{\rho}$, the amount of dislocations that cross-slip in a time step δt can be written as:

$$\delta \boldsymbol{\rho}^{\mathrm{CS}} = (\dot{p}_{\mathrm{CS}} \delta t) \boldsymbol{\rho} = \delta p_{\mathrm{CS}} \boldsymbol{\rho}, \tag{10}$$

where $\dot{p}_{\mathrm{CS}}$ is the cross-slip probability rate. Equation (7) can thus be rewritten into two parts:

$$\begin{aligned} \dot{\boldsymbol{\rho}}^{\mathrm{G}} &= \nabla \times (\mathbf{v}^{\mathrm{G}} \times \boldsymbol{\rho}^{\mathrm{G}}) \\ \delta \dot{\boldsymbol{\rho}}^{\mathrm{CS}} &= \nabla \times (\mathbf{v}^{\mathrm{CS}} \times \delta \boldsymbol{\rho}^{\mathrm{CS}}), \end{aligned} \tag{11}$$

where the velocity fields $\mathbf{v}^{\mathrm{G}}$ and $\mathbf{v}^{\mathrm{CS}}$ are dislocation velocities on slip plane and cross-slip plane, respectively. In a similar fashion, the plastic shear rate on glide and cross-slip planes are updated by the following equation:

$$\begin{aligned} \dot{\gamma}^{\mathrm{G}} \mathbf{n}^{\mathrm{G}} / b &= -\mathbf{v}^{\mathrm{G}} \times \boldsymbol{\rho}^{\mathrm{G}} \\ \dot{\gamma}^{\mathrm{CS}} \mathbf{n}^{\mathrm{CS}} / b &= -\mathbf{v}^{\mathrm{CS}} \times \delta \boldsymbol{\rho}^{\mathrm{CS}}. \end{aligned} \tag{12}$$

where $\dot{\gamma}^{\mathrm{G}}$ and $\dot{\gamma}^{\mathrm{CS}}$ represent the plastic shear strain on glide and cross-slip planes, respectively.

In both equation (11) and equation (12), $\delta \boldsymbol{\rho}^{\mathrm{CS}}$ depends on the cross-slip probability, δp_{CS}, which is based on a number of conditions. It is expressed in the form:

$$\delta p_{\mathrm{CS}} = \iota_{\theta} \iota_{\tau} \iota_{\mathrm{M}} \dot{p} \delta t. \tag{13}$$

The indicator functions ι_{θ}, ι_{τ} and ι_{M} are used to check for the cross-slip conditions. An indicator is set to be 1 if the condition is satisfied and 0 otherwise. ι_{θ} checks whether the dislocation orientation is within a small range of orientation deviation (15° [14]) from the screw orientation; ι_{τ} checks whether the resolved shear stress (RSS) on the cross-slip plane is larger than its value on the glide plane; ι_{M} is evaluated by a Monte Carlo scheme, in which a random number between 0 and 1 is generated. If δp_{CS} is larger than the random number ι_{M} is set equal to 1.

During numerical solution of the hyperbolic equations (7) numerical errors accumulate and the density ceases to satisfy the divergence free condition,

$$\nabla \cdot \boldsymbol{\rho} = 0. \tag{14}$$

One way to enforce this condition at all times during the solution is to enforce it numerically by adding it to the equations to be solved.

To close the kinetic equations (7) and (12) the velocities $\mathbf{v}^{\mathrm{G}}$ and $\mathbf{v}^{\mathrm{CS}}$ are required. The dislocation velocity vector takes on the form:

$$\mathbf{v}^{\mathrm{G}} = v^{\mathrm{G}}\boldsymbol{\zeta}^{\mathrm{G}} \tag{15}$$

on the glide plane and:

$$\mathbf{v}^{\mathrm{CS}} = v^{\mathrm{CS}}\boldsymbol{\zeta}^{\mathrm{CS}} \tag{16}$$

on the cross slip plane, with v^{G} and v^{CS} being the magnitude of the velocities and $\boldsymbol{\zeta}^{\mathrm{G}}$ and $\boldsymbol{\zeta}^{\mathrm{CS}}$ the velocity directions, which are perpendicular to the dislocation lines.

The stress field $\boldsymbol{\sigma}$ that acts as a driving force for dislocations satisfies the equilibrium condition:

$$\nabla\cdot\boldsymbol{\sigma} = 0. \tag{17}$$

where the stress tensor $\boldsymbol{\sigma}$ is linearly related to the elastic strain tensor, which is the total strain minus the plastic strain. As such, the equilibrium equation can be rewritten in the form:

$$\nabla\cdot[\mathbf{C}:(\boldsymbol{\varepsilon}^{\mathrm{tot}} - \boldsymbol{\varepsilon}^{\mathrm{P}})] = 0, \tag{18}$$

where $\mathbf{C}$ is the elastic tensor in Hooke's law, $\boldsymbol{\varepsilon}^{\mathrm{tot}}$ the total strain, and $\boldsymbol{\varepsilon}^{\mathrm{P}}$ the plastic strain. Now by assuming that dislocations exist on all slip systems, the plastic distortion tensor, when summed over all slip systems, provides the desired plastic strain tensor in equation (18):

$$\boldsymbol{\varepsilon}^{\mathrm{P}} = \mathrm{sym}\,\boldsymbol{\beta}^{\mathrm{P}} = \mathrm{sym}\sum_{\varphi=1}^{N_{\mathrm{ss}}}\boldsymbol{\beta}^{\mathrm{P}}_{\varphi} = \mathrm{sym}\sum_{\varphi=1}^{N_{\mathrm{ss}}}\gamma_{\varphi}\mathbf{S}_{\varphi}, \tag{19}$$

where N_{ss} is the number of slip systems, and $\boldsymbol{\beta}^{\mathrm{P}}_{\varphi}$, γ_{φ} and $\mathbf{S}_{\varphi}$ are the plastic distortion, plastic shear and Schmidt tensor on slip system φ, respectively. In the current formulation, $\boldsymbol{\varepsilon}^{\mathrm{tot}}$ is decomposed into two parts, a perturbation $\tilde{\boldsymbol{\varepsilon}}$ superposed on an average field $\bar{\boldsymbol{\varepsilon}}$. The latter is the control parameter for loading. These two strains correspond to displacement of different characteristics. The gradient of an average displacement $\bar{\mathbf{u}}$ gives the average strain while the gradient of the perturbation displacement $\tilde{\mathbf{u}}$ is the perturbation strain. With this in mind, equation (18) is further rewritten as:

$$\nabla\cdot[\mathbf{C}:(\mathrm{sym}(\nabla\tilde{\mathbf{u}}) + \bar{\boldsymbol{\varepsilon}} - \mathrm{sym}(\boldsymbol{\beta}^{\mathrm{P}}))] = 0, \tag{20}$$

where the perturbation displacement $\tilde{\mathbf{u}}$ is the unknown fields to be solved for. The stress field obtained from the solution of equation (20) is used to determine the magnitude of velocity on each slip system via the resolved shear stress expression,

$$\tau_{\varphi} = \boldsymbol{\sigma}:\mathbf{S}_{\varphi} = (\boldsymbol{\sigma}\mathbf{n}_{\varphi})\cdot\mathbf{b}_{\varphi}/b_{\varphi} \tag{21}$$

and the mobility law,

$$v_{\varphi} = M\tau_{\varphi}. \tag{22}$$

with M being the mobility coefficient. The numerical value of M shown in table 1 includes b already.

3. Numerical schemes

The solution proceeds as follows at a given time step: with a specified velocity field, equation (11) is solved to evolve dislocations on each slip system. The area swept by dislocations over their glide planes is integrated by Orowan's law (12) giving the plastic distortion. Equation (19) gives the plastic strain, which is subsequently used in the equilibrium equation (20). The solution of equation (20) provides the stress field which is resolved on each slip systems to generate a velocity field for the next times step. The kinetic equations (11) are hyperbolic and describe the evolution of the dislocation density field through the transport of dislocation lines. We choose a first order implicit time discretization to handle the time derivatives.

A least square finite element method (LSFEM), which suits well for the first order differential operators [15], is selected to solve the equation after time discretization. The equilibrium equation (20) is solved using a Galerkin finite element method [16]. During the simulation, the stiffness matrix of the final linear systems for equation (20) remains unchanged and only the loading vector varies. So an LU decomposition of the stiffness matrix can be implemented at the beginning of the simulation and reused afterwards. Equations (12) are ODEs which will be solved using a two-step Adam-Bashforth method. The finite element mesh used in the solution is customized to form a superlattice of the FCC crystal. The space is first partitioned into octahedral and tetrahedral elements and each octahedron is split into two pyramids. The faces of such an octahedron are parallel to slip planes and edges are parallel to Burgers vectors; see [12] for details. The mesh size in the current case is on the order of 20 nm, which is chosen in the range of annihilation distance of dislocations of opposite Burgers vectors. The time step for the simulation is adaptive and its value is determined according to the magnitude of the dislocation velocity. The average time step is around 6 ns which is about 10 times larger than the one used in discrete simulation [17].

4. Results and discussions

In this section some sample results from the simulation of a bulk of copper single crystal under tensile loading in the [001] direction are presented. The crystal is first populated with random dislocation loops and relaxed. The loops size is on the order of the simulation box or larger and so they appear as large curved segments arranged so as to conserve the Burgers vector for each slip system. Periodic boundary conditions are enforced. The loading is applied on the crystal by updating the applied strain $\bar{\boldsymbol{\varepsilon}}$ in z [001] direction at a strain rate of 30/s. All of the 12 slip systems are activated. The parameters used in the simulation are listed on table 1. We note that the cross slip probability value is relatively high so as to accelerate the effect of cross slip. This value can be calibrated using DDD simulations of cross slip and analysing the resulting rates in a continuous fashion [18].

Table 1. Simulation parameters.

Parameters	Values
Cross-slip probability over a time step (δp)	0.1
Initial dislocation density	$1.10\times10^{12}/\mathrm{m}^2$
Number of slip systems	12
Magnitude of Burgers vector	0.256 nm
Dislocation mobility constant (M)	2.56×10^{-6} m/(Pa·s)
Dimensions of the bulk	5μm×5.13μm×5.30μm

The total dislocation density ρ_{tot} at a point is given by:

$$\rho_{\mathrm{tot}} = \sum_{N_{\mathrm{BV}}} \left\| \boldsymbol{\rho}^{\mathrm{BV}} \right\| , \tag{23}$$

where $\boldsymbol{\rho}^{\mathrm{BV}}$ is the dislocation density vector corresponding to a given Burgers vector and the total number of Burgers vector is $N_{\mathrm{BV}} = 6$. The volumetric average of ρ_{tot}, denoted here by $<\rho_{\mathrm{tot}}>$, defines the average dislocation density at a given strain. The average dislocation density for various

Burgers vectors is denoted by $<\rho^i_{tot}>$ (i=1,2,..., 6). Using a similar notation, the average stress along the loading direction, $<\sigma_{33}>$, is obtained by the volumetric average of stress component σ_{33}.

The stress-strain curve and average dislocation density are plotted in figure 1(a). The characteristic hardening regime starts right after yielding which happens at about 100 MPa. The value of the yield strength is sensitive to the details of cross slip and short range reactions. Not all such reactions are included in the model at this point. Glissile junction formation, for example, is still missing in the model. The flow stress, both the initial yield point value and the instantaneous values in the hardening regime, is above the experimental values. Several factors might contribute to this difference. For example, the strain rate implemented here is much higher than that in the real experiment, which is usually less than or close to 10^{-3}/s. Secondly, the material is treated as isotropic in the simulation, which is not true for single FCC crystal. An investigation shows that the anisotropic treatment causes the yielding point to drop about one third of the value calculated by the isotropic treatment [19]. In addition, the value of cross slip probability is high, which leads to frequent cross slip events that lead to rapid dislocation density multiplication. Lastly, glissile junction formation, which provide alternate paths for crystal shearing, is absent in the model.

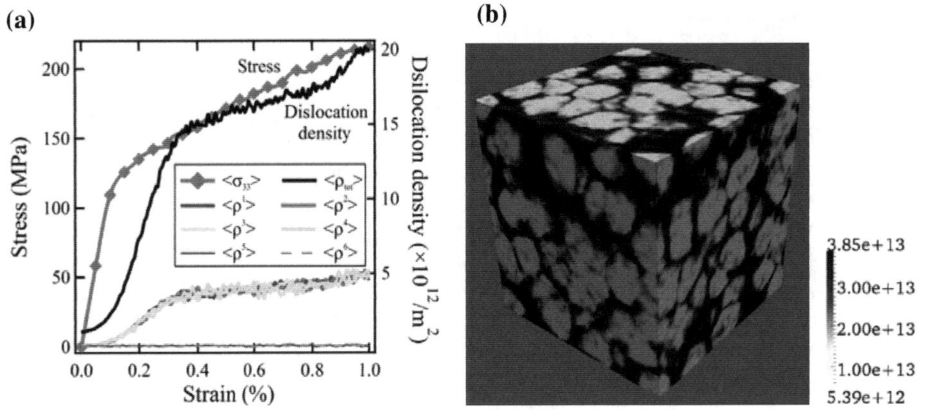

Figure 1. (a) Applied stress and average dislocation density versus strain. (b) Cell structure at 1% strain. The dislocation density on the colour bar is in units of m^{-2}.

The dislocation density evolution shown in figure 1(a) indicates that the density evolves mainly on the 8 slip systems oriented for slip. On the other hand, the density does not evolve on the slip systems with Burgers vectors normal to the loading axis. This result is consistent with the DDD simulation of copper. It has been observed that the dislocation density distribution in the crystal gradually changes from the initial random distribution to a cell structure following the initial yielding of the crystal. Figure 1(b) shows the spatial distribution of the dislocation density in the crystal at 1% strain. The cell pattern is already obvious at this level. As mentioned earlier, 3D cells form once cross slip is activated. The partial density patterns corresponding to different Burgers vectors look different from the one shown in figure 1 (b). Namely, the cells associated with dislocations of a given Burgers vector do not appear to be completely surrounded by dense dislocation walls. This implies that cells form by cooperation of dislocations on multiple slip systems.

Several deformation factors are important in the context of recrystallization [20,21]. These include the local lattice misorientation, the distribution of the dislocation density shown above and its geometrically necessary density (GND) distribution, and the distribution of elastic strain energy in the crystal. Misorientation determines the likelihood of a subgrain becoming a nucleus and influences the speed of its boundary during growth – it is known that the stored elastic energy, which is relieved by recrystallization, is a part of the thermodynamic driving force. A robust simulation of crystallization should thus start from the end of the deformation stage with a 3D map of the deformed state of the crystal with the stored energy distribution. One of the goals of the current work is to demonstrate that such deformed state maps are possible to obtain once the deformation itself is properly modelled. As such, some of the relevant fields are shown here.

The elastic lattice rotation under deformation fixes the lattice misorientation. The lattice rotation vector, Ω_k, is defined by

$$\Omega_k = (e_{ijk}\omega_{ij})/2, \tag{24}$$

where ω_{ij} is the rotation tensor. The latter is the skew symmetric part of the elastic distortion. The direction of vector Ω_k is the axis about which a lattice point rotates. In order to visualize this vector, each component of it is linearly mapped to the RGB value which ranges from 0 to 255. A colour map is formed based on the final RGB values for all components of Ω_k. In this scheme, if the rotational axis of a point is mostly toward $+x$ direction, the colour at the point is purely red. Likewise, purely green colour and blue colour represent $+y$ and $+z$ directions, respectively.

The colour map of lattice rotation is plotted in figure 2 both for the 3D body (figure 2 (a) and for the top surface of the crystal (figure 2(b)). In figure 2(b), the norm of the dislocation density tensor, which measures the geometrically necessary dislocations (GND) density, is superposed in black and white on the RGB map. It is seen that, by crossing the GND wall, the lattice rotation changes its direction (the gradual transition of colour around dark lines in figure 2) and that, within GND cells, the lattice rotation is almost the same. It is also noted that the GND density is mostly concentrated in the cell walls. The current preliminary analysis thus shows that it is the GND distribution that is tied with the lattice misorientation [22], and that capturing the GND distribution is an important mesoscale aspect of both deformation and recrystallization.

It is to be noted here that GND is responsible for the elastic strain energy distribution in crystals. The elastic strain energy density is plotted in figure 3. In doing so, the applied stress is first subtracted from the stress tensor with the remaining part referred to as internal stress tensor, σ_{ij}^{I}. The internal elastic energy density, E^{I}, is then estimated from:

$$E^{\mathrm{I}} = \frac{1+\nu}{2E}\sigma_{ij}^{\mathrm{I}}\sigma_{ij}^{\mathrm{I}} - \frac{\nu}{2E}\sigma_{kk}^{\mathrm{I}}\sigma_{jj}^{\mathrm{I}}, \tag{25}$$

Just as the distribution of total dislocations, the internal energy distribution is highly inhomogeneous, which can be seen clearly from figure 3 (a). Although the pattern of E^{I} differs from that of ρ_{tot}, the former quantity tends to be at high values around dislocation walls (figure 3 (b)). The reason for this is obvious – the dislocation walls represent the regions where the GND density is significant and the elastic strain (stress) is highest where GND is highest. The stress (elastic strain) state of the walls is thus important if we are to use this information in the context of post-deformation behaviour of metals. A distinctive difference between E^{I} and ρ_{tot} patterns is that the elastic energy decays much slower away from dislocation-dense area, which is due to the long range nature of dislocation elastic fields.

5. Conclusions

A continuum dislocation dynamics model capable of capturing dislocation cell structure has been presented. The model is based on coupled dislocation kinetics and crystal mechanics equations, and it is able to capture the dislocation density evolution, cell structure formation, stress-strain behaviour, and all other internal fields required to understand, say, recrystallization of deformed metals. The preliminary results show that cross slip is a necessary mechanism for cell structure formation and that the cells form and become smaller in size according to the empirical similitude rule. It is also shown that the GND distribution in the crystal is concentrated in the dislocation walls and that the GND density clearly marks the lattice misorientation in the crystal. The elastic strain energy distribution in the crystal is found to be highest within and near the cell walls.

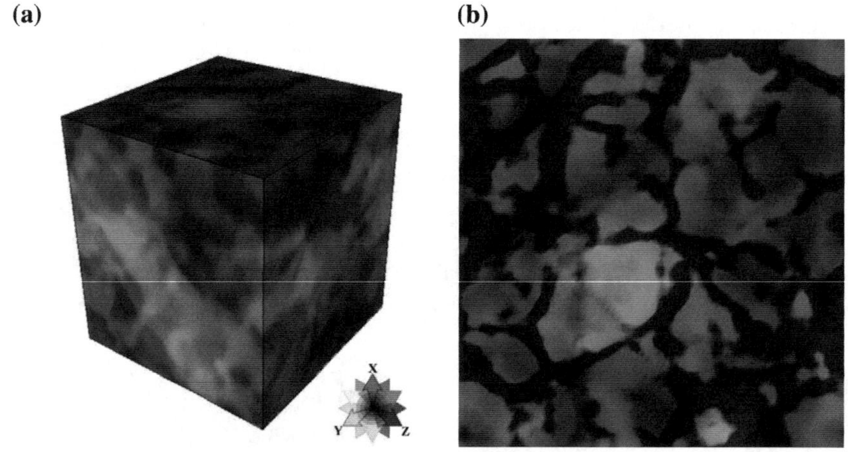

Figure 2. RGB colour representation of lattice rotation. A red colour means the lattice rotation vector along the *x* direction, green in the direction of *y*, and blue in the direction of *z*. A body visualization is shown in (a) and a (001) plane cross section in (b). The latter is plotted together with the GND pattern represented by the norm of the dislocation density tensor. It is clear that the non-trivial values of GND are associated with the points of significant lattice misorientation.

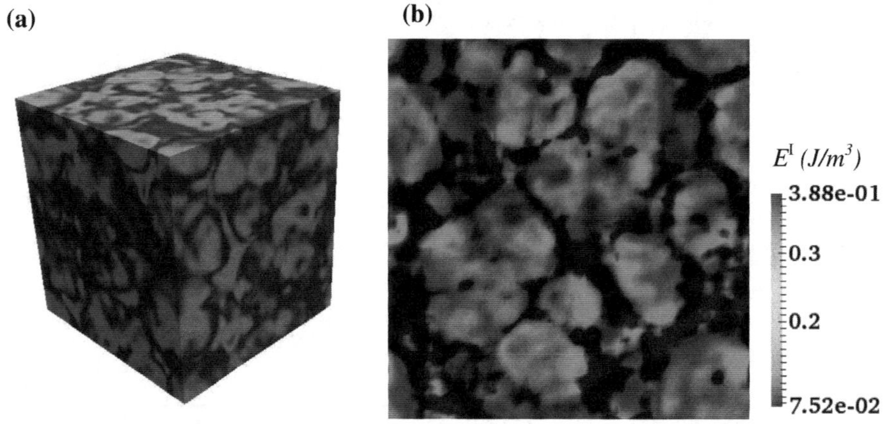

Figure 3. Internal elastic energy distribution at strain = 0.5% plotted in 3D cube (a) and on (001) plane (b). The pattern of total dislocation density ρ_{tot} is shown along with the energy distribution in figure (b). It is seen that the elastic energy distribution is highly inhomogeneous and tends to take on larger values near dislocation walls.

We remark here that unloading simulations should be performed to estimate the residual stress tensor in order to come up with the right value of elastic energy density. For now thus, expression (25) should be considered an approximate, albeit reasonably accurate, estimate of the residual elastic energy density. We further remark that the model currently applies to small deformation cases (few %

strain), and that an extension of the model to finite deformation situations is required to simulate realistic deformation conditions encountered in real experiments. Finally, the model still requires a calibration of parameters used to represent cross slip and short range reaction rates, which can be accomplished using discrete dislocation dynamics simulations.

Acknowledgments
This work was supported by in part the U.S. DOE Office of Basic Energy Sciences, Division of Materials Science & Engineering via contract # DE-FG02-08ER46494 at Florida State University and by internal funding from the School of Nuclear Engineering at Purdue University.

References

[1] Kröner E 1958 *Kontinuumscheorie der versetzungen und eigenspannungen* (Berlin: Springer)
[2] Nye J F 1953 *Acta Metall.* **1** 153
[3] Amodeo R J and Ghoniem N M 1990 *Phys. Rev. B* **41** 6958
[4] Acharya A and Roy A 2006 *J. Mech. Phys. Solids* **54** 1687
[5] Roy A and Acharya A 2006 *J. Mech. Phys. Solids* **54** 1711
[6] Chen Y S, Choi W, Papanikolaou S, Bierbaum M and Sethna J P 2013 *Int. J. Plast.* **46** 94
[7] Groma I 1997 *Phys. Rev. B* **56** 5807
[8] Zaiser M, Miguel M C and Groma I 2001 *Phys. Rev. B* **64** 224102
[9] El-Azab A 2000 *Phys. Rev. B* **61** 11956
[10] El-Azab A 2006 *Scr. Mater.* **54** 723
[11] Sandfeld S, Hochrainer T, Gumbsch P and Zaiser M 2010 *Philos. Mag.* **90** 3697
[12] Xia S and El-Azab A 2015 *Model. Simul. Mater. Sci. Eng.* In the press
[13] Mura T 1987 *Micromechanics of Defects in Solids* (Tokyo: Springer)
[14] Wang Z Q, Beyerlein I J and LeSar R 2007 *Model. Simul. Mater. Sci. Eng.* **15** 675
[15] Jiang B 1998 *The Least-Squares Finite Element Method* (Springer)
[16] Zienkiewicz O C, Taylor R L and Zhu J Z 2013 *The Finite Element Method: Its Basis and Fundamentals* (Butterworth-Heinemann)
[17] Deng J and El-Azab A 2009 *Model. Simul. Mater. Sci. Eng.* **17** 075010
[18] Deng J and El-Azab A 2010 *Philos. Mag.* **90** 3651
[19] Han X, Ghoniem N M and Wang Z 2003 *Philos. Mag.* **83** 3705
[20] Doherty R D, Hughes D A, Humphreys F J, Jonas J J, Juul Jensen D, Kassner M E, King W E, McNelley T R, McQueen H J and Rollett A D 1997 *Mater. Sci. Eng. A* **238** 219
[21] Humphreys F J and Hatherly M 2004 *Recrystallization and related annealing phenomena* (Oxford: Pergamon)
[22] Mohamed M S, Larson B C, Tischler J Z and El-Azab A 2015 *J. Mech. Phys. Solids* **82** 32

IOP Conf. Series: Materials Science and Engineering **89** (2015) 012054 doi:10.1088/1757-899X/89/1/012054

Nucleation at hardness indentations in cold rolled Al

C L Xu[1, 2], Y B Zhang[2], G L Wu[1], Q Liu[1], D Juul Jensen[2]

[1] College of Material Science and Engineering, Chongqing University, Chongqing 400044, China

[2] Danish-Chinese Center for Nanometals, Section for Materials Science and Advanced Characterization, Department of Wind Energy, Technical University of Denmark, Risø Campus, DK-4000 Roskilde, Denmark

Email: chaoxu@dtu.dk

Abstract. Nucleation of recrystallization near hardness indentations has been investigated in slightly cold rolled high purity aluminium. Samples were cold rolled to 12% and 20% reductions in thickness and indentations were done with two different loads (500 g and 2000 g). The samples were annealed at 300 °C for 1 h and nuclei were identified. It is found that the indentations are preferential nucleation sites. With EBSD maps around indentation tips, the orientation relationship between nuclei and matrix is analyzed. Finally, effects of rolling reduction and indentation load on local misorientations and stored energy distributions and thus on nucleation are discussed.

1. Introduction

Indentation hardness testing of materials is a simple and quick method for obtaining information about the mechanical properties of materials [1], such as yield stress [2, 3] and elastic modulus [4, 5]. In order to understand the mechanisms occurring during indentation better, lots of research of the deformation zones below indentations have been done by different authors [6-10]. A typical early investigation was conducted by Hill et al. [11], who explored the deformation of a plane surface with a wedge and showed that the average strain caused by wedge indentation increases with increasing wedge angle. Further studies investigated the misorientations underneath indents and the dislocation mechanisms explaining the changes within the plastic zone at different indentation depth [12-14] and explored indentation size effects [15-17].

Indentations are also known to provide additional driving force for the nucleation and growth of new grains during annealing. The aim of the present work is to investigate where and how nucleation happens near hardness indents and what orientations the nuclei have. The focus is on the influence of indents load, in cold rolled high purity Al with an initial microstructure coarse enough to do at least one 500 g indentation and one 2000 g indentation within each grain. In this paper, a statistical analysis is made based on all the data obtained from all the grains after annealing. This analysis is referred to as sample scale observation. Furthermore, in order to investigate orientation relationships in more detail, one of the grains is characterized near the indentation tip in the deformed state. This investigation is referred to as local scale observation in the following.

2. Experimental

Two high purity (99.996%) aluminium samples 94 × 46 × 3 mm^3 were used. Initially, they had a coarse and inhomogeneous microstructure with an average grain size of 300 μm. This grain size is too fine for the planned experiment, therefore all surfaces were ground using 4000# SiC paper and electro polished to prevent nucleation at surface scratches and finally the samples were annealed at 590 °C for 7 days. After annealing, grains with sizes ranging from 500 μm to 7 mm were obtained. These grains are large enough to do several microhardness indents away from the grain boundaries. The samples were cold rolled to 12% and 20% reduction in thickness. Seven samples were cut from them and were carefully ground and electro polished. Microhardness indentations were done on the RD-TD planes using a Vickers diamond indenter of pyramidal shape with loads of 500 g and 2000 g. The distance between each indentation was larger than $3d$, where d is the length of the diagonal lines of the indentations, thereby avoiding deformation zones overlapping. To initiate nucleation, the samples were annealed at 300 °C for 1 h and characterized using electron channeling contrast (ECC) and electron backscattered diffraction (EBSD) with a step size of 2 μm. It should be noted that because of the change in surface position at the indentation, only some scattered pixels can be indexed within the indentation. In order to reveal the indentation-affected zone, one of the cold rolled and indented samples was ground first to reveal the tip of a 500 g indentation and then the sample was ground again to reveal the tip of a 2000 g indentation. After each grinding, the sample was mechanical polished using 40 nm oxide polishing suspension and characterized using EBSD. In all EBSD maps shown in this paper, fine white lines are low angle boundaries (LABs) (≥ 2° and <15°) and coarse black lines are high angle boundaries (HABs) (≥15°).

3. Results and discussion

3.1. Sample scale observations

3.1.1. Nucleation potentials. In total, 57 microhardness indentations were done on six samples and 54 new grains are detected around 28 indentations based on EBSD maps. It should be noted that twins are not counted as recrystallized grains here. For the indentations where we could not be sure if there are nuclei because of only a few indexed pixels at the indents in the EBSD maps, ECC was used to characterize the indentations and an additional 5 nuclei were thereby observed around an additional 4 indentations. Figure 1(a) is an example of using ECC to detect a nucleus that could not be seen by EBSD. Of all the nucleated indentations, most stimulate 1 nucleus except that 8 are found to form 2 nuclei and 8 have 3 or more nuclei. All nuclei have grown into relatively large sizes ranging from tens of micrometres to 300 μm. It should be noticed that no nuclei are detected away from the indentations. This result clearly shows the potentials of indentations to act as nucleation sites, due to the extra driving force provided by the indentation deformation, as suggested in previous investigations [e.g. 13, 14].

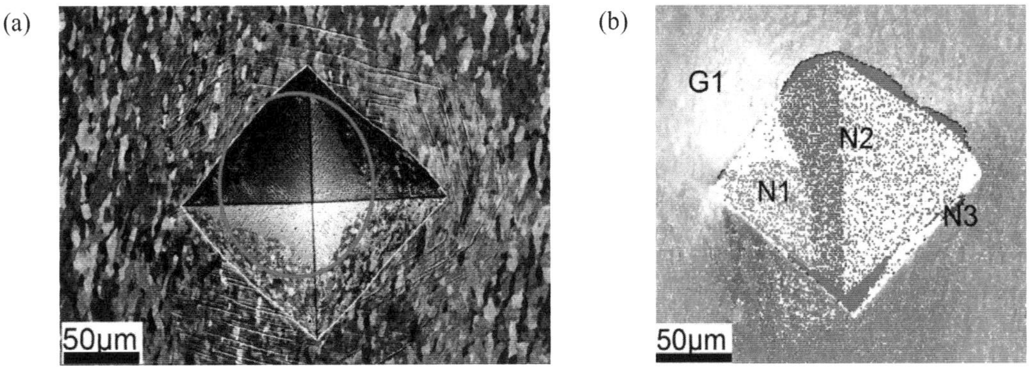

Figure 1. Examples of the nucleation in samples 20% cold rolled and 2000 g indented: (a) an ECC picture showing one nucleus (marked with a red circle) and (b) an EBSD map (without any noise reduction) showing 3 nuclei at another indentation.

The indentations are classified into four types (A, B, C and D) according to the different rolling reductions and different indentation loads, as shown in table 1. The table also reports the total number of indents, and those that have stimulated nucleation. Compared with Type B indentations (20%, 500 g), Type A indentations (20%, 2000 g) have higher nucleation potentials. Similarly, Type D indentations (12%, 500 g) lead to fewer nuclei than Type C indentations (12%, 2000 g). When the samples with 20% rolling reduction (Type A and B) are compared with those less heavily rolled (Type C and D), it is also clear that more nuclei are seen at the indents in the 20% rolled samples (see table 1). These observations clearly show that a higher/larger deformation leads to more nucleation after annealing at 300 °C for 1 hour. It is worth noting that even the most weakly deformed samples (12% rolling and 500 g indentation) can nucleate as was observed when these samples were annealed at 311 °C for 1 h.

3.1.2. Orientation relationships. The microstructures around the 28 nucleated indentations observed using EBSD were analyzed, and the misorientations between each nucleus and the pixel in the neighboring recovered matrix with the smallest misorientation were measured. 11 out of 41 (27%) around Type A indentations have misorientation angles lower than 15° to their neighboring recovered matrix, while for Type B indentations the fraction of the nuclei that have misorientation angles lower than 15° is 33%. One example is shown in figure 1(b). Although only some scattered pixels are detected by EBSD at the indentation surface, we still can in this case clearly distinguish the nuclei, which are numbered as N*i* (*i*=1-3) and the recovered matrix is marked as G1. The boundaries between N1 and G1 are composed of both LABs and HABs. Since some of the boundaries are low angle, N1 is among the 27% of nuclei with orientations near the matrix.

The remaining nuclei, i.e. the 73% nuclei around Type A indentations, 67% nuclei around Type B indentations and the nucleus at Type C indentations, only form HABs with the recovered matrix. For example, both nuclei N2 and N3 only have HABs to the surrounding recovered matrix and no LABs are detected (figure 1(b)). Therefore, one may speculate if this type of nuclei (as N2 and N3) have orientations different from the matrix.

Table 1. Deformation modes and corresponding nucleation at 4 types of indentations

Type	Cold rolling reductions	Loads of indenting	Number of indents	Number of indents with nuclei	Percentage
A	20%	2000 g	18	18	100%
B	20%	500 g	18	13	72%
C	12%	2000 g	10	1	10%
D	12%	500 g	11	0	0

The potentials for growth of nuclei depend on the misorientations between the nuclei and the entire surrounding matrix. Figure 2 shows the distribution of the maximum misorientation angles between the nuclei, which are formed in the 20% cold rolled samples, and their surrounding recovered matrix. It can be observed that the matrix misorientation angles for Type B indentations (load: 500 g) concentrate in the range of 20°~ 30° and do not exceed 35°, while for Type A indentations, the misorientation angles spread widely from 10° to 50°.

3.2. Local scale observations

Although a large fraction of the nuclei as reported above after some growth has orientations different from the observed recovered matrix around the indentations, it does not necessarily mean that these nuclei have been formed with new orientations different from those at the nucleation sites in the deformed matrix. To investigate that, one grain within a 20% cold rolled sample, having the same initial orientation as the example shown in figure 1(b), was chosen to check the deformation microstructure around a 500 g indentation tip and a 2000 g indentation tip in the deformed state.

3.2.1. Local orientation relationship. The microstructures within an area of 85×85 μm² around both indentation tips were characterized by EBSD. The {111} pole figure for the microstructures at the indentation of 2000 g is shown in figure 3, where a large orientation spread in the deformed matrix around the indentation tip is revealed as grey clouds. As the selected deformed grain has the same orientation as that annealed and shown in figure 1(b), the orientations of the nuclei N1-N3 are also plotted in figure 3 as large crosses, circles and diamonds, respectively. When comparing these orientations with the orientation spread at the indentation tip in the deformed state (grey cloud in figure 3), it is clear that N1 (marked by diamonds) has an orientation well within the deformed orientation spread, while N2 (marked by circles) and N3 (marked by crosses) both are at the outskirts but yet within the orientation spread in the deformed state. It is thus very likely that these 3 nuclei had nucleated from the sites in the deformed matrix with the actual orientation around the indentation tip, i.e. they formed with an orientation within the spread of the deformed matrix and grew to meet new areas with different orientations, thereby forming high angle boundaries (and as N2 and N3, they grew to belong to the large fraction of nuclei which had a minimum misorientation to the surrounding recovered matrix of above 15°).

It is clear that the indentations lead to significant extra deformation and consequently extra orientational scatter which has to be considered when studying nucleation. This is investigated in detail for both the 500 g and the 2000 g indentation in the following.

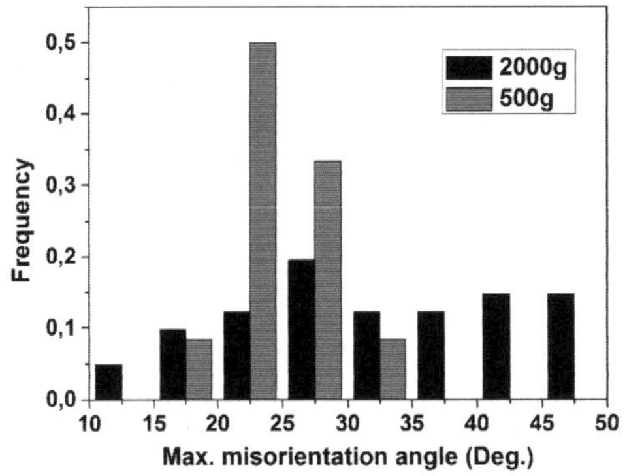

Figure 2. Maximum misorientation between nuclei and the surrounding pixels in the recovered matrix. The red and black histograms represent the data for the 500 g and the 2000 g indentations in the 20% cold rolled samples, respectively.

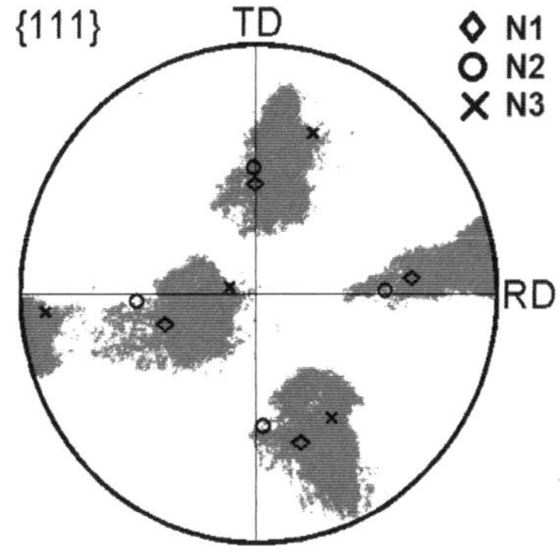

Figure 3. {111} pole figure showing the orientations within an area of 85×85 μm² around the tip of a 2000 g indentation in the deformed state of the 20% cold rolled sample. The grey cloud represents the orientations of the deformed matrix.

3.2.2 Microstructure and stored energy. In figure 4(a) and (b), the measured orientations are shown and the figures reveal the largely symmetric distributions of orientations given by the pyramidal shape of the indenter. The orientation distributions caused by the indenter may also be visualized with respect to the orientation in the 20% rolled matrix away from the indentation. This is shown in figure 4(c) and (d). In these figures, the matrix away from the indentation has more or less the same orientation and thus is shown in blue i.e. with misorientations below 15°. Most of the indentation zone both after 500 g and 2000 g load is rotated between 10° and 30° away from the matrix, and shown in green in figure 4(c) and (d). Small local zones near the tip and along a symmetry axis, however, have misorientations to the matrix above 35°; most pronounced after the 2000 g indentation where a series of misorientation peaks above 40° are seen along a symmetry axis (see figure 4(d)). Such peaks are expected to be powerful nucleation sites, leading to nuclei with orientations far away from the matrix orientations. The largest rotation angles of ~50° and ~35° under the 2000 g and 500 g indentations, respectively, agree well with the observed maximum misorientations between nuclei and the surrounding matrix shown in figure 2.

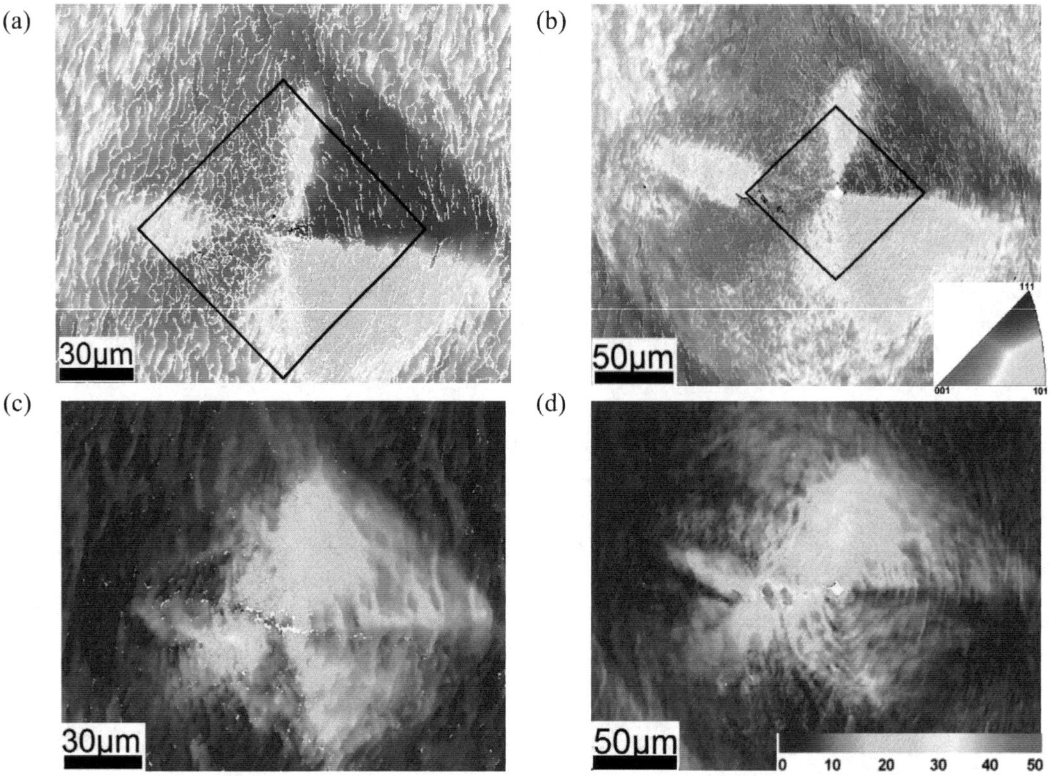

Figure 4. The deformation microstructures around a 500 g indentation (a, c) and a 2000 g indentation (b, d) in a 20% cold rolled sample. The color in (a) and (b) is defined by inverse pole figure color and the color in (c) and (d) is based on the misorientation angle between each pixel and the orientation in the 20% rolled matrix away from the indentation. The two diamonds in (a) and (b), which have the same size of 120 µm along the diagonals are used for direct comparison of stored energies. For the 500 g indentation, the size of the diamond is chosen to almost cover the region with a large misorientation (the green area in (c)).

The extra driving force introduced by the indentations may be estimated as stored energy (SE) from the measurements shown in figure 4(a) and (b). Such estimation is based on the measured misorientations across deformation-induced boundaries which can be obtained from EBSD data using a method described in [18], including all boundaries with misorientation ≥2°.

For this estimation an area near the tip is used (marked by the diamond in figure 4(a) and (b)). The size of the area is chosen to be the same for the 500 g and 2000 g indents to allow a direct comparison of SE at this central indentation zone in the two cases. The calculated results are 0.12 MJ/m^3 and 0.15 MJ/m^3 for the 500 g and 2000 g indentations, respectively. The SE in an area of the same size away from the indentation is only 0.04 MJ/m^3.

It has to be noted that the present analysis is based on 2D measurements which leads to uncertainties in the calculation of SE. Furthermore, with EBSD and ECC techniques it is not possible to follow the nucleation in-situ away from the polished surface. Therefore it is not possible to pin point the exact nucleation sites in the deformed matrix. Only by non-destructive 3D measurements, e.g. by synchrotron X-rays, it would be possible to overcome these limitations.

4. Conclusions

High purity very coarse grained Al was used to investigate effects of indenting loads and rolling reductions on recrystallization nucleation. It was found that the indentations lead to substantial additional grain subdivision of the deformed microstructure and large orientation rotations near the indentation tips. The higher the indentation loads the larger are the subdivision and additional SE. For indentations with a 500 g load, high angle boundaries with up to 25° misorientation are formed, local zones with misorientation up to 35° away from the matrix are observed and the additional SE near the tip is 0.08 MJ/m^3. The numbers for a 2000 g load are 25°, 50° and 0.11 MJ/m^3 respectively. It is thus not surprising that nucleation occurs near the indentation. No nuclei are observed away from the indentations. The higher the indentation load, the more nuclei. However, also the rolling reduction contributes here. Whereas the majority (>70%) of the indentations in the 20% cold rolled samples lead to nucleation, only 1 indentation out of 21 causes detectable nucleation in the 12% cold rolled samples at the chosen annealing condition (300 °C for 1 h). By local orientation measurements near the indentation tip in the deformed state, it is shown that the orientational spread observed there covers the orientations of the nuclei observed in the annealed state of an identical sample. It is thus suggested that the nuclei form at local sites in the indentation zone with orientations as those present there and will have large misorientations to the rolled matrix, into which they grow after some annealing.

Acknowledgements

CLX, GLW and QL gratefully acknowledge the support from the NSFC of China (Grant No. 51471039, 51327805 and 51421001) and Fundamental Research Fund of Central Universities (Grant No. 10611205CDJZXZ138802). All Authors wish to thank the support from the Danish National Research Foundation (Grant No.DNRF86-5) and the NSFC of China (Grant No. 51261130091) to the Danish-Chinese Center for Nanometals. Helpful discussions with and suggestions from Jun Sun and Xuehao Zheng are grateful acknowledged.

References

[1] Tabor D 1951 *The hardness of Metals*, Oxford, UK, Oxford University Press
[2] Clausner A and Richter F 2015 *Eur. J. Mech. A/Solid* **51** 11-20
[3] Kramer D, Huang H, Kriese M, Robach J, Nelson J, Wright, Bahr D and Gerberich WW 1999 *Acta Mater.* **47** 333-43
[4] Hu Z, Lynne K J, Markondapatnaikuni S P and Delfanian F 2013 *Mater. Sci. and Eng. A* **587** 268-82
[5] Vandeperre L J and Clegg W J 2005 *Acta Mater.* **492-493** 555-60
[6] Mulhearn T O 1959 *J. Mech. Phys.* **7** 85-96
[7] Keh A S 1960 *J. Appl. Phys.* **31** 1538-45
[8] Inkson B J, Steer T, Mobus G, Wabner T 2000 *J. Microsc-oxford* **201** 256-69
[9] Mogilecsky P 2005 *Philos. Mag.* **85** 3511-39
[10] Rester M, Motz C and Pippan R 2007 *Acta Mater.* **55** 6427-35
[11] Hill R, Lee E H and Tupper S J 1947 *Proc. R. Soc. A.* **188** 273-89
[12] Kiener D, Pippan R, Motz C and Kreuzer H 2006 *Acta Mater.* **54** 2801-11
[13] Xie G, Wang L, Zhang J and Lou L H 2012 *Scripta Mater.* **66** 378-81
[14] Zambaldi C, Rosters F, Raabe D and Glatzel U 2006 *Mater. Sci. Eng. A* **454-455** 433-40
[15] Demir E, Raabe D, Zaafarani N and Zaefferer S 2009 *Acta Mater.* **57** 559-69
[16] Nix W D and Gao H 1998 *J. Mech. Phys. Sol.* **46** 411-25
[17] Begley M R and Hutchinson J W 1998 *J. Mech. Phys. Sol.* **46** 2049-68
[18] Godfrey A, Cao W Q, Liu Q and Hansen N 2005 *Metall. Mater. Tran. A* **35** 2371-78

Effects of severe plastic deformation on grain refinement and martensitic transformation in a metastable β Ti alloy

A Zafari[1], X S Wei[1,2], W Xu[1,3] and K Xia[1]
[1] Department of Mechanical Engineering, University of Melbourne, Victoria 3010, Australia
[2] School of Materials Science and Engineering, Tongji University, Shanghai 201804, P. R. China
[3] Centre for Additive Manufacturing, School of Aerospace, Mechanical and Manufacturing Engineering, RMIT University, Victoria 3083, Australia

E-mail: k.xia@unimelb.edu.au

Abstract. A Ti-5553 metastable β alloy was subjected to high pressure torsion (HPT) to obtain a nano β grain structure. The β grains were dramatically refined from an initial size of ~1 mm to < 50 nm during HPT. The stress induced martensitic α" plates divided β grains into smaller domains and enabled subgrains to form even in very fine grains. The reverse α" to β transformation occurred when the thickness of the α" laths reached < ~10 nm, leading to pure β nano-grains. Further deformation to an equivalent strain of 242 resulted in grain growth and the formation of well defined grain boundaries.

1. Introduction

The introduction of severe plastic deformation (SPD) methods to metallurgical engineering has enabled researchers to produce bulk nanocrystalline metallic alloys with superior mechanical properties such as high strength and good fatigue resistance [1, 2]. The production of nano-grained titanium (Ti) alloys has recently drawn the attention of scientists due to their low density and compatibility with the human body, making them suitable candidates for aerospace and biomedical applications. However, the majority of studies has focused on the influences of SPD on the α-phase dominated materials, such as pure titanium and α/β alloys [3-6] although the lack of ductility in such alloys has made SPD processing difficult, preventing the formation of uniform nano-grained Ti alloys.

On the other hand, the β Ti alloys have great potential to be successfully SPD processed thanks to their better ductility as compared to the α and α/β Ti alloys. Recent investigations have revealed that the stability of the β phase could play a vital role in the formation of β nano-grains. A stress induced martensitic phase, known as α", forms in metastable β-alloys during plastic deformation, contributing significantly to β grain refinement [7-10]. It has also been claimed that α" is changed back to β during SPD [7-10]. The reverse phase transformation of α" → β is attributed to the generation of heat either at the α"/β interfaces [7-9] or in adiabatic shear bands [10] as a result of mechanical work, increasing local temperatures over the austenitic starting temperature (A_s). There is, however, little direct experimental evidence to support the speculated mechanisms and explanations. In the present investigation, a metastable β Ti alloy (Ti-5553) was subjected to high pressure torsion (HPT) to shed light on the role of the stress induced martensitic phase transformation in the microstructural evolution. Extensive transmission electron microscopy (TEM) revealed that the α" martensitic plates broke down large β grains into fine blocks, accelerating the grain refinement process, and that grain boundary migration at a very high level of strain could be responsible for the growth of nano-sized grains.

2. Experimental materials and procedures

A commercial cast Ti-5Al-5V-5Mo-3Cr (wt.%) metastable β alloy was used. Discs of 1.5 mm in thickness and 10 mm in diameter were cut from cylindrical rods of 10 mm in diameter after solution treatment at 1000 °C for 1 hour followed by water quenching, obtaining β grain sizes of ~1 mm. HPT was conducted at room temperature under a pressure of 6 GPa and at a rotation speed of 1 rpm for 0.5, 1, 5 and 20 turns. The equivalent strain is calculated using [1]

$$\varepsilon_{eq} = \frac{2\pi N r}{t\sqrt{3}} \quad (1)$$

where r is the distance from the disc centre, N the number of turns, and t the thickness of the disc.

Materials characterisation was carried out by transmission electron microscopy (TEM, FEI Tecnai F20). A scanning electron microscope with focused ion beam (SEM/FIB, FEI Nova 200 Nanolab DualBeam) was used to prepare TEM samples at r of 0.5, 2.5 and 5 mm, respectively.

3. Results

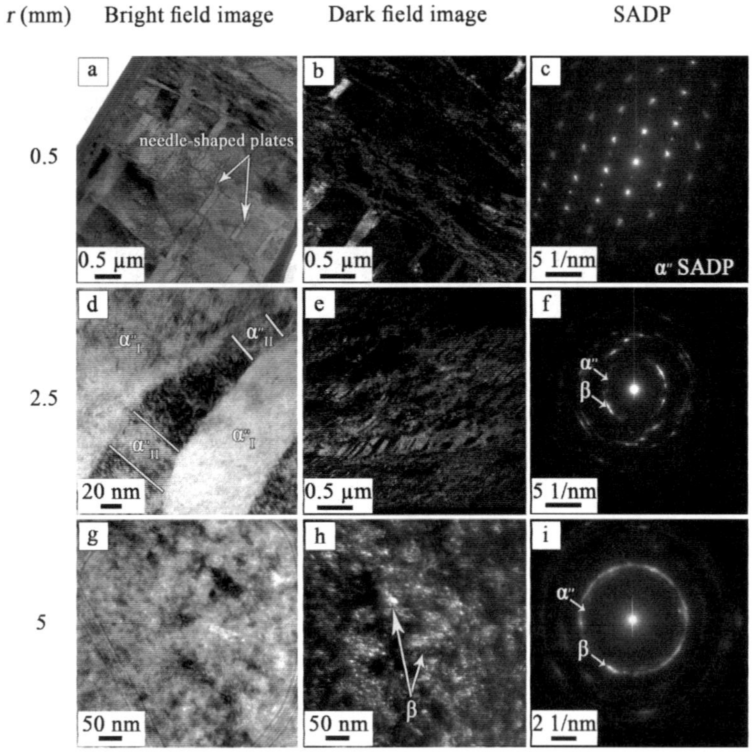

Figure 1. TEM microstructures and corresponding SADPs after 0.5 turn of HPT at r = 0.5 (a-c), 2.5 (d-f) and 5 mm (g-i).

Figures 1a-c show the TEM results at r = 0.5 mm after HPT for half a turn (ε_{eq} = 0.6). Some needle-shaped plates are visible in figures 1a and 1b with an average thickness and length of ~100 nm and ~1 μm, respectively. The selected area diffraction pattern (SADP) shown in figure 1c identifies the plates as the stress induced α" phase. Figures 1a and 1b also indicate that the α" plates were fragmented into smaller pieces as a result of shearing during HPT. Figures 1d and 1e show TEM microstructures at a greater distance from the centre (i.e. r = 2.5 mm and ε_{eq} = 3). As seen in figure 1d, the secondary α" (α''_{II}) formed between the primary α" (α''_{I}) plates. Figure 1e also shows parallel α" and β layers, indicating a twin structure consistent with what was reported earlier [11-13]. The length and thickness of a large number of the α''_{I} plates were measured to be ~200 and ~90 nm, respectively. The broken ring in the SADP (figure 1f) suggests the formation of a very fine grain structure after straining to ε_{eq} = 3. At the edge of the disc (i.e. r = 5 mm and ε_{eq} = 6), as shown in figures 1g and 1h, no α" was observed, although the α" ring was detected in the corresponding SADP (figure 1i). The more complete rings in figure 1i indicates the formation of nanocrystalline β and α".

Figures 2a and 2b show the TEM microstructures at r = 2.5 mm after 1 turn of HPT (ε_{eq} = 6). Although the equivalent strain is the same as that in the specimen shown in figures 1g-i, martensitic plates of 10-20 nm in thickness are visible. The SADP shown in figure 2c indicates a sharper and more complete α" ring compared to that in figure 1i, indicating the presence of a higher amount of the martensitic phase. Further straining to 5 turns at r = 5 mm (ε_{eq} = 30), however, contributed to the apparent disappearance of the α" plates in the bright field (figure 3a) although a faint, broken α" ring in SADP (figure 3b) suggests the martensitic phase still exists.

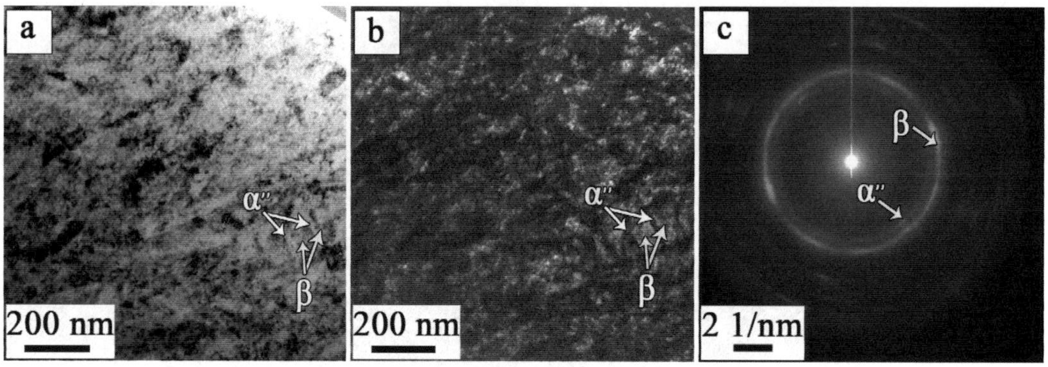

Figure 2. (a) Bright field, (b) dark field and (c) SADP at r = 2.5 mm after HPT for 1 turn.

Figures 4a and 4b show the dark field TEM image and SADP, respectively, at r = 2.5 mm after HPT for 20 turns (ε_{eq} = 242). The average β grain size was measured to be 30-50 nm (figure 4a) and no α" was detectable in the SADP (figure 4b). The β grain boundaries in figure 4a are much better-defined compared to those in figures 2 and 3.

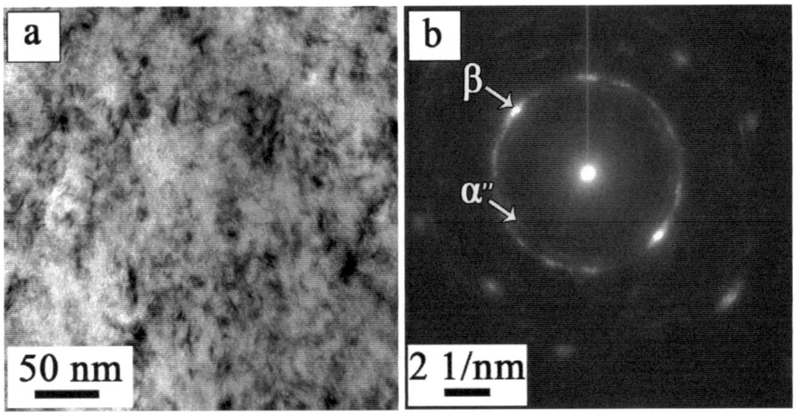

Figure 3. (a) Bright field and (b) SADP at r = 2.5 mm after HPT for 5 turns.

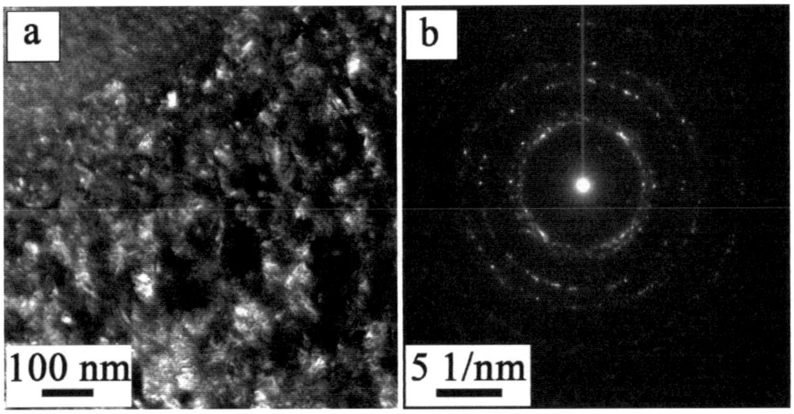

Figure 4. (a) Dark field and (b) SADP for β at r = 5 mm after HPT for 20 turns.

4. Discussion

4.1. Grain refinement mechanism

Molybdenum equivalency (Mo_{eq}) is one of the most commonly used criteria for determining the stability of the β phase in Ti alloys. It is reported that the alloys with Mo_{eq} < 10 wt.% are not stable enough to prevent the formation of the stress induced α" martensitic phase during deformation [14]. Mo_{eq} is calculated to be 8.1 wt.% for Ti-5553 and α" is expected to form during HPT as observed in figures 1-3. According to previous studies on such metastable β-Ti alloys [7-9], the formation of the martensitic plates would divide large β grains into smaller blocks, facilitating grain refinement. The results of the present work provide consistent evidence. As seen in figures 1a and 1b, the α" plates broke down the β grains into smaller blocks and as the deformation continued the secondary α" (α''_{II}) formed (figure 1d), dividing β into even finer domains. Further, the sizes of the α" plates became smaller with increasing strain (figures 1d-i, 2 and 3) due to their fragmentation (figures 1a and 1b), leading to plates of 10-20 nm at ε_{eq} = 6 (figure 2). However, the division of β grains by itself cannot produce a nanocrystalline β structure since the divided β blocks all have the same orientation. The generation of a high density of dislocations inside these small domains during HPT would lead to fine subgrains and eventually grains with high angle boundaries from the rotation of the subgrains. The dividing by α" would repeat in these smaller β grains, accelerating the grain refinement process.

Figures 1g-i, 3 and 4 show that the amount of α" decreases with increasing strain and it completely disappeared at sizes of < ~10 nm at ε_{eq} = 242. The disappearance of the α" phase is attributed to a reverse α"→ β transformation. A local increase in temperature above A_s has been suggested as the cause of the reverse transformation [7-10]. However, calculations based on a model by Pereira *et al.* [15] revealed that the temperature of the HPT discs may only reach a maximum value of 66 °C, much lower than A_s = 166 °C [16]. Therefore, temperature rise could not explain the reverse transformation observed in the current study. Instead, the martensitic plate's size is viewed as a crucial factor in the reverse transformation. A model has been developed, taking the Gibbs-Thomson effect into consideration [17]. The analysis results in that the α" plates with sizes of < ~10 nm become unstable because of a drastic increase in the Gibbs free energy. As a result of the reverse phase transformation, all α" plates eventually change back to β and the final microstructure consists of entirely β grains of 30-50 nm in size (figure 4).

In comparison, the finest grain size attained in a stable Ti-20Mo alloy is in the range of 100-150 nm [18]. This is largely attributed to the presence of the stress induced α" plates in Ti-5553, which not only divide the β grains but also act as obstacles to dislocation movement, reducing the rate of dislocation absorption into grain boundaries. Therefore, subgrains and high angle grain boundaries could continue to form even inside the β grains smaller than 100 nm.

4.2. Differences between microstructures at r = 2.5 mm after 1 turn and r = 5 mm after half a turn

The equivalent strain is apparently the same (ε_{eq} = 6) both at *r* = 2.5 mm after 1 turn and at *r* = 5 mm after half a turn. Nevertheless, larger α" plates are observed in the former (figure 2) whereas α" is only identified by SADP in the latter (figure 1i). However, the equality of strains for these two different experimental conditions is true only if the thickness of the disc is uniform across its diameter. Considering the fact that HPT discs are normally thinner at the edge, the ε_{eq} would be greater than 6 (Eq. 1). Therefore, the fragmentation and refinement of the martensitic plates at the edge of the sample (*r* = 5 mm) would be more severe than those at *r* = 2.5 mm after one turn of HPT, leading to the formation of invisible fine α" plates and probably the reverse phase transformation. In addition, the hydrostatic pressure is small at the edge and torsion is mostly responsible for the deformation [19]. In other words, the portion of torsion at the edge is greater than that closer to the centre. This may lead to more fragmentation of the α" plates and reduction of their sizes at the edge so that they are not observable in bright and dark field images (figures 1g and 1h).

4.3. HPT induced grain growth

Figure 4a shows β grains of 30-50 nm with well-defined grain boundaries formed after HPT for 20 turns, whereas the β grain boundaries are not clear after 1 and 5 turns of HPT (figures 2 and 3). HPT induced grain growth may be responsible for this phenomenon. It is believed that deformation induced grain growth occurs in nanocrystalline materials as a result of grain boundary-mediated deformation mechanisms such as grain rotation and grain boundary sliding or migration. It is possible that grains have rotated and coalesced during HPT [20]. Ni *et al.* [21] studied this phenomenon in a Ni-Fe alloy. They showed that three-dimensional low angle subgrain boundaries, containing a high density of dislocations, formed while grains were rotating during HPT. Further straining led to the elimination of the subgrain boundaries from which dislocations glide away on different {111} planes. A similar mechanism could occur in the HPT of the Ti-5553 alloy. In other words, the β grains obtained after 1 to 5 turns of HPT could be smaller than those formed after 20 turns owing to deformation induced grain growth.

5. Conclusions

(1) Grains of < 50 nm were obtained in a commercial Ti-5553 alloy by HPT. The grain refinement is attributed to the formation of the stress induced martensitic phase, α'', which divide large β grains into small blocks and enable subgrains to form even in nano-sized β grains.

(2) Fragmentation of the α'' plates led to their decreasing thickness with increasing strain. Theoretical calculations reveal that the reverse α'' to β transformation is triggered when the size of the α'' plates is reduced to < ~10 nm owing to significantly rising free energy from the Gibbs-Thomson effect.

(3) Even though the equivalent strain at r = 2.5 mm after 1 turn of HPT is calculated to be the same as that at r = 5 mm after half a turn, the size of the α'' plates in the latter was much smaller. This is because the calculation assumes the same disc thicknesses at r = 2.5 and 5 mm, but in reality, the disc was usually thinner at the edge (r = 5 mm), giving rise to a larger strain and consequently smaller martensitic plates.

(4) Deformation induced grain growth may have occurred after 20 turns of HPT, making the β grain boundaries well-defined.

Acknowledgement
We are grateful to Dr. Darren Edwards of the Defence Science and Technology Organisation for conducting the HPT processing. AZ acknowledges the receipt of a University of Melbourne postgraduate scholarship. This investigation is partially supported by the *ARC Centre of Excellence for Design in Light Metals* and the *Australia-China Research Centre for Light Metals*.

References
[1] Estrin Y and Vinogradov A 2013 *Acta Mater.* **61** 782-817
[2] Verlinden B 2005 *Metalurgija* **11** 165-82
[3] Semiatin S L and DeLo D P 2000 *Mater. Des.* **21** 311-22
[4] Zhu Y T and Huang J Y 2003 *J. Mater. Res.* **18** 1908-17
[5] Ko Y G, Lee C S and Shin D H 2008 *Scr. Mater.* **58** 1094-7
[6] Zherebtsov S, Salishchev G, Galeyev R and Maekawa K 2005 *Mater. Trans. JIM* **46** 2020-5
[7] Xu W, Kim K B, Das J, Calin M, Rellinghaus B and Eckert J 2006 *Appl. Phys. Lett.* **89** 031906
[8] Xu W, Wu X, Calin M, Stoica M, Eckert J and Xia K 2009 *Scr. Mater.* **60** 1012-5
[9] Xu W, Wu X, Figueiredo R B, Stoica M, Calin M, Eckert J, Langdon T G and Xia K 2009 *Inter. J. Mater. Res.* **100** 1662-7
[10] Chen W, Sun Q, Xiao L and Sun J 2012 *Metall. Mater. Trans.* A **43** 316-26
[11] Ping D H, Yamabe-Mitarai Y, Cui C Y, Yin F X and Choudhry M A 2008 *Appl. Phys. Lett.* **93** 151911
[12] Grosdidier T and Philippe M J 2000 *Mater. Sci. Eng.* A **291** 218-223
[13] Inamura T, Hosoda H, Kim H Y and Miyazaki S 2010 *Phil. Mag.* **90** 3475-98
[14] Doraiswamy D and Ankem S 2003 *Acta Mater.* **51** 1607-19
[15] Pereira P H R, Figueiredo R B, Huang Y, Cetlin P R and Langdon T G 2014 *Mater. Sci. Eng.* A **593** 185-8

[16] Duerig T W, Albrecht J, Richter D and Fischer P 1982 *Acta Metall.* **30** 2161-72
[17] Zafari A, Wei X, Xu W and Xia K 2015 *Submitted to Acta Mater.*
[18] Xu W, Edwards D P, Wu X, Stoica M, Calin M, Kühn U, Eckert J and Xia K 2013 *Scr. Mater.* **68** 67-70
[19] Pereira P H R, Figueiredo R B, Cetlin P R and Langdon T G 2014 *IOP Conf. Series: Mater. Sci. Eng.* **63** 012041
[20] Wang Y B, Li B Q, Sui M L and Mao S X 2008 *Appl. Phys. Lett.* **92** 011903
[21] Ni S, Wang Y B, Liao X Z, Alhajeri S N, Li H Q, Zhao Y H, Lavernia E J, Ringer S P, Langdon T G and Zhu Y T 2011 *Scr. Mater.* **64** 327–30

Influence of structural heterogeneity on the structural coarsening during annealing of polycrystalline Ni subjected to dynamic plastic deformation

H W Zhang[1,*], Z P Luo[1], N Hansen[2] and K Lu[1,*]
[1]Shenyang National Laboratory for Materials Science, Institute of Metal Research, CAS, 110016 Shenyang, China
[2]Danish-Chinese Center for Nanometals, Section for Materials Science and Advanced Characterization, Department of Wind Energy, Technical University of Denmark, Risø Campus, DK-4000 Roskilde, Denmark

E-mail: hwzhang@imr.ac.cn; lu@imr.ac.cn.

Abstract. The structural heterogeneity of a polycrystalline Ni subjected to dynamic plastic deformation to a strain of 2.3 was characterized, and its influence on the structural coarsening behaviour during post annealing was investigated. Structural heterogeneity on the large scale manifests itself by formation of two types of layers: low misoriented regions (LMRs) and highly misoriented regions (HMRs). On the small scale, the heterogeneity was characterized by different distributions of boundaries and textures in each layer. LMRs contain only low angle boundaries and one dominating crystallographic orientation. In contrast HMRs contain both low and high angle boundaries (>15°) and the texture is mixed with <011> close to the compression axis. During annealing, LMRs coarsen uniformly and recrystallization nucleation is difficult to form. In HMRs, the structural evolution is heterogeneous and recrystallization nuclei are readily formed. The importance of structural heterogeneity during structural design for high performance nanostructure was highlighted.

1. Introduction

Plastic deformation of metals is intrinsically heterogeneous, and different grains or different parts of one grain are usually subjected to different deformations and in turn distinct microstructures and textures are formed [1, 2]. As is the main theme of this symposium, structural heterogeneity is of greatly scientific and technological importance in terms of the structural adjustment for property optimization in particular for nanometals with high strength but low ductility and low thermal stability.

Recently, high strain rate deformation, as has been demonstrated by dynamic plastic deformation (DPD) [3-6], has drawn increasing interest owing to the significant grain refinement down to nanometer regime as well as the novel final structures that are distinct from the low strain rate deformed counterparts. Taking polycrystalline Ni for consideration, DPD reduces the structural scale down to ~120 nm at a substantially low strain of 2.3, forming typical lamellar structure composed of 70% low angle boundaries. Comparatively, for low strain rate deformation a similar refinement effect generally requires a strain as larger as >12, and the final product is characterized by an ultrafine grained structure containing >70% high angle boundaries [7, 8]. Besides, DPD fabricates heterogeneous deformation microstructures, distinct regions are characterized by different microstructures and textures [9]. The structural coarsening behaviours of DPD Ni during post annealing as well as the influence by the structural heterogeneity will be investigated in the present study.

2. Experimental

Table 1. Chemical compositions of the Ni samples (wt.%)

C	Si	Mn	P	S	Cr	Fe	Al	Co	Cu	Ti	Mg	Ni
0.002	0.007	0.002	0.004	0.001	0.001	0.011	0.001	0.002	0.001	0.005	0.003	99.96

A polycrystalline Ni with a purity of 99.96 wt.% (Table 1) was used as the experimental material. The starting material was fully recrystallized, with grain size ranging from 30-300 μm (an average of 100 μm), Fig. 1. Following our previous investigation [6], cylindrical Ni samples with a diameter of 20 mm and a length of 20 mm was subjected to DPD at room temperature to a strain of ε=2.3, hereafter denoted as DPD Ni. The strain is defined as the logarithm ratio of the initial height L_0 to final height L_f : $\varepsilon = \ln(L_0/L_f)$. Annealing was carried out at 100-700 °C for 1h in a quartz tube filled with Ar gas, followed by air cooling to room temperature. The microstructure at each annealing temperature was quantitatively characterized.

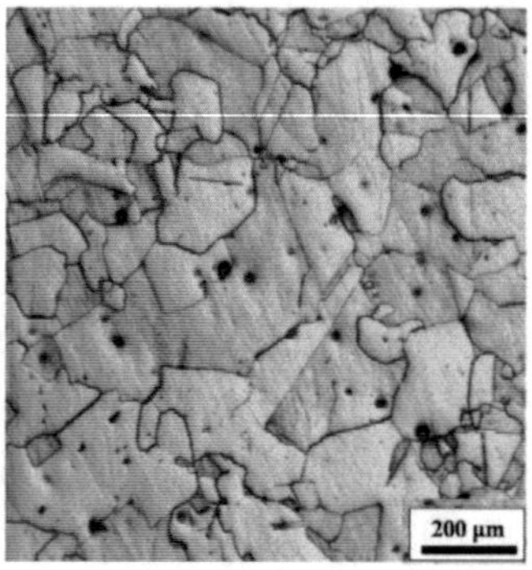

Figure 1. Optical microscopic image of the starting polycrystalline Ni sample.

Structural characterization was carried out on the longitudinal section that is parallel to the loading direction by a JEOL 2010 transmission electron microscope (TEM) and a FEI Nova NanoSEM 430 field emission gun scanning electron microscope (FEG-SEM). The boundary spacing of the deformation microstructure was determined by measuring, on the TEM micrographs, the interception length along lines perpendicular (D_T) and parallel (D_L) to the compression axis, respectively. The random boundary spacing (D_r) was calculated according to: $D_r=2/(1/D_T+\pi/(2D_L))$ [10]. Their boundary misorientation angles and the microtexture were measured via CBED in TEM and EBSD in SEM. The structure and texture of the annealed samples were characterized only by EBSD. The EBSD data were collected with a scanning step size of 0.1 μm and were analyzed with a HKL Channel 5 system, where boundaries that are misoriented less than 2° were excluded for the statistical analysis. The Vickers microhardness (H_V) of the as-deformed and as-annealed DPD Ni samples was measured on the longitudinal section with a load of 0.98 N and a loading time of 10 s. Each data point was averaged at least 10 measurements.

3. Results

3.1. Deformation microstructures

A boundary map reconstructed from EBSD data reveals two distinct regions in the DPD Ni, Fig. 2a. One region termed as highly misoriented regions (HMRs) is a mixture of low angle boundaries (LABs, indicated by thin green lines) and high angle boundaries (HABs, indicated by bold black lines), and the other named as low misoriented regions (LMRs) contains solely low angle boundaries. As indicated by the <110> pole figures in Fig. 2, LMRs show one single orientation, whereas HMRs contain many orientations with <110> close to the compression axis (CA), i.e. <110>fiber texture.

TEM observation, Fig. 2b, shows that the microstructure inside both HMRs and LMRs is a lamellar structure, which is characterized by straight and flat lamellar boundaries (LBs) bordering volumes composed of interconnecting dislocation boundaries and low densities of loose dislocations. Such microstructure is the typical microstructure developed in metals of high stacking fault energy during monotonic deformation to low-to-medium strains, where plastic deformation was accommodated by dislocation slip [1, 2, 7, 10]. However, the length of these lamellar boundaries reaches several to tens of microns, whereas the width (D_T in Table 2) is in the nanometer regime, which gives rise to an aspect ratio significantly larger than 100. D_T and D_L for HMRs (114.2 nm and 332.8 nm) are slightly smaller than those of LMRs (136.1±6.1 nm and 349.8±11.9 nm). Besides, LMRs include only LABs, whereas HMRs contain 29.8% HABs.

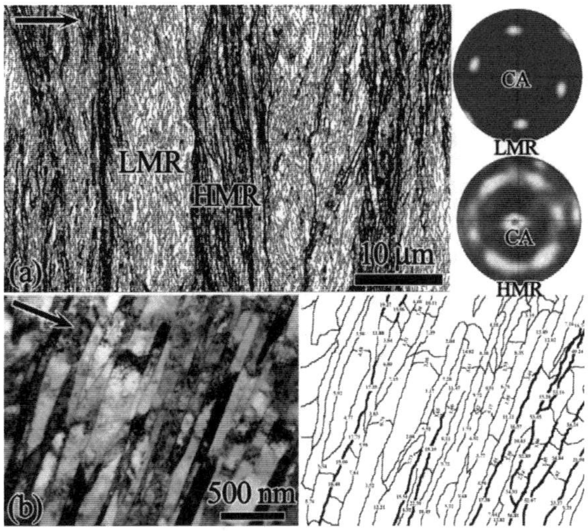

Figure 2. (a) Boundary map reconstructed from EBSD data showing two types of regions and the corresponding pole figures in DPD Ni: LMRs and HMRs. (b) TEM observation together with the boundary sketch of the lamellar structure formed in the DPD Ni. Thin green line in (a) and thin black line in (b) indicates low angle boundaries, whereas high angle boundaries are indicated by bold black lines. The compression axis (CA) was indicated by arrow.

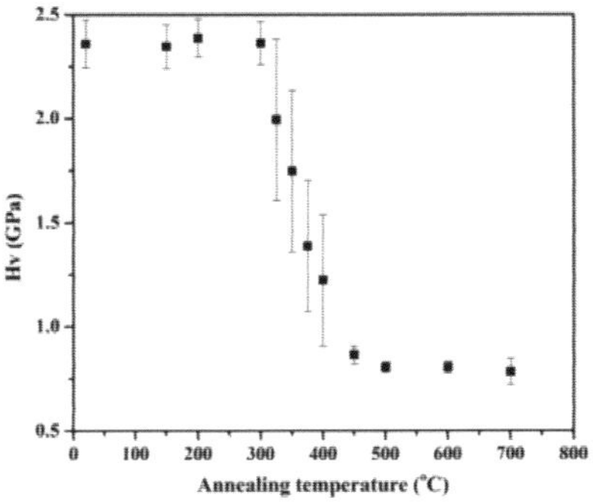

Figure 3. Variation of hardness as a function of annealing temperature of DPD Ni.

Table 2. Boundary parameters of HMRs and LMRs

Structure	D_T (nm)	D_L (nm)	D_{av} (nm)	θ_{av} (<15°)	f_{HAB} (>15°)
HMRs	114.2	332.8	148.4	5.5	29.8%
LMRs	136.1	349.8	169.0	4.9	0

3.2. Annealing behavior

3.2.1. Hardness-Temperature

During annealing at temperature from 100-700°C for 1h, the hardness of DPD Ni decreases with an increase of annealing temperature, Fig. 3, exhibiting three stages: i) a slight hardness change as temperature increases up to 300°C, corresponding to a low temperature recovery; ii) a sharp drop of hardness from 2.4 to 0.9 GPa in the temperature range from 300 to 450 °C, implying a recrystallization at medium temperatures; and iii) a small decrease of hardness from 0.9 to 0.8 GPa in the temperature span of 450-700 °C, i.e. a grain growth after recrystallization at high temperature. Such hardness change with annealing temperature was also observed in a 99.967% Ni subjected to HPT to a strain of 100 [8]. However, compared with HPT Ni, the hardness change in the recovery stage for DPD Ni is less significant and the recrystallization temperature corresponding to 50% hardness drop is higher by ~200°C.

3.2.2. Microstructure-temperature

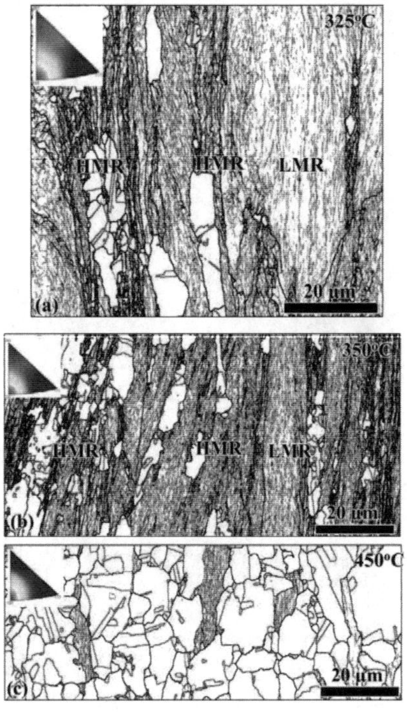

Figure 4. Boundary maps reconstructed from EBSD data of DPD Ni during annealing at different temperatures for 1h. Red lines indicate the annealing twin boundaries. Inverse pole figures are inserted.

The microstructure of the DPD Ni annealed at temperatures lower than 300°C resembles the deformation microstructure. No apparent changes in the structural pattern and boundary parameters were detected for both HMRs and LMRs. As temperature increases up to 325°C, Fig. 4a, LMRs remain the deformed state, however, high angle boundaries in the HMRs becomes wavy and rough. Small (nuclei) and large recrystallization grains are also observed with clean interiors bordered by high angle boundaries. These grains frequently contain annealing twins. Some grains are formed in the regions close to the interface between LMRs and HMRs, others are in the HMRs interiors. For both cases, the nucleation positions are composed of high density of high angle boundaries. Besides, the recrystallization grains are not equiaxed but extended along LBs. The longitudinal size (parallel to LB, D_L) is larger than the transversal size (perpendicular to LBs, D_L) by a factor of 1-8. As temperature increases to 350°C, Fig. 4b, LMRs remain unchanged, while more areas in HMRs recrystallize. Nucleation takes place continuously in the clusters of high angle boundaries, which is supplemented with the concurrently growing along LBs leading to banded recrystallization regions. After annealing at 450°C, most of the deformation matrix has been recrystallized, leaving small amount of LMRs embedded in the recrystallization grains, Fig. 4c. These recrystallization grains, however, are roughly equiaxed, though extended morphologies can still be detected in particular for small grains. It should be noted that as evidenced from the inserted inverse pole figures (IPF) in Fig. 4, <110> fiber texture was well retained even fully recrystallization has taken place.

The variation of the boundary spacing as a function of annealing temperature for both HMRs and LMRs gives rise to "S"-shape, Fig. 5. Two slightly coarsening stages observed at low temperatures (stage I) and high temperatures (stage III) are bridged by a stage with a fast increase of boundary spacing at medium temperatures (stage II). The intersection of the tangent lines of stage I and stage II points to the onset temperature for structural coarsening (T_{on}). For LMRs, the T_{on} is 410°C, whereas for HMRs it is 360°C. It is also found that the structural coarsening during stage II for LMRs is faster, and the final grain size is also relatively bigger.

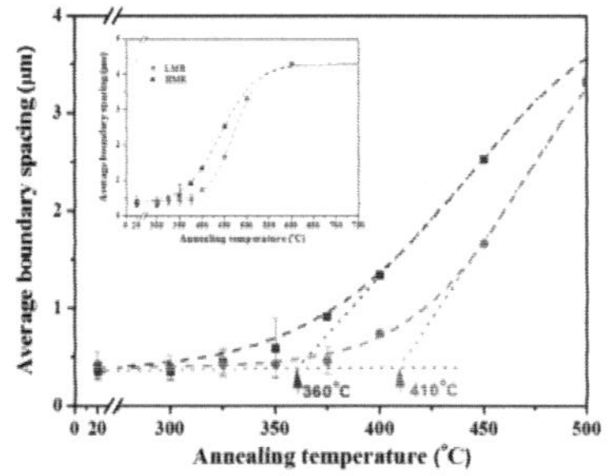

Figure 5. Variation of boundary spacing as a function of annealing temperature for HMRs and LMRs in DPD Ni.

4. Discussion

4.1. Structural heterogeneity

The deformation microstructures of DPD Ni as evidenced by both TEM and EBSD are heterogeneous. Such heterogeneity exhibits on both a fairly large scale and a small scale. Large scale heterogeneity manifests itself by the formation of two different layers defined as HMRs and LMRs, whereas the small scale heterogeneity characterizes by the difference in the structural parameters and textural components. Such structural heterogeneity has its cause partially related to the macroscopic subdivision, which has recently been found in polycrystalline copper subjected to DPD, where it was suggested that a large initial grain size (~200 μm) and different crystallographic orientations of the individual grain can result in different deformation patterns [11]. A similar dependency might result in a correlation between the scale of the macroscopic subdivision and the original grain size distribution. Assuming that the coarse grains with the grain size of 30-300 μm (average of 100 μm) are subjected to a total strain of ε=2.3 (thickness reduction of 90%), the grains would be compacted to 3-30 μm. This thickness is in accord with the layer thickness of HMRs and LMRs as shown in Fig. 2. This implies that HMRs and LMRs may be corresponding to the different original grains and their interfaces are most likely the original grain boundaries. However, it should be noted that lamellar boundary spacing for both HMRs and LMRs (136 nm for LMRs and 114 nm for HMRs) is tens to hundreds of times smaller than that of the compacted original grain boundaries (3-30 μm). This means that the original grain boundaries only account for a small fraction, whereas most of the HABs are produced by plastic deformation. That is to say HMRs and LMRs were predominately formed through the orientation splitting into different parts with distinct deformation behaviours.

The difference in structural scale and boundary distribution may lead to the difference in the stored energy that acts as the driving force for structural coarsening during post annealing. The energy can be estimated according to the structural parameters listed in Table 2. Generally, stored energy to a first approximate is contributed from elastic energy from dislocation structure and boundary energy from high angle boundaries, neglecting the contributions from vacancies and long range stress. The elastic energy (E_d) is approximated the energy per unit length of dislocation line multiplied by the dislocation density (ρ) [12]:

$$E_d = \rho \frac{Gb^2 f(\nu)}{4\pi} ln(\frac{R}{b}) \tag{1}$$

where G is the shear modulus (79 GPa), b is the Burgers vector (0.249 nm), υ is the Poisson's ratio (0.31) and $f(\upsilon)$=(1-υ/2)/(1-υ), i.e. $f(\upsilon)$~1.225. R is the upper cut-off radius and is approximated by the average dislocation spacing, i.e. $R\sim1/\rho^{1/2}$. Generally, the loose dislocations presented in between boundaries are of the order of 10^{14}-10^{15}m^{-2}, which gives rise to a stored energy of ~0.2-2 MPa. For LABs, the stored energy can be viewed as contribution from the dislocations present in them, following *eq.* (1). The dislocation density (ρ_{LAB}) can be estimated based on the misorientation angle (θ): $\rho_{LAB}=1.5\theta S_v/b$, assuming that the boundaries are mixed boundaries [1].

The contribution from HABs (E_b) can be the product of surface area per unit volume ($S_v=1/D_T+\pi/(2D_L)$) [10] and the boundary energy per unit area (γ) [13]:

$$E_b = \gamma \cdot S_v \cdot f_{HAB} \tag{2}$$

where γ for Ni is 0.866 J/m^2 [14] and f_{HAB} is the fraction of high angle boundaries. The total contribution can be a linear additivity of E_b and Ed.

By taking the structural parameters in Table 2, the stored energy for LMRs (E_{LMR}) and HMRs (E_{HMR}) are 11.2 MJm^{-3} and 13.7 MJm^{-3}, respectively. The calculation does reflect the heterogeneous distribution of stored energy, however, the difference is not very large (~23%), which should be partially attributed to the fact that ~70% boundaries in HMRs are LABs.

Consequently, the above analysis underpins the structural heterogeneity presented in the deformation microstructures of DPD Ni, which in turn influences the structural behaviours during the subsequently annealing.

4.2. Structural coarsening

The structural coarsening of DPD Ni subjected to annealing at different temperatures shows different rates, i.e. slow coarsening during low and high temperature annealing and fast structural coarsening at medium temperatures. This is reflected by the hardness change with an increase of annealing temperature (Fig. 3), which depicts three annealing stages: recovery, recrystallization and grain growth. Such a correlation between hardness and annealing temperature has been well documented in literature [13]. However, the structural coarsening behaviours of HMR and LMR differ significantly owing to their distinct structural heterogeneity.

4.2.1 HMRs

The microstructure of HMRs coarsens in a heterogeneous manner. First of all, recrystallization nucleation takes place heterogeneously. In the sample annealed at 325°C for 1h (Fig. 4a), nucleation occurs locally in HMRs while the rest of sample remains in the deformation state. Such heterogeneous nucleation should have its cause related to its structural features. HMR contains ~30% high angle boundaries and 70% low angle boundaries. High angle boundaries, owing to the higher energy, high mobility and low activation energy, are more mobile [13]. This could be verified by the observation that local wavy and roughness appeared in the high angle boundaries in the HMRs (Fig. 4a). As triple junction migration has been proposed as an important recovery mechanism for lamellar structure composed of mainly high angle boundaries [15, 16], the presence of high angle boundaries in HMRs should have the chance of forming mobile Y-type tripe junction that is composed of three lamellar boundaries with medium to high misorientation angle [15]. Such triple junction was observed in the boundary sketch of Figure 2b, which is composed of three boundaries that are misoriented by 53.45°, 15.2° and 53.6°, respectively. As a result, the readily migration of high angle boundaries together with the high mobile triple junction may be responsible for the nucleation process, which is supported by the observation that recrystallization nuclei are only observed in the places with high density of high angle boundaries.

The heterogeneous structural coarsening manifests itself also by the directional growth. As evidenced by Fig. 4, the recrystallization grains inside HMRs are hardly observed as equiaxed in shape but frequently extended along LBs. It seems that the parallel growth is more easily than perpendicular growth. Such directional growth is less apparent in the lamellar structure fabricated by cold rolling [17] and HPT [8] to large strains. The cause should be related to the heterogeneous distribution of boundary and texture in DPD Ni. DPD-induced lamellar structure in HMRs contains around 70% low angle boundaries which include both LBs and interconnecting dislocation boundaries. The LABs usually form clusters that are adjacent to those with high angle boundaries. Comparatively, for large strain cold rolled Ni [10] and high pressure torsion (HPT) deformed Ni [7, 8], lamellar structure was also fabricated but the HABs accounts for >50-60%. HABs in these samples are the lamellar boundaries but LABs are mainly the interconnecting dislocation boundaries in between LBs, and the formation of HAB clusters or LAB clusters in these samples are not observed. Besides, HMRs show <110> fiber texture, which is contributed from LAB clusters and HAB clusters. The former shows only single orientation, but the latter contains many orientations with <110> close to compression axis. The distribution of HAB clusters and LAB clusters leads to the alternatively distributed textures. On the contrary CR Ni forms rolling textures that are composed of different textural component, while HPT Ni produces shear texture that consists of different shear component. In both cases, banded regions with single orientation were not observed.

4.2.2 LMRs

With an increase of annealing temperature, the microstructures of LMRs coarsen with a uniform manner, and recrystallization nuclei were not observed even after they were consumed by the growth of recrystallization grains formed in HMRs. The uniform structural coarsening could be reflected by the roughly linear changes of boundary spacing with annealing temperature in the temperatures lower than 375°C, where the growth of recrystallization grains into LMRs is less significant (see Fig. 5). As recrystallization grains formed in HMRs start to consume LMRs at temperature > 375°C, the linear relationship will not hold. Both the uniform structural coarsening and the difficulty in recrystallization nucleation should have their causes related to the structural and textural character. As has been evidenced by the structural characterization, boundaries including LBs and interconnecting dislocation boundaries in LMRs are low angle boundaries. Owing to the low energy, low mobility and high activation energy for boundary migration [13], the migration of such boundaries requires higher temperatures. Since all the boundaries are low angle boundaries, triple junction migration should be reduced to minimum as mobile Y-type junctions that are composed of three high angle boundaries are actually not involved in this regions. Low angle boundaries also lead to low stored energy, i.e. low driving force for boundary migration. Besides, LMRs show single orientation with small orientation spread (see Fig. 2). Under such condition, even if nucleation takes place inside LMRs, the growth into deformed matrix with similar orientation should be expected difficult owing to the orientation pinning effect [18].

4.3. Structural design

The present investigation underpins the significant influence of structural heterogeneity presented in the deformation microstructure on the structural coarsening behaviours during post annealing. Structural coarsening preferentially takes place in regions composed of different boundaries and different orientations, i.e. more heterogeneous regions. On the contrary, the regions consisting of identical structural and texture, i.e. less heterogeneous regions, are thermally more stable. With respect to the high strength but low ductility of nanometals, the necessary structural adjusting to produces so-called bi-model or multi-modal grain size distribution has significantly optimized the strength and ductility combination [19, 20]. The present investigation may provide with potential structural optimization strategy through adjusting the distribution and volume fraction of hard and soft regions by proper annealing treatment. As the structural heterogeneity has the cause related to the stability of the starting orientation, the choice of single crystallites with unstable orientations to subject to high strain rate deformation would produce also such heterogeneous structure. Under such condition, it might be possible to control the orientation splitting through adjusting the processing parameters, which in turn control the structural heterogeneity and the structural coarsening behaviour during annealing.

Besides, owing to fact that microstructures with less heterogeneity are thermally more stable. Structural design by tailoring the microstructure through such microstructure would simultaneously enhance the strength and thermal stability of nanometals. For the present case, lamellar structure composed of only low angle boundaries can be a potentially option for such purpose. Actually, recent investigation has support this hypothesis [21], where nanometre scale laminated structure with mainly low angle boundaries was fabricated in a Ni sample by means of surface mechanical grinding treatment, the hardness can be as high as 6.4 GPa that is 3 times higher than the UFG counterparts, more interestingly, the onset temperature for structural coarsening is also higher by 40K. With this purpose, single crystals with relatively stable orientation to subject to high strain rate deformation might be a choice, which is one of our on-going projects.

5. Conclusions

A polycrystalline Ni was subjected to dynamic plastic deformation to a strain of 2.3. The structural heterogeneity was examined and the influence on the structural coarsening behaviour during post annealing was investigated. The following conclusions can be reached:

1. DPD introduces a heterogeneous deformation microstructure in Ni. Low misoriented regions (LMRs) composed of only low angle boundaries coexisted with highly misoriented regions (HMRs) with a mixture of both low and high angle boundaries. LMRs show roughly a single orientation, whereas HMRs contain many orientations but with <011> close to the compression axis.

2. The structural coarsening is greatly influenced by structural heterogeneity. LMRs coarsen uniformly and formation of recrystallization nuclei is difficult. HMRs, on the contrary, show heterogeneous structural coarsening. Nucleation preferentially takes place in regions with high angle boundaries and the grain growth along LBs is faster than that perpendicular to the LBs.
3. Structural heterogeneity, being a result of plastic deformation, is shown of great importance in terms of the structural design for high performance nanostructure due to the close relationship between deformation microstructure and the nucleation and recrystallization behaviour.

Acknowledgments

The authors gratefully acknowledge the financial support of the Ministry of Science and Technology of China (Grant 2012CB932201), the National Natural Science Foundation (Grant No. 51231006, 51171182), and Danish-Chinese Center for Nanometals (Grant 51261130091, DNRF86-5).

Reference

[1] Hansen N 2001 *Metall. Mater. Trans.* A **32** 2917
[2] Huang X and Winther G 2007 *Phil. Mag.* **87** 5189
[3] Tao NR and Lu K 2007 *J. Mater. Sci. Technol.* 23 771
[4] Li YS, Tao NR and Lu K 2008 *Acta Mater.* **56** 230
[5] Yan FK, Tao NR and Lu K 2012 *Acta Mater.* **60** 1059
[6] Luo ZP, Zhang HW, Hansen N and Lu K 2012 *Acta Mater.* **60** 1322
[7] Zhang HW, Huang X and Hansen N 2008 *Acta Mater.* **56** 5451
[8] Zhang HW, Huang X, Pippan R and Hansen N 2010 *Acta Mater.* **58** 1698
[9] Luo ZP, Mishin OV, Zhang YB, Zhang HW and Lu K 2012 *Scripta Mater.* **66** 335
[10] Hughes DA and Hansen N 2000 *Acta Mater.* **48** 2985
[11] Lin FX, Zhang YB, Tao NR, Pantleon W and Juul Jensen D 2014 *Acta Mater* **72** 252
[12] Kuhlmann-Wilsdorf D 1989 *Mater. Sci. Eng.* A **113** 1
[13] Humphreys FJ and Hatherly M 2014 in *Recrystallization and related annealing phenomena* 2nd ed. New York: Elsvier
[14] Murr LE 1975 *Interfacial phenomena in metals and alloys.* Addison-Wesley Publishing Company
[15] Yu TB, Hansen N and Huang XX 2011 *Proc. Royal. Soc.* A **457** 2135
[16] Yu TB, Hughes DA, Hansen N and Huang X 2015 *Acta Mater.* **86** 269
[17] Li XL, Liu W, Godfrey A, Juul Jensen D and Liu Q 2007 *Acta Mater.* **55** 3531
[18] Juul Jensen D 1995 *Acta Metall. Mater.* **43** 4117
[19] Wang YM, Chen MW, Zhou FH and Ma E 2002 *Nature* **419** 912
[20] Zhao YH, Zhu YT and Lavernia EJ 2010 *Adv. Eng. Mater.* **12** 759
[21] Zhao YH, Zhu YT and Lavernia EJ 2010 *Adv. Eng. Mater.* **12** 759
[21] Liu XC Zhang HW and Lu K 2013 *Science* **342** 337.

Recrystallization behaviour of AA6063 extrusions

K Zhang[1], T Pettersen[2], C O Paulsen[1], K Marthinsen[1], B Holmedal[1] and A Segatori[3]

[1]Department of Materials Science and Engineering, Norwegian University of Science and Technology, NO-7491 Trondheim, Norway
[2]SINTEF Materials and Chemistry, box 4760 Sluppen, NO-7465 Trondheim, Norway
[3]SAPA Technology, SE-61281 Finspång, Sweden

Email: kai.zhang@ntnu.no

Abstract. Cylindrical profiles of an AA6063 aluminium alloy were produced in a lab-scale direct extrusion set-up. The extrusion was performed at 300 °C, 450 °C and 550 °C, respectively, with the same ram speed. Immediate water quenching was applied to the profiles and the end of billet (butt-end) after extrusion. Microstructure and texture of the material in different states were measured by electron back-scattered diffraction. Only the profile extruded at 300 °C, was found in the deformed state after extrusion, featuring a fibrous grain structure and a strong <111> and weak <100> double fibre texture. Post-extrusion annealing of this profile at 450 °C resulted in an almost fully recrystallized structure (recrystallized fraction of 87%) and with a texture similar to that of the as-deformed state. The profile extruded at 450 °C was almost fully recrystallized (recrystallization fraction 91%) already after quenching, and with a texture characterized by a weak <111> and strong <100> double fibre. The profile extruded at 550 °C showed a partially recrystallized grain structure with recrystallization fraction of 71%, and with a texture dominated by a <100> fibre. The influence of the deformation conditions on the recrystallization behaviour, in terms of recrystallization kinetics and mechanisms, are discussed in view of these results.

1. Introduction

Due to aluminium's extremely good formability, aluminium profiles with very complex shapes can be extruded. However, productivity and for some high demand application areas consistent properties may be a challenge. One main issue is grain structure control and the ability to produce extrusions with a homogenous grain size and texture through the thickness of the profile. The microstructure of the extrusions is strongly dependent on the extrusion parameters [1-3] and on the microstructure of the billet material itself [4, 5]. This work aims to investigate the influence of extrusion temperature on the microstructure evolution and the recrystallization (RX) behaviour of an AA6063 aluminium alloy after axisymmetric extrusion. The motivation is to establish a better understanding of the recrystallization behaviour as basis for further development of numerical models for the recrystallized grain size and texture in this and similar alloys after extrusion.

2. Material and experiments

2.1. Material

The material investigated was an AA6063 alloy (chemical composition in wt% Si 0.4; Mg 0.5; Fe 0.096; Mn 0.017; Ti 0.01; Ga 0.012; Al remaining). After direct chill (DC) casting, the material was homogenized at 580 °C for 135 min, with a heating rate of 250 °C/h and a cooling rate of 400 °C/h. The grain structure and texture of the billet material were measured by electron back-scattered diffraction (EBSD) in a field emission scanning electron microscope (FESEM) equipped with the TSL OIM software, while the back scatter electron mode (BSE) was used to characterize the particles.

It was found that the initial grains were of equiaxed shape with an average diameter of 90 μm. The texture of the as-cast material was a random texture. The average diameter of the constituent particles was 1.4 μm, whereas the volume fraction was about 0.4%. There were no fine second-phase particles which could retard the RX process.

2.2. Lab-scale extrusion

Billets of 20 mm diameter and 22 mm length were machined from the homogenized material. Cylindrical profiles of 3 mm diameter were produced by direct extrusion in a laboratory scale direct extrusion set-up. Hence, the average accumulated strain at the end of extrusion was about 3.8. The container, billet, die and ram were heated together to the testing temperature before the start of extrusion. A thermocouple of type K with a diameter of 0.5 mm was inserted in the die in such a way that it ended at the circumference of the bearing in the middle of the 3 mm long bearing channel. The thermocouple did have contact with the flowing metal and recorded the temperature during extrusion.

The results reported in this paper refer to extrusions performed at 300 °C, 450 °C and 550 °C, respectively, with the same ram speed of 4.5 mm/s. The ram stroke was 9 mm, which accounted for an extrusion time of 2 s. At the end of extrusion, the extruded material and the die were immediately pushed into cold water while the press ring was retracted. The end of the billet and the profile were quenched below 300 °C within 2 s after the end of deformation, reading from the recorded temperature history by the installed thermocouple in the bearing.

2.3. Post-extrusion annealing treatment

The profile extruded at 300 °C was cut into small pieces along the extrusion direction (ED). A sample close to the end of the billet was annealed at 450 °C for 10 s in a salt-bath, followed by quick water quenching.

2.4. Microstructure characterization

The microstructure of the three as-extruded profiles and the post-extrusion annealed sample was measured by EBSD. The scan was made in the TD - ED plane of the cylindrical profiles (TD: transverse (radial) direction). The scanned areas were positioned very close to the butt-end, where the quenching was fastest and the material flow was close to steady state. The scanned area for the profile extruded at 300 °C was 0.5 mm along ED and 1.5 mm along TD, covering the region from the centre to the surface. The scanned areas for the other three samples covered the whole diameter region, i.e. 3 mm along TD. The step size in EBSD for the profile extruded at 300 °C and its post-extrusion annealed counterpart was 1 μm, whereas the step size was 3 μm for the other two profiles.

Recrystallized grains were identified as those being partly or fully surrounded by high-angle boundaries (defined as misorientation angles larger than 15°), with grain size larger than 6 times the scanning step size and a grain orientation spread (GOS) below 2°. The GOS was calculated by first averaging the orientation for each grain. The spread was then obtained by averaging the deviation between the orientation of each point in the grain and the average orientation for the grain. The GOS parameter has been found to be a successful approach with respect to EBSD data to discriminate between recrystallized and deformed grains [6]. Once the recrystallized grains were identified, the RX fraction could be evaluated. The recrystallized grain size is given in terms of the equivalent circle diameter.

The crystallographic texture was analyzed by using all scanning points. ED inverse pole figures (IPF) were employed to illustrate the texture.

3. Results and discussion

3.1. Results

The microstructures of the profiles extruded at three different temperatures are shown in Figure 1. Figure 2 illustrates the microstructure of the profile extruded at 300 °C and then annealed at 450 °C for 10 s. The colours in Figure 1 and 2 were coded according to ED IPF, where grains of the <111> and the <100> fibre texture are in blue and red colours, respectively. The corresponding textures of the three profiles and the post-extrusion annealed sample are shown in Figure 3 by means of ED IPFs. The RX fraction, average recrystallized grain size and the maximum intensities of the <100> fibre and of the <111> fibre are compiled in Table 1.

The profile extruded at 300 °C features an elongated, fibrous grain structure, as shown in Figure 1(a). Grains of the <111> fibre and of the <100> fibre appear alternating in the micrograph along TD. The area fraction of grains of the <111> fibre texture is larger than that of the <100> fibre. The difference in area fractions of the two texture fibres is also reflected by the texture strength, where a strong <111> and weak <100> double fibre is shown, see Figure 3(a). The <111> fibre is nearly 4 times stronger than the <100> texture fibre. Such a strong <111> and weak <100> double fibre has previously been reported as the deformation texture for axisymmetric extrusion of face-centred-cubic (FCC) metals [1, 7, 8]. The RX fraction is 15% by the criteria of GOS < 2°. However, no large equiaxed grains can be found in Figure 1(a). The recrystallized grains may be attributed to experimental error (low EBSD pattern quality for this profile) and to RX 'nuclei' which were too small to be well resolved in the current scan. The profile extruded at 300 °C was regarded to be in the deformed state.

The profile extruded at 450 °C features an equiaxed grain structure, as shown in Figure 1(b). The RX fraction is 91%. Hence, the profile is near fully recrystallized. The average grain size was 27.8 μm. The texture is a very weak <111> and strong <100> double fibre, where the <100> fibre is about four times stronger than the <111> fibre.

The profile extruded at 550 °C has a RX fraction of 71%, and can thus be considered as partially recrystallized. It is observed in Figure 1(c) that the recrystallized grains are of equiaxed shape. The average size of those recrystallized grains was 40 μm. There is a gradient of grain size along TD (radial direction), i.e. the size of recrystallized grains increases from the surface towards the centre, until the un-recrystallized region in the very centre of the profile. The texture is a single strong <100> fibre, as shown in Figure 3(c) and Table 1.

As described above, the as-extruded profile produced at 300 °C was in the deformed state. The microstructure after post-extrusion annealing at 450 °C for 10 s is shown in Figure 2. The RX fraction after post-extrusion annealing is 87%, similar to the RX fraction for the 450 °C extrusion, while the average grain size was 12.3 μm. The texture of the post-extrusion annealed sample is shown in Figure 3(c). The texture is a <111> and <100> double fibre, and both fibres are of similar intensities.

3.2. Discussion

The extrusion temperature has a great influence on the extrusion microstructure. Extrusion at 300 °C resulted in a deformed microstructure in the profile, whereas the profiles were fully or partially recrystallized when extrusion took place at 450 °C and 550 °C, respectively. The texture of the deformed profile was a typical strong <111> and weak <100> double fibre. At higher extrusion temperatures, the <100> fibre got stronger at the expense of the <111> fibre. The maximum intensity of the <100> fibre was four times stronger than that of the <111> fibre for extrusion at 450 °C. At 550 °C, the <100> fibre was dominant in the texture, while the <111> fibre was less than random. Due to fast dynamic recovery, discontinuous dynamic recrystallization will be suppressed for the current material [9]. Since the recrystallized texture is different from the deformed one, geometrical dynamic recrystallization or continuous dynamic recrystallization are also excluded [9]. Static recrystallization (sRX) is the responsible RX mechanism for the RX behaviour of extrusions at high temperature. However, the sRX kinetics at 450 °C is very fast; recrystallization is nearly completed (RX fraction of 91%) within the delay time 2 s from the end of deformation to quenching. It is noted though that the kinetics is slower at the higher extrusion temperature of 550 °C. This can be explained by a slower sRX kinetics with a lower Zener-Hollomon parameter (less driving force for recrystallization) [10]. For the post-extrusion annealed sample at 450 °C, the RX fraction was 87% after 10 s. The post-extrusion annealing texture is a near-equally strong <111> and <100> fibre. This texture is clearly different from that of the high temperature extrusions, where the <100> fibre is dominant.

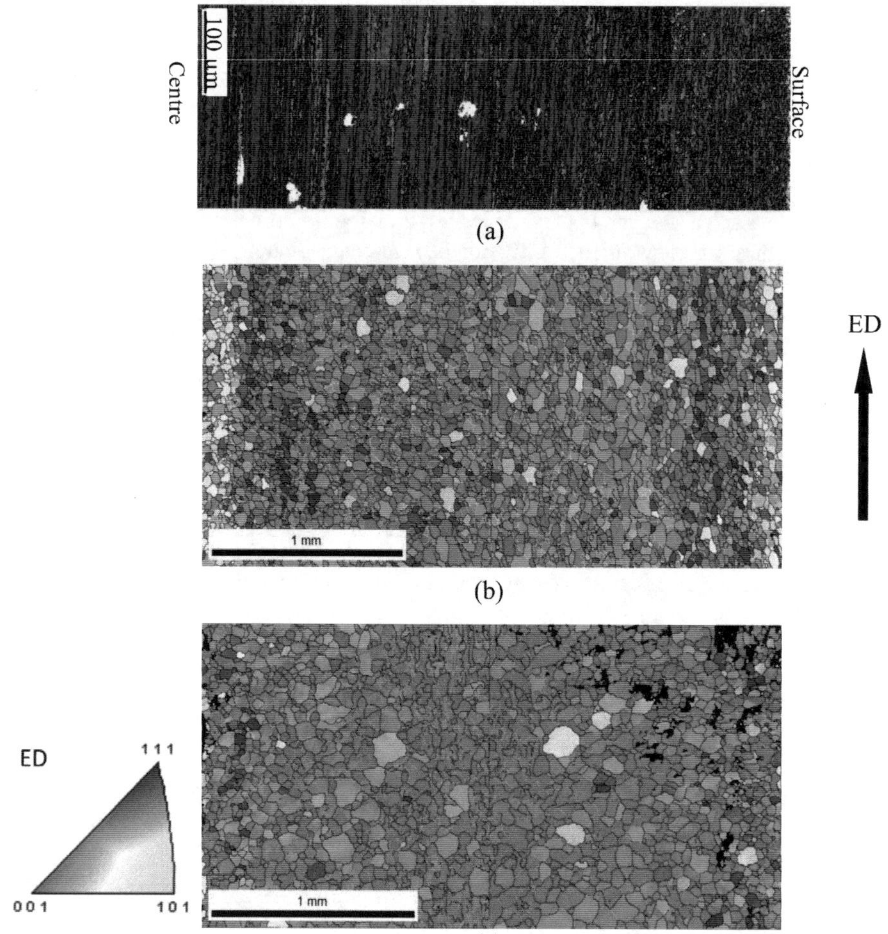

Figure 1. EBSD micrographs of the microstructure of profiles extruded at (a) 300 °C, (b) 450 °C and (c) 550 °C. Dark lines represent boundaries of misorientation larger than 15°.

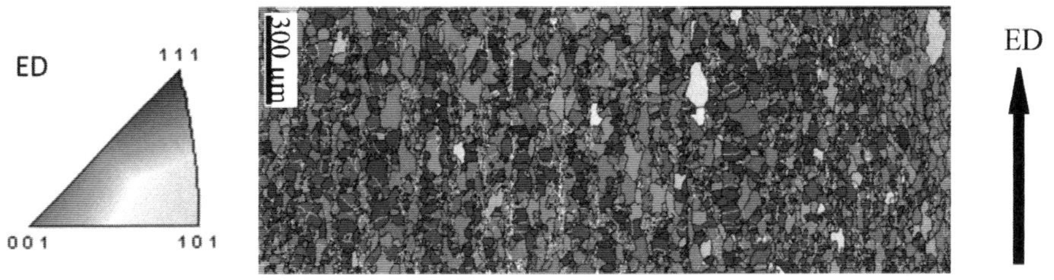

Figure 2. EBSD micrographs of the microstructure of the profile extruded at 300 °C and post-extrusion annealed at 450 °C for 10 s. Dark lines represent boundaries of misorientation larger than 15°.

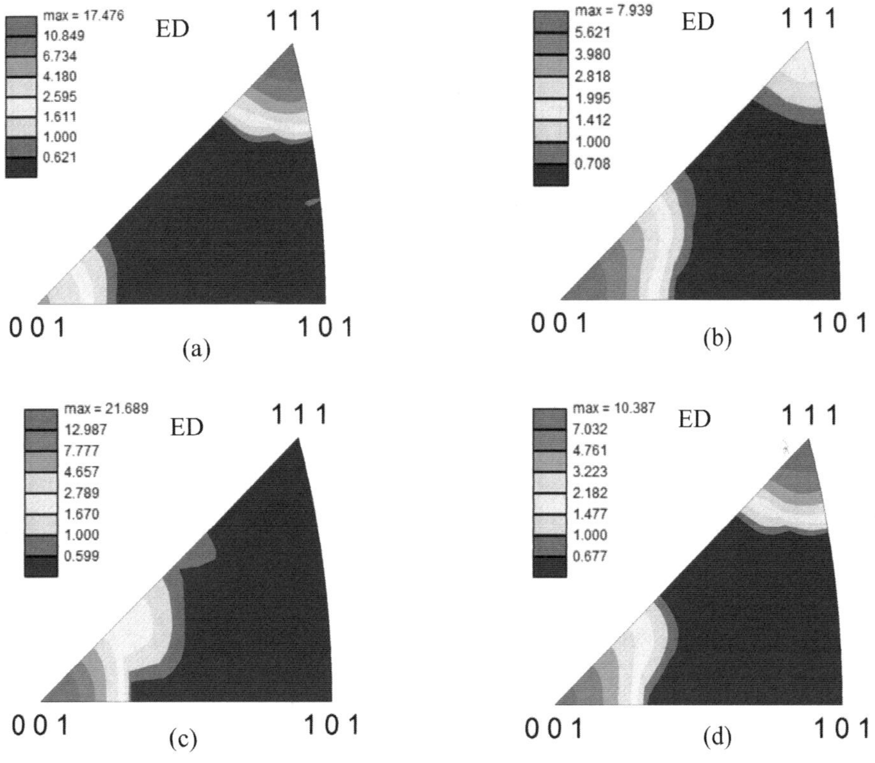

Figure 3. The ED IPFs of the profiles extruded at (a) 300 °C, (b) 450 °C, (c) 550 °C and (d) the profile extruded at 300 °C and post-extrusion annealed at 450 °C for 10 s.

Some preliminary considerations with respect to the different RX textures, based on the 'oriented nucleation' theory [11], can be made. It has been reported in the literature [7] that the <100> fibre grains have a lower stored energy than the <111> fibre grains, due to their faster recovery rates. Hence, the <100> fibre grains are potent nuclei for the recrystallization [7]. Both profiles extruded at 450 °C and 550 °C have a very strong <100> fibre. The post-extrusion annealed sample also has a stronger <100> fibre compared to that of the deformed state. All these RX textures give supports for the <100> fibre nucleation assumption for the current material. For the post-extrusion annealed sample significant <111> fibre nucleation has also taken place, given the relatively strong <111> fibre present in the RX texture. However, with increasing deformation temperatures, <100> fibre grain nucleation clearly becomes the preferred nucleation mechanism, at the expense of the <111> fibre grains which get less favourable. The reason, however, is not clear, and more thorough investigations should be made to explain these different RX behaviours.

Table 1. Summary of texture, RX fraction and average diameter of recrystallized grains for the three extruded profiles and the post-extrusion annealed sample.

Extrusion temperature	300 °C	450 °C	550 °C	300 °C
Post-extrusion annealing	-	-	-	450 °C, 10 s
RX fraction	15%	91%	71%	87%
Max. <100> fibre intensity (times random)	4.6	7.9	21.7	8.1
Max. <111> fibre intensity (times random)	17.5	1.9	<1	10.4
Average recrystallized grain size (μm)	-	27.8	40	12.3

4. Conclusion

The extrusion temperature is shown to have a great influence on the RX behaviour of the AA6063 material. For high temperature extrusion (above 450 °C for the alloy and processing conditions in the present work) the extruded profiles recrystallize directly after extrusion due to fast sRX, and the <100> fibre is the dominant RX texture component. The kinetics, however, of sRX decreases again at the highest temperature, due to less driving force for recrystallization. For warm temperature extrusion (300 °C in this work) the profile is in the deformed state after extrusion, featuring a strong <111> and weak <100> double fibre texture. After post-extrusion annealing at higher temperature, the texture is an equally-strong <111> and <100> double fibre texture, which is clearly different from the textures after high temperature extrusion. This difference may be explained by orientated nucleation under different extrusion temperatures, i.e. <100> fibre grain nucleation is the preferred nucleation mechanism and outweigh the <111> fibre grains, which become even less favourable for nucleation at higher extrusion temperatures. Further investigations should be made to fully understand the different RX textures and behaviours at high temperature extrusion and the post-extrusion annealing after warm extrusion.

Acknowledgement

This research work has been supported by the IPN project COSMETEX (228918/O30) in Norway. The financial support by the Research Council of Norway and the industrial partner SAPA AS is gratefully acknowledged.

References

[1] Inakazu I, Kaneno Y and Inoue H 1994 *Mater. Sci. Forum* **157-162** 715-720.
[2] Donati L, Segatori A, El Mehtedi M and Tomesani L 2013 *Int. J. Plast.* **46** 70-81.
[3] Peng Z and Sheppard T 2004 *Modell. Simul. Mater. Sci. Eng.* **12** 43.
[4] Eivani A R, Zhou J and Duszczyk J 2014 *Comput. Mater. Sci.* **86** 193-199.
[5] Chen J Q, Poole W J and Parson N C 2014 *Mater. Sci. Forum* **794-796** 1127-1132.
[6] Mitsche S, Poelt P and Sommitsch C 2007 *J. of Microscopy* **227** 267-274.

[7] Vatne H E, Pedersen K, Lohne O and Jenssen G 1997 *Textures Microstruct.* **30** 81-95.
[8] Calnan E A 1954 *Acta Metall.* **2** 865-874.
[9] Sakai T, Belyakov A, Kaibyshev R, Miura H and Jonas J J 2014 *Prog. Mater Sci.* **60** 130-207.
[10] Gutierrez I, Castro F R, Urcola J J and Fuentes M 1988 Mater. Sci. Eng., A 102 77-84.
[11] Humphreys F J and Hatherly M 2004 Chapter 12 - Recrystallization Textures, in: F.J.H. Hatherly (Ed.) Recrystallization and Related Annealing Phenomena (Second Edition), Elsevier, Oxford, 2004, pp. 379-413.

Low temperature annealing of cold-drawn pearlitic steel wire

X Zhang[1], J I Bech[2], N Hansen[1]

[1]Danish-Chinese Center for Nanometals, Section for Materials Science and Advanced Characterization, Department of Wind Energy, Technical University of Denmark, DK-4000 Roskilde, Denmark
[2]Section for Composite Materials, Department of Wind Energy, Technical University of Denmark, DK-4000 Roskilde, Denmark

Email: xzha@dtu.dk

Abstract. Cold-drawn pearlitic steel wires are nanostructured and the flow stress at room temperature can reach values above 6 GPa. A typical characteristic of the nanostructured metals, is the low ductility and thermal stability. In order to optimize both the processing and application of the wires, the thermal behaviour is of interest. This has been studied by annealing the wires for 1h at temperatures from ambient temperature to 300 ℃ (573 K). It is expected that a raising temperature may lead to structural changes and a reduction in strength. The change in strength is however not expected to be large. For this reason we have applied a very precise technique to measure the tensile properties of the wires from a strain of 10^{-4} to the maximum strain of about 1-2%. The structural changes have also been followed to estimate and relate strength changes to changes in structural parameters and morphology.

1. Introduction

High carbon cold drawn pearlitic steel wires have the highest strength of mass-produced steel products. Nowadays the wires with strength ranging from 2 to 4 GPa have been widely used in a variety of applications including cables for suspension bridges, steel cords for automobile tyres and springs, for their high elastic modulus, reliable and reproducible mechanical properties and low cost. Much effort has been put into improving the tensile strength of such wires to a maximum experimental value of 6~7 GPa (E/33) [1, 2], much higher than the maximum flow stresses of polycrystalline bcc metals which range from E/100 to E/50 [3]. The mechanical properties including strength and ductility can be adjusted by low temperature annealing which is also used to reduce a hazardous macro tensile residual stress introduced during the drawing process. The wires have limited ductility (normally a total elongation around few percent), and the uniform elongation is small at the order of 0.3%~0.5%, which makes it difficult to determine accurately the stress-strain dependency of importance for the various applications of the wires.

The influence of temperature on the mechanical property is of interest for both scientific and technological communities and many investigations have covered the effect of annealing on the mechanical properties of drawn wires [2, 4-6]. However, these investigations focus mainly on temperatures above 200 ℃ and limited work has been done on the quantitative stress and strain relationship due to the difficulty of precise strain measurement as the wire diameters are small of the order of 0.1 mm (wire diameter range: 0.175 - 0.5 mm), also the high sensitivity of total elongation measurements to the extensometer knife/tensile axis alignment. We have therefore developed an in-house testing method for precise strain measurements, but the present study will focus on the changes in the mechanical response of wires annealed up to 300 ℃, and the mechanisms behind.

2. Experimental methods

A high-strength near-eutectoid steel with a carbon content of 0.8 wt.% supplied by NV Bekaert SA (Zwevegem, Belgium) Technology Center Laboratory was used in this study. The specimens were taken at the intermediate step of the overall wire drawing process from the as-patented wire (1.26 mm) to the final drawn wire (0.20 mm). The wire diameter is 0.332 mm with the drawing strain of 2.67. The wires were annealed at 100, 150, 200, 260 and 300 ℃ (373 - 573 K) for 1 hour and tested using in-house developed capstan grips. The tensile test method for wires was according to ISO 6892-1:2009. An Instron electro-mechanic test machine was used with a 1000 N load cell and a Fiedler laser scanner extensometer, as shown in figure 1. Capstan type grips were used for holding the wire (figure 1). The capstan grips were designed in house for static and fatigue testing of high strength steel wires. The wires were wound on the capstan grips and clamped at about 250 degrees from the wires entering point. Thus a part of the tensile load is transferred by friction over a long distance before the stress concentration at the clamp. The capstan diameter was ø90 mm, and the wire runs in a groove on the surface of the grip. The initial center-to-center grip distance was 300 mm.

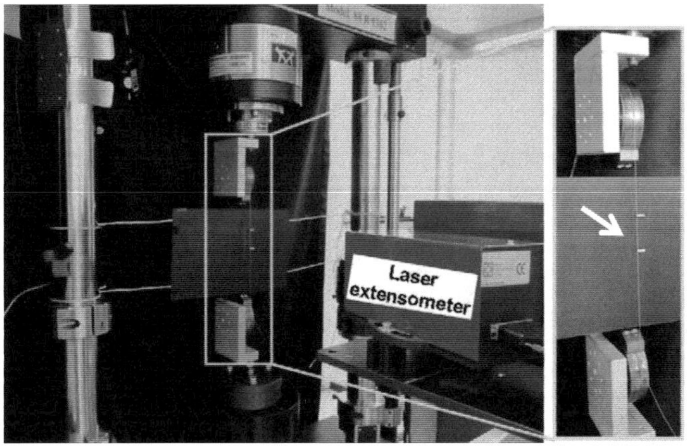

Figure 1. Tensile test setup showing loading capstans, strain labels and laser extensometer for the strain measurement of wire indicated by the white arrow

The strain was measured with a Fiedler P50 parallel scanner laser extensometer. A laser beam driven by a rotating prism, scans continuously along the axis of the sample length. Diffuse scattering of the laser light occurred at the strain labels. The scattered laser light was observed by a receiver. As the laser crosses the strain labels, the intensity of scattered light changes. The time of these changes was measured by a specially developed multi-stop counter. Because the speed of the rotating prism is known, the measured time can be used to calculate the position of the strain labels in the observation area, see figure 2a. Figures 2b and 2c show how the strain is defined. The laser extensometer scanned the distance between two adhesive strain labels attached to the wire (see figure 2b), and the initial gauge length L_0 is 20 mm ±1 mm. Since the extensometer measures the relative change in time between passing the two strain labels, the accuracy of initial gauge length is not critical for measuring relative elongation.

The tensile test setup allowed the proof stress at different offsets (0.05, 0.1 and 0.2%) to be determined accurately and thereby the strain hardening behavior. Other properties of interest were the ultimate tensile strength, the uniform/total elongation and the elastic modulus, see figure 3.

For the observation in a JEOL3000 field emission gun transmission electron microscope at 300 kV, the wires were ground, electro-polished with a double-jet sample-making method and then ion-cleaned in a low angle ion milling and polishing system (Fischione instruments) to remove the oxidation layer introduced during electro-polishing.

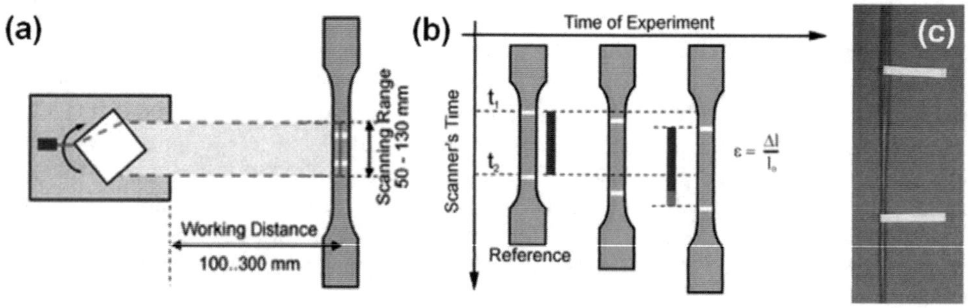

Figure 2. Principle of laser scanning extensometer (a), strain measurement by laser scanning and adhesive labels (b) attached to the wire (c).

3. Results

Nominal stress-strain curves for annealing at temperatures in the range from 20 ℃ to 300 ℃ (293 K - 573 K) are shown in figure 4. These curves can be divided into three groups based on the annealing temperature: (i) RT - 100 ℃ where the temperature effect is small; (ii) 100 – 200 ℃ showing a significant increase in flow stresses but only a small change in UTS and total elongation; (iii) 200 – 300 ℃ showing a decrease in strength with increasing temperature and no change in elongation to 260 ℃ and a significant increase to a total elongation of 5.5% reached at 300 ℃. The observed changes in mechanical properties with increasing annealing temperature are summarized in figure 5. The microstructural evolution with increasing annealing temperature has been followed by transmission electron microscopy focusing on the interlamellar spacing (ILS) between the cementite lamellae and the density and arrangement of dislocations. These parameters are chosen as they determine the contribution to the flow stress from boundary and dislocation strengthening, respectively. In the original wire, the interlamellar spacing: 28 nm, which is the sum of thickness of ferrite (23 nm) and thickness of cementite lamellae (5 nm) [7, 8]. The dislocation structures in the ferrite lamellae are single dislocations and dislocation cell boundaries, with a dislocation density ~ 10^{16} m^{-2} at the drawing strain of 2.67. The dislocation structure is shown in figure 6. The dislocations are spread in the ferrite lamellae, with the two ends of the line located at the ferrite/cementite boundaries [8], i.e., threading dislocations. During the low-temperature annealing process below 200 ℃, the morphology and structural parameters including ferrite thickness and cementite thickness are almost the same as those in the original wire. Further increase of the annealing temperature gradually changes the microstructure and the differences can be observed in the annealed wire at 300 ℃: the slight re-precipitation/spheroidization of cementite, a little change in the interlamellar spacing and a relatively small decrease in dislocation density.

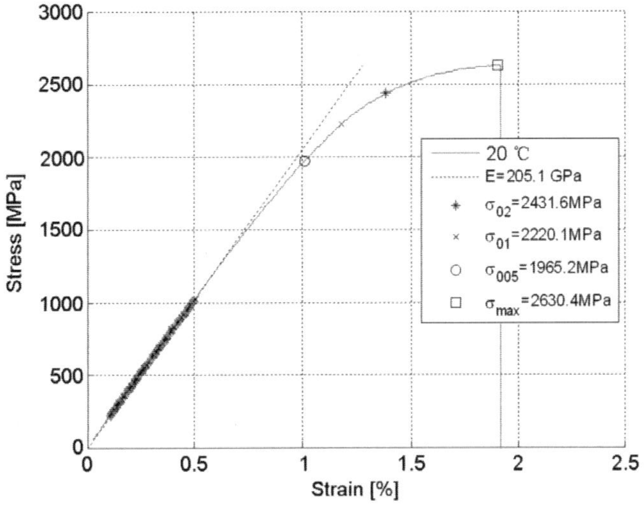

Figure 3. Demonstration of the determination of flow stress at 0.05% (σ_{005}), 0.1% (σ_{01}), 0.2% (σ_{02}) and UTS (σ_{max}).

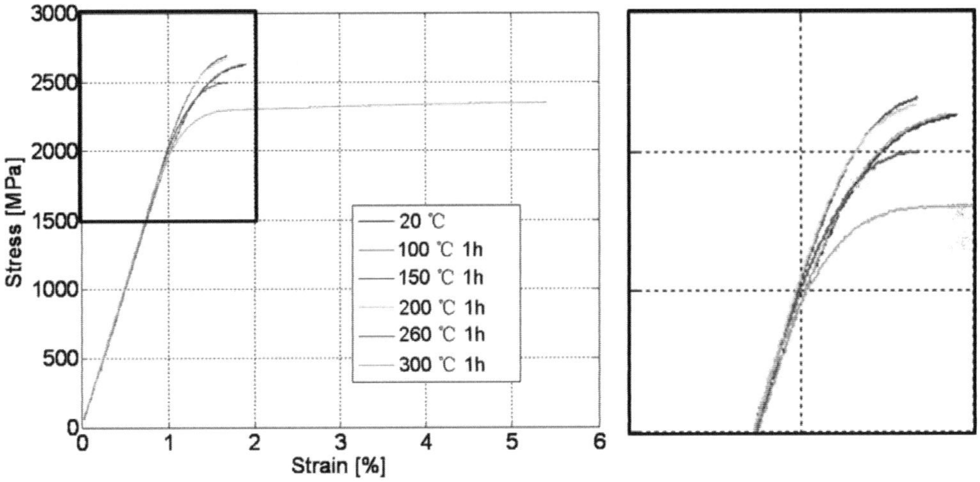

Figure 4. Tensile stress-strain curves of the wires. The area in the black frame is magnified to the right.

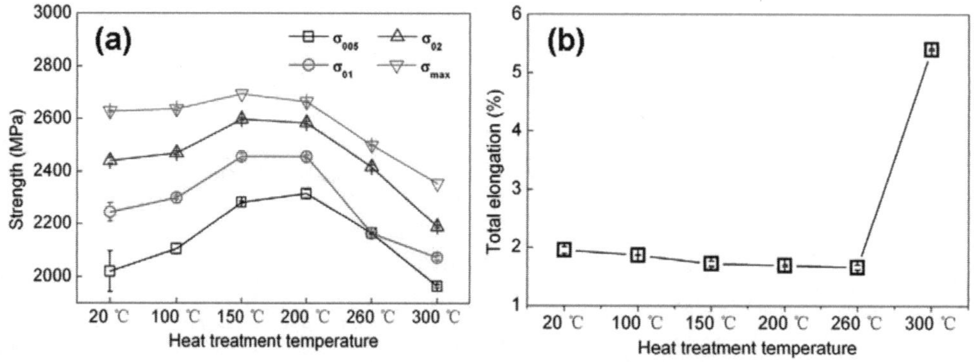

Figure 5. 0.05, 0.1 and 0.2% and maximum stress (a) and total elongation of the wires (b) versus the annealing temperature. The annealing time is 1 hour.

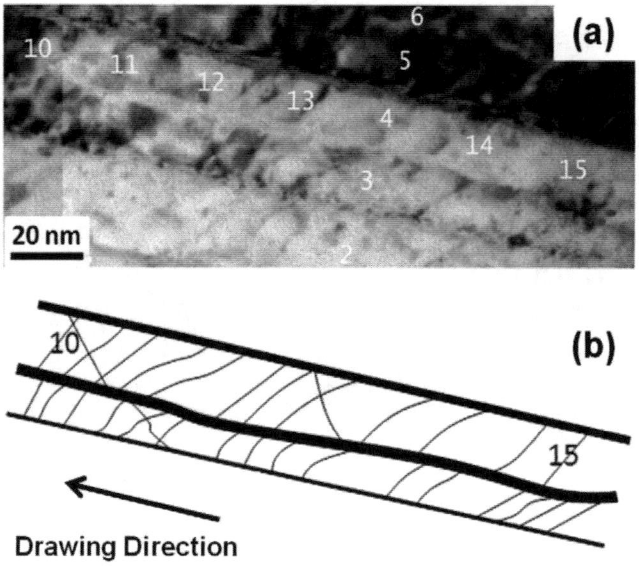

Figure 6. TEM micrograph of the wire of a drawing strain of 2.67 (a) and a sketch of the dislocation structure in the ferrite lamellae (b) showing threading dislocations [8].

4. Discussion

The mechanical data show a significant effect of annealing temperature: a strength increase in the temperature range lower than 200 ℃ and a strength decrease in the temperature range higher than 200 ℃. These changes will be analyzed based on the three strengthening mechanisms we proposed before: boundary strengthening related to the distance between cementite lamellae, dislocation strengthening related to the dislocation density in the ferrite lamellae and solid solution hardening related to the carbon concentration in the ferrite lamellae based on the change of thickness of cementite lamellae [8].

During the low temperature annealing process, C and N atoms dissolved in ferrite will migrate to the dislocations. To analyze their influences on the strength, it is necessary to examine the interaction energies between C and N atoms and the dislocations in steel. From the values reported, 0.78 eV/atom for C [9] and 0.47 eV/atom for N [10], it is reasonable to assume that the effect of dislocation pinning of C atoms is more intense than that of N atoms and in the present study we only consider the dislocation pinning of C atoms.

For the strength increase in the temperature range lower than 200 ℃, it is most notable for the proof stress at small strains, and relatively small in UTS. Since the low-temperature annealing below 200 ℃ for 1h does not bring enough detectable changes in the structural parameters in the TEM to cause the strength increase based on the above three strengthening mechanisms; this strength increase therefore points to the pinning of mobile dislocations/dislocation sources by C atoms/precipitates, as the other possibilities such as removal of mobile dislocations seems less plausible. This is because for steels the starting temperature for the dislocation movement is estimated to be 250 ℃ and the relief of macro residual stress introduced by quenching in steels starts from around 250 ℃ [11]. The pinning of the mobile dislocations/dislocation sources brings more carbon into the ferrite and this has been analyzed by the three dimensional atom probe tomography (3D-APT) [12]. When the dislocations and dislocation boundaries are taken into account, the maximum solubility of carbon in the ferrite also depends on the dislocations/dislocation boundaries in the ferrite lamellae, as it is well known that pipe diffusion of carbon occurs along the edge dislocation line and that dislocation boundaries are easy paths compared with interstitial diffusion in the pure ferrite matrix. Also the rate of dislocation boundary diffusion increases with the misorientation angle. It is reported that in the iron containing 0.011 wt. % carbon in solution, carbide precipitation occurs around 160℃, with the continuous growth at higher temperatures [13]. This is possible in the present case but difficult to detect by the technique we are using because the precipitation location is preferably the original cementite which decompose during the drawing processes and also taken into account the small spacing of ferrite lamellae. This means the thickness of the cementite lamellae may not change but the carbon content in the deformed cementite lamellae may increase. This process may be described as cementite decomposition/precipitation. In the future, the chemical composition change of the cementite lamellae with the annealing temperature will be analyzed by the advanced analytical TEM.

For the strength decrease of the annealed wires at temperatures 200℃ and 300℃, taken into account the three strengthening mechanisms, the key factors appears to be the reduction of carbon content in the ferrite by cementite precipitation/growth/spheroidization and the decrease of dislocation density in the ferrite lamellae, as little change in the distance between cementite lamellae has been observed. The dislocation density decrease may speed up by the increase of dislocation mobility with temperature and by the unpinning of the carbon atoms. In the future, the dislocation density will be quantified in the annealed wires which will contribute to the quantification of other mechanisms for the strength contribution.

Although the modern techniques such as 3D atom probe can provide the detailed information of the carbon atom distribution, the microstructure in the drawn pearlitic steel wire is so heterogeneous [14-25] and the measured carbon concentration varies greatly from one place to another. It is thus very difficult to conclude whether the carbon concentration in ferrite lamellae increases or not during the annealing up to 200 ℃ [4] and how the carbon concentration in ferrite lamellae influences the dislocation density during the annealing up to 300 ℃. For the future quantitative investigation of the carbon atom distribution and its redistribution during the low temperature annealing, the corresponding microstructural observation together with the 3D atom probe characterization of the same areas is demanded. This kind of quantitative investigation also needs several reproducible samples as the 3D atom probe characterization is destructive. To pick up the reproducible samples from the heterogeneous microstructure in the drawn wire is very difficult but becomes possible based on our recent findings that the structural evolution during the drawing process is hierarchical as the structural variations have their cause in a different macroscopic orientation of the cementite in the initial patented structure with respect to the wire axis and that the through-diameter variations subdivide the lamellar structure into two distinctly different types: one (called A_A) has a smaller interlamellar spacing and smaller dislocation density than the other (called A_BC) [8].

5. Conclusion

The changes in mechanical behavior and microstructure have been investigated for pearlitic steel wires annealed for 1h in the temperature range from 20 to 300 ℃ (293 – 573 K). The conclusions are as follows:

- By applying a very precise tensile test setup it is possible to measure the proof stress and the strain hardening from the beginning of plastic deformation to the maximum stress and strain.
- The proof stress at small offsets increases significantly with temperature in the range of 20 - 200 ℃ and decreases at temperatures above 200 ℃. These changes are related to interactions between carbon atoms in solution and the dislocation structure as the changes in strengthening parameters such as the interlamellar spacing and the dislocation density are small.
- The evolution of mechanical properties shows that annealing at 300 ℃ significantly improves the strain hardening behaviour and the total elongation with a decrease in strength of ~ 10%.

Acknowledgements

The authors thank NV Bekaert SA Technology Center Laboratory (Zwevegem, Belgium) for the supply of the pearlitic steel wires used in this investigation and Dr. X Huang for his valuable comments during preparation of the manuscript. X. Zhang and N. Hansen gratefully acknowledge the support from the Danish National Research Foundation (Grant No. DNRF86-5) and the National Natural Science Foundation of China (Grant No. 51261130091) to the Danish–Chinese Center for Nanometals, within which this work has been performed.

References

[1] Zhang X, Hansen N, Godfrey A and Huang X 2014 *35th Risø International Symposium on Materials Science: New Frontiers of Nanometals (Risø, Denmark, 1-5 September 2014)* pp 153–170
[2] Li Y J, Choi P, Goto S, Borchers C, Raabe D and Kirchheim R 2012 *Acta Mater.* **60** 4005
[3] Gil Sevillano J 1986 *Acta Mater.* **34** 1473
[4] Hong M H, Reynolds W T, Tarui T and Hono K 1999 *Metall. Mater. Trans. A* **30** 717
[5] Zelin M 2002 *Acta Mater.* **50** 4431
[6] Nam W J, Song H R and Park D B *Mater. Sci. Forum* **539-543** 4632
[7] Zhang X, Godfrey A, Huang X, Hansen N, Liu W and Liu Q 2009 *30th Risø international symposium on materials science: nanostructured metals – fundamentals to applications (Risø, Denmark, 7-11 September 2009)* 409
[8] Zhang X, Godfrey A, Huang X, Hansen N and Liu Q 2011 *Acta Mater.* **59** 3422
[9] Gridnev V N and Gavriliyuk V G 1982 *Phys. Met.* **14** 531
[10] Petarra D P and Beshers D N 1965 *Acta Metall.* **15** 791
[11] Kelly A 1966 *Strong solids* (Clarendon Press, Oxford) p 104
[12] Li Y J, Choi P, Borchers C, Westerkamp S, Goto S, Raabe D and Kirchheim R 2011 *Acta Mater.* **59** 3965
[13] Wagenblast H, Fujita F E and Damask A C 1964 *Acta Metall.* **12** 347
[14] Zhang X, Hansen N, Godfrey A and Huang X 2012 *33rd Risø international symposium on materials science: Nanometals – status and perspectives (Risø, Denmark, 3-7 September 2012)* 407
[15] Zhang X, Godfrey A, Hansen N and Huang X 2013 *Acta Mater.* **61**, 4898
[16] Zhang X 2009 *Quantitative investigation of microstructural evolution during the cold wire-drawing of a pearlitic steel wire and its relationship with mechanical properties* (Beijing: Tsinghua University)
[17] Hosford W F 1964 *Trans. Met. Soc. AIME* **230** 12
[18] Langford G 1977 *Metall. Trans. A* **8** 861
[19] Zhang X, Godfrey A, Liu W and Liu Q 2010 *Acta Metall Sin.* **46** 141

[20] Zhang X, Godfrey A, Hansen N, Huang X, Liu W and Liu Q 2010 *Mater. Charac.* **61** 65
[21] Toribio J and Ovejero E 1998 *Scripta Mater.* **39** 323
[22] Zhang X D, Godfrey A, Liu W and Liu Q 2011 *Mater. Sci. Techno.* **27** 562
[23] Gridnev V N, Gavrilyuk V G, Dekhtyar I Y, Meshkov Y Y, Nizin P S and Prokopenko V G 1972 *Phys. stat. sol. A* **14** 689
[24] Languillaume J, Kapelski G and Baudelet B 1997 *Acta Mater.* **45** 1201
[25] Zhang X, Andrew G, Liu W and Liu Q 2009 *Heat Treatment of Metals* **9** 1

Orientation-dependent recrystallization in an oxide dispersion strengthened steel after dynamic plastic deformation

Z B Zhang[1,4], N R Tao[2,4], O V Mishin[1,4] and W Pantleon[3,4]

[1] Section for Materials Science and Advanced Characterization, Department of Wind Energy, Technical University of Denmark, Risø Campus, 4000 Roskilde, Denmark
[2] Institute of Metal Research, Chinese Academy of Science, Shenyang 110016, China
[3] Section for Materials and Surface Engineering, Department of Mechanical Engineering, Technical University of Denmark, 2800 Kgs. Lyngby, Denmark
[4] Sino-Danish Center for Education and Research

E-mail: zhenbo.zhang@manchester.ac.uk

Abstract. The microstructure of the oxide dispersion strengthened ferritic steel PM2000 has been investigated after compression by dynamic plastic deformation to a strain of 2.1 and after subsequent annealing at 715 °C. Nanoscale lamellae, exhibiting a strong ⟨100⟩ + ⟨111⟩ duplex fibre texture, form during dynamic plastic deformation. Different boundary spacings and different stored energy densities for regions belonging to either of the two fibre texture components result in a quite heterogeneous deformation microstructure. Upon annealing, preferential recovery and preferential nucleation of recrystallization are found in the ⟨111⟩-oriented lamellae, which had a higher stored energy density in the as-deformed condition. In the course of recrystallization, the initial duplex fibre texture is replaced by a strong ⟨111⟩ fibre recrystallization texture.

1. Introduction

Oxide dispersion strengthened (ODS) steels are considered to be promising structural materials for the next-generation fission and fusion reactors because of their excellent resistance to both irradiation damage and high-temperature creep [1,2]. It has been suggested that irradiation induced swelling might be reduced in materials with well-refined microstructures, and that these materials may therefore exhibit improved irradiation tolerance compared to their coarse-grained counterparts [3,4]. One way to refine the microstructure is via plastic deformation when original grains are subdivided by deformation-induced dislocation boundaries [5].

In previous studies [6,7], dynamic plastic deformation (DPD) at high strain rates (10^2-10^3 s^{-1}) [8] has been demonstrated to be more efficient in refining the microstructure of a ferritic/martensitic steel than deformation at low strain rates. In the present work, the microstructure and the texture of a ferritic ODS steel PM2000 are investigated after DPD to a strain of 2.1. As the service temperatures of ODS steels in reactors are expected to be rather high, the thermal stability of the microstructure obtained by DPD is analysed focusing on orientation-dependent recovery and recrystallization taking place during annealing.

2. Experimental

A ferritic steel PM2000 (see Table 1) with oxide nanodispersoids was received in the form of a hot-extruded rod with a diameter of 13 mm [9]. Cylindrical specimens with a diameter of 6 mm and height of 9 mm were machined with their cylinder axis along the extrusion direction. The specimens were compressed by 5 passes at room temperature by DPD at a strain rate of $10^2 - 10^3$ s^{-1} to an equivalent strain of 2.1 [10]. The deformed material was then annealed isothermally at 715 °C for different durations.

Table 1. Nominal chemical composition (wt. %) of PM2000 [10].

Cr	Al	Ti	Y_2O_3	Fe
20	5.5	0.5	0.5	Bal.

The microstructure of the deformed and annealed samples was investigated by electron backscatter diffraction (EBSD) in a Zeiss Supra 35 FEGSEM equipped with a Channel 5 system. Due to the limited angular resolution of the EBSD technique [11,12], misorientation angles θ less than 2° were not considered in the analysis. Boundaries with misorientation angles θ between 2° and 15° were defined as low angle boundaries (LABs), while boundaries with msiorientation angles θ larger than 15° were classified as high angle boundaries (HABs). The volume fraction of the different texture components were calculated allowing a 10° orientation deviation from the exact ⟨100⟩ and ⟨111⟩ fibres.

The recrystallized volume fraction was determined by identifying individual recrystallized grains as regions at least partly surrounded by HABs, having an equivalent circular diameter (ECD) above 3 μm and an internal point-to-point misorientation below 2° [13]. The boundary area density S_V was determined from the EBSD data [14] and the energy density stored in the form of boundaries in the deformed or recovered microstructure was calculated as $u = \gamma S_V$. The surface energy of HABs was assumed to be $\gamma_{HAB} = 617$ MJ/m^2 [15], whereas the surface energy of LABs with misorientation angles $\theta < \theta_{cr} = 15°$ was calculated using the Read-Shockley equation [16]:

$$\gamma_{LAB} = \gamma_{HAB} \frac{\theta}{\theta_{cr}} \left(1 - \ln \left(\frac{\theta}{\theta_{cr}} \right) \right) \quad (1)$$

3. Results

The orientation map obtained by EBSD (figure 1a) from PM2000 after DPD to a strain of 2.1 shows a nanoscale deformation structure with lamellar boundaries almost perpendicular to the compression axis (CA). The microstructure is substantially refined by DPD as is evident from the spacing $d_{\theta>2°}$ measured along the CA between boundaries with $\theta > 2°$, which decreased from 580 nm in the as-received material to 76 nm after DPD (see table 2). The fraction of LABs (f_{LAB}) in this deformed microstructure is 52%. It is apparent that the map contains mainly two types of regions which appear either in red or in blue, corresponding to ⟨100⟩ and ⟨111⟩ fibre texture components, respectively.

The microstructural evolution during annealing after DPD is illustrated by orientation maps for the samples annealed at 715 °C for 10, 20, and 80 min (figure 1b-d). It is seen that the lamellar morphology observed in the as-deformed material is retained after 10 min of annealing (figure 1b). The fraction of LABs after 10 min of annealing is similar to that in the as-deformed condition (see table 2). However, during annealing the microstructure coarsens and the average spacing $d_{\theta>2°}$ increases from 76 nm in the DPD sample to 176 nm after 10 min at 715 °C. The material is still mainly in the recovered state, with only a single recrystallization nucleus found in the characterized area, which accounts for 1% of the volume fraction. After 20 min of annealing, the volume fraction of recrystallized material (f_{rex}) is 68% and the average boundary spacing $d_{\theta>2°}$ is 750 nm. After annealing for 80 min, the material is almost fully recrystallized (f_{rex}=97%) with $d_{\theta>2°}$ = 3750 nm. The fraction of LABs in this almost fully recrystallized microstructure is only 19% (cf. table 2).

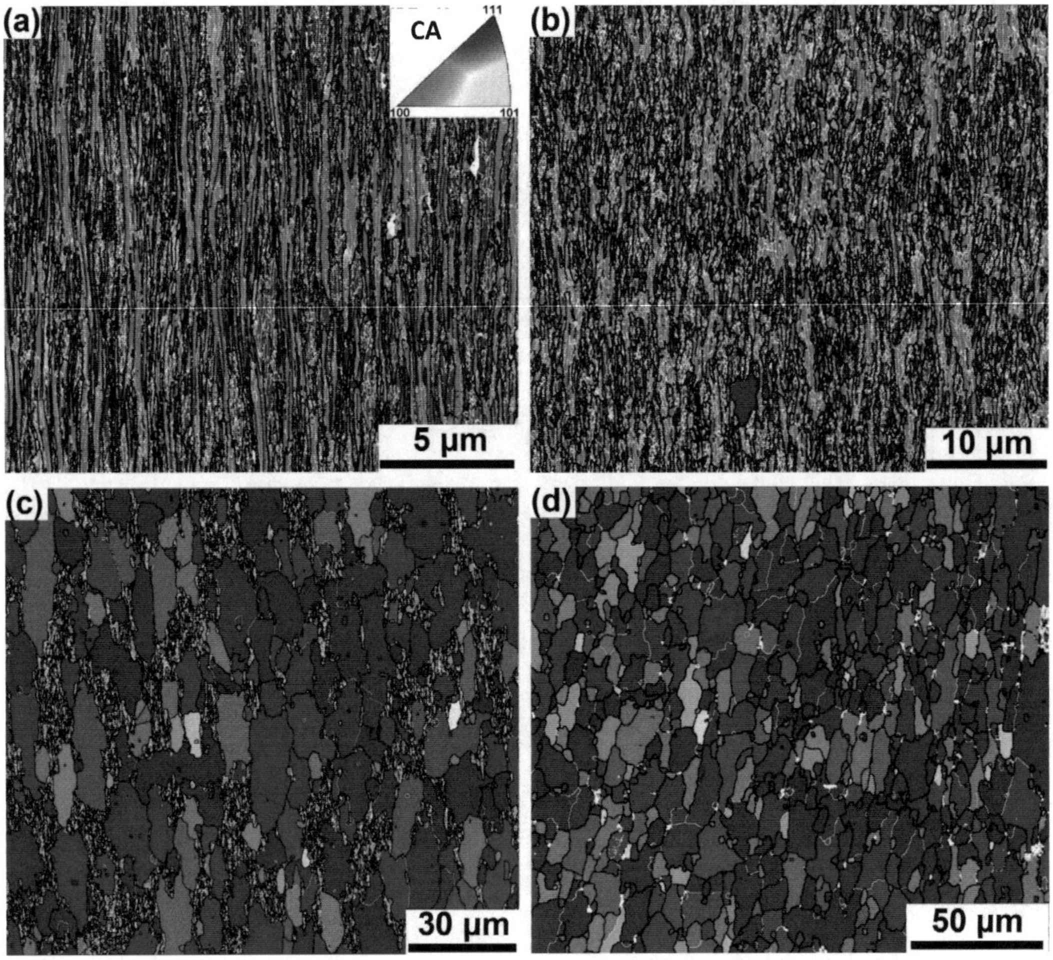

Figure 1. Orientation map obtained by EBSD from the longitudinal section of PM2000: (a) after compression by DPD to a strain of 2.1, and after subsequent isothermal annealing at 715 °C for (b) 10 min, (c) 20 min and (d) 80 min. The maps are coloured according to the crystallographic direction along the CA (see the inset). Black and white lines represent HABs and LABs, respectively. The CA is horizontal.

Table 2. Microstructural parameters of PM2000 after DPD and subsequent annealing at 715 °C.

Annealing duration (min)	$d_{\theta>2°}$(nm)	f_{rex} (%)	f_{LAB} (%)
0	76	0	52
10	176	1	54
20	750	68	47
80	3750	97	19

The orientation maps provide evidence that the majority of recrystallized grains have a ⟨111⟩ direction along the CA (blue in figure 1b-d). In the sample annealed for 80 min (figure 1d), such ⟨111⟩-oriented grains occupy 67% of the area. The evolution of crystallographic texture is further illustrated by the inverse pole figures in figure 2. It is seen that the ⟨100⟩ + ⟨111⟩ duplex fibre texture in the as-deformed PM2000 evolves during annealing into a well-defined ⟨111⟩ fibre texture, whereas the strength of the ⟨100⟩ fibre texture component decreases gradually. From the volume fractions of the individual texture components in figure 3 it is apparent that during recrystallization the ⟨111⟩ fibre strengthens primarily at the expense of the ⟨100⟩ fibre, i.e. recrystallization in PM2000 after compression by DPD is orientation-dependent.

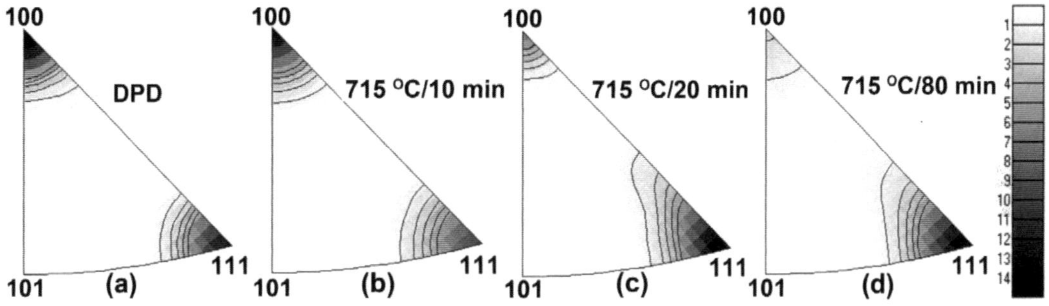

Figure 2. Inverse pole figures representing the crystallographic directions parallel to the CA in PM2000: (a) after compression by DPD to a strain of 2.1, and after subsequent isothermal annealing at 715 °C for (b) 10 min, (c) 20 min and (d) 80 min.

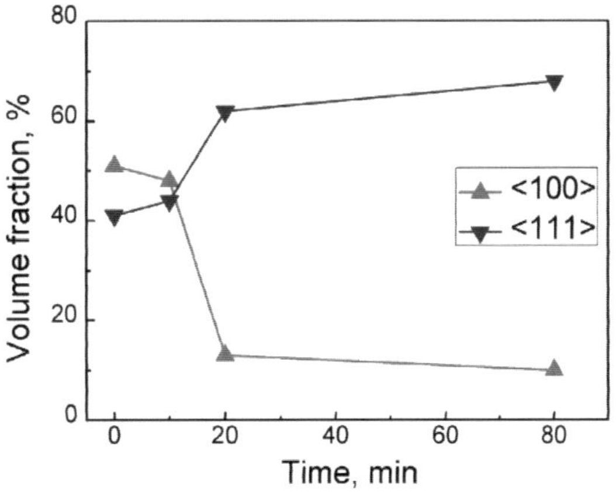

Figure 3. Volume fractions of the ⟨100⟩ and ⟨111⟩ fibre texture components in PM2000 after compression by DPD to a strain of 2.1 and subsequent isothermal annealing at 715 °C.

4. Discussion

Compression of PM2000 by DPD to a strain of 2.1 results in nanoscale lamellar structures with a boundary spacing of 76 nm as determined by EBSD along the CA. This spacing is considerably smaller than the value of 110 nm obtained in a modified 9Cr-1Mo steel compressed by DPD to a similar strain [7]. The smaller spacing in PM2000, as compared to the modified 9Cr-1Mo steel, is attributed to the finer starting microstructure as well as the enhanced structural refinement due to the oxide nanodispersoids in PM2000.

The lamellar structure shows a strong ⟨100⟩ + ⟨111⟩ duplex fibre texture, which is in agreement with earlier reports on textures developing during free compression in body centred cubic materials [6,7,17-19]. The development of such a duplex fibre texture can be rationalized in terms of Taylor-type models based on ⟨111⟩-pencil glide of dislocations in body centred cubic materials [20].

During annealing, lamellae of the different fibre texture components exhibit very different recrystallization behaviour, with the majority of the recrystallized grains belonging to the ⟨111⟩ fibre. This orientation-dependent recrystallization leads to a dominant ⟨111⟩ fibre texture in the recrystallized material. This texture evolution during recrystallization of PM2000 is similar to that previously observed during annealing of heavily compressed iron [18,19].

To understand the reason for the observed differences in the annealing behaviour of regions having different orientations, structural parameters of the deformed and recovered material in the present PM2000 sample were analysed separately for the two different texture components (see figure 4). Figure 4a demonstrates that after DPD the boundary spacing $d_{\theta>2^\circ}$ within the ⟨100⟩-oriented lamellae is larger (80 nm) than that within the ⟨111⟩-oriented lamellae (68 nm). After 10 min of annealing, however, this difference becomes smaller, while after 20 min of annealing, $d_{\theta>2^\circ}$ of the ⟨111⟩-oriented lamellae is even larger than that of the ⟨100⟩-oriented lamellae (note that for partially recrystallized samples, the reported spacings $d_{\theta>2^\circ}$ represent only recovered regions). Figure 4b shows the stored energy density u calculated from the EBSD data for the ⟨100⟩- and ⟨111⟩-oriented regions in the deformed and annealed samples. It is seen that in the deformed microstructure $u_{\langle 111\rangle}$ is considerably higher than $u_{\langle 100\rangle}$. This difference, however, becomes very small after 10 min at 715 °C (see figure 4b). After annealing at 715 °C for 20 min $u_{\langle 111\rangle}$ is even somewhat lower than $u_{\langle 100\rangle}$.

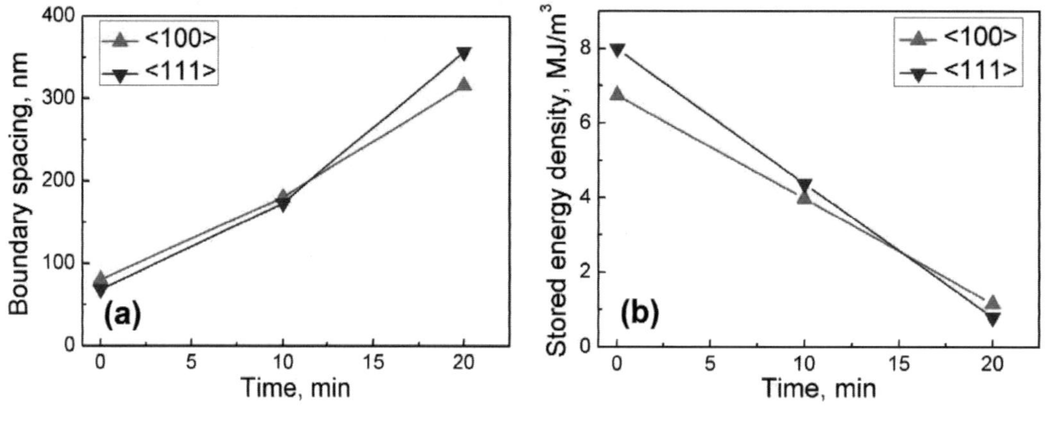

Figure 4. Evolution of structural parameters during recovery of PM2000 compressed by DPD to a strain of 2.1: (a) the average boundary spacing between boundaries with misorientation angles θ above 2° measured along the CA; (b) the stored energy density for the ⟨100⟩ and ⟨111⟩ fibre texture components. Both parameters are determined from the EBSD data for the deformed and recovered regions only.

The differences between the ⟨100⟩- and ⟨111⟩-oriented lamellae in the deformed microstructure are attributed to the difference in their Taylor factors during compression. Assuming pencil glide, the Taylor factor for body centred cubic crystals compressed along ⟨100⟩ is 2.1, while for compression along ⟨111⟩, the Taylor factor is 3.2 [21] which are respectively the lowest and highest values among all directions. This considerable difference between the two fibre texture components indicates that during compression, regions of the ⟨111⟩ component experience appreciably larger plastic slip than regions of the ⟨100⟩ component. Consequently, the work done to deform individual grains with a higher Taylor factor is larger than that to deform grains with a lower Taylor factor. Therefore, the stored energy density can also be expected to be larger in regions with a higher Taylor factor. It is believed that this higher stored energy in ⟨111⟩-oriented lamellae results in faster recovery (as seen in the faster reduction of the stored energy in figure 4b), also leading to preferential nucleation in these ⟨111⟩-oriented regions.

5. Conclusion

Compression by DPD to a strain of 2.1 significantly refines the initial microstructure of PM2000, resulting in nanoscale lamellae with a ⟨100⟩ + ⟨111⟩ duplex fibre texture. Orientation-dependent recrystallization takes place during annealing at 715 °C leading to strengthening of the ⟨111⟩ fibre texture component primarily at the expense of the ⟨100⟩ fibre texture component. As a result, a strong ⟨111⟩ fibre texture is present when recrystallization is almost complete. It is suggested that this orientation-dependent recrystallization is caused by a higher stored energy in the ⟨111⟩-oriented lamellae after DPD. During annealing these lamellae recover faster than lamellae of the ⟨100⟩ fibre texture component, which possibly causes the observed preferential nucleation of ⟨111⟩-oriented grains.

Acknowledgements

Financial support from the Sino-Danish Center for Education and Research is gratefully acknowledged. The authors are grateful to Prof. M. Heilmaier for providing the initial PM2000 sample. OVM also gratefully acknowledges the support from the Danish National Research Foundation (Grant No. DNRF86-5) and the National Natural Science Foundation of China (Grant No. 51261130091) to the Danish-Chinese Center for Nanometals.

References

[1] Odette G R, Alinger M J and Wirth B D 2008 *Ann. Rev. Mater. Res.* **38** 471
[2] Chant I and Murty K L 2010 *JOM* **62** 67
[3] Bai X M, Voter A F, Hoagland R G, Nastasi M and Uberuaga B P 2010 *Science* **327** 1631
[4] Song M, Wu Y D, Chen D, Wang X M, Sun C, Yu K Y, Chen Y, Shao L, Yang Y, Hartwig K T and Zhang X 2014 *Acta Mater.* 74 285
[5] Hansen N 2006 *Metall. Mater. Trans. A* **32** 2917
[6] Zhang Z B, Mishin O V, Tao N R and Pantleon W 2015 *Mater. Sci. Technol.* **31** 715
[7] Zhang Z B, Mishin O V, Tao N R and Pantleon W 2015 *J. Nucl. Mater.* **458** 64
[8] Li Y S, Tao N R and Lu K 2008 Acta Mater. **56** 230
[9] Schneibel J H, Heilmaier M, Blum W, Hasemann G and Shanmugasundaram T 2011 *Acta Mater.* 59 1300
[10] Zhang Z B, Mishin O V, Tao N R and Pantleon W 2014 *IOP Conf. Ser: Mater. Sci. Eng.* 63 012065
[11] Humphreys F J 2006 *J. Mater. Sci.* 36 3833
[12] Mishin O V, Östensson L and Godfrey A 2006 *Metall. Mater. Trans. A* **37** 489
[13] Wu G L and Jensen D Juul 2008 *Mater. Charact.* **59** 794
[14] Godfrey A, Hansen N and Jensen D Juul 2007 *Metall. Mater. Trans. A* **38** 2329
[15] Humphreys F J and Hatherly M 2004 Recrystallization and related annealing phenomena (second edition) (Oxford: *Elsevier*)
[16] Read W T and Shockley W 1950 *Phys. Rev.* **78** 275
[17] Hu H 1974 *Texture* **1** 233
[18] Dillamore I L, Katoh H and Haslam K 1974 *Texture* **1** 151
[19] Khatirkar R, Krishna K V M, Kestens L A I, Petrov R, Pant P and Samajdar I 2011 *ISIJ Int.* 51 849
[20] Hosford W F 2005 *Mechanical Behavior of Materials* (Cambridge University Press)
[21] Rosenber J M and Piehler H R 1971 *Metall. Trans.* **2** 257

IOP Conf. Series: Materials Science and Engineering **89** (2015) 012060 doi:10.1088/1757-899X/89/1/012060

The kinetics of grain growth in near-micrometre grain size copper produced by spark plasma sintering

K N Zhu[1], Q Ruan[1], A Godfrey[1]

[1]Key Laboratory of Advanced Materials (MOE), School of Material Science and Engineering, Tsinghua University, Beijing 100084, P.R. China

E-mail: awgodfrey@mail.tsinghua.edu.cn

Abstract. Samples of copper with a grain size in the near-micrometre regime and in a nearly fully recrystallized condition have been prepared by a spark plasma sintering (SPS) process. The thermal stability of these samples in the temperature range from 950 to 1050 °C has been investigated by an analysis of the kinetics of grain growth, based on microstructural measurements using electron backscatter diffraction. In the temperature range from 950 to 1000 °C the activation energy for grain growth is approx. 300 kJ mol^{-1}, corresponding to a value one and a half times the activation energy for self-diffusion in pure copper. In the temperature range from 1000 to 1050 °C the activation energy for grain growth is found to decrease, although annealing in this temperature range additionally results in non-uniform grain growth. For preparation of samples with a desired uniform grain size at present the best method appears to be via control of the temperature during the SPS process.

1. Introduction

It has been reported that metal materials with a microstructural scale in the near-micrometre regime have interesting mechanical properties, such as the presence of a yield point phenomena [1], annealing hardening and deformation softening [2], and enhanced mechanical strength [3]. The traditional route for the preparation of bulk samples with a near- or sub-micrometre scale involves severe plastic deformation followed by an annealing treatment. A complication in the analysis of the mechanical properties of such samples though is the presence in the microstructure of low angle deformation boundaries, and of a strong deformation texture. An alternative approach is the use of a spark plasma sintering (SPS) process [4,5] where it has been found that samples with a nearly fully recrystallized structure and almost random texture can be prepared.

For aluminium, grain size control of samples prepared by SPS can be achieved by selecting the initial powder size, as the native alumina film results in a dispersion of fine oxides around each original powder particle, thereby strongly pinning grain growth [4]. In copper such an approach is not possible, and instead it has been shown that grain size control in the near-micrometre regime can be achieved by the use of a fine initial powder size combined with sintering at different temperatures [6]. As an alternative to this annealing of as-sintered samples can also be considered as a method to achieve a desired grain size. It is known, however, that nanocrystalline materials may exhibit a remarkable resistance to grain growth [7] and hence investigation of the grain growth kinetics of SPS is therefore of interest for the control of grain size in near-micrometre grain size samples prepared using the SPS process.

2. Experimental

Based on previous work [6] the samples used in this investigation were prepared using a sintering temperature of 800 °C for a duration of 180 s, using an applied load of 50MPa. All samples were prepared using a SPS-1050 machine (Sumitomo Heavy Industries Ltd., Japan), starting with spherical copper powder (99.5% purity) with an initial particle size of approx. 0.5 μm. The density of each sample was measured using the Archimedes method. In each case the final density was typically above 98%. For the annealing treatments each sample was first sealed in a vacuum tube and evacuated to a pressure of 10^{-5} Pa. Annealing was carried out at either 950, 1000 or 1050 °C for different annealing durations in an air furnace (up to 14400 s), followed by cooling to room temperature in the furnace.

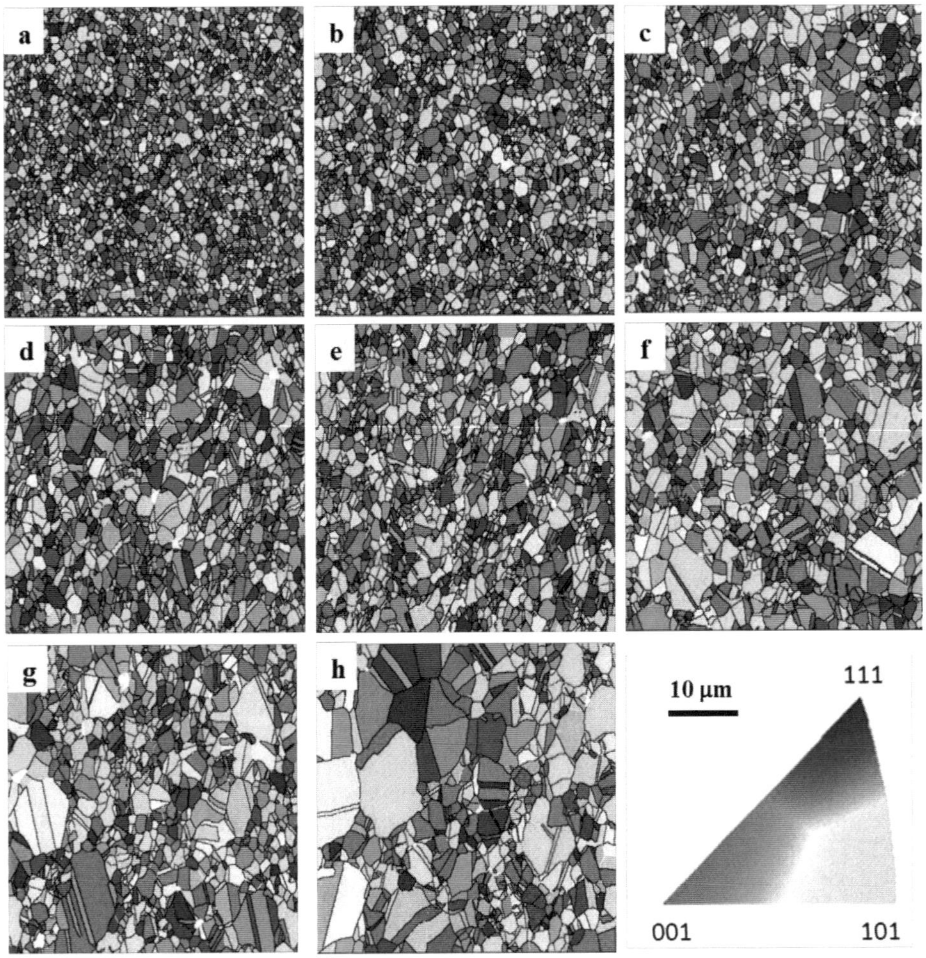

Figure 1. EBSD images of copper SPS samples processed under different annealing conditions: a) as-sintered sample, (b) 950 °C for 14400 s; (c) 1000 °C for 3600 s; (d) 1000 °C for 14400 s; (e) 1050 °C for 1800 s; (f) 1050 °C for 3600 s; (g) 1050 °C for 7200 s; (h) 1050 °C for 14400 s. In each case the colouring corresponds to the crystal direction parallel to the applied load during the SPS preparation of the samples.

For metallographic examination the annealed samples were first sectioned in a mid-plane, and then ground on SiC paper to 1000 grit. This was followed by a polishing treatment using argon gas in a cross-sectional polisher machine (JEOL IB-09020 CP, Japan). The microstructure and texture of the sintered samples were studied in the scanning electron microscope (SEM) and using electron backscatter diffraction (EBSD). All investigations were carried out in a plane containing the SPS-loading direction. For each annealing condition (combination of time and temperature) between five and seven EBSD maps were collected. The average grain size of each case D, was measured from the EBSD data using a line intercept method, taking a grain boundary misorientation angle definition of 15°. Note that in this study twin boundaries were excluded from the calculations of the grain size (i.e. were not counted as grain boundaries).

3. Results and discussion

A summary of the evolution of the microstructure in samples annealed under different conditions is given in figure 1. It is seen that the grain size increases as expected with annealing time and temperature. In all case the samples have a nearly random texture. The average grain size, D_0, for the as-sintered SPS sample (figure 1a) is approx. 0.7 μm. Only a small amount of grain growth is seen in the sample annealed at 950 °C for 14400 s (figure 1b). Annealing at a temperature of 1000 °C results in increased grain growth, with a clear increase in grain size seen after 3600 s annealing at this temperature (figure 1c). Annealing at a temperature of 1050 °C for 14400 s results in the development of a non-uniform grain size (figure 1h), with evidence of this already seen after 7200 s annealing at this temperature (figure 1f).

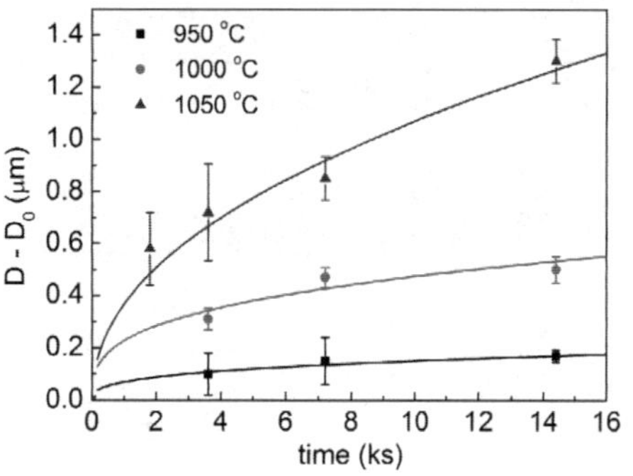

Figure 2. Evolution of the average grain size as a function of different annealing time at temperatures of 950, 1000, and 1050 °C.

The evolution of the average grain size under different conditions of annealing time and temperature from 950 to 1050 °C is shown in figure 2. Also shown in the figure are error bars corresponding to the standard deviation in the grain size measurements. It can be clearly seen that the grain size increases with annealing time and temperature. The graph confirms the visual observation that at 950 °C grain growth occurs very slowly, with the average grain size increasing by less than 0.2 μm after 14400 s annealing. At an annealing temperature of 1050 °C grain growth occurs more rapidly, with the average size increasing by more than 0.5 μm during the first 1800 s of annealing.

In order to determine the grain growth kinetics the data have been fitted at each temperature to the equation for classical grain growth kinetics [7-10] where the average grain size and annealing time are related by the equation:

$$D - D_0 = Kt^{1/n}, \tag{1}$$

where D is the mean grain size, D_0 is the initial grain size, K is a temperature dependent constant, t is the annealing time and n is the grain growth coefficient representing the grain growth behaviour (typically with values of between 2 and 4 for the annealing of metals).

The results are shown as the solid lines in figure 2. Values of n representing the grain growth kinetics at temperatures 950, 1000 and 1050 °C are found from the fitting to be 3.0, 3.1 and 2.2, respectively. Theoretically a grain growth exponent of n = 2 represents a classical uniform grain growth process driven by grain boundary curvature. In the present case the value of $n = 2$ at 1050 °C corresponds however to a process where non-uniform coarsening takes place.

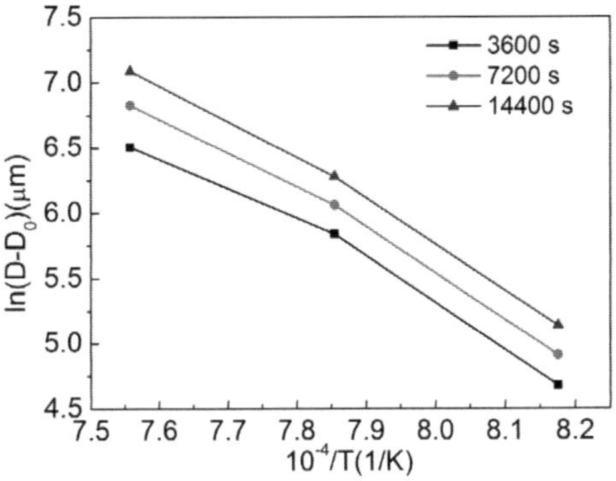

Figure 3. Plot of ln(D-D$_0$) versus 1/T for the Cu SPS samples annealed at durations of 3600, 7200, and 14400 s at 950, 1000, and 1050 °C.

For a constant value of n the activation energy, Q, for the grain growth process can be calculated using the standard assumption that K can be expressed in Arrhenius form according to [11,12]

$$K = K_0 \exp(-Q/RT), \tag{2}$$

where Q is the activation energy, T is the temperature in Kelvin, K_0 is the pre-exponential value, which we assumed here to be constant over the temperature range investigated, and R is the gas constant. Accordingly the value of Q can be calculated from a plot of ln(D - D_0) versus $1/T$ at fixed annealing durations, where the slope of the plot is $-Q/R$. The data for the present experiment are plotted in figure 3. Not surprisingly, given the change in n between 1000 and 1050 °C, the points do not lie on a straight line. Accordingly the activation energy has been calculated separately for the ranges of 950 to 1000 °C and 1000 to 1050 °C (values given in table 1). The value of approx. 300 kJ mol^{-1} for the range 950 to 1000 °C can be regarded as a reasonable estimate, as the grain growth exponent, n, is approximately constant in this range, whereas the lower value of 200 kJ mol^{-1} for the range 1000 to 1050 °C should be regarded as an upper bound, given the increase in n over this range.

Table 1. Activation energies for grain growth during annealing of SPS samples calculated from figure 3.

Annealing time	Activation energy (kJ mol^{-1})	
	950 – 1000 °C	1000 – 1050 °C
3600 s	301	187
7200 s	298	215
14400 s	295	228

The activation energy can also be calculated by determining the annealing time required to achieve the same increase in grain size (i.e., for fixed values of D - D_o), in which case the effect of the change in the grain growth exponent, n, at the highest temperature can be explicitly taken into account. For a fixed change in grain size we can write

$$\exp(-Q/RT_2)t_2^{1/n_2} = \exp(-Q/RT_1)t_1^{1/n_1}, \tag{3}$$

from which the activation energy in the range T$_1$ to T$_2$ can be calculated as

$$Q_{T_1-T_2} = R\left[\frac{\ln t_1}{n_1} - \frac{\ln t_2}{n_2}\right] \Big/ \left[\frac{1}{T_1} - \frac{1}{T_2}\right]. \tag{4}$$

Results for this second method are presented in table 2, where is seen that the activation energy decreases from approx. 300 kJ mol^{-1} to approx. 140 kJ mol^{-1} as the annealing temperature is increased from 1000 to 1050 °C. It should be remembered, however, that the latter is still only an apparent activation energy, as the value of n changes at some unknown temperature in the range 1000 to 1050 °C. It is nevertheless instructive to compare these values with those for grain boundary and self-diffusion in high purity copper, which are approx. 120 and 200 kJ mol^{-1}, respectively [13]. The activation energy for grain growth in the lower temperature range is much higher than either of these values, with the activation energy between these value in the higher temperature range. In either case an activation energy near that for grain boundary diffusion would be expected for a normal grain growth process in a pure metal.

Table 2. Different annealing times to achieve the same increase in grain size and correspondingly calculated activation energies.

D-D_0 (μm)	Time to achieve fixed increase in grain size (s)			Activation energy (kJ mol^{-1})	
	950 °C	1000 °C	1050 °C	950-1000 °C	1000-1050 °C
0.5	-	11818	1970	-	134
0.4	-	5909	1212	-	135
0.3	-	2424	606	-	144
0.17	14384	485	-	288	-
0.1	2909	75	-	306	-

The high value of the activation energy at 950 and 1000 °C, together with calculated values for the grain growth exponent and the microstructural observations of non-uniform coarsening at 1050 °C suggest the possibility of either an unknown impurity or micro-porosity in the SPS samples [14], resulting in either pinning or drag of the grain boundaries below a certain temperature. Preliminary microstructural examination in a field-emission gun SEM has not revealed the presence of fine oxides or pores, though it is possible such features, if present, can only be seen using higher resolution imaging.

It should be pointed out that non-uniform coarsening seen at 1050 °C implies that some grains do indeed grow at a faster rate at this temperature, with a correspondingly lower activation energy for grain growth. This again suggests some pinning or solute breakaway transition, as non-uniform grain growth in pure metals is usually texture-related i.e., the rapid growth of an isolated with a very different orientation inside a strong texture [15]. The SPS samples, however, have a nearly random texture so such an explanation is unlikely to account for the onset of the non-uniform coarsening.

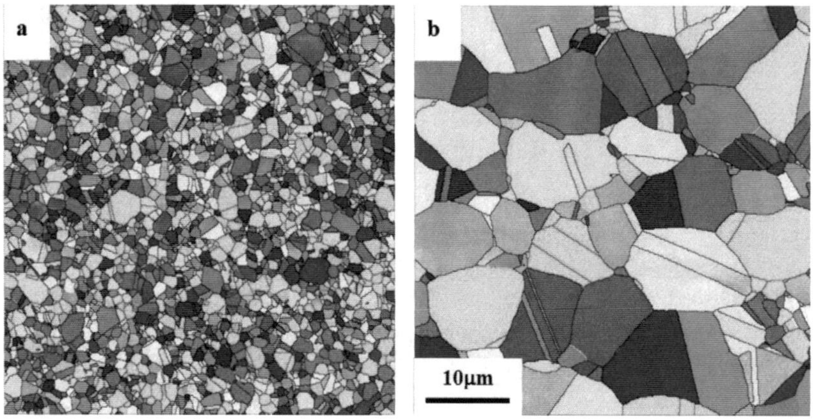

Fig. 4. EBSD images of Cu SPS samples prepared at different sintering temperatures: (a) 900 °C; (b) 1000 °C. In each case a sintering time of 180 s and an applied sintering load of 50 MPa were used.

The microstructure of samples sintered at 800 °C and then annealed can also be compared with those directly sintered at a higher temperature. Figure 4 shows the microstructure of samples prepared by SPS at temperatures of 900 and 1000 °C using a sintering time of 180 s and an applied pressure of 50 MPa. It is clearly seen that increasing the sintering temperature from 900 and 1000 °C results in a large change in grain size even with the short sintering duration of 180 s. Moreover the grains in the sample sintered at 1000 °C for 180 s have a similar size to those in the sample sintered at 800 °C and then annealed at 1050 °C for 14400 s, but with a more uniform size distribution. It will be interesting in future work to explore whether this is related to the effect of the sintering pressure on recrystallization during the early stages of sintering or whether the sintering pressure also affects the grain growth kinetics.

4. Summary and conclusions

Copper samples with a grain size in the near-micrometre regime and in a nearly fully recrystallized condition have been prepared by use of the SPS process, and the kinetics of grain growth in these samples has been studied. A high activation energy of approx. 300 $kJmol^{-1}$ (one and a half times larger than the activation energy for self-diffusion in copper) is found for grain growth in the temperature range from 950 to 1000 °C, confirming the relatively high thermal stability of the SPS samples. In the temperature range from 1000 to 1050 °C the activation energy for grain growth decreases and annealing in this temperature range results in non-uniform grain growth. For preparation of samples with a desired uniform grain size at present the best method appears to be via control of the temperature during the SPS process.

Acknowledgements

The authors gratefully acknowledge the support from the Danish National Research Foundation (Grant No DNRF86-5) and the National Natural Science Foundation of China (Grant Nos. 51261130091 and 50971074) to the Danish-Chinese Center for Nanometals, within which this work was performed.

References

[1] Tsuji N, Ito Y, Saito Y and Minamino Y 2002 *Scripta Mater.* **47** 893
[2] Huang X, Hansen N and Tsuji N 2006 *Science* **312** 249
[3] Kamikawa N, Huang X, Tsuji N and Hansen N 2009 *Acta Mater.* **57** 4198
[4] Le G M, Godfrey A, Hansen N, Liu W, Winther G and Huang X 2013 *Acta Mater.* **61** 7072
[5] Le G M, Godfrey A and Hansen N 2013 *Mater. Design* **49** 360
[6] Zhu K N, Godfrey A 2014 *Proc. 33rd Risø International Symposium on Material Science:New Frontiers of Nanometals* S Fæster et al. (Roskilde: Risø-DTU) p 537
[7] Lu K 1996 *Mater. Sci. Eng. R* **16** 161
[8] Gleiter H 1989 *Prog. Mater Sci.* **33** 223
[9] Gil F J, Picas J A, Manero J M, Forn A and Planell J A 1997 *J. Alloys Compd.* **260** 147
[10] Humphreys F J and Hatherly M 2004 *Recrystallization and Related Annealing Phenomena* (Oxford: Elsevier Science) p 122
[11] Ganapathi S K, Owen D M and Chokshi A H 1991 *Scripta Metall. et Mater.* **25** 2699
[12] Lu L, Tao N R, Wang L B, Ding B Z and Lu K 2001 *J. Appl. Phys.* **89** 6408
[13] Edalati K and Horita Z 2011 *Scripta Mater.* **64** 161
[14] Vandermeer R A and Hu H 1994 *Acta Metall.* **42** 3071
[15] Abbruzzese G and Lücke K 1986 *Acta Metall.* **34** 905

Studying the influence of triple junction energy and mobility on annealing processes

D Zöllner and P Streitenberger
Institute for Experimental Physics, Faculty for Natural Sciences, Otto von Guericke University Magdeburg, Germany

E-mail: dana.zoellner@ovgu.de

Abstract. The kinetics of grain growth in nanocrystalline two-dimensional polycrystals and thin films can be controlled by the triple junctions of the microstructure. In the present work the influence of energy and mobility of grain boundaries and their junctions on annealing is investigated by analytic theories and modified Monte Carlo Potts model simulations. To that aim each structural feature of a polygonal grain is assigned its own specific energy and finite mobility, which results in four different limiting cases of self-similar growth kinetics each characterized by its own metrical and topological properties.

1. Introduction

Grain growth is based on the fact that polycrystalline materials (Figure 1a) are thermodynamically unstable and undergo continuous grain coarsening driven by a reduction of the total Gibbs free energy of the system. Conventionally, it is assumed that only the grain boundaries contribute to the reduction of the Gibbs free energy via the decrease of the total grain boundary area. In particular, for the simplest case of normal grain growth it follows that only two characteristic parameters enter the theory namely the specific energy and mobility of the grain boundaries [1-3]. However, in recent years it has been shown that at very small average grain sizes that are typical for nanocrystalline materials, triple junction controlled motion has been observed during recovery [4] and triple lines as well as quadruple points of polyhedral grain networks may have a likewise significant effect on grain growth kinetics. In particular, the effect of finite junction mobilities on grain growth has been studied by means of experimental, theoretical and simulation methods (compare [5-9] and the literature within).

This is, in particular, also true for coatings, thin films and layers, where the thermal stability is a subject of great technological importance [10, 11]. In such systems it is often observed that upon annealing grain growth almost stagnates when the average grain size and the film thickness are of nanocrystalline character [3, 5]. In an outstanding paper, Barmak et al. [12] observed that during annealing of nanocrystalline Al and Cu thin films a stage of very slow grain growth is reached that is characterized by grain size distributions showing a large population of very small grains. As a possible cause of grain growth stagnation they investigated amongst other forces triple junction drag, which is, according to Gottstein, Shvindlerman and co-workers [3, 6, 8, 9], an effective rate limiting process especially at very small grain sizes. For this purpose the real 3D film was approximated as an ideal two-dimensional polycrystal, which was subjected to triple junction controlled grain growth by means of a multi-vertex front tracking simulation method (compare [12] and the references within).

However, two-dimensional grain growth under the condition of triple-junction drag has already been studied by means of vertex dynamics simulation methods earlier by Weygand et al. [13] and

Novikov [7]. Other two-dimensional and also three-dimensional simulation methods of triple-junction controlled grain growth have been developed in recent years. While Barrales-Mora [14] investigated 2D triple junction limited grain growth by network models, Zöllner and Streitenberger [15, 16] extended the MC Potts model in order to simulate triple and quadruple junction mobility controlled grain growth. On the latter basis, two-dimensional simulations of triple-junction mobility limited coarsening have been performed by the present authors in [17-19].

Based on our work on three-dimensional nanocrystalline grain growth [15], in the present work we attribute to each type of grain boundary junction of a 2D polycrystal its own specific energy and mobility (Figure 1b). Looking for self-similar, quasi-stationary scaling states of two-dimensional grain growth we find four types of growth kinetics, which are characterized by average growth laws obeying power law behavior and associated self-similar grain size distributions. The theoretical considerations are substantiated by simulation results using a modified Monte Carlo Potts model [16, 17].

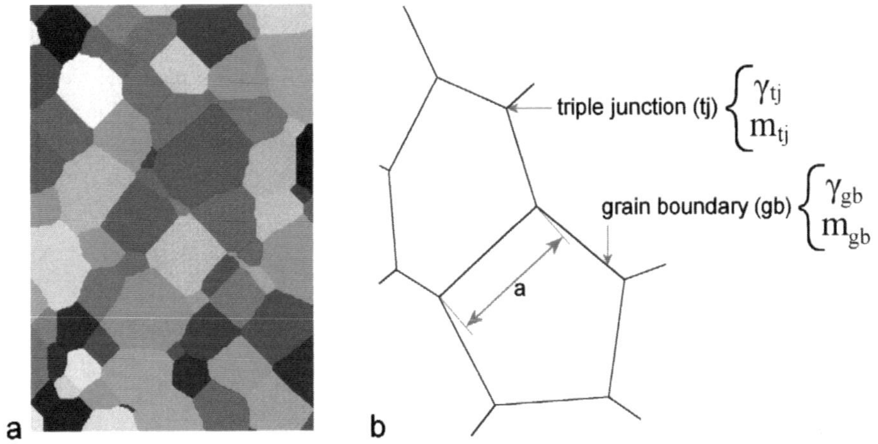

Figure 1. a – Detail of a grain microstructure obtained by grain growth under triple junction mobility and energy control; b – Sketch of two polyhedral grains highlighting the grain features and parameters of interest.

2. Theoretical background

In the following we assume that a two-dimensional polycrystalline grain ensemble can be represented by a polygonal network, where each grain may be approximated by an average N-gon as defined by Rios and Glicksman [20] (compare Figure 1b). Average N-gons, where N is the number of sides or connecting neighbors of a grain, are two-dimensional proxies for the average metric, energetic and kinetic behavior of real network grains. They have only two types of grain junctions, the grain boundaries (*gb*) and the triple junctions (*tj*), where we assume that each of them is characterized by its own specific Gibbs free energy γ_{gb} and γ_{tj}, respectively. Taking further into account that two grains share a grain boundary and three grains a triple point the total free energy of a polygonal network of N_g grains can be written as

$$G = \sum_i^{N_g} \left(\frac{\gamma_{gb}}{2} L_i + \frac{\gamma_{tj}}{3} N_i \right) = \sum_i^{N_g} \left(\pi \gamma_{gb} R_i + \frac{1}{3} \gamma_{tj} N_i \right). \tag{1}$$

Here γ_{gb} is the usual grain boundary energy, γ_{tj} the triple point energy or line tension per unit length [3, 8], L_i is the total boundary length, and N_i the number of triple points of the i-th grain. In particular, L_i is expressed by the grain area (A_i) equivalent radius using the relation $L_i = 2\pi\kappa \cdot R_i$. For sake of simplicity and without essential loss of general validity the form factor κ will be mostly set equal to one.

An estimation on the basis of the average N-gon defined by Rios and Glicksman [20] yields $\kappa = 1.050$. In Eq. (1) each grain is now characterized by the grain size R_i as a metrical property and the number of sides or neighbors N_i as a topological property.

Considering grain growth as a non-equilibrium dissipative process the reduction of the Gibbs free energy per unit time, that is $\dot{G} = \mathrm{d}G/\mathrm{d}t$, must be related to the generated dissipated heat per unit time Q. Following the approach developed by the present authors in [15] the total dissipated heat per unit time of all grains is given by the expression

$$Q = \sum_i^{N_g} \left(\frac{L_i v_{gbi}^2}{2m_{gb}} + \frac{N_i v_{tji}^2}{3m_{tj}} \right) = \sum_i^{N_g} \left(\frac{\pi}{m_{gb}} R_i + \frac{1}{3m_{tj}} N_i \right) \dot{R}_i^2 , \tag{2}$$

where m_{gb} and m_{tj} are the finite mobilities, v_{gbi} and v_{tji} the corresponding velocities, and the simplifying assumption $v_{gbi} = v_{tji} = \dot{R}_i$ has been used.

In order to derive the evolution equation $\dot{R}_i = \mathrm{d}R_i/\mathrm{d}t$ of grain i we apply a thermodynamic approach on the basis of Onsager's principle of constrained maximum dissipation and as we have applied it to three-dimensional nanocrystalline grain growth in [15] according to which the total dissipated energy per unit time Q acquires an extremum with respect to the growth velocities $\dot{R}_i$,

$$\frac{\partial}{\partial \dot{R}_i} \left(Q + \eta \left(\dot{G} + Q \right) + \mu \dot{A} \right) = 0 , \tag{3}$$

subject to the conditions of energy conservation, $\dot{G} + Q = 0$, and total area conservation, $\dot{A} = 0$. For the variation with respect to $\dot{R}_i$ in Eq. (3) it has to be taken into consideration that the number of faces N_i and the grain size R_i are not independent of each other. Similar as in our statistical theory of normal [22] and 3D nanocrystalline grain growth [15] this grain topology-grain size-correlation is taken into account by considering N_i as a continuous function of the reduced grain size $x_i = R_i/\langle R \rangle$ such that $\mathrm{d}N_i/\mathrm{d}R_i = (\mathrm{d}N_i/\mathrm{d}x_i)/\langle R \rangle$, where $\langle R \rangle$ is the average grain size of the ensemble. With this assumption we obtain from Eqs. (1) to (3) the following evolution equation of an effective polygonal grain in a two-dimensional polycrystal (suppressing from now on the index i),

$$\dot{R} = \frac{\mu - \dfrac{\gamma_{gb}}{\langle R \rangle x} - \dfrac{\gamma_{tj} N'}{3\pi \langle R \rangle^2 x}}{\dfrac{1}{m_{gb}} + \dfrac{N}{3\pi m_{tj} \langle R \rangle x}} , \tag{4}$$

where we have used $R = \langle R \rangle x$ and $N' = \mathrm{d}N/\mathrm{d}x$.

In general, power law behavior of the average growth law, $\langle R \rangle \propto t^n$, and self-similar scaling of the grain size distribution function, $F(R,t) = g(t) \cdot f(x)$, are observed only in the case that the evolution equation is separable (cf. [1, 15] and the literature within), that is $\dot{R} = \langle \dot{R} \rangle \cdot \tilde{G}(x)$, where $\langle \dot{R} \rangle$ is the average and $\tilde{G}(x)$ the dimensionless size-dependent part of the growth law. In particular, for Eq. (4) this is only true if the time evolution is controlled only by one particular specific energy and mobility (see illustration in Fig. 1b), where the evolution equation $\dot{R}$ as given by Eq. (4) reduces to the following separated growth laws

$$\dot{R} = \begin{pmatrix} \dfrac{\mu - \dfrac{\gamma_{gb}}{\langle R \rangle x}}{\dfrac{1}{m_{gb}}} & \dfrac{\mu - \dfrac{\gamma_{gb}}{\langle R \rangle x}}{\dfrac{N}{3\pi m_{tj} \langle R \rangle x}} \\ \dfrac{\mu - \dfrac{\gamma_{tj} N'}{3\pi \langle R \rangle^2 x}}{\dfrac{1}{m_{gb}}} & \dfrac{\mu - \dfrac{\gamma_{tj} N'}{3\pi \langle R \rangle^2 x}}{\dfrac{N}{3\pi m_{tj} \langle R \rangle x}} \end{pmatrix}. \tag{5}$$

It can be noted that in Eq. (5) only the relevant parameters are remaining compared to Eq. (4). The top row of the matrix form Eq. (5) represents two types of growth laws that are controlled by the grain boundary energy γ_{gb} in conjunction with the finite mobilities m_{gb} and m_{tj}. The lower row represents two further types of possible growth laws, this time controlled by the specific energy of triple junctions γ_{tj} in conjunction with one of the finite mobilities m_{gb} and m_{tj}. All these growth laws are separable and represent, therefore, different quasi-stationary self-similar growth regimes. Applying the requirement of total area conservation [19], we obtain four different types of average growth laws, from which after integration the following power law behavior for the average grain size follows,

$$\begin{pmatrix} \langle R \rangle^2 & \langle R \rangle \\ \langle R \rangle^3 & \langle R \rangle^2 \end{pmatrix} \propto t\,. \tag{6}$$

In particular, the top row represents here also the two types of growth kinetics that are controlled by the grain boundary energy γ_{gb} in conjunction with the finite mobility of grain boundaries and triple junctions, respectively, as they have been derived in [3, 6]. In addition, the second row represents two further types of possible growth kinetics controlled by the specific energy of the triple junctions γ_{tj} in conjunction with one of the finite mobilities of grain boundaries and triple junctions m_{gb} and m_{tj}, respectively. The time exponents n in the power laws of Eq. (6) vary in the range of 1/3 to 1 (respectively $1/n$ varies between 1 and 3). This agrees with the corresponding kinetics controlled by triple lines in the case of three-dimensional nanocrystalline grain growth considered in Ref. [15], which has partly been observed experimentally [3-6].

3. Grain growth under boundary and junction control

According to the above Eq. (6) the average grain size of the ensemble shows power law behavior for all four kinetics of grain growth controlled by the grain boundary or triple junction energy in conjunction with one of the finite mobilities of grain boundaries or triple junctions. These four cases have been simulated using the modified Potts model according to Refs. [16-19], where details regarding the implementation and particularly regarding the parameters can be found, yielding:

- $\langle R \rangle^2 \propto t$ for grain growth controlled by the energy and mobility of the grain boundaries (subsequently named grain boundary-grain boundary-control; Figure 2a),
- $\langle R \rangle \propto t$ for grain growth controlled by the energy of grain boundaries and mobility of triple points (grain boundary-triple junction-control; Figure 2b),
- $\langle R \rangle^3 \propto t$ for grain growth controlled by the energy of triple points and mobility of grain boundaries (triple junction-grain boundary-control; Figure 2c) and also
- $\langle R \rangle^2 \propto t$ for grain growth controlled by the energy and mobility of triple points (triple junction-triple junction-control; Figure 2d)

just as expected according to Eq. (6). While these four kinetics hold mostly for longer annealing times (sections II), we can clearly take notice of the early transient growth regime showing deviations (sections I), of which the duration seems to depend on the growth kinetics itself and on the initial grain microstructure.

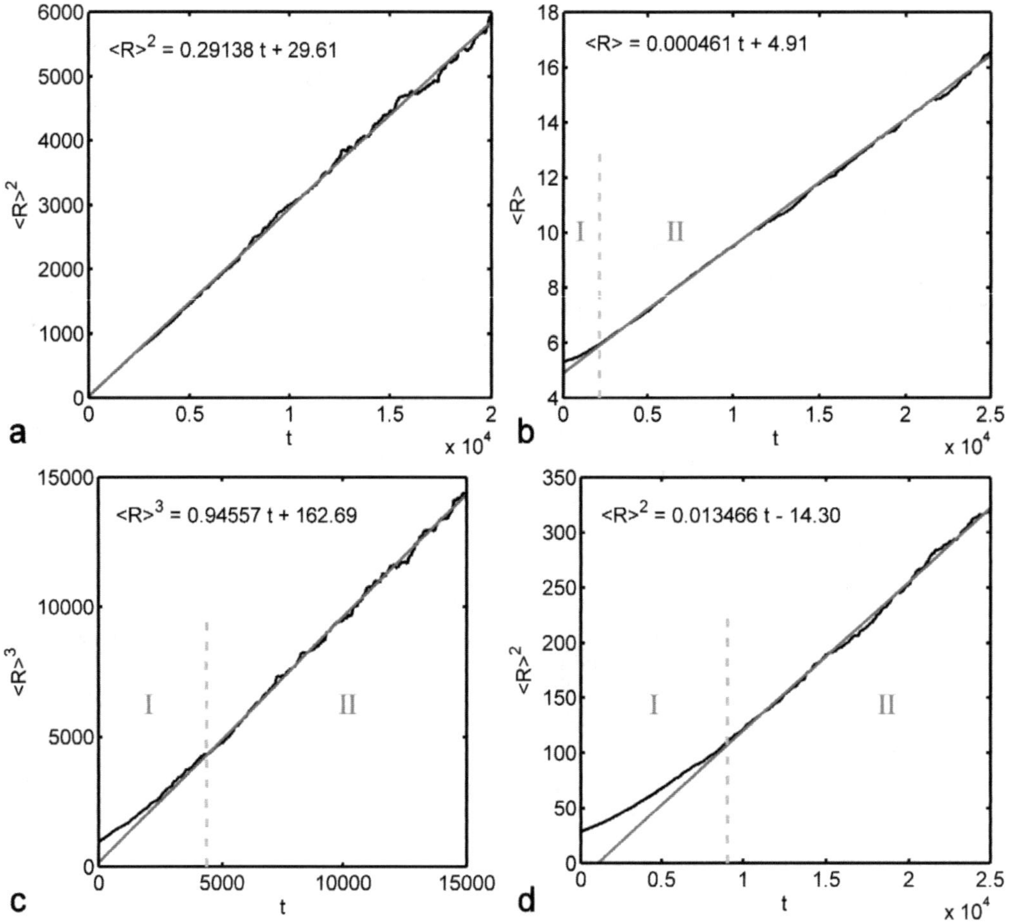

Figure 2. Average growth laws for grain growth under grain boundary and/or triple junction control for: a – grain boundary-grain boundary-control; b – grain boundary-triple junction-control; c – triple junction-grain boundary-control; d – triple junction-triple junction-control. Black: simulation results. Grey: late time behavior according to the analytical fit, Eq. (6).

However, such temporal developments of the average grain size characterize microstructures only partly. The corresponding counterpart is the grain size distribution, in which also the width of the distribution and the standard deviation, respectively, are of importance. The latter is shown for all four kinetics in Figure 3 as the standard deviation *sd* of the scaled grain size *x*. Again different growth regimes (I, II and III) can be found in the temporal development of *sd*[*x*]. We find it particularly true that the growth kinetics in Eq. (6) as seen in Figure 2 are apparently associated with a quasi-stationary, self-similar state, where the standard deviation and therewith also the scaled grain size distribution is time-independent, although *sd*[*x*] shows clearly visible fluctuations. Nevertheless, each quasi-stationary state is characterized by its own value of *sd*[*x*](*t*), which indicates that each growth kinetics is characterized by a different scaled size distribution.

One may notice that the durations of the quasi-stationary states for the different kinetics in Figure 3, where particularly Fig. 3c and d show the transition (III) from the temporarily limited self-similar growth regime to the true self-similar regime of normal grain growth for large grains and long annealing times, are of different lengths compared to those in Figure 2, of which the origin is a subject of future interest.

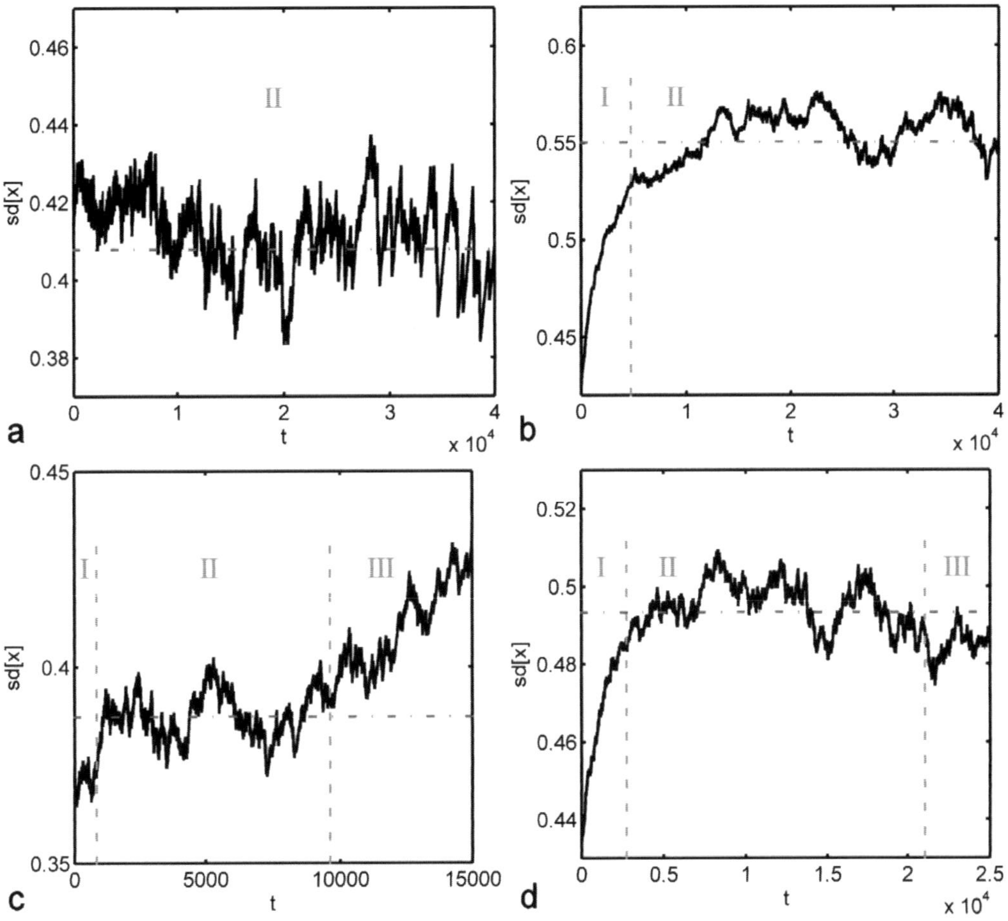

Figure 3. Temporal development of the standard deviation of the scaled grain size for: a – grain boundary-grain boundary-control; b – grain boundary-triple junction-control; c – triple junction-grain boundary-control; d – triple junction-triple junction-control.

Self-similarity can alternatively also be determined through an analysis of the topology of the microstructures, e.g., by finding a time-independent relation between the scaled grain size x and the number of sides or neighboring grains N, where N characterizes also the number of triple junctions in a two-dimensional grain network. This relation is shown in Figure 4 for each kinetics at three different annealing times, which have been taken from the self-similar regimes (II) according to Fig. 3. In particular, the data have been plotted by categorizing them in size resp. neighbor classes because of the variation of this relation for individual grains (compare Ref. [18]). It can be seen that the results are indeed time-independent in all four cases.

These topological relations, which are also needed in the evolution equations of Eq. (5), show the following characteristic features. In the cases of grain boundary mobility controlled grain growth of Figs. 4a and 4c, that is for $m_{gb} << m_{tj}$ in both cases, the course of $N(x)$ exhibits a pronounced curvature, which can be described by the quadratic function

$$N(x) = c_2 x^2 + c_1 x + c_0 . \tag{7}$$

This non-linear behavior of $N(x)$ has already been proven as a characteristic property of 2-D normal grain growth (where $\gamma_{tj} = 0$), which is important for a consistent mean-field theoretical description of normal grain growth in agreement with the simulation results [21, 22]. The results in Fig. 4c show that also for triple-junction energy driven grain growth (where $\gamma_{gb} = 0$) with finite grain boundary mobility gives a similar non-linear behavior of the topological function, which is also well described by Eq. (7). In contrast, the two cases of triple-junction mobility controlled grain growth of Figs. 4b and 4d, that is for $m_{gb} >> m_{tj}$ in both cases, show clearly a linear dependence of the topological function on the relative grain size described by

$$N(x) = c_1 x + c_0 . \tag{8}$$

or alternatively by $x(N) = a_1 N + a_0$ as for triple junction energy-triple junction mobility control in Fig. 4d. A linear behavior of $N(x)$ resp. $x(N)$ is in agreement with theoretical results based on N-sided polygonal grain models with straight grain boundaries (cf. [20, 23]), an assumption which is in agreement with the observed grain microstructure under triple junction mobility control in Fig. 1a.

Figure 5 shows the corresponding grain size distributions, where the symbols represent the simulation results and the solid lines our analytical expression [19]:

$$f(x) = \frac{2x_c\tilde{\Gamma}(x+x_1)}{\left(x^2+(x_1-\tilde{\Gamma})x+x_c\tilde{\Gamma}\right)^2}\exp\left[-2\frac{x_1+\tilde{\Gamma}}{\sqrt{\Delta}}\left[\arctan\left(\frac{2x+x_1-\tilde{\Gamma}}{\sqrt{\Delta}}\right)-\arctan\left(\frac{x_1-\tilde{\Gamma}}{\sqrt{\Delta}}\right)\right]\right] \tag{9}$$

with

$$\Delta = 4x_c\tilde{\Gamma} - (x_1 - \tilde{\Gamma})^2 \geq 0 .$$

Eq. (9) is derived from the solution of the continuity equation for the grain size distribution function $F(R,t) = g(t) \cdot f(x)$, where $\dot{R}$ is given by one of the expressions in Eq. (5) (for the procedure see Ref. [19]). $f(x)$ is normalized to one. For given values of x_1 and $\tilde{\Gamma}$, the scaled critical grain size x_c is determined from the scaling condition $\langle x \rangle = \int x f(x) dx = 1$.

For purely triple junction drag—the two cases in the right column in Eq. (5)—the parameter x_1 is given by $x_1 = c_0 / c_1 = -a_1$ following from the linear topological function Eq. (8) represented in Figs. 4b and 4d. For the two cases of grain boundary mobility dominated grain growth—the left hand column in Eq. 5—$N(x)$ is given by the quadratic polynomial Eq. (7) represented in Figs. 4a and 4c. One can show [24] that under this condition Eq. (9) is also a solution for these two cases, when the parameter x_1 is formally set equal to zero, that is $x_1 = 0$.

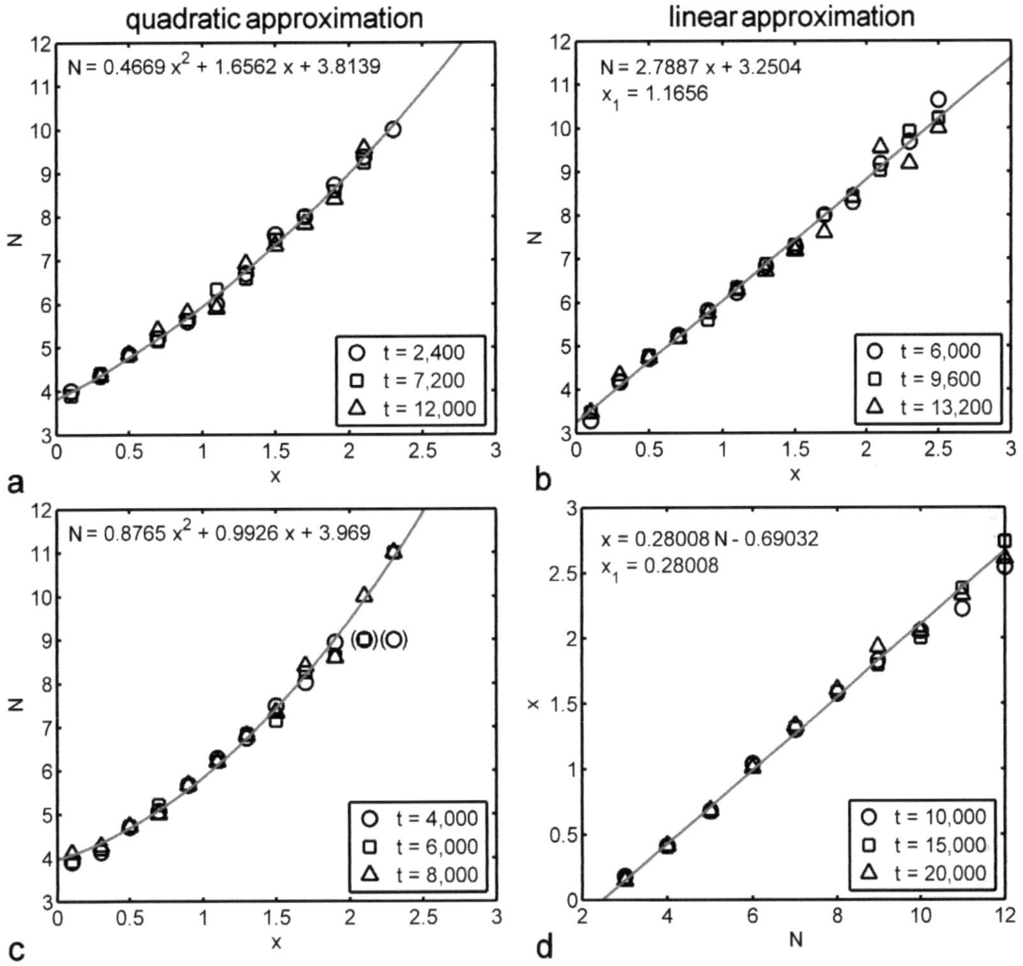

Figure 4. Topological correlation showing $N(x)$ for: a – grain boundary-grain boundary-control; b – grain boundary-triple junction-control; c – triple junction-grain boundary-control; d – triple junction-triple junction-control.

The comparison of the analytical grain size distributions $f(x)$, where $\tilde{\Gamma}$ is used as a free fit parameter, with the numerical data resulting from the MC Potts model simulations shows in Figure 5 that the scaled analytical grain size distribution function Eq. (9) reflects the essential features of the grain size distributions of all the four considered cases of self-similar grain growth controlled by the respective pair of grain boundary and triple point energy and mobility, where self-similarity is shown again for the same three different annealing times as in Fig. 4.

A prominent feature of the two grain size distributions for triple junction mobility controlled grain growth in Figs. 5b and 5d is the non-zero value of the size distribution at $x = 0$. This property reflects the influence of the finite triple junction mobility on the grain growth kinetics leading to an accumulation of grains having very small sizes, as is typical for nanocrystalline materials [15, 19].

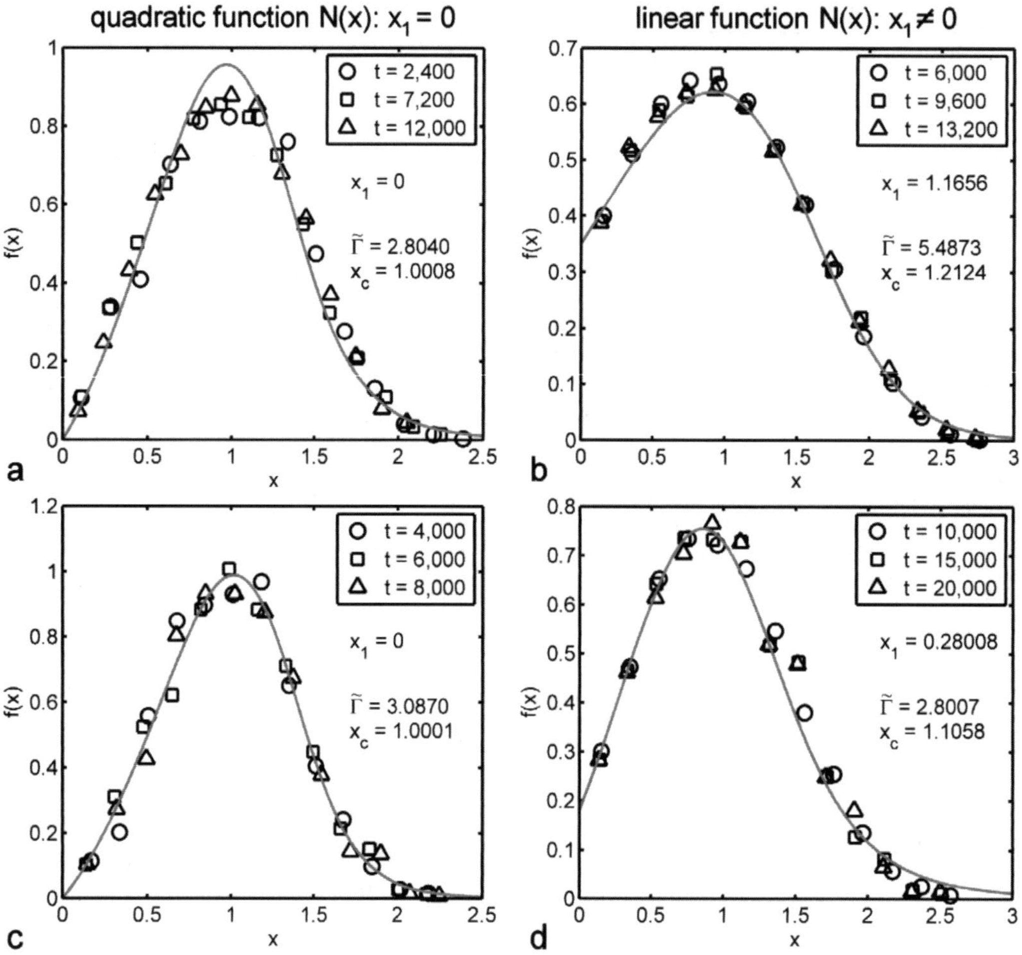

Figure 5. Scaled and self-similar grain size distributions for: a – grain boundary-grain boundary-control; b – grain boundary-triple junction-control; c – triple junction-grain boundary-control; d – triple junction-triple junction-control.

4. Conclusions

In the present paper we have investigated the influence of the energy and mobility of grain boundaries and their triple junctions in nanocrystalline two-dimensional polycrystals and thin films on annealing by analytic theories and modified Monte Carlo Potts model simulations. To that aim each structural feature of a polygonal grain has been assigned its own specific energy and finite mobility.

As a result we have derived four different limiting cases of self-similar growth kinetics each characterized by its own:

- average growth law following Eq. (6) as shown in Figure 2, where also the influence of the initial microstructure on annealing can be seen,
- standard deviation of the scaled grain size (Figure 3), which indicates that each kinetics is characterized by
- scaled grain size distribution, which we find to be in good agreement with our analytical solution, Eq. (9), as presented in Figure 5, and

- topological relation between number of neighboring grains and scaled grain size as can be seen in Figure 4.

It is particularly noteworthy that while grain growth under triple junction mobility control has been in the focus of investigations in recent years [6-9, 15-19], the present paper is to our knowledge the first record of grain growth kinetics under triple junction energy control in two-dimensional polycrystalline grain microstructures.

References

[1] Atkinson HV. 1988 *Acta Metall.* **36** 469
[2] Humphreys FJ, Hatherly M. 2004 *Recrystallization and Related Annealing Phenomena* (Oxford: Elsevier)
[3] Gottstein G, Shvindlerman LS. 2010 *Grain Boundary Migration in Metals* (Boca Raton: CRC Press)
[4] Yu T, Hansen N, Huang X. 2011 *Proc. R. Soc.* **A 467** 3039
[5] Koch CC, Ovid'ko IA, Seal S, Veprek S. 2007 *Structural Nanocrystalline Materials* (Cambridge: Cambridge University Press)
[6] Gottstein G, Shvindlerman LS. 2006 *Scripta Mater.* **54** 1065
[7] Novikov VYu. 2008 *Materials Letters* **62** 2071
[8] Barrales Mora LA, Mohles V, Shvindlerman LS, Gottstein G. 2008 *Acta Mater.* **56** 1151
[9] Gottstein G, Shvindlerman LS, Zhao B. 2010 *Scripta M*ater. **62** 9140
[10] Thompson CV. 1990 *Annu. Rev. Mater. Sci.* **20** 245
[11] Thompson CV. 2001 *Solid State Phys.* **55** 269
[12] Barmak K, Eggeling E, Kinderlehrer D, Sharp R, Ta'asan S, Rollett AD, Coffey KR. 2013 *Progr. Mater. Sci.* **58** 987
[13] Weygand D, Bréchet Y, Lépinoux J. 1999 *Interface Sci.* **7** 285
[14] Barrales-Mora LA. 2010 *Math. Comp. Simul.* **80** 1411
[15] Streitenberger P, Zöllner D. 2011 *Acta Mater.* **59** 4235
[16] Zöllner D. 2011 *Comput. Mater. Sci.* **50** 2712
[17] Zöllner D. 2012 *Scripta Mater.* **67** 41
[18] Zöllner D. 2013 *Comput. Mater. Sci.* **79** 759
[19] Streitenberger P, Zöllner D. 2014 *Acta Mater.* **78** 114
[20] Rios PR, Glicksman ME. 2000 *Scripta Mater.* **51** 629
[21] Streitenberger P, Zöllner D. 2006 *Scripta Mater.* **55** 461
[22] Zöllner D, Streitenberger P. 2006 *Mater Sci Forum* **467-470** 1129.
[23] Zöllner D, Rios PR. 2014 Acta Mater **70** 290.
[24] Streitenberger P, Zöllner D. to be published